T5-CVD-889

Crystal Growth from
High-Temperature Solutions

Crystal Growth from High-Temperature Solutions

D. ELWELL

Department of Physics
Portsmouth Polytechnic
Portsmouth, England

AND

H. J. SCHEEL

IBM Zurich Research Laboratory
8803 Rüschlikon, Switzerland

1975

ACADEMIC PRESS

London   New York   San Francisco

A Subsidiary of Harcourt Brace Jovanovich, Publishers

ACADEMIC PRESS INC. (LONDON) LTD.
24/28 Oval Road
London NW1

United States Edition published by
ACADEMIC PRESS INC.
111 Fifth Avenue
New York, New York 10003

Library of Congress Catalog Card Number: 74-18000
ISBN: 0-12-237550-8

Printed in Great Britain by
Robert MacLehose and Co. Ltd
Printers to the University of Glasgow

Contents

Acknowledgements

The book has benefited considerably from the criticism and comments of many colleagues who have read one or more chapters on their particular, specialized topics. The authors are indebted to the following for their contributions:

H. Arend, A. Armington, A. Authier, W. Bardsley, P. Bennema, S. L. Blank, R. W. Brander, J. C. Brice, B. Cleaver, B. Cockayne, T. F. Connolly, J. Felsche, R. Ghez, E. A. Giess, A. R. Goodwin, H. G. B. Hicks, B. Honigmann, D. T. J. Hurle, E. Kaldis, B. Lewis, R. C. Linares, D. J. Marshall, K. A. Müller, H. Müller-Krumbhaar, J. W. Mullin, J. W. Nielsen, D. Pohl, H. Posen, A. Preisinger, J. P. Remeika, J. M. Robertson, C. J. M. Rooymans, L. Rybach, C. S. Sahagian, M. Schieber, M. B. Small, S. H. Smith, W. Tolksdorf, B. M. Wanklyn.

The advice of P. Bennema, C. S. Sahagian and M. Schieber on important aspects is most appreciated. Valuable discussions with the following are also acknowledged with gratitude: S. L. Blank, C. H. L. Goodman, R. C. Linares, K. A. Müller, J. W. Nielsen, T. S. Plaskett, J. P. Remeika, J. M. Robertson, S. H. Smith, W. Tolksdorf, B. M. Wanklyn.

In addition, many colleagues, too numerous to mention, contributed preprints or other information. We are grateful to H. V. Alexandru, N. P. Luzhnaja, H. Sasaki and V. A. Timofeeva for their assistance with references, particularly those not easily accessible in languages familiar to the authors, to Dr. L. Bohaty for his translation from Russian work, and to Mrs. M. Scheel for collecting the data of Table 2.2.

Further comments and suggestions from readers would be highly appreciated by the authors.

Photographs or diagrams were kindly contributed by: A. Authier, Balzers AG, Liechtenstein (Dr. H. D. Dannöhl), W. T. Stacy, C. F. Cook, H. Frederiksson, S. E. R. Hiscocks, B. Isherwood, H. Klapper, D. E. Lepore, E. O. Schulz-DuBois, W. Tolksdorf, A. M. Vergnoux, B. M. Wanklyn, R. H. Wentorf, J. M. Woodall.

D.E.

H.J.S.

1. Introduction to High-Temperature Solution Growth

1.1. The Importance of Materials Research

"We have for some time labelled civilizations by the main material which they have used: The Stone Age, the Bronze Age and the Iron Age . . . a civilization is both developed and limited by the materials at its disposal. . . . Today, man lives on the boundary between the Iron Age and a New Materials Age." These words were used by Sir George Thomson, Nobel prizewinner in 1937, and were quoted in a review of the evolution of materials science and technology by Promisel (1971).

It is more valid today to say that the age of new materials is well and truly upon us. Novel metals such as titanium are increasingly replacing steel for several applications, and metals generally are experiencing competition from ceramics, plastics and composite materials. The rate of innovation is, of course, now so rapid that it has become erroneous to speak of an "age"; the term is still widely used and the particular choice of emphasis—Atomic Age, Electronic Age, Computer Age and so on—depends on the writer's personal viewpoint. Nevertheless it cannot be denied that the development of new materials has contributed greatly to the changes which have characterized the 20th century, particularly in aviation, communications and packaging.

The rapid development of new materials has led to the creation of *materials science* as a discipline in its own right. Materials science is an integration of physics, chemistry, crystallography, metallurgy, ceramics and glass technology and the need for new materials has resulted in a breakdown of many traditional barriers. The new concept is that a "consumer" will specify the properties of the material required rather than stipulating that it should be metallic, ceramic or plastic. As a result, scientific societies have tended to widen their scope with, for example, metallurgical societies including plastics and non-metallic composites in their conferences and discussions.

Materials scientists tend to have varied backgrounds, and are very rarely trained in materials as such because of the slow rate of development of degree courses in this discipline. The normal training of materials scientists is chemistry, physics or metallurgy but it is not uncommon for electrical engineers, mineralogists or geologists to move into materials science or technology.

Crystal growth is a relatively small but important area of materials science. It is clearly more difficult to prepare single crystals than poly-crystalline material and the extra effort is justified only if single crystals have outstanding advantages. Their chief merits are the anisotropy, uniformity of composition and the absence of boundaries between in-dividual grains which are inevitably present in polycrystalline materials and which lead, for example, to optical absorption or scattering or to trapping of conduction electrons. Grain boundaries are also absent in glasses, which can be used for laser generation and some other applications, but the absence of an ordered structure often has severe disadvantages. In addition, there are many experiments which can be performed only on single crystals, such as measurements of magnetocrystalline anisotropy, and which are vital for the detailed understanding of the properties even of the polycrystalline material. The relationship between crystal growth (or materials preparation), characterization and physical properties is shown in Fig. 1.1. Careful characterization should be made of all crystals used and the resulting information used by the crystal grower in order to

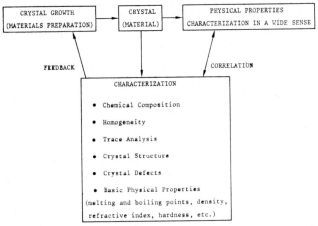

FIG. 1.1. Interrelation between crystal characterization, growth and properties.

optimize the crystal growth technique. On the other hand, characterization is necessary in order to validate any physical measurements.

The production by Stockbarger (1936, 1949) of 1500 fluorite crystals, up to 15 cm in diameter, for military application in high-quality lenses and the successful synthesis of quartz crystals by Nacken during the second world war initiated systematic crystal-growth activity, and further stimulation was given by the development of the germanium transistor, by the growth of ferroelectric $BaTiO_3$ (Blattner et al., 1947; Remeika, 1954), and of magnetic garnets (Bertaut and Forrat, 1956; Geller and Gilleo, 1957; Nielsen and Dearborn, 1958).

The widest use of single crystals is probably that of ruby for bearings (also as a laser material and as a gemstone) and that of silicon in electronic devices such as the transistor. Silicon is normally preferred to germanium because of its larger band gap (1.1 eV compared with 0.72 eV for Ge), which results in a lower sensitivity to changes in temperature, but was initially considered relatively difficult to prepare in single crystal form. The increasing use of silicon, particularly in MOS (metal-oxide-semiconductor) and in LSI (large-scale integration) technology is favoured by its outstanding crystallization properties and by the passivation by an oxide layer formed directly on the silicon surface. It should not be forgotten that the degree of perfection now possible has been attained only as a result of considerable research and development investment of money and effort. The III–V semiconductors have important advantages for certain applications; for example the high mobility of carriers in GaAs favours the use of oscillators, and GaP and its solid solutions with GaAs are now used as light-emitting diodes.

This book describes one method by which single crystals and thin film layers may be obtained, namely by the use of solutions at high temperatures. In the various chapters theoretical and practical considerations will be outlined for the growth of large crystals and of crystalline films having a high degree of perfection. It is hoped that this discussion will stimulate interest in crystal growth from high-temperature solutions and will encourage more crystal growers to adopt solution techniques.

I.2. High-temperature Solution Growth

In the growth of crystals from high-temperature solutions (the abbreviation HTS will be used extensively) the constituents of the material to be crystallized are dissolved in a suitable solvent and crystallization occurs as the solution becomes critically supersaturated. The supersaturation may be promoted by evaporation of the solvent, by cooling the solution or by a transport process in which the solute is made to flow from a hotter to a cooler region.

The principal advantage of using a solvent is that crystal growth occurs at a lower temperature than that required for growth from the pure melt. A reduction in the temperature is desirable or even essential for many materials, in particular those in the following categories.

a. Those which are *incongruently melting*, that is which decompose before melting so that crystallization from the melt results in some other phase.
b. Materials which undergo a phase transition which results in severe strain or even fracture; crystals of such materials should be grown at temperatures below this transition.
c. Materials which have a very high vapour pressure at the melting point.
d. Materials which have a very volatile constituent and whose chemical composition may therefore change on heating close to the melting point.
e. Highly refractory materials which require difficult or expensive techniques for crystallization from the melt.

Other advantages of crystal growth from solutions are based on the fact that the growing crystal is not exposed to steep temperature gradients and that the crystal can grow in an unconstrained fashion, that is the crystal can grow free from mechanical or thermal constraints into the solution and so develop facets. This, in combination with the relatively low growth temperature compared with the melting point of the solute, often results in a better crystal quality with respect to point defects, dislocation densities and low-angle grain boundaries, compared to crystals grown directly from their own melt.

The disadvantages of the method are substitutional or interstitial incorporation of solvent ions into the crystal, microscopic or macroscopic inclusions of solvent or impurities, non-uniform doping, a slow growth rate and container problems.

Until recently it was comparatively rare for HTS techniques to be preferred over all possible alternatives on the grounds of crystal quality. One of the principal aims of this book is to propose that these solution techniques should be much more widely used because crystals and crystalline layers of high quality and purity and large size can be produced if careful attention is paid to such questions as the choice of a solvent and the experimental conditions for stable growth. Also homogeneous doping and solid solutions can be achieved under adequately controlled conditions.

The subject of crystal growth from high-temperature solutions includes a number of related techniques which are often treated separately. The most widely used is *flux growth* in which the solvent is a molten salt or oxide, or a mixture such as $PbO + PbF_2$. The use of the word "flux" for the solvent is derived from the more general use of this term for a substance used to reduce the melting temperature, or to dissolve oxides as in soldering.

The term *growth from non-stoichiometric melts* is sometimes used to include all cases where crystals are grown from a liquid containing only the crystal constituents but of a composition which differs from that of the crystal. The scope of this book includes this technique along with "flux growth" and also growth from metallic solvents.

Liquid phase epitaxy (LPE) is a technologically important process in which a thin layer of crystalline material is deposited from solution onto a substrate of similar composition or surface structure.

The literature of high-temperature solution growth, as for all aspects of crystal growth, is very widespread. In addition to the more specialized journals such as the Journal of Crystal Growth, Materials Research Bulletin, Journal of Materials Science, and Kristall und Technik, a very large number of journals on crystallography, physics, chemistry, metallurgy, ceramics, mineralogy and electronic technology may contain articles dealing partly or wholly with HTS growth.

Although this book is the first on the subject an extensive number of review articles have appeared over the last ten years. Reference is made to Laudise (1963), White (1965), Schroeder and Linares (1966), Cobb and Wallis (1967), Roy and White (1968), Luzhnaya (1968), Laurent (1969), Elwell and Neate (1971), Chase (1971), Nielsen (1972), Brice (1973), Elwell (1975) and Wanklyn (1975). Although review articles have been used in the compilation of this book, the subject matter has been synthesized from the original articles and any interpretation of theory and of observations is our own or that of the original authors.

The terms "high temperature" and "solution" are not exactly defined in the context of this book. The experiments described are normally in the range from 300°–1800°C and the solute concentration is usually between 1% and 30% in terms of either molar or weight fraction. Some examples will, however, be discussed where the solute concentration is as high as 80% or 90%. Crystals grown from HTS are normally facetted, but again exceptions will be occasionally encountered in the text. Clearly the boundaries between HTS and other growth methods may be only roughly specified; the principal alternative growth methods are described briefly in the next section.

The boundary between solution growth and growth from a doped melt is particularly difficult to define. The distinction which will be made here is that in the latter case the dopant is added in order to change some property of the crystal by virtue of its presence *in the lattice*. In solution growth the "dopant" is added to the melt in order to lower the crystallization temperature and is rejected by the growing crystal under ideal conditions.

The most popular materials grown from high-temperature solution are oxides crystallized from molten salts and this emphasis is reflected in the subject matter of this book. However, all types of solute and solvent will be considered, with the important exception of *hydrothermal growth* in which the solvent is water at high temperatures and pressures (the hydrothermal technique would justify an independent treatment in the form of a book).

We have included the *vapour→liquid→solid* (VLS) process in which solute constituents are supplied by transport in the vapour phase. Examples of crystal growth under extreme conditions (e.g. high pressure) will be considered, including the technologically important crystallization of diamonds. In both the VLS process and the growth of diamonds the amount of solvent relative to solute is generally small but the inclusion of these two processes is justified here because of the vital role of the solution stage.

1.3. Alternative Methods of Crystal Growth

Single crystals may be produced by the transport of crystal constituents in the solid, liquid or vapour phase. An introduction to the general problems and classification of crystal growth methods has been given by Laudise (1970), who refers to some earlier reviews. The various methods may be classified as outlined below.

1. *Crystal growth from the solid state.* Polycrystalline solids contain a very large number of crystallites and crystals of useful size may be obtained if a few of these can be made to grow preferentially with the elimination of their neighbours within the material. Such recrystallization may be achieved by straining the material and subsequently annealing it, or by sintering, and is possible only in those materials which are stable at the high temperatures where appreciable diffusion can occur. Large crystals of several materials, particularly metals, have been obtained by this method which probably deserves more attention than it receives at present. It is, however, unlikely that any process which is carried out in the solid state can reproducibly result in crystals with the necessary high quality required for many applications.

2. *Growth from the pure melt.* Crystallization by fusion and resolidification of the pure material has many advantages from both the theoretical and experimental viewpoints. Apart from possible contamination from the crucible material and the surrounding atmosphere, no impurities are introduced in the growth process and the rate of growth is normally much higher than that possible by other methods. Mainly for the latter reason, melt growth is commercially the most important method of crystal growth. With favourable materials the crystal quality can be extremely good and the commercial production of dislocation-free silicon is one of the major advances in the field of crystal growth. There are, however, a large number

of materials which cannot be crystallized from their pure melts for the reasons outlined in the previous section. Even when melt growth is possible, as in the case of alumina Al_2O_3 and spinel $MgAl_2O_4$, a solution method may be preferred on the grounds of crystal quality. Growth from the melt is the subject of a book by Brice (1965, 1973).

3. *Growth from the vapour.* A wide variety of materials may be grown from the vapour phase and the reaction equilibria are now well understood for many systems. With certain materials growth may occur using a single component system, by sublimation-condensation or sputtering. Molecular beam techniques have also been applied recently to crystal growth problems.

The most frequently used method for the growth of bulk crystals utilizes *chemical transport reactions* in which a reversible reaction is used to transport the source material as a volatile species to the crystallization region. The book by Schäfer (1962, 1964) and the review of Nitsche (1967) are strongly recommended as an introduction to this subject, and Kaldis (1974) discusses the principles of crystal growth from the vapour phase.

The limitations of chemical transport are partly of chemical origin, especially the problem of finding a suitable transporting agent, and formidable difficulties arise if it is necessary to produce uniformly doped materials or solid solutions. In addition, it is rarely possible to grow large crystals because of multinucleation, although exceptions such as Al_2O_3, CdS, ZnSe, GaP, GaAs and Cd_4GeS_6 are known. The commercial importance of vapour growth is in the production of thin layers by *chemical vapour deposition* (CVD) where usually irreversible reactions (e.g. decomposition of silicon halides or of organic compounds) are used to deposit material epitaxially on a substrate. CVD may often be used as an alternative to liquid phase epitaxy.

4. *Growth from aqueous solution.* The simplest and oldest method of crystal growth is that from aqueous solutions. Enormous quantities of materials such as sugar, salt and of inorganic and organic chemicals are crystallized from solution in water, and excellent quality crystals of such ferroelectric and piezoelectric materials as ammonium dihydrogen phosphate ADP, potassium dihydrogen phosphate KDP and triglycine sulphate TGS are commercially grown for use in electronic devices. Chemical instability or the requirement of an appreciable solubility in water excludes a number of classes of materials from being crystallized from aqueous solutions, a low solubility arising either from a different type of bonding (metals, semiconductors, covalent compounds) or from a high lattice energy (many metal oxides).

Substances with a low solubility in water can be grown by the *gel technique* which has been reviewed by Henisch (1970). The principle relies on the slow migration of crystal constituents (ions) through an inorganic ($SiO_2 \times H_2O$) or organic gel so that a very slow reaction occurs with the formation of the sparingly soluble compound. When the concentration of this compound exceeds the solubility limit crystals will be formed, the main function of the gel being to control the slow flow of the reacting ions. From the gel large crystals have not yet been obtained but the gel technique is a useful tool for exploratory research as well as a fascinating hobby for crystal growers who may observe the slow crystallization process in a test tube.

5. *Growth from nonaqueous solvents.* There are quite a number of ionizing solvents which show similar solvent characteristics in some respects to water, for example the liquids HF, SO_2, H_2S, NH_3, $SbBr_3$ and $POCl_3$. However, these solvents also show remarkable differences from water and are therefore capable of dissolving a number of elements and compounds insoluble in or chemically reactive with water. Systematic crystal growth from the above ionizing solvents has not yet started.

The growth from solution (in ionizing or covalent solvents) of polymers and crystals of biochemical interest is also a little-explored but very important field. Little crystal growth of organic compounds from covalent solvents is performed although the organic chemist uses recrystallization from solvents or solvent mixtures as a standard purifying process.

6. *Hydrothermal growth.* A number of metals, metal oxides and other compounds practically insoluble in water up to its boiling point show an appreciable solubility when the temperature and pressure are increased well above 100°C and one atmosphere, respectively. Therefore these materials can be grown by the *hydrothermal method* reviewed by Ballman and Laudise (1963), oxides generally being grown from alkaline and metals from acid solutions (Rau and Rabenau, 1968). The requirement of high pressures presents practical difficulties and there are only a few crystals of good quality and large dimensions grown by this technique. The outstanding example of industrial hydrothermal crystallization is quartz (SiO_2) which is produced in large vessels for the manufacture of commercial piezoelectric crystals. Hydrothermal growth might in several other cases be an alternative to HTS growth when large high-quality crystals or layers are needed. One serious disadvantage of this technique is the frequent incorporation of OH^- ions into the crystals which makes them useless for many applications.

7. *Comparison with high-temperature solution growth.* Although no rigorous rules may be established for the general evaluation of a growth method, the following criteria will be of importance in the assessment of any method.

a. The universality, that is the number of materials to which the method may be applied.
b. The size and quality of the crystals grown.
c. The requirements on apparatus and chemicals.
d The requirements on experience, theory and time.
e. Particularly in industrial applications, the crucial factor is the cost per acceptable crystal or layer.

The greatest general advantage of HTS growth is its wide applicability. Except for those organics, hydrates and similar compounds which decompose at relatively low temperatures, there seems to be no limitation to the type of crystal which can be grown from high-temperature solution. In most cases a practical solvent can be found and also growth from non-stoichiometric melts is often possible. The use of HTS growth for the crystallization of oxides is well known but it may not be widely appreciated that such materials as nitrides, borides and carbides may often best be grown from solution.

Until comparatively recently, nucleation was a major problem and the majority of crystals grown from HTS were rather small. This problem has now been largely solved by general advances in techniques and by greater understanding of the growth mechanisms. Large crystals, weighing in some cases over 200 g, have now been produced of a variety of materials ranging from yttrium iron garnet ($Y_3Fe_5O_{12}$), lithium ferrite ($LiFe_5O_8$), gadolinium aluminate ($GdAlO_3$) and barium titanate ($BaTiO_3$) to $Ba_2MgGe_2O_7$, ruby and emerald (see Chapter 10).

The quality of HTS-grown crystals may vary considerably but generally it is a function of the effort (experience, cost, time) devoted to the growth of high-quality crystals. Normally the crystals will contain a higher concentration of impurities than crystals grown from the melt, but will have a lower concentration of equilibrium defects such as vacancies and frequently a lower dislocation density. As an example, the dislocation density of Al_2O_3 crystals grown by a variety of crystal growth techniques is shown in Table 1.1. In this comparison, which looks very favourable for HTS growth, it should be remembered that a number of defects do not depend only on the growth technique but originate from the cooling procedure of the crystal after growth and from any mechanical cutting, grinding and polishing processes. Evidence for the high quality which may be achieved in HTS growth is provided by the observation by Nelson and Remeika (1964) of laser action in an as-grown crystal of ruby a few mm in length, which was silver plated on its natural growth faces.

Apart from the problem of chemical contamination by substitution and inclusions which may occur as the crystals are grown in the presence of a large mass of impurity (the solvent), the main handicap of HTS growth is that the maximum stable growth rate is relatively slow (see Chapters 6, 7). This limitation can be a severe disadvantage compared with melt growth for the production of large crystals.

TABLE 1.1. Measured Dislocation Densities of Alumina Al_2O_3 and Ruby Al_2O_3:Cr Crystals Grown by Various Techniques

Technique	Range of dislocation or etch pit densities per cm^2	Reference
High-temperature solution growth	$0-<10^2$	Belt (1967b)
	$0-<10^2$	Janowski *et al.* (1965)
	$0-\sim 10^2$	Linares (1965)
	10^2-10^4	Champion and Clemence (1967)
	10^3-10^7	Sahagian and Schieber (1969)
	10^2	Stephens and Alford (1964)
	10^2-10^4	White and Brightwell (1965)
Flame fusion growth	$6.10^4-5.10^6$	Sahagian and Schieber (1969)
	10^6-10^8	Champion and Clemence (1967)
	$3.10^5-2.10^6$	Alford and Bauer (1967)
Plasma fusion growth	$4.10^5-5.10^6$	Sahagian and Schieber (1969)
	$7.10^5-7.10^6$	Alford and Bauer (1967)
Czochralski growth	$2.10^3-8.10^5$	Sahagian and Schieber (1969)
	10^2-10^4	Belt (1967a)
	10^2-10^3	Cockayne *et al.* (1967)
Bridgman-Stockbarger type	10^5	Schmid and Viechnicki (1970)
Electron beam technique	$1.10^2-5.10^4$	Sahagian and Schieber (1969)
Hydrothermal growth	$8.10^4-4.10^5$	Sahagian and Schieber (1969)
Vapour growth	$2.10^4-1.10^6$	Sahagian and Schieber (1969)
	$\sim 10^2$	Schaffer (1964)
Natural crystals	$1-5.10^3$	Sahagian and Schieber (1969)

Although the cost of the furnace and other ancillary equipment is relatively low, a further disadvantage is encountered in the majority of experiments where the only satisfactory crucible material is platinum. This material is very expensive and crucibles often form the major item of capital expenditure. This problem is serious only in the initial stages of development since the cost of reclamation and refabrication of damaged crucibles is low. As with all crystal growth experiments, the cost of materials can be extremely high, but this depends on the amount of contamination which can be tolerated in the crystals grown and therefore on the necessary purity of the starting materials.

The problems mentioned will be examined, together with the advantages of HTS growth, at various points in the text.

1.4. Applications

Crystals grown from high-temperature solutions are used both for academic research and by industry, and we shall consider these catagories separately.

Research use

One of the most important uses of HTS growth is in the synthesis of new materials. Several examples are known of materials which were first formed either by accident or during systematic studies of novel solute-solvent systems. This type of study has been mainly performed on compounds of the transition metals and the rare earths, especially complex oxides. There remains a large area for further study, for example of compounds of the palladium and platinum transition groups, and it is certain that there are many materials with interesting magnetic properties which await discovery.

Of even greater importance is the vital role particularly of flux growth in the preparation of materials in single crystal form for the first time. Many examples may be quoted of materials which were first crystallized from solution prior to the development of now more familiar methods. The synthesis of crystals of new materials has led to important advances, particularly in magnetism, ferro- and piezoelectricity and laser research.

The rapid growth in the number of materials available in single crystal form has led to the establishment of materials information centres in a number of countries,† and the excellent service operated by T. F. Connolly has made a valuable contribution to materials research and has benefitted crystal growers and consumers in many countries. The main value of the information centres is that they prevent much duplication of effort and encourage the maximum benefit to the scientific community of centres specializing in the preparation of materials. Where duplication cannot be

† *United States*: Research Materials Information Centre, P.O. Box X, Oak Ridge, Tennessee 37830 (T. F. Connolly).
France: Centre de Documentation sur les Synthèses Cristallines, Laboratoire de Physique Moléculaire et Cristalline, Faculté des Sciences, Chemin des Brusses, 34—Montpellier (Mlle A. M. Vergnoux).
Britain: Electronic Materials Unit, Royal Radar Establishment, Malvern, Worcs. (O. C. Jones).
Germany: Prof. R. Nitsche, Institut f. Kristallographie d. Universität, 78 Freiburg i. Br., and Dr. A. Räuber, Institut für Angewandte Festkörperphysik der Fraunhofer-Gesellschaft, 78 Freiburg i. Br., Eckerstrasse 4.

avoided it is desirable that original crystal growth experiments be repeated with a minimum of effort and therefore that the original publication should contain sufficient information to allow easy repetition.

Exploratory materials research cannot be expected to yield crystals which are large or of very high quality and much further effort may be required to produce crystals which are suitable for more demanding experiments, such as electro-optic modulation or acoustic experiments. The production of large crystals of good quality normally requires more sophisticated techniques and a better understanding of the phase diagram and the growth mechanism.

Increasing attention is being given to the deposition of thin films by liquid phase epitaxy. Provided that the lattice parameter match between the film and the substrate is good and the substrate itself is relatively free from defects, the quality of the films can be extremely high. If the experiment can be performed on such films, their use has the great advantage that relatively little time is required for the growth process, and so one of the major disadvantages of solution methods can be overcome.

Industrial materials

The materials which have been crystallized from high-temperature solution in industry may be divided into the categories listed below.

1. *Magnetic oxides.* Insulating magnetic materials, especially yttrium iron garnet or YIG ($Y_3Fe_5O_{12}$), find several applications in the communications field. As an example, small spheres of single crystal YIG are used in the Telstar satellite system. D. A. Lepore of Airtron, the major commercial suppliers of magnetic garnets, has estimated that the current market for such crystals is in the region of $250–300,000. The plant used at Airtron for their garnet production is shown in Fig. 1.2.

Marriott (1970) describes the design and performance of tunable bandpass filters, which utilize ferromagnetic resonance in YIG, and of power limiters which are based upon the nonlinear processes which occur at high microwave power levels. The same author also discusses the use of a YIG sphere to control the frequency of a Gunn oscillator; the basis of this device is that the oscillator is coupled to a resonant "cavity" provided by the YIG sphere in an applied magnetic field.

Some applications of YIG crystals are also discussed by Schmitt and Winkler (1970), who describe filters, delay lines and magneto-optic modulators. The delay lines are based on phonon propagation in YIG at microwave frequencies and can introduce a variable delay time with low insertion loss. Magneto-optic modulation has been proposed as a means of transmitting information on a laser beam, using the Faraday effect. A device for short-range communication by magneto-optic modulation of an

FIG. 1.2. Plant used for large-scale production of crystals from fluxed melts (courtesy D. A. Lepore, Airtron Division of Litton Industries Inc.).

infra-red beam has been produced. Large YIG crystals (with relatively large regions of inclusions) have been prepared from large (4–8l) crucibles by Adams and Nielsen (1966) and by Grodkiewicz *et al.* (1967). Large inclusion-free garnet crystals were produced by Tolksdorf and Welz (1972) for some of the applications mentioned, using a stirring technique developed by Scheel (1972).

The most lively interest in magnetic materials at present concerns the development of devices based on thin garnet films which exhibit cylindrical domains with the axis of magnetization normal to the plane of the film. These cylindrical or "bubble" domains are highly mobile and their use has been proposed for memory devices. Such devices are still in the development stage but the films which have been prepared by the LPE process appear at present to be of greater promise than those grown by chemical vapour deposition, hydrothermal epitaxy or by sputtering and pilot plants for LPE production of bubble films have recently been constructed.

2. *Lasers*. Mention has already been made of the observation by Nelson and Remeika (1964) of laser action in as-grown ruby. Comparative studies of the quality of ruby crystals for laser applications such as that of Bradford *et al.* (1964) show that HTS-grown crystals have important advantages over those grown by other methods. However, crystals grown by the Czochralski method are normally preferred in practice since crystals of adequate quality can be prepared relatively rapidly.

3. *Semiconductor devices*. The development of the LPE process by Nelson (1963) has ensured an important role for solution growth in semi-conductor device technology. Bulk crystals of III–V semiconductors are normally grown from the melt, and LPE is used to form $p–n$ junctions by deposition of a thin layer of doped material. Semiconductor diodes are used as lasers, microwave generators and particularly for light-emitting devices. LPE-grown materials are normally preferred because of a higher efficiency than can be obtained by alternative methods. Examples are reported by Rupprecht (1966) for GaAs lasers and by Panish *et al.* (1971) for $Al_xGa_{1-x}As$ room temperature continuous lasers, by Kang and Greene (1967) and Solomon (1968) on high resistivity—high mobility GaAs layers with excellent photoluminescent properties, and in a review by Casey and Trumbore (1970) on light-emitting diodes. The latter devices are now produced commercially by LPE and the value of the annual production is probably much higher than that of magnetic garnet crystals.

Thin film microcircuits require insulating substrates, preferably of high thermal conductivity. Comparisons of $MgAl_2O_4$ substrate crystals grown by various methods were performed by Wang and McFarlane (1968) and by Wang and Zanzucchi (1971) who found that the flux-grown material had much the lowest dislocation density. It seems clear that solution methods would be used for the production of $MgAl_2O_4$ substrates if large crystals could be grown quickly and reproducibly. The material which is potentially the most attractive for substrates is BeO, on account of its high thermal conductivity. Further work is necessary on establishing the optimum conditions for growth of this material, and it presents hazards because of its high toxicity, but the work which has been done to date has shown that suitable crystals could be grown from HTS with little further development effort.

4. *Nonlinear optical materials*. Many dielectric materials have been crystallized from high-temperature solutions and several applications have been envisaged, particularly of materials which are piezo-, pyro- or ferroelectric. Potentially the most important application appears at present

to be as nonlinear optical materials for modulation and second-harmonic generation of laser light as a means of communication with a high density of information transmission.

5. *Gemstones.* The high value placed on gems for use in jewellery has led to many attempts to synthesize and imitate natural gemstones, and the importance of early experiments in the development of high-temperature solution growth will be discussed in the next chapter. Synthetic emeralds and rubies are marketed on quite a large scale, the most important producers being Union Carbide (Linde Division) and C. F. Chatham in the U.S.A., P. Zerfass in Germany and P. Gilson in France. The wholesale value of synthetic emerald production may be in the region of $1–2 million.

Chatham and Gilson use a flux method, since the stones produced most closely resemble the natural gems in colour and in the nature of the inclusions. This is one area of solution growth where the solvent inclusions are considered beneficial! Chatham claims that his emeralds take over a year to "mature" but it is unlikely that such a period is really necessary since the process is almost certainly seeded growth from a flux (Flanigen *et al.*, 1967), probably lithium molybdate or tungstate, vanadium oxide or lithium vanadate. The high-quality emeralds grown by Linde are produced hydrothermally. Chatham also produces rubies for gemmological use and flux-grown rubies are also marketed under the name "Kashan". The large difference in price, which can be a factor 10^3 or more, between natural and synthetic gemstones naturally encourages attempts to pass synthetic stones as natural products. It is perhaps remarkable that natural emeralds and rubies have retained their high value when the differences between these and the flux-grown stones are marginal and take an expert to detect, and the gemmological trade need to be extremely vigilant in their tests to distinguish natural from synthetic stones.

Diamonds are now produced synthetically in several countries and synthetic stones account for about a fifth of the world supply of industrial diamonds. The method requires extremely high temperatures and pressures, but the graphite-diamond transition occurs as an exsolution process from a metallic solvent as will be discussed in Chapter 7. Stones weighing several carats (1 carat = 0.2 g) have now been produced (up to 1 carat of gem quality) but the sole application of synthetic diamond at present is in cutting, grinding and polishing in industry. Approximately 20 tons of graphite have been transformed to diamond by the General Electric Company alone.

6. *Electro-crystallization.* Electrolysis of fused salts is normally used for the commercial production of metals such as aluminium, and has great technological importance. The process of crystal growth from fused salts is analogous in many respects, except for the requirement of electron

transfer in deposition of the metal. Fused salt electrolysis has been used to grow crystals of oxides in reduced valence states (Kunnmann, 1971) and it has also been proposed to use this method for crystal growth of compound semiconductors (Cuomo and Gambino, 1970).

1.5. Current Problems and Future Trends

High-temperature solution growth is progressing gradually from an art to a science. At present, experimenters would find it difficult to give a convincing explanation for their choice of several of the operating conditions and other variables. Crystallization from solution is a complex process and only a few experiments are understood quantitatively.

A major problem which awaits clarification concerns the detailed nature of the processes which occur in the solution. Little information is available on the nature of the solute-solvent interaction, the association and dissociation of ions or on the formation of complexes in the high-temperature solutions. It is particularly difficult to give any confident statement regarding the processes in the vicinity of the crystal-solution interface. The questions of prearrangement or clustering of the ions or molecules in the diffusion boundary layer or at the crystal surface and of the nature of the desolvation stage are still largely open. The crystal itself is, of course, relatively easy to examine since it is comparatively stable, but the structure of the solution is a very wide field for future research.

It is fascinating to picture crystallization from solution on an atomic scale. Although this process is slow in comparison with growth from the melt, it does involve the ordered arrangement of 10 to 100 or more molecular layers on the crystal surface every second, with the rejection of an equivalent amount of solvent. One must hope that the latest techniques of physical chemistry will soon be applied to the clarification of the detailed stages involved, as discussed in Chapters 3 to 6.

The theory of crystal growth generally is at best understood only semi-quantitatively, since there remain several parameters which appear in any detailed theoretical treatment which cannot be assigned a value any more reliable than an order-of-magnitude estimate. At present most progress is hoped from computer simulation of nucleation and crystallization processes and from comparison of the results obtained with surface observations or with measured nucleation and crystal growth data. More quantitative studies are required of crystal growth under carefully controlled conditions, and measurements of such parameters as the viscosity of the solution and the diffusion coefficient of the solute are normally lacking for systems of interest in crystal growth.

Many more materials could be crystallized from solution than at present. Until now, flux growth has been used almost exclusively for oxides and the

LPE process has been mainly concentrated on III–V semiconductors and on magnetic garnets. There has been, for example, little work on carbides (except for SiC), borides, sulphides, selenides and tellurides, pnictides, intermetallic compounds and on some complex oxides such as germanates.

In only very few cases have intensive efforts been made to grow crystals weighing more than a few grams. It is clear that many more crystals could be grown to 100 or 200 g if systematic attempts were made to find a good solvent and to optimize the conditions of growth.

As industrial uses of more sophisticated materials grow, a stage could be envisaged where high-temperature solution growth will be used for industrial crystallization with batches of say 100 kg of crystals. Criteria for the design of such crystallizers are available for some systems, at least to approximately the level of confidence now used for crystallization from aqueous solutions. Production on such a scale might be necessary if solution methods are to be competitive for a wide range of materials on economic grounds.

In summary, we believe that the subject of crystal growth from high-temperature solution is still relatively unexplored. Major advances are likely in the next few years in the understanding particularly of the nature of solutions and of the interface kinetics. Other problems such as nucleation and the effects of impurities are also little understood at present. On the practical side there remain many interesting materials which have still not been crystallized and many more of which crystals are not available in good size or quality. Increasing use of technologically important materials such as yttrium iron garnet may require the development of apparatus for bulk crystallization from high-temperature solutions.

References

Adams, I. and Nielsen, J. W. (1966) *J. Appl. Phys.* **37**, 812.

Alford, W. J. and Bauer, W. H. (1967) In "Crystal Growth" (H. S. Peiser, ed.), p. 71. Pergamon, Oxford.

Ballman, A. A. and Laudise, R. A. (1963) In "The Art and Science of Growing Crystals" (J. J. Gilman, ed.), p. 231. Wiley, New York.

Belt, R. F. (1967a) *J. Am. Ceram. Soc.* **50**, 588.

Belt, R. F. (1967b) In "Advances in X-Ray Analysis" (J. B. Newkirk and G. R. Mallett, eds.), Vol. 10, 159. Plenum Press, New York.

Bertaut, E. F. and Forrat, F. (1956) *Compt. Rend.* **243**, 382.

Blattner, H., Matthias, B. and Merz, W. (1947) *Helv. Phys. Acta* **20**, 225.

Bradford, J. N., Eckardt, R. C. and Tucker, J. W. (1964) NRL Report 6080.

Brice, J. C. (1965) "The Growth of Crystals from the Melt" North-Holland, Amsterdam.

Brice J. C. (1973) "The Growth of Crystals from the Liquid" North-Holland, Amsterdam.

Casey, H. C., jr. and Trumbore, F. A. (1970) *Mat. Sci. and Engng.* **6**, 69.

Champion, J. A. and Clemence, M. A. (1967) *J. Mat. Sci.* **2**, 153.

Chase, A. B. (1971) In "Preparation and Properties of Solid State Materials" (R. A. Lefever, ed.), p. 183. Dekker, New York.

Cobb, C. M. and Wallis, E. B. (1967) Report AD 655388.

Cockayne, B., Chesswas, M. and Gasson, D. B. (1967) *J. Mat. Sci.* **2**, 7.

Cuomo, J. J. and Gambino, R. J. (1970) U.S. Patent 3.498.894.

Elwell, D. (1975) In "Crystal Growth" (B. R. Pamplin, ed.) Pergamon, Oxford.

Elwell, D. and Neate, B. W. (1971) *J. Mat. Sci.* **6**, 1499.

Flanigen, E. M., Breck, D. W., Numbach, N. R. and Taylor, A. M. (1967) *Am. Min.* **52**, 744.

Geller, S. and Gilleo, M. A. (1957) *Acta Cryst.* **10**, 239.

Grodkiewicz, W., Dearborn, E. F. and Van Uitert, L. G. (1967) In "Crystal Growth" (H. S. Peiser, ed.), p. 441. Pergamon, Oxford.

Henisch, H. K. (1970) "Crystal Growth in Gels" Pennsylvania State University Press.

Janowski, K. R., Chase, A. B. and Stofel, E. J. (1965) *Trans. AIME* **233**, 2087.

Kaldis, E. (1974) In "Crystal Growth, Theory and Techniques" (C. H. L. Goodman, ed.) Plenum Press, New York.

Kang, C. S. and Greene, P. E. (1967) *Appl. Phys. Lett.* **11**, 171.

Kunnmann, W. (1971) In "Preparation and Properties of Solid State Materials" (R. A. Lefever, ed.), p.1. Dekker, New York.

Laudise, R. A. (1963) In "The Art and Science of Growing Crystals" (J. J. Gilman, ed.), p. 252. Wiley, New York.

Laudise, R. A. (1970) "The Growth of Single Crystals" Prentice-Hall, Englewood Cliffs, New York.

Laurent, Y. (1969) *Rev. Chim. Min.* **6**, 1145.

Linares, R. C. (1965) *J. Phys. Chem. Solids* **26**, 1817.

Luzhnaya, N. P. (1968) *J. Crystal Growth* **3/4**, 97.

Marriott, S. P. A. (1970) *Marconi Rev.* 1st Quarter, 79.

Nelson, H. (1963) *RCA Review* **24**, 603.

Nelson, D. F. and Remeika, J. P. (1964) *J. Appl. Phys.* **35**, 522.

Nielsen, J. W. (1972) Paper at IVth All-Union Conf. on Crystal Growth, Tsakhkadzor USSR (to be published in "Growth of Crystals").

Nielsen, J. W. and Dearborn, E. F. (1958) *J. Phys. Chem. Solids* **5**, 202.

Nitsche, R. (1967) *Fortschr. d. Mineral.* **44**, 231.

Panish, M. B., Sumski, S. and Hayashi, I. (1971) *Metall. Transact.* **2**, 795.

Promisel, N. E. (1971) *J. Electrochem. Soc.* **118**, 163C.

Rau, H. and Rabenau, A. (1968) *J. Crystal Growth* **3/4**, 417.

Remeika, J. P. (1954) *J. Amer. Chem Soc.* **76**, 940.

Roy, R. and White, W. B. (1968) *J. Crystal Growth* **3/4**, 33.

Rupprecht, H. (1966) Symposium GaAs, Reading, p. 57.

Sahagian, C. S. and Schieber, M. (1969) In "Growth of Crystals" (N. N. Sheftal, ed.) Consultants Bureau, New York, **7**, 183.

Schaffer, P. S. (1964) Vapor Phase Growth of Single Crystals, U.S. Air Force Contract No. AF 19(628)—2383, Final Report, Lexington Laboratories Inc. (Jan. 1964).

Schäfer, H. (1962) "Chemische Transportreaktionen" Verlag Chemie, Weinheim/ Bergstr.

Schäfer, H. (1964) "Chemical Transport Reactions" Academic Press, New York (new edition planned).

Scheel, H. J. (1972) *J. Crystal Growth* **13/14,** 560.

Schmid, F. and Viechnicki, D. (1970) *J. Am Ceram. Soc.* **53,** 528.

Schmitt, H. J. and Winkler, G. (1970) *Elecktro-Anzeiger* **5** and **9.**

Schroeder, J. B. and Linares, R. C. (1966) *Progr. Ceram. Sci.* **4,** 195.

Solomon, R. (1968), Proc. Int. Symp. GaAs (C. I. Pederson, ed.), p. 11. Phys. Soc. Conf. Ser. No. 7, Inst of Phys., London.

Stephens, D. L. and Alford, W. J. (1964) *J. Am. Ceram. Soc.* **47,** 81.

Stockbarger, D. C. (1936) *Rev. Sci. Instr.* **7,** 133.

Stockbarger, D. C. (1949) *Disc. Faraday Soc.* **5,** 294, 299.

Tolksdorf, W. and Welz, F. (1972) *J. Crystal Growth* **13/14,** 566.

Wang, C. C. and McFarlane, S. H. (1968) *J. Crystal Growth* **3/4,** 485.

Wang, C. C. and Zanzucchi, P. J. (1971) *J. Electrochem. Soc.* **118,** 586.

Wanklyn, B. M. (1975) In "Crystal Growth" (B. R. Pamplin, ed.) Pergamon, Oxford.

White, E. A. D. (1965) *Techn. Inorg. Chem.* **4,** 31.

White, E. A. D. and Brightwell, J. W. (1965) *Chem. and Ind.* 1662.

2. History of Crystal Growth from Solutions

2.1. Development of Ideas on Crystals and Crystallization

Gems and crystals have always attracted mankind, and the belief in the virtues of gems and some minerals dates back at least two thousand years. The use of gems for jewellery originates from even earlier times, but in the early stages the colour and transparency were the main attractive factors. Beads were clearly considered of value by Bronze-age chieftains and un-facetted gems of many colours were used in the crowns of emperors and kings. The facetting of gems began only in the 15th century and cutting, cleaving and polishing techniques were subsequently developed in order to enhance the "fire" of gems, that is the inherent optical effects which arise from their high refractive index and dispersion.

Crystals as such were not recognized until comparatively recently although their forms were frequently admired and described. Accordingly, there was no general word for crystals, only specific names for gems and other minerals and for certain salts. According to Marx (1825), Homer (8th century B.C.) in his works "Ilias" and "Odysseus" was the first to use the word "crystallos" in the sense of "ice", and only since the time of Plato (427–347 B.C.) was "crystallos" also used for "rock crystal". The transparency of quartz crystals led Diodorus Siculus (around 30 B.C.) to the idea that they are formed from pure water which solidifies through the power of a godlike fire. Seneca (about A.D. 0–65) and Pliny (A.D. 23–79) believed that rock crystals were formed by "condensation" of water during a period of cold lasting up to several centuries. This belief survived seventeen centuries and was mentioned in an encyclopedia by Gleditsch in 1741. Boyle (1672) and Hottinger (1698) were among the first to prove that rock crystals were not formed from ice.

"Crystal" was for several hundred years identified with quartz = rock crystal (crystallus montius), but the dissertations of Hottinger (1698) on

"Krystallologia" and of Capeller (1723) on "Prodromus crystallographia" initiated the generalization of the word crystal. Bartholinus (1669) had previously used "crystal" for other minerals (crystallus islandicus = calcite) and Hagendorn in 1671 had used the term for organic crystals like "crystalli benzoes". "Quartz" (first mentioned by Agricola, 1530) became the scientific name for rock crystal, see Tomkeieff (1942), but it took about 200 years before the general term "crystal" gained wide acceptance, and several authors of the 17th and 18th centuries still used expressions like "corpora angula" and "figured stones", while "icicles" was used to describe needle-like crystals.

"Crystallisatio" was used for crystallization from the 17th century, whereas in earlier times expressions like condensation or coagulation were used for the process of crystal formation. It was difficult for the scientists of that time to understand how hard crystals could be formed from a clean and "soft" liquid. The concept of solutions was also not developed although Anselmus Boetius de Boodt (1609) distinguished between the evaporation and slow cooling techniques for crystallization from solution.

According to Schoen and colleagues (1956) mankind has obtained salt from natural deposits as well as by artificial preparation since the earliest recorded history, and Caldwell (1935) reports that artificial evaporation and crystallization of salt is shown on a Chinese print of 2700 B.C. Crystallization of salt was also described on the Egyptian "Papyrus Ebers" of about 1500 B.C. and by Aristoteles (384–322 B.C.). Other early crystallization processes are the preparation around 300 B.C. of sugar from sugarcane syrup in India mentioned by Ray (1956) and the crystallization of cupric sulphate (blue vitriol) and of a few other salts described by Pliny in his work "Naturalis Historia". Similar crystallizations were also reported by Arabian alchemists of the 9th to the 11th century. Birringuccio (1540) described in detail the preparation and crystallization of saltpeter, and Agricola (1546) described the crystallization of salt and stressed the importance of this material for mankind. The knowledge of his time was set down by Libavius in his textbook of alchemy (1597).

The alchemist Geber (whose works according to Darmstaedter (1922) were written in the 12th or 13th century, possibly in South Italy or Spain and who is frequently confused with the Arabian alchemist Dschabir Ibn Haijan of the 9th century) mentioned the purification of several salts by recrystallization, and also described sublimation and distillation. Geber believed in the transformation or "ennobling" of metals and since he applied many high-temperature processes it is probable that he occasionally grew crystals from high-temperature solutions. In his work "Summa Perfectionis Magisterii" he discussed the various opinions of that time on alchemical processes and concluded that special conditions such as a

B

"medium heat" are required to prepare (crystallize) metals and that the quality of the products depends also on the arrangement of the stars. The belief in the influence of the stars goes back to the Babylonians and was also mentioned by Plato. Furthermore, Geber argues that it must be easier to prepare the minerals and metals despite their dense structure than to imitate animals like oxen and goats because the latter have souls and somehow a higher degree of perfection!

This argument shows clearly that there was a very vague conception of the structure of crystals and thus of the crystallization process until the 17th century. Although Democritos (about 470–400 B.C.) had already expressed his ideas on indivisible particles of matter, the atoms, it was not until the year 1611 that Kepler concluded from the regular hexagonal form of snow crystals (De nive sexangula) that they are built up by a regular arrangement of small spherical particles. Thus Kepler was probably the first to describe the principle of order in crystals, caused by attractive forces or by pressure from outside. His ideas were extended by Robert Hooke who wrote in his work "Micrographia" in the year 1665:

I think, had I time and opportunity, I could make probable, that all these regular Figures . . . arise only from three or four several positions or postures of *globular* particles, and those the most plain, obvious, and necessary conjunctions of such figur'd particles that are possible. . . . And this I have *ad oculum* demonstrated with a company of bullets, and some few other very simple bodies; so that there was not any regular Figure, which I have hitherto met withall, . . . that I could not with the composition of bullets or globules, and one or two other bodies, imitate, even almost by shaking them together.

Huygens in 1960 postulated flattened spheroid particles of calcite (Niggli, 1946) and Newton (1730) believed that crystals consist of small indivisible particles of "several sizes and figures" (Burke, 1966) which might even be irregular.

In contrast to this regular packing of spheres or similar molecular forms, another point of view on crystalline arrangement was that of Bartholinus (1669), Bergman (1773) and Haüy (1784), who postulated that crystals are built of small geometrical units, cubes in the case of salt and rhombohedra in the case of calcite. This geometrical view of the structure originated not only from corresponding forms of certain as-grown crystals but also from the cleavage characteristics of many crystals. According to Smith (1960), Grignon in 1775 was the first to recognize that metals consist of small crystallites and his dendrite drawing is often reproduced.

The rule of the constancy of angles between crystal faces introduced by Steno (1669) and Bartholinus (1669) became fundamental for the correct recognition of crystal habit (as discussed further in Chapter 5) and thus for the development of the crystallographic and mineralogical sciences, although it was Guglielmini (1688) who exactly defined the law of constant angles and the relation of the habit to the chemical species. The latter topic was amplified by Cappeller (1723), Leeuwenhoek (1685, 1705) and by Romé de l'Isle (1783). Linnaeus (1768) and especially Werner (1774) contributed much to the classification of minerals by careful observation and description of their external characteristics, not only of their forms but also of their other properties such as colour, manner of cohesion, coldness to the touch, weight, smell and taste, and according to Werner blowpipe tests could be useful for qualitative chemical analysis.

Corresponding to the vague opinions on crystals it is no wonder that a variety of strange ideas on the mechanism of crystallization was held by alchemists (Adams, 1954). The vegetative growth of crystals was especially popular among many chemists and was believed by Paracelse (1493–1541) and still in the 18th century by the French botanist Tournefort in 1702 and by Robinet in 1761. The latter deduced his hypothesis from the concept that nature is uniform in the treatment of all her productions, and consequently there had to be a uniform process for the generation of animal, vegetable and mineral species (see Burke, 1966). Thus it was a major step for Steno (1669) and for Hottinger (1698) to recognize that a crystal does not grow from the interior like a plant or animal (intussusception) but grows by deposition of material on the external faces (juxtaposition). Another question, in connection with the formation of minerals, was whether or not water played a role, and this quarrel between the *plutonists* and the *neptunists* was not settled until the end of the last century.

In the early stages of chemistry, mineralogy and crystallography these sciences were descriptive, and as late as 1822 Mohs expressed the idea that "natural history observes natural products as they are, not how they were formed" (after Michel, 1926).

On the other hand Boyle (1672) believed that gems and minerals were formed from liquids impregnated with various substances by processes analogous to those occurring in his experiments on crystal growth from aqueous solution. Boyle appears to have been the first scientist to apply the techniques of experimental chemistry to the solution of mineralogical problems. In his work "An Essay About the Origine and Virtues of Gems" he proposed the hypothesis that many gems were once fluid bodies or are in part made up of fluid substances. In particular he claimed that gems were formed from solutions and presented the following five "proofs".

1. The transparency of gems: "it is unlikely that bodies that were never fluid should have the arrangement of their constituent parts that is requisite to transparency, which permits easy passage of light through them".

2. The external "figuration" (crystal habit) of gems, because "corpuscles of various substances will coagulate in liquids and yield crystals similar to those of gems".
3. The internal textures of gems which resemble those he observed in crystals that were once fluid (e.g. common salt and silver). The interior grain of certain gems and other minerals is visible, and muscovite and diamond, for example, are easily cleaved along the visible plates (cleavages) and along the grain.
4. The fourth "proof", based on the variety of colours in gems, is not convincing, in contrast to the next one:
5. Solid gems may contain heterogeneous matter, and Boyle describes a crystal "in the midst of which there was a drop of water", as evidence that gems grow from solutions.

It is interesting to note that Boyle did not believe in the "architectonick" explanation of crystal habits, although he observed that a solution of "stony stiriae" resulted in a "coagulated mass" when the solvent was quickly evaporated, whereas slow evaporation yielded "distinct crystals, long, transparent, and curiously shaped" (similarly Rouelle described in 1745 the crystallization of salt in truncated pyramids or squares during fast evaporation of the water, whereas slow evaporation resulted in cubes: precursors of experiments on morphological stability!).

The mineral- and gem-forming solutions had, according to Boyle, some special virtues. Although he never saw any great feats performed by those hard and costly stones (such as diamonds, rubies, sapphires), he believed in the testimonies of physicians and patients and tried to explain the medicinal virtues of gems in terms of "effluvia" or "steams", originating from earth fluids and present in the gems. To the objections of other critical scientists that gems will not "part with effluvia or portions of themselves, since they lose not of their weight" Boyle replied that mercury kept in water for a day or two would kill worms without any sensible loss of weight of the mercury. Therefore "effluvia may be of so small specific gravity, as not to make the gem at all heavier in specie than crystal itself". In Boyles' view, before gems solidify the "petrescent" substance unites with a solution, exhalation, or metallic substance, and upon crystallization the gems retain the "virtues" that were added. The medicinal powers are liberated from gems as minute corpuscular "effluvia" in rapid motion. If scientists like Boyle believed in the virtues of gems, it is not surprising that many recipes against illnesses like cholera or tuberculosis contained powdered quartz and other minerals or gems, in addition to saltpeter, red corals, prepared crab eyes and celestial "theriacs" (Wallbergen, 1760). Hottinger (1698) also shared this belief in the virtues of gems and gave several references on this topic. Even nowadays the belief in some "higher forces" of natural gems is not uncommon (see for instance Adams, 1954), and the value of natural gems in the jewellery industry has continued to rise although synthetic gems like hydrothermal ruby or emerald are available in superior quality.

It is not the purpose here to outline in detail the history of crystallography, mineralogy and crystal growth because there are several excellent works on the former two topics, such as that of Burke (1966), Niggli (1946), Lenz (1861), Kobell (1864), Marx (1825), Sohncke (1879) and Agricola (1546).

2.2. Early Crystallizations from High-temperature Solution

The discussion in Section 2.1 of the historical development of the concepts of crystals and crystallization is included to give a background to early experiments on crystal growth from high-temperature solution. Mention must also be made of flux growth in nature which started during the solidification of the earth's crust. Crystals are constantly growing in the crust, up to 30 km or so below the surface. Water and other volatile compounds are present in the "flux", which is called magma, but they merely act as flux modifiers, not as hydrothermal systems. (For the oldest granitic rocks, which are early precambrian amphibolite facies gneisses from West Greenland, a rubidium-strontium age of 3980 ± 170 million years was reported by Black et al., 1971.)

Nature has not only produced the very high quality crystals used as gems but occasionally single crystals of enormous size, which have grown mainly in pegmatites or hydrothermally in rock fissures. According to Bauer (1932), an aquamarine crystal of good quality weighing 110.5 kg was found in 1910 in a pegmatite quarry in Minas Geraes in Brazil. Topaz crystals up to 135 kg were also found in Brazil, and the largest high-quality topaz "Braganza" of 1680 carats was believed to be diamond and used in the Portuguese crown. Other large gemstones are a star sapphire of 563 carats (Star of India), a ruby of 100 carat (De Longstar ruby), an emerald made into a container for ointments weighing 2680 carats (now in Vienna), and, according to Kunz (1890), an almandine garnet crystal of 4.4 kg was found in 1885 in 35th Street, near Broadway in New York!

Artificial crystallization from high-temperature solutions probably dates from the time when man could produce high temperatures, by the use of fire. Incidental flux growth probably occurred during the preparation of metals and alloys, and of ceramics and glasses, since these crafts were practised by several ancient civilizations. Crystalline products were probably grown from high-temperature solutions during the middle ages when alchemists attempted the synthesis of the "elixir of life" and the "philosopher's stone" and the "transmutation" of "base" metals into gold, and although little or no attention was given at that time to the resulting crystals.

The alchemists tried to keep for themselves their secrets on "universal processes" by using special symbols as are shown in Fig. 2.1. The process described was discovered by Magister Heinrich Eschenreuter in a hiding place in the monastery Schwartzbach on May 6, 1403, and was hidden again for religious reasons in a hole in the wall of the monastery of St. Marienzell in Thuringia on October 10, 1489. There it was rediscovered by the Benedictine monk Basilius Valentini in the year 1762. This monk, who edited several books on alchemy and the philosopher's stone, published the recipe given in Fig. 2.1 in 1769, during the time of enlightenment, when the Church began to tolerate science. The English translation runs approximately as follows.

> *Take in the name of God Christ Jesus the son of Mary "minera mercurii" and prepare from this a "blood-red extract" with "spiritus vini", "spiritus nitri" and "spiritus salis", and from this "blood-red extract" make a "vinegar"† with "spiritus salis ammoniaci" and "spiritus tartari" with a strong fire. This "vinegar" add again to the residue "caput mortuum" and treat it until it is transformed into a "deep-red oil or sulphur of the materia philosophorum": Take this "deep-red oil" and put into a container and seal this with Lutum (loam), and let the Lutum dry completely, and when dry put it into a "melt furnace" and apply "moderate fire", primus gradus ignis (of the first grade of heat), and let it stand for forty days that it does not "melt". After these forty days let it stand in "melting" again for forty days so that it will become and look quite black by effervescing, and let it stand in the "melting" until it becomes white, as crystals which look milky, and when it has stood as long as before then you will see that it looks like a glass, and will appear quite dark red transparent, and keep it so long in the "liquid as a water" until there is no more change, then you have prepared the philosophers' tinctura.*

More attention was paid to synthetic crystals only from the late 18th century on, when their character and their importance were recognized and when chemical analyses could be performed. Realistic speculations regarding the formation of natural crystals could only then be expressed. Although already at the time of Agricola, in the 16th century, over 100 different salts had been synthesized and crystallized from aqueous solutions, it was not until the early 19th century that the first crystals were deliberately grown from high-temperature solutions.

Around 1800 the crystallization behaviour of lavas from the region of Edinburgh and from Etna and Vesuvius was studied by Hall (1798) and by Watt (1804) who used melts up to 330 kg. Watt found crystals up to 1 mm size on slow cooling (during eight days) and obtained glasses when

---

† probably "essence" is meant.

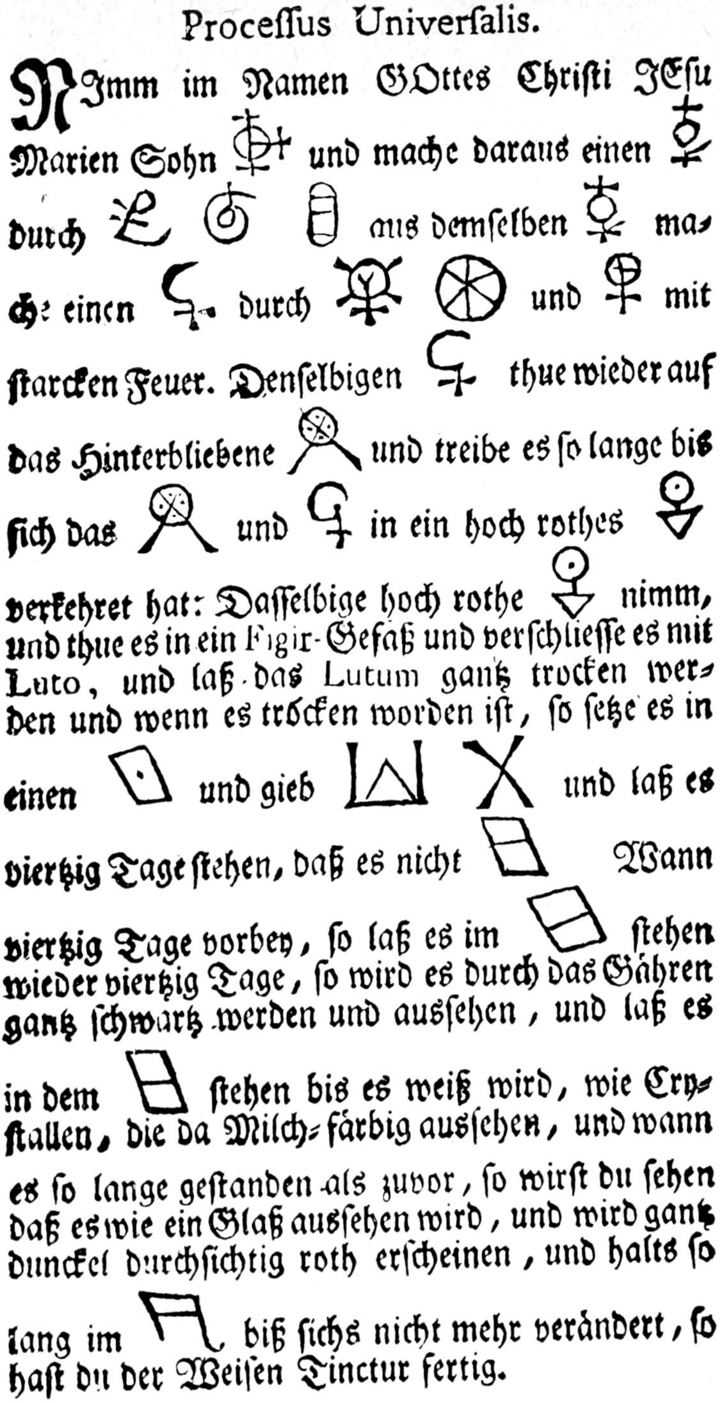

Proceſſus Univerſalis.

Nimm im Namen GOttes Chriſti JEſu Marien Sohn ⚥☿ und mache daraus einen ☿ durch ♃ ☉ ☋ aus demſelben ☿ ma-che einen ♄ durch ♅ ⊕ und ♀ mit ſtarcken Feuer. Denſelbigen ♄ thue wieder auf das Hinterbliebene ⚼ und treibe es ſo lange biß ſich das ⚼ und ♄ in ein hoch rothes ☿ verkehret hat: Daſſelbige hoch rothe ☿ nimm, und thue es in ein Figir-Gefäß und verſchlieſſe es mit Luto, und laß das Lutum gantz trocken wer-den und wenn es trocken worden iſt, ſo ſetze es in einen ◇ und gieb △ ✕ und laß es vierzig Tage ſtehen, daß es nicht ▱ Wann vierzig Tage vorbey, ſo laß es im ▱ ſtehen wieder vierzig Tage, ſo wird es durch das Gähren gantz ſchwartz werden und ausſehen, und laß es in dem ▱ ſtehen bis es weiß wird, wie Cry-ſtallen, die da Milch-färbig ausſehen, und wann es ſo lange geſtanden als zuvor, ſo wirſt du ſehen daß es wie ein Glaß ausſehen wird, und wird gantz dunckel durchſichtig roth erſcheinen, und halts ſo lang im ⊓ biß ſichs nicht mehr verändert, ſo haſt du der Weiſen Tinctur fertig.

Fig. 2.1. Alchemical recipe for preparation of a universal medicine, discovered by Eschenreuter in 1403 and published by Valentini (1769).

the melts were quenched. The crystals were magnetic and had relative densities ranging from 2.743 to 2.949—a good characterization for that time. Among the first syntheses from high-temperature solutions, Wöhler (1823) reported the preparation of tungsten bronze crystals from sodium tungstate flux, and other mineral syntheses (although not by typical flux techniques) were reported by Berthier (1823) and by Mitscherlich (1823). According to Haüy in 1822 apatite was prepared by Saussure from gypsum and phosphoric acid.

The synthesis by Gaudin (1837) of ruby rhombohedra of up to 0.187 g weight by melting potassium alum with potassium chromate attracted many mineralogists and led Böttger (1839) and Elsner (1839) to repeat the experiment. From this time the activity and interest in mineral synthesis began to increase, especially in France, but also in Germany.

Supersaturation in high-temperature solutions was achieved in these early experiments not only by the slow-cooling and evaporation techniques,

TABLE 2.1. Early Reports of Crystals Grown by the Flux-Reaction Technique

Solvent	Reactant	Result	Reference
Na_2WO_4	H_2 (vapour)	Na_xWO_3	Wöhler (1823)
$CdCl_2$	H_2S (vapour)	CdS	Durocher (1851a)
$K_2Cr_2O_7$	SiO_2 (from crucible)	Cr_2O_3	Svanberg (1854)
Al	$SiCl_4$ (vapour)	Si	Sénarmont (1856)
AlF_3	B_2O_3 (vapour)	Al_2O_3	Deville and Caron (1858)
$NaCl + NaAlO_2$	HCl (vapour)	Al_2O_3	Debray (1861a)
$AlPO_4 + CaCO_3$, formation of $Ca_3(PO_4)_2 + CaCl_2$ during the reaction	HCl (vapour)	Al_2O_3	Debray (1861a)
$MgCl_2$	H_2O (vapour)	MgO	Deville (1861b)
$Ba(NO_3)_2$	NO_2 (decomposition)	BaO	Brügelmann (1877, 1878)
$Ca(NO_3)_2$	NO_2 (decomposition)	CaO	
$Sr(NO_3)_2$	NO_2 (decomposition)	SrO	
$NiCl_2$	H_2O (vapour)	NiO	Ferrières and Dupont, see Bourgeois (1884)
$CoCl_2$	H_2O (vapour)	CoO	Bourgeois (1884)
$KF + ZnF_2$	H_2O (vapour)	ZnO	Gorgeu (1887c)
Na_2SO_4 or $K_2SO_4(+ ZnSO_4)$	H_2O (vapour)	ZnO	Gorgeu (1887c)
$Na_2SO_4 + ZnSO_4$	SiF_4 (vapour)	Zn_2SiO_4	Gorgeu (1887c)
$NaCl + ZnCl_2 + SiO_2$	H_2O (vapour)	Zn_2SiO_4	Gorgeu (1887c)
$Na_2SO_4(+ ZnSO_4 + FeSO_4)$	SO_3 (decomposition)	$ZnFe_2O_4$	Gorgeu (1887d)
BaF_2, CaF_2 or AlF_3	H_2O (through porous ceramic crucible)	Al_2O_3	Frémy and Verneuil (1888, 1890)
$FeCl_2$	H_2 (vapour)	Fe	Osmond and Cartaud (1900)
$SnS(+ CdCl_2)$	$CdCl_2$ (vapour)	CdS	Viard (1903)
$SnS(+ ZnCl_2)$	$ZnCl_2$ (vapour)	ZnS	Viard (1903)

but also by flux-reaction techniques (see Chapter 7). In fact, in one of the first flux-growth experiments Wöhler (1823) passed hydrogen over sodium tungstate melts and obtained crystals of tungsten bronzes. In this case the supersaturation was achieved and maintained by the reduction of Na_2WO_4 to Na_xWO_3 by hydrogen. Further examples of the flux-reaction technique are given in Table 2.1 and further discussion is superfluous except to mention that in most cases constituents of the crystal are transported via the vapour phase to the liquid from which the crystals grow as a result of some reaction. Thus these examples belong to the VLRS technique as discussed in Chapter 7.

Reference to the 19th-century literature also reveals a number of reports of the use of mineralizers. It is interesting that Aristoteles is said to have claimed that solid materials react only in the presence of a liquid. We know today that this statement is only partially true (solid-state reactions), but it was established in ancient technology that the addition of small amounts of fluxes greatly accelerates ceramic, glass-forming and metallurgical processes. The expression "agent minéralisateur" or "mineralizer" was introduced by Beaumont (1849) and by Deville, and, according to Moroze-wicz (1899), it had a significant influence on the development of mineral synthesis. Vapour and liquids were termed mineralizers when they had an appreciable effect on lowering the melting point, although Friedel (1880) accepted only gaseous mineralizers such as SiF_4, HCl and HF. Other scientists defined mineralizers as catalysts for crystallization. This view is still shared by several mineralogists today, although in 1899 Morozewicz wrote that mineralizers are nothing else but solvents and that one should abandon the term "mineralizer" in chemical and physical sciences.

When growth is unstable, solvent becomes included in the crystals, and these inclusions play a major role in the identification of natural crystals (see Gübelin, 1953) as well as in the deduction of the growth history of minerals. Material can be included only during growth, either as a liquid initially (from the solution) or as solid material which is captured by the growing crystal; an exception would be the phenomenon of eutectic crystallization but this is rarely found in natural stones. According to Gübelin, the importance of inclusions as a means of investigating the genesis and growth of minerals was emphasised by Sir David Brewster in the early 19th century.

Resistance heating and the use of electric power became popular only in the 20th century, and prior to this many high-temperature experiments were conducted in glass-making or pottery furnaces. For example, Ebelmen made his many syntheses of crystals in the famous porcelain factory at Sèvres in France, Fremy collaborated with the glass factory at St. Gobain and Morozewicz made his experimental investigations on crystallization of

B2

minerals from (synthetic) magmas in a Siemens furnace in the glass factory at Targowerk near Warsaw. A diagram of the furnace used by Morozewicz is shown in Fig. 2.2.

For large melts like the 330 kg mentioned above and the 50 kg melts of Morozewicz, large ceramic crucibles were used as a complement to the small platinum crucibles. Early experiments on flux growth were reviewed

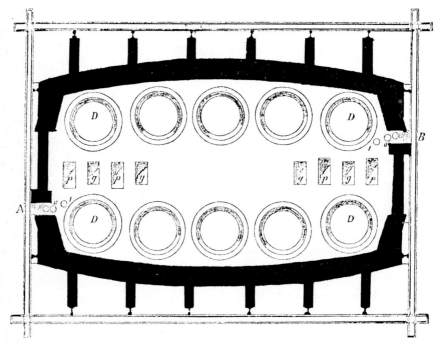

FIG. 2.2. Arrangement of the crucibles containing up to 50 kg melts in a Siemens furnace as used by Morozoewicz (1899), with large ceramic crucibles at D, air and gas channels at p and g, and with exhaust gas channels at A and B.

by Fremy and Verneuil (1888, 1890) and in the book of Fremy (1891). These authors used crucibles up to 50 litre in size and made from porous ceramics in order to facilitate the intrusion of humid air into the fluoride solutions used for crystallization of oxides, mainly of ruby. Large graphite crucibles, of 15 cm height and more than 10 cm diameter, were also used by Hampe for the preparation of aluminium borides, and 1000 kg of coke was required to heat the furnace for six days.

Although such large crucibles were used, and Goethe had already recognized that larger crystals could be grown from larger crucibles, the mineralogists and chemists of the past century did not generally obtain crystals larger than a few millimetres in diameter. This is due to various factors which are described later, and in particular to the use of impure chemicals, poor temperature control and programming, lack of knowledge of solubility and supersolubility curves and of the stable growth rates, and to uncontrolled evaporation of solvent. As described by Nassau and Nassau (1971) one experiment of Fremy and co-workers yielded up to 24,000 ruby crystals weighing a total of 1200 g from a 12 litre batch. It is not surprising that those mineralogists who were interested in the synthesis of large gem crystals tried other techniques.

Several attempts were made in the second half of the 19th century to crystallize ruby directly in a flame before Verneuil discovered the conditions of crucible-free flame-fusion growth which allowed reproducible growth of large crystals. The Verneuil technique became popular for growth of ruby for watch bearings and gemstones from 1904 onwards, and nowadays about 200 tons are produced per year. The success of the flame-fusion technique led to a decrease in the activity in flux growth during the following forty years and only occasionally was the flux technique used for the preparation of materials. Examples of its use are the crystallization of emerald at IG Farben in the thirties and the synthesis of mica and substitutes for quartz for military applications during World War II.

Renewed interest in crystal growth from high-temperature solutions began only after the second world war, when the importance of materials research and solid-state physics became obvious. The electronic and the computer industries could not have developed without the discovery and production of transistors for which high-quality crystals were required in large quantities. Flux growth was particularly stimulated by the fabrication of solid-state lasers and by the synthesis and crystallization of ferroelectric barium titanate (see Chapter 1) and of the rare-earth garnets. These recent developments are described elsewhere in this book.

2.3. Comments on the Table of Early References

In view of the large volume of early work on high-temperature solution growth a detailed description of the findings of the various scientists active in this field would be prohibitively long although it would be interesting from a historical viewpoint. A summary of this work is therefore given in the form of a Table. This Table 2.2 is probably incomplete and complex solid solutions or unidentified phases have in general been omitted. For example, the many silicates and solid solutions described in Doelter's extensive work on silicate melts (1904–1908) have not been included, nor have the results described by Vogt (1892–1904) and by Morozewicz (1899). In addition, amalgams which were prepared by Geber around A.D. 1200 (see Darmstaedter, 1922–1969) and studied extensively in the past centuries, have also been omitted together with the metallic crystals which were prepared by cooling their solutions in mercury (Puschin, 1903). The numerous crystalline phases observed during the investigations of phase diagrams (see Chapter 3) have also been omitted, although mention must be made of the enormous value in crystal growth of the careful phase diagram investigations at the US National Bureau of Standards, the US Geological Survey and elsewhere, from the beginning of the 20th century on.

Some errors may be present in the Table since the contemporary techniques for crystal identification were poorly developed. Characterization relied mainly on morphology, particularly by optical measurement of the angles between faces and hence deduction by stereographic projection of the ratio of the axes and the angles between them in the elementary cell. Additional observations of the optical indicatrix of cleavage, hardness, specific gravity and colour were sometimes reported in addition to chemical analysis of the major constituents.

The data of Table 2.2 were collected from the works of Doelter (1912–1931), Groth (1906), Fouqué and Lévy (1882) and Chirvinskii (1903–1906), which are abbreviated as D (with the volume number), G, FL and C, respectively. Comparison has been made with the reviews of Favre (1856), Gurlt (1857), Fuchs (1872), Daubrée (1879), Friedel (1880), Bourgeois (1884), Meunier (1891a), Michel (1914, 1926) and Grossmann and Neuburger (1918), in addition to reference to the original literature. The Table should be useful in demonstrating the large number of crystals prepared and of solvents used, especially since the early literature is often rather inaccessible.

The first column of Table 2.2 contains the chemical formula of the material crystallized and the second the mineral name, where appropriate. Column three lists the solvent, column four the reference, and column five the source using the abbreviations given above and the volume and page numbers.

TABLE 2.2.

Formula	Name	Solvent	Reference	Source
Ag	Silver	AgI (electrolysis)	Groth (1906)	G 5
AgCl	Chlorargyrite, Cerargyrite	AgI	Lehmann (1877)	G 200
Ag$_2$S		AgCl, SbCl$_3$	Doelter (1886a)	DIV/1, 247
		Bi	Rössler (1895, 1898)	G 145
Ag$_2$Se	Naumannite	Bi	Rössler (1895, 1898)	G 145
Ag$_3$AsS$_3$	Proustite	S, As$_2$S$_3$	Margottet (1877), see also Friedel (1880)	DIV/1, 253
Ag$_3$SbS$_3$	Pyrargyrite	S	Margottet (1877)	DIV/1, 247
AlB$_2$		B$_2$O$_3$	Wöhler and Deville (1857)	
		B$_2$O$_3$, Na$_3$AlF$_6$	Hampe (1876)	G 50
AlB$_{12}$		B$_2$O$_3$	Wöhler and Deville (1857)	
		B$_2$O$_3$	Joly (1883)	G 50
Al$_3$B$_{48}$C$_2$		B$_2$O$_3$	Hampe (1876)	
AlBO$_3$	Jeremejewite	CdO + B$_2$O$_3$	Ebelmen (1848b, 1888)	DIII/2, 425,
			Mallard (1888a)	DIII/2, G 101
Al$_2$BeO$_4$	Chrysoberyl	see BeAl$_2$O$_4$	Meunier (1890)	
Al$_2$Be$_3$Si$_6$O$_{18}$	Beryl, emerald	see Be$_3$Al$_2$Si$_6$O$_{18}$	Gaudin (1837, 1869)	
Al$_2$SiO$_5$	Sillimanite	K$_2$CO$_3$ + AlF$_3$	Böttger (1839)	
Al$_2$O$_3$	Corundum	K$_2$S, K$_2$SO$_4$	Ebelmen (1848b, 1851a, b)	DIII/2, 446
Al$_2$O$_3$:Cr	Ruby	Na$_2$B$_4$O$_7$, BaCO$_3$ or Na$_2$CO$_3$	Debray (1861a, b)	DIII/2, G 101
		K$_2$SO$_3$, Ca-phosphate	Elsner (1839)	DIII/2, 446
		K$_2$SO$_3$, Ca-phosphate	Parmentier (1882)	DIII/2, 446
		K$_2$Cr$_2$O$_7$, K$_2$MoO$_4$	Fremy and Feil (1877),	DIII/2, 446
		PbO	Fremy (1891)	DIII/2, G 102

Table 2.2 cont.

Formula	Name	Solvent	Reference	Source
		BaF_2, CaF_2, AlF_3	Fremy and Verneuil (1888, 1890)	DIII/2, 446
		AlF_3	Deville and Caron (1858)	DIII/2, 446
		Na_3AlF_6	Meunier (1887/1888)	DIII/2, 446
		Na_3AlF_6	Gmelin-Kraut II (1911) 617	DIII/2, 446
		$Na_2B_4O_7$	Hönigschmidt (1907, 1910)	DIII/2, 446
		Silicate melts	Morozewicz (1895, 1899)	DIII/2, 446
		$Na_2S(+NaAlSiO_4)$	Hautefeuille and Perrey (1890a)	G 101
As_2S_2	Realgar	S, $NaHCO_3$	Sénarmont (1851a)	G 154
Au	Gold	Hg	Knaffl (1863)	G 6
			Lang (1863)	G 6
B_6C		B_2O_3	Joly (1883)	G 50
$BaSO_4$	Barite	KCl, NaCl, $BaCl_2$	Gorgeu (1887a)	DIV/2, 245
		KCl	Manross (1852)	DIV/2, 245
		$MnCl_2$	Gorgeu (1885)	DIV/2, 196
$BaSeO_4$		$NaNO_3 + KNO_3$	Diacon (1888)	C 90
BaO		$Ba(NO_3)_2$	Brügelmann (1877, 1878)	G 74
Be		$BeCl_2$	Nilson and Petterson (1883)	G 8
		$BeCl_2$	Brögger and Flink (1884)	G 8
		$KBeF_3$ (electrolysis)	Lebeau (1899)	G 8
$BeAl_2O_4$	Chrysoberyl	$Ca_2B_2O_7$, B_2O_3	Ebelmen (1845, 1848a, 1851b, d)	DIII/2, 513 ; Gurlt (1857)
		BeF_2, AlF_3, B_2O_3	Deville and Caron (1858)	DIII/2, 513
		Na_3AlF_6	Lacroix (1887)	DIII/2, 513
		Na_2S, K_2S	Hautefeuille and Perrey (1888a, 1890a)	DIII/2, 513
$Be_3Al_2Si_6O_{18}$	Beryl, emerald	B_2O_3	Ebelmen (1848b)	DII/2, 598

Formula	Mineral	Solvent	Author	Reference
		$Li_2Mo_2O_7$	Hautefeuille and Perrey (1888b, 1890b)	DII/2, 598
		$LiVO_3$	Hautefeuille and Perrey (1888b, 1893)	DII/2, 598
$BeNaPO_4$	Berylonite	$Na_2SiO_3 + B_2O_3$	Traube (1894)	DII/2, 598
BeO	Bromellite	Na-phosphate	Ouvrard (1890)	DIII/1, 315
		$K_2CO_3 + SiO_2$, B_2O_3 $Na_2B_4O_7 + SiO_2$	Ebelmen (1848b, 1851a, 1851c, 1861)	G 69
		Na_2S, K_2S Silicate melts	Hautefeuille and Perrey (1888a, 1890a, 1891, 1893)	
Be_2SiO_4	Phenacite	$LiVO_3$	Hautefeuille and Perrey (1888b)	G 69
Bi		KNO_3 (decomp.)	Quesneville (1830)	DIII/1, 811
		Hg	Puschin (1903)	G 22
Bi_2O_3	Bismite	KOH	Nordenskiöld (1860, 1861a)	G 109
Bi_2S_3	Bismuthinite	Bi	Werther (1842), Rössler (1895, 1898)	DIV/1, 65; G 160
		Bi or S	Schneider (1854)	DIV/1, 65; G 160
		K_2S	de Sénarmont (1851b)	DIV/1, 65; G 160
			Lagerhjelm (1816)	DIV/1, 65; G 160
			Heintz (1844)	DIV/1, 65; G 160
C	Carbon, graphite, (diamond?)	$BiCl_3$	Durocher (1851a, 1854)	Gurlt (1857)
		Fe	Deville (1856a)	G 13
		Fe	Moissan (1896, 1905)	G 12
		Olivine	Friedländer (1898, 1900)	G 12
		Silicate melts	Hasslinger (1902, 1903, 1917)	G 12

Table 2.2 cont.

Formula	Name	Solvent	Reference	Source
$CaCr_2O_4$	see under Cr			
CaF_2	Fluorite	$CaCl_2$, KCl, NaCl	Scheerer and Drechsel (1873)	DIV/3, 203; G 206
CaI_2O_6	Lautarite	$NaNO_3$	de Schulten (1898a)	DIV/3, 448
$CaMoO_4$	Powellite	NaCl	Hjortdahl (1887)	DIV/2, 783
$Ca_2Nb_2O_6F$	Pyrochlore	$CaCl_2$ + NaF	Holmquist (1896, 1899)	DIII/1, 97
		CaF_2	Joly (1876)	DIII/1, 97
$Ca_2(Ta, Fe)_2O_7$	Pyrochlore	K_2CO_3	Ebelmen (1851c)	DIII/1, 97
$Ca_5(PO_4)_3(F, Cl)$	Apatite	$CaCl_2$, CaF_2	Manross (1852)	DIII/1, 345
		CaF_2	Briegleb (1856)	DIII/1, 345
		NaCl	Forchhammer (1854)	DIII/1, 345
		$CaCl_2$, CaF_2	Deville and Caron (1863b)	DIII/1, 345
		NaCl, KCl	Ditte (1883, 1884)	DIII/1, 345
		$CaCl_2$, CaF_2	Nacken (1912)	DIII/1, 345
$CaSnB_2O_6$	Nordenskjöldin	$CaCl_2(?)$	Ouvrard (1906)	DIII/2, 426
$CaSO_4$	Anhydrite	KCl, $MgSO_4$ + B_2O_3	Manross (1852)	DIV/2, 196
		NaCl	Simmler (1859)	DIV/2, 196
		$MnCl_2$	Gorgeu (1885)	DIV/2, 196
$CaSeO_4$		$NaNO_3$ + KNO_3	Diacon (1888)	C 90
$CaTiO_3$	Perovskite	Na_2CO_3, K_2CO_3	Ebelmen (1851a, c)	DIII/1, 40
		$CaCl_2$ + SiO_2	Hautefeuille (1864a)	DIII/1, 40
		$BaCl_2$	Bourgeois (1886)	DIII/1, 40
$CaWO_4$	Scheelite	NaCl + $CaCl_2$	Manross (1852)	DIV/2, 821; FL 186
		NaCl + $CaCl_2$	Michel (1879a)	DIV/2, 821; FL 186
		$CaCl_2(?)$	Debray (1862, 1863)	DIV/2, 821; FL 186
$CaMgSi_2O_6$	Diopside	CaF_2, $CaCl_2$	Berthier (1835)	
		$CaCl_2$	Lechartier (1868)	DII/1, 530; FL 106
		$Na_2B_4O_7$	Doelter (1890)	DII/1, 642

Formula	Mineral	Medium	Author (year)	Reference
$CaFeSi_2O_6$	Hedenbergite	B_2O_3, Na-phosphate, CaF_2	Bauer (1899)	DII/1, 642
$3Na_3AlSiO_4 \cdot CaSO_4$	Hauynite	$Na_2B_4O_7$	Doelter (1890)	DII/1, 642
		Na_2SO_4	Lemberg (1885)	DII/2, 255
		Salts and silicates	Morozewicz (1892, 1893, 1899)	DII/2, 255
$CaSiO_3$	Wollastonite	$CaCl_2$	Lechartier (1868)	FL 112
			Velain (1878)	
$Ca_3Al_2Si_3O_{12}$	Grossularite	$CaCl_2$	Gorgeu (1883b)	DII/2, 891
$CaTiSiO_5$	Sphene, Titanite	$CaCl_2$	Hautefeuille (1865)	DII/1, 62, 65, FL 178
		$CaSO_4$	Michel (1892)	DII/1, 62, 65, FL 178
$CaAl_2Si_2O_8$	Anorthite	NaF, KF, CaF_2 + diorite	Velain (1878), Bauer (1899)	DII/2, 996
		$Na_2HPO_4 + B_2O_3 + SnCl_2$	Medanich (1903)	DII/2, 996
$(CaAl_2Si_2O_8,$ $NaAlSi_3O_8$ s.s.)	Oligoclase	$CaF_2 + NaCl$	Petrasch (1903)	DII/3, 226, FL 140
		$CaF_2 + MgF_2$	Medanich (1903)	DII/3, 226, FL 140
		$LiCl + Na_2MoO_4$		
$(CaAl_2Si_2O_8,$ $NaAlSi_3O_8$ s.s.)	Labradorite	AlF_3	Meunier (1890)	DII/3, 284
		Silicates	Petrasch (1903)	DII/3, 284
		Augite	Fouqué and Lévy (1878)	DII/3, FL 61, 139
$(Ca, Na)_2(Al, Mg)$ $[(Si, Al)_2O_7]$	Melilite	Na_2CO_3	Bourgeois (1883)	DII/2, 963
		$NaF + CaF_2 + LiF$	Bauer (1899)	DII/2, 963
		$MgCl_2 + CaCl_2 + NH_4Cl$		
		$Na_3PO_4 + NaVO_3 + LiCl$		
CaO		$Ca(NO_3)_2$	Medanich (1903)	DII/2, 963
CdO	Monteponite	$Na_2B_4O_7$	Brügelmann (1877, 1878)	G 70
			Florence (1898)	DIII/2, 305
CdS	Greenockite	$CaCl_2$, CaF, BaS	Deville and Troost (1861)	DIV/1, 347; FL 303, G 150
			Dupont and Ferrières (1882)	FL 304
		$K_2CO_3 + S$	Schüler (1853)	DIV/1, 347; FL 304

Table 2.2 cont.

Formula	Name	Solvent	Reference	Source
CdSe	Cadmoselite	$CdCl_2$	Durocher (1851a)	DIV/1, 347
		Metal halides	Viard (1903)	DIV/1, 348
$CePO_4$	Monazite	$CdCl_2$	Fouzes-Diakon (1900)	G 150
CeO_2		$CeCl_3$	Radominsky (1875)	DIII/1, 558
		$Na_2B_4O_7$	Nordenskiöld (1860, 1861a)	G 85
		$NaCl$, $Na_2B_4O_7$	Sterba (1901)	G 85
CoS		$BaS + NaCl$	Hjortdahl (1867)	DIV/1, 680; G 138
Co_3O_4		$CoCl_2$	Schwarzenberg, see Gorgeu and Bertrand (1887)	C 80
Co_2SiO_4		$CoCl_2$	Bourgeois (1889)	C 88
CoO		B_2O_3, $Na_2B_4O_7$	Ebelmen (1851b, c)	G 71
		$CoCl_2(+H_2O)$	Ferrières and Dupont, after Bourgeois (1884)	G 71
Cr_2O_3	Eskolaite	$Co_3(PO_4)_2 + K_2SO_4$	Grandeau (1886)	
		Ca-borate	Ebelmen (1847, 1851)	DIV/2, 721; G 103
		$K_2Cr_2O_7 + SiO_2$	Svanberg (1854)	DIV/2, 721; G 103
$CaCr_2O_4$		K_2CrO_4	Weyberg (1906)	DIV/2, 705
$Li_2Cr_2O_4(?)$		Li_2CrO_4	Weyberg (1906)	DIV/2, 705
$FeCr_2O_4$	Chromite	Na_3AlF_6	Meunier (1887a, 1888)	DIV/2, 705
		B_2O_3	Ebelmen (1851b, d)	DIV/2, 705
$MgCr_2O_4$		B_2O_3	Ebelmen (1851b, d)	DIV/2, 705
$MnCr_2O_4$		B_2O_3	Ebelmen (1851b, d)	DIV/2, 705
			Meunier (1887a, 1888)	DIV/2, 705
$ZnCr_2O_4$		B_2O_3	Ebelmen (1851b)	DIV/2, 705
			Meunier (1887a, 1888)	DIV/2, 705
$NaCrS_2$	Daubreelite	Na_2CO_3, Na_2CO_3—K_2CO_3	Schneider (1873b, 1897)	DIV/2, 656
$FeCr_2S_4$		$FeCl_2$, S	Meunier (1891b)	
Cu	Copper		Behrens (1894)	G 3

Formula	Mineral	Solvent	Author	Reference
Cu_5FeS_4	Bornite	Silicates	Vogt (1892)	G 3
		Silicates	Washington (1894)	G 3
		NaCl	Böcking (1855)	DIV/1, 161
		$Na_2B_4O_3$	Marigny (1864)	DIV/1, 161
Cu_2S	Chalcocite	Pb	Rössler (1895, 1898)	G 143
CuS	Covellite	$ZnSO_4$(?)	Rogers (1911, 1912)	DIV/1, 98
Cu_2Se	Berzelianite	CuCl	Fouzes-Diakon (1900)	G 145
$CuWO_4$	Cuprotungstate	NaCl	Michel (1879b)	FL 186
$CuZn_2$		Zn	Charpy (1896)	G 43
Fe	Iron	$FeCl_2$	Haidinger (1850)	G 41
		$FeCl_2$	Osmond and Cartaud (1900)	G 41
		Hg	Nerad (1932)	
$FeAs_2$	Arsenoferrite	$Na_2B_4O_7$	Berthier (1836, 1837)	DIV/1, 604
Fe_3As_2(?FeAs)		$Na_2B_4O_7$	Descamp (1878)	DIV/1, 604
FeAsS	Arsenomarcasite	Na_2S, $Na_2S + Na_2CO_3$	de Sénarmont (1851a, b)	DIV/1, 634
$FeCr_2O_4$	see under Cr			
$FeCr_2S_4$	see under Cr			
Fe_2O_3	Hematite	$Na_2B_4O_7$	Hauer (1854)	DIII/2, 635
		$Na_2B_4O_7$	Rose (1867a, 1869a)	DIII/2, 635
		$Na_2B_4O_7$	Florence (1898)	DIII/2, 635
Fe_3O_4	Magnetite	Silicate melts	Ebelmen (1851a)	DIII/2, 649
		$FeCl_2$	Durocher (1851b)	FL 240
		$FeF_2 + B_2O_3$	Deville and Caron (1858)	DIII/2, 649
		$Na_2B_4O_7 + PbO$	Florence (1898)	DIII/2, 661
		$Na_2B_4O_7$	Hooslef (1856)	DIII/2, 823
Fe_3P	Schreibersite	Na-phosphate	Faye (1863)	DIII/2, 823
$(Fe, Ni)_3P$		Sb_2S_3	Morel (1888)	G 146–147
FeS	Pyrrhotite	$FeCl_2$	Doelter (1886b)	G 146–147, T 101

Table 2.2 cont.

Formula	Name	Solvent	Reference	Source
FeS_2	Pyrite	K_2S_x	Deville, see Friedel (1880)	DII/1, 719
Fe_2SiO_4	Fayalite	$FeCl_2$	Gorgeu (1883a, 1887b)	
$(Fe, Mg)_2SiO_4$	see Mg			
$NaFeSi_2O_6$	see Na			
$FeWO_4$	Ferberite	$NaCl$	Geuther and Forsberg (1861)	DIV/2, 848
$(Fe, Mn)WO_4$	Wolframite	$NaCl$	Michel (1879a)	DIV/2, 848
$(Fe, Mn)Nb_2O_6$	Columbite	FeF_2, MnF_2	Michel (1879a)	DIV/2, 848
$FeAl_3$		Al	Joly (1876)	FL 182
HgS	Cinnabar	$HgCl_2$	Guillet (1902)	G 45
$CsHgCl_3$		$CsCl$	Durocher (unpublished)	DIV/1, 361
Ir	Iridium	$FeS_2 + Na_2B_4O_7$	Wells (1892)	G 367
		Ag	Debray (1882)	G 38
			Rössler (1900)	G 38
$K_2Pb(SO_4)_2$	see Pb			
$KAlSiO_4$	Kaliophilite? Potash nepheline?	KCl	Gorgeu (1887a)	DII/2, 410
		KCl, K_2SiO_4	Weyberg (1908)	DII/2, 410
		KCl, K_2SO_4	Ginsberg (1912)	DII/2, 410
		KF	Duboin (1892a)	E 16
$(Na, K)AlSiO_4$	Nepheline	Na-vanadate Na_2CO_3	Hautefeuille (1880a)	DII/2, 220; FL 155; E 20, FL 156
$KAlSi_2O_6$	Leucite	K-vanadate	Hautefeuille (1880b, c, 1888)	DII/2, 470; E 9
		KVO_3	Hautefeuille and Perrey (1888c)	DII/2, 470, E 9–11
		KF	Duboin (1892a, b)	DII/2, 470
		NaF	Doelter (1897)	DII/2, 470
K-Zn-Silicate		KF, KCl	Duboin (1905)	DII/1, 790

Formula	Mineral	Solvent/compounds	Author (year)	References
$KAlSi_3O_8$	Orthoclase†	WO_3, Na_3PO_4, K_3PO_4, K_2SiF_6	Hautefeuille (1876, 1877, 1880f), see also Friedel (1880)	DII/2, 551, FL 132, 141
$KAl_3Si_3O_{10}F_2$	Fluor-muscovite	$KF + K_2SiF_6$	Doelter (1888)	DII/2, 438, 718
		$KF + NaF$	Doelter (1888)	DII/2, 438, 718
$KMg_3AlSi_3O_{10}F_2$	Fluor-phlogopite	$MgF_2 + KF$	Doelter (1888)	DII/2, 438, 718
$K(Mg, Fe)_3AlSi_3O_{10}F_2$	Fluor-biotite	$K_2SiF_6 + NaF + MgF_2$	Chroustchoff (1887)	DII/2, 718
		K_2SiF_6	Hautefeuille and St. Gilles (1887)	DII/2, 718
$KFeSi_3O_8$	Iron-feldspar(?)	KVO_3	Hautefeuille and Perrey (1888d)	C 86
$LiAlSiO_4$	Eucryptite	$LiVO_3$	Hautefeuille and Perrey (1890a)	DII/2, 195
$LiAlSi_4O_{10}$	Petalite	$LiVO_3$	Hautefeuille (1880d)	FL 159, T 100
$Li_2Cr_2O_4$	see under Cr			
MgF_2	Sellaite	KCl, NaCl	Röder (1863), Cossa (1876, 1877)	DIV/3, 188; G 205
MgO	Periclase	Ca-borate, B_2O_3	Ebelmen (1851b, c)	DIII/2, 287; G 69
		$MgCl_2$	Daubrée (1854), Dumas (1859)	DIII/2, 287
		$MgCl_2(+H_2O)$	Deville (1861b)	G 69
		$MgSO_4$, K_2SO_4	Debray (1861c)	DIII/2, 287; 689
		$MgCl_2(+SiO_2+ZrO_2)$	Hjortdahl (1866)	G 70
		KOH	de Schulten (1898b)	DIII/2, 287
$MgAl_2O_4$	Spinel	B_2O_3	Ebelmen (1848b)	DIII/2, 287
		$AlCl_3 + Na_3AlF_6$	Meunier (1887b)	DIII/2, 522
		Silicates	Morozewicz (1895, 1899)	

† The potassium feldspar crystallizing at high temperatures should be called *sanidine* according to the nomenclature of Laves (1960).

Table 2.2 cont.

Formula	Name	Solvent	Reference	Source
$MgCr_2O_4$	Magnesiachromite	B_2O_3	Ebelmen (1851b, d)	DIV/2, 705; Gurlt (1857)
$MgTiO_3$	Geikielite	$MgCl_2$	Ebelmen (1863)	DIII/1, 42
		$MgCl_2$	Hautefeuille (1865)	DIII/1, 42; FL 178
		$MgCl_2$	Bourgeois (1892)	DIII/1, 42
Mg_2PO_4F	Wagnerite	$MgCl_2$	Deville and Caron (1863a)	DIII/1, 319; FL 107
$MgSiO_2$	Enstatite	B_2O_3, KOH	Ebelmen (1851b, 1855a)	DIII/1, 319; DII/1, 329
		$CaCl_2$, $MgCl_2$	Hautefeuille (1865)	DIII/1, 319; FL 98
		$MgCl_2$	Lechartier (1868)	DIII/1, 319
$Mg_2Al_4Si_5O_{18}$	Cordierite	$MgF_2 + WO_3$	Doelter (unpublished)	DII/2, 626
$CaMgSi_2O_6$	see under Ca			
$(Mg, Fe)_2SiO_4$	Olivine	Silicates	Berthier (1823),	FL 102, 103
			Reiter (1906)	DII/1, 308
$Mg_3B_7O_{13}Cl$	Boracite	NaCl, $MgCl_2$, B_2O_3	Heintz and Richter (1860)	see Friedel (1880)
$MnCr_2O_4$		B_2O_3	Ebelmen (1851b, d)	DIV/2, 705; Gurlt (1857)
Mn_3O_4	Hausmannite	$Na_2B_4O_7$	Nordenskiöld (1861a)	DIII/2, 893
		$CaCl_2$	Kuhlmann (1861)	DIII/2, 893
		Silicates	Ebell (1876)	DIII/2, 893
$(Mn, Fe)WO_4$	Wolframite	NaCl	Michel (1879a)	DIV/2, 851
		NaCl	Geuther and Forsberg (1861)	DIV/2, 851
$MnWO_4$	Hübnerite	NaCl	Geuther and Forsberg (1861)	DIV/2, 847
$MnSiO_3$	Rhodonite	$MnCl_2$	Bourgeois (1885)	DII/1, 730
			Gorgeu (1883a, 1887b)	DII/1, 730
			Doelter and Hussak (1884)	DII/3, 362
Mn_2SiO_4	Tephroite	$MnCl_2$	Doelter and Hussak (1884)	DII/1, 712
			Gorgeu (1883b)	DII/3, 362
$Mn_3Al_2Si_3O_{12}$	Spessartite	$MnCl_2$	Doelter and Hussak (1884)	DII/3, 362

Formula	Mineral	Reagent	Author	Reference
MnO				
MoS_2	Molybdenite	B_2O_3, $Na_2B_4O_7$	Ebelmen (1851b, c)	G 70
		Molybdates	Debray (1904)	DIV/1, 70; G 157
		K-molybdate	de Schulten (1889a, b, 1891)	DIV/1, 70
		MoO_3	Meunier (1887b, 1890)	DIV/1, 70
		K-molybdate	Guichard (1901)	DIV/1, 70
MoO_2		$K_2CO_3 + B_2O_3 + MoO_3$	Mauro and Panebianco (1881)	G 93–94
NaF	Villiaumite	K_2MoO_4	Stevanović (1903)	G 93–94
$NaBePO_4$	Berylonite	Na_2SO_4—$CaCO_3$	Doelter, see p. 32	DIV/2, 1426
$(Na, K)AlSiO_4$	Nepheline	see Be		
$NaFeSi_2O_6$	Acmite, Aegirite	see K		
$Na_6Al_6O_{24}\cdot Na_2SO_4$	Nosean	$NaCl$	Weyberg (1905a, 1907)	DII/2, 339
		Na_2SO_4	Lemberg (1883)	DII/2, 253
		Na_2SO_4	Morozewicz (1899)	DII/2, 253
		Na_2SO_4	Weyberg (1911, 1914)	DII/2, 253
$Na_3Al_3Si_3O_{12}\cdot NaCl$	Sodalite	$NaCl$	Lemberg (1876)	DII/2, 239
		$NaCl$	Morozewicz (1899)	DII/2, 239
		$NaCl$	Mügge (1892)	DII/2, 239
		$LiBr$	Weyberg (1905b)	DII/2, 239
		$(\rightarrow Li_7Al_7Si_7O_{28}\cdot LiBr?)$		
$NaNbO_3$		$Na_2B_4O_7$	Mallard (1887a, b, 1888a, b, 1889), Knop (1871), Nordenskiöld (1860, 1861a)	G 112
$NaAlSi_3O_8$	Albite†	WO_3, Na_2WO_4, P_2O_5, V_2O_5	Hautefeuille (1874), see also Friedel (1880)	DII/2, 401; FL 136
		Na_2WO_4	Wallace (1909)	DII/2, 401
		Na_2WO_4	Day and Allen (1905)	DII/2, 401

† The sodium feldspar crystallizing at high temperatures should be called *analbite* according to the nomenclature of Laves (1960).

Table 2.2 cont.

Formula	Name	Solvent	Reference	Source
Na_xWO_3		$Na_2WO_4(+H_2)$	Wöhler (1823)	DII/2, 401
Nb_2O_5		B_2O_3	Mallard (1887a, b, 1888a, b, 1889)	G 112
NiAs	Nicollite	B_2O_3	Holmquist (1896, 1899)	G 112
Ni_3P		$NiCl_2$	Durocher (1851b)	DIV/1, 709; FL 277
(Ni_5P)		Ca-phosphate	Garnier, after Jannetaz (1882)	DIII/2, 823
NiS	Millerite	NiF_2	Poulenc (1882)	DIV/1, 696
Ni_2SiO_4		$NiCl_2$	Bourgeois (1889)	C 88
NiO		B_2O_3, $Na_2B_4O_7$	Ebelmen (1851b, c)	G 70, Gurlt (1857)
		$NiCl_2(+H_2O)$	Ferrières and Dupont, after Bourgeois (1884)	G 70
Os		$Ni_3(PO_4)+K_2SO_4$	Grandeau (1886)	G 70
		$Na_2B_4O_7+FeS_2$	Debray (1882)	G 38
P		Pb	Hittorf (1865)	G 17, 19
		P+S	Mitscherlich (1822–1823)	G 17, 19
PbO	Litharge	KOH	Michel (1890)	DIII/1, 209; G 77
(Pb_3O_4, PbO_2)	Massicotite	$NaNO_3$, KNO_3		
	Minium ?	KOH	Geuther (1883), Nordenskiöld (1861a)	DIII/1, 209; G 77, G 76
	Plattnerite	KOH(?)	Becquerel (1857)	DIII/1, 211
PbS	Galena	$Na_2B_4O_7$	Marigny (1864)	DIV/1, 420, G 152
		PbO, NH_4Cl	Weinschenk (1890)	DIV/1, 420, G 152
		Pb	Rössler (1895, 1898)	DIV/1, 420, G 152
			Manross, see Friedel (1880)	DIV/1, 420, G 152
$PbSO_4$	Anglesite	$CaCl_2$	Michel (1888)	DIV/1, 842
PbSe	Clausthalite	$PbCl_2$, NaCl	Michel (1888), Diacon (1888)	DIV/1, T 90
$(PbSeO_4?)$		$NaNO_3$, KNO_3		

PbSe	Clausthalite	Pb	Rössler (1895, 1898)	G 153
PbCrO₄	Crocoite	KCl	Manross (1852)	DIV/2, 736
Pb₅(AsO₄)₃Cl	Mimetite	PbCl₂	Deville and Caron (1859, 1863b)	DIII/1, 708
		PbCl₂	Lechartier (1867)	DIII/1, 708
		PbCl₂	Michel (1887)	DIII/1, 707
Pb₅(PO₄)₃Cl	Pyromorphite	NaCl(+ PbCl₂)	Manross (1852, 1858)	DIII/1, 454
		NaCl(+ PbCl₂)	Deville and Caron (1863a)	DIII/1, 454
		PbCl₂	Michel (1887)	DIII/1, 454
Pb₅(VO₄)₃Cl	Vanadinite	PbCl₂	Hautefeuille (1871, 1873, 1883)	DIII/1, 837, FL 183
PbK₂(SO₄)₂	Palmierite	Na₂SO₄	Ditte (1883)	DIII/1, 837
PbMoO₄	Wulfenite	NaCl + PbCl₂	Zambonini (1920, 1924)	DIV/2, 634
		NaCl + PbCl₂	Manross (1852)	DIV/2, 797
		NaCl	Schultze (1863)	DIV/2, 797
		NaCl	Michel (1881)	FL 193
PbMoO₄: 9 wt% Bi₂Mo₃O₁₂			see Doelter	DIV/2, 805
PbMoO₄ : Y₂Mo₃O₁₂		NaCl	see Doelter	DIV/2, 805
PbWO₄	Stolzite	NaCl + PbCl₂	Manross (1852)	FL 191
		NaCl	Geuther and Forsberg (1861)	
Pt	Platinum	Silicates	Ebelmen (1851a)	G 41
		NaNO₃	Köttig (1857)	G 41
PtAs₂		Na₂CO₃	Murray (1875)	DIV/1, 788
PtTe₂		Te	Rössler (1897)	G 155
Rh		Pb, Bi	Rössler (1900)	G 38
Ru		Na₂B₄O₇ + FeS₂	Deville and Debray (1879)	DIV/1, 790; G 38

Table 2.2 cont.

Formula	Name	Solvent	Reference	Source
RuS_2	Laurite	Pb	Rössler (1900)	G 38
Sb_2S_3	Stibnite	$Na_2B_4O_7 + FeS_2$	Deville and Debray (1879)	DIV/1, 790; G 158
		$Sb_2O_3 + S + I$	Jannasch and Remmler (1893)	DIV/1, 59
Se		$SbCl_3$	Durocher (1851a)	Gurlt (1857)
Si		Na_2Se_x	Mitscherlich (1855)	G 33
		$Al(+SiCl_4)$	Deville (1856b), after Sénarmont (1856)	G 13
		?	Henry, after Miller (1858)	G 13
		?	Percy, after Miller (1866–1867)	G 13
SiO_2	Quartz	$Na_2Si_2O_5 + K_2SiF_6$	Friedel (1888)	G 87
		$Na_2WO_4, Li_2WO_4(+B_2O_3)$	Hautefeuille (1880a)	G 87
		LiCl	Hautefeuille and Margottet (1881)	G 87
SiO_2	Tridymite	Alkali phosphates $Na_2WO_4, CaCl_2$	Hautefeuille (1880a)	G 89
SiO_2	Tridymite	LiCl	Hautefeuille and Margottet (1881)	G 87, 89
		Alkali phosphates, Silicates	Rose (1869b)	DIII/1, 192; G 89
		Na_2WO_4	Hautefeuille (1870, 1878)	
		$Na_2WO_4, NaVO_3, Na_3PO_4$	Hautefeuille, see Friedel (1880)	
SiO_2	Cristobalite	$Na_2B_4O_7(+BeO)$	Ebelmen, see Mallard (1887)	DII/1, 278, 199
SiP_2O_7		HPO_3	Hautefeuille and Margottet (1883)	G 112–113
Sn	Tin	Pb, Bi	Miller (1843)	DIII/1, 176

Formula	Mineral	Flux	Reference	Code
SnO_2	Cassiterite	Hg	Puschin (1903)	G 14-15
		Slags	Abel (1859)	DIII/1, 184
		Slags	Bourgeois (1888)	DIII/1, 184; G 95
		Sn	Vogt (1899)	DIII/1, 184; G 95
		$Na_2B_4O_7$	Wunder (1870b)	DIII/1, 184; G 96
SnS	Herzenbergite	$SnCl_2$	Schneider (1855)	G 150
$SnSe$		$SnCl_2$	Schneider (1866)	G 140
SnP_2O_7		HPO_3	Hautefeuille and Margottet (1886, 1888)	G 113
$CaSnB_2O_6$	Nordenskjöldin	$CaCl_2$(?)	Ouvrard (1906)	DIII/2, 426
$SnSb$		Sn	Behrens (1898), Stead (1897)	G 61
$SrSO_4$	Celestite	$NaCl, SrCl_2, KCl$	Gorgeu (1883b, 1887b)	DIV/2, 221, 245
		KCl	Manross, see Friedel (1880)	
$SrSeO_4$		$NaNO_3 + KNO_3$	Diacon (1888)	C 90
SrO		$Sr(NO_3)_2$	Brügelmann (1877, 1878)	G 74
Ta_2O_5		$B_2O_3(+BaCO_3)$	Mallard (1887a, b, 1888a, b, 1889)	G 112
ThO_2	Thorianite	B_2O_3	Holmquist (1896, 1899)	G 112
		$B_2O_3, Na_2B_4O_7$	Hillebrand (1893, 1896)	DIII/1, 227
		$Na_2B_4O_7$	Nordenskiöld and Chydenius (1860)	DIII/1, G 86
			Rammelsberg (1873)	
		$Na_2B_4O_7$	Bahr (1862, 1863)	
		K_3PO_4	Troost and Ouvrard (1889)	
		$Na_2CO_3—ThF_4$	Duboin (1908)	
ThP_2O_7		HPO_3	Troost (1885)	G 113
TiO_2	Rutile	B_2O_3, phosphates, alkali silicates	Ebelmen (1851b, d)	DIII/1, 24; G 91

Table 2.2 cont.

Formula	Name	Solvent	Reference	Source
	Rutile	$Na_2B_4O_7$	Rose (1867a, b, c)	DIII/1, 24; G 91
	Anatase	Phosphates	Doss (1894)	DIII/1, 24; G 91
	Rutile	Phosphates	Doss (1894)	DIII/1, 24; G 91
	Rutile ($BaTiO_3$?)	Na_2WO_4	Hautefeuille (1880e)	DIII/1, 24; G 91
	Rutile	$BaCl_2$	Bourgeois (1886)	DIII/1, 24; G 91
	Brookite	SnO_2, SiO_2	Deville and Caron (1861)	G 91
		$CaF_2 + KCl + SiF_4 + HCl$	Hautefeuille (1864b, 1865)	DIII/1, 36, FL 171
		$K_2SiF_6 + HCl$	Hautefeuille (1864b, 1865)	DIII/1, 36, FL 171
$FeTiO_3$	Ilmenite	$FeCl_2$	Bourgeois (1892)	DIII/1, 43, 51
		$Na_2B_4O_7$	Rose (1867a)	DIII/1, 51
$MgTiO_3$	see Mg			
$CaTiO_3$	see Ca			
TiP_2O_7		HPO_3	Hautefeuille and Margottet (1886)	G 113
$CaTiSiO_5$	see Ca			
Tl_2O_3		$TlNO_3$	Thomas (1904)	G 604
UO_2	Uraninite	$Na_2B_4O_7$	Hillebrand (1893)	DIV/2, 934; G 86
$Pb_5(VO_4)_3Cl$	Vanadinite	see Pb		
WO_3		$Na_2B_4O_7$, $Na_2WO_4 + Na_2CO_3 + HCl$	Nordenskiöld (1861b, 1867)	DIV/2, 811; G 110
$Y_2Mo_3O_{12}$		$Na_2MoO_4 + PbCl_2 + Bi_2(MoO_4)_3$	Doelter	DIV/2, 805
YPO_4	Xenotime	YCl_3	Radominsky (1875)	
ZnO	Zincite	$Na_2B_4O_7$	Florence (1898)	DIII/2, 299
K-Zn-Silicate	see K			
ZnS	Sphalerite, Wurtzite	$ZnCl_2$	Durocher (1851a)	Gurlt (1857)
		$CaF_2 + BaSO_4$	Deville and Troost (1861, 1865)	DIV/1, 341, G 148

Formula	Mineral	Composition	Reference	Source
Zn_2SiO_4	Willemite	$K_2CO_3 + S$	Schneider (1873a)	G 148
$ZnAl_2O_4$	Gahnite	$SnCl_2$	Viard (1903)	DIV/1, 342
		ZnF_2	Deville (1861a)	DII/1, 783
		B_2O_3	Ebelmen (1848b)	DIII/2, 530
		$CaCl_2 + AlCl_3 + ZnCl_2$	Daubrée (1854)	DIII/2, 530
		$B_2O_3 + AlF_3 + ZnF_2$	Deville and Caron (1858)	DIII/2, 530
$ZnCr_2O_4$	Zinc chromite	B_2O_3	Ebelmen (1851b, d)	DIV/2, 705; Gurlt (1857)
$ZnFe_2O_4$	Franklinite	$Na_2B_4O_7 + PbO$	Florence (1898)	DIII/2, 661
$ZnSiO_3$		B_2O_3	Ebelmen (1851b)	DIII/2, 661; Gurlt (1857)
		B_2O_3	Ebelmen (1855b)	DII/1, 786
		B_2O_3	Traube (1894)	DII/1, 786
$ZnSeO_4$		$NaNO_3—KNO_3$	Diacon (1890)	C 90
$ZnSe$		$ZnCl_2$	Fouzes-Diakon (1900)	G 149
ZrO_2	Baddeleyite (and tetragonal zirconia)	$Na_2B_4O_7$	Nordenskiöld (1860, 1861a)	DIII/1, 132; G 83, 93
		$Na_2B_4O_7$	Knop (1871)	DIII/1, 132; G 83, 93
		Phosphates, $Na_2B_4O_7$	Wunder (1870a, b)	DIII/1, 132; G 83, 93
		Na_2CO_3	Lévy and Bourgeois (1882)	DIII/1, 132; G 83, 93
		$Na_2B_4O_7$, Phosphates	Florence (1898)	DIII/1, 132
		Na_2CO_3	Morozewicz (1909)	DIII/1, 132
$ZrSiO_4$	Zircon	Li_2MoO_4	Hautefeuille and Perrey (1888c)	DIII/1, 148; G 92
		Li_2MoO_4	Stevanović (1903)	DIII/1, 148; G 92
		Na-borate	Gürtler (1907)	DIII/1, 148; G 92
ZrP_2O_7		HPO_3	Hautefeuille and Margottet (1886)	G 113

References

Abel, F. A. (1859) *Neues Jahrb. Min.* 815.

Adams, F. D. (1954) "The Birth and Development of Geological Sciences" Dover, New York.

Agricola, G. (1530) "Bermannus, sive de re metallica" Basle.

Agricola, G. (1546) "Ausgewählte Werke" III, VEB Deutscher Verlag der Wissenschaften Berlin, 1956.

Bahr, J. F. (1862) *Efvers. Svenska Akad. Förh.* 415.

Bahr, J. F. (1863) *Pogg. Ann. d. Phys.* **119**, 572.

Bartholinus, E. (1669) "Experimenta Crystalli Islandici disdiaclastici, quibus mira et insolita Refractio detegitur" Hafniae, Copenhagen.

Bauer, K. (1899) *Neues Jahrb. Min.* Beil.-Bd. **12**, 535.

Bauer, M. (1932) "Edelsteinkunde" 3rd Edn, Tauchnitz, Leipzig.

Beaumont, E. de (1849, 1853) "Emanations volcaniques et métallifères".

Becquerel, A. C. (1857) *Comp. Rend.* **44**, 38; *Ann. Chim. Phys.* **51**, 105.

Behrens, H. (1894) "Das mikroskopische Gefüge der Metalle und Legierungen", p. 68. Hamburg, Leipzig.

Behrens, H. (1898) Verhandl. K. Akad. Wet. Amsterdam 35.

Bergman, T. (1773) Nov. Acta Soc. Sci., Uppsala I.

Berthier, P. (1823) *Ann. Chim. Phys.* **24**, 374, 396.

Berthier, P. (1835) *Journ. prakt. Chem.* **4**, 457.

Berthier, P. (1836) *Ann. Chim. Phys.* **62**, 113.

Berthier, P. (1837) *Journ. prakt. Chem.* **10**, 13.

Birringuccio (1540) Pirotechnica.

Black, L. P., Gate, N. H., Moorbath, S., Pankhurst, R. J. and McGregor, U. R. (1971) *Earth and Planetary Science Letters* **12**, 245.

Böcking (1855) "Dissert" p. 29. Göttingen.

Boodt, Anselmus Böetius de (1609) "Gemmarum et lapidum historia" Hanau.

Böttger, R. (1839) *Ann. d. Pharm.* **29**, 75, 85.

Bourgeois, L. (1883) *Ann. Chim. Phys.* **29**, 450.

Bourgeois, L. (1884) "Réproduction artificielle des minéraux", p. 51. Paris (*Encycl. Chim.* **2**, append.)

Bourgeois, L. (1885) *Bull. Soc. Min.* **6**, 64.

Bourgeois, L. (1886) *Compt. Rend.* **103**, 141.

Bourgeois, L. (1888) *Bull. Soc. Min.* **11**, 58.

Bourgeois, L. (1889) *Compt. Rend.* **108**, 1177.

Bourgeois, L. (1892) *Bull. Soc. Min.* **15**, 194.

Boyle, R. (1672) "An Essay about the Origine and Virtues of Gems" (Reprint 1972 by Hafner, New York).

Briegleb, H. (1856) *Ann. Chem. Pharm.* **97**, 95.

Brögger, W. C. and Flink, G. (1884) *Z. Kryst.* **9**, 228.

Brügelmann, G. (1877) *Wiedemanns Ann. d. Phys.* **2**, 466.

Brügelmann, G. (1878) *Wiedemanns Ann. d. Phys.* **4**, 277.

Burke, J. G. (1966) "Origins of the Science of Crystals" Univ. of Calif. Press, Berkeley and Los Angeles.

Caldwell, H. B. (1935) *Chem. & Met. Eng.* **42**, 213.

Cappeller, M. A. (1723) "Prodromus crystallographiae".

Charpy, G. (1896) *Compt. Rend.* **122**, 670; *Bull. Soc. d'Encour.* (*3*) **1**, 180

Chirvinskii, P. N. (1903/1906) "Artificial Production of Minerals in the 19th Century" (Russian), Kiev.

Chroustchoff, K. von (1887) *Tschermaks Min. Mitt.* **9,** 55.
Cossa, A. (1876) *Acc. Linc.* **1** and (1877) *Z. Kryst.* **1,** 207.
Darmstaedter, E. (1922) "Die Alchemie des Geber" Springer, Berlin. (Reprint 1969 by Sändig, Wiesbaden.)
Daubrée, G. A. (1879) "Etudes Synthétiques de Géologie Expérimentale" Paris.
Daubrée, M. (1854) *Compt. Rend.* **39,** 135.
Day, A. and Allen, E. T. (1905) *Z. Phys. Chem.* **54,** 1; *Am. Journ. Sci.* **29,** 127.
Debray, M. (1861a) *Compt. Rend.* **52,** 955, 985.
Debray, M. (1861b) *Neues Jahrb. Min.* 702.
Debray, M. (1861c) *Compt. Rend.* **53,** 985.
Debray, M. (1862) *Compt. Rend.* **55,** 287, 262; *Bull. Soc. Chim.* 95; *Ann. Chim. Phys.* **125,** 95.
Debray, M. (1863) *Chem. Centr.* 208.
Debray, M. (1882) *Compt. Rend.* **95,** 879.
Debray, M. (1904) In "Handbuch der Mineralogie" (C. C. Hintze, ed.) Leipzig (1) **1,** 418.
Descamp, M. (1878) *Compt. Rend.* **86,** 1066.
Deville, H. St. Claire (1856a) *Compt. Rend.* **42,** 49.
Deville, H. St. Claire (1856b) *Ann. Chim. Phys.* (*3*) **47,** 169.
Deville, H. St. Claire (1861a) *Compt. Rend.* **52,** 1304.
Deville, H. St. Claire (1861b) *Compt. Rend.* **53,** 199.
Deville, H. St. Claire and Caron, H. (1858) *Compt. Rend.* **46,** 754.
Deville, H. St. Claire and Caron, H. (1859) *Compt. Rend.* **47,** 985.
Deville, H. St. Claire and Caron, H. (1861) *Compt. Rend.* **53,** 163.
Deville, H. St. Claire and Caron, H. (1863a) *Ann. Chim. Phys.* (*3*) **68,** 443.
Deville, H. St. Claire and Caron, H. (1863b) *Ann. Chim. Phys.* (*3*) **67,** 447.
Deville, H. St. Claire and Debray, M. (1879) *Compt. Rend.* **89,** 587; *Bull. Soc. Min.* **2,** 185.
Deville, H. St. Claire and Troost, L. (1861) *Compt. Rend.* **52,** 920, 1304.
Deville, H. St. Claire and Troost, L. (1865) *Ann. Chim. Phys.* (*5*) **5,** 118.
Diacon, M. H. (1888) *Compt. Rend.* **106,** 878.
Diacon, M. H. (1890) *Compt. Rend.* **110,** 832.
Ditte, A. (1883) *Compt. Rend.* **96,** 575, 846, 1048, 1226.
Ditte, A. (1884) *Compt. Rend.* **99,** 792, 967.
Doelter, C. (1886a) *Z. Kryst.* **11,** 29, 40.
Doelter, C. (1886b) *Tschermaks Min. Mitt.* **7,** 535.
Doelter, C. (1888) *Tschermaks Min. Mitt,* **10,** 67, 76.
Doelter, C. (1890) *Chem. Min.*
Doelter, C. (1897) *Neues Jahrb. Min. II,* 17.
Doelter, C. (1904–1908), *Sitz. Ber. Kaiserl. Akad. Wiss. Wien, Math.-naturwiss. Klasse* **113,** 177, 495; **114,** 529; **115,** 617; **116,** 1243; **117,** 299.
Doelter, C. and Hussak, E. (1884) *Neues Jahrb. Min. I,* 169.
Doelter, C. (and Leitmeier, H.) (1912–1931) "Handbuch der Mineralchemie", Volumes I–IV, Steinkopff, Dresden and Leipzig.
Doss, B. (1894) *Neues Jahrb. Min. II,* 147.
Duboin, A. (1892a) *Bull. Soc. Min* **15,** 191.
Duboin, A. (1892b) *Compt. Rend.* **114,** 1361.
Duboin, A. (1905) *Compt. Rend.* **141,** 254.
Duboin, A. (1908) *Compt. Rend.* **146,** 490.
Dumas, J. B. (1859) *Ann. Chim. Phys.* (*3*) **55,** 189.

Dupont, E. and Ferrières (1882) In "Synthèse des Minéraux et des Roches" (F. Fouqué and M. Lévy, Eds.), p. 304. Masson, Paris.

Durocher, J. (1851a) *Compt. Rend.* **32**, 823.

Durocher, J. (1851b) Unpublished experiments, after Fouqué and Lévy (1882), pp. 240, 277.

Durocher, J. (1854) *Pogg. Ann. d. Phys.* **91**, 401.

Ebell, P. (1847) *Compt. Rend.* **24**, 279.

Ebell, P. (1876) *Dinglers Polytechn. Journ.* **220**, 155.

Ebelmen, J. (1845) *Compt. Rend.* **25**, 279.

Ebelmen, J. (1847) *Ann. Chim. Phys.* (*3*) **22**, 236.

Ebelmen, J. (1848a) *Journ. prakt. Chem* **43**, 472.

Ebelmen, J. (1848b) *Ann. Chim. Phys.* (*3*) **22**, 211, 213.

Ebelmen, J. (1851a) *Compt. Rend.* **32**, 710.

Ebelmen, J. (1851b) *Ann. Chim. Phys.* (*3*) **33**, 34, 58.

Ebelmen, J. (1851c) *Compt. Rend.* **33**, 525.

Ebelmen, J. (1851d) *Compt. Rend.* **32**, 230, 332.

Ebelmen, J. (1855a) *Traveaux Scientif. I* 182.

Ebelmen, J. (1855b) *Traveaux Scientif. I* 186.

Ebelmen, J. (1861) *Cimie, Céram. etc.* (Paris) **1**, 207.

Ebelmen, J. (1863) *Bull. Soc. Chim.* **5**, 202.

Ebelmen, J. (1888) *Bull. Soc. Min.* **11**, 308.

Elsner, L. (1839) *Journ. prakt. Chem.* **17**, 175.

Favre, A. (1856) *Bull. Soc. Géol.* **13**, 307.

Faye, A. (1863) *Compt. Rend.* **57**, 803.

Florence, W. (1898) *Neues Jahrb. Min. II*, 127–144.

Forchhammer, J. G. (1854) *Ann. Chem. Pharm.* **90**, 77.

Foullon, H. von (1884) *J. K.K. geol. Reichsanst.* **34**, 367–382.

Fouzes-Diakon, M. H. (1900) *Compt. Rend.* **130**, 832; **131**, 1207.

Fouqué, F. and Lévy. M. (1878) *Compt. Rend.* **87**, 700.

Fouqué, F. and Lévy, M. (1881) *Compt. Rend.* **92**, 890.

Fouqué, F. and Lévy, M. (1882) "Synthèse des Minéraux et des Roches" Masson, Paris.

Frémy, E. (1891) "Synthèse du rubin" Dunod, Paris.

Frémy, E. and Feil, C. (1877) *Compt. Rend.* **85**, 1029.

Frémy, E. and Verneuil, A. (1888) *Compt. Rend.* **106**, 565.

Frémy, E. and Verneuil, A. (1890) *Compt. Rend.* **111**, 667.

Friedel, C. (1880) Revue Scientifique, Sept. 1880, 242–248.

Friedel, C. (1888) *Bull. Soc. Franc. Min.* **11**, 29.

Friedländer, J. (1898) "Verh. d. Ver. z. Beförd. d. Gewerbefleisses" Berlin.

Friedländer, J. (1900) *Z. Kryst.* **33**, 490.

Fuchs, C. W. (1872) "Die künstlich dargestellten Mineralien" Haarlem.

Gaudin, A. (1837) *Compt. Rend.* **4**, 999.

Gaudin, A. (1869) *Compt. Rend.* **69**, 1342.

Geuther, A. (1883) *Ann. d. Chem.* **219**, 56.

Geuther, A. and Forsberg, E. (1861) *Gött. gel. Anzeiger*, 225; *Ann. Chem. Pharm.* **120**, 268; *Liebigs Ann.* **120**, 270.

Ginsberg, A. S. (1912) *Z. anorg. Chem.* **73**, 277.

Gleditsch, J. F. (1741) "Curieuses und Reales Natur-Kunst-Berg-Gewerk-und Handlungs-Lexikon" Gleditsch, place of print not given.

Gmelin-Kraut, K. (1911) "Handbuch der anorgan. Chemie" II, 617, IV, 50 and 311.

Gorgeu, A. (1883a) *Compt. Rend.* **97**, 323.
Gorgeu, A. (1883b) *Bull. Soc. Min.* **6**, 136, 281.
Gorgeu, A. (1885) *Ann. Chim. Phys.* (6) **4**.
Gorgeu, A. (1887a) *Ann. Chim. Phys.* (4) **10**, 145.
Gorgeu, A. (1887b) *Bull. Soc. Min.* **10**, 263, 284.
Gorgeu, A. (1887c) *Compt. Rend.* **104**, 120.
Gorgeu, A. (1887d) *Compt. Rend.* **104**, 580.
Gorgeu, A. and Bertrand, E. (1887) *Bull. Soc. Min.* **10**, 261.
Grandeau, L. (1886) *Ann. Chim. Phys.* (6) **8**, 216, 219.
Grossmann, H. and Neuburger, A. (1918) "Die synthetischen Edelsteine" 2nd Edition, 1928.
Groth, P. (1906) "Chemische Krystallographie" First part, Engelmann, Leipzig.
Gübelin, E. J. (1953) "Inclusions as a Means of Gemstone Identification" Geolog. Inst. of America, Los Angeles.
Guglielmini, D. (1688) "Riflessioni filosofiche dedotte dalle figure de sali" Bologna; see also "De salibus dissertatio epistolaris", Venice (1705).
Guichard, M. (1901) *Ann. Chim. Phys.* (7) **23**, 552.
Guillet, L. (1902) *Compt. Rend.* **134**, 236.
Gurlt, A. (1857) "Uebersicht der Pyrogenneten Künstlichen Mineralien namentlich der Krystallisierten Hüttenerzeugnisse" Engelhardt, Freiberg.
Gürtler, W. E. (1907) *Chem, Z.B. 1518*; German Patent DRP 18200.
Haidinger, W. (1850) *Jahrb. geol. Reichsanstalt Wien* **1**, 151.
Hall, J. (1798) *Edinburgh Roy. Soc. Trans.* **5** (1805), 8, 56.
Hampe, W. (1876) *Ann. Chem.* **183**, 75.
Hasslinger, R. von (1902) *Sitz. Ber. Kaiserl. Akad. Wiss. Wien* **111** (IIb), 619.
Hasslinger, R. von (1903) *Monatsh. Chemie* **23**, 817.
Hasslinger, R. von (1917) *Z. Kryst.* **40**, 643.
Hauer, C. von (1854) *Sitz. Ber. Wiener Akad.* **13**, 456.
Hautefeuille, P. (1864a) *Compt. Rend.* **59**, 732.
Hautefeuille, P. (1864b) *Compt. Rend.* **62**, 148.
Hautefeuille, P. (1865) *Ann. Chem. Phys.* (4) **4**, 129, 140, 154, 167.
Hautefeuille, P. (1870) *Bull. Soc. Min.* **1**, 2.
Hautefeuille, P. (1871) *Compt. Rend.* **72**, 869.
Hautefeuille, P. (1873) *Compt. Rend.* **77**, 896.
Hautefeuille, P. (1874) *Compt. Rend.* **84**, 1301.
Hautefeuille, P. (1876) *Compt. Rend.* **83**, 616.
Hautefeuille, P. (1877) *Compt. Rend.* **85**, 952.
Hautefeuille, P. (1878) *Compt. Rend.* **86**, 1133.
Hautefeuille, P. (1880a) *Ann. de l'Ecole norm. sup.* **9**, 370.
Hautefeuille, P. (1880b) *Compt. Rend.* **90**, 313.
Hautefeuille, P. (1880c) *Ann. de l'Ecole norm. sup.* **2**, 4.
Hautefeuille, P. (1880d) *Compt. Rend.* **90**, 541.
Hautefeuille, P. (1880e) *Compt. Rend,* **90**, 868.
Hautefeuille, P. (1880f) *Compt. Rend.* **90**, 830.
Hautefeuille, P. (1883) *Compt. Rend.* **96**, 1048.
Hautefeuille, P. (1888) *Compt. Rend.* **107**, 786.
Hautefeuille, P. and Margottet, J. (1881) *Bull. Soc. Franc. Min.* **4**, 241.
Hautefeuille, P. and Margottet, J. (1883) *Compt. Rend.* **96**, 1052.
Hautefeuille, P. and Margottet, J. (1886) *Compt. Rend.* **102**, 1017.
Hautefeuille, P. and Margottet, J. (1888) *Z. Kryst.* **13**, 424.

Hautefeuille, P. and Perrey, A. (1888a) *Compt. Rend.* **106,** 487.
Hautefeuille, P. and Perrey, A. (1888b) *Compt. Rend.* **106,** 1800.
Hautefeuille, P. and Perrey, A. (1888c) *Compt. Rend.* **107,** 786, 1000.
Hautefeuille, P. and Perrey, A. (1888d) *Compt. Rend.* **108,** 1150.
Hautefeuille, P. and Perrey, A. (1890a) *Bull. Soc. Franc. Min.* **13,** 145.
Hautefeuille, P. and Perrey, A. (1890b) *Ann. Chim. Phys.* (*6*) **20,** 447.
Hautefeuille, P. and Perrey, A. (1891) *Z. Kryst.* **18,** 322.
Hautefeuille, P. and Perrey, A. (1893) *Z. Kryst.* **21,** 306.
Hautefeuille, P. and St. Gilles, P. de (1887) *Compt. Rend.* **104,** 508.
Haüy, R. J. (1784) "Essai d'une théorie sur la structure des cristaux" Paris; see also "Traité de Cristallographie", 2 volumes, Paris, 1822.
Heintz, W. (1844) *Pogg. Ann. d. Phys.* **63,** 57.
Heintz, W. and Richter, G. E. (1860) *Pogg. Ann. d. Phys.* **110,** 613.
Hillebrand, W. F. (1893) *Bull. geol. Surv. U.S.* **113,** 41; *Z. anorg. Chem.* **3,** 234.
Hillebrand, W. F. (1896) *Z. Kryst.* **25,** 283.
Hittorf, W. (1865) *Pogg. Ann. d. Phys.* **126,** 214.
Hjortdahl, T. (1866) *Ann. Chem.* **137,** 236.
Hjortdahl, T. (1867) *Compt. Rend.* **65,** 75.
Hjortdahl, T. (1887) *Z. Kryst.* **12,** 411.
Holmquist, P. J. (1896) *Bull. geol. Inst. Uppsala* **3,** (5).
Holmquist, P. J. (1899) *Z. Kryst.* **31,** 305.
Hönigschimdt, O. (1907) *Monatshefte Chem.* **28,** 1107; *Sitzber. Wiener Akad.* IIb, 116.
Hönigschmidt, O. (1910) *Z. Kryst.* **47,** 697.
Hooke, R. (1665) "Micrographia" London.
Hooslef, H. (1856) *Ann. Chem. Pharm.* **100,** 99.
Hottinger, J. H. (1698) "Krystallologia, dissertatio de crystallis, harum naturam, admentem veterum" Bodmer, Zurich.
Huygens, C. C. (1690) "Traité de la lumière" Leyden.
l'Isle, Romé de (1783) "Essai de crystallographie" Paris.
Jannasch, P. and Remmler, W. (1893) *Ber. Dt. Chem. Ges.* **26,** II, 1422.
Jannetaz, E. (1882) *Bull. Soc. Min.* **5,** 17.
Joly, M. (1876) Thèse, p. 63.
Joly, M. (1883), *Compt. Rend.* **97,** 456.
Kepler, J. (1611) "De Nive Sexangula" Tampach, Frankfurt a/Main.
Knaffl, E. (1863) *Dinglers Polytechn. Journ.* **168,** 282; *Chem. Zentralblatt* **8,** 705.
Knop, A. (1871) *Ann. Chem.* **159,** 33, 56.
Kobell, F. von (1864) "Geschichte der Mineralogie von 1650 bis 1860" Cottaschen, Munich.
Köttig, O. (1857) *Journ. prakt. Chem.* **71,** 190.
Kuhlmann, F. (1861) *Compt. Rend.* **52,** 1286.
Kunz, G. F. (1890) "Gems and precious stones of North America" The Scientific Publishing Co., New York.
Lacroix, A. (1887) *Bull. Soc. Min.* **10,** 157.
Lagerhjelm, M. P. (1816) *Schweigg. Journ.* **17,** 416.
Lang, W. R. (1863) *Phil. Mag. London* (*4*) **25,** 435.
Laves, F. (1960) *Z. Krist.* **113,** 265.
Lebeau, P. (1899) *Ann. Chim. Phys.* (*7*) **16,** 457.
Lechartier, G. (1867) *Compt. Rend.* **65,** 172.
Lechartier, G. (1868) *Compt. Rend.* **67,** 41.

Leeuwenhoek, A. van (1685) *Phil. Trans. Roy. Soc.* **15**, 1073.
Leeuwenhoek, A. van (1705) *Phil. Trans. Roy. Soc.* **24**, 1906.
Lehmann, O. (1877) *Z. Kryst.* **1**, 458.
Lemberg, J. (1876) *Z. Dt. geol. Ges.* **28**, 603.
Lemberg, J. (1883) *Z. Dt. geol. Ges.* **35**, 590.
Lemberg, J. (1885) *Z. Dt. geol. Ges,* **37**, 963.
Lenz, H. O. (1861) "Mineralogie der alten Griechen und Römer" (Reprint 1966 by Sändig, Wiesbaden).
Lévy, A. M. and Bourgeois, L. (1882) *Bull. Soc. Min.* **5**, 136.
Libavius, A. (1597) "Die Alchemie des Andreas Libavius" Verlag Chemie, Weinheim (Reprint 1964).
Linnaeus, C. (1768) "Systema naturae" III, Holmiae.
Mallard, E. (1887) *Compt. Rend.* **105**, 227.
Mallard, E. (1887a) *Compt. Rend.* **105**, 1260.
Mallard, E. (1887b) *Ann. d. Mines* **12**, 427.
Mallard, E. (1888a) *Bull. Soc. Min.* **11**, 305, 308.
Mallard, E. (1888b) *Z. Kryst.* **14**, 605.
Mallard, E. (1889) *Z. Kryst.* **15**, 650.
Manross, N. S. (1852) *Ann. Chem. Pharm.* **81**, 219, 243, **82**, 338, 348, 358; *Journ. prakt. Chem.* **58**, 55.
Manross, N. S. (1858) *Compt. Rend.* **47**, 885.
Margottet, J. (1877) *Compt. Rend.* **85**, 1142.
Marigny, F. (1864) *Compt. Rend.* **58**, 967.
Marx, C. (1825) "Geschichte der Krystallkunde", Karlsruhe.
Mauro, F. and Panebianco, R. (1881) *Mem. Accad. Lincei, Roma* **9**, 418.
Medanich, G. (1903) *Tschermaks Min. Mitt. II*, 20.
Meunier, S. (1887a) *Bull. Soc. Min.* **10**, 196.
Meunier, S. (1887b) *Compt. Rend.* **104**, 411,'1111.
Meunier, S. (1888) *Compt. Rend.* **107**, 1153.
Meunier, S. (1890) *Compt. Rend.* **111**, 509; *La Nature* **36**, 32, 13.
Meunier, S. (1891a) "Méthods de Synthèse en Minéralogie" Paris.
Meunier, S. (1891b) *Compt. Rend.* **112**, 818.
Michel, H. (1914) "Die künstlichen Edelsteine" Teubner, Leipzig. (2nd Edition 1926).
Michel, L. (1879a) *Bull. Soc. Min.* **2**, 142.
Michel, L. (1879b) Lab. de M. Friedel, see FL p. 186.
Michel, L. (1881) Lab. de M. Friedel, see FL p. 193.
Michel, L. (1887) *Bull. Soc. Min.* **10**, 133.
Michel, L. (1888) *Bull. Soc. Min.* **11**, 186.
Michel, L. (1890) *Bull. Soc. Min.* **13**, 56.
Michel, L. (1892) *Bull. Soc. Min.* **15**, 254.
Miller, W. W. (1843) *Phil. Mag.* **22**, 263; *Pogg. Ann.* **58**, 660.
Miller, W. W. (1858) *Phil. Mag.* **31**, 397.
Miller, W. W. (1866/1867) *Proc. Roy. Soc. London* **15**, 11.
Mitscherlich, E. (1822/1823) *Abhandlung Akad. Berlin*, 43 (see Mitscherlich's gesammelte Werke, p. 193).
Mitscherlich, E. (1855) *Monatsber. Berl. Akad.* 409 (Werke p. 626).
Moissan, H. (1896) *Ann. Chim. Phys.* (7) **8**, 289, 306, 466.
Moissan, H. (1905) *Traité de chim. min.* **2**, 206.
Morel, M. (1888) *Bull. Trav. Univ. Lyon* 284.

Morozewicz, J. (1892) *Neues Jahrb. Min. II*, 42.

Morozewicz, J. (1893) *Neues Jahrb. Min. II*, 139.

Morozewicz, J. (1895) *Z. Kryst.* **24**, 285.

Morozewicz, J. (1899) *Tschermaks Min. Mitt.* **18**, 1, 105, 225.

Morozewicz, J. (1909) *AnzeigerKrakauer Akad.* 207.

Mügge, O. (1892) Mikr. Phys. (Stuttgart) I, 323.

Murray, A. (1875) *Edinburgh Phil. Journ.* **4**, 202; *Gmelin-Kraut*, "Handbuch der anorg. Chem." **3**, 1192.

Nacken R. (1912) *Zentralblatt Min.* etc. 545.

Nassau, K. and Nassau, J. (1971) *Lapidary Journ.* **24**, 1284, 1442, 1524.

Nerad, A. J. (1932) *Trans. Am. Inst. Chem. Eng.* **28**, 12.

Newton, I. (1730) "Opticks" London; reprinted New York, 1952.

Niggli, P. (1946), "Die Krystallologia von Johann Heinrich Hottinger" (1698), Sauerländer, Aarau.

Nilson, L. F. and Petterson, O. (1883) see Humpidge, *Phil. Trans. Roy. Soc. London* **174**, 605 and Brögger and Flink (1884).

Nordenskiöld, A. E. (1860) *Oefv. K. Vet. Ak. Förh. Stockholm* **17**, 448.

Nordenskiöld, A. E. (1861a) *Pogg. Ann. d. Phys.* **114**, 616.

Nordenskiöld, A. E. (1861b) *Pogg. Ann. d. Phys.* **112**, 160.

Nordenskiöld, A. E. (1867) *Pogg. Ann. d. Phys.* **130**, 412, 614.

Nordenskiöld, A. E. and Chydenius, J. J. (1860) *Pogg. Ann. d. Phys.* **110**, 642.

Osmond, F. and Cartaud, G. (1900) *Ann. Mines Paris* (*9*) **18**, 113.

Ouvrard, L. (1890) *Compt. Rend.* **110**, 1333.

Ouvrard, L. (1906) *Compt. Rend.* **143**, 345.

Parmentier, F. (1882) *Compt. Rend.* **94**, 1713.

Petrasch, K. (1903) *Neues Jahrb. Min.* Beil. Bd. **17**, 498.

Poulenc, M. (1882) *Compt. Rend.* **114**, 1426.

Puschin, N. A. (1903) *Z. f. anorg. Chem.* **36**, 201, 243.

Quesneville, G. (1830) *Schweigg. Journ. Chem. Phys.* **60**, 378.

Radominsky, J. (1875) *Compt. Rend.* **80**, 304.

Rammelsberg, C. F. (1873) *Ann. d. Phys.* **150**, 219.

Ray, P. (1956) "History of Chemistry in Ancient and Medieval India" Indian Chem. Soc., Calcutta.

Reiter, H. H. (1906) *Neues Jahrb. Min.* Beil. Bd. **22**, 183.

Röder, F. (1863) Dissertation, Göttingen.

Rogers, A. F. (1911) *School of Mines Quart.* **32**, 298.

Rogers, A. F. (1912) *Neues Jahrb. Min. I*, 219.

Rose, G. (1825) *Ann. d. Phys.* **5**, 533.

Rose, G. (1867a) *Mon.-Ber. Berliner Akad.* 129, 450.

Rose, G. (1867b) *Journ. prakt. Chem.* **101**, 217.

Rose, G. (1867c) *Journ. prakt. Chem.* **102**, 385.

Rose, G. (1869a) *Chem. Zentralbl.* 1.

Rose, G. (1869b) *Mon.-Ber. Berliner Akad.* 449; *Ber. Chem. Ges.* **2**, 388.

Rössler, H. (1895) *Z. anorg. Chem.* **9**, 50.

Rössler, H. (1897) *Z. anorg. Chem.* **15**, 407.

Rössler, H. (1898) *Z. Kryst.* **29**, 299.

Rössler, H. (1900) *Chem. Ztg.* (No. 69), 733.

Rouelle, G. T. (1745) *Mém. Acad. Roy. Sci.* 57–59.

Scheerer, T. and Drechsel, E. (1873) *Neues Jahrb. Min.* 755; *Journ. prakt. Chem.* **7**, 63.

Schneider, R. (1854) *Pogg. Ann. d. Phys.* **91**, 414.
Schneider, R. (1855) *Pogg. Ann. d. Phys.* **95**, 169.
Schneider, R. (1866) *Pogg. Ann. d. Phys.* **127**, 626.
Schneider, R. (1873a) *Pogg. Ann. d. Phys.* **149**, 386.
Schneider, R. (1873b) *Journ. prakt. Chem.* **8**, 38.
Schneider, R. (1897) *Journ. prakt. Chem.* **56**, 415.
Schoen, H. M., Grove, C. S. and Palermo, J. A. (1956) *J. Chem. Education* **33**, 373.
Schüler, E. (1853) Inaug.-Dissert. Göttingen; *Ann. Chim. Pharm.* **87**, 41; *Journ. prakt. Chem.* **60**, 249.
Schulten, A. de (1889a) *Bull. Soc. Min.* **12**, 545.
Schulten, A. de (1889b) *Geol. For. Förh.* **11**, 401.
Schulten, A. de (1891) *Z. Kryst.* **19**, 108.
Schulten, A. de (1898a) *Bull. Soc. Min.* **21**, 144.
Schulten, A. de (1898b) *Bull. Soc. Min.* **21**, 87.
Schultze, H. (1863) *Ann. Chem. Pharm.* **126**, 49.
Sénarmont, H. de (1851a) *Compt. Rend.* **32**, 409.
Sénarmont, H. de (1851b) *Ann. Chim. Phys.* (*3*) **32**, 129, 157, 170.
Sénarmont, H. de (1856) *Ann. Chim. Phys.* (*3*) **47**, 169.
Simmler, R. T. (1859) *Journ. prakt. Chem.* **76**, 428.
Smith, C. S. (1960) "A History of Metallography" Univ. of Chicago Press.
Sohncke, L. (1879) "Entwicklung einer Theorie der Kristallstruktur" Leipzig.
Stead, J. E. (1897) *Berg- und Hüttenmänn. Ztg.* 333; *Journ. Soc. Chem. Ind.* **16**, 506.
Steno(nis), N. (1669) "De solido intra solidum naturaliter contento dissertationis prodromus" Florence.
Sterba, J. (1901) *Compt. Rend.* **133**, 294.
Stevanovič, S. (1903) *Z. Kryst.* **37**, 253.
Stillman, J. M. (1960) "The Story of Alchemy and Early Chemistry" Dover, New York.
Svanberg, A. N. (1854) *Journ. prakt. Chem.* **54**, 188.
Thomas, V. (1904) *Compt. Rend.* **138**, 1697.
Tomkeieff, S. I. (1942) *Min. Mag.* **26**, 172.
Traube, S. (1894) *Neues Jahrb. Min. I,* 275.
Troost, G. (1885) *Compt. Rend.* **101**, 210.
Troost, L. and Ouvrard, L. (1889) *Ann. Chim. Phys.* **17**, 273.
Valentini, Basilii (1769) "Chymischer Schriften anderer Theil" Krauss, Vienna and Leipzig.
Velain, C. (1878) *Bull. Soc. Franc. Min.* **1**, 113.
Viard, H. (1903) *Compt. Rend.* **136**, 892.
Vogt, J. H. L. (1892) "Mineralbildung in Schmelzmassen I", p. 237. Kristiania.
Vogt, J. H. L. (1899) *Z. Kryst.* **31**, 279.
Vogt, J. H. L. (1903/1904) "Die Silikatschmelzlösungen", Vols I and II. Kristiania.
Wallace, R. (1909) *Z. anorg. Chem.* **63**, 45.
Wallbergen, J. C. (1760) "Sammlung natürlicher Zauberkünste oder aufrichtige Entdeckung verschiedener bewährter, lustiger und nützlicher Geheimnisse" Mezler, Stuttgart.
Washington, H. S. (1894) *Amer. Journ. Science* **48**, 411.
Watt, G. (1804) *Phil. Trans.* 279.
Weinschenk, E. (1890) *Z. Kryst.* **17**, 498.

Wells, H. L. (1892) *Amer. Journ. Science* **44**, 221.

Werner, A. G. (1774) "Aeusserliche Kennzeichen der Fossilien" Leipzig.

Werther, G. (1842) *Journ. prakt. Chem.* **27**, 65.

Weyberg, Z. (1905a) *Ann. Univ. Varsovic* **3**, 1.

Weyberg, Z. (1905b) *Zentralbl. Min.* 653.

Weyberg, Z. (1906) *Zentralbl. Min.* 646.

Weyberg, Z. (1907) *Z. Kryst.* **44**, 86.

Weyberg, Z. (1908) *Zentralbl. Min.* 395.

Weyberg, Z. (1911) *Tr. Mus. geol. Pierre le Grand Acad. Sc. St. Petersburg* **5**, 118.

Weyberg, Z. (1914) *Z. Kryst.* **53**, 612.

Wöhler, F. (1823) *Ann. Chim. Phys.* (*2*) **29**, 43.

Wöhler, F. and Deville, H. St. Claire (1857) *Ann. Chem. Pharm.* **101**, 113, 347.

Wunder, G. (1870a) *Journ. prakt. Chem.* **1**, 475.

Wunder, G. (1870b) *Journ. prakt. Chem.* **2**, 206, 211.

Zambonini, F. (1920) *Boll. Com. Geol. Ital.* **49**, 1.

Zambonini, F. (1924), see Millosevich, *Neues Jahrb. Min.* II, 181.

3. Solvents and Solutions

3.1. Introduction

Certain problems of the natural sciences survive all generations and demand more interest with increasing data and with increasing penetration into the knowledge of nature. The solutions represent such a fundamental problem: their role is likewise important in pure chemistry, in technology and in the processes in animated and unanimated nature, but their explanation belongs to the most difficult and challenging problems of chemistry.

These sentences have not yet lost their validity, although they appeared in a treatise on the historical development of solution theories by P. Walden (1910) which is recommended for anyone interested in the history of science. Solutions are certainly understood much less than are gases or crystalline solids, although simple liquids are now partially understood. Their structure is determined to a large extent by repulsive forces amongst the molecules, and the attractive forces (analysed by van der Waals and, using perturbation treatments, by Barker and Henderson, 1967, 1968) have a less important effect on the structure (Rowlinson, 1969, 1970; Bernal, 1968). Much further work, theoretical and experimental, is necessary to gain an understanding of polyatomic liquids and of solutions, and here especially systematic experiments and large-scale computer simulation studies promise to increase greatly the knowledge in this complex field (see Neece and Widom, 1969; Eyring and Jhon, 1969; Barker and Henderson, 1972).

In the case of high-temperature solutions the state of knowledge is even smaller. A review of equilibrium theory of pure fused salts has been given by Stillinger (1964) and of thermodynamic properties of high-temperature solutions by Blander (1964). Some results of structural investigations of molten salts by neutron and X-ray diffraction, by electronic spectroscopy and by vibrational spectroscopy were reviewed by Levy and Danford (1964), Smith (1964) and James (1964), respectively. Reviews on all practical aspects of molten salts and molten-salt solutions are given by Belyaev and co-workers (1964), Bloom and Hastie (1965), Bloom (1967), Janz (1967), Mamantov (1969) and Janz and Tomkins (1972, 1973). A modern and comprehensive survey of theoretical and experimental aspects of ionic inter-actions in electrolyte solutions and in fused salts has been compiled by Petrucci (1971) and deals particularly with physicochemical properties of ionic high-temperature solutions and with crystal field and molecular orbit theories applied to fused-salt systems. Bloom and Bockris (1959) summarize the information necessary for a rough estimate of the structure of a molten salt as follows.

1. The type of entities present, i.e. ions, molecules, complex ions, etc.
2. The nature and effect of holes or vacancies.
3. Distribution functions relating to the relative positions of structural entities and holes.
4. The nature of the bonds or interionic forces amongst the various entities in the melt.

Of great importance is the presence of holes, free space, in the liquids. The presence of such holes is proved by the volume expansion when solids melt and by conductivity measurements. It is the presence of free space in liquids which favours transport processes.

Krasnov (1968) pointed out that the properties of solutions are closely related to the radii of the ions. These radii depend on the concentration: in highly concentrated solutions the ionic radii are similar to those known from corresponding crystal structures (Samoilov, 1957), whereas in dilute solutions they approximate to the radii of Böttcher (1943, 1952) which were derived from undeformed gaseous ions.

In the following a brief description of solution concepts, of solute-solvent interactions and solubilities and of equilibria in solutions will be given, followed by sections on "ideal" high-temperature solvents, on phase diagrams and on special solvents.

A rather detailed general discussion of solutions seems necessary with respect to the understanding of the processes occurring during crystalliza-tion from high-temperature solutions. The usual tendency is to consider only the reactions at the solid interface and to neglect the nature of the solution itself. From the following sections it will become clear that the type of solution and of the solute-solvent interactions may play a decisive role in the crystal growth process. In certain cases a systematic optimization of the solvent question may enable the experimentalist to decrease the tendency of unstable growth and to increase the width of the metastable region in order to prevent inclusions and multinucleation, respectively. The choice of solvent also determines the degree of solvent ion incorporation and therefore the purity of the grown crystals. It is interesting to note that many solvents chosen by experience are often optimum in many aspects, especially in flux growth of oxides, whereas for other compounds general rules are given in Section 3.6.1.

Preliminary experiments to synthesize new materials in crystalline form, as small crystals, do not require an exact knowledge of the corresponding phase diagram. For the systematic growth of large crystals and for the achievement of a high yield, the phase diagram has to be known to a sufficient degree. Therefore a discussion of phase-diagram determination as well as selected solubility data are given in the last part of this chapter.

3.2. General Solution Concepts

There is no fundamental difference between a solution and a mixture of liquids but it is convenient in a description of high-temperature solutions to consider the less refractory phase as the solvent while the solute is the phase which crystallizes first on cooling. The solute may have several components and solution behaviour may be extremely complex but the treatment given here will be restricted to basic considerations.

Since the total volume, entropy, enthalpy, etc. of the solution are, in general, not equal to the sum of the individual quantities prior to mixing it is necessary to use *partial molar* quantities. For example, in attempting to assign a fraction of the total volume to the separate constituents it is convenient to consider the partial molar volumes $\bar{V}_1$ and $\bar{V}_2$ which represent respectively the changes in volume of the solution on adding one mole of components 1 and 2 at constant temperature T and pressure p.

It may be readily shown that the Gibbs free energy of an ideal mixture (as of ideal gases) exceeds the value for the same unmixed material at constant T and p by an amount

$$\Delta G = RT \sum_A n_A \ln x_A,$$

where n_A is the amount of species A and x_A its fractional concentration $(x_A = n_A / \sum_i n_i)$. Thus the chemical potential $\mu_A = \partial(\Delta G)/\partial n_A$ of species A is given by

$$\mu_A = \mu_A^0 + RT \ln x_A, \tag{3.1}$$

where the superscript ^{0} refers to the pure liquid at the same temperature

c2

and pressure. If a mixture (or solution) exhibits non-ideal behaviour, the departure from ideality may be expressed in terms of the relative activity a_A which is defined (see Guggenheim, 1967) by

$$\mu_A = \mu_A{}^0 + RT \ln a_A. \tag{3.2}$$

Comparison of Eqns (3.1) and (3.2) shows that, for an ideal mixture,

$$a_A = x_A. \tag{3.3}$$

The condition for a solution of material A to be ideal is that its chemical potential in the solution is the same as that of the pure substance in the solid state, so that

$$\mu_A(s) = \mu_A{}^0 + RT \ln a_A. \tag{3.4}$$

Division by T and differentiation with respect to temperature gives

$$\frac{\partial}{\partial T}\left[\frac{\mu_A(s)}{T}\right] - \frac{\partial}{\partial T}\left(\frac{\mu_A{}^0}{T}\right) = R\frac{\partial}{\partial T}\ln a_A. \tag{3.5}$$

Now it is a familiar relation that

$$\frac{\partial}{\partial T}\left(\frac{\mu}{T}\right) = \frac{-H}{T^2},$$

where H is the enthalpy so that substitution in Eqn (3.5) gives

$$\frac{\Delta H_f}{T^2} = R\frac{\partial}{\partial T}\ln a_A,$$

where $\Delta H_f = H_A(s) - H_A{}^0$ is the enthalpy of fusion. Thus in an ideal solution where Eqn (3.3) holds,

$$\frac{\partial}{\partial T}\ln x_A = \frac{\Delta H_f}{RT^2}$$

or

$$x_A = x_\infty \exp\left(-\Delta H_f/RT\right). \tag{3.6}$$

If $\ln x_A$ is plotted against reciprocal temperature for an ideal solution, a straight line of slope $(-\Delta H_f/R)$ is obtained with an upper limit of $\ln x_A = 0$ at the melting point T_M, as in Fig. 3.1. Thus if the heat of fusion and melting point are known the solubility curve can be plotted and will not depend on the solvent. Since the entropy of fusion of similar compounds is approximately constant, a high melting point is associated with a low solubility. This latter relation was recognized by Lavoisier (1794).

The assumption of an ideal solution also requires that $p_A \propto x_A$ (Henry's law) where p_A is the *fugacity* or vapour pressure of A. If the constant of

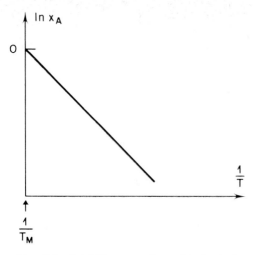

FIG. 3.1. Solubility curve for an ideal solution.

proportionality is the saturated vapour pressure $p_A{}^0$ of the pure liquid, the resulting relation

$$p_A = p_A{}^0 x_A \qquad (3.7)$$

is known as Raoult's law.

Real solutions exhibit departures from Eqns (3.6) and (3.7) and these departures are often expressed by the deviation from unity of the activity coefficient a_A/x_A. According to Hildebrand and Scott (1970) non-ideal solutions may be divided into athermal, associated and regular solutions. In *associated solutions* departures from regularity are caused primarily by bonding between the molecules of one component in the solution, whereas

TABLE 3.1. Classification of Solutions

Type of solution	Enthalpy of mixing $\bar{H}_1 - H_1{}^0$	Entropy of mixing $\bar{S}_1 - S_1{}^0$	Remarks
Ideal	0	$-R \ln x_1$	$a_1 = x_1$ $V_1 \simeq V_2$
Regular	+	$-R \ln x_1$	$a_1 > x_1$ $V_1 \simeq V_2$
Non-ideal, athermal	0	$< -R \ln x_1$	$a_1 < x_1$ $V_2 > V_1$
Irregular, associated	+	$> -R \ln x_1$	$a_1 > x_1$
Irregular, solvated	−	$< -R \ln x_1$	$a_1 < x_1$

in *solvated solutions* the principal bonding is between solute and solvent molecules. *Athermal solutions* represent a particular case in which the enthalpy of mixing is zero but the activity does not have the ideal value. This classification of solution types is indicated in Table 3.1.

Regular solutions approximately obey Raoult's law up to a solute concentration of about 15 m%. It has, however, been argued by Haase (1956) that "regular solutions" are somewhat hypothetical, or at least extremely rare (the possible exceptions being certain metallic solutions), and several definitions of a regular solution are in use.

Departures from ideal behaviour may also be displayed graphically on a plot of the Gibbs free energy of mixing versus composition, as shown in Fig. 3.2 (Darken and Gurry, 1953).

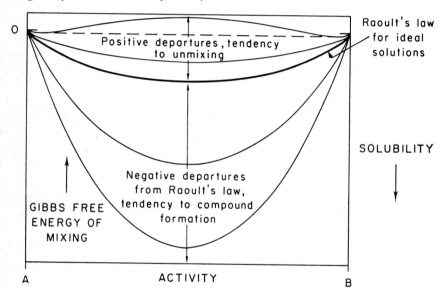

FIG. 3.2. Gibbs free energy of solutions with departures from ideal behaviour.

A positive deviation from Raoult's law corresponds to a smaller solubility for a solid and a negative deviation corresponds to a greater solubility than would be calculated from the melting point and the heat of fusion of the solid. The simplest possible model of a condensed system is based on the assumption that the total cohesive energy is the sum of interactions or "bonds" between neighbouring entities. The solution theory based on this model has been developed mainly by Rushbrooke (1938) and by Guggenheim (1952). A brief discussion has been given by Kleppa (1958). In general, the interaction energy W_{12} when species 1 and 2 form a solution will differ from the arithmetic mean of the individual bonding energies W_{11} and W_{22} so that there will be departure from Raoult's law.

Hildebrand has suggested that a useful parameter for characterizing the attraction between molecules is the solubility parameter δ, defined as

$$\delta = \Delta\omega_{\text{vap}}^{1/2}, \tag{3.8}$$

where $\Delta\omega_{\text{vap}}$ is the energy of vaporization per unit volume of the solution.

The significance of this parameter is that the energy of mixing of a regular solution may be written in the form

$$\Delta W^M = V(\delta_1 - \delta_2)^2 \phi_1 \phi_2, \tag{3.9}$$

where ϕ_1 and ϕ_2 are the volume fractions of the components of the binary solution

$$\phi_1 = \frac{N_1 V_1}{N_1 V_1 + N_2 V_2} \qquad \phi_2 = \frac{N_2 V_2}{N_1 V_1 + N_2 V_2}$$

with V_1 and V_2 the molar volumes of the pure substances. The usefulness of the solubility parameter is not generally accepted and other semi-empirical descriptions of non-ideal solutions have been proposed (see Haase, 1956).

The condition for complete miscibility of metals according to Hildebrand is that

$$\frac{V_1 + V_2}{2}(\delta_1 - \delta_2)^2 < 2RT.$$

However, Mott (1968) has shown that a better prediction of immiscibility can be obtained if the Pauling electronegativity difference W_B is included, so that

$$\frac{V_1 + V_2}{2}(\delta_1 - \delta_2)^2 - W_B \lesssim 2RT. \tag{3.10}$$

The value of W_B in calories is roughly $23{,}060\,\eta(\chi_1 - \chi_2)^2$, with χ_1 and χ_2 the electronegativities of metals 1 and 2 in volts and η the appropriate number of bonds between molecules of 1 and 2. Mott suggests that $\eta \simeq \text{C.N.}/2$, where C.N. is the coordination number in the solution. Binary metals systems exhibiting liquid immiscibility are listed in Hildebrand and Scott (1964).

Immiscibility in oxide systems has been described by the screening concept of Weyl (1956), the electrostatic bond concept of Warren and Pincus (1940), the ionic field-strength concept of Dietzel (1942, 1948/1949) and related crystal chemical relations as reviewed by Levin (1970). A theory which predicts the extent of liquid immiscibility gaps in simple ionic systems has been derived by Blander and Topol (1965).

An obvious and well-known fact is the higher solubility at higher temperatures for most systems according to Eqn (3.4) for an endothermic dissolving reaction. The rare phenomenon of *retrograde solubility* (with the solubility decreasing with increased temperature) is connected with an exothermic dissolution process.

The effect of pressure on the solubility of solids can be discussed in terms of deviations from Raoult's law: ideal solutions and solutions with positive deviation will show a lowering of the solubility with higher pressure, whereas a negative deviation from Raoult's law is frequently associated with an increase in solubility with pressure.

For further details regarding the principles of solutions reference may be made to Garrels and Christ (1965), Darken and Gurry (1953), Kleppa (1958), Hildebrand and Scott (1970), Robinson and Stokes (1968), Rowlinson (1969, 1970), Prigogine (1957) and Petrucci (1971) and especially to Reisman (1970).

Ionic Solutions

According to Blander (in Mamantov, 1969) the models for molten salts are simpler than for most other liquids, since the structure of molten salts can as a first approximation be considered as a dense mass of ionic spheres with a strong tendency towards alternation of positive and negative ions. Thus, the total energy of ionic molten-salt systems is equal to the sum of the pair interactions between cations, anions, and each other, mainly given by the Coulomb interactions. The potential for a cation-anion pair is given by

$$U = -\frac{e^2}{r} + \frac{k}{r^n},$$

where n is of the order of 8 to 12. For a gaseous dipole $k = (e^2 r_0^{n-1}/n)$ with r_0 the internuclear separation. For a solid of the NaCl type,

$$\frac{U}{n} = -\frac{1.7e^2}{r} + \frac{6k}{r^n}$$

therefore the equilibrium value of r for the solid is about 19% larger than r_0 for the vapour molecule if $n = 8$. Lowering the coordination leads to significant foreshortening and occurs when highly polarizable ions are stabilized in regions of high field intensity. A simple example is the ion-induced dipole interaction with $W = \frac{1}{2}\alpha E^2$ for weak fields where W is the energy, α the polarizability, and E the field intensity. Particular symmetries of anions around transition-metal cations are generated by ligand field effects which are dependent on interactions of nearest-neighbour anions as well as on next nearest-neighbour cations.

Cobb and Wallis (1969) discuss approaches to solubility with special emphasis on ionic high-temperature (non-ideal) solutions. However their treatment, which follows standard thermodynamic theory and which is similar to the discussions of Garrels and Christ (1965), will not be given here since the necessary knowledge of the thermodynamic quantities is generally not available. The desired relationship between solubility and the change in Gibbs free energy is based on the partial molar free energy of mixing of the solute in the solvent, which itself is a function of the solute concentration. The calculation of the partial free energy of mixing from the solubility product concept is only possible when either the solutions are completely dissociated (which normally is not the case) or when the degree of dissociation (and association) are known as a function of temperature for the given system. For partially ionic systems the easiest approach is to determine the solubility at various temperatures experimentally and inter- and extrapolate the solubility curve. Some conclusions regarding the thermodynamic functions, the degree of dissociation and association might then be drawn from such experimental values.

In order to estimate oxide solubilities the Temkin (1945) model is frequently used. Concentrations are expressed in ionic fractions, and the ionic activity coefficients are assumed to be unity. Then the solubility product K' of a completely dissociated oxide is defined by $\ln K'(T) = \Delta G^F / RT$ and the solubility changes with temperature according to

$$\frac{\partial}{\partial T}[\ln K'(T)] \approx -\frac{\Delta H_m^F}{RT^2} + \frac{\Delta C_p}{RT^2}(T_m - T).$$

Here ΔC_p is the difference in the heat capacities of the solid and liquid oxide solute, ΔH_m^F its enthalpy of fusion and T_m its melting point.

The greatly simplified approach of Temkin assumes that the heat of mixing of the oxide in the flux is zero. Nevertheless this model often predicts thermodynamic properties of unknown flux systems with considerable accuracy.

Of the necessary data generally only T_m is known. Published thermodynamic data are summarized in Table 3.2. For oxides of which the enthalpies of melting and heat capacity changes are not known, some rules are available for estimating these quantities. For completely ionized salts ΔS_m^F might be estimated by *Richard's rule* where the entropy of melting is taken as roughly R, the gas constant (Darken and Gurry, 1953). Brice (1973), who gives expressions which may be used to estimate various parameters, quotes $\Delta S_m^F = 8.8 \pm 3.8 \; \mathcal{J}(\text{g atom})^{-1} \, (^\circ C)^{-1}$. Non-ionic solids can have much lower entropies of fusion, and also phase transitions in the solid near T_m will lead to low estimated enthalpies of fusion when the latter are determined using

$$\Delta H_m^F = T_m \, \Delta S_m^F.$$

TABLE 3.2. Thermodynamic Properties of Oxides
(Table 2, p. 26, Cobb and Wallis, 1969)

Oxide	$T_m[K]$	ΔH^F Kcal/mole	ΔS^F cal/degree mole	ΔC_p cal/degree mole
MgO	3173	18.5	5.8	
CaO	2873	12	4.2	
Al_2O_3	2313	26	11	
TiO_2	2106	11.4	5.4	
ZrO_2	2950	20.8	7.0	
$VO_2(V_2O_4)$	1815	27.21	14.99	10.1
Nb_2O_5	1785	24.59	13.78	
WO_3	1745	17.55	10.06	
Cu_2O	1502	13.4	8.9	
SiO_2	2001	1.84	0.92	
SnO_2	1898	11	5.8	

Cobb and Wallis (1969) estimated using Richard's rule the thermodynamic data for such oxides which are, according to the rules of Pauling, more than 70% ionic (and which are supposed to dissociate almost completely in solution). These data are given in Table 3.3. The difference in the heat capacities of the solid and liquid oxide are on average 1 cal deg^{-1} gram atom^{-1} of the oxide.

With these thermodynamic data and with Temkin's model the solubility behaviour of ionic systems can be estimated. However, the estimate is

TABLE 3.3. Estimated Thermodynamic Properties of Pure Oxides
(Table 3, p. 28, Cobb and Wallis, 1969)

Oxide	$T_m[K]$	ΔH_m^F Kcal/mole	ΔS_m^F cal/degree mole
Li_2O	2000	12.0	6.00
SrO	2688	10.8	4.00
BaO	2196	8.8	4.00
Y_2O_3	2690	26.9	10.0
La_2O_3	2590	19.6	10.0
Ce_2O_3	1960	19.6	10.0
Eu_2O_3	2323	23.2	10.0
Dy_2O_3	2613	26.1	10.0
MnO	2058	8.2	4.0
CoO	2078	8.3	4.0
NiO	2230	8.9	4.0

very approximate as long as one is unable to take into account the solute-solvent interactions and the degree of covalency.

The solubility model described above for ionic oxides will also hold approximately for mixed oxides or other complex ionic compounds. The solubilities of incongruently melting compounds in novel solvents can be estimated (for the region below the incongruent melting point) by extrapolating from a known phase diagram or solubility curve to the hypothetical melting point and by estimating the heat of fusion of the compound by the use of Richard's rule.

Finally the theoretical approach to solutions should be mentioned. Although statistical mechanical treatments are potentially of great practical value this field has still remained a domain of theoretical physical chemists and has not found application in crystal growth from HTS, mainly because of the lack of understanding of solutions in general and especially at high temperatures. Future progress in this field should allow the calculation of solubilities from basic atomic and molecular data (Blander, 1964; Sundheim, 1964; Kaufman and Bernstein, 1970; Petrucci, 1971; Reisman, 1970).

3.3. Solute-solvent Interactions

The characteristic feature of a solvent is its ability to bring into a single phase (the liquid solution) one or more other (solid) compounds, the solutes. This dissolution process involves interactions of solvent and solute molecules which differ according to the types of bonding between solvent molecules and between solute molecules.

These various interactions and correspondingly the solubility cannot be connected with the thermal effects observed during dissolution. The heat of solution is the difference between the lattice energy and the solvation (hydration) energy, but this difference is small compared with each of the latter energies. As an example the lattice energy of KCl (corrected for 298°K) is 169 Kcal/mole, the heat of solution (in water) is $\Delta H = +4.1$ Kcal/mole, and therefore the total heat of hydration is the difference of 164.9 Kcal/mole.

The hydration energy can be roughly estimated for the case where an ion is brought from vacuum into water, which is taken as a continuous medium with a relative permittivity, ϵ, equal to 80. From electrostatics the amount of binding energy is $W_H = -\frac{1}{2}(e^2/r)(1 - 1/\epsilon)$, the energy of a charged sphere in a medium of relative permittivity ϵ being $\frac{1}{2}(e^2/r\epsilon)$ with r the ionic radius. For K^+ with $r = 1.33$ Å the hydration energy would be 122 Kcal/gram atom, and for Cl^- 89 Kcal/gram atom. Thus the sum for the idealized estimate of 211 Kcal/mole is much higher than the value of 164.9 Kcal/mole quoted above. The relations between the enthalpy of hydration of ions and the ionic radii have been reviewed by Morris (1968), whereas Fajans (1921) and Fajans and Karagunis (1930) had already established the correlation between solubility, lattice energy, hydration energy and the tendency to form solid hydrates.

The silver halides would have a hydration energy similar to KCl but the lattice energy is much higher by the contribution of the Van der Waals-London attraction between the readily polarizable cations and anions, which explains the low solubility in water. However the silver halides are readily soluble in the more polarizable solvent liquid ammonia which forms donor bonds with the silver atoms.

It is also the large lattice energy which is responsible for the low solubility of the mainly ionic oxides (e.g. MgO 940 Kcal/mole, Al_2O_3 3618 Kcal/mole), sulphides (e.g. CuS, CdS), sulphates, phosphates, nitrides, etc. in water.

In high-temperature solutions, especially in those of oxides and oxide compounds in molten salts, the type of bonding is mainly ionic. For the case where both solvent and solute are ionic a separation (lowering the Coulomb attraction forces of the solute) and a solvation of the solute ions will occur. This solvation may be weak, irregular and constantly varying as in many aqueous solutions, and this case would correspond to the regular solution defined above. If the solvation action is strong, as in irregular solutions, there will be some additional short-range order, and relatively stable complexes or compounds may be formed.

The solvating or complexing action of an ionizing solvent involves electron or ion transfer processes. The acid-base concept of Lewis (1923) and the solvent-system concept of acids and bases developed by Gutmann and Lindquist (1954) enable us in simple cases to describe the above transfer processes. Lewis acids are defined as acceptors for electrons (or negative ions) and bases are electron donors. Gutmann and Lindquist distinguish cationtropic and aniontropic reactions depending on the charge of the transferred ion. In a cationtropic system an acid is a cation donor and a base a cation acceptor, whereas in an aniontropic system an acid acts as anion acceptor and the base as anion donor. For example, halide solvents are aniontropic favouring the transfer of fluoride or chloride ions. However, Payne (1965) raises the question of whether the terms "acid" and "base" are necessary for non-protonic solvents. Often there are uncertainties in the interpretation of the species present in the solution and so it is proposed to use the terms "acid" and "base" only with protonic solvents (e.g. in aqueous systems) and in such cases where there is a clear advantage.

The acid-base concept with respect to oxides and oxysalts has been discussed by Flood and Förland (1947) who apply the definition of Lux (1939)

$$base = acid + O^{2-}$$

3. SOLVENTS AND SOLUTIONS

to a variety of reactions in high-temperature chemistry of oxides. The oxygen activity corresponds to the hydrogen activity of the classical protonic acid-base concept of Brönsted (1923, 1934). The acid-base relationship based on oxygen seems to be useful in glasses and in a discussion of high-temperature reactions and stabilities of oxides (Flood and Knapp, 1963). Thus the relative concentrations of the various silicate anions in the system $PbO—SiO_2$ could be estimated as a function of composition (Flood and Knapp), and phase diagrams could be calculated (Förland, 1955).

For oxide systems an earlier approach for finding relationships between properties and composition is that of the cationic field strengths (Weyl, 1932; Dietzel, 1942, 1948/1949). Compound formation, glass-forming and immiscible regions in phase diagrams, melting points and stabilities of oxide compounds, and other properties in oxide systems, especially silicate systems, were empirically related to the field strength, the attractive force $Z_c Z_O/a^2$ between cation and oxygen, with Z_c and Z_O being the charges of the cation and oxygen, respectively, and a the distance between cation and oxygen. A revised form of the field-strength concept, the screening theory, and the acid-base relationships in aqueous solutions, fused salts and in glass systems was proposed by Weyl (1956) and Weyl and Marboe (1962). The field-strength concept was also used by Stoch (1968) to discuss the liquidus of oxide systems. He showed that the solubility of an oxide in a binary melt is approximately proportional to the difference in the reciprocals of the ionic potentials of the cations of the two components.

The higher the value of the field strength, the less ionic is the bonding. Thus the field strength is related to the ionicity of Pauling (1960) which is expressed in terms of the electronegativity differences χ_A and χ_B of cation and anion by $1 - \exp\left[-\frac{1}{4}(\chi_A - \chi_B)^2\right]$. According to the electrostatic valence rule of Pauling, the electrostatic bond strength of stable ionic compounds is defined as $Z/C.N.$, where Z is the charge of the cation and C.N. is its coordination number. The concept of the electrostatic bond-strength has been widely applied to the study of ionic crystals in order to limit the number of possible structures and has been used, for example, by Block and Levin (1957) and by Levin (1970) to interpret immiscibility in oxide (glass) systems. Brown and Shannon (1973) reviewed recent bond-strength concepts and derived empirical bond-strength/bond-length curves for oxides.

These empirical or semi-empirical relationships are convenient for the experimentalist to get some idea of solute-solvent interactions in solutions, since exact theoretical treatments and thermodynamic data are normally missing. The initial approaches have been made in the application of crystal-field theory to ionic melts and of molecular orbital theory to partially covalent fused systems. In addition, several experimental techniques such as ultraviolet and visible electronic spectroscopy, infrared and Raman spectroscopy, nuclear magnetic resonance and X-ray and neutron diffraction techniques have now been applied to obtain information on structural details of ion pairing and complexing and on association and dissociation in molten salts. An excellent compilation of recent approaches to the investigation of ionic solutions and melts has been given by Petrucci (1971). In addition Mamantov (1969) edited a book on experimental techniques for the investigation of molten salts.

Depending on the type of solute and solvent there are other forces between entities in the solutions in place of or in addition to the Coulomb forces between ions mentioned above. Large ions are polarized by small ions of high polarizing strength, thus deforming the electron shell of the former in such a way that electron pairs are formed (another way of describing the field-strength concept). The bond type is then partially covalent. In high-temperature solutions the forces between permanent dipoles (the origin of hydrogen bonding) and between permanent and induced dipoles (P. Debye) are not important. In non-ionic solutions, covalent bonds or the Van der Waals-London type of bonding between nonpolar molecules dominate. Metal and alloy solutions generally are mainly metallic with partially ionic character and many metallic solutions behave as regular solutions (Kleppa, 1958). Good agreement may be obtained between fairly simple theoretical models and experimental data, as shown in the case of solutions of the III–V semiconductors by Panish and Ilegems (1972).

3.4. Chemical Equilibria and Complexes in Solutions

The general principles of equilibria in solutions and the law of mass action, etc., which are applicable in principle also to high-temperature solutions, are treated in many textbooks on chemical thermodynamics and physical chemistry, and therefore will not be discussed here. The books of Garrels and Christ (1965), Smith (1963), Reisman (1970), Kubaschewski, Evans and Alcock (1967), Blander (1964), Petrucci (1971), Davies (1962), Nancollas (1966), Harned and Owen (1959), Friedman (1962), Darken and Gurry (1953), Lumsden (1966) and Mamantov (1969) deal especially with high-temperature solutions.

The principles of acid-base equilibria have been discussed in the previous section. In the following discussion on equilibria in solutions we shall merely discuss some selected topics in reaction and redox equilibria which are significant for crystal growth from high-temperature solutions. Complex formation in solutions needs a somewhat more detailed treatment because of its consequences in the crystal-growth processes. Further information on reaction and redox equilibria may be obtained from the many phase diagrams, of which the most important compilations are mentioned in Section 3.9.

3.4.1. Redox equilibria

Redox equilibria become important when oxides or other compounds of a lower valence state of the metal than that normally stable in air at the applied growth temperatures have to be grown. Examples discussed here are EuO and the other europium (II) chalcogenides, V_2O_3, and the TiO_{2-x} phases. Such phases as well as complex compounds prepared from them often show interesting semiconducting or magneto-optic properties and occasionally other interesting effects such as metal-insulator transitions.

This type of compound, however, has generally been little explored, mainly because of the extreme difficulty encountered in the preparation of homogeneous crystals. This difficulty is encountered even for the cases where the "*suboxides*" have a relatively broad range of stability, as with EuO and V_2O_3, but the difficulties are extreme for the suboxides in the systems Ti—O, Nb—O, and Ta—O. Bartholomew and White (1970) showed that Ti_3O_5 and the Magneli phases in the system Ti—O (see Fig. 3.3) can be grown from $Na_2B_4O_7$—B_2O_3 fluxes under equilibrium conditions if the partial pressure of oxygen as well as the growth temperature are controlled within narrow limits.

Generally the oxygen partial pressure of the gas phase is achieved at the appropriate temperature by gas mixtures containing oxygen (for oxidizing conditions) and by H_2O—H_2, CO_2—CO, and CO—H_2 mixtures according to Richardson and Jeffes (1948) as shown in Fig. 3.4. For sulphides the sulphur partial pressures of a variety of sulphides and the corresponding H_2S—S ratios as a function of temperature are shown in Fig. 3.5.

More recent thermodynamic data have been compiled by Kubaschewski, Evans and Alcock (1967), in the JANAF Thermo-chemical Tables, by Kelley (1961), by Reed (1972) (the latter with practical overlay charts for the various thermodynamic parameters—standard free energy, partial pressure, etc.), by Janz (1967) and by Mills (1974).

The characterization of the many possible phases, for example the determination of the exact stoichiometry, represents a major problem since the accuracy of chemical analytical methods is generally not sufficient to differentiate, for instance, between adjacent oxides in the Ti—O system and to determine the degree of solid solubility. An exact formula, however, can be derived from a careful X-ray structure determination, and later identification is then possible by comparison with the X-ray powder diagram taken with an instrument of good resolution (see Chapter 9).

If the compound (e.g. suboxide) forms a eutectic with its metal and has

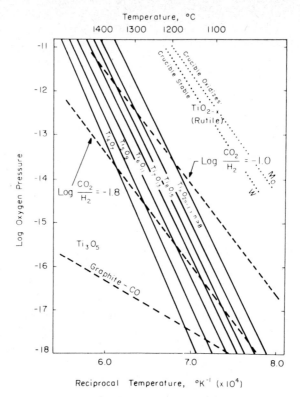

FIG. 3.3. Stability regions of titanium oxides (Bartholomew and White, 1970).

only a limited solid solubility range then it may be grown from the excess metal. This method was used in the case of the europium chalcogenides (EuO, EuS, EuSe, EuTe) which were grown from excess europium by Eick *et al.* (1956), Shafer (1965) and by Reed and Fahey (1966). The stability range of EuO was estimated by McCarthy and White (1967) and recently the phase diagram of Eu—O was redetermined by Shafer *et al.* (1972), and is shown in Fig. 3.6.

On the other hand, however, low-melting oxides of a high valence state of the metal act as solvent for growth of the suboxide, as was demonstrated by the growth of VO_2 from V_2O_5 by Sasaki and Watanabe (1964) and by Sobon and Greene (1966). The phase diagram was obtained by Burdese (1957) and is shown in Fig. 3.7. In addition the epitaxial growth of CrO_2 from CrO_3 at high pressure was reported by DeVries (1966).

A special redox equilibrium has given problems to a number of crystal growers who required to grow chromium-doped compounds from lead-containing fluxes in order to find candidates for replacement of the ruby

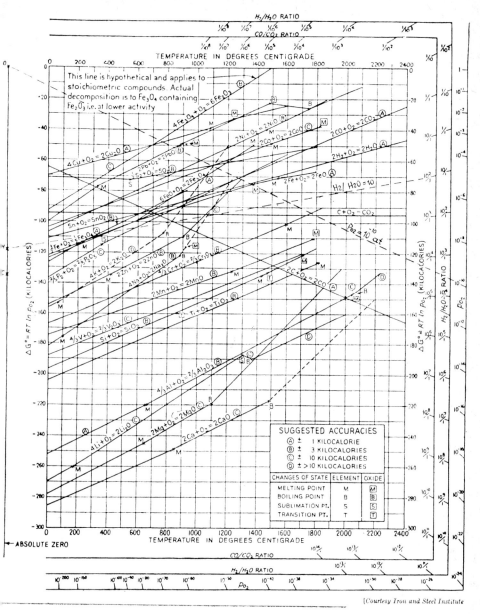

FIG. 3.4. Standard free energy of formation of metals oxides (Richardson and Jeffes, 1948; from Darken and Gurry, 1953).

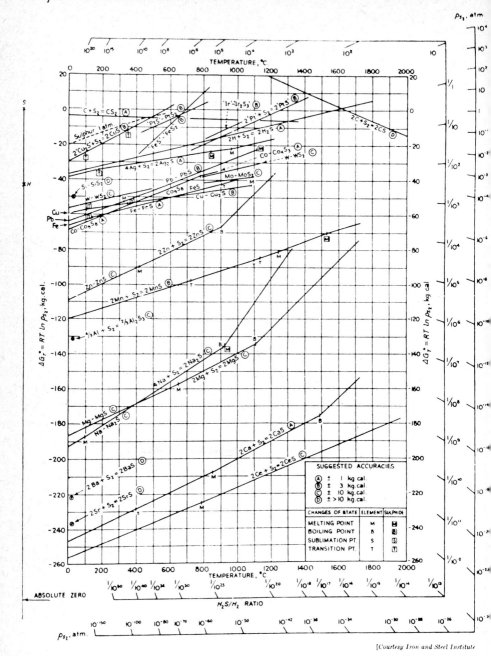

FIG. 3.5. Standard free energy of formation of metal sulphides (Richardson and Jeffes, 1952; from Darken and Gurry, 1953).

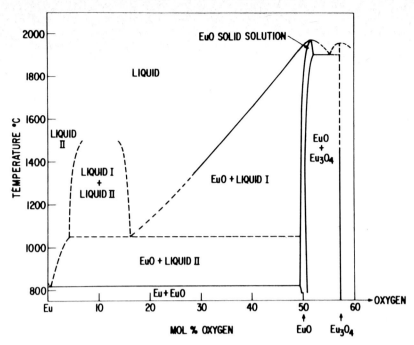

FIG. 3.6. Phase diagram of part of the Eu—O system (Shafer *et al.*, 1972).

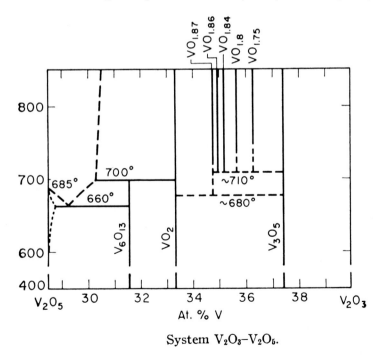

System $V_2O_3–V_2O_5$.

FIG. 3.7. Phase diagram of V_2O_5—V_2O (Burdese, 1957).

laser (the only laser based on chromium transitions), and to grow rare-earth orthochromites:

$$Cr_2O_3 + 1.5O_2 \underset{> \sim 1000°C}{\overset{< \sim 1000°C}{\rightleftarrows}} 2CrO_3 \, (-5.3 \text{ Kcal}).$$

The oxidation of Cr^{3+} to Cr^{6+} in air in lead-containing solutions probably starts well above 1000°C as is evident from experiments to grow chromium-doped Al_2O_3, $GdAlO_3$, $MgAl_2O_4$, $ZnAl_2O_4$, etc. by the slow-cooling technique: the centres of the larger crystals (grown between 1300° and about 1100°C) have the typical reddish ruby-colour of Cr^{3+}, whereas the outer region of the crystals (grown below 1050°C) and the small crystals (nucleated below 1050°C) are colourless—in other words, free from chromium.

In the determination of the system PbO-chromium oxide in air, Negas (1968) found the congruently melting (at 918°C) compound Pb_2CrO_5, which gives a eutectic of 908°C with Cr_2O_3 and of 807° with Pb_5CrO_8 on the PbO-rich side. However, Fukunaga and Saito (1968) indicted that in the pure system Cr—O, CrO_2 is the only stable phase below approximately 300°C. This problem of the stability of six- or four-valent chromium at low temperatures has still to be clarified.

In this connection it should be mentioned that the preparation of the interesting (for magnetic recording) material CrO_2 is quite difficult since it seems to be stable only at low temperatures and since its solubility is very low at such temperatures due to its high lattice energy. Therefore the most promising method for such difficult compounds involves the use of high-pressure solutions (DeVries, 1966).

Six-valent chromium probably could not be incorporated into the above oxide compounds because of the difficulty of charge compensation. The reaction equilibrium and the formation of $PbCr_2O_7$ (Grodkiewicz and Nitti, 1966) are unlikely in view of the phase diagram mentioned above. In order to incorporate chromium homogeneously into crystals and to prepare chromium-containing compounds various methods have been proposed. Remeika (1956) proposed the use of Bi_2O_3 as solvent instead of lead-containing fluxes, and Grodkiewicz and Nitti (1966) proposed the solvent evaporation technique at high temperatures (for a detailed discussion of homogeneous doping, see Section 7.1.3).

The previously mentioned chromium oxide redox equilibrium, when modified for the case of alkali chromates according to Flood and Muan (1950):

$$2M_2Cr_2O_7 \rightleftarrows 2M_2CrO_4 + Cr_2O_3 \text{ (solid)} + 1.5O_2$$

for $M = $ Li, Na, K, and Tl could be used to grow Cr_2O_3 by slowly increasing the temperature of a dichromate melt. The same principle of shifting the redox equilibrium has been used by McWhan and Remeika (1970) and by Foguel and Grajower (1971) in order to grow V_2O_3 crystals by reduction of V_2O_5 in the flux.

In several cases acid-base relationships influence the oxidation state of transition metals in such a way that an increase in concentration of a basic oxide favours a higher valence state of the transition metal ion. This was demonstrated, for example, for glass systems by Paul and Lahari (1966) and applied to growth of nickel ferrite (spinel) crystals from BaO—B_2O_3 fluxes by Smith and Elwell (1968): BaO-rich solutions produced $NiFe_2O_4$ crystals with less Fe^{2+} than B_2O_3-rich fluxes. Redox equilibria in glasses have been discussed by Weyl (1951), Tress (1960, 1962), Holmquist (1966), Johnston (1966) and by Paul and Douglas (1966).

An important redox equilibrium is that of

$$2Fe^{3+} \rightleftarrows Fe^{2+} + Fe^{4+}$$

since it affects optical absorption and other properties of magnetic garnets of the type $R_3Fe_5O_{12}$. This equilibrium is shifted not only by the temperature but also by Ca and Si addition to the PbO—B_2O_3 solution. A model proposed by Nassau (1968) cannot explain all the experimental facts satisfactorily.

An appreciable number of redox equilibria have been studied in connection with phase-diagram determinations (see Section 3.8), and a classical paper on this topic has been published by Muan (1958) on the systems Fe—O, FeO—Fe_2O_3—SiO_2 and MgO—FeO—Fe_2O_3—SiO_2. The study of redox and other equilibria is one of the main fields of modern experimental and theoretical petrology, as discussed for example by Garrels and Christ (1965).

3.4.2. Reaction equilibria

Instead of a general discussion on phase equilibria and phase diagrams we wish to restrict the discussion to a few special groups of reactions, which are important for the experimentalist (and sometimes overlooked).

Temperature-dependent chemical reactions of the type (Barrer, 1949)

$$Al_2O_3 + 3CaF_2 \rightleftarrows 3CaO + 2AlF_3$$

are worth mentioning since they can lead to an extreme loss by evaporation of the most volatile component (here AlF_3). A similar equilibrium was proposed by Morozewicz (1899):

$$SiO_2 + Na_2WO_4 \rightleftarrows Na_2SiO_3 + WO_3$$

with WO_3 as the volatile compound. The shifting of the above equilibria is obvious from the law of mass action and the principle of Le Chatelier. Other temperature-dependent equilibria are significant for the preparation of simple oxides or oxide compounds, respectively, for instance the reactions postulated by Grodkiewicz and Nitti (1966):

$$PbHfO_3 \underset{\longleftarrow}{\overset{1200°C}{\longrightarrow}} PbO + HfO_2$$

$$PbTiO_3 \underset{\longleftarrow}{\overset{1200°C}{\longrightarrow}} PbO + TiO_2.$$

These reactions suggest that HfO_2 and TiO_2 can be prepared from lead-containing fluxes only by solvent evaporation above 1200°C, and $PbTiO_3$ and $PbHfO_3$ have to be grown at much lower temperatures.

These few examples of equilibria demonstrate their significance for crystal growth from high-temperature solutions, and the principles outlined in this chapter on solvents and solutions should be kept in mind when "inexplicable" results are obtained.

A special case of reaction equilibria in solutions is shown by the *reciprocal salt systems*. These are mixtures of salts containing at least two cations and two anions, which in the simplest case undergo the reciprocal reaction

$$AX + BY \rightleftarrows AY + BX.$$

The thermodynamics of reciprocal systems, their deviation from ideal solution behaviour and its explanation, the reciprocal Coulomb effect, and some theoretical approaches with experimental evidence have been reviewed by Blander (1964) and Blander and Topol (1965). Several aspects of phase diagrams of reciprocal systems have been discussed by Ricci (1964).

Instead of losing reactants by evaporation, reactants can be transported to the system. As an example, for the growth of GaAs crystals or layers $AsCl_3$ is transported to the gallium solution. GaP is produced according to (Plaskett, 1969) $2Ga + 2PH_3 \rightarrow 2GaP + 3H_2$. This growth mechanism is similar to the *VLS* mechanism (Section 7.1.2.C) where crystal constituents are transported by vapour, enter the liquid solution in order to be transported to the growing (solid) crystal.

Another type of reaction involves reactive gases which decompose constituents of the solvent in order to form the required phase (Brixner and Babcock, 1968), for example,

$$BaCl_2 + 6Fe_2O_3 + H_2O \rightarrow BaFe_{12}O_{19} + 2HCl.$$

Further examples of reactions in high-temperature solutions are given in Section 7.1.2.C.

3.4.3. Complex formation in solutions

The solubility is increased by solvation and by complex formation in the solution as well as by other solute-solvent interactions. Complex formation and other forms of association generally result in maxima or minima when the following properties are measured and plotted *versus* composition: density, liquidus temperature, vapour pressure, viscosity, specific conductance, surface tension, etc. However, such maxima and minima are not necessarily indicative of complex formation and results must be interpreted with care. The deviation from the ideal solubility curve may indicate the degree of association as shown in Section 3.2. Valuable information on the coordination of many elements in molten salts or solutions is obtained by various other experimental and theoretical techniques as discussed by Petrucci (1971) and by Mamantov (1969).

Spectroscopic evidence for groups of ions of well-defined symmetry has often been established, especially of configurations of transition metal ions. Such spectroscopic studies are performed on a time scale orders of magnitude faster than the diffusion time. Therefore the observed local structure is often described as "configuration", "centre" or "group". The term "complex" or "species" should be used only for such groups which have a lifetime long with respect to the characteristic diffusion times. The lifetime is important with respect to the method of investigation, but also with respect to transport phenomena. The "holes" in liquids are small compared with the size of the species, and generally the quasi-lattice model gives a more realistic and preferable description of a liquid. Therefore, the motion of complexes as well as of other species is connected with a cooperative rearrangement in a volume (of the liquid) large compared to that of the complex. In the absence of applied fields this results in random diffusive displacements. However, the presence of an electric field in electrolysis, or of a concentration gradient ahead of the crystal growing from a supersaturated solution, imposes a directional bias so that large complexes can have an appreciable mobility. The presence of the field implies that the moving complex has to move against the stream of other particles (of opposite or neutral sign in electrolysis, of solvent in the case of solution growth) in such a way that no (large) holes are formed.

Vibrational spectroscopy, especially Raman spectra, can give more information on complexes in high-temperature solutions than electronic spectroscopy. However, the observed number and frequency of the Raman bands are dependent on the force constants, masses and geometrical arrangement of all the ions in the quasilattice of the solution. Only when the force constants of the complexes are large compared with those of the rest of the solution is it possible to relate the number of observed bands by group theory to the symmetry of the complex in question. In several cases unambiguous information on the geometry of complexes has been obtained from vibrational spectra which in other cases are prone to misinterpretation.

In several solution systems which can be quenched as glasses, evidence on the type of complexes and on the coordination can be obtained from optical spectra of the glasses, as discussed by Berkes and White (1966), or by other techniques such as electron paramagnetic resonance or the Mössbauer effect.

However the difficulties of most of the above techniques increase at high temperatures and often make measurements impossible. Therefore little information on complex formation in high-temperature solutions is available for temperatures above 800°C which is a lower limit for most crystal growth. Reviews on complex ions in molten salts have been given by Van Artsdalen (1959), Blander (1964), Bloom (1967), in the review "Molten Salts as Solvents" by Bloom and Hastie (1965) and in the books of Mamantov (1969) and Petrucci (1971).

As an example of the dependence of viscosity on complex formation, Fig. 3.8 shows viscosity isotherms for the systems KCl—MgCl$_2$ (a) and NaF—AlF$_3$ (b). In both cases the complex formation as a precursor of compound formation (KMgCl$_3$, Na$_3$AlF$_6$) is indicated by singularities in the viscosity isotherms. Also the decreasing tendency of complex formation at higher temperatures is obvious from Fig. 3.8. Complex formation is to be expected frequently in molten compounds having a congruent melting

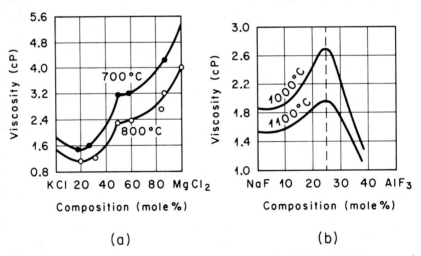

FIG. 3.8. Viscosity isotherms of (a) KCl—MgCl$_2$, (b) NaF—AlF$_3$ (Belyaev et al., 1964).

point as was shown above. Phase diagrams are therefore expected to give indications of complex formation in the melts in which, at least at not too high temperatures, the short-range order or the corresponding crystalline phase is preserved. Hence the many phase-diagram determinations play a major role in the study of complex formation. Cryoscopic measurements also indicate the degree of association as was demonstrated by a number of measurements by Rolin (1951) on the effect of a variety of oxides on the melting of cryolite (Na_3AlF_6). A review of cryoscopy and phase relations in dilute molten-salt solutions was given by Kozlowski (1970).

The formation of complexes in melts of congruently melting compounds increases significantly as the liquidus temperature is approached. This is proved by measurement of the specific conductance (Bloom, 1963). A mixture of $PbCl_2$ and NaCl shows the normal relation between specific conductance and temperature, i.e. an almost linear dependence on approaching the liquidus temperature to within a few degrees as shown in Fig. 3.9. The $CdCl_2$—KCl mixture (with the congruently melting compound $KCdCl_3$), however, shows a strong decrease of conductance at about 30–40° above the liquidus temperature, which may indicate a tendency to complex formation.

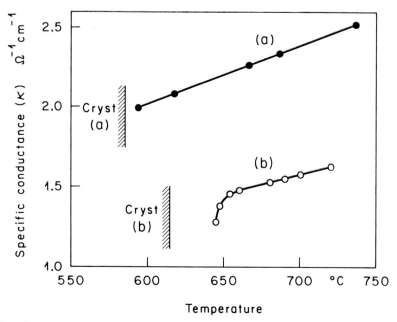

FIG. 3.9. Specific conductance near the crystallization front, which is indicated by the shaded line. (a) 50% NaCl + 50% $PbCl_2$ (molar), (b) 79.6% KCl + 20.4% $CdCl_2$ (Bloom, 1963).

Another factor favouring complex and compound formation is the difference between solute and solvent with regard to their acid-base character according to Lewis (1923, 1938). As an example, the difference in compound formation in molten-salt mixtures of $AlCl_3$ and of $SbCl_3$ is shown in Fig. 3.10 (Kendall *et al.*, 1923). In the little-reactive $SbCl_3$ the Cl as well as the Sb ions have a noble gas electron configuration, whereas in the case of $AlCl_3$ the aluminium is lacking two electrons from a noble gas electron configuration and therefore tends to react with other salts.

Molecules of high polarizability show a larger tendency to chemical combination and complex formation in the solution than those of low polarizability. This fact is well established from the phase diagrams of

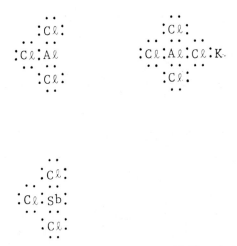

FIG. 3.10. Electron configuration of $AlCl_3$, $SbCl_3$ and of $KAlCl_4$.

lead chloride with the alkali halides which show an increasing degree of compound formation with increasing radius of the alkali ion. Thus $PbCl_2$—NaCl forms a simple eutectic, while in $PbCl_2$—KCl one congruently and one incongruently melting compound are formed, and in $PbCl_2$—RbCl three congruent compounds exist.

An interesting case of the utilization of complex formation was reported by Kunnmann *et al.* (1965) who described the crystallization of transition metal oxide compounds from sodium tungstate-sodium pyrotungstate solutions as will be discussed in Section 3.6.

Also discussed later (Section 3.6.2 and Chapter 7) will be the role of complex formation in determining not only the solubility but also the width of the metastable range. The effect of complex formation on mass transport and, through its effect on desolvation and interface kinetics, on the stability of growth may also be appreciable. There are cases where complex formation is desirable in order to achieve a sufficient width of the metastable region in order to prevent multinucleation (as in growth of ZrO_2, ThO_2, $SrTiO_3$, TiO_2, etc. from alkali fluoride fluxes). On the other hand, complex formation should be prevented in viscous solutions such as in borate fluxes.

The success of several complex solvent mixtures such as PbO—PbF_2 or PbO—PbF_2—B_2O_3 is due not only to their high dissolving power but also because these excellent solvent systems appear to show an optimum degree of complex formation and an optimum viscosity, therefore a sufficiently broad metastable region. On the other hand, the complexes formed are not stable enough to retard the interface kinetics.

Such statements are all qualitative. Further systematic experimental and theoretical work is necessary to find quantitative information on complex formation in solution and its effect on the crystal-growth processes. The role of complex formation ahead of the growing crystals and its influence on the habit is discussed in Section 3.6.3, in Chapter 5 and in Section 7.1.1.

3.5. High-temperature Solvents and Solutions

It is generally desirable that the aspects outlined in the first part of this chapter should be systematically applied to high-temperature solvents and solutions. Very little work has been done to date on the solubility of oxides in molten salts and on the structural interpretation of solutions of oxides (and chalcogenides generally) in molten salts. Other high-temperature solutions like metallic-ionic systems have also not been studied in detail. This may partially be attributed to the fact that the conditions (high temperatures, high volatility) are extremely unfavourable for experimental investigations. Another reason is that crystal growth from high-temperature solutions has become really active only in the past 15 years. Therefore it is no surprise that in most reviews and books on molten salts little (or no) attention has been given to solutions of oxides in molten salts. Crystal growers normally are unable to find the time for such basic investigations. However, information on such systems is urgently needed for the selection of the optimum solvents and growth conditions. What kind of interaction of solute and solvent is desirable for a high solubility, a reasonably broad metastable region and a small tendency to inclusion formation? This important question can only be answered by systematic studies of the structure and type of solution (deviation from ideality, degree of solvation) on a variety of molten salt-oxide systems and by comparison with crystal-growth experiments.

Compilations of solvents used for crystal growth from high-temperature

D

solutions have been given by Laurent (1969), Wanklyn (1975) and in Chapters 2 and 10, and several aspects of HTS are discussed by Elwell (1975).

Fortunately, since about 1966 a new field of physico-chemical investigation has been directed towards ionic, covalent, and metallic melts at "high" temperatures. But the "high" temperatures of such investigations rarely reach 1000°C due to the principle of the method or to experimental difficulties, and so there is little direct relevance to crystal growth from high-temperature solutions which is mostly carried out above 1000°C. Nevertheless the results obtained in research on high-temperature melts are very valuable for crystal growers, and it would be of great interest to increase the research on high-temperature solutions to include systems of interest for crystal growth at temperatures above 1000°. There is a lack of basic information on high-temperature solutions, and a systematic investigation of such properties as density, expansion coefficient, viscosity, thermal and electrical conductivity, heat capacity and on structural properties (association, dissociation) needs to be carried out so that the data obtained can be correlated with crystal-growth phenomena. The few measurements which have been carried out will be discussed in Section 3.7.

3.6. Choice of Solvents

3.6.1. Properties of ideal high-temperature solvents
The wide variety of substances crystallized from high-temperature solutions make a generalized discussion of an ideal solvent difficult. The first indications of requirements for high-temperature solvents were given by Anikin (1958), and additional aspects have been listed in the several reviews on flux growth which were cited in Chapter 1. In the following we shall discuss the published criteria for the choice of solvents and add some of our own observations.

The main requirement of a solvent is that it will dissolve the solute to an appreciable amount, a practical minimum being of the order of one per cent. Generally one can say that crystal growth is facilitated by a high solute concentration if the required temperature is not too high and if the solution is not too viscous. A favourable viscosity of the solution is 1 centipoise to 1 poise, while the maximum practical viscosity would be about 10 poise. The solvent must not react with the solute to form solid compounds, or if it does, these compounds must not be stable in the crystallization temperature range of the phase required. Furthermore the range of solid solutions between solute and solvent should be as small as possible, and often it is the incorporation of solvent ions into the grown crystals which prevents the application of a particular solvent. The desirable properties of an ideal solvent are listed in Table 3.4.

TABLE 3.4. Properties of an Ideal Solvent

1. High solubility for the crystal constituents.
2. The crystal phase required should be the only stable solid phase.
3. Appreciable change of solubility with temperature.
4. Viscosity in the range of 1 to 10 centipoise.
5. Low melting point.
6. Low volatility at the highest applied temperature (except when the solvent evaporation technique is used).
7. Low reactivity with the container material (see Chapter 7).
8. Absence of elements which are incorporated into the crystal.
9. Ready availability in high purity at low cost.
10. Density appropriate for the mode of growth (see Chapter 7).
11. Ease of separation from the grown crystal by chemical or physical means (see Chapter 7).
12. Low tendency of the solvent to "creep" out of the crucible (see Chapter 7).
13. Low toxicity.

There is no solvent which fulfills all these ideal properties and a compromise is always necessary, depending on the type of the crystal and on the requirement of size and quality which makes some "ideal" property particularly important. The first three properties of Table 3.4 have to be fulfilled if a high yield of crystals is to be achieved by any of the normal techniques used to obtain supersaturation. A low viscosity is desirable to facilitate the achievement of a sufficiently homogeneous solution and of a fast stable growth rate as discussed in Chapter 6, so that crystal growth occurs uniformly at low supersaturation. A low melting point implies that crystal growth may occur at a convenient temperature in order to prevent destructive phase transitions, volatilization of solute or solvent constituents, and a high concentration of equilibrium defects. Often incongruent melting of the phase to be crystallized necessitates a low growth temperature. In addition, high temperatures are inconvenient with respect to furnace and power requirements. The other points will be discussed in the next sections except for the last "ideal" property, the low toxity, which is obviously desirable but very rarely fulfilled. For instance, the most popular solvents for oxides and oxide compounds contain lead and fluorine and are therefore very poisonous and most high-temperature solvent chemicals are poisonous to some extent. Therefore care must always be taken in handling chemicals, especially if they are in fine powder form, during the weighing and crucible-filling procedure. Poisonous vapours from the hot furnaces are also dangerous and may cause acute or chronic illnesses.

3.6.2. Crystal chemical aspects of choice of solvents

The various types of interaction between solute and solvent have been
mentioned above (Sections 3.3 and 3.4). From the solubility point of view
one would like to have a great chemical and crystal-chemical similarity
between solute and solvent, but that generally favours mutual solid
solubility. Therefore the optimum choice is a solvent which is chemically
similar in the type of bonding to the solute, but crystal-chemical differences
should exist in order to prevent solid solubility between solute and solvent.
Most experience has been gained from flux growth of oxides which are
therefore taken as examples for the discussion of many questions. Obvious
crystal-chemical differences are established by differences in ionic size
or in the stable valency states at the growth temperature. If these differences
are large enough and if they are not compensated by solvent constituents

TABLE 3.5. Examples of Solvent-solute Pairs with Crystal-chemical Differences

Difference of Cationic and Anionic Radii
CeO_2, ThO_2 from Li_2WO_4
CoO, TiO_2 from $NaCl$

Difference of Cationic Radii
MgO, $MgAl_2O_4$ from PbF_2, $Be_3Al_2Si_3O_{18}$ from Li_2MoO_4, Na_2MoO_4, etc.
$Y_3Fe_5O_{12}$ from $BaO.0.6B_2O_3$
Al_2O_3 from LaF_3

Difference of Anionic Radii
$BaTiO_3$ from $BaCl_2$
$PbZrO_3$ from $PbCl_2$
$BaWO_4$ from $BaCl_2$
CaO, CoO, CuO from $NaCl$

Difference of Valency States of both Cation and Anion
$PbTiO_3$, $BaTiO_3$ from KF
Al_2O_3 from PbF_2
$Y_4Si_3O_{12}$ from KF
TiO_2 from Na_3AlF_6, $NaCl$

Difference of Valency State of Cation
VO_2 from V_2O_5
$ZrSiO_4$ from Li_2O-MoO_3, Na_2O-MoO_3
Al_2O_3 from PbO

Difference of Valency State of Anion
Al_2O_3 from LaF_3, PbF_2
$MgAl_2O_4$, MgO, TiO_2, MnO from PbF_2
$BaTiO_3$ from $BaCl_2$

or by impurities, little incorporation of solvent ions into the crystals grown can be expected.

Examples of crystal-chemical differences between solute and solvent ions are given in Table 3.5.

A clear correlation between crystal-chemical differences between solute and solvent and the concentration of incorporated impurity is not possible since most chemical analyses will be influenced by solvent inclusions due to unstable growth. Therefore only such analyses can be discussed which show low impurity concentrations, whereas high impurity concentrations of ions of different (from the solute) valence states indicate flux inclusions (not incorporation of flux ions into the crystal lattice) except for such cases where charge compensation is possible. As an example of the increase in impurity concentration in the presence of charge-compensating impurity, Barczak (1965) reported that the cosolubility of V_2O_5 and NiO in haematite, Fe_2O_3, was much higher than the separate solubilities of these compounds. Some examples of high impurity concentrations have been reported and these are plausible when the crystal-chemical differences are small. For example, it may be expected that rare-earth compounds grown from bismuth-containing fluxes, and niobates and tantalates grown from vanadate fluxes contain large amounts of bismuth and vanadium, respectively, as shown in Table 3.6.

Table 3.7 lists examples of low impurity contents of crystals grown from solvents which have large differences in ionic radii and/or valence state compared with the solute.

Care is necessary for those cases where either the solute or the solvent contains ions of variable valency state, especially the transition metal ions. Cations of V, Nb, Ta, Cr, Mo, W, Ti, Fe and the other iron-group elements and several rare-earth elements readily change their valency state and can therefore reduce or eliminate crystal-chemical differences between solute and solvent which arise from differences in the valency states (and ionic radii). In such cases unexpected incorporation of impurities (direct, or indirect through charge compensation) might occur and should be mini-

TABLE 3.6. Substitutional Bi^{3+} Concentration in Crystals Grown from Bi-containing Solvents

Crystal	Solvent	Impurity	Reference
$GdAlO_3$	Bi_2O_3—B_2O_3	5.5% Bi	Wanklyn (1969)
$TbAlO_3$	Bi_2O_3—B_2O_3	1.4% Bi	Wanklyn (1969)
$TbNbO_4$	Bi_2O_3—V_2O_5	2% V, 4% Bi	Garton et al. (1972)
$GdVO_4$	Bi_2O_3—V_2O_5	7% Bi	Garton et al. (1972)
$Dy_3Fe_5O_{12}$	Bi_2O_3—V_2O_5	20% Bi	Garton et al. (1972)

TABLE 3.7. Impurity Concentration in Crystals Grown from Solvents with a Large Difference in Ionic Radii and/or Valence State

Crystal	Solvent	Impurity	Reference
$LiFe_5O_8$	PbO	0.2% Pb	Folen (1960)
Al_2O_3	PbF_2	0.05% Pb	Giess (1964)
Al_2O_3	$PbO—PbF_2$	0.05–0.1% Pb	Chase and Osmer (1964)
NiO	PbF_2	0.1% Pb	Hill and Wanklyn (1968)
$\beta\text{-}Ga_2O_3$	$PbO—PbF_2$	0.07–0.2% Pb	Katz and Roy (1966)
$MgAl_2O_4$	PbF_2	<10 ppm Pb	Cloete et al. (1969)
$MgAl_2O_4$	PbF_2	0.1% Pb	Wang and McFarlane (1968)
$MgGa_2O_4$	$PbO—PbF_2$	0.02% Pb	Giess (1962)
$GdFeO_3$	$PbO—PbF_2$	0.05% Pb	Wanklyn (1969)
$Y_3Al_5O_{12}$	$PbO—PbF_2—B_2O_3$ with excess Al_2O_3	0.01% Pb	Van Uitert et al. (1965)
NpO_2	$PbF_2—B_2O_3$	0.02% B, 0.1% Pb	Finch and Clark (1970)
NpO_2	$Li_2O—MoO_3$	10 ppm Mo	Finch and Clark (1970)
$GdVO_4$	$PbO—V_2O_5$	0.03% Pb	Garton et al. (1972)
ZnS	Na_2S_X ($X=2$–5)	35 ppm Na	Scheel (1974a)
CdS	Na_2S_X ($X=2$–5)	30–50 ppm Na	Scheel (1973, 1974a)
FeS_2	Na_2S_X ($X=2$–5)	65 ppm Na	Scheel (1974a)
$CdCr_2Se_4$	$CdCl_2—PbCl_2$	<50 ppm Pb	Kuse (1970)

mized by proper control of the experimental conditions (temperature, oxygen partial pressure).

A few examples are given in Table 3.8 of high impurity concentrations in flux-grown crystals which could only be explained by:

a. flux inclusions (which can be too small to be detected by the naked eye or which are difficult to detect in the opaque orthoferrites),
b. unusual valence state of a cation (e.g. Fe^{4+}, Pb^{4+}),
c. a high concentration of vacancies,
d. a high concentration of charge-compensating impurities, or
e. by a combination of these effects.

In case of HfO_2 a charge-compensated replacement of $Hf^{4+} + O^{2-}$ by $Bi^{3+} + F^-$ would be possible if the ionic radii of Hf^{4+} and Bi^{3+} were not too different. Such questions could be answered if complete impurity levels were determined and published. The possibility cannot be excluded that flux inclusions are responsible for some of the high impurity concentrations listed in Table 3.8. This is also indicated by the low lead incorporation (0.05% Pb) reported for $GdFeO_3$ grown from $PbO—PbF_2$ flux (see Table 3.7) compared to the 0·9% Pb for the same compound listed in Table 3.8. It is questionable whether such a difference could be accounted for by different growth conditions. Remeika and Kometani (1968) suggested the com-

TABLE 3.8. High Impurity Concentrations in Some Flux-grown Crystals

Crystal	Solvent	Impurity	Reference
1. HfO_2	BiF_3—B_2O_3	5% Bi	Grodkiewicz and Nitti (1966)
2. Gd_2TiO_5	PbO	4% Pb	Garton and Wanklyn (1968)
3. $LaFeO_3$	PbO—B_2O_3	13% Pb	Remeika and Kometani (1968)
4. $RFeO_3$ (R = La—Lu)	PbO—B_2O_3	0.08% to 13.4% Pb	Remeika and Kometani (1968)
5. $GdFeO_3$	PbO—B_2O_3	0.9% Pb	Remeika and Kometani (1968)
6. $Gd_{0.9}Tm_{0.1}FeO_3$	PbO—PbF_2—B_2O_3	0.72% Pb	Giess et al. (1970)

TABLE 3.9. Examples of Solvents Chosen for their Similarity with the Solute

Crystal	Solvent	Reference
1. Common cations and common anions:		
$BaTiO_3$, $SrTiO_3$	TiO_2	Belruss et al. (1971)
$Ca_5(PO_4)_3Cl$	$CaCl_2$	Prener (1967)
$Pb_3MgNb_2O_9$	PbO	Mylnikova and Bokov (1962)
Pb_2CoWO_6	PbO—$PbWO_4$	Bokov et al. (1965)
$NaCrS_2$, $NaInS_2$	Na_2S_X (X = 2–5)	Scheel (1974a)
$KFeS_2$	K_2S_X (X = 2–6)	Scheel (1974a)
ZnS	Ba_2ZnS_3	Malur (1966)
$K(Nb, Ta)O_3$	K_2CO_3	for example: Wilcox and Fullmer (1966)
2. Common cations:		
$BaTiO_3$	$BaCl_2$	Blattner et al. (1947)
CdS	$CdCl_2$	Bidnaya et al. (1962)
FeS_2	$FeBr_3$	Wilke et al. (1967)
$PbTiO_3$	$PbCl_2$	Nomura and Sawada (1952)
$PbZrO_3$	PbF_2	Jona et al. (1955)
$CdCr_2Se_4$	$CdCl_2$	Berger and Pinch (1967)
3. Common anions:		
$CaWO_4$	Na_2WO_4	Robertson and Cockayne (1966)
Oxides	PbO, Bi_2O_3, V_2O_5 etc.	Many examples (see Chapter 10)
RVO_4	$Pb_2V_2O_7$	Feigelson (1968)
RPO_4	$Pb_2P_2O_7$	Feigelson (1964)
RVO_4	V_2O_5	Brixner and Abramson (1965)

pensation of the Pb^{2+} incorporated into $RFeO_3$ by four-valent iron, whereas Lefever *et al.* (1961) proposed dendrite formation and trapping of flux.

If unusually high impurity concentrations are observed and if there is indication of the formation of solid solutions (for instance from the absence of inclusions) this should be checked by appropriate techniques (see Chapter 9) and in many cases an exact determination of the lattice constants clarifies the situation.

Another approach for the choice of good solvents is to look for those with anions or cations in common with the solute. The examples in Table 3.9 illustrate this possibility.

In many cases little solid solubility is to be expected (at not too high growth temperatures) when there are large differences in the type of bonding between solute and solvent. Normally this advantage has the great handicap of a low solubility, but in several cases this choice of solvent has been applied successfully as in the preparation of metals from ionic melts or in the crystallization of BaO using Ba as solvent (Sproull *et al.*, 1951; Libowitz, 1953).

In summarizing the choice of solvents by crystal-chemical principles it may be stated that, by a proper choice, impurity levels of less than 0.1% (often in the ppm range) can be achieved as long as stable growth (see Chapter 6) takes place and solvent inclusions are prevented, and high purity chemicals are used.

3.6.3. Additional criteria for choice of solvents

Several authors (J. W. Nielsen, E. A. D. White, and others) attributed the excellent solvent properties of PbO, PbF_2, Bi_2O_3, KF and BaO $\times$ $0.6B_2O_3$ to the high polarizability of the respective cations. But a number of excellent solvents are known which do not contain readily polarizable cations, for example Na_3AlF_6, LiF, $Na_2B_4O_7$, whereas several others show a medium degree of polarizability, such as the vanadates, molybdates and tungstates. One possibility seems to be that solute and solvent ions should not have too large a difference in their polarizabilities according to the similarity rule. On the other hand, large differences in polarizability, the solvent being readily polarizable (large ions) and the solute having a high polarizing strength (small ions), or *vice versa*, may have the effect of strong solvent-solute interactions and thus lead to higher solubilities unless stable compounds are formed (see Section 3.3).

Quite often the materials with good solvent properties are those which form compounds with the solute at lower temperature or in a different concentration range, or which can be considered as reciprocal salt systems (Scheel, 1974b). The relation between solute-solvent compound formation and solubility is analogous to the effect of hydration on solubility in water (see Section 3.3). Cases of this type are listed in Tables 3.10 and 3.11. In Table 3.10 only simple solute compounds are listed but the same arguments which hold, for instance, for Fe_2O_3 are also applicable to Fe_2O_3-containing compounds ($R_3Fe_5O_{12}$, $M^{2+}Fe_2O_4$, $RFeO_3$, etc.). Complex solvents are excluded from the Table for simplicity. Table 3.10 shows that

TABLE 3.10. Examples of Solute-solvent Pairs which Form Compounds
(C = Congruent Melting, I = Incongruent Melting)

Solute	Solvent	Compound	Melting point	Phase diagram†
Al_2O_3	PbO	$PbAl_2O_4$	980°C I	280
Fe_2O_3	PbO	$PbFe_{12}O_{19}$	1315°C I	282
Cr_2O_3	PbO	Pb_2CrO_5	918°C C	2134
TiO_2	PbO	$PbTiO_3$	1295°C C	2561
ZrO_2	PbO	$PbZrO_3$	1570°C I	2330
$La_2O_3(R_2O_3)$	PbO	$La_2Pb_4O_7$	1220°C C	2328
Al_2O_3	B_2O_3	$Al_4B_2O_9$	1035°C I	2339
ZnO	B_2O_3	$Zn_5B_4O_{11}$	1045°C I	300
La_2O_3	B_2O_3	$La_6B_2O_{12}$	1386°C I	321
MgO	B_2O_3	$Mg_3B_2O_6$	1356°C C	261
Cr_2O_3	V_2O_5	$CrVO_4$	810°C I	333
Al_2O_3	V_2O_5	$AlVO_4$	695°C I	320
NiO	V_2O_5	$Ni_3V_2O_8$	1210°C I	279
ZnO	V_2O_5	$Zn_3V_2O_8$	890°C C	2338
ZrO_2	V_2O_5	ZrV_2O_7	750°C I	2405
Fe_2O_3	Bi_2O_3	$Bi_2Fe_4O_9$	920°/960°C I	2358/2357

† Number of the phase diagram from the compilation of Levin et al. (1964, 1969).

TABLE 3.11. Examples of Solute-solvent Pairs which Tentatively can be Described as Reciprocal Systems

Solute	Solvent	Reaction
CuO (or CaO, CoO, NiO)	NaCl	$CuO + 2NaCl \rightleftarrows CuCl_2 + Na_2O$
$CaWO_4$	NaCl	$CaWO_4 + 2NaCl \rightleftarrows CaCl_2 + Na_2WO_4$
$BaSO_4$	NaCl	$BaSO_4 + 2NaCl \rightleftarrows BaCl_2 + Na_2SO_4$
$BaTiO_3$	KF	$BaTiO_3 + 2KF \rightleftarrows BaF_2 + K_2TiO_3$
ZnS	$PbCl_2$	$ZnS + PbCl_2 \rightleftarrows ZnCl_2 + PbS$
Al_2O_3	PbF_2	$Al_2O_3 + 3PbF_2 \rightleftarrows 2AlF_3 + 3PbO$
Y_2O_3	PbF_2	$Y_2O_3 + PbF_2 \rightleftarrows 2YOF + PbO$
$SrTiO_3$	PbF_2	$SrTiO_3 + PbF_2 \rightleftarrows SrF_2 + PbTiO_3$
Zn_2SiO_4	$Li_2Mo_2O_7$	$Zn_2SiO_4 + Li_2Mo_2O_7 \rightleftarrows 2ZnMoO_4 + Li_2SiO_3$
$CaSiO_3$	Na_2WO_4	$CaSiO_3 + Na_2WO_4 \rightleftarrows CaWO_4 + Na_2SiO_3$
$Y_3Fe_5O_{12}$	PbB_4O_7	$12Y_3Fe_5O_{12} + 9PbB_4O_7 \rightleftarrows 36YBO_3 + 5PbFe_{12}O_{19} + 4PbO$

D2

compound formation of solute-solvent pairs is very common, but a few exceptions are listed in Table 3.12.

Compound formation of a special type occurs in reciprocal salt systems in which equilibria of the type $AX + BY \rightleftarrows AY + BX$ occur. The interactions in many solute-solvent systems can be formulated as reciprocal reactions, and supersaturation can be achieved by shifting the

TABLE 3.12. Examples of Solute-solvent Pairs which do not Form Compounds and which cannot be Described as Reciprocal

Solute	Solvent	No. of phase diagram (Levin et al., 1964, 1969)
BeO	WO_3	2294
BeO	$Li_2O.2.25MoO_3$	2425
ZnO	PbO	2326
SiO_2	V_2O_5	2401

equilibrium in the required direction, for example by a change in temperature. Examples of such reciprocal crystal-growth systems are given in Table 3.11.

Aspects of the choice of solvents which have to be found out by experiment are the effects of the solvent on the width of the metastable region (prevention of uncontrolled nucleation) and on stable growth.

Complex solvent compositions have several advantages with respect to a low melting point and to lowering the vapour pressure of the most volatile solvent (or crystal) component. On the other hand, certain additions may lead to formation of undesired volatile compounds like SiF_4, AlF_3, etc.

Impurities in the solutions are generally undesirable since they may become incorporated in the crystal, or promote enhanced solvent ion incorporation by charge compensation, and frequently impurities cause deleterious growth. For example, Austerman (1965) reported that addition of 0.03 to 0.1 wt% SiO_2 to the lithium molybdate solvent caused a severe degradation of the perfection of the BeO crystals grown.

However, the addition of certain ions or compounds may sometimes have very favourable effects on growth or crystal habit and examples are listed by Wanklyn (1975). The most frequently used additive is B_2O_3 (1–5%) which is said to increase the width of the metastable region, to increase the solubility and to decrease the vapour pressure of the solvent, although high B_2O_3 additions are unfavourable due to an increase in the viscosity. In many cases, complex formation in front of the growing crystals is expected to be the reason for favourable growth. In the case of

B_2O_3, unstable complexes such as $FeBO_3$ and YBO_3 would be possible. Remeika (1970) reported a dramatic reduction in the number of crystals nucleated on adding 0.5–1% of V_2O_5 to PbO—PbF_2 solutions used for the growth of ferrites and garnets. Scheel (see Kjems *et al.*, 1973) observed a drastic increase in the size of inclusion-free $LaAlO_3$ crystals grown from PbO—PbF_2—B_2O_3 solvent when 0.7 wt% V_2O_5 was added to the flux, probably due to $LaVO_4$ complex formation. Other examples where complex formation by additives plays a favourable role are Al_2O_3 grown from Bi_2O_3—PbF_2 with the addition of La_2O_3 ($\rightarrow LaAlO_3$, Chase, 1967), $Bi_4Ti_3O_{12}$ grown from Bi_2O_3 with the addition of GeO_2 or MoO_3 (Epstein, 1970) and BeO grown from Li_2O—MoO_3 with the addition of $LiPO_3$ or $LiBO_2$ (Austerman, 1964).

The addition of monovalent ions is always advantageous when viscous fluxes like $Na_2B_4O_7$, $BaO \times 0.6B_2O_3$ and PbO—B_2O_3 are used because of the network-breaking action of alkali and halogen ions. As examples, LiF was reported by Anikin *et al.* (1965) and NaF by Baker *et al.* (1965) to be favourable for the growth of TiO_2 and ThO_2, respectively, from sodium borate flux. Similar arguments hold for the addition of Li or fluoride to very concentrated solutions.

The formation of large and high-quality crystals in nature from magmatic (especially from pegmatite) melts could not be understood without the presence of so-called "mineralizers" (Vogt, 1903/1904) like OH^-, F^-, Cl^-, CO_2 and SO_2. Buerger (1948) has semiquantitatively described the role of such mineralizers which act by breaking Si—O—Si bridges in the silicate network of the mineral melts and thus increase the fluidity of the crystallizing magma. Both OH^- and F^- are most effective in doing this and, due to their small atomic weights, small weight fractions correspond to large atomic fraction and are thus very effective.

For several examples of the beneficial role of certain additives (Al_2O_3 to $Y_3Al_5O_{12}$, CaO and SiO_2 to $Y_3Fe_5O_{12}$) no good explanation has been found. A careful analysis of crystalline perfection by X-ray topography, of impurities and of microinclusions, of the habit, and of the surface morphology might help to find an explanation for the action of such additives.

3.7. Practical Solvents and their Properties

3.7.1. Solvents for growth of oxides, oxide compounds, silicates

Most of the common solvents like PbO, halides, borates, molybdates, tungstates and vanadates have been used for mineral synthesis in the last century (see Chapter 2). In the past twenty years many systems have been tried as high-temperature solvents, yet the few types of compound

TABLE 3.13. Properties of Solvents used for Growth of Oxides and Oxide Compounds

Type of solvent	Solvent	Melting point °C	Boiling point °C	Density at 20°C	Viscosity (at °C, in cP)	Solubility of solvent	Examples of applications	Remarks
Lead- and bismuth-compounds	PbO	888°	1472°	9.53	Fig. 3.11	Hot HNO_3—H_2O	Perovskites, garnets, spinels	Corrodes Pt above 1300°C
	PbF_2	855° (824°)	1293°	8.24	Fig. 3.11	Hot HNO_3—H_2O	Al_2O_3, MgO, ZnO, $MgAl_2O_4$	Corrodes Pt above 1300°C
	$PbCl_2$	498°	954°	5.8	Fig. 3.11	Hot H_2O or acid	$PbZrO_3$, $PbTiO_3$	Corrodes Pt
	PbO—PbF_2	~500° Eut.		~9	Fig. 3.11	Hot HNO_3—H_2O	Garnets, perovskites, spinels	Corrodes Pt above 1300°C
	PbB_2O_4	~500°		5.6	Fig. 3.11	Hot HNO_3—H_2O	Al_2O_3, Fe_2O_3, BeO, YIG, $PbTiO_3$	Corrodes Pt above 1300°C
	$Pb_2P_2O_7$	824°		5.8		Hot HNO_3—H_2O	RPO_4, Fe_2O_3, $MgFe_2O_4$	
	$Pb_2V_2O_7$	720°		~6		Hot HNO_3—H_2O	RVO_4, TiO_2, Fe_2O_3, Ga_2O_3	
	Bi_2O_3	820°	1890° dec.	8.9		Hot HNO_3—H_2O	$Bi_4Ti_3O_{12}$, $Bi_2Fe_4O_9$, Fe_2O_3	Relatively corrosive for Pt
	Bi_2O_3—B_2O_3	620–720°				Hot HNO_3—H_2O	Fe_2O_3, $FeBO_3$, $GaFeO_3$	Relatively corrosive for Pt
	BiF_3	727°	1027°	5.32		Hot HNO_3—H_2O	HfO_2, $MnCr_2O_4$	Relatively very volatile
Borates	B_2O_3	~460°	1860°	2.46	Fig. 3.11	Hot H_2O	$LiFeO_2$, $LiFe_5O_8$	
	$NaBO_2$	966°	1434°	2.464		H_2O	$CdTiO_3$	
	$Na_2B_4O_7$	741°	1575° dec.	2.367	Fig. 3.11	H_2O, hot H_2O-acid	BeO, Al_2O_3, Fe_2O_3, TiO_2	
	KBO_2	950°				H_2O	$RFeO_3$, $CdTiO_3$	
	$K_2B_4O_7$	815°		1.74		H_2O, hot H_2O-acid	Cr_2O_3, Fe_2O_3, TiO_2	
	BaB_4O_7	910°				Hot HNO_3—H_2O	YIG, YAG, $RFeO_3$, $BaTiO_3$	
	BaB_2O_4	1105°			Fig. 3.11	Hot HNO_3—H_2O	YIG, $Ba_2Zn_2Fe_{12}O_{22}$	
	$Ba_2B_2O_5$	~915° Eut.				Hot HNO_3—H_2O		
	$LiBO_2$	845°		1.4		Hot HNO_3—H_2O	Cr_2O_3, Fe_2O_3	Slightly soluble in hot H_2O
	$Li_2B_4O_7$	930°				Hot HNO_3—H_2O		Slightly soluble in hot H_2O
Vanadates, molybdates, tungstates	V_2O_5	690°	1750° dec.	3.36		Acids, alkali	VO_2, V_2O_3, RVO_4	
	$LiVO_3$	616°						
	$NaVO_3$	630°				H_2O, hot acid	YVO_4, (Y, Eu)VO_4	
	MoO_3	795°	1151° subl.	4.69		Acid		
	Li_2MoO_4	705°		2.66	Fig. 3.11	Hot alkali	$ZrSiO_4$, Zn_2SiO_4, TiO_2	
	$Li_2Mo_2O_7$	600°			Fig. 3.11	Hot alkali	BeO, TiO_2, $ZrSiO_4$	

Category	Compound			Density		Solvent	Crystals grown	Remarks
Vanadates, molybdates, tungstates (cont.)	Na_2MoO_4	687°		3.28	Fig. 3.11	H_2O	TiO_2, Be_2SiO_4, $ZrSiO_4$	
	$Na_2Mo_2O_7$	612°			Fig. 3.11	Hot acid	$SrMoO_4$, $NiFe_2O_4$	
	$K_2Mo_2O_7$	500°			Fig. 3.11	Hot alkali	TiO_2	
	Li_2WO_4	742°		3.71	Fig. 3.11	Hot alkali	CeO_2, ThO_2, TiO_2	
	$Li_2W_2O_7$	720°			Fig. 3.11	Hot alkali	$LiR(WO_4)_2$, ThO_2	
	Na_2WO_4	698°			Fig. 3.11	H_2O	Al_2O_3, SiO_2, Fe_2O_3	
	$Na_2W_2O_7$	730°		4.18	Fig. 3.11	Hot alkali	(Ca, Sr, Ba, Cd etc.) WO_4	
	$K_2W_2O_7$	650°			Fig. 3.11			
Sulphates, phosphates, hydroxides	Na_2SO_4	884°		2.68		H_2O	NiO:Li	
	$Zn_3(PO_4)_2$	900°		4.0		acid	$(Zn, Sb)_3O_4$	
	NaOH	318°	1390°	2.13	Fig. 3.11	H_2O	α-Ga_2O_3 (high pressure)	
	KOH	360°	1320°	2.0	450:1.7, 600:0.8	H_2O	$KNbO_3$	
Halides	LiF	842°	1676°	2.64		HF—H_2O		Also soluble in methanol
	LiCl	610°	1382°	2.1	650:1.6, 850:0.7	H_2O	$LiFePO_4$, $LiNiPO_4$	
	NaF	988°	1695°	2.56		H_2O, HF—H_2O	$BaTiO_3$, $Cd_2Nb_2O_7$	Soluble in glycerine
	NaCl	801°	1413°	2.2	Fig. 3.11	H_2O	TiO_2, $CaWO_4$, $BaSO_4$	
	KF	856°	1502°	2.5		H_2O, HF—H_2O	$BaTiO_3$, $KNbO_3$, CeO_2	Soluble in methanol
	KCl	772°	1407°	1.9	800:1.1, 900:0.9	H_2O	$KNbO_3$	
	$CaCl_2$	782°	1627°	2.2	800:4.25, 850:3.65	H_2O	$Ca_5(PO_4)_3Cl$, $Ca_2Nb_2O_7$	
	$SrCl_2$	873°	1250°	3.05		H_2O	Sr_2NiWO_6, $SrSnO_3$	
	BaF_2	1280° / 1354°	2137°	4.9		Hot acid	$MgAl_2O_4$, $BaFe_{12}O_{19}$	
	$BaCl_2$	962°	1560°	3.9	Fig. 3.11	H_2O	$BaTiO_3$, $BaWO_4$, $BaTi_3O_7$	
	ZnF_2	872°	1502°	4.9		Hot H_2O or acids		
	$CdCl_2$	568°	960°	4.05	Fig. 3.11	H_2O		Slightly soluble in methanol
	LaF_3	1493°		5.94		Hot acid	Al_2O_3	
	$LaCl_3$	860°	1862°	3.84		H_2O	LaOCl	
	Na_3AlF_6	1000°		2.9	Fig. 3.11	Hot acid (slightly)	Al_2O_3, TiO_2, WO_3	

commonly utilized have remained the same and can be classified as follows:

a. Lead and bismuth compounds.
b. Borates.
c. Vanadates, molybdates and tungstates, particularly of the alkali metals.
d. Alkali halides, carbonates, etc.

This classification is not exact and rarely used solvents are not included; many solvent systems are combinations of two categories, for example, PbO—B_2O_3, PbO—V_2O_5, etc. We now consider some examples in detail since their use illustrates the factors influencing the choice of a solvent. The important properties of these fluxes are listed in Table 3.13. The literature data of the examples given in Table 3.13 and further applications

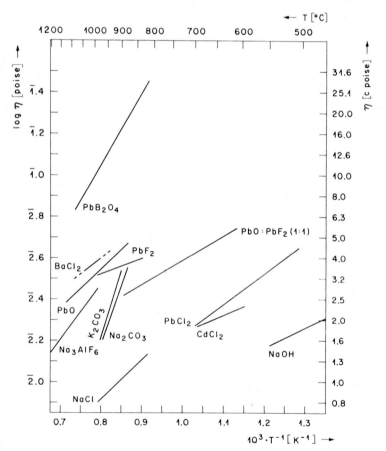

FIG. 3.11(a). Viscosity of high-temperature solvents: (a) low viscosity solvents, (b) high viscosity solvents, (c) alkali molybdates and tungstates.

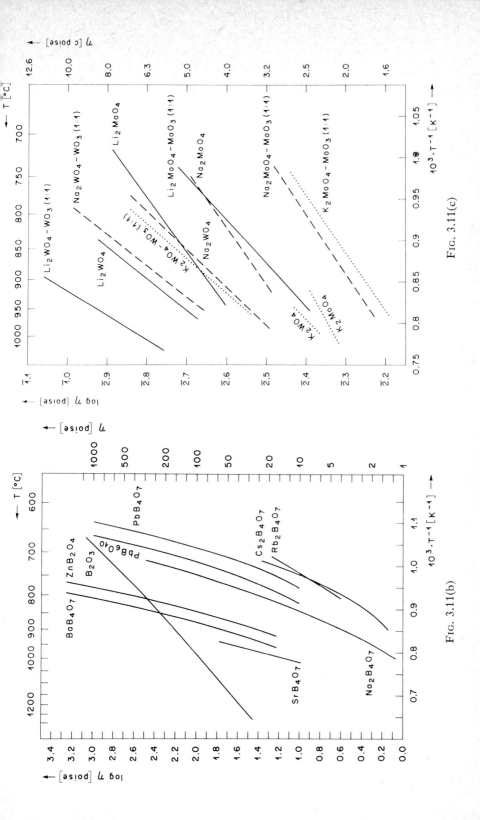

FIG. 3.11(c)

FIG. 3.11(b)

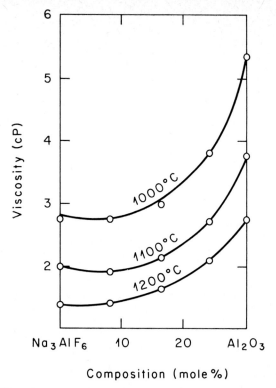

FIG. 3.12. Viscosity isotherms for Na_3AlF_6/Al_2O_3 solutions (Belyaev *et al.*, 1964).

of the common as well as of unusual solvents may be obtained from the Tables in Chapters 2 and 10. The viscosities are given in Figs 3.11(a) and 3.11(b) for solvents of low and high viscosities, respectively, and in Fig. 3.11(c) for alkali molybdates and tungstates. The viscosities of the solutions are quite different and generally higher, as is shown with the example cryolite-Al_2O_3 in Fig. 3.12. Cobb and Wallis (1967) determined the viscosity of a solution of 20 mole % Al_2O_3 in LaF_3 to vary between 10.6 cp at 1360°C and 9.3 cp at 1500°C, while Bruton and White (1973) measured the viscosity of an 8 mole % $PbTa_2O_6$ solution in $Pb_2V_2O_7$ solution as about 2 cp at 1200°C. Tolksdorf *et al.* (1972) reported that the viscosity of a solution of 5.91 wt% $Fe_2O_3 + 0.70$ wt% R_2O_3 in 91.37 $PbO + 2.02B_2O_3$ (wt%) varies from 83 cp at 900°C to 63 cp at 1100°C.

Lead Oxide PbO

Lead oxide was already used as a solvent in 1877 by Fremy and Feil and was rediscovered by Remeika (1956) and applied to the growth of crystals of ferrites and several other compounds. It is still widely used mainly as a constituent of a binary or ternary flux composition, and such lead-based fluxes have been the most widely used of all fluxes. The high solubility of refractory oxides in lead-containing melts is attributed to the strong polarizability of the Pb^{2+} ions, and dissolution is presumably effected by the formation of complex ionic species between solute and solvent. The ionic nature of molten PbO is indicated by its high electrical conductivity, which has a value of about 1Ω cm at its melting point (McKenzie, 1962).

Disadvantages of PbO are its toxicity, its tendency to attack platinum under reducing conditions or at temperatures above 1300° and, for certain techniques, its high volatility. Lead-based solvents are not practical for compounds containing large divalent cations (Ca^{2+}, Sr^{2+}, Ba^{2+}) since solid solutions will be formed.

Lead Fluoride PbF_2

Lead fluoride has a lower melting point and generally a higher solvent power than PbO. Some high-quality crystals have been grown from PbF_2 solution by the evaporation technique [for example, Al_2O_3 : Cr by White and Brightwell (1965) and $MgAl_2O_4$ by Wood and White (1968), Robertson and Taylor (1968), Wang and McFarlane (1968)]. The high vapour pressure of PbF_2, useful in the evaporation technique, has severe disadvantages in crystal growth by slow cooling since it causes uncontrolled supersaturation if the crucibles are not completely sealed. Therefore in the latter technique, cobinations of PbF_2 with oxides to decrease the volatility are often used.

$PbO—PbF_2$

The solvent system $PbO—PbF_2$ offers improvements over the individual components due to its low eutectic temperature (at around 500°C). Nielsen (1960) found that better-quality crystals of yttrium iron garnet were formed from this system than from PbO alone, and attributed this improvement to the lower melting point and viscosity. However, other factors such as complex formation (for instance of YOF) as discussed in Section 3.4 might also contribute to the improved growth.

The lower viscosity of mixed $PbO—PbF_2$ melts compared to the components has been confirmed by Oliver (1965). The viscosity data for PbO, PbF_2, $PbO—PbF_2$-mixtures and for $Pb(BO_2)_2$ are included in Figs 3.11(a) and (b). These data may not be very accurate but they do give some idea of the viscosities of these popular solvent systems. The viscosities of solutions of oxides in these solvents will generally be much higher than those of the solvents, but few viscosity determinations of saturated high-temperature solutions have been published. Such measurements are necessary for a full understanding of the hydrodynamics in the solutions and of the growth mechanism. Evaporation losses from this mixed system are mainly of PbF_2. This causes a change in the PbO—PbF_2 ratio (and therefore in the solute-PbO and solute-PbF_2 ratios) so that frequently inhomogeneities may occur in the crystals grown. As was shown by Nielsen *et al.* (1967) the preferred evaporation of PbF_2 is especially harmful for the growth of $Y_3Fe_5O_{12}$—$Y_3Ga_5O_{12}$ and relatively homogeneous solid solutions can be prepared if the PbF_2 evaporation is suppressed by sealing the crucibles.

PbO—PbF_2—B_2O_3

As discussed previously (Sections 3.4.3 and 3.6.3) the addition of relatively small amounts of B_2O_3 to PbO—PbF_2 generally improves the quality of the crystals grown and decreases the tendency to uncontrolled nucleation, probably by increasing the width of the metastable region and by complex formation in front of the growing crystal. Also B_2O_3 is thought to decrease the volatility of the PbF_2 in the solution. The effect of small additions of B_2O_3 on the viscosity of PbO—PbF_2 and of solutions (including 15–30% solute) is complex (Coe and Elwell, 1974).

This complex solvent system has been used for growth of a large variety of oxides and oxide compounds.

PbO—B_2O_3

Fluxes in this system are clearly worthy of investigation since they combine the advantages of borate and of lead-containing solvents. There are four compounds of low melting point in this system (Levin *et al.*, 1964, Fig. 281) of which PbB_4O_7 has the highest melting point of 768°C. On the B_2O_3-rich side a large liquid immiscibility region exists up to 800°C. PbO—B_2O_3 might be used for seeded growth (for example, top-seeded solution growth) and for the working temperature a compromise has to be found between the high viscosity at low temperatures (below 1000°) and the high volatility at high temperatures (above 1200°). Linares (1962a) demonstrated that the crystallization field of $Y_3Al_5O_{12}$ is extended appreciably with increasing B_2O_3 concentration (relative to the stability fields of Al_2O_3 and $YAlO_3$). Schieber (1967) proposed PbO—B_2O_3 as a fluorine-free solvent for rare-earth garnets, and for several magnetic rare-earth garnets of the type $R_3Fe_5O_{12}$, the optimum mole ratio of $Fe_2O_3 : B_2O_3$ was said to be $3 : 1$.

$Pb_2P_2O_7$, PbO—As_2O_3, PbO—V_2O_5

$Pb_2P_2O_7$ was proposed as a solvent by Wickham (1962) since its volatility is less than that of PbO or PbF_2. It is particularly useful for the growth of phosphate crystals but has quite a high solubility for Fe_2O_3 and was used to grow ferrite crystals. Accordingly, lead arsenates and lead vanadates are useful solvents for growth of rare-earth arsenates and vanadates, respectively. However, other rare-earth compounds cannot be grown from solvents rich in P_2O_5, As_2O_3 or V_2O_5 because of the stability of the corresponding rare-earth compounds. For growth of TiO_2 the solvent PbO—V_2O_5 was used by Belyaev and Chodakov (1952), and Linares (1967) used this flux successfully for growth of emeralds. PbO—V_2O_5 fluxes are found to have a very low volatility and have been used to grow a wide range of crystals. One special advantage of this solvent is that V^{5+} prevents reduction of the PbO to metallic lead by its reduction to V^{4+} or V^{3+} (Wanklyn, 1970).

Bi_2O_3-based Solvents

In many respects bismuth compounds show similarity to lead compounds with respect to solvent properties although solubilities in the bismuth compounds are generally lower. One reason that Bi_2O_3-based solvents have not found so much use as the PbO fluxes is the valence state of Bi^{3+}: it cannot be used for the many cases where compounds containing large trivalent cations have to be grown, such as the rare-earth compounds. Other disadvantages of Bi_2O_3-based fluxes are their relatively high viscosity and corrosiveness. As proposed by Remeika (1956) PbO should be replaced when Cr^{3+}-containing compounds or chromium-doped crystals have to be grown by the slow-cooling technique. Of the many Bi-containing compounds, Bi_2O_3 fluxes are an obvious choice. The addition of Bi_2O_3 to BaO—B_2O_3 fluxes (see "BaO—B_2O_3") can produce a marked decrease in the liquidus temperature (down to $\sim 600°$ compared to approximately $900°$ for the BaO—B_2O_3 eutectics) and compositions around the eutectic composition 23.4% BaO : 62.4% Bi_2O_3 : 14.2% B_2O_3 have been found to be preferable to BaO—B_2O_3 solvents for the growth of spinel ferrites (Elwell et al., 1972). This nonvolatile system can be used with advantage for seeded growth, for instance, by the top-seeded solution-growth technique.

Alkali Borates

B_2O_3 is not suitable as a solvent due to its high viscosity (see Fig. 3.11b). The viscosity is greatly lowered when the glassy network is broken up by the addition of monovalent metal ions such as alkali ions, and the alkaline earth ions decrease the viscosity of B_2O_3 to a lesser degree. The alkali borates and especially molten borax, $Na_2B_4O_7$, have been well known as fluxes for over a century and have found applications in various fields.

An investigation of the solvent behaviour of the alkali borates was undertaken by Berkes and White (1969a). Accurate liquidus curves were determined and the departures from ideality were attributed to clustering of the boron and oxygen atoms. This model led to the conclusion that, in

$Na_2B_4O_7$, for example, $8.5B_2O_3$ units formed the average-sized cluster at 820°C, decreasing to $2B_2O_3$ units at 1120°C. A similar tendency was exhibited by the $K_2B_4O_7$ and $Rb_2B_4O_7$ solvents. The B_2O_3 clusters exclude the Ni^{2+} ions which are weakly bonded in large interstitial sites in the boron-oxygen network. The weak interaction of Ni^{2+} was confirmed by optical spectroscopy of the borate liquids after quenching to glasses. The crystal-field splitting of the Ni^{2+} ions was found to be small, and to decrease as the atomic weight of the alkali increases.

Solutions in $Na_2B_4O_7$, $K_2B_4O_7$ and $Rb_2B_4O_7$ of NiO show a positive departure from ideality and therefore a tendency to immiscibility. In the system $Li_2B_4O_7$—NiO a negative departure from ideality was observed, that is, the solubility of NiO was remarkably greater than that of an ideal solution and therefore indicates compound (complex) formation.

Although NiO has not been grown as good-quality crystals from alkali borates, similar materials such as Fe_2O_3 have been grown by Barks and Roy (1967). Berkes and White conclude that a good high-temperature solvent is one in which the solute does not form a tightly bound complex with the solvent but is excluded by clusters or complexes formed by the solvent atoms. However, we feel that this is not a generally valid rule, on the contrary it was proposed in Sections 3.4.3 and 3.6.3 that complex formation in the solutions and in front of the growing crystals generally has a beneficial effect. The conclusion of Berkes and White holds for the case where the solvent or the solution has a high viscosity.

Sodium tetraborate has been used as a solvent for the growth of a wide variety of crystals and it is a useful solvent for the rapid growth of small crystals. However, larger high-quality crystals have not been grown from this flux which is very viscous and volatile.

Potassium borate KBO_2 with a melting point of 950°C was used by Marezio *et al.* (1970) for growth of small crystals of $RFeO_3$ and by Sholokhovich *et al.* (1970) for $CdTiO_3$, $CdTiO_3$—$SrTiO_3$ solid solutions and for $PbTiO_3$, but probably KBO_2 is not the best solvent for growth of large crystals due to its high viscosity.

BaO—B_2O_3

The use of barium borate fluxes was advocated by Linares (1962a) who used this system to grow several iron-containing oxides as well as CeO_2, TiO_2, ZnO and a variety of other oxides. The advantages of BaO—B_2O_3 fluxes are a very low volatility, low rate of attack on platinum crucibles and low density compared with the lead fluxes. The latter property means that crystals tend to grow at the base of the crucible rather than the surface (in an appropriate temperature gradient), where excessive nucleation may result from foreign bodies or evaporation.

The viscosity of these fluxes is high and spontaneous nucleation often yields a large number of tiny crystals. Barium borates are of particular value for seeded-growth methods in which the crystal grows just below the surface of a solution in an open crucible. The typical composition range used is from $BaO \times 0.5B_2O_3$ to $BaO \times 0.6B_2O_3$.

The low volatility permits the use of these fluxes at temperatures higher than those normally used, and a considerable reduction in viscosity might be achieved by growth at $1400°–1500°C$ provided that this is not compensated by the increase in solute concentration. Lowering of the viscosity may also be achieved by addition of Bi_2O_3 as discussed earlier, or of BaF_2 (Burmeister, 1972), and the ternary $BaO—BaF_2—B_2O_3$ has been proposed as a solvent for rare-earth garnets and orthoferrites by Hiskes et al. (1972) and its properties are discussed by Elwell et al. (1974) and Hiskes (1975).

Vanadates, Molybdates, Tungstates

The oxides V_2O_5, MoO_3 and WO_3 have all been used as fluxes but are very volatile. Therefore the alkali vanadates, molybdates and tungstates are generally preferred and a wide variety of crystals has been prepared from them. In particular, many silicates and germanates are preferably grown from these solvents. A well known example is emerald which was grown from lithium molybdate and lithium vanadate as early as 1888, and in the recent commercial production by various companies of emerald (for gem purposes) this group of solvents has been used with success. The high valence state of vanadium, molybdenum and tungsten prevents significant incorporation into most compounds, although the easy reducibility of these three ions might cause problems in certain cases. For growth of vanadates, molybdates and tungstates the corresponding solvents are obviously well suited as solvents containing the same anions and as can be seen from the Tables of Chapter 2 and Chapter 10. Many properties of the alkali molybdates and tungstates have recently been published by Gossink (1971), and the viscosity data from this work are given in Fig. 3.11(c).

A systematic study of the alkali vanadates, molybdates and tungstates for the growth of BeO was carried out by Newkirk and Smith (1965). The solubility is highest for the lithium compounds, and the solubility in the $Li_2O—MoO_3$ system increases regularly with the MoO_3 concentration. Dissolution was considered to be due to the formation of $(BeO)_x(MoO_3)_y$ complexes. This complexity is indicated, for example, by the observation of crystal growth at $1400°C$, which is above the melting point of pure MoO_3. For a given alkali metal, the solubility increases in the order $V \rightarrow Mo \rightarrow W$. However, the usefulness of the alkali-vanadate fluxes for growth of BeO is obviated by a very strong tendency to creep.

Kunnmann et al. (1965) described the solvent action of the sodium-tungstate fluxes in terms of Lewis acid-base theory. The Lewis acid WO_3 (an electron pair acceptor) dissolves the basic oxides which form the crystal, and crystallization could be promoted by the addition of the Lewis base Na_2WO_4 (electron donor). The dissolution and growth of a metal oxide MO in sodium pyrotungstate can be represented by the equation

$$x\text{MO} + y\text{Na}_2\text{W}_2\text{O}_7 \underset{\text{cooling}}{\overset{\text{heating}}{\rightleftharpoons}} x(\text{MO} \times n\text{WO}_3) + (y - nx)\text{Na}_2\text{W}_2\text{O}_7 + nx\text{Na}_2\text{WO}_4.$$

The $MO \times nWO_3$ complex must be stable only in the liquid phase so that the oxide is crystallized on cooling. The principles proposed have been verified by the application of these fluxes to grow a range of vanadium spinels such as $Co_{1+\delta}V_{2-\delta}O_2$ by an electrolytic method (Rogers et al., 1966).

The system K_2O—V_2O_5 has been studied by Holtzberg et al. (1956) and according to Shannon (private communication) the low-melting KVO_3 (mp 520°C) is a suitable flux not only for many vanadates but also for other compounds.

Miscellaneous Solvents for Growth of Oxides

Of the alkali halides only the fluorides dissolve an appreciable amount of oxides whereas, for instance, NaCl dissolves generally less than 1% oxide at 1000°C, at which temperature it already has a relatively high vapour pressure. Occasionally, small crystals of oxide compounds have been prepared from NaCl.

The most popular alkali halides are KF and Na_3AlF_6. The latter, cryolite, has been used for a long time as a flux, and because of its techno-logical importance its properties as a solvent have been studied in some detail. The solubilities of oxides at 1000°C in cryolite after Belyaev et al. (1964) are given in Table 3.14 and show interesting regularities. The solubility of the alkaline earth oxides increases from BeO to BaO according to the increasing degree of ionicity, and the low solubility of the transition metal oxides may be attributed to their partial covalent character. B_2O_3 and WO_3 are exceptions probably due to some form of complex formation. In Table 3.14 the solubilities of the oxides in a cryolite melt containing 5% Al_2O_3 are also given because they indicate the decrease in the oxide solubilities. Potassium fluoride KF, known as solvent already in the last century, was reintroduced by Remeika (1954) for the growth of the first large $BaTiO_3$ "butterfly" crystals and since then it has often been used (occasionally in combination with other halides) for the growth of titanate perovskites, and also for several other compounds. KF is volatile at high temperatures and has a very low viscosity. Generally, small crystals are obtained from KF-based fluxes due to uncontrolled spontaneous nucleation.

LaF_3 has been investigated systematically by Cobb and Wallis (1967)

TABLE 3.14. Solubilities of Oxides in Molten Cryolite and Cryolite
Containing 5% Al_2O_3 at 1000°C (mole %)

Oxide	Solubility in Na_3AlF_6	Solubility in $Na_3AlF_6 + 5\% Al_2O_3$
B_2O_3	∞	∞
WO_3	87.72	86.14
BaO	35.75	22.34
Al_2O_3	19.77	—
CaO	13.12	8.46
MgO	11.65	7.02
BeO	8.95	6.43
SiO_2	8.82	—
TiO_2	4.87	4.15
Mn_3O_4	2.19	1.22
CuO	1.13	0.68
CdO	0.98	0.26
V_2O_5	0.05	0.21
ZnO	0.51	0.004
NiO	0.32	0.180
Co_3O_4	0.24	0.140
Fe_2O_3	0.18	0.003
Cr_2O_3	0.13	0.050
SnO_2	0.08	0.010

as a potential solvent for the preparation of homogeneous ruby-laser crystals, but it seems that LaF_3 has not found a wider application since it needs a protective atmosphere.

The alkali carbonates have been successfully used for the growth of the corresponding niobates, tantalates and solid solutions of potassium tantalate-niobate (KTN), sometimes by the top-seeded solution-growth technique. With this technique (for which solvents of low volatility are a necessity) large $SrTiO_3$ and $BaTiO_3$ crystals were grown from excess TiO_2, and excess GeO_2 was used as solvent for seeded growth of complex germanates (Belruss et al., 1971).

Only the most popular and useful fluxes for the growth of oxides have been mentioned. Other solvents might be found in the Table (Chapter 10), from Laurent (1969), Wanklyn (1975) or, of course, by reasoning according to the criteria for the choice of solvents described in Section 3.6. Other solvents might be chosen with the help of phase diagrams, of which for the oxide field the most complete compilation is that of Levin et al. (1964, 1969) and for the field of metals and semiconductors those of Hansen and Anderko (1958), Elliott (1965) and Shunk (1969).

3.7.2. Solvents for growth of metals, alloys, semiconductors

Molten salts are not only common solvents for ionic compounds like many oxides but can also act as solvents for metals. In particular, the solubility behaviour of metal halides has been investigated, especially for their parent metals.

When there is no appreciable chemical interaction between the salt and the metal, the solution becomes partially metallic, and particularly the solutions of the alkali, the alkaline-earth and the rare-earth metals in their respective halides are prone to this behaviour. Solutions with strong metal-salt interactions show relatively small changes in conductivity upon addition of metal to the molten salts, and this type of behaviour is frequently found in the systems of the transition and post-transition metals.

In industrial electrolysis another type of metal solutions is of importance, namely, solutions of metals in salts of another metal. According to the electrochemical series of metals in molten salts (see Delimarskii and Markov, 1961) the one metal displaces the other according to the example

$$Cd + PbCl_2 \rightleftarrows CdCl_2 + Pb.$$

The equilibrium constant for this reaction may be obtained from the solubility behaviour of the metals in unlike salts. In the above example cadmium will displace lead below 650°C and lead will displace cadmium above 650°C due to a change in the equilibrium constant with temperature. The principles of solutions of metals in molten salts have been reviewed by Fischer (1954), Delimarskii and Markov (1961) and by Bredig (1963).

Metallic solutions as the medium for crystal growth of elements and a variety of compounds have been reviewed by Luzhnaya (1968). Metallic solvents can be used for the growth of many metals and alloys which do not form stable compounds or solid solutions and show miscibility in the liquid state. Liquid immiscibility in metallic systems has been discussed in Section 3.2.

Comparitively few metals and intermetallic compounds have been systematically grown from metallic solutions, notable exceptions being the growth of silicon from gallium and indium solutions by Keck (1953) and by Keck and Broder (1953), from tin solutions by Goss (1953), from silver and zinc solutions by Hartenberg (1951) and from gold by Carman *et al.* (1954). Germanium and silicon crystals have been grown from Al, Ga, Sn, Cd, Sb, Ag and Zn solutions by Faust and John (1964) and the superconductor Nb_3Sn has been crystallized by Hanak and Johnson (1969) from tin solution. However, many more binary metallic systems are simple eutectics (Reisman, 1970, p. 158) and are thus suitable for crystal growth of either component.

Intermetallic compounds can often be prepared from one of the components in the appropriate temperature range. For example, the III–V semiconductors GaP, GaAs, GaSb have been grown from gallium solutions in many laboratories either as bulk crystals or as thin layers, and intermetallic compounds such as V_2Ga_5, VGa_5, $MnGa_{5.2}$, $NbGa_3$, Ta_5Ga_3, $CrGa_4$, $FeGa_3$ have been prepared as small crystals by R. Reinmann (unpublished). It is obvious from the phase diagrams (of which many are still unknown) that $AuGa_2$, $PrGa_2$, $MgGa_2$, UGa_3 and several other compounds could also be crystallized from gallium solutions. Several metals and many of their intermetallic compounds can be grown from low-melting metals such as gallium, aluminium, bismuth, silver and gold. From bismuth solutions the following compounds have been crystallized by Teitel et al. (1954) and by Barton and Greenwood (1958): $BaBi_3$, $CaBi_3$, $CeBi_2$, Mg_3Bi_2, $BiSe$, $SrBi_3$, Bi_2Te_3, Th_3Bi_5, UBi_2 and $ZrBi_2$, and large crystals of MnBi have been obtained by Ellis et al. (1958).

Solvents for III–V *and* II–VI *Semiconductors and for Chalcogenides generally*
For the growth of III–V compounds the most popular solvents are the corresponding metals (e.g. gallium, indium), especially for layer growth and device fabrication by liquid phase epitaxy. However, other metals like zinc, tin, lead, mercury and cadmium have been used by several authors for III–V and the isoelectronic $II–IV–V_2$ compounds as reviewed by Luzhnaya (1968), by Faust and John (1964) and by Spring-Thorpe and Pamplin (1968). Faust et al. (1968) discussed the effect of solvents (Ag, Al, Au, Cd, Ga, In, Pb, Sn, Zn) and of added impurities on the habit of the semiconductor crystals grown (Si, Ge, III–V compounds). II–VI compounds (the sulphides, selenides, tellurides) are rarely grown from the constituent elements; they are generally grown either from metallic solutions (gallium, indium, bismuth, thallium, tin), from halides or from chalcogenides as solvents. For example, ZnS has been grown from ZnF_2 and $ZnCl_2$, and $CdCl_2$ acts as solvent for CdS, CdSe and CdTe. A high solubility (at moderate temperatures) for several sulphides is shown by lead chloride (Linares, 1968; Wilke et al., 1967), thus $PbCl_2$ could play the same important role in flux growth of sulphides as does PbF_2 for growth of oxides. Koutaissoff (1964) has prepared europium sulphide EuS crystals using LiCl—KCl eutectic as solvent for the reaction

$$EuCl_2 + Li_2S \rightarrow EuS + 2LiCl.$$

Many chalcogenides and pnictides have been prepared as crystalline products with the addition of the corresponding halides (Kweestroo, 1972).

Although several chalcogenides have been used for the growth of II–VI compounds, for example Ba_2ZnS_3 for the growth of ZnS by Malur (1966), they have not found wide application except for alkali polysulphides (for example, Na_2S_2—Na_2S_5) which have been used by Garner and White (1970) for the growth of cinnabar (HgS) and by Scheel (1974a) for the growth of many sulphides: ZnS, CdS, CuS, α-MnS, FeS_2, NiS_2, CoS_2, PbS, Cu_3VS_4, $KFeS_2$, $NaInS_2$, $NaCrS_2$, etc. Sodium polysulphides have low melting temperatures and eutectics (200° to 450°C), a common anion, are easily dissolved in water or alcohols, and alkali ions are incorporated into the crystals only as ppm traces (Scheel, 1974a). Information on the sodium polysulphide melts was obtained by Cleaver and Davies (1973), South *et al.* (1972) and Bell and Flengas (1966); and Garbee and Flengas (1972) studied the structural and electrical properties of PbS—$PbCl_2$ and of Cu_2S—CuCl and FeS—$FeCl_2$ solutions, respectively.

Solvents for Borides, Carbides, Pnictides

Owing to the extremely high melting points of many of the borides, carbides, etc., crystal growth by direct techniques (Czochralski, Bridgman, sublimation) at those high temperatures is extremely difficult due to container problems and precise growth control. In the crucible-free electron-beam zone melting, arc-imaging and plasma flame-fusion techniques, the stoichiometry as well as crystal quality (strain, grain boundaries, dislocation densities) present still unsolved problems.

For crystal growth of these classes of compounds often with very high melting points (e.g. TiB_2 2800°C, $TaC_{0.98}$ 3983°C) those transition metals which form compounds of relatively "low" melting point with the corresponding non-metal (B, C, etc.) may act as solvents.

Rowcliffe and Warren (1970) were able to grow 2 mm crystals of $TaC_{0.96}$ by the technique of Robins (1959) using solutions of Ta and C in iron which were slowly cooled from 2200°C to 1450°C, and Gerk and Gilman (1968) grew WC crystals of 1 cm diameter from cobalt solutions by the top-seeded solution-growth technique (see Chapter 7). Silicon carbide, one of the most intensively studied crystals with potential applications in many fields, has been prepared from several metal solutions which are listed in Table 3.15. The solubility of carbon in silicon at 1600°C is only 0.03 at %, but it can be enhanced by the addition of several transition elements. Of the transition metals, cobalt, nickel and chromium have been applied as thin molten layers for epitaxial vapour-liquid-solid (*VLS*) growth. The *VLS* growth mechanism often acts in the growth of whiskers (see Chapter 7), and a variety of droplets on the top of whiskers of many compounds have been analysed. These droplets act as a solvent for thin-layer transport growth and even for growth of bulk crystals of the whisker

TABLE 3.15. Solvents used for Growth of Silicon Carbide (m.p. $>2700°C$)

Solvent	Melting point of metal	Melting points of compounds	References of SiC growth
Silicon	1410°C		Hall (1958), Halden (1960), Beckmann (1963), Nelson et al. (1966), Bartlett (1969)
Silicon + Chromium	1900°C	$Cr_3Si \sim 1750°$, $Cr_2Si\ 1600°$, $CrSi\ 1640°$, $CrSi_2\ 1540°$, $Cr_3C_2\ 1890°$, $Cr_7C_3\ 1780°$, $Cr_4C\ 1520°$	Griffiths and Mlavsky (1964), Wright (1965), Knippenberg and Verspui (1966), Silva et al. (1967), Comer and Berman (1970), Kalnin and Tairov (1966), Kunagawa et al. (1970)
Silicon + Cobalt	1492°	$Co_2Si\ 1330°$, $CoSi\ 1400°$, $CoSi_3\ 1306°$	Marshall (1969), Comer and Berman (1970)
Silicon + Iron	1536°	$FeSi\ 1420°$, $Fe_3C\ 1227°$	Halden (1960), Ellis (1960)
Silicon + Nickel	1455°	$Ni_2Si\ 1318°$, $NiSi\ 992°$, Ni_3C	Baumann (1952), Ellis (1960), Berman and Comer (1969), Comer and Berman (1970)
Silicon + Silver	961°	no compound formation	Pickar (1967)
Silicon + Gold	1063°	no compound formation	Berman and Comer (1969)

material. A variety of carbides and carbide solid solutions have been prepared from metallic solvents by Jangg et al. (1968), and this technique has found industrial applications.

As for the carbides, transition metal elements or compounds have been used for the preparation and growth of borides and the pnictides. For example, B_6P has been formed from Ni solution (Burmeister and Greene, 1967), AlB_2 from Al solution (Horn et al., 1952) and BP and BAs have been grown from Cu_3P solution by various groups (see Chapter 10). Silicon diphosphide has been crystallized from Sn and Sn—Mg solutions (Spring-Thorpe, 1969). A variety of borides, carbides, silicides, germanides, phosphides, arsenides, etc., have been prepared by electrolytic reduction of the corresponding salts dissolved in a molten salt stable at the preparation conditions (see for instance Aronsson et al., 1965; Delimarskii and Markov, 1961; Wold and Bellarance, 1972; Kunnmann, 1971).

Tin was found to be a powerful solvent for the growth of II–IV–V$_2$ compounds as reported by Rubenstein and Ure (1968) for ZnSnP$_2$, and by Faust and John (1964) for ZnSnAs$_2$ and as reviewed by Spring-Thorpe and Pamplin (1968) for several II–IV–V$_2$ compounds. The latter authors claim that indium, lead, bismuth, cadmium and zinc could also be used for the growth of ternary phosphides and arsenides.

3.7.3. Solvents for miscellaneous elements and compounds

A peculiar effect of the role of solvent in the crystallization of diamond and graphite was described by Wentorf (1966). In the stability field of diamond (60 kb, 1600°C) carbon crystallizes in its high-pressure modification from solutions in the transition metals of groups VI, VII and VIII (best from Ni, Fe and Fe—Ni alloys, see Chapter 7) and as graphite from oxides, sulphides, halides and silicates such as Cu$_2$O, CuCl, Cu$_2$S, AgCl, AlCl$_3$, ZnO, ZnS, CdO, FeS and silicates containing hydroxyl such as serpentinite, biotite, muscovite and hydrous alkaline alumino-silicates (Wentorf, 1966). This fact was explained by the formation of positively charged carbon ions in solvents of the first group (Fe, Ni), whereas the second group shows a smaller solubility for carbon and dissolves it probably as a nearly neutral species. In Li$_2$C$_2$ and Ca$_2$C$_2$ the carbon is negatively charged. Wentorf attributes the effect of the charge of the dissolved carbon species (having different partial atomic volumes) on the crystallization of either diamond or graphite to the rates of nucleation of the two forms which is mainly determination by the change (with pressure) of the free-energy content of the crystalline phase relative to its state in solution, which depends on the differences in partial atomic volumes. When crystallizing graphite or diamond from iron solutions the addition of a small amount of silicon or aluminium is helpful in suppressing the formation of the iron carbide Fe$_3$C (Strong and Chrenko, 1971; Sumiyoshi et al., 1968). The solubility behaviour of carbon in nickel at high pressure has been discussed in detail by Strong and Hanneman (1967). In the diamond synthesis by the Swedish group, iron carbide was used as solvent (Liander and Lundblad, 1960). It is still not certain under which conditions crystallization of the large high-quality diamonds has occurred in nature. Borazon BN, the III–V analogue of diamond, has been prepared by Wentorf (1957, 1965) and by DeVries and Fleischer (1972) from Li$_3$BN$_2$(Mg$_3$N$_2$, Ca$_3$N$_2$) solutions at pressures of 40 to 60 kb and temperatures of about 1600 and 1900°C. Wentorf (1957) states that with increasing atomic number of the metal component of the solvent the process of crystallization of cubic boron nitride requires higher pressures, and DeVries and Fleischer assume that the structure of the Li_3BN_2—BN solutions consists of B—N chains cross-linked by lithium ions.

3.8. Determination of Solubility Curves and Phase Diagrams

A large number of high-temperature solution growth experiments have been performed with little or no knowledge of the appropriate phase diagram. The normal procedure for the growth of a new material is to select a suitable solvent on the basis of past experience, taking into account any special requirements of the material. A number of trial experiments using small quantities of chemicals and fairly rapid cooling rates are then carried out, with solute concentrations normally in the range 10–25%. If one or more compositions are found to yield crystals of the required phase, the experiment is repeated with such compositions on a larger scale and with much slower cooling rates in order to obtain crystals of the desired size.

However, if the conditions of any crystal-growth experiment are to be optimized and crystals of high quality grown, some knowledge of the phase diagram is necessary. It will be clear from the previous discussion that many high-temperature solutions contain five or more components, and the determination of a complete phase diagram in such cases is totally impractical. As stressed by Roy and White (1968) the solution is treated as a pseudo-binary system with the phase to be crystallized as one component (solute) and the solvent as the other. The information required for crystal growth is the solubility curve of the phase to be crystallized and the stability field of this phase—the range of composition and temperature over which it is stable.

A purely theoretical approach for determination of phase diagrams is possible only for cases where the solutions fulfill the conditions of a simple model and where the appropriate thermodynamic data of the components are known. Generally, the experimental determination of the phase diagrams is necessary and in the following the various approaches to determine the solubility curves are briefly presented. However, the role of phase diagrams on solid solubility and in the type of dopant incorporation as reviewed by Reisman (1970) and by Panish (1970a) will not be discussed here; the liquidus curves will be the only aspect covered.

Knowledge of the phase diagrams is not only helpful or necessary for planning crystal-growth experiments, it is also indicative of the type of solution, and of the solute-solvent interactions as outlined in Sections 3.2–3.4 and as discussed in detail by Reisman (1970). A review of thermo-analytical investigations of crystal-growth problems has recently been presented by Schultze (1972).

3.8.1. Direct techniques

Phase-diagram investigations should be performed under equilibrium conditions and it is important to allow sufficient time for equilibration prior to measurement. This time depends on the mobility of the components (thus on diffusion constants, viscosity and convection) and can be shortened by continuous mixing and by stirring.

The simplest method for determination of a solubility curve is to introduce excess crystals or even a polycrystalline mass of the solute into the solvent at a controlled temperature. After dissolution has proceeded for several hours, the undissolved material is separated from the saturated solution and weighed, and the loss in weight due to dissolution represents the solubility at that temperature. Solubility values are normally quoted as a function of the whole solution, either in mole (atomic) % or in weight %.

Some investigators have recovered the undissolved material by removal of the solution after equilibration by dissolution in some reagent after cooling to room temperature. This method is clearly very time-consuming and the construction of a solubility curve can occupy several weeks. Separation of the saturated solution from excess solute may be done by careful decanting or by using sieve arrangements, and Hall (1963) has removed excess floating compounds from the saturated melts by a quartz glass loop in his determination of solubilities of III–V compounds in gallium and indium.

An alternative but related approach is to remove a sample of the solution after equilibration and to determine the solubility by chemical analysis of this solution. If the quantity of solution removed is small compared with the total mass of solution, a solubility curve may be constructed fairly quickly by taking samples over a range of equilibrium temperatures. In this way Ray (1969) determined the solubilities of NiO, CoO, FeO and Cu_2O in neutral silicate melts.

Seed crystals may be introduced into the approximately saturated solution. If the seed is put into an unsaturated solution it will dissolve and if put into a supersaturated one it will grow (Timofeeva and Kvapil, 1966; Manzel, 1967). Instead of facetted seed crystals the use of spheres has the advantage that growth (supersaturation) can be more easily detected by the formation of facets (Timofeeva and Konkova, 1968). By this technique the solubility curves can be determined as accurately as ± 10°C if care is taken of the experimental parameters (homogeneous solution, dipping time, no interfering heat transfer, etc.) and if the volatility of any components of the solution is negligible.

Wagner and Lorenz (1966) simply used the disappearance and the formation of solid ZnSe and ZnTe on the surface of highly reflecting gallium and indium melts for solubility determination, and by temperature cycling, the liquidus temperatures could be measured very precisely. However, due to thermal gradients in the furnace with a full length view-port, the accuracy is estimated to be only $\pm 10°C$.

Quenching methods. The main disadvantage of the above dissolution-extraction methods is that they give no direct information on changes in the phase which is crystallized, and are therefore unsuitable for investigations of novel systems or for stability field determinations.

The most popular method of performing more detailed investigations involves quenching the solutions to room temperature followed by examination of the quenched sample or a polished section of it by optical microscopy. This method relies on the ability to detect undissolved particles, but optical microscopy is an extremely sensitive means of detection, and a second phase can normally be distinguished at concentrations as low as 0.05%.

The solubility of a material at a known temperature is determined by quenching samples containing a range of different solvent-solute ratios following equilibration at that temperature. If the solubility is below the solubility limit, no undissolved material will be present on quenching and the accuracy of determination of any point on a solubility curve may be increased by the use of samples having progressively smaller differences in solute concentration. Solidus temperatures may be determined by a similar method. Since several samples are required for the determination of a single point of the liquidus curve, it is desirable to quench several samples at the same time. These samples may either be of different compositions quenched from the same temperature, or of the same composition equilibrated at different temperatures in a furnace of which the temperature gradient is accurately known. A multiple-sample apparatus has been described by Cobb and Wallis (1969).

3.8.2. Hot-stage microscopy

The methods described above, though simple, are all very laborious and lengthy. Alternative methods normally involve the application of a varying temperature and so do not yield equilibrium values. However, since crystallization does not occur under equilibrium conditions, the values obtained are normally sufficiently accurate for the determination of conditions for crystal growth, and any inaccuracy is more than compensated by the great saving in time and possibly materials and the additional information which may be obtained.

The ideal method, in principle, is the visual polythermal hot-stage microscopic method in which the heated sample is located near the objective of a microscope by means of which it may be observed continuously. The amount of heat supplied to the sample must necessarily be small to prevent damage to the objective and it is normally found preferable to locate the sample at the junction of a thermocouple which is used for temperature determination. The region of the thermocouple in the immediate vicinity of the junction is made of much finer wire than the rest of the thermocouple so that heating is localized to this region. Heating may be performed by passing an alternating current through the thermocouple wire, with temperature sensing through the same thermocouple.

After a period of dissolution the sample is slowly cooled, and the formation of small crystals observed through the microscope when the solution becomes supersaturated. By temperature cycling the precision might be increased. In practice the observation of crystallization may be extremely difficult particularly if the melt is opaque or if crystallization occurs below the surface of the melt or if the crystals possess optical properties not substantially different from the solution. The main source of inaccuracy is, however, preferential volatilization of one component of the solution which can cause rapid changes in composition. Even with "non-volatile" fluxes such as $BaO \times 0.6B_2O_3$, the increase in solute concentration with time due to evaporation is very marked, due to the high area-to-volume ratio of the sample.

The solubility of sodium niobate in a borax flux was determined by this method by Burnett *et al.* (1968) but the liquidus temperatures quoted could be in error by 100°C due to solvent evaporation. Hot-stage microscopy is, however, valuable for rapid determination of the phase diagram of novel solvent systems (Elwell *et al.*, 1972).

3.8.3. Differential thermal analysis

In differential thermal analysis (DTA) two samples of similar thermal capacity are located in a furnace heated or cooled at a uniform rate. Thermocouples are inserted into the two samples and are connected in opposition so that the temperature difference ΔT between the samples is measured. If neither sample undergoes a phase change in the temperature range investigated, the value of ΔT will ideally be zero throughout the whole range. However, if one sample is inert while the other undergoes a phase change, the evolution or absorption of latent heat will result in a temperature difference between the samples. If ΔT is recorded for an unknown sample, any thermal effect such as crystallization or dissolution may be detected.

Several problems are encountered in the application of DTA to crystallization from high-temperature solutions, and these have been discussed by Elwell *et al.* (1969). Apart from the corrosive nature of the fluxes, the main problem is that crystallization occurs rather slowly and the solute concentration does not normally exceed 25%, so the values of ΔT are rather low. Also, at temperatures above about 1000°C, electrical conduction in the ceramic specimen holders may become appreciable and this can provide the main contribution to electrical noise on the ΔT signal as the temperature is increased. In the apparatus of Elwell *et al.* (1971) the

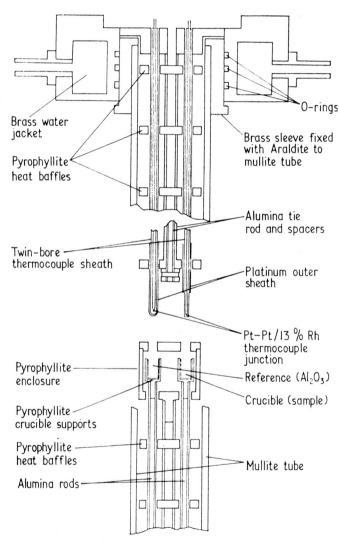

Brass water jacket

Pyrophyllite heat baffles

O-rings

Brass sleeve fixed with Araldite to mullite tube

Alumina tie rod and spacers

Twin-bore thermocouple sheath

Platinum outer sheath

Pt–Pt/13 % Rh thermocouple junction

Pyrophyllite enclosure

Reference (Al_2O_3)

Crucible (sample)

Pyrophyllite crucible supports

Pyrophyllite heat baffles

Alumina rods

Mullite tube

Fig. 3.13. DTA apparatus for fluxed melts (Elwell *et al.*, 1971).

E

noise due to conduction in ceramic components is reduced by mounting each crucible containing a sample or reference material (normally α-Al_2O_3 powder) on an independent support. The thermocouples are also mounted independently of each other and the junctions are immersed just below the surface of the solutions. In order to protect the thermocouple junctions from flux attack, they are enclosed in a sheath of very fine platinum foil. This platinum sheath is electrically earthed and so acts as a screen against stray electric fields. The signal-to-noise ratio with this apparatus is higher by an order of magnitude than that obtained from the more usual arrangement in which the thermocouple is inserted in a well in the base of the crucible. Figure 3.13 shows the layout of the apparatus in diagrammatic form. A typical record of ΔT versus T for a crystallizing solution is shown in Fig. 3.14. The signal due to crystallization persists throughout the crystallization range and a very large signal normally appears on solidification. In any apparatus the signal will be accompanied by a gradual

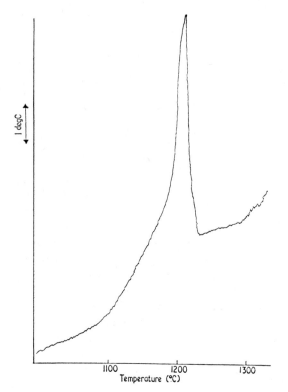

FIG. 3.14. DTA thermogram for 18 wt% $NiFe_2O_4$ crystallizing from a BaO—Bi_2O_3—B_2O_3 flux (Elwell *et al.*, 1971).

drift of the base line due to mismatch between the thermal capacities of the sample and the reference material, and particularly to changes in the thermal properties of the sample due to the formation of a layer of solute around the thermocouple sheath during cooling.

Since crystallization proceeds as the temperature is lowered rather rapidly, some supercooling inevitably occurs and so the liquidus temperatures indicated on cooling will be low. The magnitude of the error may be estimated by comparison of the curves obtained during heating and cooling; in the crystallization of nickel ferrite from barium borate, this was normally 10–20°C at a cooling rate of 2–3°C/minute. Karan and Skinner (1953), in their DTA investigation of the phase diagram $BaTiO_3$—KF, found a difference of 65 ± 5°C for the heating and cooling curves, respectively, with a cooling rate of 3–4°C/minute.

One advantage of DTA is that the crystals which form during cooling are normally large enough for identification by X-ray crystallographic or other means. It is, in principle, a good technique for volatile samples, which may be contained in sealed crucibles to which the thermocouple junctions are welded.

3.8.4. Thermogravimetry and other methods

Thermogravimetry involves a determination of the changes in weight in a sample during heating or cooling. Application of this method to crystal-growth studies from aqueous solutions was reported by Bennema (1966), and apparatus for the application of thermogravimetry to high-temperature solutions was described by Smith and Elwell (1967) and by Nielsen (1969).

In the determination of crystallization temperatures, a long platinum wire is suspended so that its tip is immersed just below the surface of the solution as shown in Fig. 3.15. As the solution is cooled below the liquidus temperature, the end of the wire acts as a nucleation centre for the growth of crystals. Provided that the density of the crystals is different from that of the solution, crystallization is then indicated by a change in weight of the wire. Usually this weight is measured on an electrobalance and is indicated on a chart recorder simultaneously with the temperature, measured as near as possible to the tip of the wire.

This method is more accurate than DTA since the rate of cooling can be reduced to an arbitrarily low value. The main source of error is that of the temperature measurement but this can be minimized by replacing the wire by a thermocouple so that crystals are formed directly on the junction.

If the solution is cooled slowly below the crystallization temperature, crystals typically up to 1–2 mm in diameter can be obtained for identification of the stable phase. The stability field can be determined readily from the weight-change record and any changes in the crystal phase can be

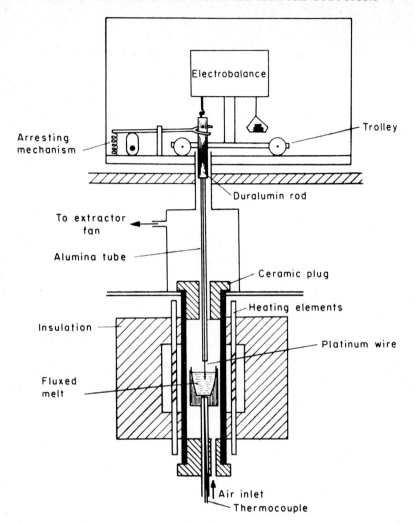

FIG. 3.15. Thermogravimetric apparatus for crystallization temperature determination (Smith and Elwell, 1967).

confirmed by periodic removal of the wire for inspection. The main limitations of the method are that it is difficult to homogenize the solution by stirring and that it is inaccurate when applied to volatile solvents since the solution must be contained in an open crucible.

Resistivity. In general, the electrical resistivity of a crystal is higher than that of its solutions and this difference could be used as a means of measuring crystallization temperatures. This method does not appear to be very accurate in practice (Smith, 1970) since the fractional area of even a fine wire electrode which is covered by a crystal immediately after nucleation is small. Optimization of this geometrical problem might produce a feasible method which would have the advantage of cheapness and ease of recording. Resistivity changes may be conveniently used to measure the melting points of novel solvent compositions.

Identification of crystalline phases will be discussed in Section 9.3, but frequently morphology and colour suffice to identify crystals of simple or well-known systems.

3.8.5. Solid phase metastability

A discussion of phase-diagram determination would be incomplete without mention of the possibility that identical systems under nominally identical conditions may not always yield the same crystal phase. Perhaps the best example of this phenomenon is encountered in the growth of garnets from PbO/PbF_2 fluxes with and without B_2O_3 addition. An attempt to ascertain the stability field of yttrium iron garnet, $Y_3Fe_5O_{12}$, has been made by Timofeeva and Lukyanova (1970). Several other phases can crystallize from this system, notably $YFeO_3$, Fe_2O_3 and $PbFe_{12}O_{19}$, and van Uitert et al. (1970), after a great deal of experience, reported that the probability of garnet formation from a favourable composition was only 50%, and this statement was re-emphasized by Tolksdorf and Welz (1972). However, by systematic studies necessary for reproducible growth of garnet films for magnetic bubble domain devices, this situation was clarified. It was shown by Levinstein et al. (1971) that supercoolings up to $120°C$ can be achieved, and Blank and Nielsen (1972) reported that a garnet—PbO—B_2O_3 solution can be saturated with respect to orthoferrite, and supersaturated with respect to garnet, simultaneously. If no seeds of either phase are introduced and nucleation occurs spontaneously, minor effects such as trace impurities or a slight volatilization of one of the solution components might accidentally determine the crystallizing phase which could thus grow in a metastable field.

Another example of metastable growth has been discussed in Section 3.7.3 where the crystallization of graphite in the stability field of diamond was explained by solute-solvent interactions; at high pressures and temperatures graphite or diamond could be crystallized depending on the type of solvent.

According to Ostwald's law of successive transformation a metastable phase may transform step by step (Ostwald'sche Stufenregel) to the stable phase. Thus metastable phases are frequently observed as low-temperature crystallization products in glasses (metastable viscous solutions of which the supersaturation is frozen in), and as shown by Scheel (1968) it is possible even to obtain relatively large crystals of metastable quartz solid solutions.

Other phases which are not the stable phases of the pure compounds can be stabilized by traces of impurities. Thus anatase and brookite may be obtained in the stability field of rutile (TiO_2) and tridymite stabilized by traces of alkali crystallizes instead of quartz (Flörke, 1955, 1956). However, one cannot speak of metastability in the latter cases since the trace-stabilized phases are thermodynamically stable. All the examples of growth of metastable phases mentioned above are exceptional, and one would expect that in almost all cases the thermodynamically stable phases crystallize from high-temperature solutions. However, Roy and White (1968) pointed out that the highest probability for growth of metastable phases is obtained in high-pressure solution growth.

3.9. Selected Solubility Data and Phase Diagrams
For the convenience of experimentalists, literature data are listed in Tables 3.16, 3.17 and 3.18 which contain solubility data and phase diagrams. However, phase diagrams which are contained in one of the following works are generally not listed:

Levin *et al.* (1964, 1969): Oxides, salts.
Landolt-Börnstein (1956): Metals, oxides, salts, etc.
Hansen and Anderko (1958): Binary alloys.
Elliott (1965): Binary alloys.
Shunk (1969): Binary alloys.
Smithells (1962): Alloys.
Delimarskii and Markov (1961): Salts and metals.
Stephen and Stephen (1963): Solubilities of inorganic and organic compounds.
Kaufman and Bernstein (1970): Computer calculations of phase diagrams.
Alper (1970): Many aspects of phase diagrams.
Panish and Ilegems (1972): Ternary III–V systems.
Gschneidner (1961): Rare earth alloys.

Table 3.16 gives solubility data of oxides and oxide compounds, Table 3.17 of metals, alloys and semiconductors, and Table 3.18 of miscellaneous elements and compounds. Further solubility data may be found in several of the original publications which are listed in the Table of Chapter 10. However, Tables 3.16–3.18 are not complete, and any published or unpublished solubility data submitted to the authors would be appreciated.

TABLE 3.16. Solubilities (Phase Diagrams) of Oxides
and Oxide Compounds

Solute	Solvent	Exper.	Theor.	References
Al_2O_3 and	Na_3AlF_6	X		Foster (1960)
Al_2O_3:Cr	PbF_2	X		Giess (1964), Airtron (1963), Wilke (1968), Adams et al. (1966)
	PbF_2		X	Timofeeva and Lukyanova (1967), White and Brightwell (1965)
	PbO	X		Torkar et al. (1966), Wilke (1968)
	$PbO-B_2O_3$	X		Timofeeva and Konkova (1968), Timofeeva and Lukyanova (1967), Timofeeva (1968)
	Li-, Na-, K-, Sr- and Ba-tungstates	X		Voronkova et al. (1967)
$BaTiO_3$	KF	X		Karan and Skinner (1953), Karan (1954)
	TiO_2	X		Rase and Roy (1955)
	$BaCl_2$	X		Rase and Roy (1957)
$(Ba, Pb)TiO_3$	$(Ba, Pb)B_2O_4$	X		Perry (1967)
$BaWO_4$	Na_2SO_4	X		Wilke (1968)
	NaCl	X		Wilke (1968), Voigt and Neels (1971)
	WO_3	X		Chang et al. (1966)
$Ba_5(Y,Ta)_4O_{15}$	BaB_2O_4	X		Layden and Darby (1966)
$BaTa_2O_6$	B_2O_3, BaB_8O_{13}	X		Layden and Darby (1966)
$Ba_5Ta_4O_{15}$	BaB_2O_4	X		Layden and Darby (1966)
BeO	$Li_2MoO_4 \cdot 1.25MoO_3$	X		Austerman (1963)
	PbO	X		Newkirk and Smith (1965)
	$PbO-PbF_2$	X		Newkirk and Smith (1965)
	SiO_2	X		Morgan and Hummel (1949)
	GeO_2	X		Cobb and Wallis (1969)
	WO_3	X		Chang et al. (1966)
Be_2SiO_4	Li_2O-MoO_3	X		Ballman and Laudise (1965)
	Na_2O-MoO_3	X		Ballman and Laudise (1965)
$CaWO_4$	$CaCl_2-LiCl$ Eut.	X		Barta et al. (1968)
	Na_2SO_4		X	Schultze et al. (1967), Wilke (1968)
	Na_2WO_4		X	Schultze et al. (1967), Cobb and Wallis (1967a)
	LiCl	X		Wilke (1962)
	NaCl	X		Voigt and Neels (1971)
	WO_3	X		Chang et al. (1966)
$CaSO_4$	Na_2SO_4	X		Wilke (1968)
	NaCl	X		Wilke (1968)
$CaMoO_4$	Li_2SO_4	X		Parker and Brower (1967)
Cu_2O	$(Na, K)_2SiO_3$	X		Ray (1969)
CoO	$(Na, K)_2SiO_3$	X		Ray (1969)

TABLE 3.16 cont.

Solute	Solvent	Exper.	Theor.	References
$CoFe_2O_4$	$PbO—PbF_2—B_2O_3$	X		Kvapil et al. (1969)
Cr_2O_3	$PbO \cdot PbF_2 \cdot 0.35B_2O_3$	X		Timofeeva and Konkova (1968),
		X		Timofeeva et al. (1969),
		X		Timofeeva (1968)
	PbO	X		Negas (1968)
EuO	Eu	X		Shafer et al. (1972)
Fe_2O_3	NaCl	X		Wilke (1964, 1968)
	$PbO \cdot PbF_2 \cdot 0.35B_2O_3$	X		Timofeeva and Konkova (1968),
		X		Timofeeva (1968)
	PbO	X		Mountvala and Ravitz (1962)
	$Na_2B_4O_7$, $K_2B_4O_7$	X		Barks and Roy (1967)
	$Na_2B_4O_7$	X		Nielsen (1969)
	B_2O_3	X		Makram et al. (1972)
FeO	$(Na, K)_2SiO_3$	X		Ray (1969)
$FeBO_3$	B_2O_3	X		Makram et al. (1972)
$KNbO_3$	$K_2CO_3(K_2O)$	X		Reisman and Holtzberg (1955)
$KTaO_3$	$K_2CO_3(K_2O)$	X		Reisman et al. (1956)
$LiFe_5O_8$	B_2O_3	X		Anderson and Schieber (1963)
MgO	PbF_2	X		Webster and White (1969)
	$MgWO_4$	X		Chang et al. (1966)
$MgFe_2O_4$	$PbMoO_4$	X		Viting and Khomyakov (1965)
$MgWO_4$	WO_3	X		Chang et al. (1966)
Mn_3O_4	$Na_2B_4O_7$	X		Nielsen (1969)
Nd_2O_3	$PbO \cdot PbF_2 \cdot 0.35B_2O_3$	X		Timofeeva and Konkova (1968),
		X		Timofeeva et al. (1969),
		X		Timofeeva (1968)
NiO	$4Li_2O \cdot 5B_2O_3$	X		Berkes and White (1969a)
	B_2O_3, $LiBO_2$	X		Berkes and White (1969a)
	$Li_2B_4O_7$, $Li_4B_6O_{11}$	X		Berkes and White (1969a)
	$(Na, K)SiO_3$	X		Ray (1969)
	$Na_2B_4O_7$, $K_2B_4O_7$	X	X	Berkes and White (1969b)
	$Rb_2B_4O_7$	X	X	Berkes and White (1969b)
$NiFe_2O_4$	PbO	X		Remeika (1955), Manzel (1967)
	$PbO—PbF_2—B_2O_3$	X		Kvapil et al. (1969)
	$BaO \cdot 0.62B_2O_3$	X		Smith and Elwell (1968)
$NiWO_4$	Na_2SO_4	X		Schultze et al. (1967)
	Na_2WO_4, $Na_2W_2O_7$	X		Schultze et al. (1967)
$PbTiO_3$	$PbCl_2$	X		Nomura and Sawada (1952)
	KF, $NaVO_3$, KVO_3	X		Belyaev et al. (1954)
	K_2MoO_4, NaF	X		Belyaev et al. (1954)
	Na_2MoO_4, Na_2WO_4	X		Belyaev et al. (1954)
	$Na_4P_2O_7$, $K_4P_2O_7$	X		Belyaev et al. (1954)
$PbZrO_3$	PbO	X		Fushimi and Ikeda (1967)
$SrTiO_3$	LiF—KF	X		Sugai et al. (1968)

TABLE 3.16 cont.

Solute	Solvent	Exper.	Theor.	References
$SrSO_4$	Na_2SO_4	X		Wilke (1968)
	NaCl	X		Wilke (1968)
$SrWO_4$	NaCl	X		Voigt and Neels (1971)
	WO_3	X		Chang *et al.* (1966)
TiO_2	$K_2B_4O_7$	X		Naumova and Anikin (1966)
	$Li_2B_4O_7$, $Na_2B_4O_7$, $K_2B_4O_7$	X		Anikin *et al.* (1965)
Y_2O_3	$PbO—PbF_2—0.35B_2O_3$	X		Timofeeva and Konkova (1968), Timofeeva (1968)
YBO_3	$BaO—B_2O_3$	X		Linares (1962a)
$Y_3Al_5O_{12}$	$PbO—0.35B_2O_3$	X		Timofeeva and Kvapil (1966)
	$PbO—PbF_2—0.35B_2O_3$	X		Timofeeva and Kvapil (1966),
			X	Timofeeva and Konkova (1968),
			X	Timofeeva *et al.* (1969),
			X	Timofeeva (1968)
	$PbO—PbF_2$	X		Bakradze *et al.* (1968), Timofeeva (1967)
	$BaO—B_2O_3$, $PbO—B_2O_3$, $PbO—PbF_2—B_2O_3$, etc.	X		Timofeeva (1967)
$Y_3Fe_5O_{12}$	$PbO—PbF_2—0.35B_2O_3$	X		Timofeeva and Konkova (1968),
			X	Timofeeva (1968)
	$BaO—B_2O_3$	X		Linares (1962a, 1964)
	$PbF_2—YF_3$	X		Sato and Hukudo (1963)
	PbO	X		Nielsen and Dearborn (1960)
	$PbO—PbF_2$	X		Kvapil *et al.* (1969)
	$PbO—PbF_2—B_2O_3$	X		Kvapil *et al.* (1969),
			X	Timofeeva and Lukyanova (1970)
YVO_4	$NaVO_3$, $NaVO_3—Na_2B_4O_7$	X		Phillips and Pressley (1967)
	V_2O_5	X		Levin (1967)
$ZnFe_2O_4$	PbO	X		Manzel (1967)
$ZnAl_2O_4$	PbF_2	X		Giess (1964)
ZrO_2	$Li_2Mo_2O_7$	X		Kleber *et al.* (1966)
	PbF_2	X		Anthony and Vutien (1965)
$ZrSiO_4$	$Li_2O—MoO_3$	X		Ballman and Laudise (1965)
	$Na_2O—MoO_3$	X		Ballman and Laudise (1965)

E 2

TABLE 3.17. Solubilities of Metals, Alloys and Semiconductors

Solute	Solvent	Exper.	Theor.	References
Ag_2S	AgCl	X		Garbee and Flengas (1972)
Al_3Er	Al	X		Meyer (1970)
CdS	$CdCl_2$	X		Izvekov et al. (1968)
	CdI_2—$CdCl_2$	X		Izvekov et al. (1968)
	Cd	X		Woodbury (1963)
	Cd, Sn, Bi	X		Rubenstein (1968)
CdSe	Cd, Sn, Bi	X		Rubenstein (1968)
CdTe	Cd, Sn, Bi	X		Rubenstein (1968)
(Cd, Zn)Te	Cd, Zn	X	X	Steininger et al. (1970)
Cu_2S	CuCl	X		Garbee and Flengas (1972)
FeS	$FeCl_2$	X		Garbee and Flengas (1972)
GaAs	Ga	X		Hall (1963)
	Ga, As, Ge, Sn	X		Panish (1966)
	Ga, Sn, Bi, Pb, Ge	X		Rubenstein (1966)
	Zn, Sn, Ge, Cu, Ag, Au, Te	X	X	Panish (1970a)
Ga(P,As)	Ga	X	X	Broder and Wolff (1963), Antypas (1970), see also Stringfellow and Antypas (1971)
(Ga, Al)As	Ga, Al	X	X	Panish and Sumski (1969), Woodall (1971)
(Ga, In)As	Ga, In	X		Kovaleva et al. (1968)
		X	X	Panish (1970b)
GaP	Ga	X		Hall (1963)
	Zn	X	X	Panish (1970a), Panish (1966)
	Ga—GaAs	X		Shih (1970)
(Ga, In)P	Ga, In	X	X	Marbitt (1970), Blom (1971),
		(X)	X	Stringfellow (1970)
GaSb	Ga	X		Hall (1963)
(Ga, In)Sb	Ga, In, Sb	X	X	Blom and Plaskett (1971)
Ge	Au	X		Wagner (1968)
HgS	Na_2S_4	X		Garner and White (1970)
HgTe	Hg	X		Dziuba (1971)
InAs	In	X		Hall (1963)
	CdI_2	X		Luzhnaya et al. (1966)
	In, As, Cd	X		Luzhnaya et al. (1966), Koppel et al. (1965)
	Sn, Zn, Pb	X		Koppel et al. (1967a)
	KCl, InI, CuCl, AgCl, AgI, ZnI_2, $CdCl_2$, $CdBr_2$, $SnCl_2$, $PbCl_2$, PbI_2	X		Koppel et al. (1967b)
In(As, Sb)	In, Sb, As	X	X	Stringfellow and Greene (1971)
In(As, P)	In	X	X	Antypas and Yep (1971), Ugai et al. (1968)
InP	In	X		Hall (1963), Shafer and Weiser (1957)
InSb	In	X		Hall (1963)
PbS	$PbCl_2$	X		Garbee and Flengas (1972)
		X	X	Bell and Flengas (1966)

TABLE 3.17 cont.

Solute	Solvent	Exper.	Theor.	References
SnTe	Sb	X		O'Kane and Stemple (1966)
ZnS	ZnF_2	X		Linares (1962b)
	$ZnCl_2$	X		Gashurov and Levine (1960), Garbee and Flengas (1972)
	Sn, Bi	X		Rubenstein (1968)
ZnSe	Ga, In	X		Wagner and Lorenz (1966)
	Zn, Sn, Bi			Rubenstein (1968)
$ZnSiP_2$	Sn	X		Spring-Thorpe and Pamplin (1968)
ZnTe	Ga, In	X		Wagner and Lorenz (1966)
	Zn, Sn, Bi			Rubenstein (1968)

TABLE 3.18. Solubilities of Miscellaneous Elements and Compounds

Solute	Solvent	Exper.	Theor.	References
BN	Li_3BN_2(50 kb)	X		DeVries and Fleischer (1971)
BaF_2	LiF	X		Neuhaus et al. (1967)
C (graphite, diamond)	Fe (57 kb)	X		Strong and Chrenko (1971)
	Ni (1 atm., 54 kb)	X		Strong and Hanneman (1967)
	Fe, Ni	X		Strong and Chrenko (1971)
$KMgF_3$	MgF_2	X		Neuhaus et al. (1967)
K_2MgF_4	KF	X		Neuhaus et al. (1967)
$LiBaF_3$	LiF	X		Neuhaus et al. (1967)
RF_3 (R = rare earth)	NaF	X		Thoma and Karraker (1966)
Si	Au	X		Hansen and Anderko (1958)
ThF_4	LiF—BeF_2	X		Thoma et al. (1960)
UF_4	LiF—BeF_2	X		Jones et al. (1962)
WC	Co	X		Gerk and Gilman (1968)

References

Adams, I., Nielsen, J. W. and Story, M. S. (1966) *J. Appl. Phys.* **37**, 832.

Airtron (1963) Report AD 432341.

Alper, A. M. (1970 (ed.) "Phase Diagrams", Vol. I–III. Academic Press, New York.

Anderson, J. C. and Schieber, M. (1963) *J. Phys. Chem.* **67**, 1838.

Anikin, I. N. (1958) *Growth of Crystals* **1**, 259.

Anikin, I. N., Naumova, I. I. and Rumyantseva, G. V. (1965) *Sov. Phys. Crystallogr.* **10**, 172.

Anthony, A. M. and Vutien, L. (1965) *Compt. Rend.* **260**, 1383.

Antypas, G. A. (1970) *J. Electrochem. Soc.* **117**, 700.

Antypas, G. A. and Yep, T. O. (1971) *J. Appl. Phys.* **42,** 3201.
Aronsson, B., Lundström, T. and Rundquist, S. (1965) "Borides, Silicides and Phosphides" Methuen, London.
Austerman, S. B. (1963) *J. Am. Ceram. Soc.* **46,** 6.
Austerman, S. B. (1964) *J. Nucl. Mater.* **14,** 225.
Austerman, S. B. (1965) *J. Am. Ceram. Soc.* **48,** 648.
Baker, J. M., Chadwick, J. R., Garton, G. and Hurrell, J. P. (1965) *Proc. Roy. Soc.* A **286,** 352.
Bakradze, R. V., Kuznetsova, G. P. Baryshev, G. P., Selivanova, T. N. and Bychov, V. Z. (1968) *Izv. Akad. Nauk SSSR Neorgan. Mat.* **4,** 395.
Ballman, A. A. and Laudise, R. A. (1965) *J. Am. Ceram. Soc.* **48,** 130.
Barczak, V. J. (1965) *J. Am. Ceram. Soc.* **48,** 541.
Barker, J. A. and Henderson, D. (1967) *J. Chem. Phys.* **47,** 2856, 4714.
Barker, J. A. and Henderson, D. (1968) *Mol. Phys.* **14,** 587.
Barker, J. A. and Henderson, D. (1972) *Ann. Rev. Phys. Chem.* **23,** 439.
Barks, R. E. and Roy, D. M. (1967) In "Crystal Growth" (H. S. Peiser, ed.) p. 497. Pergamon , Oxford.
Barrer, R. M. (1949) *Disc. Faraday Soc.* **5,** 333.
Barta, C., Zemlicka, J. and Stankov, S. (1968) *Kristall u. Technik* **3,** K35.
Bartholomew, R. F. and White, W. B. (1970) *J. Crystal Growth* **6,** 249.
Bartlett, R. W. (1969) In "High Temperature Technology" Butterworths, London.
Barton, P. J. and Greenwood, G. W. (1958) *J. Inst. Metals* **86,** 504.
Baumann, H. N. (1952) *J. Electrochem. Soc.* **99,** 109.
Beckmann, G. E. J. (1963) *J. Electrochem. Soc.* **110,** 84.
Bell, M. C. and Flengas, S. N. (1966) *J. Electrochem. Soc.* **113,** 27, 31.
Belruss, V., Kalnajs, J., Linz, A. and Folweiler, R. C. (1971) *Mat. Res. Bull.* **6,** 899.
Belyaev, A. I., Shemtschushima, E. A. and Firsanova, L. A. (1964) Physikalische Chemie geschmolzener Salze, VEB Leipzig.
Belyaev, I. N. and Chadokov, A. L. (1952) *J. exp. theoret. Physik (USSR)* **22,** 376.
Belyaev, I. N., Sholokhovich, M. L. and Barkova, G. V. (1954) *Zhur. Obshchei Khim.* **24,** 212 (see Phase Diagr. for Ceramists Fig. 1184, 1185).
Bennema, P. (1966) *Phys. stat. sol.* **17,** 555.
Berger, S. B. and Pinch, H. L. (1967) *J. Appl. Phys.* **38,** 949.
Berkes, J. S. and White, W. B. (1966) *Phys. and Chem. of Glasses* **7,** 191.
Berkes, J. S. and White, W. B. (1969a) *J. Crystal Growth* **6,** 29.
Berkes, J. S. and White, W. B. (1969b) *J. Am. Ceram. Soc.* **52,** 481.
Berman, I. and Comer, J. J. (1969) *Mat. Res. Bull.* **4,** S107.
Bernal, J. D. (1968) *Growth of Crystals* **5A,** 123.
Bidnaya, D. S., Obukhovskii, Y. A. and Sysoev, L. A. (1962) *Russ. J. Inorg. Chem.* **7,** 1391.
Blander, M. (1964) In "Molten Salt Chemistry" (M. Blander, ed.) 127–237. Interscience, New York.
Blander, M. and Topol, L. E. (1965) *Electrochim. Acta* **10,** 1161.
Blank, S. L. and Nielsen, J. W. (1972) *J. Crystal Growth* **17,** 302.
Blattner, H., Matthias, B. and Merz, W. (1947) *Helv. Phys. Acta* **20,** 225.
Block, S. and Levin, E. M. (1957) *J. Am. Ceram. Soc.* **40,** 113.
Blom, G. M. and Plaskett, T. S. (1971) *J. Electrochem. Soc.* **118,** 1831.
Blom, G. M. (1971) *J. Electrochem. Soc.* **118,** 1834.
Bloom, H. (1963) *Pure and Appl. Chem.* **7,** 389.

Bloom, H. (1967) "The Chemistry of Molten Salts" Benjamin, New York.
Bloom H. and Bockris, J. O. M. (1959) In "Modern Aspects of Electrochemistry Vol. II", 160–261. Butterworths, London.
Bloom, H. and Hastie, W. (1965) In "Non-Aqueous Solvent Systems" (T. C. Waddington, ed.) Academic Press, London, New York.
Bokov, V. A., Kizhaev, S. A., Mylnikova, I. E. and Tutov, A. G. (1965) Sov. Phys.—Solid State 6, 2419.
Böttcher, C. J. F. (1943) Recueil Trav. Chem. 62, 325, 503.
Böttcher, C. J. F. and Scholte, T. G. (1952) Recueil Trav. Chem. 70, 203.
Bredig, M. A. (1963) Mixtures of Metals with Molten Salts, Report ORNL 3391.
Brice, J. C. (1973) "The Growth of Crystals from Liquids" North-Holland, Amsterdam.
Brixner, L. H. and Abramson, E. (1965) J. Electrochem. Soc. 112, 70.
Brixner, L. H. and Babcock, K. (1968) Mat. Res. Bull. 3, 817.
Broder, J. D. and Wolff, G. A. (1963) J. Electrochem. Soc. 110, 1150.
Brönsted, J. N. (1923) Rec. Trav. Chim. Pays-Bas 42, 718.
Brönsted, J. N. (1934) Z. Physik. Chem. 169, 52.
Brown, I. D. and Shannon, R. D. (1973) Acta Cryst. A 29, 266.
Bruton, T. M. and White, E. A. D. (1973) J. Crystal Growth 19, 341.
Buerger, M. J. (1948) Am. Min. 33, 744.
Burdese, A. (1957) Ann. Chim. (Rome) 47, 795 (see Fig. 24 of Levin et al. 1964).
Burmeister, R. A. (1972) Paper presented at AACG Conference on Crystal Growth, Princeton.
Burmeister, R. A. and Greene, P. E. (1967) Trans. Met. Soc. AIME 239, 408.
Burnett, D., Clinton, D. and Miller, R. P. (1968) J. Mat. Sci. 3, 47.
Carman, J. N., Stello, P. E. and Bittmann, C. A. (1954) J. Appl. Phys. 25, 543.
Chang, L. L. Y., Scroger, M. G. and Phillips, B. (1966) J. Am. Ceram. Soc. 49, 385.
Chase, A. B. (1967) J. Am. Ceram. Soc. 50, 325.
Chase, A. B. and Osmer, J. A. (1964) Am. Min. 49, 469.
Cleaver, B. and Davies, A. J. (1973) Electrochim. Acta 18, 719 and 727.
Cloete, F. L. D., Ortega, R. F. and White, E. A. D. (1969) J. Mat. Sci. 4, 21.
Cobb, C. M. and Wallis, E. B. (1967) Report AD 655388.
Cobb, C. M. and Wallis, E. B. (1967a) US Govt. Res. Develop. Rep. 67, 115.
Cobb, C. M. and Wallis, E. B. (1969) Report AD 687497.
Coe, I. M. and Elwell, D. (1974) J. Crystal Growth 23, 345.
Comer, J. J. and Berman, I. (1970) Report AFCRL-70-0130.
Darken, L. S. and Gurry, R. W. (1953) "Physical Chemistry of Metals" McGraw-Hill, New York.
Davies, C. W. (1962) "Ionic Association" Butterworth, London.
Delimarskii, Yu.K. and Markov, B. F. (1961) "Electrochemistry of Fused Salts" Sigma Press, Washington.
DeVries, R. C. (1966) Mat. Res. Bull. 1, 83.
DeVries, R. C. and Fleischer, F. (1972) J. Crystal Growth 13/14, 88.
Dietzel, A. (1942) Z. Elektrochem. 48, 9.
Dietzel, A. (1948/1949) Glastechn. Ber. 22, 41–50, 81–86, 212–224.
Dziuba, E. Z. (1971) J. Crystal Growth 8, 21.
Eick, H. A., Baenziger, N. C. and Eyring, L. (1956) J. Am. Chem. Soc. 78, 5147.
Elliott, R. P. (1965) "Constitution of Binary Alloys, First Supplement" McGraw-Hill, New York.

Ellis, R. C. (1960) In "Silicon Carbide" (J. R. O'Connor and J. Smiltens, eds.) p. 124. Pergamon, New York.
Ellis, W. C., Williams, H. J. and Sherwood, R. C. (1958) *J. Appl. Phys.* **29,** 534.
Elwell, D. (1975) In Proc. 2nd. Int. Spring School Crystal Growth (to be published by North-Holland, Amsterdam).
Elwell, D., Neate, B. W. and Smith, S. H. (1969) *J. Thermal Analys.* **1,** 319.
Elwell, D., Neate, B. W., Smith, S. H. and D'Agostino, M. (1971) *J. Physics E* **4,** 775.
Elwell, D., Morris, A. W. and Neate, B. W. (1972) *J. Crystal Growth* **16,** 67.
Elwell, D., Capper, P. and Lawrence, C. M. (1974) *J. Crystal Growth* **24/25,** 651.
Epstein, D. J. (1970) Report AD 715 312.
Eyring, H. and Jhon, M. S. (1969) "Significant Liquid Structures" Wiley, New York.
Fajans, K. (1921) *Naturwiss.* **9,** 729.
Fajans, K. and Karagunis, G. (1930) *Z. Angew. Chem.* **43,** 1046.
Faust, Jr., J. W. and John, H. F. (1964) *J. Phys. Chem. Solids* **25,** 1407.
Faust Jr., J. W., John, H. F. and Pritchard, C. (1968) *J. Crystal Growth* **3/4,** 321.
Feigelson, R. (1964) *J. Am. Ceram. Soc.* **47,** 257.
Feigelson, R. (1968) *J. Am. Ceram. Soc.* **51,** 538.
Finch, C. B. and Clark, G. W. (1970) *J. Crystal Growth* **6,** 245.
Fischer, H. (1954) "Elektrokristallisation von Metallen" Springer-Verlag, Berlin.
Flood, H. and Förland, T. (1947) *Acta Chem. Scand.* **1,** 592–604, 781–789, 790–798.
Flood, H. and Muan, A. (1950) *Acta Chem. Scand.* **4,** 359, 364.
Flood, H. and Knapp, W. J. (1963) *J. Am. Ceram. Soc.* **46,** 61.
Flörke, O. W. (1955) *Ber. Dtsch. Keram. Ges.* **32,** 369.
Flörke, O. W. (1956) *Naturwiss.* **43,** 419.
Foguel, M. and Grajower, R. (1971) *J. Crystal Growth* **11,** 280.
Folen, V. J. (1960) *J. Appl. Phys.* **31,** 1665.
Förland, T. (1955) Freezing Point Depression and its Structural Interpretation, Office of Naval Research Techn. Report No. 63, NRO32-264.
Foster Jr., P. A. (1960) *J. Am. Ceram. Soc.* **43,** 66.
Fremy, E. and Feil, F. (1877) *Compt. Rend.* **85,** 1029.
Friedman, H. L. (1962) "Ionic Solution Theory" Wiley-Interscience, New York.
Fukunaga, O. and Saito, S. (1968) *J. Am. Ceram. Soc.* **51,** 362.
Fushimi, S. and Ikeda, T. (1967) *J. Am. Ceram. Soc.* **50,** 129.
Garbee, A. K. and Flengas, S. N. (1972) *J. Electrochem. Soc.* **119,** 631
Garner, R. W. and White, W. B. (1970) *J. Crystal Growth* **7,** 343.
Garrels, R. M. and Christ, C. L. (1965) "Solutions, Minerals, and Equilibria" Harper and Row, New York.
Garton, G. and Wanklyn, B. M. (1968) *J. Mat. Sci.* **3,** 395.
Garton, G., Smith, S. H. and Wanklyn, B. M. (1972) *J. Crystal Growth* **13/14,** 588.
Gashurov, G. and Levine, A. K., (1960) *J. Chem. and Engng. Data* **5,** 517.
Gerk, A. P. and Gilman, J. J. (1968) *J. Appl. Phys.* **39,** 4497.
Giess, E. A. (1962) *J. Appl. Phys.* **33,** 2143.
Giess, E. A. (1964) *J. Am. Ceram. Soc.* **47,** 388.
Giess, E. A., Cronemeyer, D. C., Rosier, L. L. and Kuptsis, J. D. (1970) *Mat. Res. Bull.* **5,** 495.
Goss, A. J. (1953) *J. Metals* **5,** 1085.
Gossink, R. G. (1971) *Philips Res. Rept.* Suppl. No. 3.

Griffiths, L. B. and Mlavsky, A. I. (1964) *J. Electrochem. Soc.* **111**, 805.

Grodkiewicz, W. H. and Nitti, D. J. (1966) *J. Am. Ceram. Soc.* **49**, 576.

Gschneidner, K. A. (1961) "Rare Earth Alloys" Van Nostrand, Princeton.

Guggenheim, E. A. (1952) "Mixtures" Oxford University Press.

Guggenheim, E. A. (1967) "Thermodynamics" 5th Edition North Holland, Amsterdam.

Gutmann, V. and Lindquist, I. (1954) *Z. Physik. Chem. (Leipzig)* **203**, 250.

Haase, R. (1956) "Thermodynamik der Mischphasen", p. 440. Springer-Verlag, Berlin.

Halden, F. A. (1960) In "Silicon Carbide" (J. R. O'Conner and J. Smiltens, eds.) p. 115. Pergamon, New York.

Hall, R. N. (1958) *J. Appl. Phys.* **29**, 914.

Hall, R. N. (1963) *J. Electrochem. Soc.* **110**, 385.

Hanak, J. J. and Johnson, D. E. (1969) ACCG Conf. Crystal Growth, NBS, Gaithersburg, Maryland.

Hansen, M. and Anderko, K. (1958) "Constitution of Binary Alloys" McGraw-Hill, New York.

Harned, S. and Owen, B. B. (1959) "The Physical Chemistry of Electrolyte Solutions" 3rd Edition. Reinhold, New York.

Hartenberg, H. (1951) *Z. Anorg. Allg. Chem.* **265**, 186.

Hildebrand, J. H. and Scott, R. L. (1964) "The Solubility of Nonelectrolytes" 3rd Edition. Dover, New York.

Hildebrand, J. H. and Scott, R. L. (1962, 1970) "Regular Solutions" 2nd Edition 1970. Prentice Hall, Englewood Cliffs, N.Y.

Hill, G. and Wanklyn, B. M. (1968) *J. Crystal Growth* **3/4**, 475.

Hiskes, R. (1975) *J. Crystal Growth* (to be published).

Hiskes, R., Felmlee, T. L. and Burmeister, R. A. (1972) *J. Electron. Mater.* **1**, 458.

Holmquist, S. B. (1966) *J. Am. Ceram. Soc.* **49**, 228.

Holtzberg, F., Reisman, A., Berry, M. and Berkenblit, M. (1956) *J. Am. Chem. Soc.* **78**, 1538.

Horn, F. H., Fullam, E. F. and Kasper, J. S. (1952) *Nature* **169**, 927.

Izvekov, V. N., Sysoev, L. A., Obukhovskii, Ya.A. and Birman, B. I. (1968) *Growth of Crystal* **6A**, 106.

James, D. W. (1964) In "Molten Salt Chemistry" (M. Blander, ed.) 507–533. Interscience, New York.

JANAF Thermochemical Tables, The Dow Chemical Company, Midland, Michigan.

Jangg, G., Kieffer, R. and Usner, L. (1968) *J. Less-Common Metals* **14**, 269.

Janz, G. J. (1967) "Molten Salts Handbook" Academic Press, New York, London.

Janz, G. J. and Tomkins, R. P. T. (1972, 1973) "Nonaqueous Electrolytes Handbook" Vol. 1, 2. Academic Press, New York, London.

Johnston, W. D. (1966) *J. Am. Ceram. Soc.* **49**, 513.

Jona, F., Shirane, G. and Pepinsky, R. (1955) *Phys. Rev.* **97**, 1584.

Jones, L. V., Etter, D. E., Hudgens, C. R., Huffmann, A. A., Rhinehammer, T.B., Rogers, N. E., Tucker, P. A. and Wittenberg, L. J. (1962) *J. Am. Ceram. Soc.* **45**, 79.

Kalnin, A. A. and Tairov, Yu.M. (1966) *Izv. Leningrad Electrotekh. Inst.* No. 61, 26.

Karan, C. (1954) *J. Chem. Phys.* **22**, 957.

Karan, C. and Skinner, B. J. (1953) *J. Chem. Phys.* **21**, 2225.

Katz, G. and Roy, R. (1966) *J. Am. Ceram. Soc.* **49**, 168.

Kaufman, L. and Bernstein, H. (1970) "Computer Calculations of Phase Diagrams" Vol. 4 of "Refractory Materials" (J. L. Margrave, ed.) Academic Press, New York and London.

Keck, P. (1953) *Phys. Rev.* **90,** 379.

Keck, P. and Broder, J. (1953) *Phys. Rev.* **90,** 521.

Kelley, K. K. (1961) Bureau of Mines Bulletin, 584, U.S. Government Printing Office, Washington, D.C.

Kendall, J., Crittendon, E. D. and Miller, H. K. (1923), *J. Am. Chem. Soc.* **45,** 963.

Kjems, J. K., Shirane, G., Müller, K. A. and Scheel, H. J. (1973) *Phys. Rev.* **38,** 1119.

Kleber, W., Ickert, L. and Doerschel, J. (1966) *Kristall u. Technik* **1,** 237.

Kleppa, O. J. (1958) "Liquid Metals and Solidification" (B. Chalmers, ed.) 56–86. Am. Soc. for Metals, Cleveland/Ohio.

Knippenberg, W. F. and Verspui, G. (1966) *Philips Res. Rept.* **21,** 113.

Koppel, H. D., Luzhnaya, N. P. and Medwedyeva, Z. S. (1965) *Russ. J. Inorg. Chem.* **10,** 2315 (Russ.).

Koppel, H. D., Medwedyeva, Z. S. and Luzhnaya, N. P. (1967a) *Zh. Neorgan. Materialy* **3,** 300 (Russ.).

Koppel, H. D., Luzhnaya, N. P. and Medwedyeva, Z. S. (1967b) *Zh. Neorgan. Materialy* **3,** 1354 (Russ.).

Koutaissoff, A. (1964) Ph.D. Thesis No. 3572, ETH Zurich.

Kovaleva, I. S., Luzhnaya, N. P. and Martikyan, S. B. (1968) *Zh. Neorgan. Khim.* **13.**

Kozlowski, T. R. (1970) In "Phase Diagrams" Vol. III (A. M. Alper, ed.) p. 271. Academic Press, New York.

Krasnov, K. S. (1968) *Growth of Crystals* **5A,** 141.

Kubaschewski, O., Evans, E. L. and Alcock, C. B. (1967) "Metallurgical Thermochemistry" 4th Edition. Pergamon Press, Oxford.

Kunagawa, M., Ozaki, M. and Yamada, S. (1970) *Jap. J. Appl. Phys.* **9,** 1422.

Kunnmann, W. (1971) In "Preparation and Properties of Solid State Materials" Vol. I (R. A. Lefever, ed.) Dekker, New York.

Kunnmann, W., Ferretti, A., Arnott, R. J. and Rogers, D. B. (1965) *J. Phys. Chem. Solids* **26,** 311.

Kuse, D. (1970) *IBM J. Res. Develop.* **14,** 315.

Kvapil, J., Jon, V. and Vichr, M. (1969) *Growth of Crystals* **7,** 233.

Kweestroo, W. (1972) In "Preparative Methods in Solid State Chemistry" (P. Hagenmüller, ed.) 563–574. Academic Press, New York, London.

Landolt-Börnstein (1956) 6th Edition, Vol. II, Part 3. Springer-Verlag, Berlin.

Laurent, Y. (1969), *Rev. Chim. Minér.* **6,** 1145.

Lavoisier, A. L. (1794) "Traité elementaire de chimie" Tom II, Partie III, p. 104, from Hildebrand and Scott (1950), p. 28.

Layden, G. K. and Darby, W. L. (1966) *Mat. Res. Bull.* **1,** 235.

Lefever, R. A., Chase, A. B. and Torpy, J. W. (1961) *J. Am. Ceram. Soc.* **44,** 141.

Levin, E. M. (1967) *J. Am. Ceram. Soc.* **50,** 381.

Levin, E. M. (1970) In "Phase Diagrams" Vol. III (A. M. Alper, ed.) p. 143. Academic Press, New York.

Levin, E. M., Robbins, C. R. and McMurdie, H. F. (1964) Phase Diagrams for Ceramists, Amer. Ceram. Soc., Columbus, Ohio.

Levin, E. M., Robbins, C. R. and McMurdie, H. F. (1969) Phase Diagrams for Ceramists, Supplement, Amer. Ceram. Soc., Columbus, Ohio.

Levinstein, H. J., Licht, S., Landorf, R. W. and Blank, S. (1971) *Appl. Phys. Letters* **19**, 486.

Levy, H. A. and Danford, M. D. (1964) In "Molten Salt Chemistry" (M. Blander, ed.) p. 109. Interscience, New York.

Lewis, G. H. (1923) "Valence and the Structure of Atoms and Molecules" New York. Reprinted by Dover Publications, New York, 1966.

Lewis, G. H. (1938) *J. Franklin Inst.* **226**, 293.

Liander, H. and Lundblad, E. (1960) *Ark. Kemi* **16**, 139.

Libowitz, G. G. (1953) *J. Chem. Soc.* **75**, 1501.

Linares, R. C. (1962a) *J. Am. Ceram. Soc.* **45**, 307; *J.Appl. Phys.* **33**, 1747.

Linares, R. C. (1962b) In "Metallurgy of Advanced Electronic Materials" (G. E. Brock, ed.) p. 329. Interscience, New York.

Linares, R. C. (1964) *J. Appl. Phys.* **35**, 433.

Linares, R. C. (1967) *Am. Min.* **52**, 1554.

Linares, R. C. (1968) *Trans. Met. Soc. AIME* **242**, 441.

Lumsdem, J. (1966) "Thermodynamics of Molten Salt Mixtures" Academic Press, New York, London.

Lux, H. (1939) *Z. Elektrochem.* **45**, 303.

Luzhnaya, N. P., Medwedyeva, Z. S. and Koppel, H. D. (1966) *Z. anorg. allg. Chem.* **344**, 323.

Luzhnaya, N. P. (1968) *J. Crystal Growth* **3/4**, 97.

Makram, H., Touron, L. and Loriers, J. (1972) *J. Crystal Growth* **13/14**, 585.

Malur, J. (1966) *Kristall u. Technik* **1**, 261.

Mamantov, G. (1969) (Ed.) "Molten Salts: Characterization and Analysis" Dekker, New York.

Manzel, M. (1967) *Kristall u. Technik* **2**, 61.

Marbitt, A. W. (1970) *J. Mater. Sci.* **5**, 1043.

Marezio, M., Remeika, J. P. and Dernier, P. D. (1970) *Acta Cryst.* B **26**, 2008.

Marshall, R. C. (1969) *Mat. Res. Bull.* **4**, S73.

McCarthy, G. J. and White, W. B. (1967) *J. Less—Common Metals* **22**, 409.

McKenzie, J. D. (1962) *Adv. Inorg. Chem. Radiochem.* **4**, 293.

McWhan, D. B. and Remeika, J. P. (1970) *Phys. Rev.* **2B**, 3734.

Meyer, A. (1970) *J. Less—Common Metals* **20**, 353.

Mills, K. C. (1974) "Thermodynamic Data for Inorganic Sulphides, Selenides and Tellurides".

Morgan, R. A. and Hummel, F. A. (1949) *J. Am. Ceram. Soc.* **32**, 255.

Morozewicz, J. (1899) *Tschermaks Min. Petrogr. Mitt.* **18**, 3, 163.

Morris, D. F. C. (1968) *Structure and Bonding* **4**, 63; **6**, 157.

Mott, B. W. (1968) *J. Mat. Sci.* **3**, 424.

Mountvala, A. J. and Ravitz, S. F. (1962) *J. Am. Ceram. Soc.* **45**, 285.

Muan, A. (1958) *Am. J. Sci.* **256**, 171.

Mylnikova, I. E. and Bokov, V. A. (1962) *Growth of Crystals* **3**, 309.

Nancollas, G. H. (1966) "Interactions in Electrolyte Solutions" Elsevier, Amsterdam.

Nassau, K. (1968) *J. Crystal Growth* **2**, 215.

Naumova, I. I. and Anikin, I. N. (1966) *Zh. Neorgan. Khim.* **11**, 1746.

Neece, G. A. and Widom, B. (1969) *Ann. Rev. Phys. Chem.* **20**, 167.

Negas, T. (1968) *J. Am. Ceram. Soc.* **51**, 716.

Nelson, D. F. and Remeika, J. P. (1964) *J. Appl. Phys.* **35**, 522.

Nelson, W. E., Rosengreen, A. and Halden, F. A. (1966) *J. Appl. Phys.* **37**, 333.

Neuhaus, A., Holz, H. G. and Klein, H. D. (1967) *Z. Phys. Chem.* (Neue Folge) **53**, 1.

Newkirk, H. W. and Smith, D. K. (1965) *Am. Min.* **50**, 44.
Nielsen, J. W. (1960) *J. Appl. Phys.* **31**, 51S.
Nielsen, O. V. (1969) *J. Crystal Growth* **5**, 398.
Nielsen, J. W. and Dearborn, E. F. (1960) *J. Phys. Chem.* **64**, 1762.
Nielsen, J. W., Lepore, D. A. and Leo, D. C. (1967) In "Crystal Growth" (H. S. Peiser, ed.) p. 457. Pergamon, Oxford.
Nomura, S. and Sawada, S. (1952) *Rep. Inst. Sci. Technol. Univ. Tokyo* **6**, 191.
O'Kane, D. F. and Stemple, N. R. (1966) *J. Electrochem. Soc.* **113**, 290.
Oliver, C. B. (1965) *J. Electrochem. Soc.* **112**, 629.
Panish, M. B. (1966a) *J. Less—Common Metals* **10**, 416.
Panish, M. B. (1966b) *J. Electrochem. Soc.* **113**, 224.
Panish, M. B. (1970a) In "Phase Diagrams" Vol. III (A. M. Alper, ed.) p. 53. Academic Press, New York and London.
Panish, M. B. (1970b) *J. Electrochem. Soc.* **117**, 1202.
Panish, M. B. and Ilegems, M. (1972) *Prog. Solid State Chem.* **7**, 39.
Panish, M. B. and Sumski, S. (1969) *J. Phys. Chem. Solids* **30**, 129.
Parker, H. S. and Brower, W. S. (1967) In "Crystal Growth" (H. S. Peiser, ed.) p. 489. Pergamon, Oxford.
Paul, A. and Douglas, R. W. (1966) *Phys. Chem. Glasses* **7**, 1.
Paul, A. and Lahari, D. (1966) *J. Am. Ceram. Soc.* **49**, 565.
Pauling, L. (1960) "The Nature of the Chemical Bond" 3rd Edition, p. 97. Cornell University Press, Ithaca, N.Y.
Payne, D. S. (1965) In "Non-Aqueous Solvent Systems" (T. C. Waddington, ed.) p. 305. Academic Press, New York, London.
Perry, F. W. (1967) In "Crystal Growth" (H. S. Peiser, ed.) p. 483. Pergamon, Oxford.
Petrucci, S. (1971) (Ed.) "Ionic Interactions" 2 volumes. Academic Press, New York, London.
Phillips, W. and Pressley, R. J. (1967) *Bull. A. Ceram. Soc.* **46**, 366.
Pickar, P. B. (1967) US Patent 3 353 914.
Plaskett, T. S. (1969) *J. Electrochem. Soc.* **116**, 1722.
Prener, J. S. (1967) *J. Electrochem. Soc.* **114**, 77.
Prigogine, I. (1957) "The Molecular Theory of Solutions" North-Holland, Amsterdam.
Rase, D. E. and Roy, R. (1955) *J. Am. Ceram. Soc.* **38**, 110.
Rase, D. E. and Roy, R. (1957) *J. Phys. Chem.* **61**, 746.
Ray, H. S. (1969) *Trans. Indian Ceram. Soc.* **28**, 82.
Reed, T. B. and Fahey, R. E. (1966) *Rev. Sci Instr.* **37**, 59.
Reed, T. B. (1972) "Free Energy of Binary Compounds", MIT Press.
Reisman, A. (1970) "Phase Equilibria" Academic Press, New York, London.
Reisman, A. and Holtzberg, F. (1955) *J. Am. Chem. Soc.* **77**, 2115.
Reisman, A., Holtzberg, F., Berkenblit, M. and Berry, M. (1956) *J. Am. Chem. Soc.* **78**, 4514.
Remeika, J. P. (1954) *J. Am. Chem. Soc.* **76**, 940.
Remeika, J. P. (1955) British Patent 783 670.
Remeika, J. P. (1956) *J. Am. Chem. Soc.* **78**, 4259.
Remeika, J. P. and Kometani, T. Y. (1968) *Mat. Res. Bull.* **3**, 895.
Remeika, J. P. (1970) Paper presented at International Conference on Ferrites, Kyoto (unpublished).
Ricci, J. E. (1964) In "Molten Salt Chemistry" (M. Blander, ed.) 239–325. Interscience, New York.

Richardson, F. D. and Jeffes, J. H. E. (1948) *J. Iron Steel Inst.* **160,** 261.
Richardson, F. D. and Jeffes, J. H. E. (1952) *J. Iron Steel Inst.* **171,** 167.
Robertson, D. S. and Cockayne, B. (1966) *J. Appl. Phys.* **37,** 927.
Robertson, J. M. and Taylor R. G. F. (1968) *J. Crystal Growth* **2,** 171.
Robins, D. A. (1959) *Nat. Phys. Lab. Gt. Brit. Proc. Symp.* **9,** 2 (quoted by Rowcliffe and Warren, 1970).
Robinson, R. A. and Stokes, R. H. (1968) "Electrolyte Solutions" 2nd Edition. Butterworth, London.
Rogers, D. B., Ferretti, A. and Kunnmann, W. (1966) *J. Phys. Chem. Solids* **27,** 1445.
Rolin, M. (1951) *Revue de Metallurgie* **48,** 182.
Rowcliffe, D. J. and Warren, W. J. (1970) *J. Mat. Sci.* **5,** 345.
Rowlinson, J. S. (1969) "Liquids and Liquid Mixtures" 2nd Edition. Butterworth, London.
Rowlinson, J. S. (1970) *Disc. Faraday Soc.* **49,** 30.
Roy, R. and White, W. B. (1968) *J. Crystal Growth* **3/4,** 33.
Rubenstein, M. (1966) *J. Electrochem. Soc.* **113,** 752.
Rubenstein, M. (1968), *J. Crystal Growth* **3/4,** 309; see also *J. Electrochem. Soc.* **113,** 623 (1966).
Rubenstein, M. and Ure, J. W. (1968) *J. Phys. Chem. Solids* **29,** N3.
Rushbrooke, G. S. (1938) *Proc. Roy. Soc.* (London) A **166,** 296.
Samoilov, O. Y. (1957) "Structures of Aqueous Electrolyte Solutions and Ion Hydration" Izv. Akad. Nauk, SSSR.
Sasaki, H. and Watanabe, A. (1964) *J. Phys. Soc. Japan* **19,** 1748.
Sato, M. and Hukudo, S. (1963) *Yogyo Kyogai Shi* **71,** 5.
Scheel, H. J. (1968) *J. Crystal Growth* **2,** 411.
Scheel, H. J. (1973) Swiss Patent 530 939 (Jan. 15, 1973).
Scheel, H. J. (1974a) *J. Crystal Growth* **24/25,** 669.
Scheel, H. J. (1974b) unpublished.
Schieber, M. (1967) *Kristall u. Technik* **2,** 55.
Schultze, D., Wilke, K.Th. and Waligora, C. (1967) *Z. anorg. allg. Chem.* **352,** 184.
Schultze, D. (1972) Registered Lecture at 1st Seminar on Crystal Chemistry, Prachovské Skaly, Sept. 19–21, 1972, pp. 2–49.
Shafer, M. W. (1965) *J. Appl. Phys.* **36,** 1145.
Shafer, M. W. and Weiser, K. (1957) *J. Phys. Chem.* **61,** 1425.
Shafer, M. W., Torrance, J. B. and Penney, T. (1972) *J. Phys. Chem. Solids* **33,** 2251.
Shih, K. K. (1970) *J. Electrochem. Soc.* **117,** 387.
Sholokhovich, M. L., Kramarov, O. P., Proskuruyakov, B. F. and Zvorykina, E. K. (1970) *Sov. Phys.-Cryst.* **14,** 884.
Shunk, F. A. (1969) "Constitution of Binary Alloys" Second Supplement. McGraw–Hill, New York.
Silva, W. J., Rosengreen, A. and Marsh, L. E. (1967) Final Report, AFML Contract F33615-67-C-1328.
Smith, F. G. (1963) "Physical Geochemistry" Addison-Wesley, Reading, Mass.
Smith, G. P. (1964) In "Molten Salt Chemistry" (M. Blander, ed) 427–505. Interscience, New York.
Smith, S. H. (1970) Ph.D. Thesis, Portsmouth Polytechnic.
Smith, S. H. and Elwell, D. (1967) *J. Mat. Sci.* **2,** 297.
Smith, S. H. and Elwell, D. (1968) *J. Crystal Growth* **3/4,** 471.

Smithells, C. J. (1962) "Metals Reference Book" 3rd Edition, 2 volumes. Butterworths, London.

Sobon, L. F. and Greene, P. E. (1966) *J. Am. Ceram. Soc.* **49**, 106.

South, K. D., Sudworth, J. L. and Gibson, J. G. (1972) *J. Electrochem. Soc.* **119**, 554.

Spring-Thorpe, A. J. (1969) *Mat. Res. Bull.* **4**, 125.

Spring-Thorpe, A. J. and Pamplin, B. R. (1968) *J. Crystal Growth* **3/4**, 313.

Sproull, R. L., Dash, W. C., Tyler, W. W. and Moore, A. R. (1951) *Rev. Sci. Instr.* **22**, 410.

Steininger, J., Strauss, A. J. and Brebrick, R. F. (1970) *J. Electrochem. Soc.* **117**, 1305.

Stephen, H. and Stephen, T. (1963) "Solubilities of Inorganic and Organic Compounds" 5 volumes. Macmillan, New York.

Stillinger Jr., F. H. (1964) In "Molten Salt Chemistry" (M. Blander, ed.) 1–108, Interscience, New York.

Stoch, L. (1968) *J. Am. Ceram. Soc.* **51**, 419.

Stringfellow, G. B. (1970) *J. Electrochem. Soc.* **117**, 1301.

Stringfellow, G. B. and Greene, P. E. (1971) *J. Electrochem. Soc.* **118**, 805.

Stringfellow, G. B. and Antypas, G. A. (1971) *J. Electrochem. Soc.* **118**, 1019.

Strong, H. M. and Hanneman, R. E. (1967) *J. Chem. Phys.* **46**, 3668.

Strong, H. M. and Chrenko, R. M. (1971) *J. Phys. Chem.* **75**, 1838.

Sugai, T., Hasegawa, S. and Ohara, G. (1968) *Jap. J. Appl. Phys.* **7**, 358.

Sumiyoshi, Y., Ito, N. and Noda, T. (1968) *J. Crystal Growth* **3/4**, 327.

Sundheim, B. R. (1964) (Ed.) "Fused Salts" McGraw-Hill, New York.

Teitel, R. J., Gurinsky, D. H. and Bryner, J. S. (1954) *Nucleonics* **12**, 7, 14.

Temkin, M. (1945) *Acta Physicochem. USSR* **20**, 411 (quoted by Cobb and Wallis, 1969).

Thoma, R. E., Insley, H., Friedman, H. A. and Weaver, C. F. (1960) *J. Phys. Chem.* **64**, 865.

Thoma, R. E. and Karraker, R. H. (1966) *Inorg. Chem.* **5**, 1222.

Timofeeva, V. A. (1967) In "Crystal Growth" (H. S. Peiser, ed.) p. 445. Pergamon, Oxford.

Timofeeva, V. A. (1968) *J. Crystal Growth* **3/4**, 496.

Timofeeva, V. A. and Konkova, T. S. (1968) *Sov. Phys.-Cryst.* **13**, 479.

Timofeeva, V. A. and Kvapil, I. (1966) *Sov. Phys.-Cryst.* **11**, 263.

Timofeeva, V. A. and Lukyanova, N. I. (1967) *Sov. Phys.-Cryst.* **12**, 77.

Timofeeva, V. A. and Lukyanova, N. I. (1970) *Inorganic Materials* **6**, 587.

Timofeeva, V. A., Lukyanova, N. I., Guseva, I. N. and Lider, V. V. (1969) *Sov. Phys.-Cryst.* **13**, 747.

Tolksdorf, W. and Welz, F. (1972) *J. Crystal Growth* **13/14**, 566.

Tolksdorf, W., Bartels, G., Espinosa, G. P., Holst, P., Mateika, D. and Welz, F. (1972) *J. Crystal Growth* **17**, 322.

Torkar, K., Kirschner, H. and Moser, H. (1966) *Ber. Dt. Keram. Ges.* **43**, 259.

Tress, H. J. (1960) *Phys. Chem. Glasses* **1**, 196.

Tress, H. J. (1962) *Phys. Chem. Glasses* **3**, 28.

Ugai, Ya.A., Goncharov, E. G., Kitina, Z. V. and Shuyreva, T. N. (1968) *Zh. Neorgan. Materialy* **4**, 348.

Van Artsdalen, E. R. (1959) In "The Structure of Electrolytic Solutions" (W. J. Hamer, ed.) 411–421. Wiley, New York.

Van Uitert, L. G., Grodkiewicz, W. H. and Dearborn, E. F. (1965) *J. Am. Ceram. Soc.* **48**, 105.

Van Uitert, L. G., Bonner, W. A., Grodkiewicz, W. H., Pietroski, L. and Zydzik, G. J. (1970) *Mater. Res. Bull.* **5**, 825.

Viting, L. M. and Khomyakov, K. G. (1965) *Vestn. Mosk. Univ. Ser. Khim.* **20**, 60.

Vogt, J. H. L. (1903/1904) "Die Silikatschmelzlösungen" II, Christiania.

Voigt, D. O. and Neels, H. (1971) *Kristall u. Technik* **6**, 651.

Voronkova, V. I., Yanovskii, V. K. and Koptsik, V. A., (1967) *Dokl. Akad. Nauk, SSSR* **117**, 571.

Wagner, P. and Lorenz, M. R. (1966) *J. Phys. Chem. Solids* **27**, 1749.

Wagner, R. S. (1968) *J. Crystal Growth* **3/4**, 159.

Walden, P. (1910) "Die Lösungstheorien in ihrer geschichtlichen Aufeinanderfolge" In F. B. Ahrens "Sammlung chemischer und chemisch-technischer Vorträge", F. Enke, Stuttgart, p. 277.

Wang, C. C. and McFarlane, S. H. (1968) *J. Crystal Growth* **3/4**, 485.

Wanklyn, B. M. (1969) *J. Crystal Growth* **5**, 219, 323.

Wanklyn, B. M. (1970) *J. Crystal Growth* **7**, 368.

Wanklyn, B. M. (1975) In "Crystal Growth" (B. R. Pamplin, ed.) Pergamon, Oxford.

Warren, B. E. and Pincus, A. G. (1940) *J. Am. Ceram. Soc.* **23**, 301.

Webster, F. W. and White, E. A. D. (1969) *J. Crystal Growth* **5**, 167.

Wentorf, R. H. (1957) *J. Chem. Phys.* **26**, 956.

Wentorf, R. H. (1965) US Patent 3.192.015.

Wentorf, R. H. (1966) *Ber. Bunsenges. Phys. Chem.* **70**, 975.

Weyl, W. A. (1932) *Glastechn. Ber.* **10**, 541.

Weyl, W. A. (1951) "Coloured Glasses" Sheffield.

Weyl, W. A. (1956) *Glass Industry* **37**, 264–269, 286, 288, 325–331, 336, 344, 346, 350.

Weyl, W. A. and Marboe, E.Ch. (1962) "The Constitution of Glasses" Interscience, New York, London.

White, E. A. D. and Brightwell, J. W. (1965) *Chem. and Industry* 1162.

Wickham, D. G. (1962) *J. Appl. Phys.* **33**, 3597.

Wilcox, W. R. and Fullmer, L. D. (1966) *J. Am. Ceram. Soc.* **49**, 415.

Wilke, K.Th. (1962) *Ber. Geolog. Ges.* **7**, 500.

Wilke, K.Th. (1964) *Z. anorg. allg. Chem.* **330**, 164.

Wilke, K.Th., Schultze, D. and Töpfer, K. (1967) *J. Crystal Growth* **1**, 41.

Wilke, K.Th. (1968) *Growth of Crystals* **6A**, 71.

Wold, A. and Bellarance, D. (1972) In "Preparative Methods in Solid State Chemistry" (P. Hagenmüller, ed.) 279–308. Academic Press, New York.

Wood, J. D. C. and White, E. A. D. (1968) *J. Crystal Growth* **3/4**, 480.

Woodall, J. M. (1971) *J. Electrochem. Soc.* **118**, 150.

Woodbury, H. H. (1963) *J. Phys. Chem. Solids* **24**, 881.

Wright, M. A. (1965) *J. Electrochem. Soc.* **112**, 1114.

4. Theory of Solution Growth

4.1. Limitations of a Theoretical Treatment

In this chapter we discuss those aspects of the theory of crystal growth from solution which relate to the growth mechanism. Reference is made where possible to experiments either on high-temperature or on aqueous solutions which support the various postulates introduced in the theory. A recent review of crystal-growth theory has been given by Parker (1970) and theoretical aspects of crystal growth from solution have also been reviewed by Bennema (1965), Khamskii (1969), Strickland-Constable (1968) and Lewis (1974).

Although the number of theoretical publications is quite extensive, reliable quantitative estimates of the growth rate under specified conditions still cannot be given for growth from solution. All the expressions for this most important parameter contain factors which cannot be assigned numerical values based on experiment. Any numerical estimates given therefore contain values which are crude approximations and so predictions from the theory are at best reliable only to the order of magnitude.

Another serious limitation mentioned in the previous chapter is that our present knowledge of the detailed atomic structure of solutions is uncertain and any model of atomistic behaviour in the neighbourhood of a crystal-solution interface is therefore highly speculative. It may be expected that the recent advances in understanding of the liquid state will lead to new experimental and theoretical studies on solutions, and there is considerable scope for original work. The content of this chapter is limited to an explanation of existing theories in order to formulate the most complete model of crystal growth from solution which can be given at present.

4.2. Nucleation

The initial stage of crystallization in a supercooled liquid is the formation of nuclei of the crystalline phase. Crystal growth, as distinct from nucleation, is the process by which these nuclei attain macroscopic dimensions.

The most important early study of nucleation was that of Tammann (1925), who determined the rate of nucleation of complex organic materials. The form of the curve he obtained is shown in Fig. 4.1. On cooling below

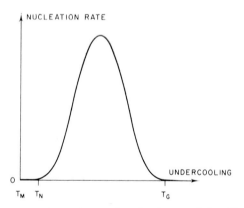

Fig. 4.1. Temperature dependence of nucleation rate (after Tammann, 1925).

the melting point T_M, the nucleation rate is low until some temperature T_N is reached at which the nucleation rate increases very rapidly. The metastable region $T_M \rightarrow T_N$ will depend on such factors as the purity of the melt and the presence of dust or other particles which may act as centres for nucleation. The maximum in the nucleation-temperature curve is due to a slowing down in the kinetics as the temperature is decreased. The fall in the nucleation rate is particularly marked in viscous melts, and will become essentially zero at some temperature T_G. If the melt is cooled to T_G without any crystallization, a glass will be formed. A similar curve to Fig. 4.1 will apply to solutions and it is possible to cool very viscous solutions to a temperature at which nucleation does not occur.

Reviews of nucleation from solution have been given by Hirth and Pound (1963), Nielsen (1964) and Zettlemoyer (1969). In most systems used for the growth of crystals, nucleation occurs heterogeneously, that is at favourable sites within the solution such as the crucible wall or the surface of the solution. Nucleation theory, however, normally describes the process of homogeneous nucleation in which the nuclei are considered to form at random throughout the solution, although estimates of heterogeneous nucleation can also be made.

Fluctuations within the supersaturated solution give rise to small clusters of molecules, known as "embryos". The probability that an embryo will grow to form a stable nucleus depends on the change in free energy associated with its growth or decay. The change in Gibbs free energy associated with the formation of a spherical embryo of radius r is given by

$$\Delta G = 4\pi r^2 \gamma - \tfrac{4}{3}\pi r^3 \Delta G_V + \Delta G_E + \Delta G_C \qquad (4.1)$$

where γ is the interfacial surface energy of the solid phase and ΔG_V the difference in the Gibbs free energy per unit volume between the solid and liquid phases. The terms ΔG_E and ΔG_C represent respectively the changes in Gibbs free energy due to the strain energy and to the configurational entropy change associated with the replacement of internal degrees of freedom of bulk crystal by rotational and translational degrees of freedom of isolated embryos (Lothe and Pound, 1962) and these are normally neglected as a first approximation.

As r increases from zero, the Gibbs free energy increases up to a critical value r^* and then decreases, so that r^* represents the minimum radius of a stable nucleus. The value of r^* is given by differentiation of Eqn (4.1) as

$$r^* = \frac{2\gamma}{\Delta G_V}. \qquad (4.2)$$

The form of Eqn (4.2) is unchanged if nuclei of nonspherical shape are considered but the numerical factor will then differ from 2.

The critical radius r^* may be related to the supersaturation in the system if the free-energy change per unit volume is written as

$$\Delta G_V = -S_V \Delta T = \frac{\phi_V}{T} \Delta T \qquad (4.3)$$

where ϕ_V is the heat of crystallization per unit volume and ΔT the magnitude of the supercooling at constant pressure. For an ideal solution, the equilibrium solute concentration is given by $n_e = n_\infty \exp(-\phi/RT)$, where $\phi \, (=\Delta H_f)$ is the molar heat of solution so that $\phi = V_M \phi_V$, with V_M the molar volume. The relative supersaturation for small values of ΔT is

$$\sigma = \frac{\Delta n}{n_e} = \frac{\phi \Delta T}{RT^2} \tag{4.4}$$

so that

$$\Delta G_V = \frac{\phi_V \Delta T}{T} = \frac{\phi \Delta T}{V_M T} = \frac{RT\sigma}{V_M}. \tag{4.5}$$

Substitution for ΔG_V in Eqn (4.2) gives the value of the critical radius as

$$r^* = \frac{2\gamma V_M}{RT\sigma} \tag{4.6}$$

so that an increase in supersaturation will decrease r^* and will therefore favour nucleation.

The value of ΔG in Eqn (4.1) for a nucleus of critical size is

$$\Delta G^* = \frac{16\pi\gamma^2}{3\Delta G_V{}^2} = \frac{16\pi\gamma^3 V_M{}^2}{3R^2 T^2 \sigma^2} \tag{4.7}$$

and, if there are n molecules per unit volume, the concentration of nuclei of critical size is

$$n^* = n \exp\left(-\Delta G^* / kT\right). \tag{4.8}$$

The nucleation rate I, being defined as the number of critical nuclei generated in unit volume per second, is given by the product of the concentration of nuclei of critical size and the rate at which molecules join such nuclei as

$$I = n^* z^* A^* = 4\pi n^* z^* r^{*2}. \tag{4.9}$$

Here z^* is the frequency of attachment of single molecules to unit area of nuclei and A^* is the area of a critical nucleus. Substitution for r^* and n^* in Eqn (4.9) gives

$$I = \frac{16\pi z^* \gamma^2 n V_M{}^2}{R^2 T^2 \sigma^2} \exp\left(-\frac{16\gamma^2 V_M{}^2}{3kR^2 T^3 \sigma^2}\right) \tag{4.10}$$

from which it is apparent that I will vary rapidly with the supersaturation σ, mainly through the exponential term.

The above treatment follows that given by Volmer and Weber (1926) who assumed that the probability of growth of the nuclei undergoes a sharp discontinuity at the critical radius r^*. Actually embryos of sub-critical size will have a finite probability of growing and those of super-critical size may shrink. A correction for such behaviour was applied by Becker and Döring (1935), but the resulting expression for I still varies rapidly with the driving force for crystal growth, which is represented by σ.

The dependence on supersaturation of the nucleation rate of potassium sulphate from aqueous solution has been measured by Mullin and Gaska (1969) and is shown in Fig. 4.2. This figure shows a comparison between the nucleation rate and the growth rate over the same supersaturation range. Nucleation is found to be extremely slow for supersaturations below

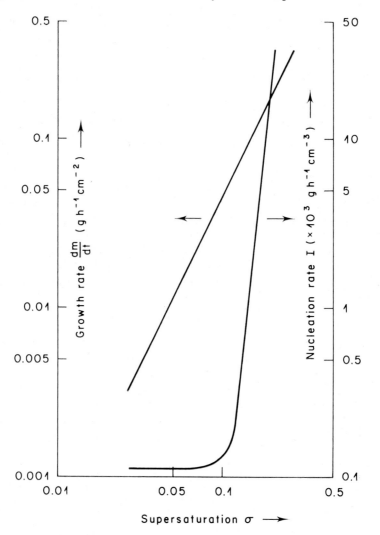

FIG. 4.2. Growth and nucleation rates of potassium sulphate (after Mullin and Gaska, 1969).

10% and so will not interfere to any appreciable extent with growth on established crystals in well-stirred solutions at supersaturations much below this value. The form of the $I(\sigma)$ curve is in quite good agreement with that of Eqn (4.10) and, in the region of supersaturation above 10%, the nucleation rate can be approximated by a power law $\sim\sigma^9$.

In the presence of a solid surface or other favourable centre, the nucleation rate increases because of a reduction in the interfacial free energy. An expression for the rate of heterogeneous nucleation may be obtained by replacing ΔG^* by some lower value, depending on the nature of the surface and the shape of the embryos. Foreign particles are well known to provide nucleation centres and the problem of achieving a really clean system makes truly homogeneous nucleation difficult to achieve experimentally.

When the conditions for nucleation are first created in a solution, a finite period is required before the steady nucleation rate is established. The rate at which the nucleation rate approaches the steady value I_0 can be described (Dunning, 1955) by a relation

$$I(t) = I_0 \exp\left(-\frac{\tau}{t}\right).$$

The time constant τ can be written as

$$\tau = \frac{N_C^2 h}{N_S^* kT} \exp\left(\frac{W_D}{kT}\right) \tag{4.11}$$

where N_C is the number of molecules in the critical nucleus and N_S^* the number of solute molecules in the layer of solution adjacent to this nucleus. Cobb and Wallis (1967) have estimated that, in the growth of Al_2O_3 from high-temperature solution, τ can have values from about 0.4 μs to 40 μs for undercoolings between 1°C and 10°C. Under normal growth conditions, therefore, this time dependence should have little effect since under-coolings are expected to be less than 10°C. Long induction periods prior to nucleation may, however, be possible in highly viscous solutions.

4.3. Rough and Smooth Interfaces

Once a crystal has nucleated in a solution, the growth process involves the transport of solute molecules from the solution to some point on the crystal surface where they become part of that surface. Of critical importance is the nature of the crystal-solution interface and we consider first the atomic models of the surfaces of crystals.

To the unaided eye, many crystals grown from solution have perfectly flat faces. The important question which will determine the growth kinetics of the crystal is whether this flatness persists down to the atomic level. Figure 4.3(a) shows a section through an idealized crystal having atomically flat faces, in which the atoms, all identical, have been represented as small cubes (this picture clearly differs very strongly from reality!). Inside the crystal any atom will have six neighbours and, if the binding energy per atom pair is W_B, the energy with which the atom is bound into the crystal is $3W_B$ since each bond is shared between two atoms. For simplicity, only nearest-neighbour interactions are considered. If a single extra atom is to be added to the crystal, it can form a bond with only one nearest neighbour and so its binding energy is only W_B. Further atoms may, of course, from extra bonds with this first additional atom (adatom) and so constitute a stable cluster, but the small energy with which the first atom is attached is clearly a major barrier to the growth of this crystal.

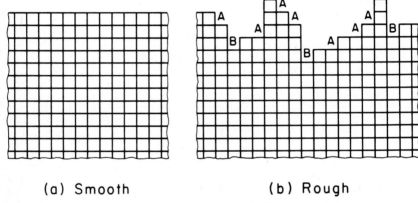

(a) Smooth   (b) Rough

FIG. 4.3. Crystal interfaces. (a) "flat", (b) "rough".

An atomically rough crystal interface will have a cross-section such as that shown diagrammatically in Fig. 4.3(b). An atom added at the sites labelled A will form bonds with two atoms in the same plane and atoms arriving at sites labelled B will form bonds with three atoms in this plane. It is clear that any atom incident on this "rough" surface will have a much greater probability of becoming part of that surface than in the case of the smooth surface. Note that this probability will depend on the binding energy W_B, $2W_B$, $3W_B$, etc., not linearly but through terms $\exp(W_B/kT)$, $\exp(2W_B/kT)$ etc., where T is the interface temperature and k is Boltzmann's constant.

From this very simple argument, we may conclude that atomically rough surfaces have a much higher rate of growth than atomically flat surfaces. Rough surfaces tend to remain rough as long as adatoms which become attached at sites such as those labelled A in Fig. 4.3 create new "corners" for the attachment of subsequent atoms. However, on a smooth surface, the rate-limiting step will be the addition of a new atom or group of atoms on that surface, since this group will form a layer with a "rough" edge at which atoms can be integrated relatively easily until the layer covers the whole crystal face and the surface is again smooth.

4.4. Models of Surface Roughness

Several calculations have been performed of the degree of roughness of a crystal surface and its variation with temperature. Burton and Cabrera (1949) used the Onsager (1944) solution of the Ising model to treat the behaviour of an array of atoms on the surface of the crystal. If U is the surface potential energy per atom of the actual surface and U_0 that of a perfectly flat surface, the surface roughness is defined as $S_r = (U - U_0)/U_0$. The parameter S_r will clearly be zero for a perfectly flat surface and so a non-zero value of S_r is a measure of the degree of roughness. A simple cubic array (such as that of Fig. 4.3) is treated and is assumed to be perfectly flat at absolute zero.

The energy required to remove an atom from the perfectly flat surface and to place it on a site in the next layer (previously empty) is $2W_B$ since four bonds must be broken. For temperatures well below a critical value T_c, $S_r = 4 \exp(-2W_B/kT)$ in which the factor $\exp(-2W_B/kT)$ is the probability of excitation of a single atom from a full to an empty layer on the surface. The variation of this function with temperature is shown in Fig. 4.4(a). It will be seen that the surface may be assumed flat provided that T is much less than $0.1\ W_B/k$. More recent treatments have predicted

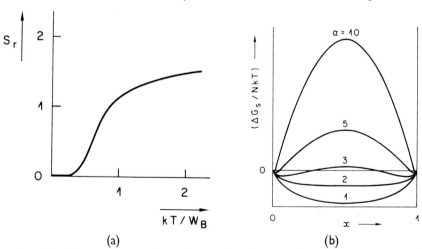

FIG. 4.4. (a) Temperature dependence of surface roughness (after Burton and Cabrera, 1949). (b) Free energy change with fractional occupation of layer (after Jackson, 1958).

curves which differ markedly from that of Fig. 4.4(a), but the trend is always from $S_r=0$ at low temperatures with the roughness increasing rapidly as T is raised above some value in the region of 0.2 W_B/k. The temperature T_c at which the surface in contact with the vapour becomes "ideally rough" is given by $W_B/k \ln (2^{1/2} - 1)^{-1}$ and is normally much higher than the melting point of the solid. For solid-liquid interfaces W_B is lower and the surface may be rough at or below the melting point.

Jackson (1958) used a rather different approach which takes into account the nature of the medium in contact with the crystal surface. His approximation involves a calculation of the change in the Gibbs free energy as atoms are added to the surface. The results are shown in graphical form in Fig. 4.4(b) as a plot of the change in free energy per atom versus the fraction x of atoms occupying a layer on the surface. The parameter $\alpha=(L/kT)f_k$, where L is the latent heat of the process, and $f_k(<1)$ is a crystallographic factor representing the fraction of all first neighbours lying in a plane parallel to the face considered. It may be seen that, for $\alpha < 2$, the free energy is a minimum when $x=0.5$, that is when the surface is rough. For $\alpha \gg 2$, the free energy is a minimum when x has a value close to 0 or 1, that is when the surface is almost smooth. For a {100} plane on a simple cubic lattice, $f_k=2/3$ and the critical condition $\alpha=2$ corresponds, for growth from a pure melt, to a melting temperature $T_M=L/3k$.

A similar problem was treated by Temkin (1966), who described the behaviour of the surface in terms of a dimensionless parameter $\gamma' = 4W/kT$, where W is the surface energy per atom. A flat surface corresponds to a high value of γ'. While the Temkin theory is related to that of Jackson, it is more general in that the number of surface layers considered is unlimited.

All the theoretical treatments such as those described suffer from the necessity to make some approximation since a rigorous solution is not possible. The most common approximations are the restriction of interactions to nearest neighbours and the so-called Bragg–Williams approximation which assumes long-range order and averages the interaction between atoms so any effects of small clusters of atoms on each other are not taken into account. Recently attempts have been made to simulate a crystal surface by computer and some results of such simulations have been reported by Binsbergen (1972) and by Bennema and Gilmer (1972). The relatively large size of the simulated surface area ($\sim 40 \times 40$ lattice units) gives more reliable results of the static surface properties like surface roughness than the presently available analytical approaches. Thus computer simulation offers considerable promise.

Hartman and Perdok (1955) proposed a treatment of crystal surfaces based on considerations of the chemical bonding within the crystal. They define periodic bond chains (PBC's) as chains running through the crystal in certain directions which contain the strongest chemical bonds. The flat (F) crystal faces are those which are parallel to at least two of these chains. Stepped or S faces are those parallel to one PBC and rough or kinked (K) faces are not parallel to any PBC. This theory gives good qualitative results for the crystal morphology of several materials but it cannot be used for quantitative work such as calculations of surface energy.

The observation of smooth, highly reflecting facets on most crystals grown from solutions suggests that these are the F faces. If a small crystal is nucleated with an approximately spherical shape in a supersaturated solution, the rough faces will have more sites available for the attachment of solute molecules and will therefore grow more rapidly. As growth proceeds, these rapidly growing faces tend to disappear and the crystal will eventually be bounded by the relatively slow-growing "habit" faces. The sequence of formation of the habit faces is illustrated in Fig. 4.5. These slow-growing faces, which form the boundaries of crystals grown under stable conditions, are of course not perfectly flat on the atomic scale. They contain vacancies and adatoms (note that the minima in Fig. 4.4(b) for $\alpha > 2$ do not occur exactly at $x = 0$ or 1), but their important property is that growth can only occur at certain sites where a new layer is nucleated.

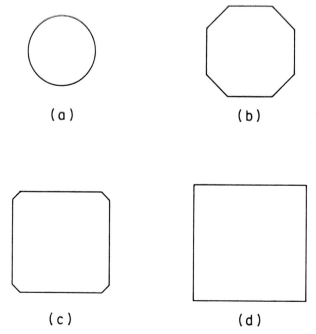

Fig. 4.5. Elimination of more rapidly growing faces during growth.

Such surfaces are referred to as "singular" and correspond to a minimum in the $\gamma(\theta)$ plot which will be discussed in Chapter 5. The mechanisms by which surface nucleation may occur are considered in Section 4.7. It may, however, be noted that the nucleation sites will often be lattice defects, although in principle growth by random two-dimensional nucleation is possible on a singular surface.

Very few observations have been reported of rough surfaces on crystals grown from high-temperature solutions. E. A. D. White (unpublished work) has noted on ruby crystals grown from solution in PbF_2 small facets which appear to be rough, but such facets are very rare and it is probable that they will be observed only when growth is terminated at a transient stage following some change in the experimental conditions which is tending to produce a habit change. Another cause of surface roughness was discussed by Scheel and Elwell (1973b) who assume a fast, unstable growth rate at the end of a crystal-growth experiment due to fast cooling when the furnace is shut off or the crucible is removed so that the remaining solution may be poured out.

The surface roughness of crystals growing in high-temperature solutions will increase with temperature and they may exhibit changes in growth rate or morphology on this account as the growth temperature is raised towards the melting point. However, we shall assume in the subsequent discussion that the faces of crystals grown by this method are atomically flat and the theory will be developed with the assumption that some surface nucleation process is necessary. The experimental evidence for this assumption will be discussed in Sections 4.11 to 4.13.

4.5. Stages in Growth from Solution

As first stressed by Kossel (1927), growth on a crystal having a flat interface requires some mechanism by which atoms (or the appropriate growth units)† will be integrated into the crystal more readily than on the remaining surface. This integration may be at the edge of a layer of monatomic thickness which spreads laterally across the crystal surface. Integration of atoms into the crystal will occur most readily at vacant sites or "kinks" along the edge of this layer since an atom entering such a kink will be able to form nearest-neighbour bonds with three atoms in the crystal. The meaning of the terms "step" and "kink" is illustrated in the diagram of a crystal surface shown in Fig. 4.6.

† Glasner (1973) has proposed that supersaturated aqueous solutions contain crystalline aggregates some 50 to 100 Å in diameter and that crystallization involves the regular arrangement of such aggregates on the crystal surface layers of unit cell height (see Section 4.12) but its confirmation would revolutionize the basic concepts of growth from solutions.

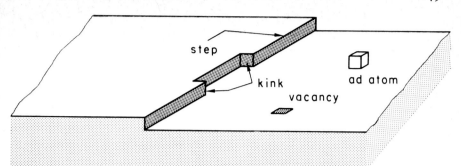

FIG. 4.6. Idealized model of "flat" crystal surface.

If a crystal which has a stepped interface is in contact with a super-saturated solution, the process of growth can be considered to occur in the following stages:

(i) Transport of solute to the neighbourhood of the crystal surface.
(ii) Diffusion through a boundary layer, adjacent to the surface, in which a gradient in the solute concentration exists because of depletion of material at the crystal-solution interface.
(iii) Adsorption on the crystal surface.
(iv) Diffusion over the surface.
(v) Attachment to a step.
(vi) Diffusion along the step.
(vii) Integration into the crystal at a kink.

The sequence (i)–(vii) is illustrated in Fig. 4.7(a). The detailed nature of the solute particles is not known but it is likely that ions of opposite sign will tend to diffuse together because of their electrostatic attraction. It is certain that some interaction between solute and solvent particles exists in the solution. Such interactions are described by the term *solvation* which is used here to include all forms of interaction. For simplicity the solute particles of Fig. 4.7(a) have been shown to be surrounded in the solution by six particles of the solvent forming a regular octahedron. Solvation may reduce the tendency of solute particles to form clusters near the crystal surface, but the importance of clustering in vapour growth has been demonstrated by Lydtin (1970) and there is a need for experiments aimed at understanding the nature of the solute particles near the crystal interface.

Stages (iii), (v) and (vii) are accompanied by partial desolvation and there will be a new flow of solvent away from the growing crystal. The solute particles may become desorbed at any stage after (iii) and the desorption process has been represented on the diagram by (iv)*. The solute does not

F

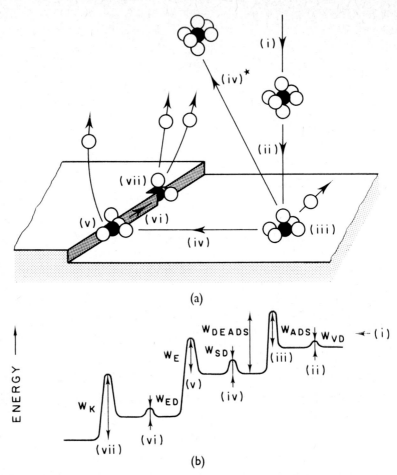

FIG. 4.7. (a) Stages in crystal growth from solution; (b) corresponding energy changes.

fully become part of the crystal until the heat of crystallization has been liberated and the desolvation process is complete.

All the stages in the growth process can be represented by relaxation times or the equivalent energy barriers and the potential energy profile for the growth process is shown schematically in Fig. 4.7(b). A similar diagram was given by Conway and Bockris (1958) for electro-crystallization and by Bennema (1967). An alternative representation would be to consider the various processes as impedances but the electrical analogue of solution growth has not been pursued, presumably because the impedances are distributed rather than discrete.

It should be noted that some of the processes (i)–(vii) occur in series but that some occur in parallel so that not all the stages are necessarily involved in the growth of a chosen material. For example, solute particles may diffuse directly to a kink site by surface migration and so eliminate the necessity for (v) and (vi). Some of the processes will normally occur so quickly (in series) and some so slowly (in parallel) that they may be neglected in comparison with the other stages. In practical crystal growth it is most important to know which process determines the rate of growth and we shall be particularly concerned in this Chapter and in Chapter 6 with the problem of deciding which step is likely to be rate-determining.

In order to discuss the growth process in more detail, it is convenient to take stages (iii)–(vii) together as the interface kinetic stage. It is also necessary to consider the origin of the steps, which has been neglected in the previous discussion. The transport process (stage (i)) by which solute is transferred to the crystal is crucial to the growth of good quality crystals but we defer discussion of this process until Chapter 6, in which the use of the theory in the design of crystal-growth experiments is considered. Stage (ii), diffusion through the boundary layer, is first considered separately for the case in which the interface kinetics are not rate determining. The interface kinetic stage (iii)–(vii) is considered separately and the general case where stages (ii) and (iii)–(iv) are combined is also treated.

4.6. The Boundary Layer

The concept of a boundary or "unstirred" layer was introduced by Noyes and Whitney (1897) and its importance in crystal growth from solution was stressed by Nernst (1904). There is often confusion between the solute diffusion boundary layer, which was introduced in the previous section, and the "hydrodynamic" boundary layer. The latter is a layer of solution which is considered stagnant because of adhesion to the crystal surface while the remainder of the solution is flowing past this surface (see Wilcox, 1969). A simple relation exists between the two layer thicknesses, and the layer referred to in the remainder of the book will be the solute diffusion boundary layer.

A boundary layer, whether diffusion or hydrodynamic, is a simplified concept in any system fluctuating with time. Its use in diffusion-limited growth can be illustrated with reference to a plane crystal surface growing uniformly in a supersaturated solution. The rate of transport of solute per unit area in the z direction, normal to this surface, is given by Fick's law as

$$\frac{dm}{dt} = -D\frac{\partial n}{\partial z} \qquad (4.12)$$

and the linear rate of growth of the crystal if its surface at $z=0$ is correspondingly, with n_0 the solute concentration at $z=0$,

$$v = \frac{D}{\rho - n_0} \left(\frac{\partial n}{\partial z}\right)_{z=0}$$

(4.13)

where ρ is the density of the crystal.† The solute concentration at the interface will approximate to the equilibrium value provided that the kinetic process is extremely rapid compared with the volume diffusion. This condition was originally assumed by Nernst (1904). If the solute gradient is uniform over the boundary layer, substitution for $(\partial n/\partial z)$ in Eqn (4.13) gives, if $\rho \gg n_0$,

$$v = \frac{D}{\rho} \frac{(n_{sn} - n_e)}{\delta}$$

(4.14)

This equation may be used to define the width δ of the diffusion boundary layer.

The existence of a boundary layer has been confirmed using optical interference methods by Berg (1938), Bunn (1949) and several other investigators, using aqueous solutions. The solute concentration is determined from the refractive index of the solution and contours of equal concentration around a growing crystal have the form shown in Fig. 4.8.

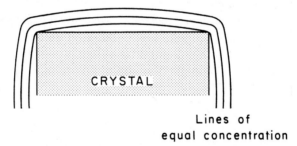

Lines of
equal concentration

Fig. 4.8. Concentration contours around a growing crystal.

The supersaturation is seen to be highest at the corners and lowest at the centre of the faces. Such a variation of the supersaturation across the face is to be expected for a polyhedral crystal and the experimental results have been explained by Seeger (1953) and by Boscher (1965), who solved the diffusion equation in three dimensions using an electrical analogue.

† The diffusion coefficient D is an effective value, since both positive and negative ions must diffuse and the requirement of local electrical neutrality must be satisfied.

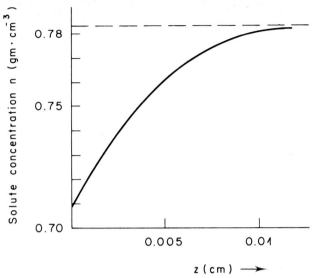

FIG. 4.9. Solute distribution adjacent to growing KBr crystal.

The fact that crystals normally grow uniformly in spite of this variation in supersaturation has been explained by Frank (1958a), who supposed that the rate of growth of any face is determined by the local value of the supersaturation at one point at which the dominant growth centre for the whole face is located. However, if solute is deposited too rapidly from the solution, it may be expected that faster growth will occur at the corners or edges of the crystal where the supersaturation is highest, and this is confirmed by experiment (Chernov, 1963; Lefever and Chase, 1962). The experimental observations of variations in solute concentration across the face of the crystal confirm the approximate nature of equations such as (4.14). The variation of the solute concentration normal to a crystal surface in aqueous solution has been measured by Goldsztaub, Itti and Mussard (1970) and their result is shown in Fig. 4.9. The equation of solute flow in one dimension is normally written in the form

$$D \frac{\partial^2 n}{\partial z^2} + u_{cf} \frac{\partial n}{\partial z} = \frac{\partial n}{\partial t}. \tag{4.15}$$

The first term represents the diffusional flow, the second growth-induced convection (Wilcox, 1972) and the third takes into account the time dependence of the solute concentration. In the steady state, $\partial n/\partial t = 0$ and so, if u is negligible,

$$\frac{\partial^2 n}{\partial z^2} = 0 \qquad \text{i.e.} \qquad \frac{\partial n}{\partial z} = \text{const} = \frac{n_{sn} - n_e}{\delta}$$

as in Eqn (4.14). The non-linearity in Fig. 4.9 is attributed to convection in the cell used by Goldsztaub *et al.* If the convection term is negligible, the time-dependent solution of Eqn (4.15) has the form

$$n(z, t) = n_e + (n_{sn} - n_e) \, \text{erf} \left(\frac{z}{2(Dt)^{1/2}} \right). \qquad (4.16)$$

The value of the boundary-layer thickness in this case will be time-dependent and integration of the growth rate over the period of the experiment is necessary if a comparison between experiment and theory is to be made.

The problem of the boundary layer was considered by Carlson (1958) who assumed laminar flow of the solution over a face of the crystal. He found that, for uniform growth of the crystal face, the concentration of solute should decrease with distance from the leading edge. As in diffusional flow, therefore, a non-uniform supersaturation over the surface is expected. Carlson derived an expression for the rate of growth of the crystal and his results give for the solute diffusion boundary-layer thickness (taking into account hydrodynamics)

$$\delta = \left\{ 0.463 \left(\frac{\eta}{\rho_{sn} D} \right)^{1/3} \left(\frac{u \rho_{sn}}{\eta l} \right)^{1/2} \right\}^{-1}. \qquad (4.17)$$

Here η is the viscosity and ρ_{sn} the density of the solution, u the flow velocity and l the length of the crystal face considered. A similar expression was used by Bennema (1967) to calculate the boundary-layer thickness and the results were found to be in agreement with his experimental values. If $\eta = 10$ cP, $\rho_{sn} = 5$ g cm^{-3}, $D = 10^{-5}$ cm^2 s^{-1}, $u = 0.1$ cm s^{-1} and $l = 5$ mm, the value of δ is calculated to be 0.055 cm and this value is probably correct to the order of magnitude for diffusion-controlled growth.

The variation of δ with the solution flow rate u may be used to explain the change in crystal-growth rate at high supersaturation when the flow rate is varied. The usual form of the variation of the crystal-growth rate in aqueous solutions with the solution flow rate is shown in Fig. 4.10. The increase in growth rate with flow rate continues until some limiting rate is reached where the growth rate becomes controlled by the interface kinetic process. Carlson's theory predicts that δ should vary as $u^{-1/2}$, and so v should depend on $u^{1/2}$. This result is in reasonable agreement with the experiments of Hixson and Knox (1951), who report $v \propto u^{0.60}$, and of Mullin and Garside (1967), whose results are described by a relation $v \propto u^{0.65}$.

A similar variation in the growth rate is observed when a crystal is

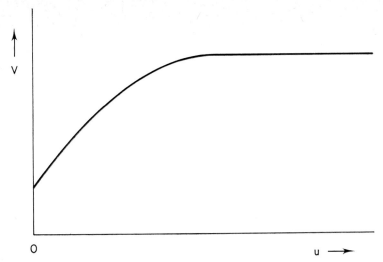

FIG. 4.10. Variation in linear growth rate with solution flow rate.

rotated in solution. The boundary-layer thickness in this case is given by Burton, Prim and Slichter (1953) as

$$\delta \simeq 2^{2/3} D^{1/3} \nu^{1/6} \omega^{-1/2} \qquad (4.18)$$

where ω is the angular velocity of rotation of the crystal and ν the kinematic viscosity of the solution. A linear dependence of the growth rate v on $\omega^{1/2}$ is found for the growth of sodium thiosulphate using the data of Coulson and Richardson (1956), for low values of ω. Laudise, Linares and Dearborn (1962) measured the variation of the growth rate of yttrium iron garnet from solution in $BaO-B_2O_3$ with crystal rotation rate. They found an increase in v for values of ω up to about 50 r.p.m., beyond which the growth rate was independent of the rotation rate. The data are insufficient to confirm an $\omega^{1/2}$ dependence at low rotation rates.

In general the observed rate of growth of a crystal will depend partly on boundary-layer diffusion and partly on the interface kinetics. Brice (1967a) has shown how the role of the boundary layer may be taken into account in order to deduce the form of the interface kinetic law. His approach is based on that of Berthoud (1912) and Valeton (1924). The solute concentration at the interface is taken as n_i and the kinetic law is assumed to have the form

$$v = A(n_i - n_e)^m \qquad (4.19)$$

where A and m are independent of the solute concentration. The growth law may also be expressed in terms of the diffusional flow by a modification of Eqn (4.14). In this case

$$v = \frac{D}{\rho} \frac{(n_{sn} - n_i)}{\delta}.\tag{4.20}$$

Elimination of n_i between Eqns (4.19) and (4.20) gives

$$\left(\frac{v}{A}\right)^{1/m} + \left(\frac{v\rho\delta}{D}\right) = n_{sn} - n_e.\tag{4.21}$$

If δ varies as $\omega^{-1/2}$ or as $u^{-1/2}$, a plot of $v^{1/m}$ versus $v\omega^{-1/2}$ or $vu^{-1/2}$ at constant supersaturation should be linear and such plots were successfully used by Brice to obtain the power m of the kinetic law. This procedure does not, however, give satisfactory results in all cases, presumably because of the simplifications introduced in assuming Eqns (4.19) and (4.20).

The variation of growth rate with boundary-layer thickness as a function of supersaturation was discussed by Scheel and Elwell (1973a) and will be treated in Chapter 6. For low and medium supersaturation Eqn (4.21) will approximately hold. However, at high supersaturation and sufficient stirring a maximum (stable) growth rate is reached which is a constant for a given solute-solvent system. Depending on n_{sn} this maximum growth rate is determined either by surface kinetics or by heat flow.

4.7. Generation of Surface Steps

We now consider interface kinetic mechanisms in detail, treating in particular crystal surfaces which are "flat" rather than "rough". The critical step in the growth of crystals having perfect or nearly perfect surfaces is the formation of a cluster of atoms sufficiently large to constitute a stable nucleus which will grow to form a new layer. The classical theory of crystal growth is analogous to the nucleation theory described in Section 4.2, with the exception that nucleation occurs on a crystal surface. In such a "two-dimensional" nucleation theory it is convenient to treat a cylindrical embryo of radius r and of height a corresponding to one growth unit (e.g. an atom or molecule). The change in Gibbs free energy on formation of such an embryo is

$$\Delta G(r) = 2\pi r\gamma_e - \pi r^2 a\Delta G_v \tag{4.22a}$$

where γ_e is the edge energy per unit length of the nucleus. The term ΔG_e of Eqn (4.1) is included (see Lewis, 1974) by putting

$$n(r) = n_0 \exp\left(-\Delta G/kT\right) \tag{4.22b}$$

where n_0 is the density of available sites.

Alternatively the free energy can be expressed in terms of the energy per growth unit (for simplicity, we shall use the term "molecule") γ_m on the edge of the cylindrical nucleus. If the length of the molecule is also a, $\gamma_m \simeq a\gamma_e$ and so

$$\Delta G \simeq \frac{2\pi r \gamma_m}{a} - \pi r^2 a \Delta G_v, \tag{4.23}$$

and differentiation gives the radius of the critical nucleus as

$$r_s = \frac{\gamma_m}{a^2 \Delta G_v} \tag{4.24a}$$

and the corresponding value of ΔG is

$$\Delta G^* = \frac{\pi \gamma_m{}^2}{a^3 \Delta G_v}. \tag{4.24b}$$

Substitution for ΔG_v from Eqn (4.5) gives

$$r_s{}^* = \frac{\gamma_m V_M}{a^2 RT\sigma}.$$

A more familiar form of this equation is obtained by putting $V_M = N_A a^3$, where N_A is Avogadro's number and the molecule is assumed to be a cube of side a. This gives

$$r_s{}^* = \frac{\gamma_m a}{kT\sigma} \tag{4.25a}$$

and correspondingly

$$\Delta G^* = \pi \gamma_m{}^2/kT\sigma. \tag{4.25b}$$

The number i^* of molecules in a critical nucleus is

$$i^* = \frac{\pi r_s{}^{*2}}{a^2} = \left(\frac{\gamma_m}{kT\sigma}\right)^2. \tag{4.25c}$$

The rate of surface nucleation, and hence of crystal growth, depends by analogy with Eqn (4.8) on $\exp(-\Delta G^*/kT)$, and it is instructive to estimate the order of magnitude of this factor as a function of the supersaturation. The energy γ_m is of the order of the binding energy W_B, introduced in Section 4.3, that is $\gamma_m \simeq \phi_m/6$, where W_B is the binding energy, ϕ_m is the heat of solution per molecule. (Strictly, the value of γ_m will be higher on low energy planes.) Using a value for $\phi = 72$ kJ mole^{-1} as found for nickel ferrite in barium borate (Elwell, Neate and Smith, 1969) so that $\phi_m \sim 2 \times 10^{-20}$ J molecule^{-1}, then, with $T = 1500$ K, $\gamma_m/kT \simeq 1$ so that $\Delta G^* \simeq \pi/\sigma$. The term $\exp(-G^*/kT)$ varies from 3×10^{-3} for $\sigma = 0.5$ and $\sim 10^{-13}$ for $\sigma = 0.1$ to $\sim 10^{-130}$ for $\sigma = 0.01$. Growth by two-dimensional nucleation therefore has a high probability except at very low supersaturation values. In the system referred to above, growth was observed experimentally at relative supersaturations down to about 1%.

F2

A discrepancy between observed growth rates from the vapour at supersaturations below 1% and the prediction from two-dimensional nucleation theory of negligible growth below 50% supersaturation (for $\phi_m/kT \simeq 12$) led Frank (1949) to propose that dislocations having a screw component act as a continuous source of layers on the surface of a crystal. The presence of the step associated with such a dislocation removes the need for surface nucleation.

Figure 4.11(a) shows the face of a crystal with a screw dislocation emerging at P. Molecules are readily integrated into the crystal at the step PQ, which is of approximately monomolecular height, and the initial growth is normal to the step as indicated by the arrow. The emergence of the screw dislocation at P fixes this point so that the rate of movement of

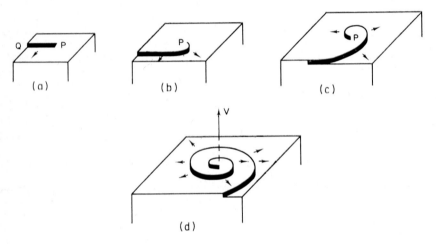

Fig. 4.11. Development of a spiral.

the layer is here zero. Elsewhere the step moves in such a way that its linear velocity is constant and angular velocity decreases with the distance from P. As the crystal grows, the step therefore winds itself up into a spiral with its centre at P. The development of the spiral is illustrated in Figs 4.11(a)–(d). In this sequence the face considered grows normal to itself at a linear rate v. The area of the face increases at the same time, due to a similar growth process on the other surfaces of the crystal. The spiral will continue to wind itself up until the separation of adjacent layers at the centre is of the order of the radius $r_s{}^*$ of the critical nucleus.

The presence of growth spirals has now been established on a large variety of crystals. These include natural crystals (Sunagawa, 1960) and synthetic crystals grown from the vapour phase (Verma, 1953) and from

100 μm

Fig. 4.12. Growth spiral on a rare-earth orthoferrite crystal (after Tolksdorf and Welz, 1972).

aqueous solution (Forty, 1951). Figure 4.12 shows a particularly beautiful example of a growth spiral on an orthoferrite crystal grown from high-temperature solution, observed by Tolksdorf and Welz (1972). The presence of such spirals provides evidence for the validity of Frank's screw-dislocation model, although the height of the steps in Fig. 4.12 is 50–150 Å rather than of monomolecular dimensions as envisaged by Frank.

Lewis (1974), in a review of two-dimensional nucleation, has pointed out that the importance of growth in solution by this mechanism has been underestimated, certainly for medium and high supersaturation. As is clear from Eqn (4.25), the probability of 2-D nucleation will depend on the factor ϕ_m/kT, which will be lower for solution growth than for growth from the vapour. Bennema et al. (1972) have confirmed by computer simulation experiments that a mechanism of growth by 2-D nucleation on growing two-dimensional nuclei can describe some experimental growth-rate data better than the screw-dislocation theory.

4.8. The Theory of Burton, Cabrera and Frank

Screw dislocations are important because they can provide a continuous source of steps which can propagate across the surface of the crystal. In order to construct a theory which will predict values for the rate of growth of the crystal, it is necessary to calculate the rate at which molecules will arrive at the steps of the spiral. A theory of crystal growth including the mechanism of step generation and of transport into the step was given by Burton, Cabrera and Frank (1951) and this BCF paper has assumed great importance since much of the content will apply to any theory of crystal growth. The theory given here was originally proposed for growth from the vapour phase but its applicability to solution growth has been strongly advocated by Bennema (1965, 1967) and by Bennema and Gilmer (1973) whose treatment we follow.

The velocity of growth will depend on the shape of the growth spiral, for which an exact expression has not been developed. BCF used the equation for an Archimedian spiral

$$r = 2r_s{}^*\theta \qquad (4.26)$$

where r and θ are the coordinates of any point on the spiral as indicated in Fig. 4.13. Equation (4.26) should be a good approximation to the behaviour of a real spiral, for positions not too close to the centre. The distance y_0 between the steps of the spiral will thus be

$$y_0 = 2r_s{}^*(\theta + 2\pi) - \theta = 4\pi r_s{}^*.$$

A more rigorous approach by Cabrera and Levine (1956) showed that a better approximation is given by

$$y_o \simeq 19 r_s{}^* = \frac{19\gamma_m\, a}{kT\sigma} \tag{4.27}$$

and this value will be used in the subsequent development.

The second part of the BCF theory is concerned with the transport of molecules from the bulk of the solution to kinks in the steps of the spiral. It is assumed that the surface-diffusion coefficient is independent of the local concentration and this, together with the neglect of surface vacancies, is the main assumption of the theory. As mentioned earlier, the nature of

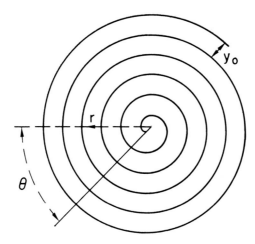

FIG. 4.13. Growth spiral.

solute particles on the crystal surface is not known but, if local electrical neutrality is assumed, it is possible to define a single relaxation time for each stage of the surface transport process in the same way that an effective volume-diffusion coefficient can be specified for the flow of ions of opposite charge.

The steps in the spiral are assumed to move negligibly slowly compared with the rate of migration of molecules on the surface. This assumption is justified since the rate at which the step moves is governed by the rate of arrival of diffusing molecules. For simplicity, the distance from the spiral centre is taken to be so large that curvature of the steps may be neglected. The net flux of particles into a strip of width dy on the surface in the region of a step will depend upon the flux j_v from the solution to the surface and on the flux j_s across the surface into the step due to the concentration gradient created by integration of molecules into the surface at the step.

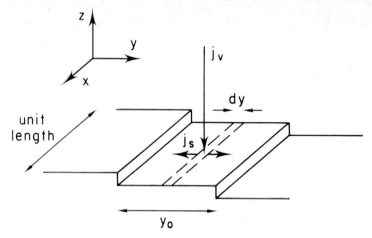

FIG. 4.14. Flow of solute to a step.

These particle fluxes are indicated in Fig. 4.14. In the steady state the two fluxes will balance and so, for unit length in the x direction,

$$\frac{dj_s(y)}{dy} - j_v = 0. \qquad (4.28)$$

The surface flux j_s can be expressed in terms of the surface-diffusion coefficient D_s and the local surface concentration n_s as

$$j_s = -D_s \frac{dn_s}{dy} = -D_s \frac{d}{dy}(n_{se}\,\sigma_s + n_{se}) = -D_s\,n_{se}\frac{d\sigma_s}{dy} \qquad (4.29)$$

where n_{se} is the equilibrium concentration at the surface far from a step and σ_s the local value of the relative supersaturation. It is convenient to introduce a variable ψ as the difference between the surface supersaturation σ_s and the supersaturation very far from a step, which is governed by the solute concentration in the bulk of the solution. Thus

$$\psi = \sigma - \sigma_s(y) \qquad (4.30)$$

and, since σ is independent of y,

$$j_s = D_s\,n_{se}\frac{d}{dy}(\sigma - \sigma_s) = D_s\,n_{se}\frac{d\psi}{dy}. \qquad (4.31)$$

The flux j_v can be written as the difference between the flux leaving the surface $n_s(y)/\tau_{\text{deads}}$ and that moving towards the surface $n_{se}/\tau_{\text{deads}}$ where τ_{deads} is the relaxation time governing deadsorption of solute from the surface (shown as (iv)* in Fig. 4.7). Thus

$$j_v = \frac{\sigma n_{se} - n_s}{\tau_{deads}} = \frac{n_{se}(\sigma - \sigma_s)}{\tau_{deads}} = \frac{n_{se}\,\psi}{\tau_{deads}}. \tag{4.32}$$

On substitution of Eqns (4.31) and (4.32) into Eqn (4.28), the differential equation of solute transport becomes

$$D_s\,\tau_{deads}\,\frac{d^2\psi}{dy^2} = \psi$$

or

$$y_s{}^2\,\frac{d^2\psi}{dy^2} = \psi \tag{4.33}$$

where $y_s = \sqrt{(D_s\,\tau_{deads})}$ is the mean distance travelled by solute molecules on the surface. Equation (4.33) has a general solution

$$\psi = A\,\exp\,(y/y_s) + B\,\exp\,(-y/y_s) \tag{4.34}$$

and it is necessary to introduce boundary conditions to obtain values for A and B. The most probable situation is that $y_s \gg x_o$, where x_o is the average distance between kinks in a step. For a set of equidistant steps of separation y_o and with the origin of y chosen to be mid-way between the steps, the boundary condition may be expressed by putting the value of ψ at a step as $\beta\sigma$, so that $\psi = \beta\sigma$ when $y = \pm\frac{1}{2}y_o$. Then, from Eqn (4.34), for $y = +\frac{1}{2}y_o$, $\psi = \beta\sigma = A\,\exp\,(y_o/2y_s) + B\,\exp\,(-y_o/2y_s)$ and for $y = -\frac{1}{2}y_o$, $\psi = \beta\sigma = A\,\exp\,(-y_o/2y_s) + B\,\exp\,(y_o/2y_s)$ from which $A = B$, and substitution in terms of $\beta\sigma$ in Eqn (4.34) gives

$$\psi = \frac{\beta\sigma\,\cosh\,(y/y_s)}{\cosh\,(y_o/2y_s)}. \tag{4.35}$$

If $x_o \gg y_s$, it is necessary to introduce an extra factor c_o into Eqn (4.35) to take into account the non-planar diffusion fields around the kinks.

From Eqn (4.31), the flux of particles towards a step may now be written as

$$j_s = D_s\,n_{se}\,\frac{d\psi}{dy} = \frac{D_s\,n_{se}\,\beta\sigma}{y_s}\,\frac{\sinh\,(y/y_s)}{\cosh\,(y_o/2y_s)}. \tag{4.36}$$

If n_{se} is measured in g cm^{-2}, j_s represents the flux in g cm^{-1} s^{-1} towards a step either of monomolecular or larger height. The linear rate of advance of the step v_{st} is obtained by multiplying j_s by the area $1/\rho a$ per unit mass of the crystal so that, for a step of monatomic height,

$$v_{st} = 2j_{s(y=y_o/2)} \cdot \frac{1}{\rho a} = \frac{2D_s\,n_{se}\,\beta\sigma}{a\rho y_s}\,\tanh\,\frac{y_o}{2y_s}. \tag{4.37}$$

The factor 2 is introduced since molecules enter the step from two sides.

In order to calculate the linear growth rate v of the crystal (in the z direction), it is necessary to multiply the flux of steps by the height of a step. For a step separation y_o, the number of steps per unit length is $1/y_o$ and so the flux of steps in the y direction will be v_{st}/y_o. If the step height is a, the rate of growth will then be

$$v = \frac{v_{st}\, a}{y_o} \tag{4.38}$$

or, on substituting for v_{st} and y_o from Eqns (4.37) and (4.27)

$$v = \frac{2 D_s\, n_{se}\, \beta \sigma^2 kT}{19 \gamma_m\, y_s\, \rho a} \tanh \frac{y_o}{2 y_s}. \tag{4.39}$$

If a parameter σ_1 is defined as

$$\sigma_1 = \frac{\sigma y_o}{2 y_s} = \frac{9.5 \gamma_m\, a}{kT y_s} \tag{4.40}$$

Eqn (4.39) may be rewritten in the form

$$v = \frac{D_s\, n_{se}\, \beta}{y_s^2 \rho} \cdot \frac{\sigma^2}{\sigma_1} \tanh \frac{\sigma_1}{\sigma} = \frac{C \sigma^2}{\sigma_1} \tanh \frac{\sigma_1}{\sigma}. \tag{4.41}$$

The variation of growth rate with supersaturation thus depends on two parameters: $C(= D_s\, n_{se}\, \beta / y_s^2 \rho)$, which determines the absolute value of v, and σ_1 which determines the shape of the $v(\sigma)$ curve. For low values of $\sigma\,(\sigma \ll \sigma_1)$ Eqn (4.41) may be approximated by

$$v \simeq \frac{C \sigma^2\, (\exp\,(2\sigma_1/\sigma) - 1)}{\sigma_1\,(\exp\,(2\sigma_1/\sigma) + 1)} \simeq \frac{C \sigma^2}{\sigma_1} \tag{4.42a}$$

while for $\sigma \gg \sigma_1$

$$v \simeq \frac{C \sigma^2\,(1 + (2\sigma_1/\sigma) + \ldots) - 1}{\sigma_1\,(1 + (2\sigma_1/\sigma) + \ldots) + 1} \simeq C\sigma. \tag{4.42b}$$

The BCF theory therefore predicts a quadratic $v(\sigma)$ curve for low values of the supersaturation with a gradual transition to a linear law as the supersaturation is increased above a critical value σ_1. A relatively large value of σ_1 for a given material should result in a quadratic growth curve while a linear $v(\sigma)$ plot should be expected according to the above theory if σ_1 is low.

Cabrera and Coleman (1963) have pointed out that at higher supersaturations, the surface supersaturation near the centre of the spiral may be lower than σ because of the depletion caused by surface diffusion to that portion of the spiral where the step spacing y_o is small. The result is that y_o decreases more slowly with σ than predicted by Eqn (4.27). This "back stress" effect makes the transition from a quadratic to a linear law occur at higher values of σ than predicted by Eqn (4.42) and a perfectly linear law is unlikely over any wide range of supersaturation values.

If a number of screw dislocations emerge at the growth centre the form of the spiral will be more complex than that shown in Fig. 4.13. In order to take into account the effect of cooperation between a number of interacting spirals, BCF introduced a factor ϵ such that

$$y_o = \frac{19r_s{}^*}{\epsilon} = \frac{19\gamma_m\, a}{\epsilon k T \sigma}. \qquad (4.43)$$

Equation (4.41) then becomes

$$v = \frac{C\epsilon\sigma^2}{\sigma_1}\tanh\frac{\sigma_1}{\epsilon\sigma}. \qquad (4.44)$$

The factor ϵ can be quite complex and some examples of cooperating dislocations will be discussed in Section 4.12.

BCF *Theory of Solution Growth*
As mentioned earlier, the BCF theory was derived for growth from the vapour. In the case of solution growth, the molecules were assumed to enter the kinks directly rather than by entering an adsorption layer and undergoing surface diffusion. The justification for this assumption was that the coefficient of volume diffusion ($\sim 10^{-5}$ cm² s^{-1}) is normally much higher than the coefficient of surface diffusion ($\sim 10^{-8}$ cm² s^{-1}) for molecules in solution so that any diffusion in a direction parallel to the crystal surface might be expected to occur in the boundary layer. If the rate of flow of solute molecules to the kinks is governed by diffusion through the boundary layer, the net flux reaching the steps, which governs their rate of advance v_{st}, will be proportional to the supersaturation σ. With $1/y_o \propto \sigma$ according to Eqn (4.27), the growth rate v will again vary as σ^2 since $v = v_{st}\, a/y_o$ [Eqn (4.38)]. BCF considered solute flow towards a kink in a hemispherical diffusion field and obtained an expression for the step velocity

$$v_{st} = \frac{Dn_e\, 2\pi\sigma}{\rho x_o}\left[1 + \frac{2\pi a(\delta - y_o)}{x_o\, y_o} + \frac{2a}{x_o}\ln\left(\frac{y_o}{x_o}\right)\right]^{-1}. \qquad (4.45)$$

For low supersaturations y_o is large and the third term in the bracket is the dominant one. In this case, $v_{st} \propto \sigma$ and a quadratic law is predicted using Eqn (4.38) since $y_o \propto 1/\sigma$. However, at high supersaturations the second term is dominant since y_o becomes small. In the latter case

$$v_{st} \simeq \frac{Dn_e \sigma y_o}{\rho a (\delta - y_o)}$$

and, neglecting y_o in comparison with δ, Eqn (4.38) gives the growth rate as

$$v = \frac{Dn_e \sigma}{\rho \delta}.$$

This case is exactly the same volume-diffusion limited situation which was considered by Nernst and described by Eqn (4.14).

4.9. Should Surface Diffusion be Included?

The difference between Bennema's treatment of solution growth and the BCF solution-growth theory rests upon whether or not surface diffusion plays an important role in the growth process. It is generally accepted that the rate of volume diffusion exceeds that of surface diffusion, but the effective area of the kink sites is small compared with the total area of the crystal face and this factor will favour a mechanism in which volume diffusion to a random point on the surface is followed by surface diffusion to a kink.

A meaningful numerical comparison between the growth rates calculated using Eqns (4.44) and (4.45) is difficult because many of the parameters in these equations are not known even to the order of magnitude. An attempted comparison is given in Fig. 4.15. In this example it has been assumed that $D = 10^{-5}$ cm^2 s^{-1}, $n_e = 1$ g cm^{-3}, $\rho = 5$ g cm^{-3}, $a = 4 \times 10^{-8}$ cm and $\gamma_m = 2 \times 10^{-20}$ J/molecule $\simeq kT$ so that, from Eqn (4.27), $y_o \simeq 10a/\sigma$. The mean separation between kinks x_o is given by BCF as

$$x_o = \frac{a}{2} \exp\left(W_B / kT\right) \tag{4.46}$$

and, with the binding energy $W_B \sim \gamma_m \sim kT$ for $T = 1500$ K, $x_o \sim a$. BCF estimate $x_o \sim 4a$ and so, for our example, we take an intermediate value of $x_o = 2a$. The boundary-layer width δ is taken to be 10^{-2} cm and the supersaturation range chosen is typical of experimental values. It is found that, with these data, the second term of Eqn (4.45) is dominant and so the growth rate in the BCF solution-growth theory is determined by volume diffusion over the whole range considered. For the surface-diffusion case we assume $D_s = 10^{-8}$ cm^2 s^{-1} and $y_s = 10^{-5}$ cm, which are typical values for aqueous solution growth according to Bennema's interpretation

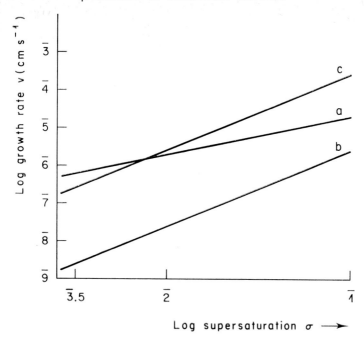

FIG. 4.15. Growth rate for BCF volume and surface-diffusion theories: (a) volume diffusion, (b) surface diffusion, $\beta = 10^{-2}$, (c) surface diffusion, $\beta = 1$.

(Bennema, 1965). The latter value gives $\sigma_1 \simeq .04$ which is within the range of supersaturation values considered. The value of the growth rate for the surface-diffusion model depends critically on the parameter β of Eqn (4.35). A value $\beta \simeq 1$ would indicate that the surface supersaturation has its maximum value and so corresponds to a maximum growth rate. Bennema's estimates of the relevant activation energies suggest a value of $\beta \sim 10^{-2}$ and the usual values are probably somewhere between these limits. There is no reason in principle why a factor β should not be included in Eqn (4.45) also. Figure 4.15 shows that v varies as σ^2 in the supersaturation range shown.

It should be emphasized that the data of Fig. 4.15 represent typical values and do not indicate the effect of surface diffusion on the system considered. Surface diffusion will always increase the growth rate, if its effect is not negligible, by increasing the probability that a solute molecule will find a kink site. Chernov (1961) also proposed a theory of crystal growth from solution based on calculation of the flow to a system of parallel steps, assuming no surface or edge diffusion. The concentration n is assumed to be described by an equation

$$D\frac{\partial n}{\partial r} = A(n - n_e)$$

where n is the concentration at a distance r from a step and A a constant which is large if the kink separation is small. Figure 4.16 shows the solute diffusion field around the steps assumed by Chernov. The solution of the diffusion equation gives for the growth rate

$$v = \frac{AakTn_e}{4\gamma} \frac{\sigma^2}{\{1 + (Aa/D) \ln (\delta\sigma_3/a\sigma) \sinh (\sigma/\sigma_3)\}} \qquad (4.47)$$

where $\sigma_3 = 4V_m\gamma/kT\delta$. Eqn (4.47) gives a rather similar result for the growth rate to that of the BCF volume-diffusion theory; at low supersaturations

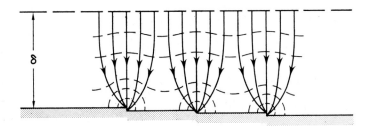

FIG. 4.16. Solute diffusion to system of steps (after Chernov, 1961, 1963).

$(\sigma < \sigma_3)$ a quadratic law is predicted and the $v(\sigma)$ curve becomes linear at high values of σ as the volume-diffusion step becomes rate controlling. Over a wide range of supersaturation values, Chernov's equation can be approximated by a law of the form

$$v \propto \sigma^{1.65}. \qquad (4.48)$$

Gilmer, Ghez and Cabrera (1971) have given a more complete treatment of the mechanism of transport of solute particles to kinks in a step, including simultaneous volume and surface diffusion. They also assume a set of equidistant parallel steps and a high density of kinks so that diffusion along the edge of a step may be neglected. A single step of height h is considered at $y = 0$, as in Fig. 4.17, and the volume and surface solute densities are related using three equations. Firstly, Fick's second law requires that, in the steady state,

$$\frac{\partial^2 n}{\partial y^2} + \frac{\partial^2 n}{\partial z^2} = 0 \qquad (4.49a)$$

since diffusion in the crystal is neglected. Secondly, the surface-diffusion

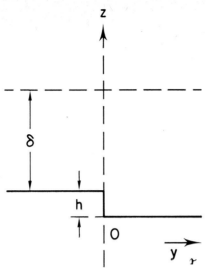

FIG. 4.17. Step at $y=0$.

process as affected by the volume to surface flow is described by the equation

$$D_s \frac{\partial^2 n_s}{\partial y^2} + D\left(\frac{\partial n}{\partial z}\right)_{z=0} = 0 \qquad (4.49b)$$

where n_s in this case is the surface concentration of solute per unit area. Finally, the exchange of solute between surface and volume is given by

$$D\left(\frac{\partial n}{\partial z}\right)_{z=0} = \frac{Dn}{\Lambda} - \frac{n_s}{\tau_{\text{deads}}}. \qquad (4.49c)$$

The factor D/Λ represents a "drift velocity" of solute molecules entering the adsorption layer from the adjacent volume such that $\Lambda = \lambda \tau_{\text{desolv}}/\tau_{v\text{diff}}$ where λ is the mean free path in the solution and the τ's are relaxation times for desolvation and volume diffusion.

The net exchange of solute at a kink is given by the net flux from neighbouring sites as

$$j = D\left(\frac{\partial n}{\partial y}\right)_{y=0} = \frac{D}{\Lambda_s}\left[(n_s)_{y=0} - n_e\right] \qquad (4.50)$$

where $\Lambda_s = \lambda \tau_{\text{kink}}/\tau_{s\text{diff}}$ is the quantity analogous to Λ for surface diffusion.

In the solution to these equations, the critical parameter is found to be $b = y_s/\Lambda$ where y_s is, as before, the mean distance travelled by an adsorbed solute molecule on the crystal surface. The growth rate in the limit $b=0$ is given by

$$\frac{\rho v}{Dn_e} = \sigma \left[\varLambda + \delta + \frac{\varLambda \varLambda_s y_o}{y_s^2} + \varLambda \left\{ \frac{y_o}{2y_s} \coth\left(\frac{y_o}{2y_s}\right) - 1 \right\} \right]^{-1}. \qquad (4.51)$$

This equation is analogous to Ohm's law in electricity, σ being the driving force for crystal growth and $\rho v / Dn_e$ a growth "current". Each of the terms in the square brackets has the character of an impedance. The first may be regarded as the impedance of the adsorption reaction and the second is that of the boundary layer. The third term represents an impedance for entering the steps and the fourth is that due to surface diffusion.

Equation (4.51) includes Chernov's theory and the BCF theory and reduces to these when the appropriate assumptions are made. The effect of a non-negligible value for b can be included only by numerical computation and examples of such calculations are given in the original paper. Results of computer simulation of crystal growth taking into account surface diffusion have been published by Gilmer and Bennema (1972).

It should be noted that, in this treatment, adsorption-controlled growth which would be expected for large values of $\varLambda$ is linear in the supersaturation. This result conflicts with that of Reich and Kahlweit (1968) which is discussed in the next Section.

4.10. The Role of Desolvation

The formation of complexes between solute and solvent is well established, and the requirement of desolvation prior to growth has been discussed briefly in Section 4.5. Desolvation must occur at the crystal surface since the surface cannot provide a driving force for desolvation at long range. If, as in the Chernov and BCF solution-growth theories, solute were to enter the kink sites directly from the solution, desolvation would have to occur at the same time as the integration process. It appears reasonable to expect that adsorption onto the surface, which permits partial desolvation and orientation of the molecules prior to entry into a kink, will be a more probable mechanism.

This latter conclusion was reached by Davies and Jones (1951) who studied the precipitation of silver chloride from aqueous solution by monitoring the electrical conductivity of the solution. They reasoned that, if the growth kinetics were determined by the reaction of Ag^+ and Cl^- ions at the interface, the rate of crystallization would be proportional to n_{sn}^2, where n_{sn} is the concentration of AgCl in the solution. Since this rate must equal the dissolution rate when $n_{sn} = n_e$, the net growth rate should be proportional to $n_{sn}^2 - n_e^2$. Experimentally they found that the rate of precipitation was proportional to $(n_{sn} - n_e)^2$, and this led them to reject a model in which adsorption was not included.

Doremus (1958) reviewed the experimental data on the precipitation of relatively insoluble salts and also stressed the importance of an adsorption layer. In experiments where ions of one constituent were added in excess of the stoichiometric ratio, the rate of precipitation was found to be substantially unchanged on adding more of the excess ions. This result is best explained by assuming the existence of an adsorption layer which is "saturated" by the excess ions since the growth rate then depends only on the minority ion concentration. Doremus extended the concept of surface reaction-controlled growth, considering both the formation of molecules on the surface prior to diffusion to a kink and the separate surface diffusion of oppositely charged ions which are integrated alternately into the crystal at the kink sites. In the first case, the precipitation rate was calculated to be proportional to $(n_{se} - n_e)^3$ for a "one–one" electrolyte AB and to $(n_{sn} - n_e)^4$ for a "two–one" electrolyte $A_2 B$. These dependences became $(n_{sn} - n_e)^2$ and $(n_{sn} - n_e)^3$ respectively for the latter model. Several examples of a cubic growth law were quoted.

Reich and Kahlweit (1968) proposed a theory which is related to the BCF volume diffusion theory but which should be applicable to those cases where desolvation at the kinks is the rate limiting kinetic process. According to their treatment, the rate of advance of steps is governed by the flux of desolvated ions to the kinks. The step velocity is given by

$$v_{st} = \frac{3 V_m a^2}{\tau_{\text{des}} x_o} (n_{sn} - n_s) \exp (W_{\text{des}}/kT) \qquad (4.52)$$

where τ_{des} is the relaxation time for desolvation at a kink and W_{des} the potential barrier for desolvation. At low supersaturations $v_{st} \propto \sigma$ through the term $(n_{sn} - n_s)$ and a parabolic $v(\sigma)$ law is expected since $y_o \propto 1/\sigma$ as in the BCF theory. At high supersaturations volume diffusion will become the rate-limiting step as predicted in all treatments of solution growth.

4.11. Comparison of Solution Growth Theory with Experiment

One spectacular success of the BCF theory is that it successfully predicted the occurrence on crystal surfaces of growth spirals, which have now been observed on a wide variety of crystals. In this section we examine the ability of this theory and its various extensions to account for experimental determinations of the variation with supersaturation of the growth rate of crystals from solution.

In interpreting experimental data, difficulty is frequently encountered in distinguishing between boundary-layer and interface-kinetic effects. Two methods are available for obtaining the form of the $v(\sigma)$ relationship for the kinetic process by experiment. The first is to measure the variation of growth rate with solution flow rate or crystal rotation rate and to extract the $v(\sigma)$ relationship using Eqn (4.21). Alternatively, high flow rates or rotation rates may be used and the assumption made that the growth rate is then controlled only by the interface kinetics. The latter assumption is often of dubious validity and experimental data may underestimate the true kinetic-controlled growth rate because no allowance is made for a desolvation or minimum diffusion stage. Unfortunately data obtained by either method are not available for growth on a habit face from high-temperature solution and we therefore consider the results of experiments on aqueous solutions. (Measurements of the growth rate in LPE experiments as a function of the substrate rotation rate will be described in Chapter 8.)

For those crystals to which Brice's method is applicable, that is for which the $v(\omega)$ or $v(u)$ data yield a straight line when plotted according to Eqn (4.21), a quadratic growth law is often found. Brice (1967a) used the experimental data on sucrose (van Hook, 1945) and $CuSO_4.5H_2O$ (McCabe and Stevens, 1951; Hixson and Knox, 1951) and found that $v \alpha \sigma^2$ except for Hixson and Knox's data above 71°C, which indicated a linear growth law. The data of Coulson and Richardson (1956) also fit a quadratic law but our attempts to apply Eqn (4.21) to the results of other investigators were not successful. For example, the data of Mullin and Gaska (1969) yield a highly non-linear plot of $vu^{-1/2}$ against $v^{1/2}$ although the growth rates at high values of u indicate a quadratic law. The extent of the discrepancy between these values and Eqn (4.21) is indicated by an increase of $vu^{-1/2}$ with v, a similar discrepancy with Eqn (4.21) being also found for citric acid using the data of Cartier *et al.* (1959). This discrepancy may be due to convective flow in the solution.

A quadratic growth law has been found for a number of materials grown under conditions of rapid flow. Examples are sodium chloride (Rumford and Bain, 1960), ammonium dihydrogen phosphate (ADP) and potassium dihydrogen phosphate (KDP) (Mullin and Amatavivadhana, 1967) and potassium sulphate (Mullin and Gaska, 1969). However, a linear growth law has been discovered by Bransom *et al.* (1949) for the growth of cyclonite, by Belyutsin and Dvoryakin (1957) for various alums and by Bennema (1966b) for potassium aluminium alum.

Discrepancies are frequently noted between the results of different investigators. For example, Mullin and Garside (1967) found that their results for potassium aluminium alum are best described by a curve of the form $v \alpha \sigma^{1.62}$, which is in agreement within experimental error with the expression given by Chernov (Eqn 4.48). The discrepancy between their results and those of Bennema may be due to the higher supersaturation range studied by Mullin and Garside. Chernov's theory is also supported by the data of Kunisaki (1957) on ethylene diamine tartrate and by

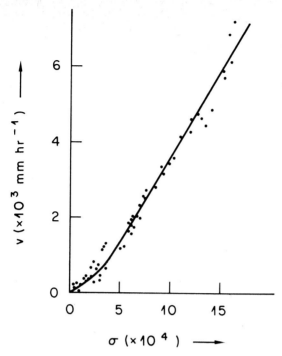

FIG. 4.18. Growth rate of sodium chlorate (after Bennema, 1967).

Garabedian and Strickland-Constable (1970), who reported a variation of the form $v \alpha \sigma^{1.73}$ for the growth of sodium chlorate.

Bennema (1967) used his own results on sodium chlorate to support his case for the inclusion of surface diffusion in crystal growth from solution. The experimental data are shown in Fig. 4.18 together with a curve plotted using the BCF surface-diffusion formula, Eqn (4.41). A similar curve would be predicted by the BCF volume-diffusion theory but in that case the linear region would be controlled by boundary-layer diffusion. Bennema found, however, that changing the stirring rate had no effect on the crystal growth rate and was therefore confident that the measured growth rate was determined by the interface kinetics. The slope in the linear region is roughly one tenth that expected for volume-diffusion control (Eqn 4.14). This discrepancy was given an alternative explanation by Gilmer *et al.* (1971) by the inclusion of the parameter Λ which appears in Eqn (4.51). Then, in the linear region,

$$\frac{dv}{d\sigma} = \frac{Dn_e}{\rho(\delta + \Lambda)} \qquad (4.53)$$

from which they estimate $\Lambda \simeq 10\delta \simeq 0.1$ cm in this case. The data of Garabedian and Strickland-Constable clearly do not agree with those of Bennema but, again, this may be due to the fact that they were obtained at much higher supersaturations.

Alexandru (1971) investigated the growth of ADP by a method similar to that used by Bennema and also found that his results were best explained by the BCF surface-diffusion theory.

Gilmer et al. (1971) used Eqn (4.51) to interpret data of Smythe (1967) on the growth of sucrose crystals. A linear dependence of v on σ was observed by Smythe at temperatures from 20°C to 70°C. The value of Λ at 21°C is estimated as 2×10^{-2} cm, which is much larger than the estimated value of $\delta = 4 \times 10^{-4}$ cm. If this interpretation is correct, the growth mechanism must involve adsorption followed by surface diffusion since Λ represents the effective impedance of the adsorption process.

When the results on precipitation, described in the last section, are included, the weight of evidence appears to favour a growth mechanism which includes a surface-diffusion process in many cases. This conclusion is supported by estimates by Conway and Bockris (1958) of the energy changes occurring during electrocrystallization. They concluded that the energy required to transfer an ion to a surface site is much less than that for direct transfer to a kink, and therefore favoured an initial surface adsorption stage. Electrocrystallization must, of course, include the transfer of an electron which is required before an ion in the solution can become a neutral atom, but the situation is otherwise identical to crystal growth from solution.

The number of $v(\sigma)$ measurements on crystals grown from high-temperature solutions is very small, and these have been made only on un-stirred solutions.† Elwell and Dawson (1972) found a linear variation for the growth of nickel ferrite from barium borate and of sodium niobate from $NaBO_2$. The data for nickel ferrite are shown in Fig. 4.19, and growth in this case is believed to be controlled by volume diffusion through the boundary layer. The value of D/δ calculated using Eqn (4.14) is found to be 5.7×10^{-4} cm s^{-1}. The value of δ estimated from Eqn (4.17) using $\eta \simeq 20$ cP, $\rho_{sn} \simeq 4.5$ g cm^{-3}, $D \simeq 10^{-5}$ cm^2 s^{-1}, $u \simeq 0.1$ cm s^{-1} and $l = 0.5$ cm is $\delta \simeq .06$ cm, which gives $D/\delta \sim 1.6 \times 10^{-4}$ cm s^{-1}. The agreement between theory and experiment is as good as can be expected in view of the uncertainties in the values of D, η and u.

A quadratic $v(\sigma)$ variation was found for the growth of barium strontium niobate $Ba_{0.5}Sr_{0.5}Nb_2O_6$ from the system $BaO—SrO—Nb_2O_5—B_2O_3$ as shown in Fig. 4.20. A remarkable feature of these results is the persistence

† Measurements on stirred solutions will be published in *J. Crystal Growth* by Elwell, Capper and D'Agostino.

of the quadratic law to supersaturations of up to 10%. A critical super-saturation σ_1 of 10% is two orders of magnitude greater than the highest value reported by Bennema (1967) for crystal growth from aqueous solution although Bennema *et al.* (1972) recently revised their estimate of σ_1 to $\sim 10^{-1}$. According to Eqn (4.40), σ_1 is given by $9.5\gamma_m\,a/kTy_s$, so that a high value of σ_1 requires either a high value of γ_m or a low value of y_s. A high value of σ_1 thus appears to be unfavourable for crystal growth since both low y_s and high γ_m will favour deadsorption of surface molecules rather than integration into the kinks, and it is found experimentally that $Ba_{0.5}Sr_{0.5}Nb_2O_6$ is a difficult material to crystallize from borate solvents. The quadratic law may also be due to a surface reaction between, say, $BaNb_2O_6$ and $SrNb_2O_6$ units, as suggested by Tiller (1971), but current knowledge of the ionic species present in the solution is insufficient to allow any firm conclusion. A quadratic $v(\sigma)$ variation was found to explain the growth-rate measurements of $NaNbO_3$ from a $NaBO_2$ flux (Dawson *et al.*, 1974) and of $KTa_{1-x}Nb_xO_3$ from a K_2CO_3 flux (Whiffin and Brice 1974).

Newkirk and Smith (1965) observed a linear variation in the growth of BeO from a number of Li_2O/MoO_3 solvents. The growth rates for this

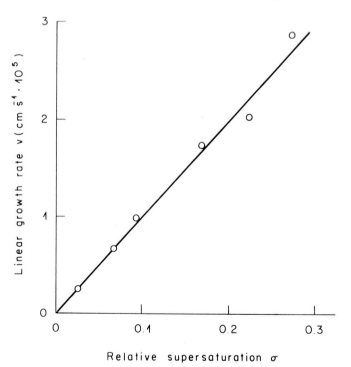

FIG. 4.19. Growth rate of nickel ferrite (Elwell and Dawson, 1972).

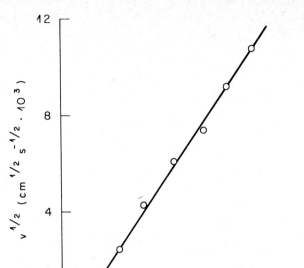

FIG. 4.20. Growth rate of barium strontium niobate (Elwell and Dawson, 1972).

material were only of the order of 10^{-8} cm s^{-1}, some 2 orders of magnitude lower than those shown in Figs 4.19 and 4.20, which are more typical of the maximum values possible in high-temperature solution growth (Scheel and Elwell, 1972, 1973a). It is unlikely that such low growth rates for BeO can be explained simply by a low coefficient of volume diffusion, and the simplest explanation would be to postulate a high value for the adsorption parameter of Gilmer et al. (1971). (It was mentioned above that a linear $v(\sigma)$ relation is difficult to explain in terms of the BCF surface-diffusion theory when the back-stress effect is included. In the next section we shall discuss the shapes of spirals which may be expected to result when the growth centre is a pair or group of spirals; one example which can result in a linear $v(\sigma)$ variation will be included.)

4.12. Non-Archimedian Spirals

In the previous discussion the growth spirals have been assumed to be of approximately Archimedian shape and to have their origin in a single dislocation with a screw component. Frequently, however, dislocations occur in pairs or groups and the spirals originating from such centres will normally have more complex shapes, and the growth mechanism may differ from the simpler case considered in Section 4.7.

If the growth centre is a pair of dislocations of like sign, separated by a distance greater than $2\pi r_s{}^*$, the shape of the resulting spiral will have the form shown in Fig. 4.21. If the crystal face is divided as shown by the heavy dashed line, which will be slightly curved, the two sections will be fed with steps from the two centres, respectively. The activity is approximately the same as that of a single spiral. When the centres are separated by less than $2\pi r_s{}^*$, the arms of both spirals reach the whole area; if the separation is much less than $r_s{}^*$, the centre effectively generates two spirals, each with the same step velocity, and so the activity of the centre will be twice that of a single dislocation.

When a pair of dislocations of opposite sign are separated by a distance greater than $2\pi r_s{}^*$, the steps join up to form closed loops, as shown in Fig. 4.22. This type of cooperation in which a screw-dislocation source generates a series of continuous layers has been observed by Forty (1951) and Griffin (1951), along with many other examples of spirals due to interacting dislocations.

If there are two similar pairs of dislocations separated by a distance large compared with the separation in each pair, the steps will combine on meeting and the number of steps passing any point on the surface will be the same as if only one pair existed. Generalizing from this statement, the growth rate of a face containing several pairs of dislocations of opposite sign will be the same as that of a face having only one such pair as the

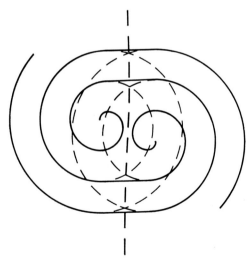

FIG. 4.21. Growth spiral due to pair of dislocations of like sign.

FIG. 4.22. Layers due to a pair of dislocations of opposite sign.

active centre. When the separation of a pair is less than $2r_s{}^*$, step motion cannot occur and so no growth will proceed from such a centre.

An interesting case arises when a group of dislocations of the same sign, all separated by the same distance smaller than $2\pi r^*$, acts as a spiral source. Such an array of dislocations may form wherever screw dislocations occurring in a group lie along some line. The type of spiral produced by this type of group is shown in Fig. 4.23. The separation y_o of the spirals generated will be determined by the separation l between the dislocations and is thus independent of the supersaturation σ. As a result, the growth rate $v(=v_{st}\,a/y_o)$ will depend on the supersaturation only through the term v_{st}. Since $v_{st}\propto\sigma$ [Eqn (4.37)], a linear $v(\sigma)$ law is expected and this may explain the experimental observation of linear kinetic laws for some materials.

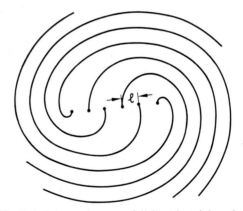

FIG. 4.23. Spiral due to a group of dislocations lying along a line.

Fig. 4.24. Spiral on barium zinc ferrite (after Cook and Nye, 1967).

Frequently the spirals which are observed experimentally do not have curved edges. If the rate of advance of a step over a crystal surface depends upon the orientation, the spiral may readily develop straight edges which are related to the slow-growing faces of the crystal. An example of such a "polygonized" spiral is shown in Fig. 4.24. This spiral was observed (using optical microscopy) by Cook and Nye (1967) on a flux-grown crystal of $Ba_2Zn_2Fe_{12}O_{22}$. The spiral is on the basal plane of the crystal and its shape clearly reflects the hexagonal symmetry normal to this plane. The height of the steps in some spirals was determined by replication electron microscopy as 14.5 Å, which corresponds to the unit-cell edge.

The spirals and growth features which are observed experimentally are often not of unit-cell dimensions but may be built up of 100–10000 unit cells. In a review by Honigmann (1958), surface studies on solution-grown crystals of eleven different materials were reported. Spirals were observed on seven of these materials and non-spiral layer growth on eight. On six materials, the steps were of one or two unit cells in height, on three they were of many unit cells in height and on two, step heights in the region of 1000 Å were observed.

The formation of "macrospirals" observable with a simple microscope

FIG. 4.25. Macrospiral formation due to periodic motion of centre.

was explained by Amelinckx, Bontinck and Dekeyser (1957) as being due to a "wobbling" of the centre of the spiral at a helicoidal screw dislocation. The effect of a periodic perturbation of the spiral centre is illustrated in Fig. 4.25, in which the regular fluctuation in the pitch of the spiral may be seen to give the impression of a spiral of greater pitch. The periodic perturbation can be included in the theory by replacing the factor ϵ in Eqn (4.44) by $\epsilon_0 \sin \omega t$, so that

$$v(t) = \frac{C\epsilon_0 \sin \omega t \, \sigma^2}{\sigma_1} \tanh \frac{\sigma_1}{\epsilon_0 \sin \omega t \, \sigma}$$

and the appearance of a macrospiral will be governed by the relative magnitudes of the frequency ω of the perturbation and the frequency of rotation of the spiral. Bennema and van Rosmalen (1972) have shown that fluctuations will always reduce the flow of steps and therefore the rate of growth.

Bennema (1969) has argued that polygonization of the macrospirals is explained more readily if surface diffusion of solute occurs than if solute enters the kink sites directly. He considered in particular the observations of Torgeson and Jackson (1965) of the macrospiral shapes on ADP crystals grown from aqueous solution. When the crystals are grown in a pure solution, the macrospirals on (100) faces are elliptical with a shorter axis in the [001] direction as shown in Fig. 4.26(a). When Cr^{3+} ions are added to the solution, the spirals become polygonized along [010] and [001] directions as shown in Fig. 4.26(b).

According to the PBC description of Hartman (1956), the {100} surfaces

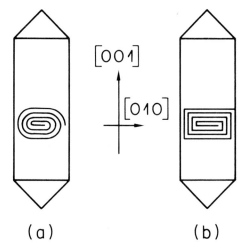

(a)   (b)

FIG. 4.26. Macrospirals on ADP, schematic (after Torgeson and Jackson, 1965).

G

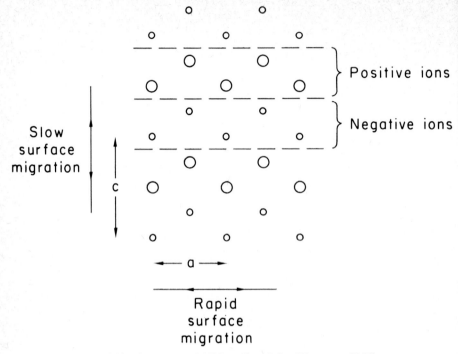

FIG. 4.27. Structure of ADP surface (after Hartman, 1956).

of ADP can be considered as narrow regions of positive ions extending in the [100] direction with a width of $c/2$, alternating with similar regions of negative ions as shown in Fig. 4.27. An ion in the surface can diffuse relatively easily along the [010] direction since it always moves past ions of the same sign. Migration along [001] is, however, relatively difficult since alternate layers are of opposite charge. This difference in surface-diffusion rates along [010] and [001] accounts for the ellipticity in the spiral of Fig. 4.26(a).

When Cr^{3+} ions are added to the solution, many of the kink sites are filled preferentially with these ions so that the number of kink sites available for growth is reduced. Polygonization results from the retardation in the rate at which steps can advance across the surface, but the anisotropy in the spiral shape is preserved since surface diffusion remains anisotropic. While alternative means of explaining these results could be considered (see Section 5), an anisotropic surface-diffusion mechanism appears to offer the simplest explanation.

Although macrospirals are observed quite frequently on crystal surfaces, a quantitative theory of their development is still lacking. A qualitative treatment of the "bunching" of steps has been given by Cabrera and Vermilyea (1958) and by Frank (1958b) based on the kinematic wave theory of Lighthill and Whitham (1955). The formation of large steps by bunching is governed by kinetics rather than by thermodynamics. The velocity of any step depends on the proximity of other steps, which will remove some of the solute. The rate of flow of steps will therefore depend on the average separation between steps and the kinematic wave theory describes the motion of macrosteps of constant separation at some rate v'_{st} which is less than the velocity v_{st} of a single step. Bunching will be particularly likely to occur if the velocity v'_{st} is increasing as crystal growth continues, since in this case newly formed steps will tend to overtake those already present on the surface. Bunching is also more probable in impure solutions, since impurity molecules which are rejected by the crystal interface tend to impede the motion of steps; highly immobile impurity ions may become incorporated into the crystal at the resulting macrosteps. Also the solution flow rate might have an effect on the average step height.

4.13. Surface Morphology of Flux-grown Crystals

Reference has been made above to the observation of growth spirals on the surfaces of orthoferrite crystals by Tolksdorf and Welz (1972) and of polygonized spirals on hexagonal ferrites by Cook and Nye (1967). These observations and the earlier ones of Sunagawa (1967) and others support the validity of Frank's screw-dislocation model. In this section we consider other observations of surface features of crystals grown from high-temperature solutions and the relation between these features and the mode of growth. A more extensive discussion of this topic has been given by Chase (1971).

When crystals nucleate in solution, the supersaturation is normally much higher than that at which the subsequent growth occurs. As a result the initial growth of spontaneously nucleated crystals tends to be highly dendritic. The dendrites grow along fast growth directions and this rapid growth reduces the supersaturation. Subsequent growth occurs more slowly but the ends of the dendrites will be located in regions of higher supersaturation than the central region, and solvent inclusions are trapped near the growth centre as the dendrite arms close. An initial dendritic growth stage has been described by several authors, for example Lefever and Chase (1962), White (1965), Chase (1968) and Scheel and Schulz-Dubois (1971). Figure 4.28(a) shows a large crystal of $GdAlO_3$ in which the central dendritic region may be clearly seen, and Fig. 4.28(b) shows the same crystal in reflected light with the large concentration of growth hillocks in the region above the dendritic core.

As growth proceeds on the dendritic core, the stepped edges of the

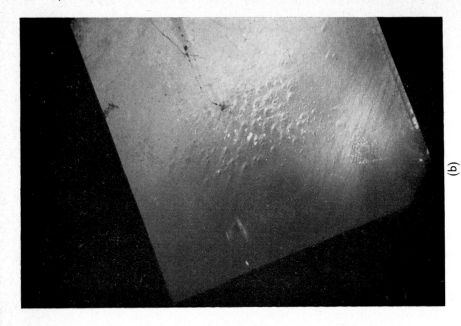

(b)

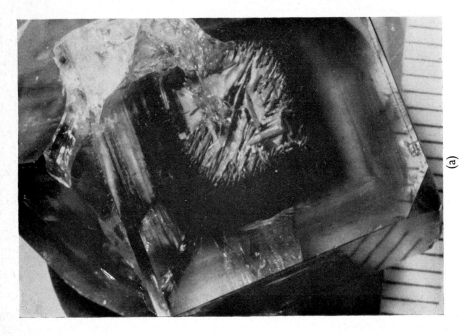

(a)

Fig. 4.28. (a) Dendritic centre of GdAlO₃ with flux inclusions; (b) effect on growth hillocks (reflection photograph.

dendrite arms provide sites for the integration of solute and a terraced structure is produced. If growth is terminated at this stage the crystals are found to exhibit a "hopper" morphology as illustrated in Fig. 4.29. The mechanism of hopper formation was discussed by Lefever and Giess (1963), who pointed out that hopper crystals will be more likely if the initial dendrites attain large dimensions and so incorporate a large fraction of the available solute.

According to Scheel and Elwell (1973a) hopper growth is assumed to be an effect of unstable growth. By increasing the supersaturation gradient, increasingly unstable growth in the following sequence will occur: flat faces→formation of inclusions→edge nucleation→hopper growth→dendritic growth.

Fig. 4.29. Hopper crystal of hematite (courtesy Mrs. B. M. Wanklyn).

An alternative mechanism of hopper formation was proposed by Amelinckx (1953). The crystals in this case were considered to grow while floating on the solution so that the centre of the face is not in contact with the supersaturated solution. Since contact with the solution occurs only at the edge, growth occurs only where a step in the growth spiral meets an edge and a narrow strip of material is deposited. This strip continues to grow along the edge of the crystal and a vertical hollow box would tend to develop except that the crystal simultaneously grows laterally. Each turn therefore appears at a greater lateral distance from the centre than the previous one and the characteristic terraced depression develops. In the extreme case of growth at the edges of a crystal, the resulting shape will be a hollow rectangular tube.

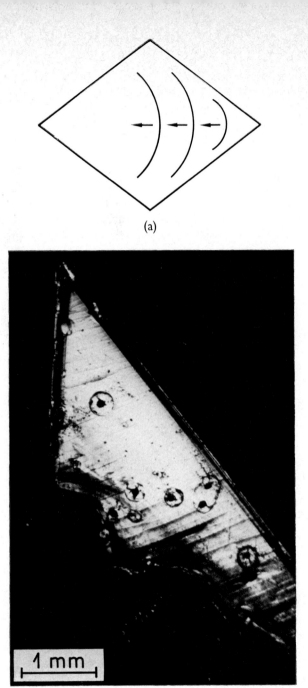

FIG. 4.30. Layer growth nucleated at edges and corners. (a) diagrammatic, (b) edge nucleation on β-eucryptite, LiAlSiO$_4$ (courtesy K. Meyer, ETH Zurich).

If all the crystal faces remain in contact with the solution, continued growth will eventually result in the establishment of the habit faces. Growth at relatively high temperatures (and presumably at rather high supersaturation) was found by Lefever and Chase (1962) to proceed by nucleation of layers at corners or edges of the garnet crystals studied. The layers in this case were normally curved in a direction concave from the point of origin, as shown in Fig. 4.30(a). This curvature arises because of the higher supersaturation at corners and edges which can lead to an increase of growth rate with distance from the centre of the face. Similar layers were observed by Chase (1968) on In_2O_3 crystals and by Quon and Sadler (1967) on yttrium iron garnet. In the latter case a similar structure made up of much finer layers was also observed. An example of edge

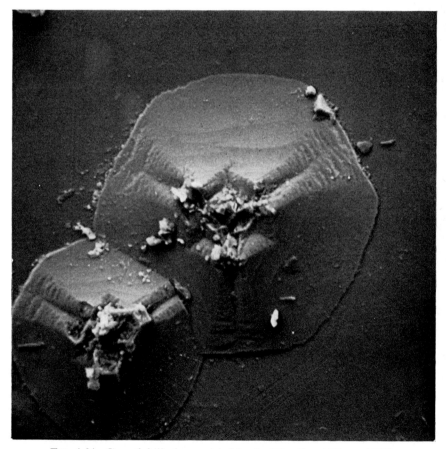

FIG. 4.31. Growth hillocks on nickel ferrite (Elwell and Neate, 1971).

nucleation on a β-eucryptite (LiAlSiO$_4$) crystal grown from a vanadate flux is shown in Fig. 4.30(b).

If the supersaturation is lowered below the value which can promote corner and edge nucleation, the characteristic features seen on most crystals are growth hillocks, consisting of layers roughly 10^{-5} cm in height. Typical hillocks are illustrated in Fig. 4.31. Growth hillocks are presumably formed by a bunching process, as described in the previous section, which gives rise to the relatively thick layers visible under the microscope. Other examples of growth hillocks have been described by Lefever and Chase (1962) and Quon and Sadler (1967) on garnets, by Chase (1968) on In$_2$O$_3$, by Sunagawa (1967) on aluminium oxide and by Scheel and Elwell (1973b) on rare-earth aluminates. Sunagawa (1967) has

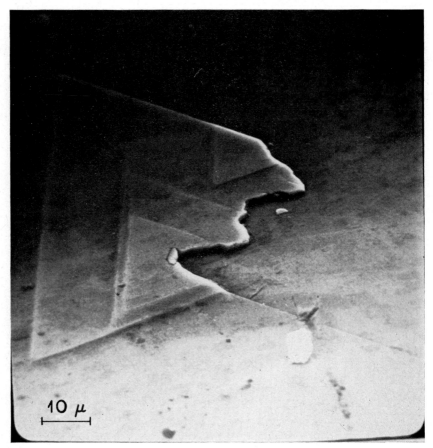

FIG. 4.32. Triangular growth layers on lithium ferrite (Elwell and Neate, 1971).

investigated a large number of flux grown crystals and has observed spirals of monomolecular step height on magnetoplumbite, $PbFe_{12}O_{19}$, on ferric oxide, alumina and yttrium iron garnet. Pyramidal layers were observed on spinel, $MgAl_2O_4$. The number of features seen on a given face appears to depend on the supersaturation and a single feature often dominates a whole face when growth occurs at low supersaturation. This decrease in the number of active centres as growth proceeds may have an influence on the maximum rate of stable growth, as is discussed in Chapter 6.

Triangular growth layers were observed by Elwell and Neate (1971) on ferrite crystals, an example being shown in Fig. 4.32. This feature appeared to be the only active growth centre on that particular face, and the layer height ($\sim 10^{-5}$ cm) is clearly determined by some bunching effect. A mechanism of crystal growth by the spreading of layers of similar height was reported by Bunn and Emmett (1949) who studied the growth of lead nitrate from aqueous solution.

As discussed earlier in the chapter, layers, hillocks and macrospirals may all have their origin in screw dislocations. Confirmation of the dislocated nature of hillock centres was reported by Lefever and Chase (1962), who found on etching the crystal surfaces that an oriented etch pit was formed at the centre of each hillock. The most likely conclusion to be drawn from these surface studies is that growth on habit faces at low supersaturation frequently occurs by the Frank screw-dislocation mechanism but that edge nucleation may be dominant at higher supersaturations.

4.14. Alternative Growth Mechanisms

Although the mechanism by which crystals grow from fluxed melts is often the BCF screw-dislocation mechanism, alternative growth mechanisms are not rare (Scheel and Elwell, 1973b).

Nucleation of surface layers at corners or edges of a crystal may be by 2-D nucleation rather than at screw dislocations. The relative ease of nucleation at corners or edges was first proposed from binding energy considerations by Stranski (1928). Corner and edge nucleation will clearly be favoured because of the relatively high concentration of solute in these regions, even if growth occurs by the screw-dislocation mechanism. Figure 4.33 shows an optical reflection micrograph of a $GdAlO_3$ crystal in which the concentration of hillocks is higher at the crystal edges due to the higher local supersaturation. As growth continues at a stable rate, the concentration of hillocks near the edges decreases and so edge growth becomes less important. The tendency of crystals to grow with raised edges is, however, favoured if growth becomes unstable, as will be discussed in Chapter 6.

A particularly powerful nucleation site may be formed when the faces

G 2

FIG. 4.33. Growth hillocks at edges and on faces of a GdAlO$_3$ crystal (Scheel and Elwell, 1973b).

of a twinned crystal meet along the twin plane at an acute angle. The resulting twin-plane re-entrant edge (TPRE) growth mechanism can be envisaged with reference to Fig. 4.34, which shows a section through a twinned crystal. The crystal grows by the propagation of layers in the directions indicated by v_L, and rapid growth may also occur in the direction of the twin plane, depending on the nature of the twin and the crystal structure.

The TPRE mechanism and its influence on the habit of crystals was described by Niggli (1920) and Spangenberg (1934), who both refer to Mügge (1911) and Becke (1911), by Wagner (1960), John and Faust (1961) and Faust and John (1964), the latter giving an extensive list of semi-

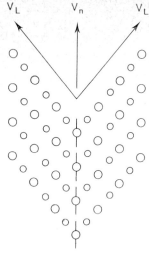

V_L V_n V_L

TWIN | PLANE

FIG. 4.34. Twin-plane re-entrant edge growth mechanism.

1 mm

FIG. 4.35. Layer spreading influenced by multidomain twinning of $NdAlO_3$ (Scheel and Elwell, 1973b).

FIG. 4.36. Detail of interactions of twin domains and growth layers on NdAlO₃ crystal (Scheel and Elwell, 1973b).

conductors grown by this mechanism. The habit of Al_2O_3, BeO and $BaTiO_3$ is controlled by the relative importance of this mechanism as will be discussed in the next Chapter.

Twin domains formed due to a phase transition during growth may affect the growth mechanism even when the angular deviation between twins is very small. Figures 4.35 and 4.36 show growth layers on the surface of neodymium-aluminate crystals. The pattern of layers is very closely related to the domain structure, although the twinning angle is less than 1° (Geller and Bala, 1956). This interrelation between growth layers and domains is not observed in crystals such as $BaTiO_3$ in which the domains are formed at temperatures well below the growth temperature.

It is not clear whether twin boundaries at very low angles act by

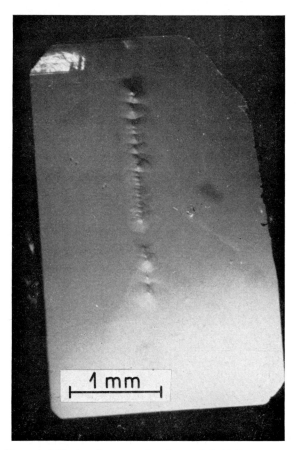

FIG. 4.37. Growth hillocks along a twin plane of $GdAlO_3$ (Scheel and Elwell, 1973b).

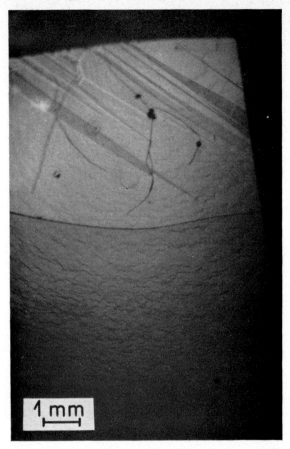

FIG. 4.38. Layers spreading from multiple twinned region of GdAlO$_3$ (Scheel and Elwell, 1973b).

providing centres for classical nucleation or because of a high concentration of screw dislocations. In some cases, the twin planes provide centres for the formation of growth hillocks as shown in Figs. 4.37 and 4.38. These photographs are of GdAlO$_3$ crystals, and examples have also been observed where twinned regions do not provide the dominant growth centres because of the presence of very active screw-dislocation sources (see Fig. 4.39).

Carlson (1958) proposed that low-angle grain boundaries may also provide more active growth centres than those due to isolated screw dislocations. Twist boundaries will give rise to screw dislocations, the separation of which is given by Nabarro (1967) as

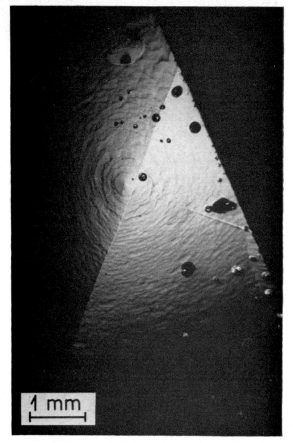

FIG. 4.39. Dominating growth centre near a twin plane of GdAlO₃ (Scheel and Elwell, 1973b).

$$d = \frac{a}{2} \operatorname{cosec} \phi/2$$

where a is the interatomic separation and ϕ the angle between the adjacent grains. Using the criterion of BCF for the cooperation between screw dislocations of like sign, that $d < 2\pi r^*$, the minimum angle for preferential growth at screw dislocations is given by

$$\sin \phi/2 > a/4\pi r^*.$$

Thus for $r^* \sim 20a$, (see Eqn (4.25a) with $\gamma_m/kT \sim 1$ and $\sigma = 0.05$), ϕ must be of the order of $1/2°$, which is typical of the values at which a twin plane acts as the dominant growth centre.

Cracks which develop in any crystal due to severe strain during growth

FIG. 4.40. Growth along crack of a GdAlO$_3$ crystal (Scheel and Elwell, 1973b).

provide many active growth centres and tend to "heal" by relatively rapid local growth. Figure 4.40 shows a GdAlO$_3$ crystal in which a crack has developed during removal of the crucible from the furnace in order to pour off the residual solution. The crack has clearly resulted in many growth centres which were more active than centres which had previously dominated growth over the whole face.

In any attempt to assess the growth mechanism of a crystal, care must be exercised to allow for the mutual influence which neighbouring faces exert upon each other. An example of this influence is shown in Fig. 4.41 which shows two faces of GdAlO$_3$ inclined at 90° to each other. Layers on the two faces run in opposite directions, and the layer-rich regions of the two faces correspond to each other. On some crystals one face had very

FIG. 4.41. Two adjacent faces (at nearly 90°) of GdAlO$_3$ (Scheel and Elwell, 1973b).

active growth centres, with adjacent faces showing hardly any features, suggesting that the latter faces grew by edge nucleation from the more active face. Such observations are contrary to the PBC concept, which treats all {100} faces of a pseudocubic perovskite as essentially equivalent, and indicate that generalizations on growth mechanisms should be expressed with care.

4.15. Summary

(i) The rate of nucleation of crystals varies rapidly with supersaturation once a critical value, of the order of 10%, is exceeded and is very low at lower supersaturations.

(ii) The initial growth following spontaneous nucleation is often dendritic, then terraced, before stable facets are established.

(iii) Crystals grown in a stable mode from high-temperature solution normally have atomically flat faces, on which growth occurs by the spreading of layers from active centres. The evidence for this statement is based on observations of growth spirals and layers and it also explains the observation that crystals preserve their shape although the supersaturation is not constant across a face.

(iv) A complete description of the growth process should include desolvation and surface diffusion of solute.

(v) In unstirred solutions, volume diffusion is the most probable rate determining step. At low supersaturations screw-dislocation growth can account for most of the experimental measurements of growth kinetics although alternative explanations are often possible. The theory of solution growth is still not quantitative since it contains several parameters which cannot be determined.

(vi) No strict generalization on the growth mechanism and the rate determining step is possible. Depending on the solute-solvent system and on the experimental parameters (supersaturation, temperature, concentration, stirring, impurity concentration, etc.) each crystal will have its individual growth history.

References

Alexandru, H. V. (1971) *J. Crystal Growth* **10,** 151.
Amelinckx, S. (1953) *Phil. Mag.* **44,** 337.
Amelinckx, S., Bontinck, W. and Dekeyser, W. (1957) *Phil. Mag.* **2,** 1264.
Becke, F. (1911) *Fortschr. Min.* **1,** 68.
Becker, R. and Döring, W. (1935) *Ann. Phys.* **24,** 719.
Belyutsin, A. V. and Dvorykin, V. F. (1957) *Growth of Crystals* (A. V. Shubnikov and N. N. Sheftal, eds.) **1,** 139.
Bennema, P. (1965), Thesis, University of Delft.
Bennema, P. (1966) *phys. stat. solidi* **17,** 563.
Bennema, P. (1967) *J. Crystal Growth* **1,** 278, 287.
Bennema, P. (1969) *J. Crystal Growth* **5,** 29.
Bennema, P. and Gilmer, G. H. (1972) *J. Crystal Growth* **13/14,** 148.
Bennema, P. and Gilmer, G. H. (1973) In "Crystal Growth" (P. Hartman, ed.) North Holland, Amsterdam.
Bennema, P. and van Rosmalen, R. (1972) *Growth of Crystals* (to be published).
Bennema, P., Boon, J. and van Leeuwen, C. (1972) Chisa Conference Report, Prague.
Berg, W. F. (1938) *Proc. Roy. Soc. A* **164,** 79.
Berthoud, A. (1912) *J. Chim. Phys.* **10,** 624.
Binsbergen, F. L. (1972) *J. Crystal Growth* **13/14,** 44.

Boscher, J. (1965) *Ann. Assoc. Int. Calc. Analog.* **4,** 117.

Bransom, S. H., Dunning, W. J. and Millard, B. (1949) *Disc. Faraday Soc.* **5,** 83.

Brice, J. C. (1967) *J. Crystal Growth* **1,** 161.

Bunn, C. W. (1949) *Disc. Faraday Soc.* **5,** 144.

Bunn, C. W. and Emmett, H. (1949) *Disc. Faraday Soc.* **5,** 119.

Burton, W. K. and Cabrera, N. (1949) *Disc. Faraday Soc.* **5,** 33.

Burton, W. K., Cabrera, N. and Frank, F. C. (1951) *Phil. Trans. A* **243,** 299.

Burton, J. A., Prim, R. C. and Slichter, W. P. (1953) *J. Chem. Phys.* **21,** 1987.

Cabrera, N. and Coleman, R. V. (1963) In "The Art and Science of Growing Crystals" (J. J. Gilman, ed.) p. 3. Wiley, New York.

Cabrera, N. and Levine, M. M. (1956) *Phil. Mag.* **1,** 450.

Cabrera, N. and Vermilyea, D. A. (1958) In "Growth and Perfection of Crystals" (R. H. Doremus, B. W. Roberts, D. Turnbull, eds.) p. 393. Wiley, New York and Chapman and Hall, London.

Carlson, A. E. (1958), Thesis, Univ. of Utah; in "Growth and Perfection of Crystals" (R. H. Doremus, B. W. Roberts, D. Turnbull, eds) p. 421. Wiley, New York and Chapman and Hall, London.

Cartier, R., Pindola, D. and Bruins, P. (1959) *Trans. Inst. Chem. Engrs.* **51,** 1409.

Chase, A. B. (1968) *J. Am. Ceram. Soc.* **51,** 507.

Chase, A. B. (1971) In "Preparation and Properties of Solid State Materials" (R. A. Lefever, ed.) p. 183. Dekker, New York.

Chernov, A. A. (1961) *Sov. Phys. Usp.* **4,** 129.

Chernov, A. A. (1963) *Sov. Phys. Cryst.* **8,** 63.

Cobb, C. M. and Wallis, E. B. (1967) Report AD 655388.

Conway, B. E. and Bockris, J. O. M. (1958) *Proc. Roy. Soc. A* **248,** 394.

Cook, C. F. and Nye, W. F. (1967) *Mat. Res. Bull.* **2,** 1.

Coulson, J. M. and Richardson, J. F. (1956) In "Chemical Engineering" Vol. 2. Pergamon Press, Oxford.

Davies, C. W. and Jones, A. L. (1951) *Trans. Faraday Soc.* **55,** 312.

Dawson, R. D., Elwell, D. and Brice, J. C. (1974) *J. Crystal Growth* **23,** 65.

Doremus, R. H. (1958) *J. Phys. Chem.* **62,** 1068.

Dunning, W. J. (1955) In "Chemistry of the Solid State" (W. E. Garner, ed.) p. 159. Butterworth, London.

Elwell, D. and Dawson, R. D. (1972) *J. Crystal Growth* **13/14,** 555.

Elwell, D. and Neate, B. W. (1971) *J. Mat Sci.* **6,** 1499.

Elwell, D., Neate, B. W. and Smith, S. H. (1969) *J. Thermal Anal.* **1,** 319.

Faust, Jr., J. W. and John, H. F. (1964) *J. Phys. Chem. Solids* **25,** 1407.

Forty, A. J. (1951) *Phil. Mag.* **42,** 670.

Frank, F. C. (1949) *Disc. Faraday Soc.* **5,** 48.

Frank, F. C. (1958a) In 'Growth and Perfection of Crystals" (R. H. Doremus, B. W. Roberts, D. Turnbull, eds.) p. 393. Wiley, New York and Chapman and Hall, London.

Frank, F. C. (1958b) In "Growth and Perfection of Crystals" (R. H. Doremus, B. W. Roberts, D. Turnbull, eds.) p. 411. Wiley, New York and Chapman and Hall, London.

Garabedian, H. and Strickland-Constable, R. F. (1970). Paper presented at BAGG Meeting, University of Bristol.

Geller, S. and Bala, V. B. (1956) *Acta Cryst.* **9,** 1019.

Gilmer, G. H. and Bennema, P. (1972) *J. Appl. Phys.* **43,** 1347.

Gilmer, G. H., Ghez, R. and Cabrera, N. (1971) *J. Crystal Growth* **8,** 79.

Glasner, A. (1973) *Mat. Res. Bull.* **8,** 413.

Goldsztaub, S., Itti, R. and Mussard, F. (1970) *J. Crystal Growth* **6,** 130.

Griffin, L. J. (1951) *Phil. Mag.* **41,** 1337.

Hartman, P. (1956) *Acta Cryst.* **9,** 721.

Hartman, P. and Perdok, W. G. (1955) *Acta Cryst.* **8,** 49.

Hirth, J. P. and Pound, G. M. (1963) "Condensation and Evaporation, Nucleation and Growth Kinetics" Pergamon, Oxford.

Hixson, A. W. and Knox, K. L. (1951) *Ind. Eng. Chem.* **43,** 2144.

Honigmann, B. (1958) "Gleichgewichts- und Wachstumsformen von Kristallen" Steinkopff, Darmstadt.

Jackson, K. A. (1958) In "Liquid Metals and Solidification" p. 174. Am. Soc. Metals, Cleveland.

John, H. F. and Faust, Jr., J. W. (1961) In "Metallurgy of Elemental and Compound Semiconductors" (R. Gruebel, ed.) Interscience, New York.

Khamskii, E. V. (1969) "Crystallization from Solutions" Consultants Bureau, New York.

Kossel, W. (1927) Nachr. Gesell. Wiss. Göttingen, Math-Phys. Kl., 135.

Kunisaki, J. (1957) *J. Chem. Soc. Japan* **60,** 987.

Laudise, R. A., Linares, R. C. and Dearborn, E. F. (1962) *J. Appl. Phys.* **33S,** 1362.

Lefever, R. A. and Chase, A. B. (1962) *J. Am. Ceram. Soc.* **45,** 32.

Lefever, R. A. and Giess, E. A. (1963) *J. Am. Ceram. Soc.* **46,** 143.

Lewis, B. (1974) *J. Crystal Growth* **21,** 29, 40.

Lighthill, M. J. and Whitham, G. B. (1955) *Proc. Roy. Soc.* **229,** 281.

Lothe, J. and Pound, G. M. (1962) *J. Chem. Phys.* **36,** 2080.

Lydtin, H. (1970) "Chemical Vapour Deposition" (J. M. Blocher and J. C Withers, eds.) 2nd Int. Conf. Electrochem. Soc., 1971.

McCabe, W. L. and Stevens, R. P. (1951) *Chem. Eng. Progr.* **47,** 168.

Mügge, O. (1911) *Fortschr. d. Mineral.* **1,** 38.

Mullin, J. W. and Amatavivadhana, A. (1967) *J. Appl. Chem.* **17,** 151.

Mullin, J. W. and Garside, J. (1967) *Trans. Inst. Chem. Engrs.* **45,** T285.

Mullin, J. W. and Gaska, G. (1969) *Can. J. Chem. Eng.* **47,** 483.

Nabarro, F. R. N. (1967) "Theory of Crystal Dislocations", Oxford University Press.

Nernst, W. (1904) *Z. Phys. Chem.* **47,** 52.

Newkirk, H. W. and Smith, D. K. (1965) *Am. Min.* **50,** 44.

Nielsen, A. E. (1964) "Kinetics of Precipitation" Pergamon Press, Oxford.

Nielsen, A. E. (1969) *Kristall u. Technik* **4,** 17.

Niggli, P. (1920) In "Lehrb. d. Mineral." p. 142. Borntraeger, Berlin.

Noyes, A. A. and Whitney, W. R. (1897) *J. Am. Chem. Soc.* **19,** 930.

Onsager, L. (1944) *Phys. Rev.* **55,** 117.

Parker, R. L. (1970) Solid State Physics (M. Ehrenreich, F. Seitz and D. Turnbull, eds.) **25,** 152.

Quon, H. and Sadler, A. G. (1967) *J. Can. Ceram. Soc.* **36,** 33.

Reich, R. and Kahlweit, M. (1968) *Ber. Bunseng. Phys. Chem.* **72,** 66.

Rumford, F. and Bain, J. (1960) *Trans. Inst. Chem. Engrs.* **38,** 10.

Scheel, H. J. and Elwell, D. (1972) *J. Crystal Growth* **12,** 153.

Scheel, H. J. and Elwell, D. (1973a) *J. Electrochem. Soc.* **120,** 818.

Scheel, H. J. and Elwell, D. (1973b) *J. Crystal Growth* **20,** 259.

Scheel, H. J. and Schulz-DuBois, E. O. (1971) *J. Crystal Growth* **8,** 304.

Seeger, A. (1953) *Phil. Mag.* **44,** 1.

Smythe, B. M. (1967) *Austr. J. Chem.* **20,** 1087.

Spangenberg, K. (1934) *Handb. der Naturwiss.* 2nd Edition, **10,** 362.

Stranski, I. N. (1928) *Z. Phys. Chem.* **136,** 259.

Strickland-Constable, R. F. (1968) "Kinetics and Mechanism of Crystallization" Academic Press, London, New York.

Sunagawa, I. (1960) *Mineral Journ.* **3,** 59.

Sunagawa, I. (1967) *J. Crystal Growth* **1,** 102.

Tammann, G. (1925) "States of Aggregation" Van Nostrand, New York.

Temkin, D. E. (1966) "Crystallization Processes" p. 15. Consultants Bureau, New York.

Tiller, W. A. (1971) Comment at ICCG3 (Marseille).

Tolksdorf, W. and Welz, A. (1972) *J. Crystal Growth* **13/14,** 566.

Torgeson, J. L. and Jackson, R. W. (1965) *Science* **148,** 952.

Valeton, J. J. P. (1924) *Z. Krist.* **59,** 135 and 335.

Van Hook, A. (1945) *Ind. Eng. Chem.* **37,** 782.

Verma, A. R. (1953) "Crystal Growth and Dislocations" Butterworth, London.

Volmer, M. and Weber, A. (1926) *Z. Phys. Chem.* **119,** 277.

Wagner, R. S. (1960) *Acta Met.* **8,** 51.

Whiffin, P. A. C. and Brice, J. C. (1974) *J. Crystal Growth* **23,** 25.

White, E. A. D. (1965) *Tech. Inorg. Chem.* **4,** 31.

Wilcox, W. R. (1969) *Mat. Res. Bull.* **4,** 265.

Wilcox, W. R. (1972) *J. Crystal Growth* **12,** 93.

Zettlemoyer, A. C. (1969) (editor) "Nucleation" Dekker, New York.

5. Crystal Habit

5.1. Historical Development

Interest in the morphology of crystals really began in 1669 when Steno described the law of constancy of angles for the faces on quartz crystals. This law was confirmed by Cappeller (1723) on a variety of crystals. The first examples of habit modification in artificially-grown crystals were described by Romé de L'Isle (1783, rocksalt octahedra in a solution to which urine was added), by Leblanc (1788, alum cubes and octahedra) and by Beudant (1817, 1818), who described the influence of impurities and of mechanical mixing on the habit.

A further stimulus to the growing interest in this field was provided by Haüy (1783, 1784), who postulated a relation between a chemical substance and its individual crystalline form, and another important development was that of Wollaston (1809) with the optical goniometer. An impressive amount of data was accumulated in the 19th and early 20th centuries on the angles between crystal faces, and an enormous variety of careful drawings was prepared. Today the possible morphologies of a crystal having a known structure may be drawn by a computer in a matter of minutes (see Figs 5.8.(II), 5.13, 5.14), but 100 years ago this was a major problem. The famous Goldschmidt atlas (1913) and the works of Groth (1906) contain a compilation of all the faces which had been observed on the known minerals. For example, for quartz 31 common and 369 rare and for calcite 148 common and 381 rare faces are listed, and the occurrence of particular crystal faces on minerals from various deposits was described by statistics.

The situation was changed by the discovery of X-ray diffraction by Friedrich, Knipping and von Laue (1912) and by the first structure determinations by W. H. and W. L. Bragg (1913). The crystal grower is now mainly interested in finding explanations for the habit modifications in artificially-prepared crystals while the crystal consumer is concerned with optimization of the shape of crystals for his particular application or measurement. If impurities are incorporated preferentially at certain faces, it is desirable to ensure that such faces are not present on the growing crystal if the highest possible purity is to be obtained.

The following discussion on the habit and its modification is, of course, only relevant when the crystal grows unconstrained in the solution, when the crystal can form equilibrium faces. This is not the case, for example, in the travelling solvent zone technique, in liquid phase epitaxy or in the pulling of crystals from solution (by a modified Czochralski technique) as will be discussed in Chapters 7 and 8. However, crystals grown by the latter techniques frequently show facets which demonstrate the strong tendency towards facet formation even in constrained crystallization processes.

This Chapter gives a brief survey of the equilibrium shape and of habit modification in crystal growth from solution. More detailed information on special topics has been given by Valeton (1915), Tertsch (1926), Burton, Cabrera and Frank (1951), Buckley (1951), Honigmann (1958), Chernov (1961), Hartman (1969), Kern (1969) and in the conference proceedings of Adsorption et Croissance Cristalline (Colloq. Intern. C.N.R.S. No. 152, Paris, 1965).

5.2. The Equilibrium Shape of a Crystal

Gibbs (1875, 1878) was the first to give a description of the equilibrium form of a crystal based on thermodynamics. The total free energy of a crystal is the sum of the free energies of the volume, of the surface and of the edges and corners. Gibbs showed that the edges and corners have an effect only when the crystals are extremely small, whereas the relative contribution of the surface free energy decreases in proportion to the linear dimensions of the crystal. For crystals of the same volume the equilibrium form is that which has a minimum surface energy. Gibbs' condition demands that $\Sigma \gamma_i A_i$ should have a minimum value, where γ_i is the specific surface free energy of the face i, A_i its area, and the summation is taken over all the faces of the crystal. This principle was also used by Curie (1885).

Wulff (1901) established the relationship between individual faces of the equilibrium shape and their individual specific surface energies by the theorem:

When a crystal is in its equilibrium shape, there exists within it a point to which the perpendicular distances from all faces are proportional to their specific surface free energies; any other possible face, not belonging to the equilibrium shape, has a surface free energy such that a plane drawn with the corresponding orientation and distance from this point would be entirely outside the crystal.

A polar diagram of the specific surface free energy is called a *Wulff plot* (or γ plot) and shows a closed surface, the distance of which from the origin is proportional to the magnitude of γ. The equilibrium shape is then found by drawing all the planes normal to the radius vectors to this surface and taking the innermost envelope. The equilibrium shape will thus be determined by the minima in the Wulff plot. If these are sharp, that is if certain faces have much lower free energy than other possible faces, the crystal will be facetted. A point on the γ surface corresponds to a part of the equilibrium surface if a sphere drawn through the origin to touch the γ surface at this point does not intersect the γ surface. A section through a Wulff plot for a simple crystal shape *CDEF* is shown in Fig. 5.1. The

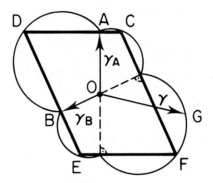

Fig. 5.1. Wulff plot for simple crystal.

diagonals *CE* and *DF* are first drawn, then circles are drawn with the diameters *OC*, *OD*, *OE*, *OF* between *O* and the corresponding corners. The normals from *O* to *CD* and *DE*, respectively, are then proportional to γ_A and γ_B and meet the minima in the γ plot. This example assumes no other minima, that is no other possible equilibrium facets. It is evident that the Wulff theorem can generally be applied to small crystals only. This is confirmed by careful experiments of Valeton (1915) and of Neuhaus (see Spangenberg, 1934) who did not find differences in solubility of various existing faces on macroscopic crystals. However there is theoretical proof by Volmer (1939), von Laue (1943), Landau (1950), Herring (1953) and Chernov (1961) that in principle the Wulff theorem should hold and that crystals should have singular faces up to their melting point. Bennema

(1973) has reviewed Herring's treatment of the equilibrium form and presented his conclusions in the form of seven theorems.

The kinetics of the microscopic growth steps at the growing surface, especially the concept of the *repeatable step*, offer an alternative possiblity of determination of the equilibrium form. Attempts to perform such calculations have been described by Kossel (1927), Stranski (1928) and by Stranski and Kaishev (1934) and were for instance successfully applied by Simon and Bienfait (1965) to the complicated structure of gypsum. However, as a rule the method of Kossel and Stranski will only be applicable to simple structures. Reviews on the Kossel-Stranski model were published by Knacke and Stranski (1952) and by Honigmann (1958).

Qualitative methods of predicting the stable crystal habit have been proposed, notably by Bravais (1866), Niggli (1919, 1920), Sohncke (1888), Donnay and Harker (1937) and by Hartman and Perdok (1955). Stable faces were connected by L. Bravais with a high lattice plane density and by Niggli with surfaces having few unsaturated bonds. Sohncke and later Donnay and Harker proposed that the faces of lowest specific surface free energy will be those of highest reticular density, that is those which have the highest density of atomic packing. Although the equilibrium faces of crystals do normally have a high reticular density, it is also necessary to consider the nature of the chemical bonds between the atoms in the crystal as was stressed by Niggli (1919, 1920).

The Hartman-Perdock PBC method, which was briefly mentioned in Chapter 4, is based on the assumption of Born (1923) that the surface energy of a crystal depends mainly on the chemical bond energies. The attachment energy of Hartman and Perdok depends on the direction (with respect to the surface) of the strong chemical bonds, which are those which release the largest amount of energy during the crystallization process. Uninterrupted chains of strong bonds are called periodic bond chains (PBC). In the example shown in Fig. 5.2, (a) has flat or *F faces* with strong bonds linking neighbouring PBC's, (b) has a stepped or *S face* since these bonds are not directed along the surface. In Fig. 5.3, the *F* and *S* faces

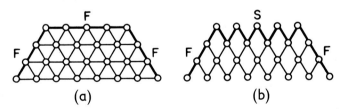

(a)   (b)

FIG. 5.2. Section of crystal with o representing a PBC. (a) Neighbouring PBC's linked by strong bonds, (b) neighbouring PBC's not linked by strong bonds along the *S* face.

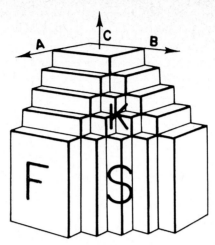

FIG. 5.3. Flat (*F*), stepped (*S*) and kinked (*K*) faces corresponding to the directions of PBC's parallel to *A*, *B* and *C*.

are depicted, together with the kinked or *K faces* which do not contain any PBC vector. Clearly the *S* and *K* faces grow very quickly and are therefore rarely, if ever, observed. The habit of a crystal is dominated by the slowly-growing *F* faces. From a knowledge of the crystal structure it is frequently possible to predict which will be the *F* faces, and many examples of the application of this method have been published. However, not all the *F* faces will be present, and it is not possible to predict with confidence which *F* faces will have the lowest rates of growth and which will be present in the crystal.

There is a certain similarity between the PBC method and the classical Kossel-Stranski theory as outlined by Honigmann (1958). The *F* and the *S* and *K* faces of Hartman and Perdok correspond approximately to the flat and rough faces, respectively, of Stranski (1932). There is, in fact, a variety of terms for the three types of faces, as listed in Table 5.1, and having the same significance. In this connection it is interesting to note that Steno described the stepped and irregular structure of intermediate faces as early as 1669.

In general, the value of theories of the equilibrium habit of crystals is limited because growth occurs under nonequilibrium conditions (except for the few observations of equilibrium forms of small crystals by Lemmlein, 1954; Klija, 1955; Bienfait and Kern 1964). The departure from equilibrium normally increases with crystal size, solute concentration, supersaturation, growth rate and impurity concentration. In addition, the type of solvent may influence the habit according to the type of

TABLE 5.1. Methods of Notation for the Three Types of Face

Kossel (1927) Stranski (1928)	glatt (flat)	vergröbert (rough)	
Stranski and Kaischev (1931)	vollständig, Gleichgewichtsform-Flächen (complete, faces of equilibrium form)	unvollständig (incomplete)	
Burton and Cabrera (1949)	close-packed	stepped	
Hartman (1953), Hartman and Perdok (1955)	F (flat)	S (stepped)	K (kinked)
Honigmann (1958)	A2 two-dimensional nucleation	A1 one-dimensional nucleation	A0 zero-dimensional nucleation
Chalmers (1958) Laudise (1970) etc.	smooth	rough	
Frank (1958), Cabrera (1959)	singular	nonsingular, vicinal	

solvent-solute interaction as discussed in Chapter 3. Frank (1958) stressed the fact, already mentioned by Gibbs (1875, 1878), that for crystals of macroscopic dimensions the energy associated with the driving force for crystallization will be larger than changes in free energy due to departures from the equilibrium shape, so that crystals will not have their equilibrium form except for the cases where the kinetically controlled habit happens to be identical with the equilibrium form. Among the other factors which might influence the habit are the crystal defects (dislocations, twin and low-angle boundaries), as discussed later. In addition the effect of surface roughening (Sections 4.2 and 4.3) will decrease the importance of the predicted equilibrium form. Normally, however, as pointed out in the review by Hartman (1969), morphological changes usually involve only F faces and are caused by changes in the relative growth rates of different F faces.

5.3. Influence of Growth Conditions on Habit

The habit of crystals growing in solution is determined by the slowest growing faces as noted in Chapter 4. The assumption in the discussion on the equilibrium form was that these faces will be the faces of lowest energy, but it is apparent that crystal habit is governed by kinetic rather than equilibrium considerations. A considerable period may, however, elapse after nucleation before the slowest growing faces dominate the habit, and a dependence of the observed habit on the duration of growth has been noted by numerous investigators (see, for instance, Buckley, 1951; Van Hook, 1961; Alexandru, 1969; Mullin, 1972).

A dependence of crystal habit on *supersaturation* is to be expected since the growth rates of different F faces often exhibit a different dependence on the supersaturation. The linear growth rate may be written as

$$v = k\sigma^m \tag{5.1}$$

where the parameters k and m depend on the face considered and on such factors as the temperature and the solution flow rate. Thus the relative growth rates of two faces denoted by 1 and 2 will be

$$\frac{v_1}{v_2} = \frac{k_1 \, \sigma^{m_1}}{k_2 \, \sigma^{m_2}}. \tag{5.2}$$

If these growth rates have the form shown in Fig. 5.4, face 2 will tend to dominate at low supersaturations where $v_2 < v_1$. Accordingly face 1 will dominate at higher supersaturations.

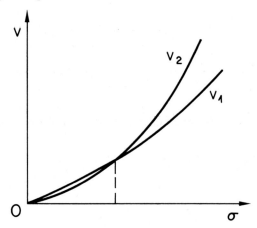

FIG. 5.4. Variation with supersaturation of growth rate of two F faces.

Kern (1969), in a review on crystal morphology and the effects of impurities, quotes potassium iodide as an example of a crystal which exhibits such a dependence of habit on supersaturation. When KI is grown from an aqueous solution at supersaturations below 10%, it is bounded by {100} faces. If the supersaturation is greater than 14%, the dominant faces are {111}. Both types of face are seen on crystals grown at supersaturations between 10 and 14%. The type of behaviour illustrated in Fig. 5.4 has been found for many crystals grown from aqueous solution, particularly by Kern and co-workers. Another example is that of Rochelle salt, which was studied by Belyustin and Dvoryakin (1958). They studied the relative growth rates of several faces and found maxima relative to {100} at different supersaturations.

The variation with *time* of crystal habit for faces inclined at different angles has been discussed by Alexandru (1969), who also considered the practical conditions for the growth of large crystals from aqueous solutions. He also pointed out the effect of the seed crystal shape on the final habit.

Since the parameter k of Eqn (5.1) depends on temperature, habit modifications will normally result from significant changes in the *growth temperature*. In general, experimental studies are not carried out at constant supersaturation and so do not distinguish between the (respective) effects of temperature and supersaturation. Gavrilova (1968) studied the effect of temperature on the morphology of magnesium sulphate $MgSO_4.7H_2O$. The changes in morphology of this substance were mainly changes in the relative elongation. The ratio of the length of the crystals along [100] to the distance between opposite {110} faces varied regularly from 7.8 at 22°C to 2.3. at 42°C. Gavrilova ascribes this effect to modifications in the structure of the solution, due to the change in concentration rather than to the effect of temperature on the growth kinetics. The $MgSO_4$—H_2O system is, however, untypical because of the large degree of solvation: crystallization above 48°C yields the hexahydrate $MgSO_4.6H_2O$. An example of separation of the effects of temperature and supersaturation on the habit for hydrothermal growth of quartz was given by Laudise (1958, 1959). The appearance of new faces, {111} and {012}, was observed on NaCl crystals by Honigmann (1952) when temperature fluctuations of 0.1 to 5°C were applied during growth, and similar observations had already been made in 1914 by Shubnikov on alum. The appearance of rough (non-equilibrium) faces is compared with the face development on spheres (Honigmann, 1958).

As discussed by Egli and Johnson (1963) a directed *flow of solution* around the growing crystal might be used to change the final shape. Alexandru (1969) found no significant difference in the habit of Rochelle salt ($NaKC_4H_4O_6.4H_2O$) crystals grown in static and in stirred solutions. Effects of solution flow have been discussed by Gülzow (1969). In the case of rapid solution flow and high supersaturation, the linear growth rate was found to vary with distance from the leading edge of the crystal. Two equivalent faces then will grow with different speeds as schematically shown in Fig. 5.5. In the "shadow" of the solution stream hopper growth might even occur, as will be discussed in Chapter 6. Gülzow also gives examples of morphology changes due to *defects* having a strong directional property, which will arise because of the increase in the rate of growth with the concentration of defects on certain crystal faces.

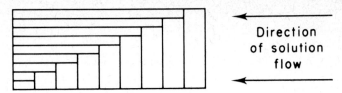

Fig. 5.5. Different growth rate of equivalent faces due to the direction of solution flow.

Twinning and especially the *twin-plane re-entrant edge* (TPRE) *mechanism* and their influence on the habit were described by Scheel and Elwell (1973) and in Section 4.14. Examples of habits of Al_2O_3, $BaTiO_3$ and BeO modified by TPRE are discussed below (Sections 5.5.1–5.5.3). Shlichta (1969) considered the effect on crystal growth of the strain energy associated with *dislocations*. This energy might be large enough to influence the step height of the layers advancing from the growth spirals. Thus crystal faces having the largest value of Burgers vector would have the highest surface energy and therefore the slowest growth rate. On the other hand, Shlichta presented evidence that the contribution of the dislocation strain energy to the morphology is at most rather small. Fordham (1949) measured the growth rates in the same crystallizer of strained and unstrained crystals of ammonium nitrate. He found that the strained crystals grew at a slightly faster rate on average, but the difference was not large compared with the scatter in the results for both sets of crystals.

Sheftal (1958) has proposed that habit changes can arise from differences in the concentration of *inclusions* near a crystal face, but this view does not appear to have received much support. Inclusions are normally seen as an effect rather than a cause of morphological changes.

Ultrasound may have an effect on the habit since the growth rate of various faces seems to be dependent on the direction, the frequency and the intensity of applied ultrasonic waves (Kapustin, 1963).

5.4. Effect of Impurities on Habit

A vast number of papers has been published on observations, largely qualitative, of morphological changes produced by the addition of small quantities of impurity to solutions. Buckley (1951) devotes a major section of his book to this topic, in particular to the effect of organic impurities on the growth of inorganic crystals from aqueous solution, and Mullin (1972) discusses in some detail the use of habit modification in the chemical industry. We shall consider here only briefly the effect of impurities on crystal morphology, and in Chapter 7 we shall discuss the distribution of impurities in solution-grown crystals (in connection with the preparation of doped crystals and of solid solutions).

An important fact to be explained in connection with impurity effects is that only very small traces, typically 0.01%, are often required to produce changes in habit. This means that many observed crystal habits may be caused by unsuspected impurity effects. The impurities normally have little effect on the dissolution rate as they are present in concentrations well below the solubility limit.

Marc (1908, 1912), Gille and Spangenberg (1927), Neuhaus (1928), and Bunn (1933) proposed that impurities form two-dimensional *complexes* on certain crystal faces and that these may become unstable three-dimensional complexes as growth proceeds. The latter complexes break up and have the effect of retarding growth on such faces through their interaction with the solute. On other faces the impurity forms stable complexes which are incorporated into the crystal and have little effect on the growth rate. The effect of the impurity on the crystal habit according to this argument arises through the reduction it causes in the supply of material to the crystal face.

More recent treatments normally consider that the impurity is adsorbed on the crystal surface and that it causes a reduction in the specific surface free energy $\Delta\gamma$. This is related to the temperature and to the packing density of adsorbed particles by the equation

$$\Delta\gamma = kTn_{S_0} \cdot \ln\left(1 - \frac{n_S}{n_{S_0}}\right) \tag{5.3}$$

which was derived by von Szyskowski from the Gibbs adsorption isotherm and the Langmuir equation. Here n_S is the actual number of adsorbed particles per square centimeter and n_{S_0} is the maximum number of adsorbed particles. Stranski (1956) has discussed this change in surface free energy with respect to the equilibrium shape regarding the Kossel crystal (see also Honigmann, 1958).

A recent experimental and theoretical study of the effect of adsorption applying the theory of Gjostein (1963) has been published by Burmeister (1971) for the case of silicon growth from the gas phase. In part these arguments can be applied to growth from solution if the pressures are replaced by the concentrations in the solution. Burmeister's equation for a rough surface is

$$\gamma_l = kTN_{ls} \cdot \ln\left(1 + \frac{p}{p_l^*}\right) \tag{5.4}$$

where γ_l is the surface (line) energy, N_{ls} the maximum number of sites available on ledges, p the actual pressure and p_l^* is a characteristic pressure. The relative density of adsorbed molecules is then

$$\frac{N_l}{N_{ls}} = 1 - \exp\left(-\frac{\gamma_S}{kTN_{ls}}\right) \tag{5.5}$$

with N_l the effective number of molecules adsorbed at ledges. The ratio of the line energy γ_l to surface energy is of the order of the lattice constant (Honigmann, 1958).

The review of Kern (1969) discusses two-dimensional adsorption compounds and the specific effects of certain impurities, such as Cd^{2+} ions for habit modification of NaCl, are ascribed to the similarity between the {111} planes of NaCl and the corresponding plane in the $CdCl_2$ structure. If an adsorption layer covers a whole surface of a crystal, growth of that face may be entirely suppressed. Cabrera and Vermilyea (1958) postulate that a fall in the growth rate will occur if the mean distance between strongly adsorbed impurity particles is comparable with the size of a critical two-dimensional nucleus for the corresponding supersaturation (see below). Chernov (1961, 1962) distinguishes between two effects of adsorbed impurities. If these are relatively small and mobile, the main effect will be to reduce the effective number of kinks. Relatively large and immobile impurities, such as organic dye molecules, act as an obstacle for the movement of surface steps. In the former case, the mean separation x_i of unoccupied kinks is increased compared with the value x_o for the same crystal in a pure solution by a relation

$$x_i = x_o + \epsilon n_i \tag{5.6}$$

where n_i is the relative concentration of impurity in the solution and ϵ is given by

$$\epsilon = \frac{a}{2V_I f^3} \cdot \left(\frac{kT}{2\pi m}\right)^{3/2} \exp \frac{W_S + W_{k_1} - W_{k_2}}{kT}. \tag{5.7}$$

Here V_I is the volume and W_S the energy of an impurity molecule, f the vibrational frequency of an impurity particle of mass m in the adsorbed state, W_{k_1} and W_{k_2} are, respectively, the energies of a free and an impurity-occupied kink. Chernov estimates $W_S + W_{k_1} + W_{k_2} \sim 40$ kJ/mole, $f \sim 3.10^{12}$ s^{-1}, so that $\epsilon \sim 10^4 a$. Thus an impurity concentration of $\sim 10^{-3}$ will increase the distance between impurity-free kinks by a substantial factor. The energy term $(W_S + W_{k_1} - W_{k_2})$ will normally depend on the orientation of the step and the resulting anisotropy of x_i will cause a corresponding anisotropy in the rate of advance of the steps. This would explain poly-gonization of the growth steps in the presence of impurities which was discussed in Section 4.12. A qualitative confirmation of Chernov's two effects of impurities as mentioned above was given by Slavnova (1958).

The effect of kink poisoning can be included in the BCF theory by introducing into Eqn (4.41) an additional factor C_o to allow for the relatively large separation between kinks in a step. The magnitude of this effect has been calculated by Chernov (1962), who found that it increased the nonlinear region of the $v(\sigma)$ variation as well as decreasing the absolute value of v compared to that for a pure system.

Sears (1958) also considered the poisoning of kink sites in a step by impurities. He proposed that the poisoning will only be effective if the impurity covers the whole step, otherwise new kinks are continuously generated by statistical fluctuations. Quantitative calculations are not presented but Sears also discusses the effects of poisons on the nucleation rate and on the spiral shape. In the example considered, potassium chloride, the growth behaviour is changed by only a few parts per million of lead chloride.

The effect of large impurity particles is to retard the growth of layers at the points of contact with the particles while these are captured into the steps. This *step pinning* was considered also by Cabrera and Vermilyea (1958). The condition for the step to re-form on the other side of the impurity particle is fulfilled when the separation between particles is greater than $2r^*$, where r^* is the radius of a critical nucleus. Cabrera and Vermilyea assume that the rate of advance v_{st} of the steps is given by

$$v_{st}/v_\infty = (1 - 2r^*d^{1/2})^{1/2} \tag{5.8}$$

where v_∞ is the step velocity in the absence of impurities and d the density of impurities on a two-dimensional lattice. If new impurity particles are flowing towards the step at a rate $\mathscr{J}_i$ then, just ahead of a step moving with a velocity v_{st}, with step density $1/y_o$,

$$d = \mathscr{J}_i y_o / v_{st}. \tag{5.9}$$

Combining (5.8) and (5.9) gives an equation in v_{st}^5 which has a solution only if $2r^*(\mathscr{J}_i y_o / v_\infty)^{1/2} < 0.54$. Thus the steps will flow only if

$$v_\infty / y_o > 14r^{*2}\mathscr{J}_i. \tag{5.10}$$

For small σ, $v_\infty / y_o \propto v$ (the linear growth rate) $\propto \sigma^2$ [Eqn (4.43)] and $r^* \propto 1/\sigma$ [Eqn (4.25)] and so growth will only occur if the supersaturation exceeds a minimum value given by

$$\sigma_{min}^3 \mathscr{J}_i = \text{constant}. \tag{5.11}$$

For high supersaturations where $v \propto \sigma$, this relation should be replaced by

$$\sigma_{min}^4 \mathscr{J}_i = \text{constant}. \tag{5.12}$$

The latter relation was confirmed experimentally by the studies of Price, Vermilyea and Webb (1958) on the electrolytic growth of silver whiskers in the presence of gelatine. Step pinning on sucrose crystals by raffinose impurities has been measured by Albon and Dunning (1962). As in the theory of Cabrera and Vermilyea, the results are interpreted by assuming

H

that pinning occurs when the distance between two impurity molecules in a step is less than $2r^*$.

Burrill (1972) has proposed a model based on the reduction in the area of crystal face available for adsorption of solute molecules due to the presence of impurity.

Mullin et al. (1970) considered that a mechanism of physical blocking of sites on the crystal surface is an oversimplification, and that impurity ions in the vicinity of the surface will retard growth by their interaction with the solvent even if they are not adsorbed on the crystal surface. The presence of impurities may reduce the effective supersaturation by a "dilution", retard diffusion, hinder aggregation of growth units and so a detailed description of their effect is likely to be highly complex.

Whatever the detail of the atomic or kinetic mechanism, impurities will clearly cause habit modification by the varying degree with which they inhibit growth on different faces. However, in some exceptional cases an increase in the growth rate of crystal faces due to impurities has been observed. Such an increase may be caused by a decrease in the surface energy which reduces the size of the critical nucleus. It is likely to occur when the increased surface nucleation rate more than compensates for the decrease in step velocity (Sears, 1958).

The effect of impurity addition is often highly beneficial to the quality of the crystals. For example, Egli and Zerfoss (1949) have pointed out that NaCl is difficult to grow from a pure aqueous solution because the super-saturation for the onset of nucleation is small. If Pb^{2+} ions are added, crystals grow very easily because the critical supersaturation for nucleation is increased and so growth can proceed at much higher supersaturations than in the pure solution. Egli and Zerfoss also noted that the impurities may not enter the lattice. High-quality ammonium dihydrogen phosphate crystals were grown from a solution containing 0.1% Fe at a rate ten times that in a pure solution, but iron could not be detected in the crystals. Wanklyn (1974) listed the cases where the presence of impurities in crystal growth from high-temperature solutions was observed to have a beneficial effect.

As mentioned in the introduction to this chapter, impurities may cause the appearance of faces which are not observed in pure solutions. Hartman (1969) has proposed that certain impurities will cause faces which are normally rough to become flat, due to adsorption of a layer of impurities over the face. This effect is illustrated in Fig. 5.6. Lateral growth is possible only at steps and the growth process is thus similar to that of a normal F face. The epitaxial layer of impurities effectively imposes its own PBC's on the face. The example discussed by Hartman is that of $Hg(CN)_2$ grown from methyl alcohol (Ledésert and Monier, 1965). Crystals grown

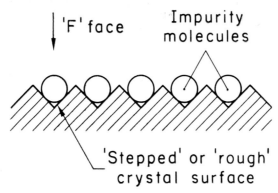

FIG. 5.6. Conversion of S and K faces by absorbed layer of impurity (schematic).

from this solvent exhibit {211} faces, which do not appear in growth by other methods and which are not F faces. This morphology is ascribed to the formation of an adsorption layer of CH_3OH molecules which give an arrangement resembling the flat {100} faces of the complex compound $Hg(CN)_2.CH_3OH$. The effect is of course due to the solvent itself and not to an impurity in this example, but the explanation could be extended to a number of impurity-dependent habit changes. However, for obvious reasons, it is extremely difficult experimentally and theoretically to describe the actual kinetic processes of the role of impurities in habit control.

5.5. Habit Changes in HTS Growth

The literature on the habit of crystals grown from HTS is too extensive for a detailed discussion on all materials. Examples of substances which have been investigated systematically or by several investigators will therefore be discussed. The examples described below illustrate the principles outlined in the previous sections, but some part of the interpretation is still speculative. An extensive tabulation of crystal habits and habit changes in flux growth has been given by Wanklyn (1974), and the influence of solvent and impurities on the habit of semiconductor crystals has been reported by Faust et al. (1968).

5.5.1. Aluminium oxide Al_2O_3

Extensive investigations of alumina, with and without the addition of Cr^{3+} to form ruby, have been made because of its application in many fields and its interesting properties. According to Timofeeva and Luky-anova (1967), the faces which are found on crystals grown from PbF_2 and from $PbO/PbF_2/B_2O_3$ solutions are listed in Table 5.2, together with their

TABLE 5.2. Relative Growth Rates of the Most Common Faces of Alumina (after Timofeeva and Lukyanova, 1967)

Face	Terminology	Estimated relative growth rates
{001}	Pinacoid	1
{101}	Main rhombohedron	10
{012}	Small rhombohedron	6
{223}	Hexagonal bipyramid	8
{110}	Prism	20

approximate relative growth rates. Since the {001} face has the slowest growth rate, alumina crystals tend to grow with a plate-like habit with the more rapid growth perpendicular to the c axis. Timofeeva and Lukyanova also examined the structure of the faces of their crystals by optical microscopy. They found growth steps on the faces of the three slowest growing faces, and low-angle boundaries on the {101} and {110} faces. No other faces appear to have been seen on Al_2O_3 crystals grown by other workers except by White and Brightwell (1965) who observed {001}, {012}, {104}, {113} and {125} faces (all hexagonally indexed). Of these the latter three are different from those in Table 5.2 and seem to be transient faces. White and Brightwell also mentioned an interesting habit change due to the influence of temperature. Below 1250°C, {001} plates with minor {012} and {104} were predominant, whereas at higher temperatures more equant and inclusion-free crystals with {001} and larger {012} faces were grown. The latter crystals were almost free from the multiple twinning effects observed in the "low-temperature" thin plates. Therefore the presence of twins seems to account for preferred nucleation sites and growth perpendicular to the c axis. This was confirmed later by Wallace and White (1967) who examined several plate-like crystals using X-ray diffraction topography. The twinning consists of a 180° rotation in the basal plane.

Linares (1965) found that his plate-like crystals were bounded only by {001} and {223} faces. He also reported that the {110} is the fastest growing face and gave the ratio of the growth rates of {223} and {001} faces as 100 : 1. The tendency of crystals to grow as thin plates is reduced when growth occurs on the walls of the crucible, well below the surface of the melt. The enhancement of growth in the direction of $\langle 001 \rangle$ was attributed to conduction of heat of crystallization through the crucible walls, which favoured growth of {001} until the {223} faces became sufficiently established to provide an appreciable area for the dissipation of heat. Increasing the cooling rate from about 0.5°C per hour to 1–5°C per hour was reported to favour the formation of more equidimensional crystals.

Nelson and Remeika (1964) had also found a tendency for crystal plates to grow near the surface and for more equidimensional crystals with fewer defects to grow near the bottom of the solution. Further confirmation is obtained from the work of Adams, Nielsen and Story (1966) who imposed a steep temperature gradient on their crucible in order to encourage growth near the base. Izvekov et al. (1968) confirmed the observation mentioned above that Al_2O_3 crystals grown at high temperatures (here above $1135°C$) have a more equidimensional habit and attributed this effect to increased surface roughness at higher temperatures. The same tendency was found by Champion (1969), who compared the habit of crystals grown by slow cooling, evaporation and gradient transport from lead flouride. Changing the method of growth had little effect apart from the trend towards thicker crystals at higher temperatures.

Janowski et al. (1965) and Chase (1966) found that the addition of La_2O_3 to alumina growing from a PbF_2—Bi_2O_3 solvent greatly reduced the incidence of plate-like crystals. Its effect is to slow down the growth rate of {101} and {102} faces so that these become predominant rather than {001}. The {101} faces are dominant at 0.1–0.2% La_2O_3, and {012} at about 1%. An increase of the impurity beyond 1.5% results in the appearance of irregular {110} faces and in deterioration in quality of the crystals. La^{3+} ions are found to enter the crystal in concentrations up to 0.9%. It is concluded by Chase (1966) that La^{2+} is incorporated into the lattice. However, the incorporation of such a high concentration of the large La^{3+} ion should shift the lattice constants by 0.01 to 0.03 Å, and it is a pity that this was not determined. It seems more probable that a large fraction of the La^{3+} and F^- ions was incorporated as flux inclusions. These can be very tiny and undetectable by the unaided eye, as was demonstrated by Linares (1965). Lanthanum additions probably act by slowing down the rate of step motion on {101} and on {012} faces but are rejected at steps on the former face and enter the lattice at steps or other sites in the latter. Crystals of high optical quality were grown with 0.5% La_2O_3 concentration in the solution.

Similar observations have been made by Scheel and Elwell (1973) during a systematic study of conditions for growth of $GdAlO_3$ and $LaAlO_3$. In Fig. 5.7. a twin of alumina with the twin plane (101) and with typical re-entrants, as grown from a solution of equal amounts (by weight) of Gd_2O_3 and Al_2O_3 in PbO—PbF_2—B_2O_3 flux, is shown.

Some typical habits of Al_2O_3 are shown in Fig. 5.8.i, and the corresponding indexed drawings in Fig. 5.8.ii. In addition to the above-mentioned habits of {001} plates and combinations of {001} with {221}, a steep

Fig. 5.7. Twinned alumina crystal with re-entrants (Scheel and Elwell, 1973).

pyramid grown from a La_2O_3-containing solution and a typical natural crystal with dominant {110} and {001} faces are shown. The latter habit is frequently found in metamorphic rocks or sediments, and often barrel-like corundum crystals with dominating {221} and minor {001} faces are found in nature. The crystal projections of Fig. 5.8.ii were drawn by computer with a program of Keester (1972).

Chase and Osmer (1970) attempted to correlate morphological changes with the composition of the solvent. Since most crystals had been grown from PbF_2 or mixtures of PbF_2 with other salts, they investigated the growth habit in solvents of different compositions in the systems PbF_2—PbO and PbF_2—MoO_3. Flat plates were found in solutions rich in PbF_2, and crystals tended to become more rhombohedral as the oxide content in the solvent was increased above 50 mole %. Chemical analysis revealed the presence of Pb^{2+} and F^- ions in the platy crystals. The habit was believed to be related to the presence of AlF_6^{3-} ions in the melt. F^- ions entering the crystal enhance the adsorption of Pb^{2+} and so slow down the growth rate, especially of the {001} face which always grows by the lateral propagation of growth layers.

Yanovskii et al. (1970) showed that alternative solvents can be found for the growth of non plate-like alumina crystals. They used various alkali and

(i)

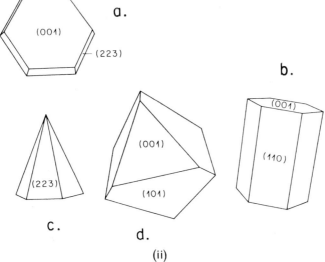

(ii)

FIG. 5.8. Habits of alumina crystals. (i) Actual crystals, (ii) corresponding computer graphs: (a) hexagonal plate {001}/{111}, (b) hexagonal prism {110}/{001}, (c) pyramid {223}, (d) typical isometric Al_2O_3 crystal with {001} and {101}.

alkaline-earth tungstates and found that the crystals grew with a bipyramidal habit with only {223} faces, occasionally modified by small {101} facets. The dominance of the {223} faces was attributed to adsorption of a layer of tungstate ions on this face, which is the most favourable for epitaxial adsorption of WO_4 tetrahedron chains. Addition of cryolite, Na_3AlF_6, to a sodium tungstate solvent resulted in the reappearance of {001} faces so that the crystals were again of plate-like habit. Thick plates with {001} and {012} faces were grown from pure cryolite by Arlett *et al.* (1967).

The influence of the growth mechanism, impurities, type of solvent, and temperature on the Al_2O_3 habit is not yet clearly understood. However, the twin growth mechanism of White and Brightwell seems to provide a plausible explanation of the platy habit. In this connection it should be mentioned that other growth mechanisms have been deduced from surface features. Although Wallace and White (1967) did not find growth spirals on a (001) plane of an alumina plate, Sunagawa (1967) observed hillocks, apparently originating from screw dislocations, on the same plane. This indicates that alumina plates often grow rapidly perpendicular to the c axis by the twin-plane reentrant-edge mechanism (see Sections 5.3 and 4.14) and slowly along $\langle 001 \rangle$ by a screw-dislocation mechanism.

The habit changes of alumina were discussed in detail since here is a very good example of the effect of the various parameters. Also it was demonstrated how by proper choice of solvent, dopant, supersaturation and growth temperature, Al_2O_3 crystals which are untwinned, free from inclusions, of equidimensional shape and low in impurity content, can be obtained from flux.

5.5.2. Barium titanate $BaTiO_3$

The perovskite-type compound barium titanate is one of the most interesting ferroelectric compounds and has been studied in great detail. Its study was particularly stimulated when Blattner, Känzig, Matthias and Merz in 1947, 1949 and Remeika in 1954 grew the first crystals suitable for physical measurements. Of the various phase transitions, that occurring at 1460°C from the high-temperature hexagonal to the cubic form is of particular importance to the crystal grower, since crystals should be grown below this temperature. The tetragonal ferroelectric phase has a Curie temperature of 120–130°C.

The dominating habits of $BaTiO_3$ crystals grown from various solvents are the pseudo-cube with {100} faces and the so-called *butterfly twin*. The latter is preferred for physical studies and is schematically shown in Fig. 5.9. These typically twinned $BaTiO_3$ crystals were first grown to a considerable size by Remeika (1954) using potassium fluoride as solvent. The

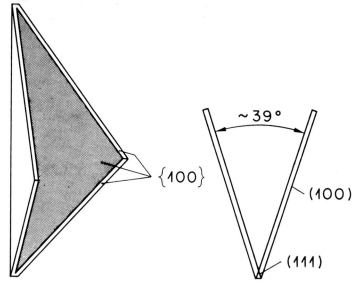

FIG. 5.9. Butterfly twin of barium titanate (after Nielsen *et al.*, 1962).

crystallographic features of the twin were described by White (1955) and by
Curien and LeCorre (1955). All faces are {100} and the twin plane is (111).
Small differences are reported in the literature for the value of the angle
between the "wings", but it is always 39° within a few minutes.

The conditions necessary for the formation of butterfly twins have been
studied by De Vries (1959) and by De Vries and Sears (1961), and have
been summarized by Nielsen, Linares and Koonce (1962). The presence of
excess $BaTiO_3$ powder at the onset of growth is most important, otherwise
cubes are formed. Nielsen *et al.* found that the yield of twins depended
strongly on the size of the undissolved particles and concluded that growth
originated on particles of micron dimensions. The nuclei for twin formation
must have {111} faces exposed and may be twinned themselves.

A high supersaturation is also necessary for the growth of the butterfly
twins. Timofeeva (1959) found that twins formed on cooling a solution of
$BaTiO_3$ in $BaCl_2$ at 20°C/h, but that more equidimensional crystals grew
when the cooling rate was 4°C/h. Sasaki (1964) reported that the rapid
lateral growth occurred mainly at temperatures above 1000°C, and that it
was difficult to grow butterfly twins below this temperature. Sasaki and
Kurokawa (1965) observed that the yield of butterfly twins depended on the
temperature gradient across the crucible and that the yield was highest with
the temperature higher at the base of the crucible.

Since it is difficult to control the concentration and size distribution of

H2

undissolved particles remaining in a solution after the initial soak period, the size and yield of butterfly twins are sensitive to minor changes in experimental procedure and to the nature of the barium titanate powder. Impurities in small concentrations have relatively little effect (Nielsen *et al.*, 1962; De Vries, 1959; Sholokhovich *et al.*, 1968) except for lanthanum oxide which is very effective in reducing the yield of twins. De Vries (1959) noted the $BaTiO_3$ faces which developed under various growth conditions ($BaTiO_3$—KF ratio, soak temperature, cooling rate) and also occasionally observed hexagonal $BaTiO_3$ crystals which seem to be stabilized by replacement of Pt^{4+}, Zr^{4+}, Au^{3+} or Fe^{4+} for Ti^{4+}. Hexagonal plates of composition $BaTi_{0.75}Pt_{0.25}O_3$ were obtained by Blattner *et al.* (1947, 1949).

A completely different morphology is obtained by the method used by Linz *et al.* (von Hippel *et al.*, 1963; Belruss *et al.*, 1971) in which $BaTiO_3$ crystals are pulled from a melt containing excess TiO_2 (see Section 7.2.7). Growth on a seed occurs at temperatures between 1396° and 1335°C at a rate of about 0.25 mm/h. The preferred orientation of the seed crystal is [110]. Figure 5.10 shows a typical habit of a $BaTiO_3$ crystal grown by this

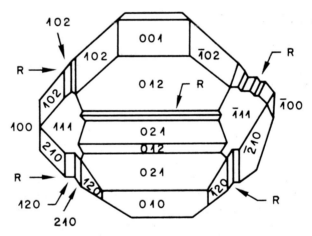

FIG. 5.10. Habit and reentrants (R) of $BaTiO_3$ grown by top-seeded solution-growth technique, growth direction {111} (von Hippel *et al.*, 1963).

top-seeded solution-growth technique. The fully grown crystals exhibit mainly {210}, {100} and {111} faces, with {210} being the most developed. The crystals also display reentrants (indicated by *R*) which are reminiscent of the (111) twinning in the butterfly twins, but are due to alternating (210) and (120) faces.

Attempts to grow butterfly twins of other perovskites such as $CaTiO_3$, $SrTiO_3$ and $PbTiO_3$ failed, and it was proposed that this fact could be correlated with the nonexistence of a hexagonal phase at high temperatures (Nielsen *et al.*, 1962).

5.5.3. Beryllium oxide BeO

Detailed studies of the growth of BeO from lithium molybdate and other solvents have been reported by Austerman (1964) and by Newkirk and Smith (1965). As in the examples described previously, BeO shows a rather complex set of habit changes which are influenced by the type of solvent, impurities, temperature and supersaturation. The effect of growth temperature and of the solvent composition in the system Li_2MoO_4—MoO_3 on the habit is shown in Fig. 5.11. The principal habits are seen to be plates, prisms and pyramids, often showing the {101} pyramid face which is seen on most crystals. The prismatic and plate crystals are bounded by {101}, {100} and generally by the (00$\bar{1}$) basal plane which by definition is the oxygen side of the polar BeO structure.

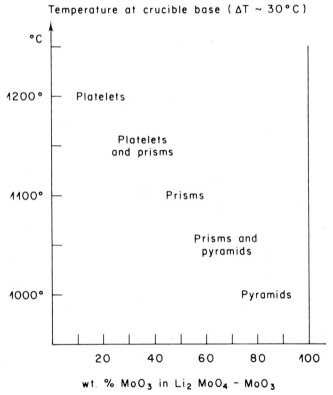

Fig. 5.11. Habits of BeO as a function of growth temperature and flux composition (after Austerman, 1964).

Many crystals seem to exhibit *twinning*, a common form being indicated by a core of reverse polarity running through a prismatic crystal. This core terminates in a small pyramid, bounded by {101} and {001} faces, which projects from the centre of the $(00\bar{1})$ face. Austerman suggests that the twinning is a discontinuity in the beryllium layers, with the oxygen layers continuing across the twin boundary. This twinning mechanism is shown in Fig. 5.12. The energy to form a twin is presumably very small, and the prevalence of twins is perhaps responsible for the large variety of growth forms.

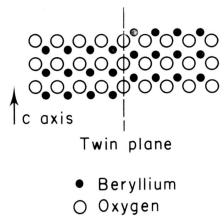

c axis

Twin plane

● Beryllium
○ Oxygen

FIG. 5.12. Twinning mechanism in BeO (schematic).

The number of crystal forms may be further increased by the addition of impurities. Austerman (1964) has noted a dozen or so different faces, and mentions in particular {111}, {102}, {20$\bar{1}$}, {12$\bar{3}$} and {11$\bar{1}$} in addition to the more common faces mentioned above. His paper gives drawings of nineteen different habits observed after adding phosphate, borate or silicate to the lithium molybdate flux or after replacing lithium by potassium. Less regular crystals were due to asymmetric twinning. Newkirk and Smith (1965) also found that borates, silicates and phosphates have the most marked effect on morphology. The addition of phosphate seems to reduce twinning and to improve the crystal quality, whereas Austerman (1965) devoted a paper to the detrimental effect of silica impurities.

Linares (1967) has grown BeO crystals from sodium borate fluxes in which the solubility is much higher than in the alkali molybdates. Prismatic crystals are formed in $NaBO_2$ and rod-shaped crystals in $Na_2B_4O_7$ when cooling rates of 1–5°C/h are used. On increasing the cooling rate to 25°C/h, hollow rods crystallize, and *whiskers* may be grown when a cooling rate of 100°C/h is applied.

Newkirk *et al.* (1967) hoped to find the actual growth mechanism by a careful electron-microscope study of the features. Flat cones (growth hillocks) were detected on the singular (001) face, and the height of the spreading layers was assumed to be below 250 Å, the resolution of the replica technique used. However, the twin boundary evidently showed a multifacetted structure and therefore offered a much larger numbei of reentrant sites. The (00$\bar{1}$) face has vicinal character and appeared highly convoluted near the twin boundary where the lateral motion and traffic of growth steps was greatest. Farther from the boundary, convolutions became more widely spaced as growth layers annihilated and reinforced one another. Thus, as in the growth of the twinned alumina plates described above, a combined twin reentrant-edge and screw-dislocation mechanism can be assumed to be the growth mechanism for the majority of the flux-grown BeO crystals.

5.5.4. Cerium oxide CeO$_2$ and thorium oxide ThO$_2$

Ceria and thoria are highly refractory materials both of which have the cubic fluorite structure. They have been grown from high-temperature solutions as crystals with similar habits and are included here to illustrate the fact that similar considerations can be applied to related groups of materials, although significant differences are often observed between members of a group.

Crystals of CeO$_2$ were probably first synthesized from high-temperature solution by Nordenskiöld (1860, 1861) whose crystals from borax solution showed a combination of {100} with {111} faces, rarely with minor {110}. These observations were confirmed by Grandeau (1886) and by Sterba (1901) who observed the same habit in CeO$_2$ crystals grown from NaCl, borax, and K$_2$SO$_4$ solutions.

Finch and Clark (1966) reported that crystals grown from Li$_2$O—WO$_3$ solvents had an octahedral habit. Linares (1967), in a systematic study of the effect of different solvents on the habit and quality of CeO$_2$ crystals, obtained octahedra from Na$_2$O—B$_2$O$_3$ solvents with Na/B$\leqslant$1.0 and from Li$_2$O—B$_2$O$_3$—MoO$_3$ with Mo/B$\leqslant$1.0. The PbO—PbF$_2$—B$_2$O$_3$ solvents always gave CeO$_2$ cubes, which are usually characterized by large regions of lamellae with flux *inclusions* between these lamellae. Despite the higher solubility of CeO$_2$ in PbO—PbF$_2$ solvents, the crystallization of inclusion-free CeO$_2$ crystals seems to necessitate the use of fluxes based on molybdates or alkali borates. CeO$_2$ grown by Wanklyn (1969) from various PbF$_2$-based solvents showed mainly {100} faces, as was also described by Zonn and Joffe (1969), who obtained cubes from a 10 PbF$_2$: 1 PbO solvent.

According to Linares (1967) crystals of thoria and of ceria grown from various solvent compositions resemble each other as regards habit. The tendency of thoria to form a cubic shape was stronger than with ceria but fewer inclusions were observed in thoria. Finch and Clark (1965) found that the addition of about 1% B_2O_3 to a Li_2O—WO_3 flux had a beneficial effect on the solubility and the growth behaviour. They do not discuss the morphology but show a crystal which appears to be mainly octahedral in shape. Chase and Osmer (1967) obtained inclusion-free thoria cubes from Bi_2O_3—PbF_2 and from PbF_2 solvents; good quality was obtained particularly by the gradient transport technique. Scheel (unpublished) has confirmed the observations of Linares (1967) and found no influence of small U-, Sb- and Bi-additions on the habit. However, from a flux containing 73% PbF_2, 21% Sb_2O_3, 3% NaF and 3% B_2O_3, yellow octahedral thoria crystals were formed together with black octahedra having the pyrochlore structure.

5.5.5. Yttrium garnets $Y_3Al_5O_{12}$, $Y_3Ga_5O_{12}$, $Y_3Fe_5O_{12}$

Since the discovery of the ferrimagnetic yttrium iron garnet (YIG) by Bertaut and Forrat (1956) and by Geller and Gilleo (1957), the interest in the garnets of yttrium and the rare earths has grown very rapidly, resulting in widespread activity on the crystal growth of these compounds. The structure of the rare-earth garnets is similar to that of natural silicate garnets. The cubic unit cell with a lattice constant of the order of 12 Å contains eight formula units. The rare-earth ions occupy irregular dodecahedral sites, the smaller Al^{3+}, Ga^3, Fe^{3+}, etc. ions being distributed on octahedral and tetrahedral sites. An extensive review of the crystal chemistry of garnets has been published by Geller (1967).

It is normally found that {110} and {211} faces are dominant on flux-grown crystals. The {100} habit reported by Timofeeva (1959, 1960) has not been confirmed since, and the crystals shown were possibly orthoferrite pseudo-cubes or hematite rhombohedra, as were found among the crystallization products of Titova (1962). However, Timofeeva (1971) claimed that garnets usually show {100} and {111} after spontaneous nucleation and only later develop the slowly growing {110} and {211} faces, especially in viscous solutions. Nielsen and Dearborn (1958) found that {110} faces of YIG were dominant when the crystals were grown slowly and that {211} became the most important faces when the cooling rate was increased to 5°/h.

Drawings of garnet habits prepared by computer (Keester, 1972) are shown in Fig. 5.13. Rhombendodecahedral faces {110} (a), ikositetrahedral faces {211} (b), and combinations of {110} with {211} (c) and of {110}, {211} and the rare {321} faces are the dominant garnet habits.

Lefever et al. (1961) showed that, for YIG, the ⟨100⟩ are the directions of rapid dendritic growth following nucleation. The ⟨111⟩ are also

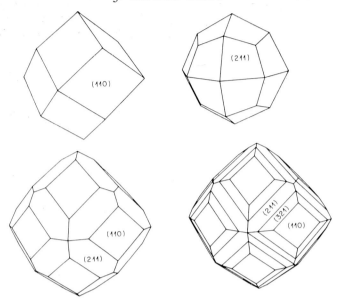

FIG. 5.13. Computer drawings of the garnet habits. (a) {100}, (b) {211}, (c) {110} + {211}, and (d) {110} + {211} + {321}.

rapid growth directions, but not so rapid as $\langle 100 \rangle$. Lefever and Chase (1962) examined the surface features of the {110} and {211} faces, which showed growth steps spreading from hillocks and which appeared to grow by a screw-dislocation mechanism. Gendelev (1963) proposed that the relative dominance of {110} and {211} faces depends mainly on the ratio of Y_2O_3 to Fe_2O_3 in the layer of solution adjacent to the growing crystals. In the (110) plane, the ratio of Y^{3+} to Fe^{3+} ions is 0.33 : 1, while in the (211) plane it is 0.60 : 1. A high Y^{3+} : Fe^{3+} ratio in the melt might therefore increase the relative growth rate of the {211}, and so the {110} would become more dominant. Similarly an increase in the Fe^{3+} concentration would favour dominance of the {211} faces. The dominance of {211} on crystals grown at higher cooling rates is explained by the lower solubility of Y_2O_3 and the lower mobility in the solution of the relatively large Y^{3+} ions. Gendelev discusses the influence of the direction of convective solution flow on the morphology, {211} faces dominating on one half of a crystal in which the supersaturation was believed to be highest, and {110} on the opposite side of the same crystal where the solute flow and hence the supersaturation were lower. He concludes that the quality of the YIG crystals depends directly on the dominance of the {110} faces, since these appear to be the slowest growing faces under conditions of low supersaturation and adequate supply of both Y^{3+} and Fe^{3+} ions. However, Tatarskii (1964) criticizes the explanation of the habit changes of YIG in terms of local differences in concentration.

Large crystals of YIG and of YIG–YGaG solid solutions grown by Scheel (unpublished) under the stirring action of the accelerated crucible rotation technique (Scheel, 1972) exhibit a dominance of the {110} faces and are free from flux inclusions. The large YIG crystals grown at the crucible base by Grodkiewicz et al. (1967) are dominated by {110} faces, but the crystals contain large regions with flux inclusions.

The morphology of yttrium aluminium garnet (YAG) is discussed by Gendelev and Titova (1968). Crystals of this substance grown from lead-based solvents are normally bounded only by {110} faces. The {211} faces, which appear only occasionally, must therefore have a relatively rapid growth rate. The development of {211} can be increased, as in the case of YIG, by high cooling rates, but crystals with dominant {211} faces usually contain a high concentration of inclusions.

Chase and Osmer (1969) found that YAG crystals grown at the bottom of the crucible have a pure {110} habit but those grown on the melt surface show additional {211} faces. Yttrium gallium garnet (YGaG) crystals have the {211} faces more strongly developed, and grow with a pure {211} habit from a melt rich in Y_2O_3 or PbO. The {110} faces are present when growth occurs in a melt rich in Ga_2O_3 or PbF_2. YAG was found to exhibit a similar dependence on flux composition, growing with a pure {110} habit from a melt rich in PbF_2 or Al_2O_3, and with {110} modified by {211} in a PbO or Y_2O_3 rich melt. The crystals grown from a melt rich in PbF_2 and Al_2O_3 or Ga_2O_3 have much more incorporated lead impurity than those grown from a melt rich in PbF_2 or Y_2O_3. This suggests that the lead, replacing yttrium in the garnet structure, changes the relative surface energies of the {110} and {211} faces, and so retards the relative growth rate of {110}. This tendency is contrary to Gendelev's explanation for YIG and some explanation other than the relative concentration of Y^{3+} in the two types of face must be considered. Chase and Osmer (1969) propose that the habit is mainly governed by the formation of a complex between PbF_2 and yttrium ions and that this complex modifies the behaviour of the surface diffusion or growth step propagation in some way leading to incorporation of the complex. This suggestion has not been substantiated by further evidence, and sytematic studies of the growth habit of garnets from other solvents have not been performed.

The habit of the yttrium garnets seems to be insensitive to the addition of impurities to the solution. Various substitutions have been made in these materials and the habit remains dominated by {110} and {211}, although {321} and other faces have been observed on naturally occurring silicate garnets. However, Wolfe et al. (1971) and van der Ziel et al. (1971) described a *facet-related site selectivity* for rare-earth ions in YAG. It is probable that such site preference accounts for the facet-related anisotropy in magnetic garnets as indicated by Callen (1971) and by Bobeck et al. (1971). These noncubic magnetic properties, introduced during growth, and therefore the habit and the growth direction in liquid phase epitaxial growth of garnets are of considerable importance in the development of magnetic bubble domain devices for logic and memory applications.

5.5.6. Zinc sulphide ZnS and cadmium sulphide CdS

Cubic and hexagonal polymorphs of zinc sulphide (zincblende and wurtzite) have been grown from high-temperature solutions. The cubic-hexagonal transition region at about 1020°C and its relationship to crystal habit in ZnS crystals grown by sublimation was studied by Hartmann (1966).

As listed in Table 5.3, cubic zinc sulphide crystallizes from flux as tetrahedra, octahedra and plates, whereas the hexagonal ZnS grows as prisms and plates. Linares (1968) found cubic {111} plates when $PbCl_2$ solutions of ZnS were cooled rather quickly (5° to 10° per hour), whereas at cooling rates of 1°C/h octahedral ZnS crystals were formed. Mita (1962) tried a number of compounds as solvents for flux growth of zinc sulphide. He obtained crystals only from NaCl, KCl, NaBr, NaI, KI and $CaCl_2$, whereas from $ZnCl_2$, K_2S and other salts no visible crystals were obtained, contrary to crystal-growth experiments of Parker and Pinnell (1968). Mita found that crystal size is closely related to the solubility of ZnS in the various solvents, and observed needles (hexagonal prisms {100} with pyramidal end faces {101}) in all solvents at crystallization temperatures of approximately 1050°C and, in addition, {001} plates of the hexagonal zinc sulphide grown from NaCl and KCl fluxes.

Parker and Pinnell (1968) reported systematic experiments intended to optimize the conditions for growth of cubic ZnS. They used a horizontal gradient transport technique and pure KCl as well as mixtures of KCl with $ZnCl_2$ and KI with $ZnCl_2$, $CdCl_2$ and $PbCl_2$ in order to obtain clear crystals up to 1 cm in size. With potassium chloride as solvent, dendritic platelets were formed at temperatures below 800°C, thin platelets from 800–830°C and thicker platelets at temperatures above 830°C. However, the crystals grown from KCl were small (~ 1 mm.). From mixed solvents (e.g. 20% KI, 80% $ZnCl_2$, $t = 845$°C, $\Delta T = 2$°) clear, large and more equidimensional crystals were obtained. These observations were attributed to complex formation in the solution.

Scheel (1974) obtained colourless tetrahedra and hollow crystals of cubic zinc sulphide by cooling sodium polysulphide solutions of ZnS.

Only the hexagonal wurtzite-type modification of CdS was obtained by

TABLE 5.3. Observed Habit Changes in Flux-grown ZnS and CdS

	Habit	Size (mm)	Solvent, exp. conditions	References
Cubic ZnS	Tetrahedra	$1 \times 1 \times 1$	KCl, 850°	Parker and Pinnell (1968)
			Na_2S	Doelter (1890, 1894)
		$1 \times 1 \times 1$	Na_2S—S, 500–600°	Scheel (1974)
	Octahedra	0.1	K_2CO_3	Schneider (1873)
		3	$PbCl_2$, 1°/hr	Linares (1968)
	Plates {111}	0.1–0.5	K_2S, Na_2S	Malur (1966)
		3×3	Ba_2ZnS_3, 1300°, 1.2°/hr	Malur (1966)
		$5 \times 5 \times 1$	$PbCl_2$, 5–10°/hr	Linares (1968)
		10	$ZnCl_2$—KI, 845°, $\Delta T = 2°$	Parker and Pinnell (1968)
Hexagonal ZnS	Prisms {100}	$10 \times 1 \times 1$	$CaF_2 + BaS + ZnSO_4$ NaCl, KCl, 1070–1200°→ 900–950°C	St. Claire-Deville and Troost (1861) Mita (1962)
		5×0.1	NaBr, NaI, KI, $CaCl_2$, 1050°C, 5°/hr	Mita (1962)
	Plates {001}	5×0.2	K_2S	Malur (1966)
			NaCl, KCl	Mita (1962)
Hexagonal CdS	Prisms {100} with minor {101} or {001}		$CaF_2 + BaS + CdO$	St. Claire-Deville and Troost (1861)
			$CdCl_2$	Bidnaya et al. (1962)
		$15 \times 1 \times 1$	Na_2S—S, 600°	Scheel (1974)
	Prisms {100} and pyramids {101}		$K_2CO_3 + S$	Schüler (1853)
			Na_2S—S, 500°, 5 at.	Scheel (1974)
			Na_2S—K_2S—S	Scheel (1974)
	Plates {001} with minor {100} or {101}		$CdO + BaS + CaF_2$	St. Claire-Deville and Troost (1861)
			$K_2CO_3 + S + C$	Schüler (1853)
			$CdCl_2$, 800–900°C	Bidnaya et al. (1962)
		$10 \times 10 \times 0.5$	$PbCl_2$, 800°→500°C	Linares (1968)
		$5 \times 5 \times 0.5$	Na_2S—S, 750–800°→ 400°	Scheel (1974)

crystallization from high-temperature solutions. Cubic CdS prepared by precipitation from aqueous solutions (Jackson, 1969) seems to transform to the hexagonal phase at about 310–370°C according to Hartmann (1966). However, Cardona et al. (1965) described the epitaxial vapour deposition of cubic CdS on the arsenic face of gallium arsenide at 710–730°C.

The morphology of hexagonal CdS grown from the vapour phase was

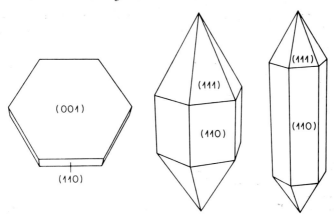

FIG. 5.14. Computer drawings of the main habits of flux-grown CdS: (a) hexagonal plates {001}/{110}, (b) and (c) prisms and pyramids.

studied in detail by Woods (1959), Bulakh (1969) and Kaldis (1969). In flux-grown CdS, a similar variety of habit change has been observed, as may be seen from the Table 5.3 and Fig. 5.14. Linares (1968) and Bidnaya *et al.* (1962) obtained {001} plates when crystallization from $PbCl_2$ and $CdCl_2$ fluxes, respectively, started at relatively high temperatures, of the order of 800°C. Scheel (1974) also obtained hexagonal plates when sodium polysulphide was used as solvent at high temperatures, whereas at lower temperatures, and with the addition of K_2S to the Na_2S—S flux, a prismatic habit with pyramidal end faces is favoured. However, this temperature relationship may not be significant, since the supersaturations are not known.

5.6. Summary on Habit Changes

Only real habit modification, that is the appearance and the change in relative importance of various types of faces on sound crystals, has been discussed in this Chapter. Unusual growth forms such as dendrites, hopper crystals, hollow crystals, whiskers, etc. will be discussed in the next Chapter. The few typical examples chosen demonstrate the influence of the various parameters on the habit. Information on other compounds may be obtained from the original literature listed in the table of flux-grown crystals in Chapter 10. However, the phenomenon of habit change should not be generalized. For instance, there are substances which do not change their habit at all such as compounds of the spinel type which generally grow as {111} octahedra.

It has been shown that habit, growth mechanism, incorporation of inclusions and impurities, and growth conditions are closely interrelated.

The study of the causes of habit modification is and will remain an interesting field for further research because, though not generally considered of great importance, habit often plays a vital role in the growth of large high-quality crystals and in their economical use. It is therefore proposed that crystal-growth publications should include observations on the morphology of flux-grown crystals for the benefit of future development in this field.

References

Adams, I., Nielsen, J. W. and Story, M. S. (1966) *J. Appl. Phys.* **37**, 832.
Albon, N. and Dunning, W. A. (1962) *Acta Cryst.* **15**, 474.
Alexandru, H. V. (1969) *J. Crystal Growth* **5**, 115.
Arlett, R. H., Robbins, M. and Herkart, P. G. (1967) *J. Am. Ceram. Soc.* **50**, 58.
Austerman, S. B. (1964) *J. Nucl. Mat.* **14**, 225.
Austerman, S. B. (1965) *J. Am. Ceram. Soc.* **48**, 648.
Belruss, V., Kalnajs, J., Linz, A. and Folweiler, R. C. (1971) *Mat. Res. Bull.* **6**, 899.
Belyustin, A. V. and Dvoryakin, V. F. (1958) *Growth of Crystals* **1**, 139.
Bennema, P. (1973) In "Crystal Growth" (P. Hartman, ed.) North-Holland, Amsterdam.
Bertaut, E. F. and Forrat, F. (1956) *Compt. Rend.* **242**, 382.
Beudant, F. S. (1817, 1818) Recherches sur les causes qui determinent les variations des formes cristallines d'une même substance minerale. See also *Ann. chim. phys.* **8**, 5; and *Ann mines* **2**, 10.
Bidnaya, D. S., Obukhovskii, Y. A. and Sysoev, L. A. (1962) *Russian J. Inorg. Chem.* **7**, 1391.
Bienfait, M. and Kern, R. (1964) *Bull. Soc. Franc. Min. Crist.* **87**, 604.
Blattner, H., Matthias, B. and Merz, W. (1947) *Helv. Phys. Acta* **20**, 225.
Blattner, H., Känzig, W. and Merz, W. (1949) *Helv. Phys. Acta* **22**, 35.
Bobeck, A. H., Smith, D. H., Spencer, E. G., Van Uitert, L. G. and Walters, E. M. (1971) *IEEE Transact. Magnet.*, INTERMAG Conf., 461.
Born, M. (1923), Atomtheorie des festen Zustandes, 2nd Edition (Leipzig, Berlin), p. 538.
Bragg, W. H. and Bragg, W. L. (1913) *Proc. Roy. Soc. London* **88**, 428; **89**, 246.
Bravais, A. (1866) Etudes Cristallographiques (Paris).
Buckley, H. E. (1951) "Crystal Growth" Wiley, New York.
Bulakh, B. M. (1969) *J. Crystal Growth* **5**, 243.
Bunn, C. W. (1933) *Proc. Roy. Soc. A* **141**, 567.
Burmeister, J. (1971) *J. Crystal Growth* **11**, 131.
Burrill, K. A. (1972) *J. Crystal Growth* **12**, 239.
Burton, W. K. and Cabrera, N. (1949) *Disc. Faraday Soc.* **5**, 33.
Burton, W. K., Cabrera, N. and Frank, F. C. (1951) *Phil. Trans. Roy. Soc.* A**243**, 299.
Cabrera, N. (1959) In "Metal and Semiconductor Surfaces" (H. C. Gates, ed.) p. 71, Wiley, New York.
Cabrera, N. and Vermilyea, D. A. (1958) In "Growth and Perfection of Crystals" (R. H. Doremus, B. W. Roberts and D. Turnbull, eds.) p. 393. Wiley, New York.

Callen, H. (1971) *Appl. Phys. Lett.* **18**, 311.
Capeller, M. A. (1723) "Prodromus Crystallographiae."
Cardona, M., Weinstein, M. and Wolff, G. A. (1965) *Phys. Rev.* **140** (A2), 633, 637.
Chalmers, B. (1958) In "Growth and Perfection of Crystals" (R. H. Doremus, B. W. Roberts and D. Turnbull, eds.) p. 291. Wiley, New York.
Champion, J. A. (1969) *Trans. Brit. Ceram. Soc.* **68**, 91.
Chase, A. B. (1966) *J. Am. Ceram. Soc.* **49**, 233.
Chase, A. B. and Osmer, J. M. (1967) *J. Am. Ceram. Soc.* **50**, 325.
Chase, A. B. and Osmer, J. M. (1969) *J. Crystal Growth* **5**, 239.
Chase, A. B. and Osmer, J. M. (1970) *J. Am. Ceram. Soc.* **53**, 343.
Chernov, A. A. (1961) *Sov. Phys.—Usp.* **4**, 116.
Chernov, A. A. (1962) *Growth of Crystals* **3**, 31.
Curie, P. (1885) *Bull. Soc. minéral. France* **8**, 145.
Curien, H. and Le Corre, Y. (1955) *Bull. Soc. Franc. Min. Crist* **78**, 604.
De Vries, R. C. (1959) *J. Am. Ceram. Soc.* **42**, 547.
De Vries, R. C. and Sears, G. W. (1961) *J. Chem. Phys.* **34**, 618.
Doelter, C. (1890) *Tschermaks Min. Petrogr. Mitt.* **11**, 324.
Doelter, C. (1894) *N. Jahrb. f. Min.* **2**, 276.
Donnay, J. D. H. and Harker, D. (1937) *Am. Min.* **22**, 446.
Egli, P. H. and Johnson, L. R. (1963) In "The Art and Science of Growing Crystals" (J. J. Gilman, ed.) p. 194. Wiley, New York.
Egli, P. H. and Zerfoss, S. (1949) *Disc. Faraday Soc.* **5**, 61.
Faust Jr., J. W., John H. R. and Pritchard, C. (1968) *J. Crystal Growth* **3/4**, 321.
Finch, C. B. and Clark, G. W. (1965) *J. Appl. Phys.* **36**, 2143.
Finch, C. B. and Clark, G. W. (1966) *J. Appl. Phys.* **37**, 3610.
Fordham, S. (1949) *Disc. Faraday Soc.* **5**, 117.
Frank, F. C. (1958) In "Growth and Perfection of Crystals" (R. H. Doremus, B. W. Roberts and D. Turnbull, eds.) pp. 3, 304. Wiley, New York.
Friedrich, W., Knipping, P. and von Laue, M. (1912) Sitz. Ber. Kgl. Bay. Akad. Wiss. p. 303.
Gavrilova, I. V. (1968) *Growth of Crystals* **5B**, 9.
Geller, S. (1967) *Z. Krist.* **125**, 1.
Geller, S. and Gilleo, M. A. (1957) *J. Phys. Chem. Sol.* **3**, 30.
Gendelev, S. S. (1963) *Soviet Phys. Crystallog.* **8**, 335.
Gendelev, S. S. and Titova, A. G. (1968) *Growth of Crystals* **6** A, 90.
Gibbs, J. W. (1875, 1878) *Trans. Connecticut Acad.* **3**. See also Gibbs, J. W. (1928) "Collected Works" Longmans, Green, New York.
Gille, F. and Spangenberg, K. (1927) *Z. Krist.* **65**, 204.
Gjostein, N. A. (1963) *Acta Met.* **11**, 957, 969.
Goldschmidt, V. (1913) "Atlas der Kristallformen" 18 volumes, Carl Winters, Heidelberg.
Grandeau, L. (1886) *Ann. Chim. Phys.* (6) **8**, 216, 219.
Grodkiewicz, W. H., Dearborn, E. F. and van Uitert, L. G. (1967) In "Crystal Growth" (H. S. Peiser, ed.) p. 441. Pergamon, Oxford.
Groth, P. (1906) "Chemische Krystallographie" Vols. 1 and 2. Engelmann, Leipzig.
Gülzow, H. (1969) *Growth of Crystals* **7**, 82.
Hartmann, H. (1966) *Kristall u. Technik* **1**, 27, 267.
Hartman, P. (1953) Thesis, Groningen.
Hartman, P. (1969) *Growth of Crystals* **7**, 3.

Hartman, P. and Perdok, W. G. (1955) *Acta Cryst.* **8,** 49, 521, 525.

Haüy, R. J. (1783, 1784) Essai d'une théorie sur la structure des cristaux.

Herring, C. (1953) In "Structure and Properties of Solid Surfaces" (R. Gomer and C. S. Smith, eds.) p. 5. Univ. of Chicago.

von Hippel, A. *et al.* (1963) *MIT Techn. Rept.* **178**.

Honigmann, B. (1952) *Z. Electrochem.* **56,** 342.

Honigmann, B. (1958) "Gleichgewichts- und Wachstumformen von Kristallen" Steinkopff, Darmstadt.

Izvekov, V. N., Sysoev, L. A., Obukhovskii, Y. A. and Birman, B. I. (1968) *Growth of Crystals 6A*, 106.

Jackson, P. A. (1969) *J. Crystal Growth* **3/4**, 395.

Janowski, K. R., Chase, A. B. and Stofel, E. J. (1965) *Trans. AIME* **233,** 2087.

Kaldis, E. (1969) *J. Crystal Growth* **5,** 376.

Kapustin, A. P. (1963) "The Effects of Ultrasound on the Kinetics of Crystallization" p. 42. Consultants Bureau, New York.

Keester, K. L. (1972) IBM San José Research Laboratory, kindly provided the computer drawings of the crystal habits.

Kern, R. (1969) *Growth of Crystals* **8,** 3.

Klija, M. D. (1955) *Dokl. Akad. Nauk SSR* **100,** 259.

Knacke, O. and Stranski, I. N. (1952) *Ergebn. exakt. Naturwiss.* **26,** 383.

Kossel, W. (1927) Nachr. Ges. Wiss. Göttingen, mathem.—phys. Klasse, p. 135.

Landau, L. D. (1950) Symposium in Commemoration of 70th Birthday of A. F. Joffe, p. 44. Acad. Sci. Press, Moscow.

Laudise, R. A. (1958) *J. Amer. Chem. Soc.* **80,** 2655.

Laudise, R. A. (1959) *J. Amer. Chem. Soc.* **81,** 562.

Laudise, R. A. (1970) "The Growth of Single Crystals" p. 81. Prentice Hall, Englewood Cliffs, N.J.

von Laue, M. (1943) *Z. Krist.* **105,** 124.

Leblanc, N. (1788) *Journ. de Phys.* **33**; *Ann. de Phys.* **23,** 375.

Ledésert, M. and Monier, J. C. (1965) In "Adsorption et Croissance Cristalline" p. 537. Intern. Colloq. CNRS No. 152, Paris.

Lefever, R. A. and Chase, A. B. (1962) *J. Am. Ceram. Soc.* **45,** 32.

Lefever, R. A., Chase, A. B. and Torpy, J. W. (1961) *J. Am. Ceram. Soc.* **44**.

Lemmlein, G. (1954) *Dokl. Akad. Nauk SSSR* **98,** 973.

Linares, R. C. (1965) *J. Phys. Chem. Solids* **26,** 1817.

Linares, R. C. (1967) *Am. Min.* **52,** 1211.

Linares, R. C. (1968) *Trans. Met. Soc. AIME* **242,** 441.

Malur, J. (1966) *Kristall u. Technik* **1,** 261.

Marc, R. (1908) *Z. physik. Chem.* **61,** 385, (1909), **67,** 470.

Marc, R. (1912) *Z. physik. Chem.* **79,** 71.

Mita, Y. (1962) *J. Phys. Soc. Japan* **17,** 784.

Mullin, J. W. (1972) "Crystallization" 2nd Edition. Butterworths, London.

Mullin, J. W., Amatavivadhana, A. and Chakraborty, M. (1970) *J. Appl. Chem.* **20,** 153.

Nelson, D. F. and Remeika, J. P. (1964) *J. Appl. Phys.* **35,** 522.

Neuhaus, A. (1928) *Z. Krist.* **68,** 15.

Newkirk, H. W. and Smith, P. S. (1965) *Am. Min.* **50,** 44.

Newkirk, H. W., Smith, D. K., Meieran, E. S. and Mattern, D. A. (1967) *J. Mat. Sci.* **2,** 194.

Nielsen, J. W. and Dearborn, E. F. (1958) *J. Phys. Chem. Solids* **5,** 202.

Nielsen, J. W., Linares, R. C. and Koonce, S. E. (1962) *J. Am. Ceram. Soc.* **45,** 12.
Niggli, P. (1919) "Geometrische Kristallographie des Diskontinuums" Leipzig.
Niggli, P. (1920) *Z. Anorg. Allg. Chem.* **110,** 55.
Nordenskiöld, A. E. (1860) Oefvers. K. Vet.-Akad. Förhandl. Stockholm, 450; (1861) *Pogg. Ann. Phys.* **114,** 616.
Parker, S. G. and Pinnell, J. E. (1968) *J. Crystal Growth* **3/4,** 490.
Price, P. B., Vermilyea, D. A. and Webb, M. B. (1958) *Acta Met.* **6,** 524.
Remeika, J. P. (1954) *J. Amer. Chem. Soc.* **76,** 940.
Romé de L'Isle, (1783) "Cristallographie" 2nd Edition (Paris) p. 379.
St. Claire-Deville, H. and Troost, L. (1861). *Compte Rend.* **52,** 920.
Sasaki, H. (1964) *J. Phys. Soc. Japan* **20,** 264.
Sasaki, H. and Kurokowa, E. (1965) *J. Am. Ceram. Soc.* **48,** 171.
Scheel, H. J. (1972) *J. Crystal Growth* **13/14,** 560.
Scheel, H. J. (1974) *J. Crystal Growth* **24/25,** 669.
Scheel, H. J. and Elwell, D. (1973) *J. Crystal Growth* **20,** 259.
Schneider, R. (1873) *Pogg. Ann. Phys.* **149,** 386.
Schüler, E. (1853) *J. prakt. Chem.* **60,** 249.
Sears, G. W. (1958) *J. Chem. Phys.* **29,** 1045.
Sheftal, N. N. (1958) *Growth of Crystals* **1,** 5.
Shlichta, P. J. (1969) *Growth of Crystals* **7,** 75.
Sholokhovich, M. L., Berberova, L. M. and Varicheva, V. I. (1968) *Growth of Crystals* **6***A*, 85.
Simon, B. and Bienfait, M. (1965) *Acta Cryst.* **19,** 750.
Slavnova, E. N. (1958) *Growth of Crystals* **1,** 117, **2,** 166.
Sohncke, L. (1888) *Z. Krist.* **13,** 214.
Spangenberg, K. (1934) In "Handwörterbuch der Naturwiss" 2nd Edition, Jena, Vol. 10, 362.
Steno, N. (1669) De solido intra solidum naturaliter contento dissertationis prodromus, Florence.
Sterba, J. (1901) *Compt. Rend.* **133,** 294.
Stranski, I. N. (1928) *Z. Physik. Chem.* **136***A*, 259.
Stranski, I. N. (1932) *Z. physik. Chem.* **B 17,** 127.
Stranski, I. N. (1956) *Bull. Soc. franç. Minéral. Crist.* **79,** 359.
Stranski, I. N. and Kaischev, R. (1931) *Z. Krist.* **78,** 373.
Stranski, I. N. and Kaischev, R. (1934) *Z. physik. Chem.* **B 26,** 100, 312.
Stranski, I. N., *et al.* (1949) *Disc. Faraday Soc.* **5,** 13–32.
Sunagawa, I. (1967) *J. Crystal Growth* **1,** 102.
Tatarskii, V. B. (1964) *Sov. Phys.—Cryst.* **9,** 113.
Tertsch, H. (1926) "Trachten der Kristalle" Borntraeger, Berlin.
Timofeeva, V. A. (1959) *Growth of Crystals* **2,** 73.
Timofeeva, V. A. (1960) *Kristallografiya* **5,** 476 (in Russian).
Timofeeva, V. A. (1971), paper presented at ICCG 3, Marseille.
Timofeeva, V. A. and Lukyanova, N. I. (1967) *Sov. Phys.—Cryst.* **12,** 77.
Titova, A. G. (1962) *Growth of Crystals* **3,** 306.
Valeton, J. J. P. (1915) *Ber. d. Math.—Phys. Klasse d. Kgl. Sächs. Ges. d. Wiss.* (Leipzig) **67,** 1.
Van Hook, A. (1961) "Crystallization" Reinhold, New York.
Volmer, M. (1939) "Kinetik der Phasenbildung" Steinkopff, Dresden, Leipzig.
Wallace, C. A. and White, E. A. D. (1967) In "Crystal Growth" (H. S. Peiser, ed.) p. 431. Pergamon, Oxford.

Wanklyn, B. M. (1969) *J. Crystal Growth* **5,** 219.
Wanklyn, B. M. (1974) In "Crystal Growth" (B. R. Pamplin, ed.) Pergamon, Oxford.
White, E. A. D. (1955) *Acta Cryst.* **8,** 845.
White, E. A. D. and Brightwell, J. W. (1965) *Chem. and Industry* 1662.
Wolfe, R., Sturge, M. D., Merritt, F. R. and Van Uitert, L. G. (1971) *Phys. Rev. Lett.* **26,** 1570.
Wollaston, W. H. (1809) *Phil. Trans.* 253.
Woods, J. (1959) *Brit. J. Appl. Phys.* **10,** 529.
Wulff, G. (1901) *Z. Krist.* **34,** 449.
Yanovskii, V. K., Voronkova, V. I. and Koptsik, V. A. (1970) *Sov. Phys.—Cryst.* **15,** 302.
van der Ziel, J. P., Sturge. M. D. and Van Uitert, L. G. (1971) *Phys. Rev. Lett.* **27,** 508.
Zonn, Z. N. and Joffe, V. A. (1969) *Growth of Crystals* **8,** 63.

6. Conditions for Stable Growth

6.1. Stability of Growth

The aim of most crystal-growth experiments is to produce crystals which are sufficiently large and perfect for some measurement or application. The crystal grower is therefore particularly concerned to establish, either by trial and error or by the application of theoretical principles, the conditions under which such large and relatively perfect crystals may be produced.

Stable growth of a crystal from solution may be defined as growth without the entrapment at any stage of solvent inclusions. Alternative definitions of stability are possible and growth-rate fluctuations will always occur on some scale. Large fluctuations may facilitate inclusion formation or compositional variations and are likely to have an adverse effect on the crystal quality.

The problem which is normally considered in a theoretical approach to the calculation of conditions for stable growth is that of morphological stability, or whether a specified shape is stable against small perturbations. This problem differs from the related question of whether a given shape or habit is preserved as the crystal grows. Both aspects are of importance and will be discussed in this chapter. Reviews on morphological stability have recently been published by Parker (1970) and by Chernov (1972).

Perturbation analyses and related studies are concerned with simple shapes but have given a number of results which are of great relevance in the design of experiments. The most important conclusions are that the stability tends to decrease as the crystal increases in size, and the concept of *constitutional supercooling*. Both these results of stability theory have been appreciated for some years but their relation to experiments on high-temperature solution has received comparatively little attention and will be stressed where possible.

Particular importance is attached in this chapter to the concept of a maximum stable growth rate and its dependence on the growth conditions. Practical considerations for the attainment of stable growth will be discussed in the later sections of the chapter.

6.1.1. Stability of the sphere and cylinder

Mullins and Sekerka (1963) used the perturbation method to examine the stability of a spherical crystal with isotropic surface kinetics growing in a supersaturated medium. They concluded that the sphere is stable against perturbations only if its radius is less than a value of $7r^*$, where r^* is the radius of the critical nucleus. For a supersaturation of 10%, the maximum stable radius was calculated to be of the order of 0.1 μm. Nichols and Mullins (1965) and Coriell and Parker (1966) studied the effect of surface diffusion on the stability of a spherical crystal and found that the maximum stable radius is increased by a large factor. In the example considered by Nichols and Mullins, this factor was of the order of 100 so that spheres were estimated to be stable to a radius of about 10^{-3} cm.

Cahn (1967) included the effect of interface kinetics and of an anisotropic surface tension. He found that the latter has no stabilizing effect but this result is not surprising for a sphere since the anisotropy is not shape preserving as it would be for a polyhedral crystal, and its main effect will be to cause an anisotropy in the instability. The inclusion of interface kinetics leads to an expression for the rate of increase of radius of the sphere given by

$$\frac{dR}{dt} = \frac{F(n_{sn} - n_e)}{1 + F\rho R/D} \qquad (6.1)$$

where F is a kinetic coefficient such that the linear growth rate $v = F(n_i - n_e)$ and the other symbols have been defined previously.

For small crystals the term $F\rho R/D$ may be neglected and the growth rate becomes $F(n_{sn} - n_e)$. Since the diffusion coefficient D does not appear in the expression for dR/dt, growth is said to be interface controlled. The interface concentration in this case is approximately the same as in the bulk solution and growth should be stable since the concentration gradient in the solution is approximately zero. At high values of R the increase of radius is given by

$$\frac{dR}{dt} = \frac{D}{\rho R} (n_{sn} - n_e)$$

and the sphere becomes unstable as shown by Mullins and Sekerka (1963).

This treatment therefore leads to the conclusion that the sphere will be stable only up to a radius such that $K\rho R/D \sim 1$. For $D = 10^{-5}$ cm^2 s^{-1}, $\rho = 5$ g cm^{-3} and $F = 10^{-4}$ cm^{-2} s^{-1} g^{-1}, the maximum stable radius is only of the order of 0.02 cm. Coriell and Parker (1967) performed a similar but more quantitative calculation for both linear and quadratic interface kinetics. In the example they quoted, of salol growing from aqueous solution, kinetic control increases the maximum size for stable growth by a factor 3000, to a value of 0.5 cm.

The stability of a cylindrical crystal has been studied by Coriell and Parker (1966) and is found to exhibit approximately the same behaviour as that predicted for the sphere. Surface diffusion was estimated to increase the maximum stable radius by a factor of about 40. Kotler and Tiller (1966) included interface kinetics and found that the maximum radius is strongly dependent on the undercooling and the kinetic coefficient.

The main conclusion to be drawn from these studies is that instability tends to occur when the crystal reaches a critical size, which will be increased by surface diffusion of solute and by interface kinetics. These predictions must, however, be treated with great caution in their application to solution growth because of the strong tendency of crystals to develop habit faces. As a result of this tendency, it has not been found possible to study the stability of spherical or cylindrical crystals in solution, as for ice crystals in water (Hardy and Coriell, 1968). Greater significance must therefore be attached to studies of polyhedral crystals and of a single plane interface, which is treated in greater detail in the following section.

6.1.2. Stability of a plane interface

A. *Constitutional supercooling/supersaturation gradient.* Perturbation treatments of the stability of a planar crystal surface growing in a doped melt were first given by Mullins and Sekerka (1964), Sekerka (1965) and Voronkov (1965). When conduction of heat through the crystal is included, the condition for stability may be written as

$$\frac{mn(1-k)v}{kD} < \frac{K_c}{K_c+K_l}\left(\frac{\mathrm{d}T}{\mathrm{d}z}\right)_c + \frac{K_l}{K_c+K_l}\left(\frac{\mathrm{d}T}{\mathrm{d}z}\right)_l. \tag{6.2}$$

Here m is the slope of the liquidus curve, n the concentration of the impurity in the bulk liquid, K the thermal conductivity and $\mathrm{d}T/\mathrm{d}z$ the temperature gradient normal to the interface, with the suffices c and l referring to the crystal and liquid, respectively. k is the partition coefficient which is defined as the ratio n_c/n_l of the impurity concentration in the crystal to that in the liquid, and which is normally less than unity. Equation (6.2) could be extended to growth from very concentrated solutions where

n_c and n_l would be the solvent concentrations in the crystal and solution, respectively.

The condition expressed by Eqn (6.2) is closely related to the *constitutional supercooling criterion*, introduced by Ivantsov (1951, 1952) as "diffusional undercooling", by Rutter and Chalmers (1953) and quantitatively by Tiller *et al.* (1953). As the crystal grows, impurities are rejected at the crystal surface and so the impurity concentration in the liquid immediately ahead of the interface becomes appreciably higher than that in the bulk of the liquid (Fig. 6.1a). This accumulation of impurities results in a depression of the equilibrium liquidus temperature T_L (according to the phase diagram) as illustrated in Fig. 6.1(b). The actual temperature distribution in the melt is as shown in the dashed line (i) of Fig. 6.1(b) and any protuberance on the interface will tend to grow (relative to the interface) since it will experience a higher supercooling.

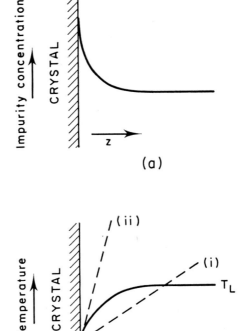

(a)

(b)

Fig. 6.1. (a) Impurity distribution and (b) equilibrium liquidus temperature T_L ahead of an advancing interface in a doped melt.

The region ahead of the interface is then unstable. Stable growth can occur if the temperature gradient at the interface is increased as in (ii) of Fig. 6.1(b), in which case the actual temperature ahead of the interface is higher than the liquidus.

The condition for constitutional supercooling may be readily derived as

$$\frac{mn(1-k)v}{kD} > \frac{dT}{dz} \qquad (6.3)$$

which is identical with Eqn (6.2) if $K_c = 0$, or if $K_c = K_L$ and the latent heat is zero. Lucid reviews of constitutional supercooling and of the modes of unstable growth which can result have been given by Tiller (1963, 1970). Experimental confirmation of the validity of the constitutional super-cooling criterion for melt growth has been provided by Walton *et al.* (1955) and Bardsley *et al.* (1961).

The similarity between growth from a doped melt and growth from solution has been pointed out by White (1965). In the latter case solvent is rejected by the growing crystal and there will inevitably be a gradient of solute ahead of the interface due to local depletion by the crystal. Tiller (1968) has proposed the application of the constitutional supercooling criterion to growth from solution by a modification of Eqn (6.3). Since the solution normally contains a number (j) of solute constituents, the condition for instability may be expressed by writing for the growth rate:

$$v > D_o \frac{dT}{dz} \bigg/ \sum_{i=1}^{j} \left\{ \frac{m_i(k_i{}^* - 1)n_i}{D_i/D_o} \right\} \qquad (6.4)$$

where D_o is the diffusion coefficient of the solvent and m_i, $k_i{}^*$, n_i and D_i refer to the solute constituent i. The effective partition coefficient is defined by

$$k_i{}^* = \left(\frac{n_c{}^i}{n_{sn}{}^i} \right)_{z=0}$$

where $n_c{}^i$ and n_{sn}^i are the concentration of i in the crystal and solution, respectively. Tiller has calculated the ratio of the maximum stable growth rate v_{max} to the gradient (dT/dz) as a function of temperature for various compound semiconductors and his results are shown in Fig. 6.2.

An alternative and much simpler calculation of the criterion for stable growth in solution under diffusion-limited conditions may be obtained by considering the condition for the appearance of a supersaturation gradient —an increase in supersaturation ahead of the interface which may be a consequence of solute diffusion (Elwell and Neate, 1971; Scheel and Elwell, 1973a). The condition for instability is that a protuberance will encounter a higher supersaturation as it advances so that at the interface,

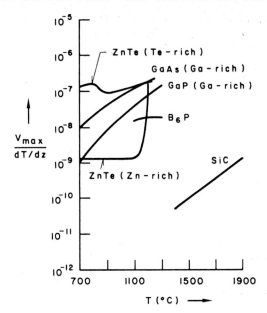

FIG. 6.2. Ratio of maximum stable growth rate to temperature gradient for solution growth of some compound semiconductors (Tiller, 1968).

$$\frac{dn}{dz} > \frac{dn_e}{dz}. \tag{6.5}$$

Now dn/dz is related to the linear growth rate v by Eqn (4.13)

$$v \simeq \frac{D}{\rho}\left(\frac{dn}{dz}\right)_{z=0}$$

and, for an ideal solution with $n_e = \text{const} \exp(-\phi/RT)$,

$$\frac{dn_e}{dz} = \frac{\phi n_e}{RT^2}\frac{dT}{dz}.$$

Substitution into Eqn (6.5) gives the condition for instability as

$$v > \frac{D\phi n_e}{\rho RT^2}\frac{dT}{dz} \tag{6.6}$$

which is substantially the same result as Eqn (6.4) for a single component.

Using typical values of $D = 10^{-5}$ cm² s⁻¹, $n_e = 1$ g cm⁻³, $\rho = 5$ g cm⁻³, $\phi = 70$ kJ mole⁻¹ and $T = 1500°$K, the maximum stable growth rate according to Eqn (6.6) will be $v_{\max} \sim 10^{-8}$ cm s⁻¹ for $dT/dz = 10$ deg cm⁻¹ or $v_{\max} \sim 10^{-9}$ cm s⁻¹ for $dT/dz = 1$ deg cm⁻¹. The value of $v_{\max}/(dT/dz) \sim 10^{-9}$ s⁻¹ deg⁻¹ is typical of the values quoted by Tiller (Fig. 6.2).

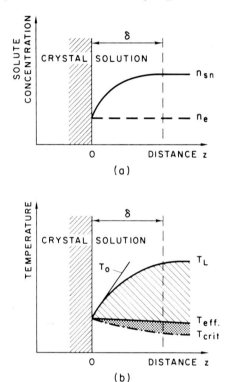

In practice crystals are grown at stable rates of the order of 10^{-6} cm s^{-1} in temperature gradients of the order of 1 deg cm^{-1}. It is therefore clear that *crystal growth from high-temperature solution normally occurs in a destabilizing supersaturation gradient.*

For stable growth to occur at rates much greater than those given by Eqn (6.6), it must be assumed that the supersaturation gradient must be insufficient for a perturbation to nucleate (Wagner, 1954; Tiller and Kang,

FIG. 6.3. (a) Solute concentration ahead of crystal growing in solution. (b) Metastable region of supersaturation gradient. (T_o corresponds to ii of Fig. 6.1.) (Scheel and Elwell, 1973a.)

1968; O'Hara *et al.*, 1968). The crystals may therefore be said to grow in a *metastable region of the supersaturation (or supercooling) gradient* as discussed by Scheel and Elwell (1973a). This region is analogous to the normal metastable or Ostwald–Miers region and is illustrated in Fig. 6.3.† In Fig. 6.3(a) is shown the actual solute concentration n_{sn} in front of the

† The width of the boundary layer is denoted approximately in the diagrams. The exact definition of δ is that of Eqn (4.14).

growing crystal for diffusion-controlled growth and the corresponding liquidus temperature T_L is shown in Fig. 6.3(b). The significance of this metastable region between T_L and T_{crit} is that a perturbation will not nucleate (or will develop at a negligibly slow rate) so long as the temperature T_{eff} ahead of the crystal exceeds the limiting value. In the example shown, growth will occur in the metastable region even if the temperature gradient at the interface is zero. If, however, the requirement of conduction of the heat of crystallization through the solution leads to a temperature distribution below that of the dashed line T_{crit} in Fig. 6.3(b), growth will be unstable and the crystal will contain an appreciable concentration of solvent inclusions.

By analogy with observations on the Ostwald-Miers region, it may be expected that the width of the metastable supersaturation (supercooling) gradient region will depend on such factors as the crystal-growth rate and the degree of disturbance, particularly thermal or mechanical shock, to which the solution is subjected.

An important question concerns the origin of the metastability and we now consider in some detail the relative importance of the various stabilizing factors which were not taken into account in the derivation of Eqns (6.4) or (6.6). We consider first the results of a perturbation approach.

B. *Perturbation analysis.* In view of the importance of the perturbation method of stability analysis, a summary is given here of a simplified treatment of the stability of a plane interface growing in solution, due to Shewmon (1965).

Consider a plane crystal surface growing in a supersaturated solution in the volume diffusion-controlled regime. Any protuberance on this surface may be analysed into a number of sine waves of different wavelength and it is convenient to discuss the stability condition in terms of such sine waves. A protuberance of the interface such as that shown in Fig. 6.4 will have components of the form

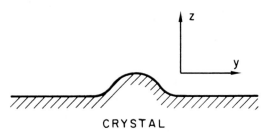

FIG. 6.4. Proturberance on plane crystal-solution interface.

$$z = \epsilon(t) \sin \omega y \qquad (6.7)$$

where ϵ is the time-dependent amplitude of the perturbation and the wavelength is specified by ω. The perturbation causes local changes in the solute concentration at the surface, which is given by the Gibbs-Thomson equation. The local curvature is assumed to be d^2z/dy^2 so that the equilibrium concentration is specified by

$$n_s = n_{se}(1 + \Gamma \epsilon(t)\omega^2 \sin \omega y) \qquad (6.8)$$

where n_{se} is the equilibrium concentration per unit volume for a flat surface, and Γ is the capillary constant $\gamma V_M/RT$. (As in Chapter 4, V_M is the molar volume of the solute and γ the surface energy per unit area.)

An approximate solution may be obtained by assuming that the interface is static, in which case the solute distribution obeys Laplace's equation $\nabla^2 n = 0$. The general solution to the latter equation for a sinusoidal interface and a static solute gradient $G = (dn/dz)_{\epsilon=0}$ is

$$n(z, y) = A + B \exp(-\omega z)\epsilon \sin \omega y + Gz. \qquad (6.9)$$

The constants A and B are chosen to make Eqn (6.9) identical with Eqn (6.8) at the interface, that is by equating coefficients with $z = \epsilon \sin \omega y$. This gives $A = n_{se}$ and $B = (n_{se}\Gamma\omega^2 - G)$, so that

$$n(z, y) = n_{se} + (n_{se}\Gamma\omega^2 - G) \exp(-\omega z)\epsilon \sin \omega y + Gz. \qquad (6.10)$$

The linear growth rate is then given by Eqn (4.13) as

$$v \simeq \frac{D}{\rho}\left(\frac{dn}{dz}\right)_{z=0} = \frac{D}{\rho}[G + (G - n_{se}\Gamma\omega^2)\omega\epsilon \sin \omega y]. \qquad (6.11)$$

The first term in Eqn (6.11) represents the growth rate v_o in the absence of any perturbation, and so the development of any perturbation relative to the mean position of the surface is given by the second term as

$$\dot{\epsilon} = \frac{D}{\rho}(G - n_{se}\Gamma\omega^2)\omega\epsilon \sin \omega y \qquad (6.12)$$

so that

$$\frac{\dot{\epsilon}}{\epsilon} = (1 - n_{se}\Gamma\omega^2/G)\omega v_o. \qquad (6.13)$$

The first term on the right-hand side of Eqn (6.13) may be interpreted physically as being due to an increase in the concentration gradient in front of the "hills" on the perturbed "valleys". The second term is due to the concentration gradients along the surface which cause solute transport and so tend to smooth out the sinusoidal disturbance. The two effects balance at a critical value of ω, denoted ω_o, such that

$$\omega_o = (G/n_{se}\Gamma)^{1/2}. \qquad (6.14)$$

I

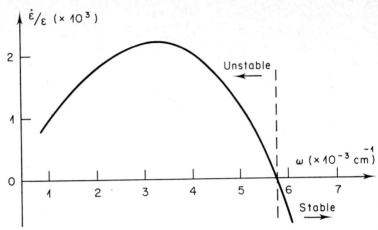

FIG. 6.5. Rate of development of instability for various values of ω ($= 2\pi/\lambda$).

Taking typical values of $G/n_{se} = 1$ cm^{-1} (1% supersaturation and $\delta = 10^{-2}$ cm) and $\Gamma = 3 \times 10^{-8}$ cm ($V_M = 40$ cm^3 mole^{-1}, $\gamma = 10^{-5}$ J cm^{-2} and $T = 1500°$K) gives a value of $\omega_o \simeq 6 \times 10^3$ cm^{-1}. A plot in Fig. 6.5 of $\dot{\epsilon}/\epsilon$ for various values of ω using the same parameters with $v_o = 10^{-6}$ cm s^{-1} shows that a particular wavelength will tend to become dominant if growth occurs under unstable conditions. In the example chosen, the maximum value of ω is about 3×10^3 cm^{-1} so that the corresponding wavelength is in the region of 20 μm. Some confirmation of this prediction is provided by the surface structure of a GdAlO$_3$ crystal shown in Fig. 6.6. The surface shows

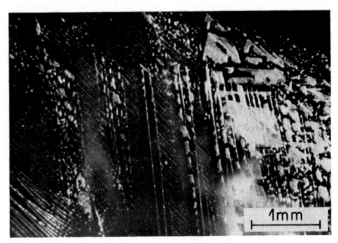

FIG. 6.6. Periodic solvent inclusions on surface of GdAlO$_3$ due to interruption of stirring at the end of growth experiment (Scheel and Elwell, 1973a).

an inclusion structure which exhibits a fine periodicity of approximately 30 μm. A transient instability in this case resulted from the cessation of stirring at the termination of growth prior to removal of the excess solution.

For values of ω greater than ω_o, $\dot{\epsilon}/\epsilon$ is negative and so the perturbation tends to decay; the interface is therefore unstable against perturbations of relatively long wavelength. The stability range is increased if the solute gradient G is small and if the capillary constant Γ is large. Absolute stability, for all values of ω, requires that $G=0$ which is incompatible with volume diffusion (and would also require $v_o=0$ under the conditions assumed!).

Additional stability is obtained if surface diffusion of the solute is included since this will tend to smooth out any disturbance by transporting solute from the "hills" to the "valleys". The inclusion of surface diffusion introduces an additional term so that Eqn (6.12) becomes

$$\dot{\epsilon}/\epsilon = \frac{D}{\rho}\left(G - n_{se}\Gamma\omega^2 - \frac{\Gamma D_s \Lambda\omega^3}{D}\right)\omega \sin \omega y \qquad (6.15)$$

where D_s is the surface-diffusion coefficient and Λ the thickness of the adsorption layer in which this diffusion occurs.

In the general case, the growth rate is determined partly by interface kinetics and Eqn (6.10) will no longer be valid. If the linear growth rate is specified in terms of the kinetic coefficient F such that $v = F(n_i - n_e)$, as in Section 6.2, an extension of the analysis outlined above leads to an expression for the growth of the protuberance

$$\dot{\epsilon} = \frac{(1 - \Gamma n_{se}\omega^2 G)\omega\epsilon v_o}{(1 + D\omega/F)} \qquad (6.16)$$

in which surface diffusion has been neglected. This equation reduces to Eqn (6.13) in the volume-diffusion regime where $D\omega/F \ll 1$.

It is seen that, according to this treatment, the condition for stability is the same as in the diffusion-controlled case since the boundary between $\dot{\epsilon} < 0$ and $\dot{\epsilon} > 0$ is still given by Eqn (6.14). As $D\omega/F$ increases, so $\dot{\epsilon}/\epsilon v_o$ will decrease and the principal effect of kinetic control will be to reduce the rate of development of the instability.

According to the above model, the instability condition applies to all values of ω below ω_o and therefore to all wavelengths above the corresponding value $\lambda_o = 2\pi/\omega_o$. However, the model must break down at long wavelengths since it requires redistribution of solute due to a perturbation of wavelength λ over a distance of the order of λ. This redistribution must occur by volume diffusion and so the model breaks down when $\lambda \gtrsim (D\tau)^{1/2}$, where $(D\tau)^{1/2}$ is the mean displacement due to diffusion. At higher values of λ, stabilization will be present because of surface tension but there will be no enhancement of the solute gradient to give instability. Since $(D\tau)^{1/2} \sim \delta$, the wavelength region over which growth is unstable will be given by

$$\delta \gtrsim \lambda \gtrsim \pi \left(\frac{n_{se} \Gamma}{G} \right)^{1/2} . \qquad (6.17a)$$

Since $G \simeq (\sigma n_e/\delta) \simeq (\sigma n_{se}/\delta)$, this condition may alternatively be expressed as

$$\delta \gtrsim \lambda \gtrsim 2\pi \left(\frac{\Gamma \delta}{\sigma} \right)^{1/2} . \qquad (6.17b)$$

An alternative stability condition is therefore that $2\pi(\Gamma\delta/\sigma)^{1/2} > \delta$ or, as is equivalent,

$$\sigma < \frac{4\pi^2 \Gamma}{\delta} \qquad (6.18)$$

The maximum supersaturation required by this condition is, however, unreasonably low. If $\Gamma \simeq 3 \times 10^{-8}$ cm and $\delta \simeq 10^{-2}$ cm as in the example above, Eqn (6.18) requires that $\sigma < 1 \times 10^{-4}$ which would correspond to an extremely slow growth rate.

In summary, Shewmon's treatment predicts that a planar crystal surface is unstable towards perturbations above some critical value of the order of 10 μm. Stability is enhanced by surface diffusion but not by surface kinetics, although the latter retards the development of the instability. The main success of this approach is in the prediction of a periodic structure of wavelength about 20 μm when growth is unstable, but it does not yield a criterion for stable growth which may be used in practice.

The effect of interface kinetics on the stability of a plane interface was considered by Tarshis and Tiller (1967). They concluded that kinetics will stabilize the interface, but only under source-limited growth conditions.

C. *Stabilization due to facetting.* The most important factor which has been neglected in the above treatment is the normal tendency of solution-grown crystals to develop habit faces. Such faces have a characteristically low energy and it may be expected that the development of a perturbation on such faces will be more difficult than on non-habit faces since protuberances will involve the formation of surfaces of relatively high energy.

The stability of habit faces has been discussed qualitatively by O'Hara *et al.* (1968), who consider both kinetic and capillarity effects. If growth on a particular facet is controlled by a single active centre which generates a growth spiral, then a perturbation which tends to increase the small angle between the resulting vicinal face and the crystallographic habit face will also increase the number of layer edges per unit area. The lateral motion of these edges will cause the surface to revert back towards the original geometry with the separation between adjacent spiral arms given by $19r^*$ according to the BCF theory (see Chapter 4). Capillarity tends to stabilize a facet when the growth is strongly anisotropic, particularly when the energy minimum in the Wulff plot is very sharp. This effect can be enhanced by non-isotropic adsorption of impurities, which could explain why the addition of impurities may sometimes result in improvement in crystal quality.

The stability of polyhedral crystals has been considered in more detail by Chernov (1972) in a review of morphological stability. Chernov explains the stability of facetted crystals in terms of the anisotropy of the surface processes. A ridge or hollow produced by some fluctuation on an anisotropic surface has along its edges much higher kinetic coefficients than at the vertex, so that it expands tangentially at a rapid rate relative to the normal growth direction. This anisotropy invalidates the use of a perturbation approach.

Of particular importance when polyhedral crystals (as distinct from a plane surface of unspecified extent) are considered is the difference in supersaturation between the corners and face centres. The variation in supersaturation across a crystal face was measured, for example, by Bunn (1949) (see Chapter 4) and was found to be about 25% in the case of sodium chlorate. According to Chernov, this supersaturation inhomogeneity is compensated by the development of vicinal faces as indicated in Fig. 6.7. The slope at the centre of the face to the crystallographic habit face must differ from that at the corners by about $2°$ if the increased kinetic coefficient at the centre is to balance the lower supersaturation. The supersaturation inhomogeneity increases as the crystal grows and the curvature

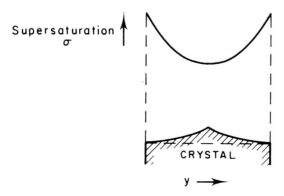

FIG. 6.7. Supersaturation inhomogeneity and compensating curvature for a crystal face (Chernov, 1972).

of the crystal must also increase if the face is to remain stable. Chernov's condition for instability is that the face centre attains some maximum deviation from the simple crystallographic orientation such that the kinetic coefficient becomes very large. This approach gives the maximum length l of a crystal having N faces as

$$l = \frac{D(p_o - p_i)\tan(\pi/N)}{2F_i\,\epsilon\theta f(\theta)} \tag{6.19}$$

where p_o and p_i are the slope of the crystal face to the crystallographic habit plane at the centre and edge, respectively. F is the kinetic coefficient, ϵ a measure of the difference in solute concentration between the edges and the face centre and θ represents the anisotropy in the growth kinetics. The complex function $f(\theta)$ depends on the anisotropy and also on the limiting value of $(p_o - p_i)$. Using typical values of the various parameters, Chernov estimates $l \sim 10^{-2}$ cm for $\theta = 2$ (a minimum value for a regular polyhedron), and $l = 10^{-1}$ cm for $\theta = 40$. These values are clearly at variance with experiment by one or two orders of magnitude.

A factor not considered by Chernov which could increase the maximum size for stable growth is the tendency for dislocations to propagate in bundles radiating either from a seed crystal or from the nucleation centre towards the centre of the crystal faces rather than towards the corners. This tendency is illustrated in Figs 4.28(a) and (b) which show the surface of a large GdAlO$_3$ crystal grown by spontaneous nucleation (Scheel and Elwell, 1973b). The high concentration of defects at the centre of the face is clearly correlated in extent with the dendritic core of the crystal, while the outer regions of the face are relatively free from defects.

The main technique which has been used to demonstrate this tendency of dislocation propagation towards face centres is that of X-ray topography, which will be discussed in detail in Chapter 9. In the topograph shown in Fig. 6.8, which is fairly typical, the dislocation bundles are revealed as white streaks and the preferential propagation towards the face centres is clearly noticeable.

If the crystal does contain a higher concentration of active sites near the face centres, an enhanced departure from the habit plane will be un-necessary and crystals will be able to grow to greater size than that predicted by Chernov without the development of excessive curvature. However, the tendency illustrated in Figs. 4.28 and 6.8 is by no means universal and alternative sources of stabilization must be considered.

Cahn (1967) also treated the stability of a habit face with growth by layer propagation but took as his stability condition the requirement that the supersaturation must not fall to zero at the face centre. By assuming that the solute is transported over the surface only by volume diffusion,

FIG. 6.8. X-ray topograph of triglycine sulphate crystal showing bundles of dislocations radiating from the seed (Vergnoux *et al.*, 1971).

Cahn arrived at an expression for the maximum size of a crystal for stable growth

$$l = \frac{D(n_{sn} - n_e)}{v\rho} .$$
(6.20)

With $D = 10^{-5} \text{ cm}^2 \text{ s}^{-1}$, $n_{sn} - n_e = 5 \times 10^{-2} \text{ g cm}^{-3}$, $v = 10^{-5} \text{ cm s}^{-1}$ and $\rho = 5 \text{ gm cm}^{-3}$, Eqn (6.20) gives $l = 2 \times 10^{-3}$ cm which is again much too small in relation to experiment.

The most likely cause of the large discrepancy between Chernov's or Cahn's treatment and experiment is in the assumption that the flow of solvent between the edges and centre of the crystal faces occurs only by volume diffusion. The principle that the difference in supersaturation between the edges and centre of a face leads to instability is likely to be correct, but convective flow must be taken into account in any realistic estimate of the maximum stable size. The importance of solution flow will be considered in the next section.

D. *Velocity gradient.* An alternative and attractively simple method of treating the effect of surface kinetics on stability was proposed by Brice (1969). If a crystal is growing in the z direction at a stable rate v, the condition for stability proposed by Brice is that a projection will grow less rapidly and a depression more rapidly than the rest of the surface. This requires that the *velocity gradient* should be negative, that is that

$$\frac{dv}{dz} < 0. \tag{6.21}$$

If it is assumed that the crystal is growing at a rate determined by the BCF formula [Eqn (4.42a)] written in the form

$$v = A\left(\frac{n_{sn} - n_e}{n_e}\right)^2 T^2 \exp\left(-B/RT\right),$$

then differentiation and substitution into Eqn (6.21) gives

$$v\left[\left(\frac{2}{T} + \frac{B}{RT^2}\right)\frac{dT}{dz} + \frac{2}{(n_{sn} - n_e)}\frac{dn}{dz} - \frac{2n\phi}{(n_{sn} - n_e)}\frac{dT}{dz}\right] < 0.$$

With $\rho v/D$ substituted for dn/dz from Eqn (4.13), the stability condition becomes

$$v < \frac{D}{\rho}\frac{dT}{dz}\left[\frac{n_{sn}\phi}{RT^2} - (n_{sn} - n_e)\left(\frac{1}{T} + \frac{B}{RT^2}\right)\right] \tag{6.22}$$

which is the same as Eqn (6.6) except for the second term in the square bracket. This term in fact reduces the maximum stable growth rate by about 35% if B is taken to have a value of 20 kJ/mole. It would be of interest to extend this model to treat the stability of a rectangular protuberance considering both its movement along and normal to the crystal surface, and the results of the above one-dimensional approach must be treated with caution. The various treatments of the effect of interface kinetics are seen to be somewhat conflicting.

6.2. Solution Flow and Stability

An increase in the rate of flow of solution past a crystal surface has two main effects. It will even out the distribution of solute over the surface and will reduce the thickness of the boundary layer. The first effect, as argued in the previous section, will lead to enhanced stability for a polyhedral crystal, but the beneficial effect of the reduced boundary-layer thickness is not so obvious and will be discussed first.

According to the concept of a metastable region of supersaturation gradient, stirring may lead to an enhancement of stability even of an infinite plane surface. The distribution of solute and the temperature

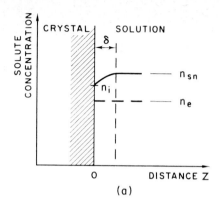

(a)

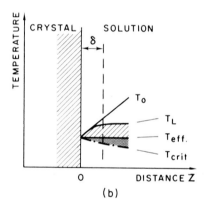

(b)

FIG. 6.9. (a) Solute concentration ahead of crystal growing in stirred solution (b) Metastable region of supersaturation gradient in stirred solution (compare Figs 6.3(a) and (b)) (Scheel and Elwell, 1973a).

relationships ahead of the interface are shown in Figs 6.9(a) and (b), which may be compared with the corresponding diagrams for an unstirred solution shown in Figs 6.3(a) and (b). The interface concentration n_i will exceed the equilibrium value as interface-kinetic control becomes dominant and the solute gradient will depend, to a good approximation, on $(n_{sn} - n_i)/\delta$. Thus, although δ decreases with stirring, there is a corresponding decrease in the width of δ_T of the thermal boundary layer. So if the temperatures of the crystal surface and the bulk solution remain constant, stabilization results since the temperarute gradient is steepened by stirring to a greater extent than the solute gradient. This additional stabilization is a result of the enhanced degree of interface control, which determines the growth rate v (see Eqn. 6.3).

Tiller (1968) reached the opposite conclusion, namely that stirring leads

12

to reduced stability in the case of an infinite plane surface. This conclusion will apply to cases where the temperature control is such that the increase in solute gradient is not compensated by an increased temperature gradient. Hurle (1961) examined the conditions for stable growth in the case of a rotating crystal and concluded that the supersaturation gradient is independent of the crystal rotation rate.

The theory of interface stability during growth from stirred melts was also considered by Delves (1968) and by Hurle (1969), who used a perturbation analysis. Delves concluded that the interface may be stabilized by fast stirring if the liquid is slightly supercooled. Under conditions near to instability, a self-excited oscillatory motion of the interface was predicted, with a wavelength of $30\,\mu m$ in the example quoted. This result is very similar to that of Shewmon which was discussed in Section 6.1.2. Hurle also concluded that the effect of stirring on the stability of a plane interface is small, and found that there is no condition of absolute stability in a stirred solution.

When crystals of finite size are considered, the important condition is that growth should be uniform over the whole surface. As discussed in the previous section, instability will result if the supersaturation falls to zero at the centre of a face, and the probability that this will occur is clearly much reduced in a flowing solution. The problem of the maximum size of a crystal face for stable growth has been considered by Carlson (1958) who assumed a region of laminar flow between the surface and the bulk solution. For the crystal to grow at a uniform rate, the concentration was assumed to vary with the distance y from the leading edge according to

$$n = n_{sn} - by^{1/2} \qquad (6.23)$$

where b is independent of n or y. The maximum length l of the face for stable growth is then determined by the condition that the surface concentration should not fall below the equilibrium value. This gives for the limiting value

$$l = 0.214\,Du/\mathrm{Sc}^{1/3}[v\rho/(n_{sn} - n_e)]^2 \qquad (6.24)$$

where Sc is the Schmidt number $\eta/\rho_{sn} D$, u the solution flow rate and v the linear growth rate. With this equation, and assuming similar values for D, v, $n_{sn} - n_e$, etc., to those in the previous examples, values of l some two orders of magnitude higher than those given by the theory of Chernov [Eqn (6.19)] or Cahn [Eqn (6.20)] are predicted. Equation (6.24) therefore appears to provide a stability criterion which may be used as a basis for practical procedures for crystal growth under stable conditions, as will be discussed further in Section 6.6.

6.3. Ultimate Limit of Stable Growth

A prediction of the constitutional-supercooling or supersaturation-gradient approach is that the maximum stable growth rate may be increased as the temperature gradient at the interface is increased. However, experimental evidence indicates that there exists for any material an ultimate rate of stable growth which cannot be exceeded even with a steep temperature gradient and a high degree of stirring.

Data for the stable growth rates from a number of typical HTS and LPE growth experiments have been listed in Table 6.1 and in no case was this rate found to exceed significantly 5×10^{-6} cm s^{-1}, or about 4 mm per day. It is probable that a limiting growth rate of this order is imposed by surface-kinetic processes such as desolvation, integration at kinks and removal of solvent molecules from the surface. In several cases spontaneous nucleation of further crystals might limit the maximum feasible growth rate.

However, faster growth rates are possible in crystal growth from the melt and it is clear that the transition from a dilute solution to a pure melt is gradual. This implies that higher stable growth rates may be achieved in solution growth if the solute concentration is relatively high. This conclusion is confirmed by the work of Belruss et al. (1971), who reported stable growth rates of 10^{-5} cm s^{-1} in top seeding experiments using a 70–90% solute concentration. At high values of the growth rate, the removal of the heat of crystallization cannot be neglected as a rate determining factor.

Wilcox (1970) has discussed the influence of a temperature gradient on crystal facetting. In high temperature gradients, crystals tend to grow without facets and it is possible that, in certain systems, even higher growth rates than those of Belruss et al. (1971) could be achieved with non-facetted crystals.†

If crystals are to remain facetted, the only possibility of faster stable growth than by the usual layer mechanism would appear to be by encouraging a high activity of hillock sources on highly dislocated faces. Figure 6.10 shows the large activity of growth hillocks on an yttrium iron garnet crystal compared with the layer mechanism. The photograph shows a surface which normally grows by spreading of layers from relatively few centres. The surface has two raised circular areas due to solution droplets which have remained after removal of the bulk of the solution by hot pouring. During cooling to room temperature, rapid growth continued on these areas and the remaining flux was subsequently removed by dissolution. One droplet shows continued layer growth with a raised rim due

† This conclusion is confirmed by results reported by Mrs. V. A. Timofeeva at ICCG 4, Tokyo, 1974.

TABLE 6.1. Experimentally Observed Growth Rates in Crystal Growth from HTS

Crystal	Solvent	Linear growth rate $\text{Å} \cdot \text{s}^{-1}$	Remarks	Reference
$Ba_2Zn_2Fe_{12}O_{22}$	$BaO—B_2O_3$	200	Pulling from solution	AuCoin et al. (1966)
$GdAlO_3$	$PbO—PbF_2—B_2O_3$	~200	Accelerated crucible rotation technique	Scheel (1972)
$NiFe_2O_4$	$NaFeO_2$	200	Pulling from solution	Kunnmann et al. (1963)
$NiFe_2O_4$	$BaO—B_2O_3$	~500	Pulling from solution	Smith and Elwell (1968)
$NiFe_2O_4$	$PbO—PbF_2$	~260	Seeded growth from solution	Kvapil et al. (1969)
$Y_3Fe_5O_{12}$	$BaO—B_2O_3$	120	Seed crystal on stirrer	Laudise et al. (1962)
$Y_3Fe_5O_{12}$	$BaO—B_2O_3$	150	Pulling from solution	Linares (1964)
$Y_3Fe_5O_{12}$	$BaO—B_2O_3$	~150	Pulling from solution	Kestigian (1967)
Growth by liquid phase epitaxy				
$Al_xGa_{1-x}As$	Ga	22	LPE, slow cooling	Blum and Shih (1971)
$GaAs$	Ga	~140	LPE, fast cooling	Kang and Greene (1967)
$GaAs$	Ga	~170	LPE, slow cooling	Kinoshita et al. (1968)
$Ga_{1-x}Al_xAs$	Ga	~10	LPE, slow cooling	Woodall (1972)
$InAs_{1-x}Sb$	In	250	LPE, gradient transport	Stringfellow and Greene (1971)
$InAs_{1-x}Sb$	In	170	LPE, gradient transport	Stringfellow and Greene (1971)
$Eu_1Er_2Fe_{4.3}Ga_{0.7}O_{12}$	$PbO—B_2O_3$ $Bi_2O_3—V_2O_5$	660	LPE, 30° supercooling	Levinstein et al. (1971)
$Eu_{0.6}Y_{2.4}Fe_{3.9}Ga_{1.1}O_{12}$	$PbO—B_2O_3$	340	LPE, slow cooling	Giess et al. (1972)
$Y_3Fe_5O_{12}$	$BaO—B_2O_3$	260	LPE, gradient transport	Linares et al. (1965)
$Y_3Fe_5O_{12}$	$BaO—B_2O_3$	8	LPE, slow cooling	Brochier et al. (1972)
$RFeO_3$	$PbO—B_2O_3$	11	LPE, slow cooling	Shick and Nielsen (1971)

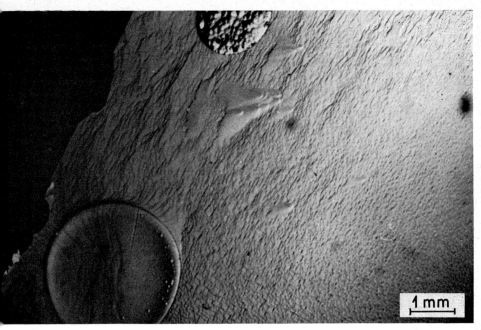

FIG. 6.10. Surface of a garnet crystal showing growth layers and two solidified solution droplets from which the solvent was dissolved. The large droplet shows continued layer growth during cooling, the smaller one the "nucleation" of growth hillocks (Scheel and Elwell, 1973b).

to faster cooling and crystallization in that region. On the other droplet many hillocks have been nucleated and the resulting region appears to be inclusion-free in spite of the rapid growth. An increase in growth rate of bulk crystals by this mechanism would clearly be at the expense of crystal quality, as measured by the dislocation density and impurity incorporation.

A method of achieving fast growth rates which has been little explored is the use of a very thin zone of solvent over the whole crystal surface, with solute supplied from the vapour phase. The advantage of a very thin zone is that supersaturation gradients would be avoided, and the use of an "ultra-thin" alloy zone for the growth of silicon has been proposed by Hurle *et al.* (1964, 1967) and by Filby and Nielsen (1966).

In view of the severe limitation imposed by the normally slow growth rates used in solution growth, any method which could permit an increase in the maximum stable growth rate by a substantial factor is worthy of investigation. The most significant contribution to fast stable growth rates is given, assuming an optimum choice of solvent and growth technique, by an adequate temperature gradient and sufficient solution flow rates at the growing crystal faces.

6.4. Experiments on Growth Stability

Several experiments have been performed, particularly with aqueous solutions, with the aim of determining the conditions for stable growth and in order to observe the effects of instability.

The existence of a maximum rate of stable growth and its dependence on crystal size was demonstrated as long ago as 1939 by Yamamoto. He measured the critical growth rate of alkali halide crystals of various sizes growing in aqueous solution. The incidence of inclusions at the higher growth rates was found by microscopic observation to depend upon the spreading of layers across the crystal face. Under stable conditions, only one layer could be seen to be advancing across a given face at any time. Unstable conditions leading to inclusion formation could be correlated with the formation of successive layers before a previous layer had reached the edge of the crystal. Yamamoto's observations led him to propose that the maximum rate of stable growth decreases in proportion to the area of the crystal face.

The decrease in the maximum stable growth rate with crystal size was also stressed by Egli and Zerfoss (1949) and by Egli (1958), although quantitative data were not given.

Detailed studies were made by Denbigh and White (1966) of the growth stability of hexamethylenetetramine. They found no inclusions in the central 65 μm of crystals and concluded that this represents a critical size below which inclusions are not formed, irrespective of the growth conditions. The incidence of inclusions in larger crystals confirmed the validity of a critical growth rate, which has a value of about 2×10^{-5} cm s^{-1} for this material. The critical growth rate was substantially independent of the stirring rate for the small crystals grown ($\sim 10^{-2}$ cm), but inclusions were not observed when the stirring rate was very high. In the batch system used, however, the main effect of stirring was to increase the nucleation rate and hence to reduce the crystal size. A quantitative investigation was also made by Alexandru (1972) of the stability of Rochelle salt. The crystals used in this case were large, up to 600 g in weight, in contrast to the relatively small crystals studied by Yamamoto (1939) and Denbigh and White (1966). The measurements were made under conditions of fairly rapid solution flow. Alexandru found that the maximum stable growth rate varies in inverse proportion to the length of the crystal face. As with Yamamoto's observations, stability was believed to be correlated with the rate of movement of layers across the crystals and was influenced by the presence of impurities in the solution. The stability condition could also be expressed in terms of a maximum supersaturation σ_{max}, which was related to the face length by an expression of the form

$$\sigma_{max} = a + \frac{b}{x}. \qquad (6.25a)$$

When a seed crystal of length x_o was used, a modified relation

$$\sigma_{max} = a + \frac{b}{x - x_o} \qquad (6.25b)$$

was found to fit the data. It is perhaps surprising that the maximum supersaturation for a given crystal size should depend on the previous history of the crystal, and these observations indicate the importance of the distribution of dislocations in the crystal.

Relatively few measurements have been made of maximum stable growth rates in high-temperature solution but Wentorf (1971) reached the same conclusion for the growth of diamond as did Alexandru for Rochelle salt, namely that the stable growth rate should decrease inversely as the diameter of the crystal. For a 1 mm crystal, the maximum growth rate was found to be about 0.2 mm hr^{-1}, decreasing to 0.04 mm hr^{-1} when the crystal reached its maximum size of 5 mm.

Bruton (1971) studied the stability of growth of lead tantalate, $PbTa_2O_6$, by top seeding from a $Pb_2V_2O_7$ flux under conditions which were believed to be turbulent. According to Carlson's criterion [Eqn(6.24)] the maximum size of crystal for stable growth was calculated to be 1.2 cm. In practice inclusion-free crystals rarely grew larger than $4 \times 2 \times 1$ mm, and larger crystals usually contained many inclusions.

Dawson et al. (1974) measured the growth rate and inclusion concentration of $NaNbO_3$ grown on a rotating seed in $NaBO_2$ as a function of the temperature difference across the melt. The results are shown in Fig. 6.11. The growth rate varies approximately as ΔT^2 and extrapolates to rather a large value at $\Delta T = 0$ because of solvent evaporation. Also shown in Fig. 6.11(a) is the line which is believed to denote the boundary between stable and unstable growth. The justification for this particular choice of stability condition is that the variation with ΔT of the inclusion concentration is very similar to a plot of the difference between the actual growth rate and the value given by this boundary line, as may be seen from Fig. 6.11(b).

The stability condition represented by the boundary in Fig. 6.11(a) is in good agreement with the supersaturation-gradient concept, if it is assumed that the temperature gradient at the crystal surface varies directly as ΔT. From Eqn (6.6), $(v/\Delta T) > (D\phi n_e/\rho RT^2 \Delta z)$ for instability, where Δz is the length over which the temperature drop occurs. If Δz is taken to be the depth of the melt (2.3 cm) with $D = 4 \times 10^{-5}$ cm^2 s^{-1}, $\phi = 59$ kJ mole^{-1}, $\rho = 4.44$ g cm^{-3}, $n_e = 1.77$ g cm^{-3}, $T = 1378$ K (all measured experimental values), the value predicted for $v/\Delta T$ is 12×10^{-9} cm s^{-1} K^{-1}.

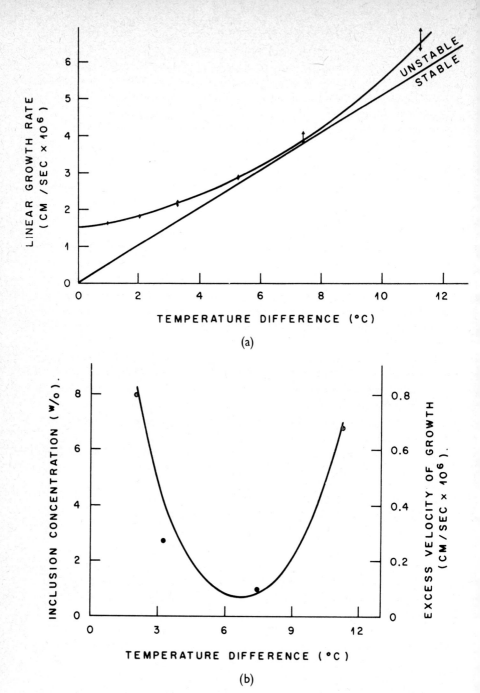

FIG. 6.11. (a) Growth rate of NaNbO$_3$ on a rotating seed in NaBO$_2$ solution, with estimated limit of stable growth. (b) Solvent inclusions concentration and excess growth rate above estimated limit for NaNbO$_3$ (Dawson *et al.*, 1974).

This is much lower than the experimental value of 5.3×10^{-7} cm s^{-1} K^{-1}. However, if the temperature is assumed, due to stirring in the bulk solution, to be dropped over a thermal boundary layer of width $\delta(Sc/Pr)^{1/2}$, where Sc is the Schmidt number and Pr the Prandtl number, a value in the region of 5×10^{-7} cm s^{-1} K^{-1} for $v/\Delta T$ is predicted by the supersaturation-gradient model. Unfortunately a direct measurement of the temperature gradient at the interface was not possible, but these results strongly support the validity of the supersaturation-gradient approach.

6.5. Results of Unstable Growth

Since stable growth has been defined in the present context as growth without solvent inclusions, it is clear that instability will result in inclusions. What is interesting is to consider the extent to which the quantative and qualitative models discussed above can account for the observed features which result from the onset of instability.

The development of a periodic disturbance has already been discussed in Section 6.1.2 and examples of theoretical treatments which predict this periodicity have been mentioned. Another observation of a crystal with periodic inclusions is illustrated in Fig. 6.12, which shows a section parallel to the growth direction of a NaNbO$_3$ crystal grown by top seeding (Dawson et al., 1974). As in Fig. 6.6, the instability has occurred at the termination

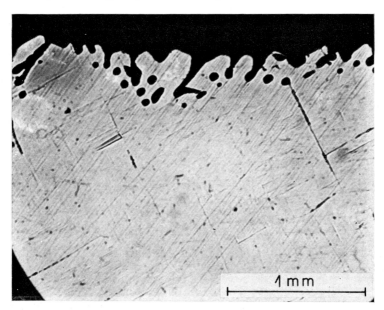

1 mm

FIG. 6.12. Section through NaNbO$_3$ showing periodic inclusion structure (Dawson et al., 1974).

of growth, possibly following removal from the solution. The period in this case is about 80 μm.

A cellular interface similar in appearance to that observed on melt-grown crystals under unstable conditions was reported by Hurle *et al.* (1962) on crystals of InSb grown from solution in a supersaturation gradient and recently by Schieber and Eidelberg (1973) on crystals of $BaFe_{12}O_{19}$. The cells in the latter case were believed to be due to platinum segregation.

Another example of a crystal with periodic inclusions, in this case $DyVO_4$ grown by slow cooling by Garton and Wanklyn (1969), is shown in Fig. 6.13. It is by no means certain that the periodicity in this case is due to excitation of the longer face since the periodicity may well be *in* the growth direction. Landau (1958) has predicted that, under constitutionally supercooled conditions, the growth rate may vary periodically, resulting in a periodic distribution of impurities. The periodicity is caused by the lowering of the degree of constitutional supercooling by an interval of unstable growth with inclusion formation, so that an interval of stable growth follows during which the instability builds up to some critical value and the cycle is re-initiated. This model could account for some of the striations and bands of inclusions which are observed in HTS-grown crystals (see Chapter 9) but has not found wide acceptance.

Particularly under diffusion-limited conditions, the higher supersatura-

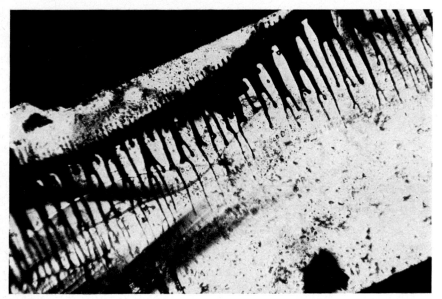

Fig. 6.13. $DyVO_4$ crystal with periodic inclusions (Garton and Wanklyn, 1969).

tion at the corners and edges of a crystal will lead to an onset of more rapid growth in these regions as the degree of constitutional supersaturation is increased. A progressive increase in the supersaturation gradient leads first to the formation of raised edges, then to the development of terraces or "hopper" crystals and finally to dendrite formation with projections in the directions of rapid growth. This sequence is illustrated in Fig. 6.14 which is taken from the paper of Fredriksson (1971), who discusses the morphology of metal crystals as a function of the growth conditions.

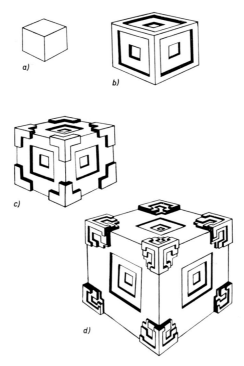

Fig. 6.14. (a)—(d) Progressive changes in shape of an ideally cubic crystal with increasing degree of supersaturation gradient (Fredriksson, 1971).

If growth occurs by a mechanism of layer spreading from corners and edges, it is very probable that the inclusions will be formed at the face centres of the crystal where the supersaturation is low. This has been confirmed by several observations by the authors, by the work of Carlson (1958) on aqueous solution growth and, for example, by Lefever and Chase (1962) on yttrium iron garnet. A more detailed description of the types of inclusion found in crystals grown from HTS will be given in

Chapter 9. Many examples of the transition from the normal habit form to skeletal and similar crystal shapes are quoted in the review by Chernov (1972).

6.6. Experimental Conditions for Stable Growth

6.6.1. Optimum programming for stable growth

The essential criterion for stable growth is that the growth rate should always lie below the maximum stable value, with a "safety margin" to allow for temperature fluctuations due to imperfect regulation or to convection overstability in the solution.

Scheel and Elwell (1972) and Pohl and Scheel (1975) presented a temperature programme for the growth of crystals by slow cooling, with the stable growth rate estimated according to Carlson's criterion which was discussed in Section 6.5. Rearrangement of Eqn (6.24) gives the maximum stable growth rate $v_{max} = dl/2dt$ for a crystal of side l as

$$v_{max} = Bn_e/\sqrt{l} \qquad (6.26)$$

with $\qquad B = (0.214\ Du\sigma^2/Sc^{1/3}\rho^2)^{1/2}. \qquad (6.27)$

Here n_e is the solubility at temperature T, and σ is the relative supersaturation $(n_{sn} - n_e)/n_e$ which is assumed to remain constant throughout the crystallization process. The crystal volume is

$$l^3 = (n_0 - n_e)V/\rho, \qquad (6.28)$$

where n_0 is the initial solubility at time $t = 0$ and V the volume of the solution. Combination of (6.26) and (6.27) yields after integration

$$n_e(T) = n_0/\cosh^2[B(n_0\rho/V)^{1/2}t]. \qquad (6.29)$$

This equation defines the temperature T as a function of the time t if the solubility curve is known and the supersaturation is given a value below some critical limit.

An example of a cooling programme based on Eqn (6.29) is shown as curve III in Fig. 6.15(a). The parameters assumed are: solution volume $V = 80$ cm³, $\rho = \rho_{sn} = 5$ g cm⁻³, $n = 15\%$ at 1600 K and 5% at 1300 K, $D = 10^{-5}$ cm² s⁻¹, $\sigma = 10^{-2}$, $Sc = 420$ and $u = 10$ cm s⁻¹. It may be seen that the deviation of the calculated programme from a constant cooling rate [II of Fig. 6.15(a)] is relatively slight except at the early stage where the crystal is very small. The corresponding growth rates are shown in Fig. 6.15(b)

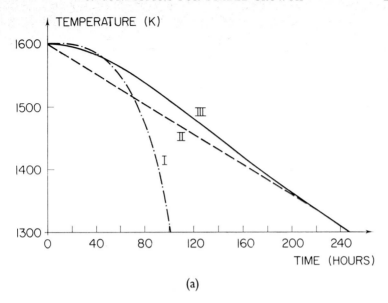

(a)

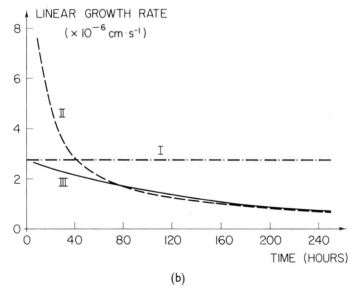

(b)

FIG. 6.15. (a) Temperature programmes for HTS growth by slow cooling. I constant linear growth rate; II constant cooling rate; III for maximum stable growth rate according to Eqn (6.29). (b) Linear growth rate for programmes I—III (Scheel and Elwell, 1972).

from which it is clear that the initial growth rate in case II is much higher than the limiting value, which explains the frequent observation of a dendritic core in large crystals grown by spontaneous nucleation.

Curve I of Fig. 6.15(a) is based on the assumption of a constant linear growth rate, in which case the temperature change from the initial value will vary as t^3. Several proposals for temperature programmes (Neuhaus and Liebertz, 1962; Koldobskaya and Gavrilova, 1962; Sasaki and Matsuo, 1963; Bibr and Kvapil, 1964; Kvapil, 1966; Cobb and Wallis, 1967; Kvapil *et al.*, 1969; Fletcher and Small, 1972; Wood and White, 1972) have been based on the use of a constant linear growth rate, but it is clear from Fig. 6.15(b) that the growth rate may exceed the maximum stable value during the later stages of growth unless the constant value is initially well below the stable limit.

Figure 6.16 shows the effect of viscosity, solution flow rate and crucible size on the temperature programme calculated using Eqn (6.29) with otherwise the same parameters as in the previous example. A total duration of 10^3 hours (about 6 weeks) is considered acceptable but twice this value would probably be prohibitive. A rapid solution flow rate can be seen to be essential if one large crystal is to be grown. Large crucibles are unlikely to result in one crystal per run but stable growth is possible if multinucleation is taken into account. The effect of flux viscosity is seen to be relatively

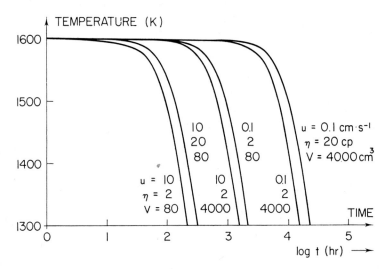

FIG. 6.16. Temperature programmes according to Eqn (6.29) for various values of solution flow rate u, viscosity η and solution volume V (Scheel and Elwell, 1972).

minor, but it is unlikely that high values of solution flow rate will be possible in viscous solvents.

The slowest cooling rates must be used in the early stage following nucleation and a considerable saving in time without the onset of unstable growth can be achieved by the use of seed crystals. The initial cooling rate with a seed crystal will be chosen to correspond to the maximum stable growth rate for the particular size of seed chosen.

The programme specified by Eqn (6.29) was calculated on the basis that the growth rate should at all times have its maximum stable value. In practice it is desirable to use a growth rate which is less than the maximum value by a sufficient margin to allow for minor temperature fluctuations within the solution. The best temperature regulation which can be obtained with commercial controllers is about $\pm 0.1°C$ and, in the example considered above, a sudden drop of $0.1°C$ would result in the deposition of about 13 mg of solute. If this drop were to occur in 10 s on a crystal of area 1 cm², the resulting growth rate would be 3×10^{-4} cm s^{-1}, which is two orders higher than the maximum stable value! In practice the super-saturation is created throughout the melt and the effect of the temperature drop is much less drastic, but sudden temperature drops of $2°C$ may occur when cooling is effected by a motor-driven helipot or similar mechanical means. This shows that excellent temperature control and programming are necessary when large inclusion-free crystals are to be grown.

Curve I of Fig. 6.17 shows the cooling rate according to the programme of Fig. 6.15 (curve III) and a less idealized practical procedure is indicated by the dotted line II. The actual values of the cooling rate proposed in this example are: $0.2°C$ hr^{-1} for the first 48 hr, $0.5°C$ hr^{-1} for the next 24 hr and $1.2°C$ hr^{-1} for the remainder of the growth period, about 220 hr. Those values are chosen to give a reasonable safety margin, except for the initial value which is selected on the basis that it is pointless to use a cooling rate which is not at least comparable with the random fluctuations (Laudise, 1963). The increase in time required by the proposed procedure is about 75 hr or 25%.

Also shown in Fig. 6.17 as curve III is the cooling rate required by the programme of Eqn (6.29) for the same conditions as for curve I but with $u = 0.1$ cm s^{-1}, a value typical of stirring by natural convection. The maximum stable value in this case is only $0.175°C$ hr^{-1} and the total time required by the programme is about 100 days. Since such a period would be unacceptable to most crystal growers we propose the use of a constant cooling rate of 0.2 or $0.3°C$ hr^{-1} for experiments using unstirred melts. Such a cooling rate will probably result in more than one crystal but should yield only a few crystals with substantial inclusion-free regions.

Temperature programming for the growth of crystals in industrial crystallizers has been discussed by Mullin and Nyvlt (1971), and Wood and White (1968) advocated the use of programming in crystal growth by flux evaporation in order to achieve a constant linear growth rate. However, the linear growth rate has to decrease according to Scheel and Elwell (1972), and the flux evaporation rate should be programmed according to the maximum stable growth rate. In certain cases a constant linear growth rate might be necessary, for instance for homogeneous doping. Constant growth conditions necessitate, according to the Burton-Prim-Slichter equation for the effective distribution coefficient, a nonvarying boundary-

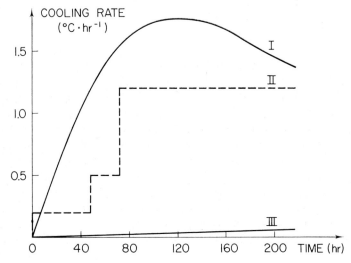

FIG. 6.17. Cooling rates for stable growth. I as for programme III of Fig. 6.16(a); II suggested practical procedure; III for an unstirred solution (Scheel and Elwell, 1972).

layer thickness and therefore a nonvarying area of the growing crystal face. These conditions are only fulfilled in liquid phase epitaxy and in such cases where the application of large seed plates is possible and crystal growth occurs mainly in the direction normal to the seed plate by proper choice of the seed orientation. In any case a value of a constant linear growth rate has to be chosen which is equal to or lower than the maximum stable growth rate for the final crystal size. From this discussion it follows that it is difficult to obtain the quasi-steady-state conditions necessary for the growth of large crystals of homogeneous dopant concentration or of homogeneous solid solutions, and the experimental conditions required are discussed in the next chapter.

6.6.2. Stirring in flux growth

The desirability of stirring for the achievement of stable growth at relatively rapid rates has been mentioned in the discussion of the examples in
Section 6.4, and stirring techniques are discussed in the next chapter.
In the great majority of experiments, some stirring action is achieved by
natural convection. The role of natural convection in crystal growth has
been reviewed by Cobb and Wallis (1967), Parker (1970), Wilcox (1971),
and Schulz-Dubois (1972).

The onset of convection is normally specified (Chandrasekhar, 1961)
by the value of the dimensionless Rayleigh number

$$R = \frac{g\alpha L^3 \Delta T}{K\nu} \tag{6.30a}$$

where α is the volume expansion coefficient, L the depth, K the thermal
diffusivity and ν the kinematic viscosity of the liquid, with ΔT the temperature difference across it. Some critical value of R, depending on an idealized
geometry, must be exceeded for convection while higher values of R may
lead to temperature oscillations, or to turbulence at even higher values.
However, in real crystal-growth systems some convection will occur below
the critical R values due to some inevitable asymmetric or reverse temperature gradients due to buoyancy.

In solutions it is also necessary to consider thermosolutal convection
due to density differences between the solute and solvent. The onset of
solutal convection may be specified by defining a solutal Rayleigh number
R_s as

$$R_s = \frac{g\beta L^3 \Delta n}{K_s \nu} \tag{6.30b}$$

where β is the rate of change of density with concentration, Δn the solute
concentration difference across the liquid and K_s the diffusivity of the
solute. Since K_s is normally lower than the thermal diffusivity K by some
orders of magnitude, convection is highly probable in solutions even if the
temperature gradient is in the "wrong" direction. Oscillations are particularly likely in solution due, for example, to "overstability" which can
occur when a destabilizing temperature gradient is opposed by a solute
gradient (Jakeman and Hurle, 1972).

Even for pure melts, it is difficult to obtain a reliable expression for the
rate of convective flow of the liquid. Cobb and Wallis (1967) derived a
simple expression for the flow rate of a liquid, unbounded in the horizontal
direction, between horizontal plates differing in temperature by ΔT. The
average flow rate u was estimated as

$$u = \frac{K}{\rho_{sn} L C_P} [0.208(R)^{1/4} - 1] \qquad (6.31)$$

where C_P is the specific heat of the liquid and R the Rayleigh number. Figure 6.18(a) shows the value of u as a function of the liquid depth L for various values of the convection parameter $a = R/L^3$ and for a fixed value of $\Delta T = 10°C$. Except for very low values of L, where u increases very rapidly, the solution flow rate is seen to remain between roughly 0.01 and 0.04 cm s^{-1}. Figure 6.18(b) shows the convection flow rate versus temperature difference ΔT for a typical value of $a = 1500$. The latter is limited in a crystal-growth experiment because of nucleation at high values of ΔT, and therefore this possibility of increasing u is also limited. Maximum rates of thermal convection flow in high-temperature ionic solutions will be of the order of 0.1 cm s^{-1} and in metallic solutions one order of magnitude faster.

From the examples illustrated in Fig. 6.16, it is clear that such values of the flow rate are too low for stable growth of a few large crystals per crucible, and so forced convection by stirring is desirable where possible. Stirring may be achieved by rotating a seed crystal as in the *top-seeded solution-growth* (TSSG) technique and the resulting flow patterns have been studied by Robertson (1966) and Carruthers and Nassau (1968) using an aqueous analogue. Stirring may be enhanced by occasional reversal of the seed rotation (Miller, 1958; Senhouse *et al.*, 1966) or of the crucible rotation direction (Nassau, 1964; Bonner *et al.*, 1965; Schroeder and Linares, 1966). This TSSG technique is restricted to solvents of low volatility and is therefore not applicable to PbO/PbF$_2$ and many other widely used solvents.

The problem of stirring a corrosive liquid at high temperatures is by no means simple. Serious problems are associated with sealing the crucible and stirring the solution in a sealed crucible at high temperatures or with the corrosive solvent vapours. The only really effective method proposed to date is the *accelerated crucible rotation technique* (Scheel and Schulz-DuBois, 1971; Scheel, 1972) in which the rate (and frequently also the sense) of crucible rotation is varied continuously (but not abruptly as in the crucible reversal mentioned above) and the inertia of the liquid used to promote mixing. Experience from aqueous solution growth indicates that flow rates of 10–50 cm s^{-1} are desirable and the practical realization of such conditions will be discussed in the next chapter.

An alternative approach to the problem of non-uniform solute flow has been proposed by Tiller (1968) and is illustrated in Fig. 6.19. He suggested the use of a convection-free cell with a seed crystal located inside a platinum tube inserted into a well-mixed solution. Convection is prevented by

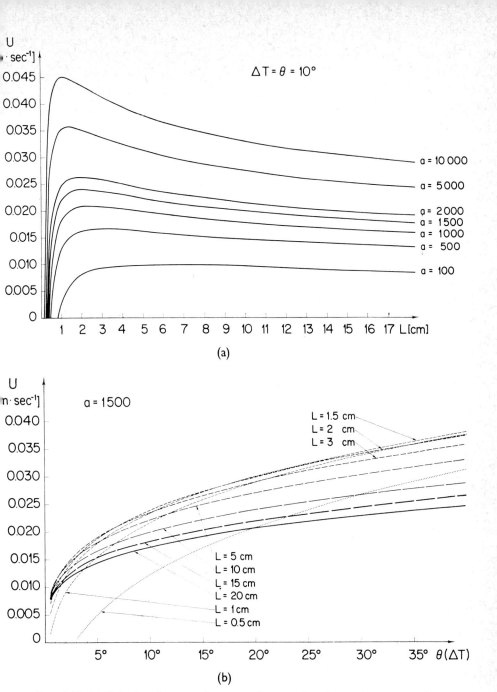

FIG. 6.18 (a) Solution flow rate by natural convection for various convection parameters a $(=R/L^3)$ and temperature difference of $10°C$ versus liquid layer height L. (b) Solution flow rate by natural convection versus temperature difference ΔT for various liquid layer heights and a constant convection parameter $a = 1500$. (Scheel and Elwell, 1973a.)

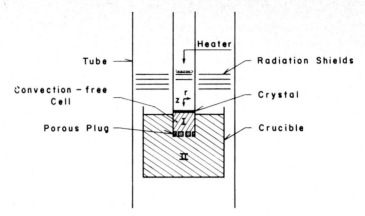

FIG. 6.19. Convection-free cell (I) in a stirred solution (II) (Tiller, 1968).

baffles across the end of the tube and a uniform flow of solute to the seed by diffusion may be realized. This method does not yet appear to have been tested in practice on Earth but could give stable growth if technical problems and problems associated with heat conduction through the platinum tube (and nucleation on the tube) could be solved. The growth rate would, however, be appreciably lower than in a stirred solution. Interesting results on convection-free growth may be expected from the Skylab experiments where convection does not occur due to the near zero gravity in space.

6.6.3. Temperature control and distribution
The importance has been stressed of using the best possible controller to regulate the temperature of the furnace. Commercial controllers using saturable core reactors or thyristors can give regulation to $\pm 0.1°C$ and their ready availability has greatly contributed to a continuing improvement in crystal size and quality. However a high degree of control is pointless if temperature oscillations due to convection overstability are present in an unstirred solution. Smith and Elwell (1968) measured oscillations of amplitude $0.5°C$ in a solution of $NiFe_2O_4$ in $BaO.0.62B_2O_3$ at $1200°C$ with a melt depth of about 3 cm. The amplitude of these oscillations was reduced to some extent by rotating the crucible (see Section 9.2.4).

By effective stirring such temperature oscillations could be prevented so that exact temperature regulation becomes meaningful. It is technically difficult to measure small temperature fluctuations in stirred high-temperature solutions but estimates of Schulz-DuBois (1972) and Scheel (1972) indicate a high degree of temperature homogenization, for instance by the accelerated crucible rotation technique, and this applies also to other stirring techniques which dominate the hydrodynamics in the high-temperature solutions.

From the discussion of the supersaturation-gradient criterion it is evident that the temperature gradient at the crystal surface is of major importance in determining the crystal quality. In theory a very large temperature gradient is desirable so that stable growth will be possible at high growth rates but in practice a compromise is always used because of an adverse effect on the crystal quality. A large gradient will result in a high degree of strain and so more dislocations will be produced in the crystal. In top-seeded growth where the temperature gradient at the crystal can be varied by changing the depth of the crystallizing interface below the liquid level and by altering the degree of coolant flow through the seed holder, a high cooling rate is often found to result in nucleation at the edges of the crystal and it is important to ensure that the radial temperature gradient across the surface of the solution is not too great.

Similar considerations apply to growth by spontaneous nucleation where localized cooling is used. Measurements of the optimum temperature distribution for the growth of garnet crystals at the base of a crucible have been reported by Tolksdorf and Welz (1972).

6.6.4. Mechanical disturbances

The concept of a metastable region of supersaturation gradient suggests that instability will be favoured if the growing crystal is subjected to mechanical shock, which will tend to nucleate any instability. This view is supported by the experience of the authors with both growth by spontaneous nucleation and by top seeding. In the former case the size of inclusion-free regions was found to be increased by a factor greater than two when the furnaces were mounted on antivibration supports inside closed metal cabinets. In top-seeded growth, a gradual deterioration in crystal quality and an increase in the nucleation of secondary crystals on the edge of a seed have been noted as the seed rotation mechanism became worn.

The practical aspects mentioned above will be discussed more fully in the next chapter.

6.7. Summary

Theoretical treatments of the growth of spherical and cylindrical crystals indicate that these will be stable only up to a critical radius, which will be increased by interface kinetics and surface diffusion.

In the case of a plane crystal surface growing in solution, unstable growth may result if there exists a supersaturation gradient ahead of the interface. A solute gradient associated with the volume-diffusion process will inevitably be present and a supersaturation gradient may be avoided only by the application of a sufficiently large temperature gradient. The experimental observation of inclusion-free growth in very small or negative temperature gradients may be explained by the assumption of a metastable region of supersaturation gradient. Stabilization is believed to result primarily from the kinetic mechanisms on the low-energy habit faces normally exhibited by solution-grown crystals.

An important factor when polyhedral crystals are considered is the difference in supersaturation between the edges and centre of any face. This supersaturation inhomogeneity may be offset by a higher kinetic coefficient at the face centres due to curvature of the vicinal face or to a higher concentration of active growth centres. Stirring the solution is desirable in order to minimize the supersaturation inhomogeneity. Even in well-stirred solutions with a large stabilizing temperature gradient, it is likely that there will exist for any material an ultimate rate of stable growth.

For the experimental attainment of stable growth, precise temperature regulation is required and mechanical shocks should be prevented. The maximum growth rate may be increased by the application of a sufficiently large temperature gradient and by stirring the solution.

A considerable body of evidence has been presented to demonstrate that the maximum stable growth rate decreases with increase in crystal size. Temperature programmes for crystal growth by slow cooling have been presented which are based on the requirement that the growth rate should never exceed its maximum stable value.

References

Alexandru, H. V. (1972) Paper presented at CHISA Congress.

AuCoin, T. R., Savage, R. O. and Tauber, A. (1966) *J. Appl. Phys.* **37**, 2908.

Bardsley, W., Callan, J. M., Chedzey, H. A. and Hurle, D. T. J. (1961) *Solid State Electronics* **3**, 142.

Belruss, V., Kalnajs, J., Linz, A. and Folweiler, R. C. (1971) *Mat. Res. Bull.* **6**, 899.

Bibr, B. and Kvapil, J. (1964) *Chem. Prumysl* (Chemical Industry), 507.

Blum, J. M. and Shih, K. K. (1971) *Proceed. IEEE* **59**, 1498.

Bonner, W. A., Dearborn, E. F. and van Uitert, L. G. (1965) *Am. Ceram. Soc. Bull.* **44**, 9.

Brice, J. C. (1969) *J. Crystal Growth* **6**, 9.

Brochier, A., Coeure, P., Ferrand, B., Gay, J. C., Joubert, J. C., Mareschal, J., Vignie, J. C., Martin-Binachon, J. C. and Spitz, J. (1972) *J. Crystal Growth* **13/14**, 571.

Bruton, T. M. (1971) Ph.D. Thesis, Imperial College, London; Bruton, T. M. and White, E. A. D. (1973) *J. Crystal Growth* **19**, 341.

Bunn, C. W. (1949) *Disc. Faraday Soc.* **5**, 132.

Cahn, J. W. (1967) In "Crystal Growth" (H. S. Peiser, ed.) p. 681. Pergamon, Oxford.

Carlson, A. E. (1958) Ph.D. Thesis, Univ. of Utah; In "Growth and Perfection of Crystals" (R. H. Doremus, B. W. Roberts and D. Turnbull, eds.) p. 421. Wiley, New York.

Carruthers, J. R. and Nassau, K. (1968) *J. Appl. Phys.* **39**, 5205.

Chandrasekhar, S. (1961) "Hydrodynamic and Hydromagnetic Stability" Oxford University Press.

Chernov, A. A. (1972) *Sov. Phys. Cryst.* **16**, 734.

Cobb, C. M. and Wallis, E.B. (1967) Rept. AD 655. 388.

Coriell, S. R. and Parker, R. L. (1966) *J. Appl. Phys.* **37**, 1548.

Coriell, S. R. and Parker, R. L. (1967) In "Crystal Growth" (H. S. Peiser, ed.) p. 703. Pergamon, Oxford.

Dawson, R. D., Elwell, D. and Brice, J. C. (1974) *J. Crystal Growth* **23**, 65.

Delves, R. T. (1968) *J. Crystal Growth* **3/4**, 562.

Denbigh, K. G. and White, E. T. (1966) *Chem. Eng. Sci.* **21**, 739.

Egli, P. H. (1958) In "Growth and Perfection of Crystals" (R. H. Doremus, B. W. Roberts and D. Turnbull, eds.) p. 408. Wiley, New York.

Egli, P. H. and Zerfoss, S. (1949) *Disc. Faraday Soc.* **5**, 61.

Elwell, D. and Neate, B. W. (1971) *J. Mat. Sci.* **6**, 1499.

Filby, J. D. and Nielsen, S. (1966) *Microelectronics and Reliability* **5**, 11.

Fletcher, R. C. and Small, M. B. (1972) *J. Crystal Growth* **15**, 159.

Fredriksson, H. (1971) *Jerkonnt. Ann.* **155**, 571.

Garton, G. and Wanklyn, B. M. (1969) *Proc. Int. Conf. Rare Earths*, Grenoble, 343.

Giess, E. A., Kuptsis, J. D. and White, E. A. D. (1972) *J. Crystal Growth* **16**, 36.

Hardy, S. C. and Coriell, S. R. (1968) *J. Appl. Phys.* **39**, 3505.

Hurle, D. T. J. (1961) *Solid State Electronics* **3**, 37.

Hurle, D. T. J. (1969) *J. Crystal Growth* **5**, 162.

Hurle, D. T. J., Jones, O. and Mullin, J. B. (1962) *Solid State Electronics* **4**, 317.

Hurle, D. T. J., Mullin, J. B. and Pike, E. R. (1964) *Phil Mag.* **9**, 423; *Solid State Comm.* **2**, 197, 201.

Hurle, D. T. J., Mullin, J. B. and Pike, E. R. (1967) *J. Mat. Sci.* **2**, 46.

Ivantsov, G. P. (1951) *Dokl. Akad. Nauk SSSR* **81**, 179.

Ivantsov, G. P. (1952) *Dokl. Akad. Nauk SSSR* **83**, 573.

Jakeman, E. and Hurle, D. T. J. (1972) *Rev. Phys. Tech.* **3**, 3.

Kang, C. S. and Greene, P. E. (1967) *Appl. Phys. Lett.* **11**, 171.

Kestigian, M. (1967) *J. Am. Ceram. Soc.* **50**, 165.

Kinoshita, J., Stein, W. W., Day, G. F. and Mooney, J. B. (1968) *Proceedings Intern. Symp. GaAs* (C. I. Pederson, ed.) p. 22. Inst. of Phys., London.

Koldobskaya, M. F. and Gavrilova, I. V. (1962) *Growth of Crystals* **3**, 199.

Kotler, G. R. and Tiller, W. A. (1967) In "Crystal Growth" (H. S. Peiser, ed.) p. 721. Pergamon, Oxford.

Kunnmann, W., Ferretti, A. and Wold, A. (1963) *J. Appl. Phys.* **34**, 1264.

Kvapil, J. (1966) *Kristall u. Technik* **1**, 97.

Kvapil, J., Jon, V. and Vichr, M. (1969) *Growth of Crystals* **7**, 233.

Landau, A. I. (1958) *Phys. Met. Metallog.* **6**, 148.

Laudise, R. A., Linares, R. C. and Dearborn E. F. (1962) *J. Appl. Phys.* **33**, 1362.

Laudise, R. A. (1963) In "The Art and Science of Growing Crystals" (J. J. Gilman, ed.) p. 252. Wiley, New York.

Lefever, R. A. and Chase, A. B. (1962) *J. Am. Ceram. Soc.* **45**, 32.

Levinstein, H. J., Licht, S., Landorf, R. W. and Blank, S. L. (1971) *Appl. Phys. Lett.* **19,** 486.

Linares, R. C. (1964) *J. Appl. Phys.* **35,** 433.

Linares, R. C., McGraw, R. B. and Schroeder, J. B. (1965) *J. Appl. Phys.* **36,** 2884

Miller, C. E. (1958) *J. Appl. Phys.* **29,** 233.

Mullin, J. W. and Nyvlt, J. (1971) *Chem. Eng. Sci.* **26,** 369.

Mullins, W. W. and Sekerka, R. F. (1963) *J. Appl. Phys.* **34,** 323.

Mullins, W. W. and Sekerka, R. F. (1964) *J. Appl. Phys.* **35,** 444.

Nassau, K. (1964) *Lapidary J.* **18,** 42.

Neuhaus, A. and Liebertz, J. (1962) *Chem. Ing. Tech.* **34,** 813.

Nichols, F. A. and Mullins, W. W. (1965) *Trans. AIME* **233,** 1840.

O'Hara, S., Tarshis, L. A., Tiller, W. A. and Hunt, J. P. (1968) *J. Crystal Growth* **3/4,** 555.

Parker, R. L., (1970) In "Solid State Physics" Vol. 25 (F. Seitz and D. Turnbull, eds.) p. 151.

Robertson, D. S. (1966) *Brit. J. Appl. Phys.* **17,** 1047.

Rutter, J. W. and Chalmers, B. (1953) *Can. J. Phys.* **31,** 51.

Sasaki, H. and Matsuo, Y. (1963) *Natl. Tech. Rept.* **9,** 116.

Scheel, H. J. (1972) *J. Crystal Growth* **13/14,** 560.

Scheel, H. J. and Elwell, D. (1972) *J. Crystal Growth* **12,** 153.

Scheel, H. J. and Elwell, D. (1973a) *J. Electrochem. Soc.* **120,** 818.

Scheel, H. J. and Elwell, D. (1973b) *J. Crystal Growth* **20,** 259.

Scheel, H. J. and Schulz-DuBois, E. O. (1971) *J. Crystal Growth* **8,** 304.

Schieber, M. and Eidelberg, J. (1973) European Research Office U.S. Army Tech. Rept. Contract No. DAJA 37-72-C2777 (see Aidelberg *et al.*, *J. Crystal Growth* **21,** 195, 1974).

Schroeder, J. B. and Linares, R. C. (1966) *Prog. Ceram. Sci.* p. 195.

Schulz-DuBois, E. O. (1972) *J. Crystal Growth* **12,** 81.

Sekerka, R. F. (1965) *J. Appl. Phys.* **36,** 264.

Senhouse, L. S., de Paolis, M. V. and Loomis, T. C. (1966) *Appl. Phys. Lett.* **8,** 173.

Shewmon, P. G. (1965) *Trans. AIME* **233,** 736.

Shick, L. K. and Nielsen, J. W. (1971) *J. Appl. Phys.* **42,** 1554.

Smith, S. H. and Elwell, D. (1968) *J. Crystal Growth* **3/4,** 471.

Stringfellow, G. B. and Greene, P. E. (1971) *J. Electrochem. Soc.* **118,** 805.

Tarshis, L. A. and Tiller, W. A. (1967) In "Crystal Growth" (H. S. Peiser, ed.) p. 709. Pergamon, Oxford.

Tiller, W. A. (1963) In "The Art and Science of Growing Crystals" (J. J. Gilman, ed.) p. 276. Wiley, New York.

Tiller, W. A. (1968) *J. Crystal Growth* **2,** 69.

Tiller, W. A. (1970) In "Phase Diagrams" Vol. 1 (A. M. Alper ed.) p. 199. Academic Press, New York.

Tiller, W. A. and Kang, C. (1968) *J. Crystal Growth* **2,** 345.

Tiller, W. A., Jackson, K. A., Rutter, J. W. and Chalmers, B. (1953) *Acta Met.* **1,** 428.

Tolksdorf, W. and Welz, F. (1972) *J. Crystal Growth* **13/14,** 566.

Vergnoux, A. M., Riera, M., Ribet, J. L. and Ribet, M. (1971) *J. Crystal Growth* **10,** 202.

Voronkov. V. V. (1965) *Sov. Phys. Solid State* **6,** 2378.

Wagner, C. (1954) *Trans. AIME*, J. Metals, 154.

Walton, D., Tiller, W. A., Rutter, J. W. and Winegard, W. C. (1955) *Trans. AIME* **203,** 1023.

Wentorf, R. H. (1971) *J. Phys. Chem.* **75,** 1833.

White, E. A. D. (1965) *Tech. Inorg. Chem.* **4,** 31.

Wilcox, W. R. (1970) *J. Crystal Growth,* **7,** 203.

Wilcox, W. R. (1971) In "Aspects of Crystal Growth" (R. A. Lefever, ed.)
    p. 37. Dekker, New York.

Wood, J. D. C. and White, E. A. D. (1968) *J. Crystal Growth* **3/4,** 480.

Wood, J. D. C. and White, E. A. D. (1972) unpublished.

Woodall, J. M. (1972) *J. Crystal Growth* **12,** 32.

Yamamoto, T. (1939) *Sci. Papers Inst. Phys. Chem. Res.* **35,** 228.

7. Experimental Techniques

It is mentioned several times in the book that the experience obtained from crystal growth from aqueous solutions can be applied to HTS† growth. The same is partially true for the experimental techniques. Therefore it is good practice for scientists and technicians entering the field of HTS growth to obtain some experience with solution growth at low temperatures where the growing crystals may be readily observed. For this reason, reference is made below to a few books and review papers which deal with experimental techniques in crystal growth from (aqueous) solutions: Brice (1973), Buckley (1951), Crystal Growth (1949), Gilman (1963), Haussühl (1964), Holden and Singer (1960), Khamskii (1969), Mullin (1972), Neuhaus (1956), Smakula (1962), Tarjan and Matrai (1972), Van Hook (1961) and Wilke (1973).

† HTS high-temperature solution(s).

This chapter deals first with the principles and with the main factors which have to be taken into account when large crystals of high quality are to be grown. However, the conditions for stable growth have been discussed in Chapter 6, and several aspects that are relevant to experiments have been mentioned in Chapters 3, 4 and 5.

The second part of this chapter treats high-temperature technology as far as is necessary for HTS growth. In particular the attainment and control of high temperatures, the crucible problems and stirring techniques are discussed in some detail.

In Section 7.3 a discussion is given of special techniques and of some relatively unusual aspects, including a speculative treatment of some possible techniques for future development.

7.I. Principles

7.I.I. Metastable region, nucleation, seeding

A typical example of a phase diagram used in crystal growth from HTS is shown diagrammatically in Fig. 7.1. The solvent can be an element, a compound or a combination of compounds. The solute is an element or a compound with a melting point generally higher than that of the solvent, but in principle one could consider growing crystals from eutectic systems

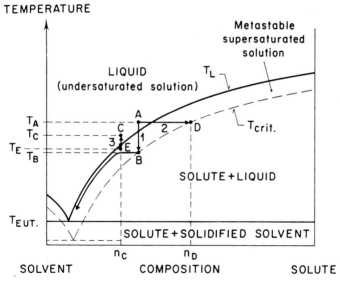

FIG. 7.1. Typical eutectic phase diagram showing the metastable Ostwald-Miers region and crystal growth by slow cooling (1), by solvent evaporation (2) and by gradient transport (3).

in which the "solvent" has a higher melting point. As described in Chapter 3 the liquidus temperature curve is given, at least approximately in the majority of cases, by the Van't Hoff equation.

A solution of composition n_A equilibrated at the temperature T_A can be cooled, in the absence of seeds and agitation, to the temperature T_B at which spontaneous nucleation occurs. In the region between the liquidus line and the dashed line intersecting B the solution is said to be undercooled or supersaturated and this region is called the *metastable* or *Ostwald-Miers* region. The metastability results because a nucleus of critical size must be formed before crystalline material is precipitated, as discussed in Section 4.2.

Metastability is a very complex phenomenon and it is still difficult to make reliable statements on the factors which critically determine the width of the metastable region. In addition, experiments are often difficult to reproduce and disagreement has often been reported between the results of workers who studied nucleation in aqueous solutions. It is, however, clear that the width of the metastable region will be greater if the build-up of high local solute concentrations can be avoided. A broad metastable region is therefore favoured by a small solution volume, a high viscosity and a low solubility. In addition the complexity of the solute and of the solution appears to have an important influence, as will be discussed later in this section. The time dependence of nucleation should also be included in any discussion of metastability (see Eqn 4.11).

The metastable Oswald-Miers region is of paramount importance for the crystal grower since for the growth of large crystals the experimental conditions must be controlled to such an extent that no unwanted nucleation can occur. Homogeneous nucleation is rather improbable in practical crystal growth experiments and will only occur in highly super saturated solutions. Generally heterogeneous nucleation occurs on the container walls or at the surface of the solution, assuming that no undissolved particles are present.

As indicated by curve 1 of Fig. 7.1, continued slow cooling of the solution from temperature T_B is accompanied by crystal growth at much lower supersaturation because of the presence of the crystals which nucleated at B. If solvent is evaporated at constant temperature (curve 2 of Fig. 7.1) the metastable region is passed and nucleation occurs at D. Alternatively, as in 3, the solvent is transported from a hotter (saturated) to a cooler (supersaturated) region.

According to Neuhaus (1956) the width of the metastable Ostwald-Miers region in growth from aqueous solutions depends on the nature of the crystallizing substance, on the degree of agitation and on additives, as demonstrated in Figs 7.2(a) and (b). Figure 7.2(a) indicates that the width of the metastable region increases with increasing degree of "complexity"

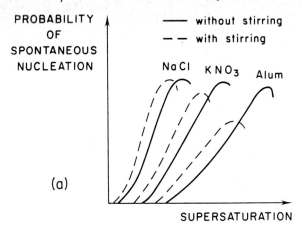

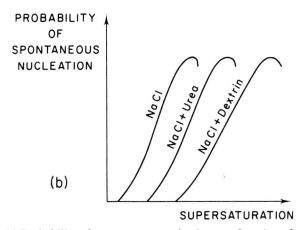

FIG. 7.2. (a) Probability of spontaneous nucleation as a function of agitation and of "complexity" of participating substance (Neuhaus, 1956). (b) Probability of spontaneous nucleation in relation to the type of additive for the example of NaCl crystallization from aqueous solution (Neuhaus, 1956).

$(NaCl \rightarrow KNO_3 \rightarrow KAl(SO_4)_2 \cdot 12H_2O)$, whereas stirring produces the opposite effect. The role of additives on the habit is discussed in Chapter 5. Here the role of additives in generally increasing the width of the metastable region is shown by Fig. 7.2(b). Neuhaus attributed the effect of urea to complex formation in front of the growing crystal, whereas dextrin is said to increase the width of the metastable region by increasing the viscosity.

It would be interesting to establish more quantitatively the relation between the width of the metastable region and such factors as the chemical and crystallographic "complexity" of the crystal, the species present in the solution, the viscosity and the degree of agitation.

In high-temperature solutions the width of the metastable region varies from 1 to 100°C, and very viscous systems (e.g. borate solutions) can be quenched as glasses. As discussed in Chapter 3, there seems to be a relationship between complex formation in the solution and the width of the metastable region. Qualitatively one is tempted to say that missing "complexity" of the crystals might be the cause of the difficulty in obtaining large crystals by flux growth of the simple compounds TiO_2, ZrO_2, BeO, whereas it is becoming easier to grow large crystals with increasing "complexity": $GdVO_4$, $SrTiO_3$, $GdAlO_3$, $Y_3Al_5O_{12}$, $Y_3Fe_5O_{12}$, $Pb_3MgNb_2O_9$. On the other hand there are many other factors which influence nucleation phenomena, for example complex formation in front of the growing crystal.

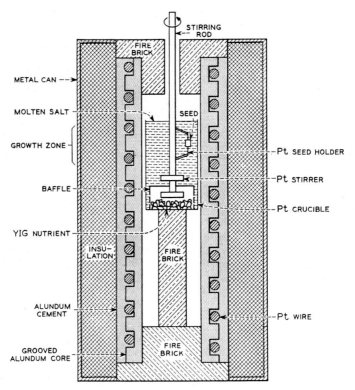

FIG. 7.3. $Y_3Fe_5O_{12}$-crystallization on a rotating seed, with nutrient in the hotter zone (Laudise, 1963).

TABLE 7.1. Crystals Prepared by Top-seeded Solution Growth by the Slow-cooling Technique†

Crystal	Solvent	Size	Reference
$KNbO_3$	K_2CO_3	12–15 g	Miller (1958)
$K(Nb, Ta)O_3$	K_2CO_3	7.5 g	Triebwasser (1959)
$BaTiO_3$	BaF_2	20 × 10 mm	Linares (1960)
$BaTiO_3$	TiO_2	1 cm³	Von Hippel et al. (1963)
$NiFe_2O_4$	$Na_2Fe_2O_4$		Kunnmann et al. (1963)
$KTaO_3$	K_2CO_3		Wemple (1964)
$K(Nb, Ta)O_3$	K_2CO_3	14 × 14 × 12 mm	Bonner et al. (1965)
$K(Nb, Ta)O_3$	K_2CO_3		Wilcox and Fullmer (1966)
$Ba_2Zn_2Fe_{12}O_{22}$	$Ba_2B_2O_5$	10 g	Aucoin et al. (1966)
$KTaO_3$	K_2CO_3		Senhouse et al. (1966)
$(Ba, Pb)TiO_3$	BaB_2O_4—PbB_2O_4	~15 × 10 mm	Perry (1967)
$(Ba, Sr)TiO_3$	TiO_2	~8 × 8 × 8 mm	Bethe and Welz (1971)
$KTaO_3$	K_2CO_3	280 g	Bonner et al. (1967)
$K(Nb, Ta)O_3$	K_2CO_3	75 × 60 × 25 mm (1300 g)	Bonner and Van Uitert (1967)
$Pb_3MgNb_2O_9$	PbO—B_2O_3	~1 cm³	Goodrum (1970)
GeO_2	Na-germanate	3–10 mm	
$K(Nb, Ta)O_3$	K_2CO_3		Epstein (1970)
$Bi_4Ti_3O_{12}$	Bi_2O_3, Bi_2O_3—GeO_2, Bi_2O_3—MoO_3		
$Gd_2(MoO_4)_3$	MoO_3		Hurst and Linz (1971)
$KNbO_3$	K_2CO_3	2 cm³	Belruss et al. (1971)
$SrTiO_3$	TiO_2	14 × 14 × 6 mm	Belruss et al. (1971)
$Y_2Ti_2O_7$	TiO_2 + Ba-borate	>1 cm³	Belruss et al. (1971)
$Ba_2MgGe_2O_7$	GeO_2	>1 cm³	Belruss et al. (1971)
$(Ca, Sr, Ba)Y_2(Zn, Mg)_2Ge_3O_{12}$	$Ca(Sr, Ba)O + GeO_2$	>1 cm³	Belruss et al. (1971)

† For experimental details see Chapter 10.

TABLE 7.2. Crystals Prepared by Top-seeded Solution Growth by the Gradient Transport Technique†

Crystal	Solvent	Size	Reference
$Y_3Fe_5O_{12}$	$BaO \times 0.6B_2O_3$	$\sim 10 \times 10$ mm	Linares (1964)
$Y_3Fe_5O_{12}$	$BaO \times 0.6B_2O_3$	$\sim 40 \times 10$ mm	Kestigian (1967)
$Y_3Fe_5O_{12}$	$BaO \times 0.6B_2O_3$	?	Bennett (1968)
$Y_3Fe_5O_{12}$	$BaO \times 0.6B_2O_3$	$20 \times 20 \times 10$ mm	Tolksdorf (1974b)
$NiFe_2O_4$	$BaO \times 0.62B_2O_3$	10×8 mm	Smith and Elwell (1968)
$K(Nb, Ta)O_3$	K_2CO_3	~ 1 cm^3	Whipps (1972)
$(Zn, Mn)_2Ba_2$ $(Fe, Mn)_{12}O_{22}$	$BaO \times 0.5B_2O_3$	$\sim 15 \times 15 \times 4$	Tolksdorf (1973)

† For experimental details see Chapter 10.

The temperature control requirements necessary for stable and homogeneous growth (see Chapter 6 and Chapter 7.1.3) are normally much more stringent than those necessary to prevent spontaneous nucleation. This implies that if spontaneous nucleation occurs during the growth of a crystal then its quality is often bad due to formation of inclusions and striations.

The growth of large crystals is favoured by the use of seed crystals, which is common practice in crystal growth from aqueous solutions. However, in high-temperature solutions technical reasons make the use of seeds very difficult. Since the seeds cannot be observed in the (platinum) crucibles and in the opaque solutions the solubility curve has to be known very exactly and the conditions carefully adjusted in order to prevent dissolution of the seed crystals.

Growth on a rotating seed immersed in the solution has been described by several authors (see Table 7.5), and an example of an experimental arrangement is shown in Fig. 7.3. Timofeeva and Kvapil (1966) fixed seed crystals near the base of an unstirred crucible and still observed multinucleation. This shows that seed crystals are only of major advantage when the solution is homogenized by forced convection or when the solution volume is small as is often the case in the travelling solvent technique and in liquid phase epitaxy.

The most popular means of nucleation control is by top-seeding as can be seen from Tables 7.1 and 7.2. The arrangement used successfully at MIT for the growth of several crystals is shown in Fig. 7.4. The position of the crystal-liquid interface is important with respect to inclusion formation, dislocation density and maximum attainable growth rate. With increasing depth of the growth face (in the solution) the temperature gradient at the interface becomes smaller and the supersaturation gradient increases, the stable growth rate decreases, but also the dislocation density

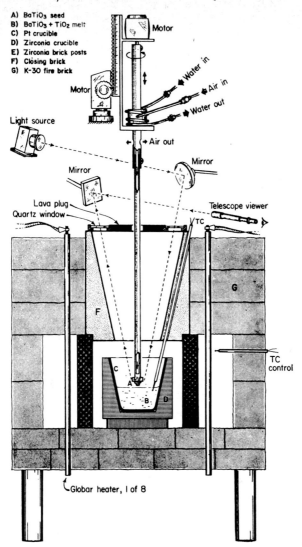

A) BaTiO₃ seed
B) BaTiO₃ + TiO₂ melt
C) Pt crucible
D) Zirconia crucible
E) Zirconia brick posts
F) Closing brick
G) K-30 fire brick

FIG. 7.4. Arrangement for crystal growth by top seeding (Von Hippel *et al.*, 1963; Belruss *et al.*, 1971).

will decrease if inclusions are not trapped. The temperature gradient also determines the degree of facetting as discussed by Wilcox (1970). The technique of growth on rotating seed crystals at the top of a solution, with or without withdrawal, has been given a variety of names (Czochralski, Kyropoulos, pulling from solution etc.), but it seems that the term *top-*

K 2

seeded solution growth (TSSG) first introduced by Linz *et al.* (1965), and described in detail by Belruss *et al.* (1971), is becoming increasingly popular and is therefore adopted in the following discussion.

The size of seeds used in HTS growth has to be a compromise between various requirements. As is discussed in Chapter 6 and may be seen from the topographs of Chapter 9, edge and spiral dislocations propagate from the seed in a direction approximately normal to the growing faces. The high-quality (nearly dislocation-free) regions between these dislocation bundles are obviously larger when small seeds are used.

Small seed crystals, on the other hand, have the disadvantage of being easily dissolved and also require an extremely low supersaturation if the maximum stable growth rate is not to be exceeded (see Section 6.6.1). Therefore one has to choose the optimum seed size in relation to the solution volume and the degree of control over the supersaturation. If large samples are available for use as seeds, it is advantageous to use seed plates of that crystallographic orientation which is optimum with respect to low impurity incorporation, high stable growth rate, and the intended application; good examples of the application of these criteria are provided by the hydrothermal growth of quartz crystals (Ballman and Laudise, 1963) and by the growth of ADP from aqueous solution (Egli and Johnson, 1963). In LPE-grown layers of magnetic garnets the orientation-dependent site preference of several ions is used to optimize the properties of the magnetic bubble domains as will be mentioned in the next chapter. As is well-known from crystal growth from aqueous solutions it is desirable to etch the seeds (to dissolve the damaged surface layer and to remove adsorbed impurities) before growth is commenced in order to obtain high quality material.

The mounting of seed crystals in flux growth of oxides normally necessitates the use of platinum wire. Since pure platinum is very soft at the temperatures used and since the wire has to be as thin as possible, it is proposed to use alloys of platinum with 1–10% rhodium or iridium. The wire thicknesses should range from 0.1 to 0.5 mm depending on the size of the seed, degree of stirring and the length of the experiment. If the seeds are immersed in the high temperature solution it is generally desirable to drill a hole into the crystal so that wire may be inserted for fixing to the seed holder. The alternative of wrapping the wire around the crystal is disadvantageous since the crystal may become loose when an outer layer is dissolved during the first stage of the experiment. Examples of seed mountings for top-seeded solution growth are shown in Figs 7.5 and 7.6. For growth from metallic solutions a seed holder may be ground from graphite, boron nitride or alumina. The seed may be conveniently held by a peg of the same material as the holder, which is inserted through a horizontal hole in the seed and the holder.

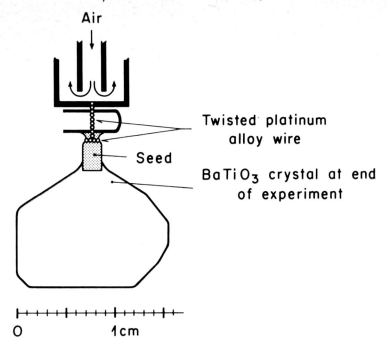

Air

Twisted platinum
alloy wire

Seed

BaTiO₃ crystal at end
of experiment

0   1 cm

FIG. 7.5. Seed mounting for growth of BaTiO₃ by top-seeded solution growth. (Courtesy A. Linz and V. Belruss, MIT.)

The difficulties mentioned above and the high-vapour pressure of many HT solvents have prevented a wide application of seeding, especially when supersaturation is achieved by slow cooling or by solvent evaporation where a number of additional crystals are generally produced by *spontaneous nucleation*. A technique to reduce the number of nuclei after their formation is based on the variation of the surface energy of a crystal with the crystal radius. If the temperature of a supersaturated solution is held constant for a long time, the small crystallites will be dissolved and the size of the larger crystallites will increase. However, the effectiveness of this process diminishes as the crystallites attain macroscopic dimensions, and it is significant only in solutions which contain a large number of crystals.

The reduction of the number of nuclei can be enhanced by the temperature cycling technique which was proposed by Schäfer (1964) for chemical transport reactions and by Hintzmann and Müller-Vogt (1969) for high-temperature solutions. The latter authors used temperature cycling during the whole cooling cycle whereas Scheel and Elwell (1972) argued that temperature oscillation is only advantageous in the first stage

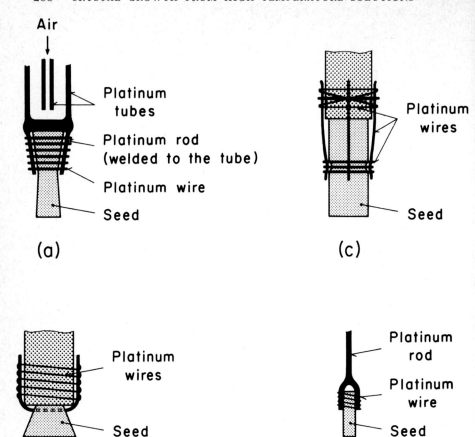

Air

Platinum tubes

Platinum rod (welded to the tube)

Platinum wire

Seed

(a)

Platinum wires

Seed

(c)

Platinum wires

Seed

(b)

Platinum rod

Platinum wire

Seed

(d)

FIG. 7.6. (a)–(d) Examples of seed mounting.

of the experiment when nucleation occurs. This is demonstrated in Fig. 7.7, which shows the procedure of Scheel and Elwell. First the mixture of solute and solvent is held at the temperature A, about 50° above the liquidus temperature T_L, for about 15 hours (the "soaking period") in order to ensure complete dissolution. Then the temperature is lowered to C, a temperature significantly below T_L and T_M, the temperature of the limit of the metastable region. Most of the crystallites formed during this initial cooling are dissolved when the melt is heated to G. This procedure is repeated ($\rightarrow C' \rightarrow E \rightarrow C'' \rightarrow J$ according to the dashed line) until a tempera-

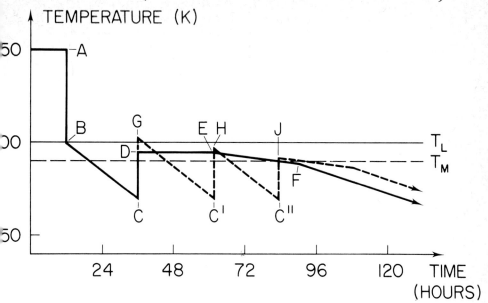

FIG. 7.7. Temperature cycling at the start of a slow cooling experiment in order to decrease the number of spontaneously nucleated crystals (Scheel and Elwell, 1972).

ture below T_L is reached such that only a few crystallites have "survived". Then the temperature programme as described in Section 6.6.1 is started from J. If the temperatures T_L and T_M are accurately known, the procedure should follow the full line ($A \rightarrow C \rightarrow D \rightarrow E \rightarrow F$) so that programmed cooling starts after the minimum period required for dissolution of the smaller crystallites.

Depending on the temperature gradients in the crucible, nucleation occurs at the surface (which is also the favoured location in the case of solvent evaporation because of high local supersaturation) or at the crucible walls. In order to provide a preferred site for nucleation *localized cooling* has been proposed in several papers and has been used in melt growth by the classical techniques of Bridgman, Kyropoulos and Czochralski.

A "cool" spot or region can be provided by an air jet (Chase and Osmer, 1967; Grodkiewicz *et al.*, 1967; Scheel, 1972), by fixing cooling fingers (heat sink, ceramic rod, heat pipe) to the crucible wall, or by placing the crucible into an appropriate temperature gradient (by suitable choice of position in the furnace with respect to the heating elements). For example nucleation at the bottom of the solution was achieved by Linares (1967) and by many other authors, but in most cases multinucleation occurred.

Kvapil (1966) and John and Kvapil (1968) used a baffle above the cooler base in order to reduce the number of nuclei as shown in Fig. 7.8. They also applied a hydraulic seal in order to prevent solvent evaporation.

However, cooling of the base is often disadvantageous in stationary crucibles: the homogenization of the solution by natural convection is minimized (see Chapter 6), and the stable growth rate is low. Therefore localized cooling of the crucible base is only effective when the solution is homogenized. With nonvolatile solutions a stirrer can be applied, and with volatile systems where the crucible has to be sealed the stirring technique of Scheel (1972) by accelerated crucible rotation (see 7.2.7) is very effective

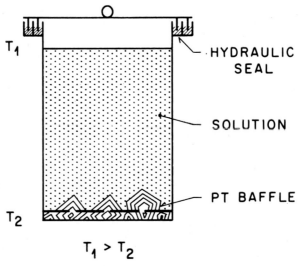

$$T_1 > T_2$$

Fig. 7.8. Reduction of the number of crystals nucleated at the cooler base by a baffle. Flux evaporation is minimized by a hydraulic seal (Kvapil, 1966; John and Kvapil, 1968).

so that by proper control of the experimental conditions one or a very few crystals are spontaneously nucleated on the cooled crucible base.

However, stirring of the solution has to be smooth; with strong agitation or by shaking of the solution the width of the metastable region is made extremely small, and multinucleation will occur due to local density fluctuations and to collision of the crystallites among themselves or with the container or stirrer. An effective technique of nucleation control was applied by Bennett (1968) and by Tolksdorf (1968) and is schematically shown in Fig. 7.9(a)–(d). The crucible is contained in a high-temperature furnace. After soaking (a) the solution is cooled until either the saturation point is exactly reached or until spontaneous nucleation has occurred (b).

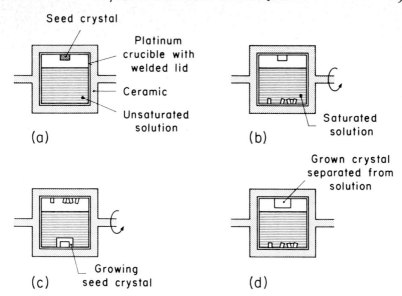

FIG. 7.9. Immersion of a seed crystal into a saturated solution. (a) Initial position at high temperature. (b) After a certain degree of cooling spontaneous nucleation occurs at the bottom, and the crucible is inverted (c) so that the saturated solution covers the crystal, which grows during continued cooling. (d) After termination of cooling the grown crystal is separated from the solution by re-inversion of the crucible into the initial position (Bennett, 1968; Tolksdorf, 1968).

The crucible is then inverted so that spontaneously nucleated crystals are separated and the seed crystal is immersed in a saturated solution and grows by continuous cooling (c). At the end of the run the crucible is inverted again, thereby separating the grown crystal from the solution (d). This technique in combination with localized cooling and with the accelerated crucible rotation technique was used by Tolksdorf and Welz (1972) to produce large inclusion-free crystals of magnetic garnets for various applications.

An alternative approach to reduce the number of crystals nucleated, which has been successfully used in a number of cases, relies upon the addition to the solution of small quantities of some material which is not incorporated into the crystal. The most popular additive is B_2O_3 which is thought to increase the width of the metastable region, probably because of the formation with the various cations of complexes corresponding to the borates which are stable at lower temperatures. The widespread use of additives such as B_2O_3 has led to the development of empirical statements of the form: "The more complex the solution, the greater the solubility and the width of the metastable region, and the fewer the number of crystals nucleated."

Remeika (1970) reported a dramatic decrease in the number of ferrite or garnet crystals nucleated in a PbO/B_2O_3 solvent when 0.1–0.5% by weight of V_2O_5 was added. Similarly Scheel (see Kjems et al., 1973) was able to prevent multinucleation of $LaAlO_3$ from a $PbO—PbF_2—B_2O_3$ flux and therefore to grow large inclusion-free crystals when 0.7 wt% V_2O_5 was added. The beneficial effect of this small V_2O_5 addition is attributed to the formation of $LaVO_4$-like complexes in front of the growing crystals. If these complexes are distributed statistically among $LaAlO_3$-like complexes, they will tend to retard the formation of critical $LaAlO_3$ nuclei and hence to reduce nucleation.

In principle the effect of addition of PbF_2 to PbO could be described in the same way since it permits the formation of complexes such as LaOF. According to this model, however, V_2O_5 will be particularly effective since it forms a larger complex.

Grodkiewicz et al. (1967) reported a beneficial effect for the growth of large garnet crystals of adding several divalent metal oxides or SiO_2. The effect of the latter was confirmed by Page (unpublished, reported by Brice, 1973) who found that the number of $Y_3Fe_5O_{12}$ crystals nucleated from a $PbO—PbF_2$ flux decreased from 50–100 when pure chemicals were used to 3 or 4 when 0.08% Si was substituted for Fe in the melt. These impurities may also act by forming complexes but additional complications arise from the tendency of iron in the crystal to exist as Fe^{2+} or Fe^{4+} ions in small concentration. One of the most remarkable examples of nucleation reduction has been reported by Robertson et al. (1973), who observed the nucleation of only one yttrium iron garnet crystal from a $PbO—PbF_2—B_2O_3$ flux when growth occurred in oxygen under a pressure of ten atmospheres. The reduction in this case was attributed to the lack of crystals of a second phase which often act as nuclei for garnets.

One point which is frequently neglected in connection with nucleation control in HTS growth is the quality of the containers. The use of new platinum containers with polished inner walls generally results in fewer crystals (by heterogeneous nucleation) than in crucibles which have been used frequently and which show strong recrystallization and a rough surface. Platinum crucibles can be re-used (as is necessary because of the high cost of re-fabrication) when they are carefully cleaned, reshaped and chemically or mechanically polished as is discussed in detail in Section 7.2.3. The same arguments apply to other types of container.

Thus control of nucleation in HTS growth is generally problematic and needs careful control of all chemical and experimental conditions. By homogenization of the solution and by use of appropriate seeds, large crystals can be grown. If spontaneous nucleation has to be chosen, smooth stirring of the solution combined with localized cooling will limit the number of nucleated crystals generally to one or very few.

7.1.2. Techniques to produce supersaturation

As in crystal growth from aqueous solutions and as is obvious from Fig. 7.1, supersaturation can be obtained by *slow cooling* (1), by *solvent evaporation* (2) or by the *temperature gradient technique* (3) in which nutrient is held in a hotter region: solution saturated in the hot region is transported to a cooler region by natural or forced convection and so becomes super-saturated. Apart from these three most popular techniques there are a few special methods to obtain supersaturation such as the *reaction technique* in which solute constituents formerly separated diffuse to the region where reaction (and crystallization) proceeds. If crystal constituents are trans-ported in the vapour phase to the liquid solution in order to precipitate the (solid) crystal, this may be termed the *VLS†* *mechanism*, which is well known as an important mode of whisker growth (Wagner and Ellis, 1964). The temperature gradient, reaction and VLS techniques, together with electrolytic growth will be referred to collectively as *transport techniques*. Reaction equilibria which are shifted by means other than by temperature change (included under "slow cooling"), solvent evaporation or tempera-ture gradients include evaporation of "reaction products" and the *salting out* effect. The techniques of obtaining supersaturation will be described according to the following classification:

    A. Slow cooling,   B. solvent evaporation,   C. transport techniques.

    The supersaturation value used in an experiment is determined by the requirement that the growth rate should not exceed the maximum stable value, as discussed in Chapter 6. It is not possible at present to give a quantitative criterion for the maximum supersaturation, apart from that specified by Eqn 6.26. If the solution, stirring rate, temperature gradient etc. are optimized, the supersaturation may have a maximum value such that the growth rate obtains its ultimate stable value. Very small super-saturations are, however, pointless since the crystal will exhibit growth and dissolution fluctuations as shown for Czochralski growth by Witt and Gatos (1968) and by Kim *et al.* (1972).

    In practice the relative supersaturation normally has a value in the region of 0.1–1%, as in aqueous solution growth, although the corresponding supercooling is appreciably higher at high temperatures. For an ideal solution with $n_e \propto \exp(-\phi/RT)$, the relative supersaturation is

               † VLS vapour-liquid-solid.

$$\sigma = \frac{\Delta n}{n_e} = \frac{\phi \Delta T}{RT^2} .$$

Thus for $\phi = 50$ kJ/mole and $T = 1200°C$, $\Delta T = 3.6°C$ for $\sigma = 1\%$.

A. Slow cooling. The most common technique for producing super-saturation in flux growth is by slow cooling (see Tables 7.1 and 7.3 and Chapter 10) where generally a linear cooling rate of 0.2°C h⁻¹ to 10°C h⁻¹ is applied. If inclusion-free crystals larger than a few millimetres have to be grown a cooling rate of less than 1°C h⁻¹ is necessary as discussed in Chapter 6.

The linear growth rate v (in cm/h) by slow cooling is related to the cooling rate (Laudise, 1963; Cobb and Wallis, 1967) according to

$$v = \frac{V}{A\rho} \left(\frac{\mathrm{d}n_e}{\mathrm{d}T}\right)\left(\frac{\mathrm{d}T}{\mathrm{d}t}\right), \tag{7.1}$$

where V is the volume of solution (cm³), A the area of the growing crystal (cm²) and ρ its density (gcm⁻³), $\mathrm{d}n_e/\mathrm{d}T$ the change in solubility per degree [g cm⁻³ °C⁻¹] and $\mathrm{d}T/\mathrm{d}t$ the cooling rate in °C/h. Obviously, Eqn 7.1 only holds when all the solute precipitated is deposited on the crystal.

The optimum cooling rate for stable growth as a function of various growth parameters is also discussed in Chapter 6. It was shown that a constant linear growth rate and therefore a cubic decrease of temperature with time will cause unstable growth. Scheel and Elwell (1972) and Pohl and Scheel (1975) have shown that the temperature regulation and the cooling rate have to be adjusted in such a way that the slope of the effective cooling curve (including any oscillations or fluctuations) never exceeds the slope of the calculated optimum cooling curve for stable growth at the corresponding temperature. The general rule is that the slower the growth rate the better and larger the crystals. However, one has to find a compromise between the slow cooling rate and the correspondingly long duration of an experiment. Also Laudise (1963) stressed the fact that cooling rates which are not at least comparable with the temperature fluctuations due to inaccurate regulation are pointless. A very slow initial cooling rate is impractical for the (normal) case where the super-solubility curve is not known with sufficient accuracy and much time is lost before nucleation starts.

The advantages of the slow cooling technique are:

1. That a closed container (sealed crucible) can be used thereby preventing evaporation of volatile solvent or solute constituents which are poisonous or corrosive and which cause uncontrolled supersaturation.
2. The technique is relatively simple for the growth of crystals up to

5–10 mm size. For larger high-quality crystals the effort on apparatus, temperature regulation and programming has to be increased.

3. The slow cooling technique is very suitable for exploratory materials research. It is usually simple to crystallize known crystals and also new phases in sizes from 2 to 5 mm, which are suitable for X-ray structure determinations and a number of physical measurements.

The slow cooling technique has several disadvantages which arise from the continuously changing growth temperature:

1. The concentration of equilibrium defects varies through the crystal (however for a typical temperature range such as 1500°K to 1200°K this is generally not critical).
2. The concentration of incorporated impurities and dopants changes according to the differences in the solubility behaviour of the solute and the dopants or impurities. Frequently this is manifested in the shifting of equilibria in the solutions as discussed in Section 3.4. For example, it is extremely difficult to produce homogeneous chromium doping in Al_2O_3 and other oxides by the slow cooling technique and the solvent evaporation or transport techniques are then preferable.
3. For the preparation of solid solutions the same arguments as above for dopants hold, only the effect on the composition is more drastic. In Section 7.1.3 the techniques to obtain homogeneous solid solutions will be discussed.
4. Frequently in slow-cooling experiments unwanted phases appear (sometimes "non-reproducibly" or depending on the crucible size: the explanation is uncontrolled evaporation of a solvent constituent as in the case of $Y_3Fe_5O_{12}$ crystals grown from PbO—PbF_2—B_2O_3 flux where magneto-plumbite, $YFeO_3$ or $Y_{20}O_{16}F_{28}$ are the unwanted phases, see Tolksdorf and Welz, 1972). Reproducibility is an absolute necessity and can in this case be achieved by sealing the crucibles by welding and by careful control of all parameters including the purity of the chemicals used.

According to the degree of nucleation control the following versions of the slow-cooling technique can be distinguished:

Spontaneous uncontrolled nucleation (Table 7.3).
Reduction in the number of crystals nucleated by an oscillatory temperature variation at the start of the experiment.
Localized cooling.
Localized cooling and ACRT stirring (see Section 7.2.7).
Bennett-Tolksdorf seeding technique (with and without ACRT).
Top-seeded solution growth.

TABLE 7.3. Examples of Crystals Grown by the Slow-cooling Technique and Spontaneous Nucleation†

Crystal	Solvent	Size (mm)	Remarks	Reference†
Al_2O_3	PbO/B_2O_3; PbF_2	30	Plates	Nelson and Remeika (1964), White and Brightwell (1965)
$BaTiO_3$	KF	$34 \times 24 \times 0.4$	Butterfly twin	Remeika (1958)
BeO	Various	10	Prisms	Newkirk and Smith (1965)
C (graphite)	Fe, Ni	$30 \times 0.5 \times 0.5$		Austerman et al. (1967)
$Cd_{1-x}Cu_xCr_2Se_4$	$CdCl_2$	3–4		Tyco (1971)
CdS	Na_2S_x	$15 \times 1 \times 1$; $5 \times 5 \times 0.2$	Prisms, plates	Scheel (1974)
$GaFeO_3$	Bi_2O_3/BiF_3	$12 \times 6 \times 6$		Linares (1962)
$GdAlO_3$	$PbO/PbF_2/B_2O_3$	$35 \times 30 \times 25$	210 g, ACRT	Scheel (1972)
$Gd_3Fe_5O_{12}$	PbO/B_2O_3	10		Remeika (1963)
$Gd_3Ga_5O_{12}$	PbO/B_2O_3	8		Remeika (1963)
In_2O_3	PbO/B_2O_3	$10 \times 10 \times 1$		Remeika and Spencer (1964)
LaB_4	La	5–8		Deacon and Hiscocks (1971)
$LiFe_5O_8$	PbO/B_2O_3	20		Pointon and Robertson (1967)
Mn_3O_4	PbF_2/PbO	10×10		Wanklyn (1972)
MnTe	Te	$50 \times 15 \times 15$	ACRT	Mateika (1972)
$NaCrS_2$	Na_2S_x	$20 \times 20 \times 0.2$	Plates	Scheel (1974)
$Y_3Al_5O_{12}$	$PbO/PbF_2/B_2O_3$	50	Large crucible	Grodkiewicz et al. (1967)
$Y_3Fe_5O_{12}$	$PbO/PbF_2/B_2O_3$	~300 g	Large crucible	Nielsen (1964)
$Y_3Fe_5O_{12}$	$PbO/PbF_2/B_2O_3$	~60	Large crucible	Grodkiewicz et al. (1967)
$Y_3Fe_5O_{12}$	$PbO/PbF_2/B_2O_3$	57 g	Bennett-Tolksdorf nucleation control	Bennett (1968)
$Y_3Fe_5O_{12}$	PbO/PbF_2	~$30 \times 25 \times 25$	(49 g) nucleation control	Tolksdorf (1968)
$Y_3Fe_5O_{12}$	$PbO/PbF_2/B_2O_3$	$30 \times 25 \times 25$	Inclusion-free, ACRT	Scheel (1972)
$Y_3Fe_5O_{12}$	PbO/PbF_2	$60 \times 50 \times 25$	250 g, ACRT	Tolksdorf (1974a)
$Y_3Fe_{3.8}Ga_{1.2}O_{12}$	$PbO/PbF_2/B_2O_3$	$30 \times 25 \times 25$	Inclusion-free ACRT	Scheel (1972)
ZnO	PbF_2	50		Nielsen and Dearborn (1960)

† For experimental details and for references see Chapter 10.

B. Evaporation of solvent. Solvent evaporation, the basis of common salt production for thousands of years, has also been frequently used in crystal growth from HTS. In this case the linear growth rate of the crystal is given by

$$v = \frac{n_e}{\rho A}\left(\frac{dV}{dt}\right) \tag{7.2}$$

with the symbols corresponding to those of Eqn 7.1, so that dV/dt is the solvent evaporation rate.

Flux evaporation has been systematically used by Tsushima (1966), Grodkiewicz and Nitti (1966), Roy (1966), Wood and White (1968) and by Webster and White (1969). The technique has been especially used for compounds which react with the solvent at lower temperatures $\left(TiO_2 + PbO \xrightarrow{<1200°C} PbTiO_3; \ HfO_2 + PbO \xrightarrow{<1200°C} PbHfO_3\right)$ or for a cation of which the valence state changes by oxidation at lower temperatures $\left(Cr^{3+} \xrightarrow{<\sim 1100°C} Cr^{6+}, \text{see Section 3.4}\right)$. Roy (1966) stressed the

TABLE 7.4. Crystals Prepared by the Solvent Evaporation Technique†

Crystal	Solvent	Size	Reference
$BaTiO_3$	$BaCl_2$	—	Timofeeva and Zalesskii (1959)
$BaTiO_3$	$BaCl_2$	—	Arend (1960)
In_2O_3	B_2O_3	few mm	Roy (1966)
TiO_{2-x}	$Na_2B_6O_{10}$	few mm	Roy (1966)
Fe_2O_3	$Na_2B_6O_{10}$	few mm	Roy (1966)
Cr_2O_3	$Na_2B_6O_{10}$	few mm	Roy (1966)
VO_2	V_2O_5	few mm	Roy (1966)
$BaAl_{12}O_{19}$	PbF_2	$2 \times 2 \times 0.1$ mm	Tsushima (1966)
Al_2O_3	PbF_2	$10 \times 10 \times 1$ mm	Tsushima (1966)
$ZnAl_2O_4$	PbF_2	$1 \times 1 \times 1$ mm	Tsushima (1966)
$ZnMn_2O_4$	PbF_2	$2 \times 2 \times 2$ mm	Tsushima (1966)
$CoMn_2O_4$	PbF_2	$2 \times 2 \times 2$ mm	Tsushima (1966)
$LaAlO_3$	$PbO + PbF_2$	$3 \times 3 \times 0.4$ mm	Tsushima (1966)
HfO_2	PbF_2—B_2O_3	~ 1 g	Grodkiewicz and Nitti (1966)
ThO_2	PbF_2—B_2O_3	~ 5 mm	Grodkiewicz and Nitti (1966)
TiO_2	PbF_2—B_2O_3	~ 2 mm	Grodkiewicz and Nitti (1966)
$Al_2O_3:Cr$	PbF_2—B_2O_3	57×2 mm	Grodkiewicz and Nitti (1966)
$YCrO_3$	PbF_2—B_2O_3	7×4 mm	Grodkiewicz and Nitti (1966)
MgO	PbF_2	—	Webster and White (1969)
$SrSO_4$	NaCl		
$BaSO_4$	NaCl	few mm	Patel and Bhat (1971)
$PbSO_4$	NaCl		

† For experimental details see Chapter 10.

advantages of the flux evaporation method especially for such cases where oxides of metals with a specified valence state have to be grown. The constant temperature and a carefully controlled oxygen partial pressure allow growth of oxides of lower valency states as was shown by Berkes *et al.* (1965), Roy (1966) and by Bartholomew and White (1970). Flux evaporation could also be used to grow crystals which show a low variation of solubility with temperature or a retrograde solubility behaviour. Several crystals prepared by solvent evaporation at high temperatures are listed in Table 7.4, and further examples are given in Chapter 10.

Since solvent evaporation may be carried out isothermally, this technique offers the advantages connected with growth at constant temperature:

1. Easy and often closer temperature control.
2. Constant concentration of equilibrium defects.
3. Approximately constant incorporation of solvent ions as impurities.
4. In the cases where the distribution coefficient of impurities or dopants is not extremely different from unity and if growth takes place either under the regime of pure diffusion control or under complete mixing, a homogeneous incorporation of dopants and impurities can be expected.

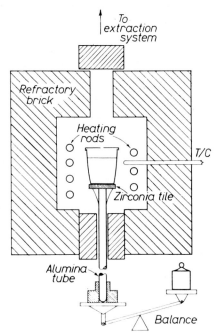

Fig. 7.10. Open system for solvent evaporation with monitoring by a balance (Wood and White, 1968).

Disadvantages of the isothermal flux evaporation technique are the difficulty of controlling the evaporation rate and thus the growth rate, and the poisonous and corrosive nature of the solvent vapours (e.g. PbF_2) if an open system such as that shown in Fig. 7.10 is used. The advantage of this system is that the balance permits a measurement of the rate of evaporation and hence of the mass deposition. The use of an air flow to remove PbF_2 vapour will, however, normally produce temperature fluctuations.

In order to prevent corrosion of furnace ceramics and heating elements by the action of the solvent vapours, closed evaporation systems have been proposed and three such designs are shown in Fig. 7.11. These systems permit control or programming of the evaporation rate through variation in the temperature T_2 at which condensation occurs.

An important feature of the flux evaporation technique is that nucleation should take place in the lower part of the crucible. The surface region of the solution has to be warmer, otherwise crystallization at the surface would diminish the free surface area of the solution and thereby decrease the rate of evaporation. Therefore, in a symmetric (to the heat source) position of the crucible, little natural thermal convection occurs.

Depending on the relative densities of solute and the solution some solutal convection can occur, the latter being favoured by a higher density of the crystal. However, any natural convection in this case of a cooler crucible base will be slow, so that growth is diffusion-controlled unless stirring is applied. This can be achieved for closed systems by the accelerated crucible rotation technique as in the apparatus shown in Fig. 7.11(c). In this apparatus, the evaporation rate will be governed by the dimensions r_1 and r_2 as well as by the temperatures T_1 and T_2 so that considerable flexibility is available to the experimeter for optimization of the growth conditions.

In principle the evaporation rate w is constant when the temperature is held constant and is given by the Arrhenius equation

$$w = w_0 \exp\left(-H_v/RT\right)$$

where w_0 is the rate constant, H_v the activation energy of vaporization, R the gas constant and T the absolute temperature. From measurement of the weight loss as a function of temperature Giess (1966) determined an activation energy of 33 kcal/mole for the evaporation of PbF_2 from a $Y_3Fe_5O_{12}$—PbO—PbF_2 melt, whereas Perry (1967) estimated an activation energy of 62 ± 5 kcal/mole for the evaporation of BaB_2O_4—PbB_2O_4 flux during crystallization of $(Ba, Pb)TiO_3$.

In order to achieve an optimum stable growth rate during the course of crystallization a program for the evaporation according to the principles of the stable growth rate of Scheel and Elwell (1972) has to be worked out,

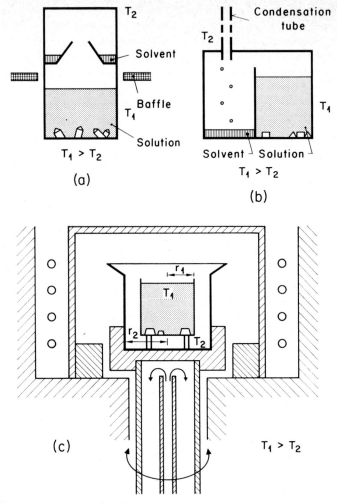

FIG. 7.11. Arrangements for flux evaporation experiments. (a) System with collection of the evaporated solvent (Hart, quoted in White, 1965). (b) Evaporation with condensation tube (Brice, 1973). (c) Flux evaporation-condensation system with ACRT stirring (Scheel, 1972).

and accordingly the rate of evaporation can be controlled by the temperature, by a gas flow to remove the solvent vapours, by baffles, by the free surface area of the solution etc.

C. Transport techniques. In principle all methods of crystal growth from solutions depend on transport of solute to the crystal, whether the supersaturation is provided by slow cooling or by solvent evaporation. The term "transport techniques" is used here for such cases where supersaturation is achieved exclusively by transport of solute or solute constituents which were initially not dissolved in the solution and are either solid (becoming gradually dissolved and transported to the growing crystal) or in the vapour phase. The following transport techniques can be distinguished:

(i) Transport by a temperature gradient (between nutrient and crystal) in bulk solutions.
(ii) Travelling solvent zone crystallization.
(iii) Diffusion of reactants (solute constituents) = flux reaction technique;
    (a) Solid source
    (b) Vapour phase source (VLRS)
    (c) Shifting equilibria.
(iv) Vapour-liquid-solid (VLS) mechanism.
(v) Electrolytic growth in high-temperature solutions.

(i) *Transport in a temperature gradient.* The principle of the gradient transport technique is shown in Fig. 7.1 (process 3). Nutrient is held at a temperature T_c in a solution of average composition n_c. If by natural or forced convection a flow of solution occurs towards a region of average temperature T_E where a seed crystal is held, the solution becomes supersaturated and crystallization occurs at the seed surface.

For natural convection the average mass transport rate and its dependence on the properties of the solution (viscosity, thermal conductivity, thermal expansion coefficient, heat capacity) is discussed in Section 6.6.2. For a given solute-solvent system the mass transport rate and hence the supersaturation and growth rate may be varied by adjusting the temperature difference, the area and form of seed and nutrient, and the depth of the solution. Alternatively the seed may be rotated so that a column of solution is drawn up from the basal region at a rate which is given by Cochran (1934). An example of the variation in crystal growth rate with rotation rate is shown in Fig. 4.10. A theoretical treatment of thermal gradient transport has been given by Dawson *et al.* (1974). If the flow of solution is assumed to be rapid compared with diffusion through the boundary layer, and neglecting solvent evaporation, the growth rate v is given by

$$v\left[\frac{\rho\delta}{D} + \frac{\rho\delta_N A}{DA_N}\right] + \left[\frac{v}{F}\right]^{1/m} = \frac{n_e\phi}{RT^2}\,\Delta T. \tag{7.3}$$

Here the subscript N refers to the nutrient, ϕ is the heat of solution, and the interface kinetics are taken to be such that $v = F(n_i - n_e)^m$. The parameter which is normally varied to adjust the growth rate is the temperature difference ΔT between crystal and nutrient.

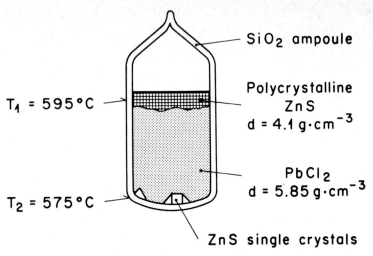

$T_1 = 595\,°C$

$T_2 = 575\,°C$

SiO$_2$ ampoule

Polycrystalline ZnS
$d = 4.1\ g \cdot cm^{-3}$

PbCl$_2$
$d = 5.85\ g \cdot cm^{-3}$

ZnS single crystals

Fig. 7.12. Apparatus used for growth of zinc sulphide by solute transport (Linares, 1968).

As an example of application of crystal growth by transport in a temperature gradient, Fig. 7.12 shows the growth of zinc sulphide where polycrystalline nutrient is floating at the hotter surface and where transport occurs downwards to the cooler end of the ampoule.

Another example of achieving supersaturation with the nutrient held in a hotter unstirred zone is shown in Fig. 7.13 where GaAs nutrient material is floating in an outer annular chamber while the slowly rotated (10 rpm) GaAs crystal grows in a slightly cooler inner chamber and is slowly withdrawn. As shown in Table 7.2 a number of authors used a gradient transport technique where nutrient is held in a hotter zone and where the rotated crystal is slowly withdrawn, and a suitable arrangement is shown in Fig. 7.14. This apparatus may incorporate an annular cooling jacket for control of the interface temperature gradient.

Growth on fully immersed seed holders as shown in Fig. 7.3 is normally more difficult if the phase diagram and parameters affecting growth conditions are not accurately known. A few examples are quoted in Table 7.5.

The advantage of a constant growth temperature is obvious for the production of homogeneous solid solutions and of homogeneously doped crystals as described in Section 7.1.3, and therefore the temperature gradient technique has been frequently used. Recently Tolksdorf and Welz (1973) described a gradient transport technique where the advantages of the Tolksdorf (1968) seeding method and the ACRT stirring technique

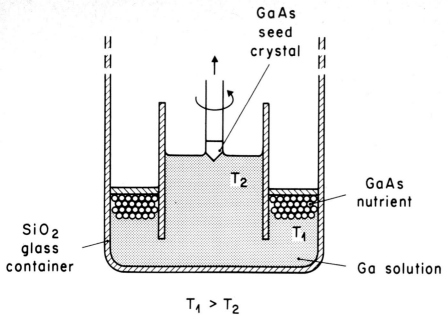

FIG. 7.13. Gradient transport technique for growth of GaAs (Lyons, 1965).

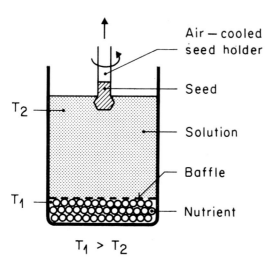

FIG. 7.14. Typical arrangement for gradient transport with top seeding.

TABLE 7.5. Crystals Prepared from Immersed Seeds by the Gradient Transport Technique[†]

Crystal	Solvent	Size	Reference
$CoFe_2O_4$	$Na_2B_4O_7$	10–15 mm	Reynolds and Guggenheim (1961)
$Y_3Fe_5O_{12}$	$BaO \times 0.6B_2O_3$	~15 mm	Laudise et al. (1962)
ThO_2	$Li_2W_2O_7$	10×3 mm	Finch and Clark (1965)

[†] For experimental details see Chapter 10.

(Scheel, 1972) are applied to isothermal growth by temperature gradient transport. The arrangement is shown in Fig. 7.15. Large homogeneous and nearly inclusion-free crystals of solid solutions $Y_3Fe_{5-x}Ga_xO_{12}$ were grown. A few inclusions were trapped at the seed crystal when the air cooling was turned on in order to initiate growth, probably due to a sudden increase in the growth rate.

In growth on a rotating seed, the thickness of the solute boundary layer is, to a good approximation, independent of the diameter of the crystal. As a result, whether or not pulling is used to maintain a constant diameter, the maximum stable growth rate should not change as growth proceeds. A reduction in the growth rate will, however, be necessary if the temperature gradient at the crystal-solution interface varies as the crystal becomes larger.

(ii) *Travelling solvent zone crystallization* (TSZC). A great variety of terminology has been used to describe the technique in which a zone of solution is made to travel through a solid in a similar manner to zone melting (Pfann, 1955, 1966): thin film solution growth, travelling heater method, thin alloy zone crystallization, temperature gradient zone melting, moving solvent method, travelling solvent method, zone melting with a temperature gradient, etc. TSZC has been reviewed by Wolff and Mlavsky (1965, 1974) and by Hemmat et al. (1970).

The principle is shown in Fig. 7.16; (a) demonstrates the technique in which the driving force for the motion of the solution is the temperature gradient so that at the hotter side single-crystalline or dense poly-crystalline feed is dissolved and then deposited at the cooler side of the solution zone. In (b) the solution zone is moved by motion of the heater, which may be a resistance heating ring or an RF coil. As the temperature profiles (c) demonstrate, both techniques are essentially equivalent. However, there are differences which determine the applicability of the two techniques, and which originate from the individual heat flow patterns as indicated in Fig. 7.16(d) and (e).

In the travelling solvent method (a) the solution moves towards an

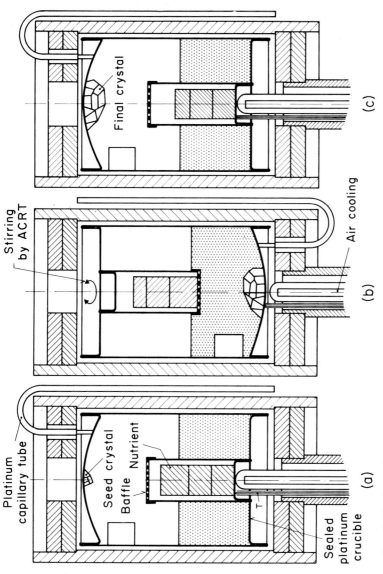

FIG. 7.15. Growth of solid solutions of $Y_3(Fe, Ga)_5O_{12}$ by gradient transport. (a) Starting position. (b) Position during crystallization. (c) Final position (after Tolksdorf and Welz, 1973).

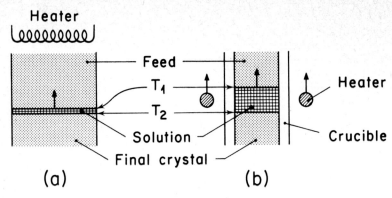

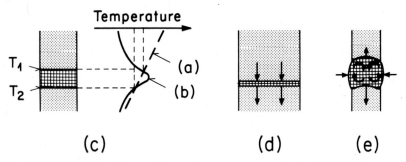

FIG. 7.16. Principle of travelling solvent zone method. (a) Solution zone travelling in a temperature gradient. (b) Solution zone travelling with a heater. (c) Temperature profiles for (a) and (b). (d) Heat flow for (a). (e) Heat flow for (b).

increasingly hotter region if the temperature gradient across the ingot or, correspondingly, if the temperature of the heater is kept constant. This would lead to a varying composition in the case of solid solutions or doped crystals, and also the rate of movement of the solution zone, and hence the growth rate, would increase with time. Therefore a temperature programme is advantageous which provides a constant growth temperature T_2 and a constant dissolution temperature T_1. According to Wolff (1965) the solution zone in the temperature gradient driven technique can be as thin as 25–100 μm and this allows the preparation of thin crystals and devices (Mlavsky and Weinstein, 1963; Griffiths and Mlavsky, 1964). In reviews on the principles of "thin alloy zone crystallization" Hurle et al. (1964, 1967) have shown that for thin solution zones the supersaturation gradient is much smaller than in bulk solutions and this allows fast growth rates without solvent inclusions. Under favourable conditions the stable growth rates can approach those used during crystallization from pure melts as discussed in Chapter 6. The migration kinetics of a solution zone in a temperature gradient have been studied by Tiller (1963), Seidensticker (1966) and by Hamaker and White (1968).

In the travelling heater method (THM) the solution zone is orders of magnitude wider so that this technique may be conveniently used to produce large crystals. This technique is especially useful for the preparation of homogeneous solid solutions and homogeneously doped crystals (see Section 7.1.3) if the temperature difference between the dissolving and the growing interfaces is kept small and when the conditions are optimized. The maximum stable growth rate, however, is much lower than in the travelling solvent method due to the thickness of the molten solution zone. Growth rates are limited to typical values for crystal growth from bulk solutions, of the order of 500 Å s^{-1} or 3 to 5 mm per day. The onset of constitutional supercooling and the maximum stable growth rates were discussed by Hemmat et al. (1970) and are also treated in Chapter 6.

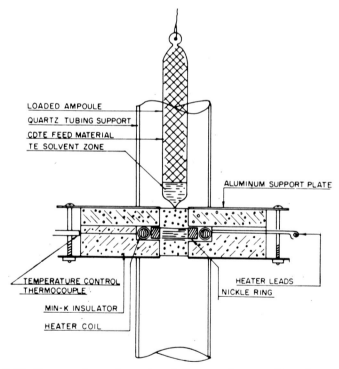

LOADED AMPOULE
QUARTZ TUBING SUPPORT
CDTE FEED MATERIAL
TE SOLVENT ZONE

ALUMINUM SUPPORT PLATE

TEMPERATURE CONTROL THERMOCOUPLE
MIN-K INSULATOR
HEATER COIL

HEATER LEADS
NICKLE RING

FIG. 7.17. Construction schematic of furnace for travelling heater method (Wald et al., 1971).

A typical furnace used for THM growth of CdTe from Te solution is shown in Fig. 7.17 and nearly single crystals of 5 cm length were obtained by Wald *et al.* (1971) and by Bell *et al.* (1970).

Table 7.6 lists some crystals which have been grown by the travelling solvent zone method and demonstrates its wide applicability.

Modifications of the travelling heater method use a heater wire to move a solution zone along a thin crystalline layer or a heating strip which moves with the solution zone through the crystal boule. The first method was used to grow $BaTiO_3$ films of about 50–125 μ thickness, as shown in

TABLE 7.6. Crystals Prepared by the Travelling Solvent Zone Method

Crystal	Solvent	Method†	Crystal size	Reference
GaP	Ga	THM		Broder and Wolff (1963), Wolff *et al.* (1958), Plaskett *et al.* (1967)
GaAs	Ga	THM		Wolff *et al.* (1958)
		THM		Hemmat *et al.* (1970)
(Ga, In)P	Ga—In	THM		Hemmat *et al.* (1970)
(Al, Ga)As	Al—Ga	THM		Hemmat *et al.* (1970)
Ga(As, P)	Ga	THM		Wolff *et al.* (1968)
Si	Au	THM		Hein (1956)
SiC	Cr	THM		Wolff *et al.* (1969)
(Zn, Hg)Te	Te	THM	4.5 cm	Wolff *et al.* (1968)
CdTe	Te	THM		Bell *et al.* (1970), Wald *et al.* (1971)
$CdIn_2Te_4$	In_2Te_3	THM		Mason and Cook (1961)
$CdCr_2Se_4$	$CdCl_2$	THM	8×35 mm	Hemmat *et al.* (1970)
ZnO	PbF_2	THM	$2 \times 3 \times 10$ mm	Wolff and LaBelle (1965)
$Y_3Fe_5O_{12}$	Fe_2O_3	THM		Abernethy *et al.* (1961)
$Y_3Fe_5O_{12}$	BaO/B_2O_3	STRIP	10×3 mm	Tolksdorf (1974b)
(Pb, Sr)TiO_3	(Pb, Sr)$_2B_2O_5$	THM		DiBenedetto and Cronan (1968)
GaP	Ga	TSM		Weinstein and Mlavsky (1964)
GaAs	Ga	TSM		Mlavsky and Weinstein (1963)
SiC	Cr	TSM		Griffiths and Mlavsky (1964)
$BaTiO_3$	BaB_4O_7	STRIP		Hemmat *et al.* (1970)
$CaCO_3$	Li_2CO_3	STRIP	15×20 mm	Brissot and Belin (1971), Belin *et al.* (1972)
InAs	In	TSM		Kleinknecht (1966)
InSb	In, Pb	TSM		Hamaker and White (1968)
$Ga_xIn_{1-x}Sb$	In, Pb	TSM		Hamaker and White (1969)
$ZnSe_xTe_{1-x}$		TSM		Steininger and England (1968)
$CaMoO_4$	Li_2SO_4	TSM	~ 1 mm	Parker and Brower (1967)

† TSM = Travelling solvent method, THM = travelling heater method, STRIP = travelling stripheater method.

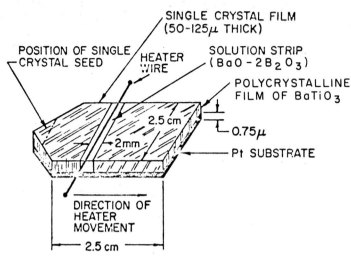

SINGLE CRYSTAL FILM
(50-125μ THICK)

POSITION OF SINGLE
CRYSTAL SEED

HEATER
WIRE

SOLUTION STRIP
(BaO - 2B₂O₃)

POLYCRYSTALLINE
FILM OF BaTiO₃

2.5 cm

0.75μ

2mm

Pt SUBSTRATE

DIRECTION OF
HEATER
MOVEMENT

2.5 cm

FIG. 7.18. Schematic of thin film growth by moving platinum wire heater (Hemmat *et al.*, 1970).

Fig. 7.18, by moving an infrared (focusing) line heater at a rate of 8.4 cm per day along with a strip of the BaB_4O_7 solution (Hemmat *et al.*, 1970). A travelling strip heater was first used by Brissot and Belin (1971) and by Belin *et al.* (1972) to grow large $CaCO_3$ crystals from a solution in Li_2CO_3 by an apparatus which is diagrammatically shown in Fig. 7.19. In this case the solution technique allows the growth at 1–2 atm. CO_2 and 700–800°C of a compound which at its melting point of about 1340°C would have a CO_2 equilibrium pressure of about 100 atm. The linear growth rate was 600 Å s⁻¹ or 5 mm per day, and the seed crystal was rotated.

Plaskett *et al.* (1967) found that in GaP crystals grown by THM in the [211] direction the twinned regions grew faster. Thus they used twinned seeds in order to obtain large inclusion-free twinned crystals from which high quality substrates for the preparation of LED† devices were cut parallel to the longitudinal twin planes.

The rate of advance of the solution, which is equal to the crystal growth rate v and to the rate of dissolution, is estimated by Wilcox (1968) to be

$$v \simeq \frac{GD\rho_c}{m(1-n_{sn})\rho_{sn}}, \tag{7.4}$$

where G is the temperature gradient in the liquid, D is the diffusion coefficient, ρ_c and ρ_{sn} the densities of the crystal and the solution, m the slope of the liquidus on the crystal-solvent phase diagram (e.g. the change

† LED = light-emitting diodes.

L

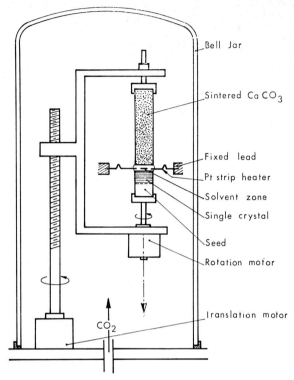

FIG. 7.19 Strip heater apparatus for growth of calcite (Brissot and Belin, 1971).

in liquidus temperature per change in weight fraction) and n_{sn} the solute concentration in the solution. This relationship is equally applicable to the movement of inclusions in a temperature gradient as discussed in Chapter 9 and shows that the solution zone travel rate increases with increasing temperature gradient, increasing solubility, increasing dependence of solubility on temperature and with increasing diffusion coefficient. The latter three properties increase with temperature, so that the temperature and the temperature gradient can be given values which allow stable growth (see discussion in Chapter 6). If the average temperature and the temperature gradient across the solution zone are kept constant, then also the maximum stable growth rate is constant because of the non-varying size of the crystal. On the other hand impurities with $k < 1$ are continuously enriched in the travelling solvent zone and may necessitate a decreasing growth rate, and also an increasing crystal perfection (decreasing number of dislocations) acts in the same direction. Equation 7.4 is approximate since it neglects the effect of interface kinetics, and applies only if the transport occurs only by diffusion.

An important aspect of the travelling solvent zone method is its purification effect. Although purification is a general effect of crystal growth from solutions (as long as growth is stable) and has been used in preparative chemistry for more than 700 years (recrystallization from solutions) the purification in TSZM is specially pronounced when the relatively small amount of solvent is considered. Even if impurities for which $k > 1$ are present in the solvent, the uptake of such impurities from the solution is clearly much lower than in growth from bulk solutions.

(iii) *Diffusion of reactants (= flux reaction technique).* In this section those techniques are discussed which differ from those in Sections i, ii and iv in that certain of the crystal constituents are present in the solution while other constituents diffuse from a solid source or enter the solution from the vapour phase. These techniques should not be confused with reactions between starting materials which may occur in the solution prior to the production of supersaturation. For example, mixtures of Y_2O_3 and Fe_2O_3 crystallize from PbO—B_2O_3, PbF_2 or $BaO \cdot 0.6B_2O_3$ solvents as $YFeO_3$, $Y_3Fe_5O_{12}$, $PbFe_{12}O_{19}$, Fe_2O_3, YBO_3 or YOF depending on the relative concentrations of starting materials but not on the degree of (or absence of!) prior solid state reaction. As another example, $BaWO_4$ crystallizes from a homogenized solution whether the starting mixture is $Na_2WO_4 + BaCl_2$ or $BaWO_4 + NaCl$. Despite the fact that reactions take place in both the above examples it would be rather inconvenient to classify all such experiments under the term "reaction technique" or "indirect flux method".

In the following we shall restrict the term "reaction technique" to those examples where by a reaction between constituents (formerly separated) supersaturation is achieved and precipitation occurs, or where by oxidation or reduction during the experiment supersaturation is continuously produced. The "reaction" is taking place during the crystal growth process, generally by transport (on a macroscopic scale) of the reactants to the crystallization region. Several examples of this "reaction technique" were reported in the last century and are listed in Table 2.1.

(a) *Solid source.* To this category belong the cases where crucible constituents become constituents of the crystal or serve as reduction or oxidation media. Reactions with crucible materials are frequently undesirable and many examples were reported in the 19th century. As a recent example lead feldspar solid solutions $PbAl_2Si_2O_8$—$KAlSi_3O_8$ crystallized in prisms up to 15 mm length on a sillimanite-type ceramic lid on which PbO and PbF_2 from a flux growth experiment condensed (Scheel, 1971).

In the thirties emerald was produced by IG Farben (Espig, 1960) by a flux reaction technique as shown diagrammatically in Fig. 7.20(a). Pieces of silica float on a solution containing beryllia and alumina in the correct

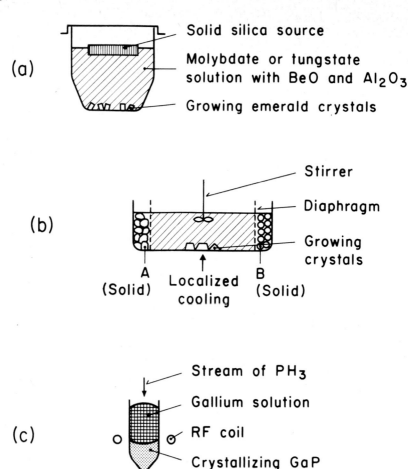

Fig. 7.20. Examples of the flux reaction technique. (a) Growth of emerald by the method of Espig (IG Farben) with silica as the solid source (after Wilke, 1956). (b) System of two reactants separated from the growth region by diaphragms. (c) Crystallization of gallium phosphide from gallium solution with PH_3 vapour as phosphorous source.

proportions and by dissolution and diffusion the reaction product emerald ($Be_3Al_2Si_6O_{18}$: Cr) is precipitated. A more sophisticated arrangement is shown in Fig. 7.20(b), where the reactants A and B are separated from the growth regions by diaphragms. Preferentially the site of nucleation is provided by localized cooling, or a seed crystal can be applied when the solution is saturated.

Examples of the use of solids which slowly dissolve and produce reduction reactions have been published by McWhan and Remeika (1970) and by Foguel and Grajower (1971). The former used slowly dissolving VN in order to grow V_2O_3 from $KF + V_2O_5$ melts in platinum crucibles, whereas Foguel and Grajower reduced V_2O_5—KF melts with the graphite crucible and grew high purity V_2O_3 crystals up to 6.5 mm length and 2 mm diameter.

(b) *Vapour phase source* (VLRS). As discussed in Chapter 2, among the earliest crystals grown from high-temperature solutions were tungsten bronzes prepared by Wöhler (1823) from sodium tungstate melts which were reduced by hydrogen to produce Na_xWO_3. Since that time many crystals have been prepared by the *vapour-liquid-reaction-solid* (VLRS) technique, and in the following only a few examples will be given. Transparent crystals of $BaFe_{12}O_{19}$, $SrFe_{12}O_{19}$, Fe_2O_3 and $LiFe_5O_8$ have been prepared from $BaCl_2$ or BaF_2, $SrCl_2$, $NaCl$ and $LiCl$ melts, respectively, containing Fe_2O_3, which were reacted with oxygen at 1250° by Brixner (1959). Similar experiments on growth by reaction with oxygen or water yielded crystals of $CaMn_2O_4$, $Ca_2Nb_2O_7$, $CaFe_2O_4$, $Ca_3Al_{10}O_{18}$, $CaCrO_4$, Ca_2SiO_4, Ca_2PO_4Cl, $Ba_5(VO_4)_3Cl$, $Ba_5(MnO_4)_3Cl$, $BaCrO_4$, $BaSb_2O_6$, $BaFe_{12}O_{19}$, $BaWO_4$, BaB_2O_4, $BaSi_2O_5$, $BaPbO_3$ and $BaTi_3O_7$ (Brixner and Babcock, 1968), Sr_2VO_4Cl and Sr_2VO_4Br (Brixner and Bouchard, 1970), Ca_2PO_4Cl, Ca_2VO_4Cl, $Sr_5(PO_4)_3Cl$, Sr_2VO_4Cl, $Ba_5(PO_4)_3Cl$ and $Ba_5(VO_4)_3Cl$ (Brixner and Weiher, 1970).

Semiconductors such as the III–V compounds may also be prepared from solutions in the Group III metals, with the Group V element supplied via the vapour phase, as in the experiment of Plaskett (1969) and of Poiblaud and Jacob (1973). The principle is shown in Fig. 7.20(c), the growth of gallium phospide being promoted by the relative motion between the ampoule and the RF coil. A similar technique was developed by Kaneko *et al.* (1973, 1974) for production of GaP crystals from gallium solution. The supply of reactant or dopant gases is also frequently used in liquid phase epitaxy as mentioned in Chapter 8.

Decomposition resulting in a volatile crystal component occasionally has been used to grow bulk crystals or crystalline layers from high-temperature solutions. As an example, DeVries (1966) was able to control the decomposition of molten CrO_3 at medium oxygen pressures in order to grow thin layers of CrO_2 epitaxially on various substrates. Another example is the crystallization of the highly refractory uranium monosulphide (M.P. 2460°C) by decomposition of uranium disulphide (M.P. 1560°) at 1700–1900° in tungsten crucibles, as reported by Van Lierde and Bressers (1966).

Complex systems have been used by Von Philipsborn (1967, 1969) and by Von Neida and Shick (1969) and Shick and Von Neida (1971) to grow a variety of chalcogenide spinels by VLRS technique. Von Philipsborn (1971) reviewed crystal growth of chalcogenide spinels of which it is difficult to obtain crystals larger than 5 mm. Another example of the VLRS mechanism is the growth of NiO whiskers from molten nickel (Ahmad and Capsimalis, 1967).

An interesting modification of the VLRS mechanism has been proposed by Wagner (1968) and named SLV growth: by a reaction $B + 2HX \rightarrow BX_2 + H_2$ volatile BX_2 is removed from the solution of A in B so that A crystallizes.

(c) *Technique of shifting chemical equilibria.* According to the law of mass action (which in its ideal form only holds for dilute solutions) of Guldberg and Waage (1867) solution equilibria are shifted when a volatile component is vaporizing. This shift in equilibrium may be used to crystallize compounds which otherwise would not precipitate. This source of supersaturation was already known in the 19th century, and Morozewicz (1899) expressed the temperature-dependent relationship

$$Na_2WO_4 + SiO_2 \rightarrow Na_2SiO_3 + WO_3 \uparrow.$$

Dugger (1966, 1967) reported on a "new hydrolysis technique" which, however, must be a shifting equilibrium technique when one analyses the experimental conditions. Large amounts of water which would be required to grow $MgAl_2O_4$ crystals by hydrolysis could not be present in a molybdenum crucible containing BaF_2, MgF_2 and Al_2O_3, which was heated at 900°C in vacuum and then heated up to 1650°C for three hours in a helium atmosphere. Thus hydrolysis cannot have taken place. It is much more probable that according to the equation

$$3MgF_2 + 4Al_2O_3 \rightarrow 3MgAl_2O_4 + 2AlF_3$$

aluminium fluoride with a boiling point of 1537°C was evaporated and the chemical equilibrium shifted to cause the precipitation of $MgAl_2O_4$.

This mechanism of shifting chemical equilibrium should be applied more often to grow other crystals of highly refractory compounds. On the other hand the growth temperature is relatively high which is disadvantageous both from the experimental point of view and because of the higher concentration of defects.

Another technique of shifting chemical equilibria which apparently has not been used in flux growth is based on the so-called "salting-out effect" where a more soluble compound dissolves and so precipitates the required phase. The isothermal solution mixing technique of Woodall (1971), which was used to grow multiple-layer films (see Section 8.4.4), is a further example of growth by shifting equilibria. As illustrated in Fig. 8.13, solutions of Ga/Al/As of different composition may be mixed isothermally to produce a supersaturated composition from which a $Ga_{1-x}Al_xAs$ solid solution precipitates.

(iv) *Vapour-liquid-solid* (VLS) *mechanism.* In general the VLS growth mechanism is one by which the solute is transported in the vapour phase prior to dissolution in the solvent and subsequent crystallization. It therefore differs from other transport techniques only in that the solute is initially transported as a vapour rather than by dissolution of nutrient material. Although this technique has been applied to the growth of bulk crystals and films, its initial application was in the growth of whiskers and the main emphasis in VLS growth has remained in this area.

The use of the VLS mechanism was first reported by Wagner and Ellis (1964, 1965) who produced whiskers of silicon up to 0.2 mm in diameter on dots of gold which were deposited on a silicon crystal. At temperatures above the eutectic, the gold dissolves the substrate and preferentially removes the relatively imperfect regions. The liquid can absorb material from the vapour readily so that the surface droplet becomes supersaturated and crystalline material is deposited. As growth proceeds, the liquid droplet remains at the end of the filament which may grow at a rate of about 1 μm/min. A solidified droplet may be seen at the end of the CaB_6 whisker (Rea and Kostiner, 1971) which is shown in Fig. 7.21. The whiskers are of high quality and are often free from dislocations. The uniformity may also be very good although thickening may occur by direct deposition from the vapour at steps on the lateral faces.

A crystalline substrate is not essential for whisker growth since super-saturation will still occur due to absorption from the vapour. As dissolution from the vapour continues, crystals will nucleate and will tend to grow as needles. In the experiments of Frosch (1967) needles were grown on the wall of the container where wet hydrogen was passed over adjacent crucibles containing GaP and Ga, respectively. The characteristic solvent droplet (in this case gallium) was found at the end of most needles. Schönherr (1971), however, disputes the effectiveness of the VLS mechanism in the growth of GaP whiskers by transport in wet hydrogen since growth was observed to cease when a whisker became covered with a Ga droplet.

Wagner (1967) has discussed the perfection of silicon whiskers grown by VLS and demonstrated that the branching and kinking, which is frequently observed, results from lateral driving forces. Temperature gradients along the substrate surface are particularly effective in producing such phenomena. Occasionally whiskers of very complex shape have been

FIG. 7.21. VLS growth demonstrated by the solidified droplet at the end of CaB$_6$ whiskers (Rea and Kostiner, 1971).

produced by the VLS mechanism, for example the continuous coils and spirals of ZnS and GaAs observed by Addamiano (1971).

Quite a wide variety of materials have now been grown by the VLS technique, either deliberately or by accident. Examples are given in Table 7.7. In addition, some interesting experiments were reported by Givargizov and Sheftal (1971) in which composite whiskers were grown, for example of silicon and lanthanum hexaboride in alternate sections.

The methods used to transport the vapour depend on the material crystallized and correspond in general to those used for chemical transport reactions and vapour phase epitaxy. Silicon, for example, may be transported by direct sublimation which has the advantage that impurities from the carrier gas are avoided. A transporting gas such as HCl or H$_2$/H$_2$O is normally preferred particularly for such materials as GaAs and GaP. The carrier gas should have a very low solubility in the solvent or, if this solubility is appreciable, should be rejected from the growing crystal. Doping of the whiskers during growth may be effected by admixtures to the vapour phase.

Few quantitative studies of whisker growth have been presented, a notable exception being that of Bootsma and Gassen (1971) who studied

TABLE 7.7. Materials Grown by the VLS Method

Material	Solvent	Form	Size	Reference
C (diamond)	Ni, Fe, Mn	Whiskers	$130 \times 50\ \mu m$	Derjaguin et al. (1968)
CaB_6	?	Whiskers	$100 \times 20\ \mu m$	Rea and Kostiner (1971)
GaAs	Au, Pd, Pt	Whiskers	—	Barns and Ellis (1965)
GaP	Ga	Whiskers	—	Holonyak et al. (1965)
GaP	Ga	Bulk crystals	20×1–2 mm	Ellis et al. (1968)
$NiBr_2$	Cu, Co, Mn, Fe	Whiskers	—	Sickafus and Barker (1967)
$Pb_{1-x}Sn_xTe$		Bulk crystals	60×9 mm	Mateika (1971)
Se	Tl	Whiskers	—	Keezer and Wood (1966)
Si	Au	Whiskers	$200\ \mu m$ dia	Wagner and Ellis (1964)
Si	Au	Film	$15 \mu m$ thick	Filby and Nielsen (1966)
SiC	Ni	Film	$20\ \mu m$ thick	Berman and Comer (1969)
SiC	Si etc.	Whiskers	—	Berman and Ryan (1971)

the growth of silicon and germanium whiskers using silane ($SiCl_4$) and germane ($GeCl_4$) decomposition, respectively. These authors found evidence to support the validity of the VLS mechanism and concluded that the decomposition at the vapour-liquid interface is rate determining rather than the solid-liquid interface mechanism.

VLS may also be used for the growth of bulk crystals but again few examples are available. Ellis et al. (1968) investigated the growth of GaP crystals in gallium metal with transport by wet hydrogen from a GaP source. Needle-shaped crystals up to 2 cm in length and 1–2 mm in cross-section were grown in an hour but many crystals exhibited twinning, branching or even curvature. Tiller (1968) proposed the use of VLS for the growth of a number of compound semiconductors in the convection-free cell which is shown in Fig. 6.19.

The largest crystals grown to date by the VLS method are probably those of $Pb_{1-x}Sn_xTe$ solid solutions reported by Mateika (1971). Crystals up to 60 mm in length and 9 mm diameter have been prepared by a specific drop technique but it should be mentioned that the liquid in this case is not a solution but a melt of the same composition as the growing crystal.

For the growth of epitaxial layers, transport of solute constituents including dopants in the vapour phase may be convenient and the technology of vapour transport, particularly of semiconductors, is well established. The use of a thin layer of solution rather than a bulk liquid is discussed in Section 6.3 where reference is made to the potential value of this arrangement for fast stable growth. If the problem of the stability of a

L2

thin surface layer can be solved, this arrangement could well become an important mode of application of the VLS technique.

Activity in whisker growth by VLS appears to have declined since 1970–71 but the versatility of the VLS technique makes it a useful tool which may be used to tackle otherwise difficult materials problems.

(v) *Electrolytic growth.* The use of electrolysis in the growth of crystals from high-temperature solution is analogous to its use for electro-deposition of metals except that in the latter case precautions are normally taken to avoid the formation of large crystals. The crystals grown by electrolysis have, however, not normally been of metals but of transition metal oxides such as MoO_2. The essential characteristic of electrolytic growth is that oxidation or reduction occurs at the electrodes and many applications of interest yield crystals in which the valence of a constituent element differs from that of its ions in the solution.

Typical experimental arrangements for electrolytic crystal growth, based on the designs of Kunnmann and Ferretti (1964) and of Rogers *et al.* (1966) are shown in Fig. 7.22(a) and (b). In the latter arrangement a central cathode carries a seed crystal (crystallization alternatively proceeds on the wire itself) which is inserted into an inner cell. The crucible wall is used as an anode so that the separation of the cathode and anode compartments prevents re-oxidation of the crystal by the gas evolved at the walls. The inner cell is mounted on a ceramic support which also provides thermal insulation and growth normally proceeds isothermally at low current densities, typically around 10 mA/cm². Holes in the cell partition permit the flow of ions in the solution but prevent the flow into the growth region of crystallites of undissolved nutrient material which may be present in the outer cell. A similar cell to that shown in Fig. 7.22(b) is described by Perloff and Wold (1967), who used an alumina crucible. The cell actually shown was used in attempts to grow Fe_3O_4 crystals from a solution of Fe_2O_3 in $BaO/0.62B_2O_3$.

According to the review of Kunnmann (1971), electrolytic crystallization may be performed in three ways—direct electrochemical decomposition of the solvent, electrochemical decomposition of a solute in an inert solvent, or by the use of electrochemical transport phenomena. The latter alternative could in principle be applied to any material which may be dissolved and subsequently recrystallized electrochemically but no examples of its use are known to the authors. The distinction between the first and second techniques is not always clear but many experiments may be considered in the first category. As an example, MoO_2 has been crystallized up to $7 \times 3 \times 2$ mm in size by reduction of K_2O/MoO_3 and Na_2O/MoO_3 melts at about 600°C (Wold *et al.*, 1964; Perloff and Wold, 1967). The process involves a reaction of the form

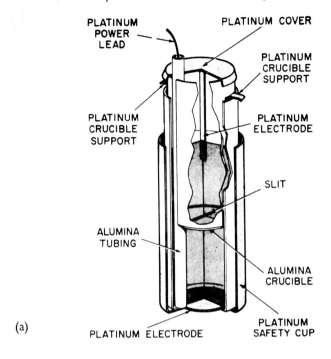

(a)

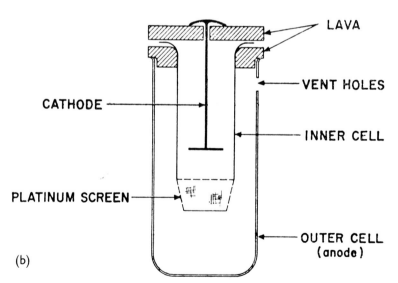

(b)

FIG. 7.22. (a) Cell for crystal growth by electrolysis of fluxed melts (Kunnmann and Ferretti, 1964). (b) Platinum double cell for electrolysis (Rogers *et al.*, 1966).

$$K_2Mo_2O_7 \rightarrow K_2MoO_4 + MoO_2 \rightarrow (\text{cathode}) + \tfrac{1}{2}O_2 \uparrow (\text{anode}).$$

The so-called sodium tungsten bronzes, which are highly conducting oxides of composition Na_xWO_3, may be similarly prepared by reduction of sodium tungstate fluxes:

$$Na_2W_2O_7 \rightarrow \frac{2-2x}{2-x} Na_2W_2O_4 + \frac{2}{2-x} Na_xWO_3 \downarrow + \frac{x}{4-2x} O_2 \uparrow$$

Reactions of this type suffer from the disadvantage that the composition of the liquid, and therefore possibly of the crystal, changes continuously as growth proceeds.

The most successful attempt to grow crystals by electrolytic decomposition from an "inert" flux was probably the crystallization of $Co_{1+x}V_{2-x}O_4$ spinels from sodium tungstate fluxes by Rogers et al. (1966). The description of the dissolution and crystallization process in terms of Lewis acid-base theory is given in Section 3.7.1. The general idea is that the basicity of $Na_2W_2O_7/yNa_2WO_4$ solvents increases directly with the value of y; if the solute can be considered as a solid solution A_zB_{1-z} of components A and B which differ in their basicity, then the value of z of the crystallizing phase will be determined by the value of y of the solvent. In the growth of $Co_{1+x}V_{2-x}O_4$ spinels, electrolytic reduction of V^{5+} to V^{4+} and V^{3+} is required and the value of x was found to depend upon the solvent composition in agreement with the principles outlined above.

Cuomo and Gambino (1970) have proposed the use of electrolysis for crystal growth and epitaxial deposition of compound semiconductors. They demonstrated that, for example, gallium phosphide could be deposited on a silicon substrate from a melt of composition

$$2NaPO_3 + 0.5NaF + 0.25Ga_2O_3$$

at 800°C. A current of 50 mA/cm² was employed and a 100 μm layer was deposited in twenty hours.

In principle a great variety of borides, carbides, germanides, etc. may be prepared by electrolytic reduction of melts containing the corresponding oxidized form—borate, carbonate, germanate, etc.

Many preparations of such compounds from halide solutions were reported by Andrieux and co-workers (Andrieux, 1929; Andrieux and Weiss, 1948; Andrieux and Marion, 1953). These syntheses are normally performed at very high current densities and may therefore yield metastable phases. Kunnmann (1971) has reported attempts to reproduce the experiments at low current densities and finds that these are successful in over half the examples considered. Although preparations of this kind are the subject of a not inconsiderable number of patent applications, further study appears necessary before the reproducible growth of crystals of good size and quality can be achieved.

7.1.3. Preparation of solid solutions and homogeneously doped crystals

Many of the materials of interest for devices or for academic studies are solid solutions or contain dopants which are added in order to produce a desired change in some property. The ability to vary the composition of a crystal is clearly an extremely powerful tool for the materials scientist and the concept has been developed of molecular engineering—tailoring a material to meet a specific set of properties by the addition of controlled quantities of dopants to some suitable host-crystal. The difference between a solid solution and a doped crystal is mainly one of degree, although the term "solid solution" is restricted to members of an isostructural series whereas dopants may differ in structure and even in valence from the host. However, the chromic oxide which is added to alumina to produce the characteristic red colour of ruby would normally be referred to as a dopant, whereas a compound $Al_{1.8}Cr_{0.2}O_3$ would be termed a solid solution between Al_2O_3 and Cr_2O_3. In the preparation of solid solution crystals it is of great importance that the properties should not vary significantly throughout the material, which implies that the composition should be uniform.

There are two types of concentration variation which normally concern the experiments: the relatively long range variation due to a change in the solution composition or growth temperature or growth rate as the crystal grows, and short range variations due to transients within the solution. Time-dependent effects have been treated by Slichter and Burton (1958) who consider the changes in concentration due to a step-function change in the melt concentration and to sinusoidal variations in the growth rate. The latter fluctuations normally lead to the periodic variations in composition termed *striations* (see Section 9.2.4). In crystals pulled from the melt the dopant concentration frequently exhibits a radial variation due to flow effects not associated with crystal rotation, such as thermal convection or crucible rotation (Carruthers, 1967). The major cause of striated dopant distributions in melt growth and probably in solution growth is oscillatory variation of the temperature (Hurle, 1966, 1967; Hurle et al., 1968). The relation between striations and composition in solid solutions is well illustrated by Fig. 7.23, in which the Ta/Nb ratio in crystals of nominal composition $KTa_{0.64}Nb_{0.36}O_3$ is compared with the optical absorption (Whiffin, 1973). In the diagram, contrast between the bands is maximized by arrangement of crossed polars, and the Ta/Nb ratio is determined by electron microprobe.

Striations may be eliminated if temperature oscillations can be suppressed, as is discussed in Section 9.2.4. The use of artificially induced

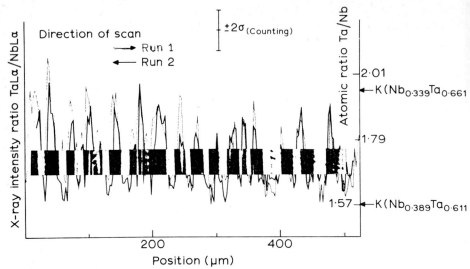

FIG. 7.23. Striations in a K(Ta, Nb)O$_3$ crystal indicated by electron microprobe measurements and an overlaid transmission photograph of the crystal taken in polarized light (Whiffin, 1973).

striations as a means of determining the instantaneous growth rate as suggested by Witt and Gatos (1969) is a potentially valuable tool for the investigation of segregation in crystals grown from high-temperature solution. The interesting technique of Mateika (1971) of producing large homogeneous crystals of solid solutions from the melt with a vapour source has been mentioned in Section 7.1.2 (C.iv). The problem of avoiding relatively long range variations in flux growth will be treated in this section.

Distribution coefficient. The incorporation of an impurity into a crystal may be characterized by a *distribution coefficient k* which is defined as the ratio of the concentration of the dopant in the solid to that in the liquid. It is possible to define three distribution coefficients:

the equilibrium coefficient $k_0 = n_s/n_{sn}$ $(v = 0)$
the interface coefficient $k^* = n_s/n_{sn}$ $(z = 0)$
the effective coefficient $k = n_s/n_{sn}$ $(v \neq 0)$

where n_{sn} is the concentration of the dopant in the solution and n_s that in the solid.

The value of k_0 may be determined from the phase diagram. Thurmond and Struthers (1957) have given an expression of the form

$$\ln k_0 = \frac{\Delta H_F' - \Delta H_M'}{RT_M} - \frac{\Delta H_F'}{RT_m'} - \frac{\Delta H_F' - \Delta N_M'}{\Delta H_F} \ln (1 - X), \qquad (7.5)$$

where ΔH_F is the enthalpy of fusion of the solvent of melting point T_m, $\Delta H_F'$ and $\Delta H_M'$ the enthalpy of fusion and mixing respectively of the solute, which melts at temperature T_M', and X the fractional concentration of solute. Values of k_0 for a wide variety of crystals and dopants have been tabulated by Kröger (1964) and by Brice (1973).

Kamenetskaya (1967, 1968) has calculated the change in composition of solid-solution crystals nucleating in a two-component liquid by considering the changes in free energy. This theory may be used to estimate the variation in the concentration of solvent molecules present substitutionally in the crystal but does not refer to the more usual situation in solution growth where the solvent is rejected by the crystal.

In general the effective value of k will differ from k_0 and a relatively simple relation which has been widely applied was derived by Burton, Prim and Slichter (1953), namely

$$k = \frac{k_0}{k_0 + (1 - k_0)\,\exp\,(-v\delta/D)}. \tag{7.6}$$

According to this equation, k will tend to k_0 at low values of v (and δ) and to unity at high values of growth rate. In practice v often varies inversely as δ, for example as the crystal rotation rate is changed, so that k will vary mainly through its dependence on k_0 and D unless growth occurs mainly under kinetically limited conditions where this inverse dependence does not apply.

Equation 7.6 is derived by solving the time-independent differential equation for solute flow in the usual boundary layer approximation. Earlier treatments of the segregation of dopants have been given by Hayes and Chipman (1939) and by Wagner (1950), the latter considering the effects of natural convection in addition to convection by the rotating crystal. A review of segregation including reference to its importance in purification has been given by Pfann (1966).

The Burton-Prim-Slichter equation is valid only so long as $k^* \simeq k_0$. If this condition is not met, a correction may be applied by considering the solute concentration in an adsorbed surface layer. Trainor and Bartlett (1961) considered the build-up of impurities due to the propagation of steps across the surface at a rate v_{st} and arrived at a relation

$$k^* = k_0[1 - v_{st}/\{\beta D_i(1 - k_0)\tau_s/\tau_i\}]^{-1}, \tag{7.7}$$

in which D_i is the surface diffusion coefficient of the impurity, τ_s/τ_i the ratio of sticking times of solvent and impurity on the surface and β a function which depends on the growth rate and the rate of impact of impurity atoms.

An alternative approach to the same problem has been used by Kröger (1964). By combining Eqn (7.6) with a theory due to Hall (1952, 1953), he arrives at an expression

$$k = \frac{k_0 + (k_{ads} - k_0) \exp{(-v_{deads}/v)}}{1 - \{1 - \exp{(-v\delta/D)}\}\{1 - k_0 - (k_{ads} - k_0) \exp{(-v_{deads}/v)}\}} \quad (7.8)$$

where v_{deads} is the rate of transfer of impurity atoms from the surface layer to the solution and k_{ads} is the distribution coefficient for adsorption into this layer. Equation (7.8) reduces to the Burton-Prim-Slichter equation if $k_{ads} = k_0$ and, while it is clearly of wider applicability, its value is reduced by the difficulty of predicting values of k_{ads} and v_{deads}.

Examples of distribution coefficient measurement. Measurements of the distribution coefficient in solution growth should be treated with caution since in some cases the dopant may be present in minute inclusions rather than substitutionally in the crystal lattice. This is particularly the case when the dopant is a constituent of the solvent, and some examples where high values of distribution coefficient may have such an origin are discussed in Chapter 9. If the dopant has a valence which differs from that of the host lattice, the distribution coefficient will depend strongly on the presence of charge compensating ions. The theory of coupled substitution of ions of unlike valence has been considered by Millett *et al.* (1967), who determined the distribution coefficients of lithium and chromium in zinc tungstate pulled from the melt, using radioactive chromium (see Section 3.6.2).

Rare earth ions in $Y_3Al_5O_{12}$: Rare-earth doping of yttrium aluminium garnet is of practical importance in connection with the application of these materials as lasers. Monchamp *et al.* (1967) found a regular dependence of the distribution coefficient on the size of the ionic radius of the rare-earth ion. The value of the distribution coefficient was found to vary with increasing rare-earth ionic radius from 1.9 for Tm to 0.25 for Pr for growth from a PbF_2/B_2O_3 flux. An alternative and closely related plot was given by Van Uitert *et al.* (1970) who plotted the distribution coefficient of Y^{3+} in $(Y, R)_3Al_5O_{12}$ (R = rare earth) solid solutions and showed that the logarithm of this coefficient varies linearly with the Espinosa ionic radii of the R^{3+} (Fig. 7.24). Also shown in Fig. 7.24 is the distribution coefficient of Pb in orthoferrites grown from PbO/B_2O_3 fluxes.

The distribution coefficient of rare-earth ions in garnets depends on the crystal facet into which substitution occurs, as was demonstrated by Wolfe *et al.* (1971).

Gallium in $Y_3Fe_5O_{12}$: The saturation magnetization of yttrium iron garnet is too high for several applications, for example in microwave devices and for bubble domain memories. Gallium substitution is

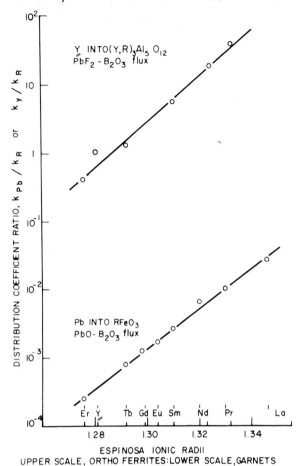

Fig. 7.24. Distribution coefficient ratios k_Y/k_R and k_{Pb}/k_R *versus* radii of rare-earth ions R, for Y entering the garnet (upper line) and for Pb entering rare-earth orthoferrites (lower line) (after Van Uitert *et al.*, 1970).

frequently employed as a means of reducing the magnetization and uniformity is often very important. The distribution coefficient of Ga in $Y_3Fe_5O_{12}$ grown from PbO/PbF_2 solvent was measured by Nielsen *et al.* (1967). These authors defined a coefficient in terms of the fractional concentration of gallium, that is

$$k' = \frac{\dfrac{\text{moles } Ga_2O_3}{\text{moles } Ga_2O_3 + \text{moles } Fe_2O_3} \text{ in crystal}}{\dfrac{\text{moles } Ga_2O_3}{\text{moles } Ga_2O_3 + \text{moles } Fe_2O_3} \text{ in solution}}.$$

The value of k' was found to be in the region of 2.0 and to vary only slowly with temperature and with the absolute Ga concentration. A large change was found to result from the preferential evaporation of PbF_2 from imperfectly sealed crucibles and the resulting variation in magnetization throughout the crystal is plotted in Fig. 9.6(a). When sealed crucibles were used a much more uniform gallium distribution was observed, since the temperature variation in k' compensates for the change in gallium concentration in the solution as growth proceeds.

Makram et al. (1968) also reported the preparation of solid solutions of $Y_3Fe_{5-x}Ga_xO_{12}$ by slow cooling from lead salt solvents. In their experiments a high pressure of oxygen was used to suppress solvent evaporation. The value of k' was found to be about 1.7. Large inclusion-free and homogeneous crystals of the same solid solution were also grown by Scheel (1972) using accelerated crucible rotation and sealed crucibles.

Distribution coefficients for Ga have also been measured in magnetic garnet films grown by liquid phase epitaxy. Giess et al. (1972) found values of 1.94–1.74 for a linear growth rate of 0.98–5.18×10^{-6} cm s^{-1} at $980°$ for solid solutions $Y_{2.4}Eu_{0.6}Fe_{5-x}Ga_xO_{12}$ grown from PbO—B_2O_3. These values are in good agreement with the 1.96–1.76 reported by Blank et al. (1973) between $1065°$ and $918°C$. Blank and Nielsen also measured the distribution coefficient in $Y_3Fe_{5-x}Ga_xO_{12}$ solid solutions and noted that their values of 2.26–2.09 between $1077°$ and $879°C$ were within 10% of the values found in bulk crystals. Data for gallium distribution coefficient as a function of temperature for $Y_{2.7}Gd_{0.3}Fe_{5-x}Ga_xO_{12}$ films are reported by Janssen et al. (1973) who used a radio-isotope labelling technique. These values range from $k' = 2.0$ at $1000°C$ to $k' = 1.4$ at $780°C$ and are therefore in general agreement with the data of other groups.

In general, however, distribution coefficients in films grown by LPE will depend upon the mismatch between the film and the substrate. This dependence is illustrated clearly for Pb and Bi incorporation into garnet films by Robertson et al. (1975). Robertson et al. (1974b) have also noted that the Pb concentration may be much higher in the initial stages of growth than in subsequent growth of the film and this variation should be taken into account when any results of distribution coefficients in epitaxially grown films are quoted.

Experimental methods. Of the methods available for the preparation of homogeneous solid solutions, the most widely used are the gradient transport techniques. These have the advantage that growth occurs isothermally and, even if the distribution coefficient differs appreciably from unity, there will be equilibrium between the growth and dissolution rates of the various constituents provided that growth does not occur too rapidly. The success of the gradient transport method in the preparation of solid solutions of composition $Y_3Fe_{5-x-y}Ga_yAl_xO_{12}$ by Linares (1965) is discussed in Section 9.2.4 (see Fig. 9.14). Other examples of the application of gradient transport to grow homogeneous solid solutions are $InAs_{1-x}Sb_x$ (Stringfellow and Greene, 1971—see Section 8.4.1) and $Y_{3-x}Nd_xAl_{5-y}Cr_yO_{12}$ (Timofeeva et al., 1969). In the latter case the improved homogeneity compared with crystals grown by slow cooling was confirmed by electron probe microanalysis.

Tolksdorf and Welz (1972) have developed apparatus in which the gradient transport method is used in combination with seeding by the Bennett–Tolksdorf technique (Section 7.1.1) and with stirring by Scheel's accelerated crucible rotation (Section 7.2.7). In the initial stage shown in Fig. 7.15(a), the solution is saturated, and spontaneous nucleation might occur. The crucible is then inverted as in Fig. 7.15(b) with the temperature maintained constant but with a "cool finger" brought into contact with the region immediately below the seed crystal. The nutrient material is supported on a perforated baffle and is in contact with the solution so that dissolution may occur, and the transport of solute across an adverse temperature gradient is enhanced by accelerated crucible rotation. The homogenizing effect of the stirring action is particularly valuable in this case. After growth has occurred for the required period, typically a few days, the crucible is reinverted and the solution runs off the crystal as in Fig. 7.15(c).

In such experiments, where solid solution crystals are grown on a seed which is an end member of the series (e.g. $Y_3Fe_{5-x}Ga_xO_{12}$ on $Y_3Fe_5O_{12}$) defects can arise because of the lattice mismatch between the seed and the crystal. A solution to this problem was suggested by Chicotka (1971) which can be applied to certain cases such as the growth of $Ga_{1-x}In_xP$ crystals on a GaP seed. If crystallization occurs initially at a relatively high temperature, the composition of the phase deposited is close to that of GaP. The temperature is then lowered with the solution in contact with nutrient material, and the crystal composition becomes progressively richer in indium as growth occurs by slow cooling. When the temperature has been lowered to that corresponding to the terminal composition, cooling is stopped and subsequent growth occurs by gradient transport. In this way the composition of the crystal is changed gradually and an abrupt change in lattice parameter is avoided.

The travelling solvent zone method (Section 7.1.2. CII) may similarly be used to prepare homogeneous crystals, provided that growth conditions are isothermal. DiBenedetto and Cronan (1968) were able to prepare very homogeneous crystals of $Pb_{1-x}Sr_xTiO_3$ by the slow passage of a $PbO/SrO/B_2O_3$ solvent zone through the source material. The solvent composition was found to be critical for the growth of homogeneous crystals.

If polythermal methods are used the resulting crystals will normally be inhomogeneous and any exceptions to this rule require rather unusual conditions. In attempting to grow $Ni_{1-x}Zn_xFe_2O_4$ crystals from PbO solution, Manzel (1967) determined the separate solubilities of $NiFe_2O_4$ and $ZnFe_2O_4$ and selected a temperature range over which the slope of the solubility curves was the same for both components. This condition will not, however, always result in homogeneity since the competing ions may have very different probabilities of incorporation into the crystal lattice.

Since according to Eqn 7.6 the interface distribution coefficient will depend upon the growth rate and the thickness of the boundary layer, it is possible in principle to prepare homogeneous solid solutions by programming growth in such a way that kinetic effects are used to compensate for the temperature dependence of the equilibrium distribution coefficient. This programming may be effected by changing the cooling rate, and hence the growth rate, or by a variation in the degree of stirring which will affect the boundary layer thickness. This type of procedure has not yet been attempted, so far as the authors are aware.

Homogeneous solid solutions may also be prepared by many of the reaction techniques described in Section 7.1.2C. Such reactions normally occur at constant temperature and the composition of the crystal grown will mainly depend upon the rate of arrival at the interface of the reactants. Since the transport rate of the various reactants may be varied independently in, for example, growth by the method illustrated in Fig. 7.20(b), the composition may be maintained constant once a steady state has been established.

Luzhnaya (1968) discusses several examples of the preparation of solid solutions from metallic solvents. The majority of the compounds listed are solid solutions of the III–V semiconductors, such as $GaSb_{1-x}P_x$. Stambaugh et al. (1961) utilized programming of the vapour pressure above the melt to prepare solid solutions of $Ga_{1-x}Al_xAs$ and $Ga_{1-x}In_xP$. This method relies on the introduction into the melt of the more volatile metallic species by transport in the vapour phase. The crystals grown in these experiments were not homogeneous but the use of the VLS mechanism does provide an alternative means of achieving homogeneity if the vapour pressure is suitably adjusted. This method was in fact used by Rodot et al. (1968) who supplied phosphorous and arsenic from separately heated ampoules to prepare homogeneous crystals of $GaAs_{1-x}P_x$ from gallium solution.

Several examples of characterization of solid solution crystals are discussed in Chapter 9 where methods of studying homogeneity are treated.

7.1.4. Preparation of special modifications and of metastable phases

In general those phases crystallize from high-temperature solutions which are thermodynamically stable. In the following a few examples will be given of the growth of special modifications stabilized by trace impurities and of the growth of metastable phases. Mineralogists are familiar with the fact that rutile is the only stable modification of pure TiO_2 and that anatase and brookite are modifications which are stabilized by trace impurities. Similarly tridymite is a form of SiO_2 which is stabilized by alkali ions and which is not a modification of pure silica as was shown by Flörke (1955).

Similarly four modifications of sodium disilicate have been prepared by Hoffmann and Scheel (1969) from $Na_2Si_2O_5$ glass on a variety of metal oxide pellets, the metal oxides having either a chemical or an epitaxial effect on the crystallization of the various $Na_2Si_2O_5$ modifications. As an example of the effect of impurities on the crystallization behaviour of alloys the compound Al_3Er may be mentioned which crystallizes in the cubic $AuCu_3$ type unless traces of silicon (less than 10^{-2} wt%) are present in which case a rhombohedral form of Al_3Er crystallizes (Meyer, 1970). Thus if a modification cannot be crystallized in its stability region, the influence of epitaxy and trace impurities is worth investigating. The effect of solvents on the crystallization of carbon as diamond or graphite is discussed in Section 3.7.3.

Metastable phases occasionally grow in the stability field of an other phase when its nucleation and growth is facilitated, for instance by providing seed crystals. Thus Roy and White (1968) claim to have grown cm-size quartz-type GeO_2 crystals on substrates of natural quartz from high-pressure solutions, and propose the use of high-pressure solution growth technique at relatively low temperatures for growth of metastable phases.

It was shown by Scheel (1968) that in cases where a system is far from equilibrium metastable phases can be grown. In the example given, crystals of up to 2 mm length of the metastable β-quartz phase of the composition $MgAl_2Si_3O_{10}$ were grown from a glass of that composition in the presence of a small amount of lithium tungstate flux in the temperature range $600°-800°$ whereas higher temperatures or long heating resulted in crystallization of the thermodynamically stable cordierite phase. According to Ostwald's step rule the less metastable quartz-phase crystallizes from a metastable glass quenched from high temperatures (and having still the high-temperature structure) until finally the stable cordierite phase is formed. The crystallization of metastable phases from glasses is a common phenomenon, and accordingly one would expect that metastable phases could also be crystallized from viscous high-temperature solutions, when the system is far from equilibrium. Similarly from systems which allow high supersaturation of the stable phase, crystals from another phase (which might be thermodynamically unstable) might form if seeds of the required phase or a substrate which allows epitaxial overgrowth are provided. As discussed in Section 3.8.5. and by Blank and Nielsen (1972) a rare-earth garnet $PbO-B_2O_3$ solution can be saturated with respect to rare-earth orthoferrite and supersaturated with respect to garnet, simultaneously, thus explaining the growth of phases in a metastable field.

The importance of seeds for the growth (reproduction) of the various polytypes of silicon carbide has been demonstrated by Knippenberg and Verspui (1966) although in this case the degree of "metastability" as deducible from differences in free energy is expected to be very small for the SiC polytypes (Knippenberg, 1963).

7.1.5. Growth of compounds with defined valence states

The great majority of HTS growth experiments yield materials in which the ions are in "normal" valence states, so that the atmospheric conditions need not be closely specified. So, for example, most crystal growth from molten salt solutions occurs in an air ambient and crystallization from metallic solutions takes place under hydrogen or an inert gas. Many examples are known, however, where materials having ions in relatively unusual valence states have been prepared from HTS or where special conditions have been used in order to produce material of high stoichiometry. The techniques which are available for such preparations may be broadly divided into three groups: (i) electrolytic growth, (ii) the use of a solvent of controlled oxidizing or reducing power, (iii) the use of controlled atmospheres.

Crystal growth by electrolysis is discussed in Section 7.1.2 where the examples are discussed of MoO_2 and the sodium tungsten bronzes. Further examples of transition metal oxides with ions in reduced valence states are given in the table of Chapter 10, and it is likely that further applications of this technique will be made, for example to produce materials of higher than normal valence state, or to grow crystals of magnetic oxides with controllable concentrations of Fe^{2+} or Fe^{4+}.

Reference is also made in Sections 7.1.2 and 3.4.1 to examples where the oxidation state of transition metals may be influenced by the basicity of the solvent. In general, additional factors such as the solubility will have a major influence on the choice of a solvent, and atmosphere control provides a simpler means by which the valence state may be varied.

The use of atmosphere control in the crystallization of phases unstable in air at the growth temperature is discussed in some detail in Section 3.4.1. In that section the Magneli phases Ti_nO_{2n-1}, europium chalcogenides, vanadium dioxide VO_2 and chromium-containing oxides are quoted as examples of materials which can be crystallized from high-temperature solution when the atmosphere is adjusted to the equilibrium range for the required phase.

Many other examples of materials, especially oxides, are known in which an atmosphere other than air is used in order to improve stoichiometry. Oxides containing iron are often difficult to prepare as single crystals with the iron in a single valence state, and considerable effort has been devoted to the preparation particularly of the ferrite spinels with negligible concentrations of Fe^{2+}. The equilibrium partial pressure of oxygen above these materials may be extremely high at temperatures near the melting point, and Ferretti et al. (1962) found that an oxygen pressure of nearly 120 atmospheres was required to prepare crystals of $CoFe_2O_4$ free from Fe^{2+} ions from a highly concentrated solution in $NaFeO_2$ at 1590°C. Such pressures require rather expensive autoclaves and the use of more easily realizable pressures at lower temperatures has been advocated by Makram and co-workers. Makram and Krishnan (1967) used an oxygen pressure of 15 atm. for the growth of garnet crystals from PbF_2/PbO by slow cooling from 1300°–900°C. The application of a gas pressure can cause major changes in the nucleation and growth characteristics, particularly from the volatile lead solvents, so that the properties of the crystals grown may be changed by factors other than those associated with the variation in stoichiometry. Some complexities of yttrium iron garnet crystal growth under pressure have been discussed by Robertson and Neate (1972) and by Robertson et al. (1973). The work of the latter authors is of particular interest since the reproducible crystallization of only one crystal per crucible is reported under oxygen pressures in excess of ten atmospheres, as discussed in Section 7.1.1. Makram et al. (1968) reported that the gallium distribution in solid solutions of composition $Y_3Fe_{5-x}Ga_xO_{12}$ was homogeneous when crystals were grown by slow cooling under an oxygen pressure of 15 atm.; they attributed this homogeneity to the absence of inclusions and voids in their crystal so that strained regions with locally high concentrations of Ga were avoided. However, the results of Nielsen et al. (1967) and Scheel (to be published) show that good homogeneity of the $Y_3Fe_{5-x}Ga_xO_{12}$ crystals is achieved by prevention of PbF_2 evaporation and seems to be relatively unrelated to the mechanism suggested by Makram and co-workers.

The growth of nickel ferrite $NiFe_2O_4$ under oxygen pressure has been reported by Makram (1966) and by Robertson et al. (1969). In the latter experiments borate solvents were used and the application of oxygen pressures of only two atmospheres was found to increase markedly the concentration of flux inclusions.

7.2. High Temperature Technology

7.2.1. Furnaces

The discussion here of furnace design is necessarily brief, and detailed descriptions of laboratory and industrial furnaces may be found in various handbooks of physics and engineering and in the books of Campbell (1959), Smakula (1962), Wilke (1963, 1973), Otto (1958) and Brice (1973).

Of the various types of heating only resistance and induction heating are used, while heating by flames, plasmas, lasers, image techniques and by electrons has not found significant application in crystal growth from high-temperature solutions. Reactions between heating elements, insulating materials, crucibles, thermocouples, etc. are important in the design and use of high-temperature furnaces, and these aspects will be discussed in Section 7.2.3.

Resistance furnaces for flux growth are generally used up to 1600°C, while at higher temperatures induction heating by R.F. generators is preferred. However, the range of *resistance heating* can be extended by use of molybdenum, tungsten or graphite heating elements in neutral or hydrogen atmospheres or in vacuum, whereas in oxidizing atmospheres rhodium (up to 1800°C), zirconia or thoria may be used. The two latter have the disadvantages that they require pre-heating and that they need high power and are sensitive to thermal shock and to steep thermal gradients. Table 7.8 summarizes the properites of a few popular heating elements, of which Kanthal† A1, silicon carbide (Globar, Crusilite, Crystolon, Silit) and molybdenum disilicide (Kanthal, Super Mosilit) are most frequently used in flux growth.

Furnaces with Kanthal A1 and similar heating elements are readily available in many shapes, and also various types of heating elements and Kanthal A1 wire are available. It should be noted that metal-based heating elements such as Kanthal A1 become brittle after the first firing. The lifetime is long if corrosion is prevented and if the temperature does not exceed 1150°C for extended periods in the case of Kanthal A1.

Silicon carbide heating elements in the form of rods, tubes or spiral tubes are suitable for the construction of simple furnaces according to individual requirements. As thermal insulation a half brick (about $4\frac{1}{2}$ inches) around the hot space is sufficient when a layer of 1 to 2 cm of ceramic wool or similar insulating material is placed between the bricks and the outer wall of, for example, asbestos. It is possible to connect the heating elements directly to the thyristors, but the latter should be adjustable according to the ageing of the SiC elements. (Usually compensation is provided for 100 to 500% increase of resistance, which tends to occur after one month

† Registerd trademark by Bulten-Kanthal AB, Sweden.

TABLE 7.8. Properties of Heating Elements Most Frequently Used in Flux Growth

Heating element	Max. working temperatures (°C)†	Atmosphere
Kanthal A1	1375° (w), 1150° (y)	air
Silicon carbide	1550° (w), 1350° (m), 1200° (y)	(dry) air
Kanthal-Super ST	1700° (w), 1600° (m), 1500° (y)	air‡
Kanthal-Super N	1700° (w), 1600° (m), 1500° (y)	air‡
Kanthal-Super 33	1800° (w), 1700° (m), 1600° (y)	air‡
Platinum	1550° (m)	air
Rhodium	1800° (w)	air
ThO_2 (with La_2O_3 or CeO_2)	2000° (w)	air
ZrO_2 (stabilized)	2400°	air
Molybdenum	1800° (w), 2400°	Vacuum or H_2
Tantalum	2600° (w)	Argon or H_2
Tungsten	2600° (w)	Vacuum
	3000° (w)	Argon
Graphite	2600° (w)	Inert gas

† For estimated or experienced durations of the order of weeks (w), months (m) or a year (y). As discussed in the text the lifetime may vary significantly.

‡ In non-oxidizing atmospheres the maximum working temperature is reduced by 100° to 350°.

to four months use.) SiC heating elements are advantageous when the duration of typical (1300°C) experiments does not exceed one or two weeks and when the furnace requires frequent cooling to room temperature although it is better to keep the furnace always above 750° even when not in use. Elmer (1953) and Bovee (1953) have discussed the properties of SiC heating elements and their lifetime. Of the latter it must be said that it depends very much on the conditions (load, ambient atmosphere and temperature, continuous or intermittent use) and on the fabrication with Globar, Crusilite and Crystolon being known for reliability. A stagnant oxidizing atmosphere is best for a long lifetime, and if corrosive vapours are produced in flux growth experiments it is preferable to blow air gently along the heating elements. Because of the so-called "silicon carbide pest" SiC elements should never be used below 750°C, otherwise the protective SiO_2 film will crack and fast oxidation will occur. The relationship between surface load and surface temperature of the SiC heating elements and the furnace temperature is shown in Fig. 7.25; for comparison the permissible surface load of Kanthal A1 wire heating elements at 1000°C is 4 W/cm² and at 1250°C 1.7 W/cm².

Molybdenum disilicide heating elements have significant advantages over SiC such as a longer lifetime, higher maximum working temperature and absence of ageing effect. Disadvantages are the higher price and low resistance, and $MoSi_2$ elements have to be handled with great care, so that

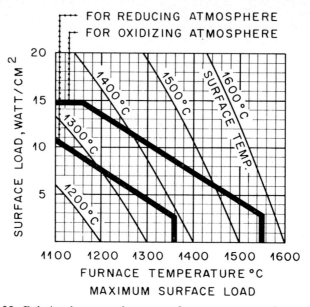

Fig. 7.25. Relation between element surface temperature, furnace temperature and surface load of silicon carbide heating elements. The maximum recommended surface load for oxidizing and reducing atmospheres is indicated by the strong lines.

furnace construction should be left to specialists unless the manufacturer's instructions† are followed very carefully.

The lifetime of Kanthal Super N elements exceeds two years when the following precautions are taken (Scheel, unpublished):

(a) The furnace is never cooled below 800°C.
(b) Dry air is gently flown into the heating element compartment continuously.
(c) Corrosive vapours cannot reach the elements.
(d) The temperature does not exceed 1500°C, and the surface load should not exceed a value of 15–20 watts per cm² (see Fig. 7.26).
(e) If used below 1450°C, the furnace should be heated to 1450°–1500°C for one to two days every six months in order to anneal the elements for strain removal.
(f) The temperature of the heating elements is regulated by continuous control.
(g) Vibrations are prevented by placing the furnace on an anti-vibration mount.

† Bulten-Kanthal AB, Hallstahammar, Sweden.

(h) The U-shaped heating elements should hang freely from the top of the furnace.

(i) The hot zone of the element should not reach into the ceramic insulation.

(j) The terminals should be kept cool by having a free distance between the roof and the contact of 4–5 cm.

The surface load of Kanthal Super elements and their temperature and current are plotted against the furnace temperature in Fig. 7.26. The shaded area shows the optimum conditions. The resistance increases

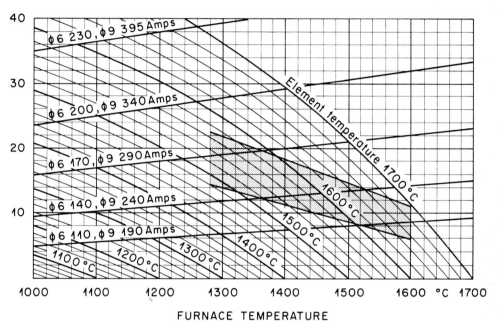

FIG. 7.26. Surface load, temperatures and current of kanthal super heating elements (of 6 and 9 mm diameter) versus furnace temperature.

steeply with temperature as shown in Fig. 7.27, where for comparison the resistance *versus* temperature change of SiC, graphite andK anthal A1 are also shown.

The oxidation behaviour of $MoSi_2$ heating elements has been studied by various groups. Fitzer (1956) was the first to describe an anomalous oxidation behaviour at temperatures below $800°$ and called it "silicide pest". On severe oxidation the "pest" oxide consisting of crystalline MoO_3 and amorphous SiO_2 particles is formed and the heating element

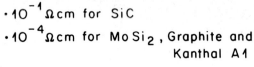

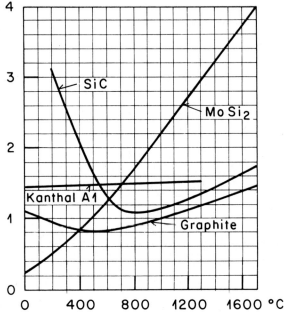

FIG. 7.27. Electrical resistivity of various heating element materials.

disintegrates. Above 800°C a protective SiO_2 layer can form according to

$$2MoSi_2 + 7O_2 \rightarrow 2MoO_3 + 4SiO_2,$$

and the molybdenum oxide evaporates (Bartlett *et al.*, 1965). As was shown by Wirkus and Wilder (1966), another oxidation reaction occurs above 1200°C and just beneath the protective SiO_2 layer. This reaction may be described by

$$5MoSi_2 + 7O_2 \rightarrow Mo_5Si_3 + 7SiO_2.$$

Thus typical $MoSi_2$ heating elements consist of interior bulk $MoSi_2$, a thin Mo_5Si_3 and an outer glassy or semicrystalline SiO_2 layer. The overall oxidation of the $MoSi_2$ elements is controlled by oxygen diffusion through the SiO_2 layer. Generally $MoSi_2$ heating elements have a protective oxide layer when delivered, and it is proposed to keep this intact by a slow flow of dry air along the heating elements.

Table 7.9 Properties of Firebricks and Insulating Ceramics

Name	Composition	Max. working temperature	Melting point	Density of compound	Density of brick/ceramic	Thermal conductivity† compound	Thermal conductivity† ceramic† brick	Lin. expansion coefficient	Thermal shock resistance	Atmosph.	Remarks
Silica	SiO_2	1100°C	1710°C	2.30	>1.5	0.012	0.003	0.5×10^{-6}/°C	good	oxid.	Corroded by fluorides
Sillimanite	$Al_2O_3 \times SiO_2$	1500°		3.25	>2.3	0.008	>0.003		medium	oxid.	Corroded by fluorides
Mullite	$3Al_2O_3 \times SiO_2$	1650°	1920°	3.16	>2.0	0.009	0.007	5.3	medium	oxid.	Corroded by fluorides
Alumina	Al_2O_3	1800°	2020°	3.97	>2.8	0.014	0.005	9.0 (8.6, 9,5)	poor	any	
Spinel	$MgAl_2O_4$	1850°	2135°	3.60	>2.2	0.013	0.005	8.8	poor	oxid.	
Zircon	$ZrSiO_4$	1700°	2420°	4.56	>3.3	0.008	0.005	4.2 (3.4, 5.6)	good	oxid.	Corroded by fluorides
Zirconia	ZrO_2‡	2300°	2700°	5.60	>4.4	0.005	0.002	10.0	poor	oxid.	
Beryllia	BeO	2300°	2570°	3.01	>2.8	0.046	0.02	8.9	good	oxid.	Extremely poisonous dust and vapours
Magnesia	MgO	2400°	2800°	3.58	>2.5	0.016	0.006	14	medium	oxid.	
Thoria	ThO_2	2400°	3050°	9.86	>6.3	0.007	0.005	9.7	poor	any	Radioactive
Urania	UO_2	2200°	2800°	10.96		0.02		10		oxid.	Radioactive
Yttria	Y_2O_3	2000°	2410°	5.01				9.3		any	
Graphite	C	3000°	> 3600°	2.25	~1.9	0.16		2–5	good	reducing or vacuum	
Boron nitride	BN (hexag.)	2000°	3000° (dec.)	2.25	~1.9	0.04–0.12	0.03–0.07	0.05–10 (~2)	good	reducing or vacuum	
Silicon carbide	SiC	1600°	2830° (dec.)	3.22	>2.6	0.10	0.02	4.4	good	oxid.	Corroded by fluorides
Silicon nitride	Si_3N_4	1500°	1900° (dec.)	3.44	>2.0	0.045		3.2 (2.47)	good	neutral	Corroded by fluorides
Silica glass wool	SiO_2	1000°	1710°	2.30		0.012	~0.001		good	oxid.	Corroded by fluorides
Ceramic wool		1200°	~1700°	3.20		0.008	~0.001		good	oxid.	Corroded by fluorides
Zirconia wool	ZrO_2‡	2300°	2550°	5.60		0.005	~0.0005		good	oxid.	Corroded by fluorides

† At 1000°C, in cal/cm°C sec. ‡ Stabilized with CaO or Y_2O_3.

Thermal insulation in resistance furnaces is generally achieved by firebricks of which the most common with their properties are listed in Table 7.9. Dense bricks can carry loads and are resistant to abrasion while the lighter (porous) bricks are better insulators. The maximum working temperature of firebricks based on mullite and sillimanite increases with the alumina content. The space containing the crucibles is preferably separated from the heating elements by dense ceramic tubes or plates. If platinum crucibles are used, care must be taken that they cannot come into contact with silicon carbide ceramic because platinum alloys with SiC at high temperatures; also borides, nitrides and metals should not contact platinum. Only dense alumina-rich firebricks or pure alumina, spinel, magnesia or zirconia ceramics are recommended as supports for platinum crucibles.

Corrosion by flux vapours shortens the lifetime of the firebricks, especially if porous bricks face the chamber containing the crucibles. The resistance to corrosion is increased if such firebricks are coated with an alumina-rich high-temperature cement. Also zirconia or SiC wall plates may be used in order to decrease the number of repairs necessary. Another application of insulating bricks and ceramics is to fill the furnace chamber as much as practical in order to reduce temperature fluctuations by air convection and to increase the thermal capacity of the whole furnace.

Induction heating is generally used when the required temperature exceeds about 1600°C. At such high temperatures difficulties arise with heating elements and their atmosphere requirements and with dense ceramics, especially when the duration of experiments is of the order of weeks or months as in many growth experiments from high-temperature solutions. For work in neutral or reducing atmospheres graphite, molybdenum, tungsten and tantalum heating elements may be used, and excessive loss of heat is prevented by the use of metallic radiation shields, generally contained in water-cooled metal containers. For oxidizing atmospheres and temperatures exceeding 1700°C stabilized zirconia or thoria tubes may be used as heating elements; however, these oxides have to be preheated to about 1600°C in order to make them sufficiently conductive.

The most reliable and practical way of heating to about 2000°C in any atmosphere for crystal growth from high-temperature solutions is by induction heating, and this source of heat has been reviewed by May (1950), Stansel (1949), Brunst (1957) and Simpson (1960). The frequency of generators is about 450 kHz, and it should be noted that typical RF generators have an output of about 50% of their power consumption. For most applications generators of 15 to 30 kW RF power are appropriate, and care should be taken in obtaining sufficient coupling between the RF coil and the conducting crucible or melt. Also the wall thickness of metal crucibles should be uniform, and the crucibles as well as insulation or radiation shields should be centred carefully in the coil.

According to Duncan *et al.* (1971) the penetration depth of the RF heating is given by

$$d = \frac{3750 \times \rho^{1/2}}{(f \times \mu)^{1/2}} \text{ cm},$$

[with ρ the resistivity in ohm-cm of the heated material, μ the relative permeability of the heated material ($=1$ for non-magnetic materials), f the induction heating frequency (Hz) and d the skin depth of the material being heated, at which the current density reaches $1/e \sim 37\%$ of the value at the surface]. The value of d is about seven times greater for 10 kHz than for 500 kHz (2.8 mm and 0.4 mm for platinum, respectively). Thus these authors propose the use of 10 kHz RF heating in order to overcome the common problems of hot spots and catastrophic failures, especially when crucibles are used near their melting points. Also 10 kHz motor-generators are said to be insensitive to line voltage fluctuations and to be very reliable.

Special heat sources might be used for specific experiments. Thus for transparent solutions the crystal growth process may be observed if furnaces with transparent regions are used. For temperatures up to 1100° quartz-glass tube furnaces with a slit in the insulation or reflectors have been used, and Lord and Moss (1970) described a furnace with water cooling between two concentric glass tubes. A furnace consisting of an inner quartz-glass tube with the heating wire wound around it and an outer concentric quartz-glass tube with a thin gold layer evaporated on it was first described by Rabenau (1963/1964) and later in detail by Reed (1973). An approximately 200–500 Å thick gold layer shows high reflectivity for the infrared but is partially transparent to visible light. Another example of transparent furnaces was reported by Wood and Van Pelt (1972) who used tin-oxide coated silica-glass tubes as transparent heating elements. A spray coating system was used to achieve a uniform coating, and such furnaces have been used up to 700°C.

Localized heating as necessary for temperature-gradient transport and travelling solution zone techniques has been discussed in Section 7.1.2. Heat pipes may be useful for localized heating as well as for localized cooling as indicated by Steininger and Reed (1972) but except for a brief mention in a paper of Steininger and Strauss (1972) heat pipes have not yet found wide application in crystal growth from high-temperature solutions. With heat pipes temperatures approaching 2000°C may be obtained, and the lifetime of the heat pipe using lithium as fluid and TZM alloy as pipe material is said to be above one year (Harbaugh and Eastman, 1970) at a temperature of 1500°C. Temperature stabilization needs special precautions

and is best achieved by pressure control of open heat-pipe systems according to Vidal and Haller (1971).

7.2.2. Temperature control (including programming)

Temperature evaluation. Of the variety of temperature measurement and
control techniques only a few are generally used because of their reliability,
accuracy and convenience, and these are indicated in Fig. 7.28, together
with their common ranges of measurement and control. Thermometers
based on thermal expansion of liquids (and gases) and of solids (bimetal
thermometers) are very common and may be used up to relatively modest
temperatures (500°C) but are not applicable or reliable at higher
temperatures. Thus, in this section only resistance thermometry, thermoelectric thermometry and optical pyrometry will be discussed. The
measurement and control of temperature have been reviewed in

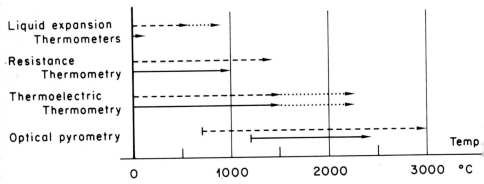

FIG. 7.28. Temperature ranges for measurement (broken lines) and reliable
control (solid lines) for various common temperature evaluation techniques.

"Temperature" (1941, 1955, 1962), by Baker *et al.* (1953, 1961), Kostkowski
(1960), Rubin and by Best (1967).

How precisely and accurately† can a temperature be measured? This
question has been discussed, in addition to above references, by Dike
(1958), Roeser and Lonberger (1958), Stimson (1961) and by Gray and
Finch (1971).

According to the International Practical Temperature Scale of 1968
(IPTS-68, 1969) the temperature is established by interpolating between
the points 100°C (boiling point of water), 419.58°C (freezing point of zinc),
961.93°C (freezing point of silver) and 1064.43°C (freezing point of gold).

† Precision is qualitatively the inverse of the (mean) deviation from a given
temperature, whereas accuracy is the inverse of the (mean) deviation from a point
on the (absolute) International Practical Temperature Scale of 1968.

In various laboratories the freezing points of antimony (630.5°C), aluminium (660.1°C) and copper (1083°C) are used for calibration, but extreme care is necessary with respect to purity (corrosion of container, influence of gases) in all direct calibration work.

Chemicals of high purity needed for calibration according to IPTS-68 may be obtained either from a national institution of standardization or from the National Bureau of Standards, Washington, D.C., U.S.A. However, it is more convenient to obtain calibrated thermocouples, resistance thermometers or pyrometers from commercial suppliers or from the above institutions. The accuracy of temperature measurement at the gold point (1064.43°C) using calibrated Pt–PtRh 10% thermocouples may be as good as 0.2°C, and using non-calibrated commercial thermo-couples the accuracy is generally better than 2.5°C.

A standard thermocouple properly used has a precision of 0.1°C around the gold point, that means that the standard deviation is 0.1° about the mean of that laboratory. At the National Bureau of Standards Pt–PtRh10% thermocouples are annealed for one hour at 1450°C before calibration.

For temperatures exceeding the melting point of gold the melting points of palladium (1552°C) and platinum (1769°C) have been proposed as standards. For temperatures up to 1300°C the platinum *versus* platinum 10% rhodium thermocouples are reliable to within 0.25% if used in a clean oxidizing atmosphere and if extended periods above 1300°C are avoided. The International Temperature Scale above the gold point is defined in terms of the Planck radiation equation and the ratio of the spectral radiance, in the visible region, of a black body at the temperature to be measured to the spectral radiance of a black body at the gold point. However, with a black body and using a calibrated optical pyrometer (with disappearing filament) the standard deviations from IPTS-68 at the gold point and at 2000°C may be as low as 0.6° and 3°, respectively. Generally in practice it is found difficult to obtain a precision of better than 6°C using an optical pyrometer.

The first part of the following section deals with temperature sensors whereas the second part briefly treats techniques for the control and programming of the temperature of furnaces.

Resistance thermometry. This topic has been reviewed in the general references mentioned previously and by Berry (1966) while Daneman and Mergner (1967) discussed the equipment necessary for measuring exact resistance ratios. Resistance thermometry is based on the change of resistance of materials with temperature, and the resistance change is measured with resistance bridges or potentiometers.

The resistance of pure metals increases with temperature whereas semi-

M

conducting materials (e.g. germanium, oxides) show a logarithmic decrease in resistance with increasing temperature which is used in the so-called *thermistors*.

Resistance thermometers allow very precise measurements of temperature as long as the resistance of the metal is not changed by recrystallization, oxidation, impurities or evaporation. Thus nickel and copper are generally used only up to 200°C, platinum up to 1000°C, and tungsten, molybdenum and tantalum (in a protective atmosphere) up to 1200°C, whereas thermistors are used up to 300°C and with special compositions (though not reliably) up to 1600°C. The platinum resistance thermometer may be calibrated if prepared from ultra-pure platinum and if mounted in a helium-filled jacket. In Fig. 7.29 the dependence on temperature of the resistance of various materials is shown.

Resistance thermometry allows the measurement of temperatures up to 1000° with a precision not achieved by other techniques (McLaren, 1957). However, the limited temperature range and lifetime have prevented its wide use in crystal growth from high-temperature solutions. An exception is the range of controllers marketed by CNS Instruments, which use

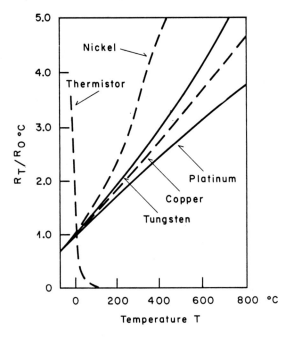

Fig. 7.29. Resistance ratios for temperatures T to the temperature of 0°C for various metals and a typical thermistor.

platinum resistance thermometers for temperature control up to 1400°C, but for temperature regulation over extended periods the resistance thermometer should not be used at temperatures higher than about 1000° because of the change of resistance due to evaporation and recrystallization of the platinum.

Thermoelectric thermometry. The most common temperature measurement and control techniques are based on thermocouples. Since the thermo-electric power of a variety of metal and alloy combinations is known, the temperature of one junction can be determined provided the other junction has a known temperature, preferably the ice point (0°C). The thermo-electric power of a variety of thermocouples has been listed as a function of temperature by Shenker *et al.* (1955) and is shown for high and low temperatures, respectively, in Fig. 7.30(a) and (b). Table 7.10 lists the sensitivities at various temperatures, the maximum temperatures of use, the necessary atmospheric conditions and the stability of a variety of thermocouples.

At temperatures up to 600° the similar thermocouples copper-constantan and iron-constantan are commonly used, whereas for temperatures up to 900°C the chromel-alumel thermocouple is popular. Pallador† and Platinel‡ (Pt, Pd, Ir, Au alloys) are thermocouples with a high sensitivity, comparable with chromel-alumel, and which can be used up to 1200 and 1300°C. Platinum *versus* platinum rhodium 10% is the most frequently used thermocouple for the range 500° to 1400°C, although other thermo-couples based on noble metals are now available which are more stable at higher temperatures and have similar sensitivity above 1000°C. As an example the platinum 94% rhodium 6% *versus* platinum 70% rhodium 30% is listed which has the advantage of having a negligibly small sensitivity around room temperature. Scheel and West (1973) use thermopiles made from several PtRh6% *versus* PtRh30% couples as sensing elements for furnace control. Such thermopiles have the advantage of very high sensitivity so that changes in e.m.f. due to variation in the furnace temperature are very large in comparison with noise or changes due to variation in the ambient temperature. The latter effects normally result in fluctuations at the input to a furnace controller of about $\pm 3\ \mu V$ which corresponds to the change in output from a Pt *versus* PtRh10% thermo-couple due to a change in the furnace temperature of $\pm 0.3°C$. A sensing device of high sensitivity to variations in the furnace temperature but insensitive to cold junction temperature changes is therefore necessary if stabilization to say $\pm 0.05°C$ is to be achieved. The PtRh6% *versus*

† Trademark of Johnson Matthey Ltd.
‡ Trademark of Engelhard Industries.

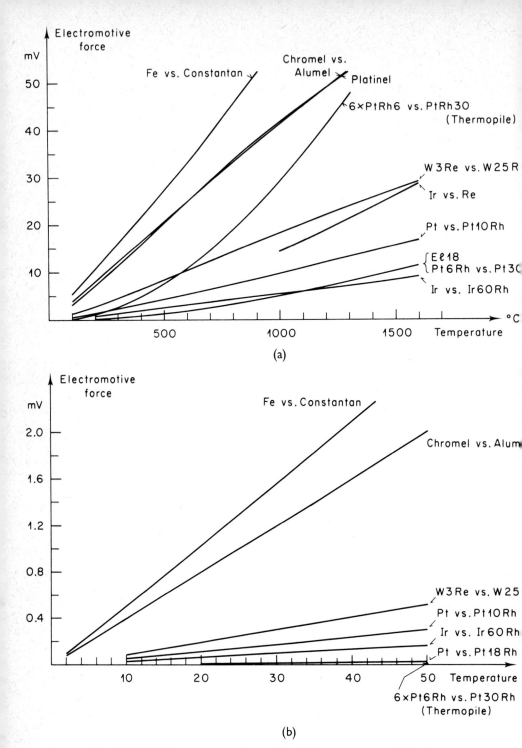

FIG. 7.30. (a) Electromotive forces (for high temperatures) of various common thermocouples and of the PtRh6–PtRh30 thermopile proposed by Scheel and West (1973). (b) Electromotive forces of various common thermocouples for the temperature range 0° to 50°C.

Thermocouple	Sensitivity: μV/°C at			Max. temp. of use		Atmosphere	Estimated temperature shift at		
	25°C	800°C	1200°C	intermittent	continuous		temp.	10 hours	1000 hours
Iron–Constantan	51	64		800°C	650°C	reducing inert, or oxidizing	650°		2.5°
Chromel–Alumel	40	40	36	1260°C	1100°C	oxidizing or neutral	1000° 1200° 1200°	0.5° 2.0° 0.5°	5.0° fails 5.0°
Pd–Pt85Ir15	19	37	45	1500°C		inert or oxidizing†			
Platinel	32	41	39	1300°C	1150°C	oxidizing, inert or reducing	1200°		3.0°
Pt–Pt90Rh10	6	10.9	12.0	1700°C	1450°C	oxidizing	1200° 1500° 1700°	0.2° 1.0° 4.0°	2.0°
Pt94Rh6–Pt70Rh30	<0.07	7.55	10.5	1800°C	1500°C	oxidizing			
Thermopile of 6PtRh6–PtRh30 thermocouples	<0.4	45.3	63	1800°C	1600°C	oxidizing			
Pt80Rh20–Pt60Rh40				1830°C	1600°C	oxidizing	1800°	6.0°	
(sensitivity below at)	25°C	1600°C	2000°C						
Ir–Rh60Ir40	3	5.8	6.2	2000°C	1800°C	inert or slightly oxidizing	2100°	10.0°	
W–Ir		25	26.3	2100°C	2000°C	inert or vacuum	2100°	10.0°	
Re–Ir				2100°C	2000°C	inert or vacuum			
W97Re3–W75Re25	10	17	14	2400°C	2200°C	inert or reducing or vacuum			
W–Re	~5	12	6.5	2400°C	2200°C	inert or reducing or vacuum			
SiC–C	~300	~300	~300	~2700°C		vacuum	Fragile, not reliable		
B₄C–C	~330	~330	~330	~2400°C		vacuum	Fragile, not reliable		

† Pd is oxidized slowly in air between 700° and 900°C; iridium above 900°C.

PtRh30% thermocouples have the additional advantage of a long lifetime at temperatures up to 1600°C.

For temperatures exceeding 1800°C thermocouples based on tungsten or rhenium are used, but the brittleness, the instability and the stringent requirements on atmosphere lead many researchers to prefer optical pyrometry where possible. The fragile nature of the carbon-based thermocouples prevent their popularity although their sensitivity is extremely high.

Some remarks on the thermocouples based on platinum alloys are appropriate as these are in common use in crystal growth from high-temperature solutions, and several of the points also hold for other thermocouple materials. Many aspects, including chemical and physical properties and stability, are discussed by Caldwell (1962) and by Vines (1941). The general features of thermoelectric temperature measurement have been discussed by Finch (1966).

The most stringent requirements are those of purity, which should be better than 99.99% of the individual metal or alloy, and of homogeneity. Any physical and chemical inhomogeneities produce spurious effects by acting as thermocouples themselves. A clean oxidizing atmosphere (e.g. air) is necessary, and traces of hydrogen, carbon monoxide, hydrocarbons, sulphur (or other chalcogenide), pnictides (except N_2) or metal-containing vapour corrode the platinum-based thermocouples or may have an effect on its thermoelectric characteristics. The oxidizing atmosphere also keeps all trace impurities oxidized so minimizing their influence. In addition, the insulating ceramic tubes should be of high purity and should not contain silica (SiO_2, mullite, sillimanite) since SiO_2 has a detrimental effect in reducing atmospheres or at high temperatures, and metals, carbides, borides etc. cannot be used in contact with platinum-based thermocouples because of alloying. Also the ceramic sheath tubes made of high-purity alumina should be impervious in order to prevent volatile flux components (e.g. PbF_2) from coming into contact with the thermocouple. The high temperature limit for a variety of thermocouples as given in Table 7.10 may be lowered by the material which it contacts (insulating or sheath tubes), and this aspect is discussed in Section 7.2.3.

The optimum diameter for thermocouples based on platinum is 0.5 mm but thinner wires of 0.1–0.2 mm may be preferred for a fast response, and thicker wires are used under harsh conditions because of their mechanical and thermal stability. Volatility is an important factor for noble metal thermocouples (and resistance thermometers) at high temperatures. For instance, the loss of platinum in vacuum is 7×10^{-20} g cm^{-2} s^{-1} at 727°C, 5×10^{-11} at 1227°C and 10^{-6} at 1727°C (Vines, 1941). According to Crookes (1912) platinum is twice as volatile as rhodium in air at 1300°C, one-third as volatile as palladium, and one-thirtieth as volatile as iridium, and these differences in volatility are partially explained by the relative ease of formation of volatile oxides.

Recrystallization is another cause of limited lifetime. Alternating cooling and heating, by its expansion effects, may cause failure of a recrystallized thermocouple junction. In order to achieve precise and reliable thermocouple output the thermocouples should be regularly replaced after a period of hours or months depending on the couple material and the conditions of use.

For exact temperature recording and control a reference junction held at a precisely controlled reference temperature is required, or the thermocouple may be connected to copper held at exactly 0°C (see Figs. 7.31a and b). A controlled 0°C junction based on Peltier cooling is convenient and the triple point cell of water with a temperature of $+0.0099° \pm 0.0001°C$ may be used if the duration of experiments is not too long (both reference systems are commercially available). Reference baths at other temperatures than 0°C have the disadvantage that the e.m.f. is not zero (or nearly zero).

Extension wires prepared from alloys with similar thermoelectric characteristics are frequently used because of their lower price and higher mechanical stability. In this way a thermocouple is extended to the reference temperature or the measuring or control unit. However, the accuracy decreases with the use of extension wires, and care must be taken that all connections of the pair of thermocouple wires or any extension wires are isothermal.

The small thermocouple signals necessitate careful shielding and grounding, otherwise stray voltages may be introduced by direct coupling or by capacitive coupling. If the thermocouple line is close to motors, transformers or power-circuits it can only be fully shielded by magnetic shielding. Non-magnetic shielding can be penetrated by magnetic fields of not too high frequency. Twisting of the thermocouples minimizes such magnetic interactions. Also a coaxial arrangement of the thermocouple line has many advantages. Grounding should be effected according to Fig. 7.31(c) or (d).

In certain cases the use of a thermopile might be necessary in which a number n of thermocouples are connected in series and produce n times the e.m.f. of a single thermocouple and thus a larger signal. If an average temperature signal is required then several thermocouples of equal resistance are connected in parallel. A special arrangement for rotating thermocouples using electrically isolated ballbearings as coupling elements in a thermocouple circuit has been described by Robertson and Scholl (1971). Such an arrangement may be advantageous as an alternative to a commutator for growth on a rotating seed or by the accelerated crucible rotation technique (see Section 7.2.7).

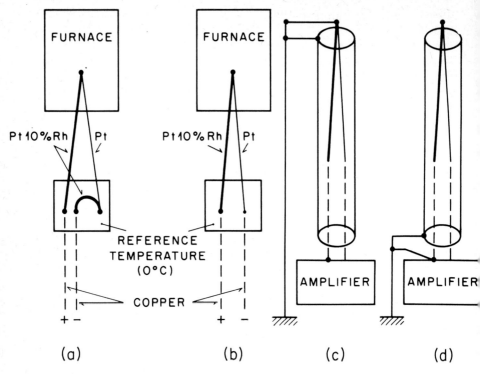

FIG. 7.31. (a)–(d) Thermocouple arrangements and methods of shielding.

Temperature measurement by *optical pyrometry* is used when thermo-electric techniques are not reliable, for example at high temperatures and in corrosive gas atmospheres, or in induction heating. However, vapours and material condensed on the observation windows might have a dramatic effect on the reliability of optical techniques. Optical thermometry has been described in "Temperature" (1941, 1955, 1962), by Harrison (1960), Dike *et al.* (1966), Euler and Ludwig (1960), Kostkowski and Lee (1962) and by Poland *et al.* (1961).

Many optical pyrometers (e.g. the disappearing filament) indicate the brightness temperature, so the spectral emissivity of the radiating source (the region of which the temperature has to be measured) as well as reflectance and absorption of the windows and lenses between object and pyrometer may lead to deviations from the true temperature. Thus the indicated temperature can be more than 10% lower than the actual temperature. Ideally the whole arrangement should be calibrated using fixed points. Emissivity is the emittance of a material having an optically smooth surface and a thickness sufficient to be opaque, and emittance is the ratio of the energy radiated by a surface (per unit area and unit time) to the energy which would be radiated by a black body at the same temperature. Applying Wien's law, the true temperature T is related to the brightness temperature T_B by

$$\frac{1}{T_B} - \frac{1}{T} = -(\lambda/C_2) \log_e \epsilon_\lambda,$$

where C_2 is the second radiation constant (1.438 cm degree) and ϵ_λ the emittance at wavelength λ. Typical values of spectral emittance are given in Table 7.11 and a few temperature corrections for various emittances and temperatures are listed in Table 7.12. Quartz-glass windows and lenses necessitate corrections of the order of 5 to 30°C. In applying temperature corrections it must be kept in mind that the surface structure of the observed hot body (e.g. crucible) has a significant effect on the emittance. The problem of the emissivity may be reduced by the use of a *two-colour* pyrometer which compares the radiation at two different wavelengths as suitable filters are introduced alternately into the radiation ("Temperature" 1962, p. 419). The output voltage from the detection system may be calibrated in terms of the source temperature.

Radiation pyrometers or photoelectric pyrometers can, in addition to indicating, also record and control the temperature and are used for example in silicon production by Czochralski-pulling.

TABLE 7.11. Spectral Emittances ϵ for Various Materials

Material	ϵ	Material	ϵ
Graphite	0.76	Al_2O_3	0.30
Graphite, powder	0.95	BeO	0.31–0.35
Gold, solid	0.04–0.16	Fe_2O_3	0.63–0.98
Gold, liquid	0.07–0.22	Fe_2O_3, liquid	0.53
Iron, solid	0.35–0.37	steel, oxidized	0.75–0.90
Iron, liquid	0.37	MgO	0.70
Molybdenum, solid	0.37–0.43	ThO_2, solid	0.50–0.57
Platinum, solid	0.30–0.38	ThO_2, liquid	0.69
Platinum, liquid	0.38	TiO_2, solid and liquid	0.51, 0.52
Tantalum	0.50	Y_2O_3	0.60
Tungsten	0.39–0.46	ZrO_2	0.40

M 2

TABLE 7.12. Temperature Corrections for Various Emittances ϵ and Temperatures, Calculated for the Effective Wavelength $\lambda = 0.655\ \mu$

ϵ	800°C	1000°	1200°	1400°	1600°	1800°	2000°	2200°C
0.1	136	196	269	355	457	575	710	864
0.2	91	131	178	233	297	371	453	546
0.3	67	95	129	169	214	265	323	387
0.4	50	71	96	125	158	196	238	284
0.5	38	53	72	93	117	145	175	209
0.6	27	39	52	68	85	105	127	151
0.7	19	27	36	47	59	72	87	103
0.8	12	17	22	29	36	45	54	64
0.9	6	8	10	13	17	21	25	30

In order to overcome the problem of absorption by vapours and windows a *light pipe* has been proposed. A sapphire single crystal is fixed near to or at the object (crucible), and the radiation transmitted by this light pipe out of the hot region is sensed by appropriate thermocouples or radiation sensitive devices.

An infrared television system has been successfully applied to Czochralski growth of refractory oxides and of copper crystals by O'Kane *et al.* (1972) and by Gärtner *et al.* (1972). Process control using a computer allowed exact control of the crystal diameter, and similar equipment could be helpful or necessary for the growth of "difficult" crystals by the top-seeded solution growth technique.

Temperature control. The signals obtained from the various temperature sensors are generally amplified and used to control the power input to the furnace. In the following we shall restrict the discussion to thermocouple sensors and resistance heating, the arrangement most typically used in crystal growth from high-temperature solutions. The regulation of RF generators has been discussed in Section 7.2.1. Mechanical temperature regulators and on-off control do not fulfil the requirements of crystal growers and have been progressively replaced by electronic controllers and by proportional or continuous (or stepless) temperature control to which the following discussion will therefore be restricted.

In a regulation system as shown diagrammatically in Fig. 7.32 pure proportional control of the furnace temperature is defined by

$$P - P_0 = K_P(W - X),$$

where P_0 is a chosen mean value of the input power. If $P - P_0$ and $W - X$ are expressed in the same units then K_P is a dimensionless constant which

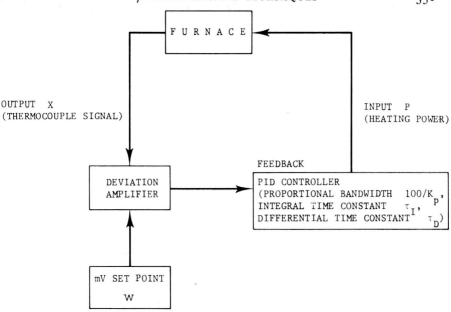

Fig. 7.32. Temperature regulation circuit (diagrammatic).

is a measure of the amplification of the feedback signal $W - X$. The
reciprocal value $100/K_P$ expressed in percent is called the proportional
band width. A purely proportional-controlled system is stable only if the
output indicates any changes in the input (heating power) immediately and
proportionally. In normal systems this is never the case, and the time
constant is a measure of the delay time caused by the heat capacity of the
furnace and of the sensing thermocouples. With pure proportional control
the steady-state temperature deviates from the set point, and the degree of
deviation changes irregularly within the limits of the proportional band.
The difference $W - X$ decreases with large K_P, but either the stability of
the system or the instrument itself limits the maximum K_P.

In order to achieve a stable temperature without a setting error in real
systems with a response time lag the so-called three-term controllers or
PID controllers should be used. These contain in addition to the adjustable
proportional band (*P*-term) the integral (*I*) and the differential (*D*) terms.
The idealized *I*-term depends on the integral of the feedback signal so that
the power delivered to the furnace includes a term

$$P_I(t) = \frac{1}{\tau_I} \int_0^t (W - X) \, dt,$$

where $(W - X)$ is again expressed in the same units as P and $1/\tau_I$ is the adjustable proportionality factor of the I-term, τ_I being called the integral time. Thus the integral term produces an increase in the heating power which is linear with time. Integral control (also named reset control) has the effect of shifting the proportional band in such a way that the deviation of the actual temperature from the set point is minimized (which is achieved neither by P- nor by $P + D$-control).

Idealized differential control or derivative (or rate) control depends on the differential of the feedback signal so that, at time t, the output of the controller has a component

$$P_D(t) = \tau_D \frac{d}{dt}(W - X).$$

The addition of differential control reduces the tendency towards oscillation when the proportional band is narrowed (i.e. K_P made large). If the chosen derivative time τ_D is too large, a small change of the actual temperature will drive the controller to its limits of zero and maximum power, so that the system tends towards on-off control with a long time constant.

The general equation for a three-term controller is

$$P(t) - P_0 = K_P(W - X) + \frac{1}{\tau_I}\int_0^t (W - X)\,dt + \tau_D \frac{d(W - X)}{dt}.$$

Although it would be possible to calculate the optimum proportional bandwidth, integral time and derivative time such a calculation would be very complicated considering the many factors which have to be taken into account (furnace design, characteristics of heating elements, power supply, thermocouples, etc.) and which are generally not known.

In practice the P, I and D terms are set by "experience" or by measuring the time constant of the system. A practical procedure to optimize the parameters of PID controllers has been described by Ziegler and Nichols (1942). Firstly, the I and D terms are adjusted in order to obtain pure proportional control by setting the integral time as large as possible and the differential time as small as possible. By increasing the amplification K_P to a value K_P(crit.) where undamped oscillations are just achieved the period τ(crit.) of these oscillations can be determined. Then the parameters of a PID controller are set as follows:

$$K_P = 0.6\ K_P(\text{crit.})$$
$$\tau_I = 0.5\ \tau(\text{crit.})$$
$$\tau_D = 0.12\ \tau(\text{crit.}).$$

If the dynamic behaviour (amplification factor F_A, time constant τ_c, dead time τ_d) of the control system is known then the parameters are set according to Oppelt (1964):

$$K_P = 1.2 \frac{\tau_c}{\tau_d} \frac{1}{F_A}$$
$$\tau_I = 2\,\tau_d$$
$$\tau_D = 0.42\,\tau_d$$

(see also Smith and Murrill, 1966).

The effect of the various control techniques for a typical furnace system is demonstrated in Fig. 7.33. As shown with examples 1 and 2 the sensing thermocouple should be near to the heating element and the same applies in *PID* control, since the dead time is thus decreased. In addition to an optimum setting of the *PID* terms, a stabilized power supply is necessary for highest precision of the temperature control, as proved by examples 8 and 9 where the temperature is shown to be unaffected by a change in the line voltage.

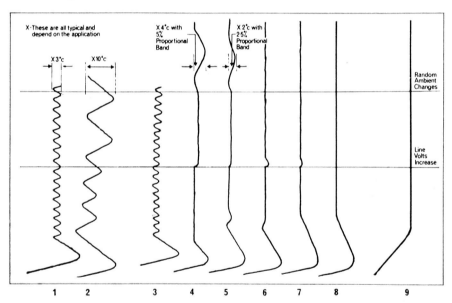

Fig. 7.33. Typical control records, recorded from the control thermocouple, using various control methods. Thermocouple $\frac{1}{2}$ inch from heater in all cases except 2 (Courtesy Eurotherm Ltd.). (1) On-off control. (2) On-off control. Thermocouple $1\frac{1}{2}$ in from heater. (3) Proportional control. Proportional band too narrow. (4) Proportional control. Proportional band wide. (5) Proportional control. Proportional band optimum. (6) Proportional + integral control. (7) Proportional + integral + derivative control. (8) Proportional + integral + derivative with power stabilization. (9) Proportional + integral + derivative with power stabilization with derivative start up.

By using a high-quality commercial *PID* controller and an arrangement similar to that shown in Fig. 7.34(a), by careful installation of the sensing thermocouple and of the signal lines, very precise control of furnace temperatures may be achieved, of the order of 0.1 to 0.5°C at temperatures around 1200°C using a normal Pt/Pt–10Rh thermocouple. The arrangement of Fig. 7.34(a) is a typical example.

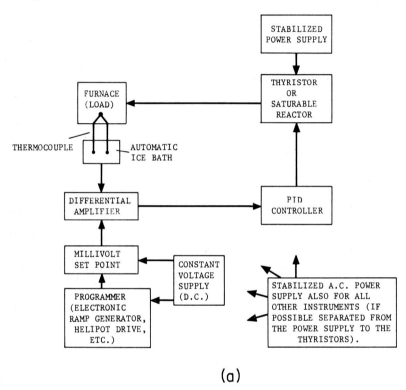

(a)

(b)   (c)

Fig. 7.34. (a) Example of furnace-sensor-controller arrangement for precise temperature regulation. (b) Continuous control by phase angle firing. (c) Continuous control by fast cycling.

The signal from the *PID* controller may regulate a saturable reactor or a thyristor which is itself connected to the heating elements directly (resistive load) or via a variac or transformer (inductive load). Saturable reactors (or magnetic amplifiers) are very reliable and not sensitive to mechanical, electrical or thermal disturbances whereas the thyristor (also called silicon-controlled rectifier S.C.R.) is faster, cheaper and more sensitive. In thyristors the power control is achieved either by *phase angle firing* or by *fast cycling*, the principles of which are schematically shown in Fig. 7.34(b) and (c). Phase angle firing obviously distorts the mains waveform and with large installations this might be serious. It can also cause radio interference, and for these reasons phase angle firing should not be used in research laboratories except where precautions are taken. Rapid cycling if synchronized with the mains does not distort the mains waveform, and there is no radiated interference. However, in both cases, specific difficulties may arise and should be discussed with specialists.

Temperature programming. According to Fig. 7.34(a) programming is achieved by changing the millivolt set point. The simplest way to do this is to change the resistance in the mV circuit by means of a motor-driven potentiometer (Helipot) or similar device. The precision is frequently not sufficient, and potentiometers are sensitive to dust and vapours thus necessitating frequent cleaning. Motors with changeable gears† attached to helipots provide a wide range of cooling rates which are, depending on the type of thermocouple, more or less linear. As was pointed out by Scheel and Elwell (1972), by Pohl and Scheel (1975), and in Section 6.6, the cooling rate should frequently be below 1°C per hour and the temperature regulation better than 0.1°C for flux growth of oxides by the slow-cooling technique, and these goals necessitate the application of fully electronic programmers‡ or of computer-assisted process control. Also the commercially available instruments with rotating cams or drums, in which the programmed value of the temperature is sensed mechanically or optically, are generally not sufficiently precise. Laudise (1963) pointed out that the precision of the temperature control should be commensurate with the slow cooling rates used, and Scheel and Elwell (1972) expressed the necessity of a continuous (stepless) cooling curve if crystals of high quality are to be grown. Further details of temperature programming in the slow-cooling technique are discussed in Sections 6.6 and 7.1.2.

---

† For instance of Halstrup, 7815 Kirchzarten, W-Germany.

‡ For example, Eurotherm Ltd. (Worthing, Sussex, England) offers electronic programmers in which capacitors are slowly discharged by field-effect transistors and thus generate an electronic ramp.

7.2.3. Crucibles

Most experiments in crystal growth from high-temperature solutions are carried out in containers, either in crucibles or in ampoules. Thus crystal growers are faced with the choice of a practical crucible material which should be resistant to corrosion by the solvent, solute and the atmosphere at the high growth temperatures. Further requirements are a high mechanical strength, ease of shaping and of cleaning, good thermal shock resistance, reasonably long lifetime (little recrystallization) and a low price.

There are two materials which fulfil most of the above requirements, namely graphite and silica glass (also called quartz glass). These two materials have many unique properties and find wide applications, graphite for several metals which have a low tendency towards carbide formation and for such compounds which are stable against reduction, especially for the preparation of semiconductor crystals and of semiconductor layers by liquid phase epitaxy. Silica-glass crucibles, boats and ampoules may also be used for the preparation of semiconductors and for those metallic melts which do not cause reduction. Obviously SiO_2 glass cannot be used to contain solvents for oxides and for oxide melts, and its upper limit of temperature is around 1100°C (1000° for extended use). For growth of most oxides and oxide compounds, platinum is the only choice because of its resistance to corrosion, and it has to be used despite its high price and softness. Thus the various aspects of the use and handling of platinum crucibles will be discussed in some detail.

As a rule suitable crucible materials are those which have a type of bonding different from that of solvent and solute provided that no decomposition reaction occurs. The difference in the chemical bond type generally leads to a sufficiently small solubility of the crucible material in the solvent, especially when this material has a relatively high melting point. Examples of the bonding difference rule are the use of platinum crucibles for ionic melts, of (ionic) oxide and of covalent graphite and nitride crucibles for metallic melts, and further examples are given in Table 7.13 which lists the most common crucible materials with their properties and their applications. Table 7.14 summarizes the corrosion resistance of various crucible materials against a variety of molten metals for 300° and 600°C, respectively, and several properties of a variety of crucible materials have already been listed in Table 7.9 (see Section 7.2.1). In Table 7.15 examples of crucibles for a variety of melts and solutions, together with the temperature range, are given [see also Janz (1967)].

A well-known rule in solid-state reactions is that the reaction between two compounds of similar bond type starts at temperatures of 0.3–0.4 T_F for metals, of 0.6 T_F for ionic compounds and 0.8–0.9 T_F for silicates.

T_F is the mean absolute melting point of the two components. According to these values, compounds may be melted in crucibles of the same bond type if the above values are not exceeded, that is, if the melting point of the solvent or solution is much lower than that of the crucible. For example, sodium chloride may be melted in alumina crucibles, and several low melting point metals may be contained in refractory metal crucibles. However, many refractory metal crucibles (except noble metals) cannot be used for ionic compounds since they are oxidized by the latter.

Table 7.16 gives the temperatures at which reactions between refractory materials start, so that mechanical contacts between two different materials (for instance crucible and support) should be held at significantly lower temperatures than those given in the table. Further information on the usefulness of crucible materials and of their reactions may be obtained from the appropriate phase diagrams (see Sections 3.7–3.9).

Much information on refractory materials, on high-temperature oxides, refractory metals, carbides, nitrides and on the corresponding phase diagrams has been collected in a book series edited by Margrave (1965–), in High Temperature Technology (1960), by Campbell (1959) and by Kingery (1959).

Composite materials, glass-ceramics and cermets find several applications in high-temperature technology due to the often unique properties originating from the combination of two or more components. However, in flux growth corrosion is generally the greatest problem, and so crucible materials are chosen mainly to minimize this problem. However, coatings of more corrosion-resistant materials on containers made of materials with high mechanical strength (and frequently lower price) may be useful for several applications in crystal growth. For instance, oxides (Al_2O_3, ZrO_2) are flame- or plasma-sprayed on to refractory metals and vice versa. Frequently one may observe that resistance to corrosion increases with higher purity of the crucible material.

Corrosion is related to the wetting angle of the melt. The larger the difference in the type of bonding between crucible and melt, the smaller the wetting and the corrosion normally observed. Although it has been proposed to determine covalency or more generally the bonding character from the measured wetting angle and thus to select optimum crucible materials, such determinations of wetting angles are not conclusive. The contact angles frequently depend strongly on the atmosphere, on the surface finish of the solid phase, on adsorbed gas and impurity layers and sometimes even on surface reaction products. However, some useful conclusions may be obtained if a series of experiments are performed as by Champion et al. (1973) on the wetting of refractory materials by molten intermetallic compounds.

TABLE 7.13. Crucible Materials, Properties and Applications†

Crucible material	Melting point (°C)	Used for	Max. temp. of use (°C)	Atmosphere	Remarks
Nickel	1453	(R.E.)F$_3$, ZrF$_4$, halides	900	oxygen-free	stable in air up to ~900°C
Niobium	2415		~1100	reducing, neutral, vacuum	Na + K 800°
			~2000	reducing, neutral, vacuum	Na + K 600°
Molybdenum	2610		1900	reducing, neutral, vacuum	Ti$_2$O$_3$ 1820–1920° (not: TiO reacts)
		Al$_2$O$_3$	2050	reducing, neutral, vacuum	Hg 600°, Na + K 700°
Tantalum	3000		2400	reducing, neutral, vacuum	Na + K 600°
		La	1700	reducing, neutral, vacuum	
Rhenium	3180		2380	reducing, neutral, vacuum	
Tungsten	3410		2500	reducing, neutral, vacuum	Eu(+EuO) 2150°, Hg, Na 600°
		Al$_2$O$_3$, La$_2$O$_2$S	2050	reducing, neutral, vacuum	
Silver	961	KOH	550		
		halides	>750		
Gold	1063	vanadates	~1000		Au–Pd alloys are stronger and resistent to higher temperature
		Na$_2$CO$_3$, etc.	900		

† Thermal shock resistance and other properties; see Table 7.9.

Material	Melting point (°C)	Compatible with	Max. temp. (°C)	Atmosphere	Remarks
Platinum	1772	silicates, earth alkali borates, LaF₃	~1500	oxidizing	
		PbO, PbF₂, Bi₂O₃, PbO—B₂O₃, V₂O₅, molybdates, tungstates	1300	oxidizing	
		NaCl	1000		
Rhodium	1960				
Iridium	2442				
Graphite	~3700 (subl.)	stable oxide compounds	2200	neutral	stable in air up to 500°C, Hg 350°, silicon 2000 (2200° for SiC)
		metals, alloys	~1000–2000	neutral, reducing	
Glassy carbon		fluorides			
		alkali halides, sulphides	1000		
Boron nitride	3000 (dec.)	similar to graphite	approx. 2000	neutral, reducing	stable in air up to 700°C
Aluminium nitride	~2200 (dec.)	Ga (+ GaP)	>1050		stable in air up to 700°C
		cryolithe, metals, alloys	1000–1600	neutral, reducing	
Silicon nitride	1900 (dec.)	metals, alloys	1500	neutral, reducing	
Silicon carbide	2830 (dec.)	metals, alloys	1500	neutral, reducing (see Section 7.2.1)	stable in air up to 1600°
Tantalum carbide	3875	metals, alloys	approx. 3000		
Oxide glasses (Pyrex, Duran, Vycor, Suppremax, etc.)	800–1200	halides (except fluorides)	500		Na + K 300°, Hg 350°
		metals, alloys, sulphides, etc.			
Silica glass	1710	halides (except fluorides)	900	oxygen-free	Te + (Zn, Hg) Te 650°
		metals, alloys	1100	oxygen-free	
		sulphides, selenides, tellurides		oxygen-free	
		silicon	1500	oxygen-free	
Mullite, Sillimanite	1850	metals, alloys	~1400	oxygen-free	Hg 350°
		alkali polysulphides	~1000	oxygen-free	

TABLE 7.13 cont.†

Crucible material	Melting point (°C)	Used for	Max. temp. of use (°C)	Atmosphere	Remarks
Alumina	2020	metals, alloys	1800	oxygen-free	Na + K 500°
		sulphides, selenides			Na$_2$CO$_3$ 900°
		halides (except fluorides)	1000	halogen or neutral	
		Ga (+ GaP)	1150		
		alkali polysulphides	1000	oxygen-free	
				oxygen-free	
				oxygen-free	
Spinel	2135	metals, alloys			
Zircon	2420	metals, alloys			
	2715				
Zirconia	2550	metals, alloys		oxygen-free	
Beryllia	2570	metals, alloys		oxygen-free	vapours and dust very poisonous!
					Na + K 600°
Calcia	2600	metals, alloys			Na + K 200°
Magnesia	2800	metals, alloys			radioactive
Thoria	3050	metals, alloys			
		Bi$_2$O$_3$	1100		
Fluorite	1360	fluorides, metals	~1150		attacked by HCl, chlorides

† Thermal shock resistance and other properties; see Table 7.9.

TABLE 7.14. Resistance of Crucible Materials to Attack by Liquid Metals at 300° and 600°C (Kelman et al., 1950). The Corrosion Resistance is Good (++), Fair (+) or Bad (−).

Material	Temp	Li	Na,K	Hg	Pb	Bi	Cd	Tl	In	Ga	Al	Sn	Zn
Pyrex Glass	600°	−	−	−	−	−	−	−	−	−	−	−	−
	300°	−	++	++	++	++	++	++	++	++	−	++	
Silica Glass	600°	−	+	+	++	++	++	++	++	++	−	++	
	300°	−	++	++	++	++	++	++	++	++	−	++	
Porcelain	600°	−	+	+	++							++	
	300°	−	−	++	++							++	
MgO (porous)	600°	+	+		++					−			
	300°	+	+		++								
Al₂O₃, BeO (dense)	600°	+	+	++	++					++	++		
	300°		+	++	++					+			
Graphite	600°	−	−	++	++	++				++	++		++
	300°	−	−		++	++				++	++		++
Cr–Ni Austen. Steel	600°	++	++	−	−	−	−	++	++	−	−	−	
	300°	+	++		++	−		++	++	+	+	+	
Nickel	600°	−	++	−	−	−	−	−	−	−	−	−	−
	300°	+	++	−	−	−	−	−	−	−	−	−	−
Ag, Au, Pt	600°	−	−	−	−	−	−	−	−	−		−	−
	300°	−	−	−(Pt+)	−	−	−	−	−	−		−	
Molybdenum	600°	++	++	++	++	++				−	−		
	300°		++	++	++	++				+	+		
Tantalum	600°		++	++	++					+	++		
	300°		++	++	++						++		
Tungsten	600°		++	++	++					++	++		
	300°		++	++	++					++	++		

TABLE 7.15. Crucible Materials Used for Various Melts and Solutions, and the Maximum Temperatures Applied

Melt or solution	Crucible	Temp. (°C)	Reference
Pb	Nb, Ta	950	
	BeO, SiO_2	1000	
	C	550	
	Porcelain	850	
	Pyrex	350	
Bi	Mo, C	1000	
	Nb	450	
	SiO_2	900	
Bi—Pb—Sn Eutectic	W	500	
	Pyrex	300	
Ga	W, Re	800	
	Ta, ZrO_2, Pyrex	500	
	Nb, Mo	400	
	C	>800	
	BeO	>1000	
	Al_2O_3	1150	
	SiO_2	1160	if oxygen-free atm.
Al	Al_2O_3, BeO	1000	
	C	1200 (1700)	
	ThO_2, ZrO_2	>1000	
Zn	C, Al_2O_3, SiC	900	
Sb	C	1300	
Ti	Y_2O_3	~1700	Helferich and Zanis (1973)
La + LaB_4	Ta	1200–1700	Deacon and Hiscocks (1971)
Eu + EuO	W	2150	Shafer et al. (1972)
US_2 + US	W	1700–1900	Van Lierde and Bressers (1966)
Ti_2O_3	Mo	1820–1920	Reed et al. (1967)
Alkali halides	Supremax glass	750	Janz (1967)
Bi_2O_3	ThO_2	1100 (4h)	Tiche and Spear (1972)
KOH + BeO	Ag	550	Levin et al. (1952)
KF + V_2O_5 + V_2O_3	Pt	950	McWhan and Remeika (1970)
V_2O_5	Pt	max. 700	Kennedy et al. (1967)
PbO + PbF_2 + Oxides	Pt	1300	Various authors
PbO + B_2O_3 + Oxides	Pt	1300	Various authors
PbO + V_2O_5 + Oxides	Pt	1350	Various authors
BaO × $0.6B_2O_3$ + Oxides	Pt	1400	Various authors
KF + $BaTiO_3$	Pt	1200	Various authors
Na_2CO_3 + $BaTiO_3$	Ni	1200	Reducing atm., Kawabe and Sawada (1957)

TABLE 7.16 Temperatures of first noticeable reaction (upper right part) and the maximum working temperatures (lower left part) in °C.

Material	m.p. / proposed max. working temp. (°C)	SiO₂	3Al₂O₃·2SiO₂	Al₂O₃	BeO	MgO	ZrO₂	ThO₂	MgAl₂O₄	ZrSiO₄	TiO₂	CaF₂	BN	SiC	AlN	Si₃N₄	C	Mo	W	Ni	Al	V	Nb	Zr	Ti	Pt	Atmosph. stock
SiO₂	1700 / 1250~			~1100	1500	1450											1400 N	1500 N	1600 N			800 1½				750 NR	ON(R) ++
3Al₂O₃·2SiO₂	1830 / 1800	1100																								750 NR	ON(R) m
Al₂O₃	2015 / 1900	1200	1200		~1800	~1800	1600	1700	>1800		~1700							2000	2000	>1800 L	>1800 L		>1800	1600	1600	1100(V) 1559(O)	ONR m
BeO	2550 / 2400 V / 2350	1600	1200	1200		~1800	1900	2100	~1800		1600						2300	1900	2100	>1800 L			1600	1600	1600	1659	ONR ++
MgO	2800 / 1600(V) / 2400(O)	1000	1100	1200	1600		<1560	2200	>1800		1500						1800	1800	2000	>1800 L			1800	1800	1400	1650(ON)	O
ZrO₂	2677 / 2100 V / 2400	1000	1200	1600	1600	1800		2200	>1800								1800	2150	2100	>1800 L			1400	1800	1600	1100	NR m
ThO₂	3300 / 2300 V / 2800	1000	1200	1500	1800	1900	2200		1800								1950	2200	2200	>1800 L			>1800	>1800	>1800	1100	RN m
MgAl₂O₄	2160 / 1950	1200	1300	1400	1600	1800																				~1600	
ZrSiO₄	2420 / 1870	1100	~1000	~1000	1200													>1800		>1800 L			1600	1600	1600	~900 (NR)	RN +
TiO₂	1150 / 700°O / 1700 N	1100	1200	~1400	~1400																					1600(O)	O
CaF₂	800 / ~800	800	~800	~1000	~1400																					1500(O)	
BN	1400	1400	1400	~1400	~1400																						N
SiC	1100																										O
AlN	2000	1400	1400	(1800) 1600	(2300) 2000	(1800) 1700	1880	1950																			NRO
Si₃N₄	1500 (1400)	1000 (1400)		(2000) 1800	(1900) 1700	(1800) 1700	2150	2200										1300									NRO
C	2100 V / 3000 N	1500 N 1400 N	1400 N	(2100) 1800	(2100) 1800	(2000) 1800	2100	2300	1600									1300	1500	>1300						<1000	RNV
Mo	2100 V	1500 N 1400 N	1400 N	(2000) 1800	(1900) 1700	(1800) 1700											1600		2300	1300						<1000	RNV
W	2400 V	1600 N 1500 N	1500 N	(2100) 1800	(2100) 1800	(2000) 1800											1500	2300		1500						<1000	RNV
Ta				(2000) 1800													2300										RNV
Ni	1200 (NR) / 1000 (O)			1800	1700	1700	~1700	1700																		800	NV
Pt	700 (R) / 1700 (O)	700	700	1500	500	1600	1100	1100		800	1500(O)																O

PROPOSED MAXIMUM WORKING TEMPERATURES

| Atmosphere | | ONV | ON | ONV | ON | ON | ON | NO | NO | NO | O | N | NRO | NRO | NRO | NRO | RNV | RNV | RNV | RNV | NV | RNV | RNV | RNV | RNV | O | |

TEMPERATURES OF FIRST NOTICEABLE REACTION

~ Estimated values; N Neutral atmosphere; R Reducing; L Liquid; O Oxidizing atmosphere; V Vacuum.

Thermodynamic data may give some general indication of the stabilities of crucible materials and their reactions with high-temperature melts, but results to better than the order of magnitude must not be expected because of the limited reliability of much thermodynamic data, of the influence of the grain size, impurities and dislocations of the crucible material, as well as of the detailed chemistry of the solution and the gas atmosphere. The data for stabilities of the crucible materials given in Tables 7.13 to 7.16 must be regarded as approximate. As an example of the estimation of interface reactions using thermodynamic data, the paper of Armstrong *et al.* (1962) may be mentioned. These authors claimed that at fairly high temperatures strong reactions with silica occur for those metals of which the oxides form low-melting eutectics with silica.

In the following paragraphs two classes of crucible materials will be discussed in detail because they are the most frequently used in flux growth, namely (*A*) ceramic crucibles, and (*B*) noble metal crucibles. The general aspects of the size of crucibles are discussed in Section B.

A. Ceramic crucibles.

Silica: Silica glass is highly corrosion-resistant and is easy to form to any desired shape. It is transparent and has a low thermal expansion coefficient, so that it is resistant to thermal shock. The resistance of silica against corrosion can be increased by deposition of a thin carbon layer obtained by decomposition of acetone or ethanol at 500 to 700°C in an atmosphere low in oxygen or by deposition of an alumina layer about 500 Å thick as reported by Widmer (1971). Silica glass crystallizes slowly around 1000°C (especially where traces of alkali, like NaCl from sweat, are present) and more quickly above 1100°, and the crystallization products, tridymite and cristobalite, undergo destructive phase transitions between 100 and 400°C so that recrystallized silica frequently breaks or explodes in this temperature range. At temperatures above 1100° softening takes place, but silica glass can be used for Czochralski growth of silicon at 1540° when its shape is maintained by an external graphite container. Chemically silica is stable against carbon, many metals, sulphides, selenides and halides up to 1000°C, but it is attacked by fluorine-containing vapours and liquids, basic melts (with formation of silicates), many oxides, hydroxides, phosphates, alkali and alkaline earth metals, aluminium and boiling phosphoric acid. Lead, tin and cadmium corrode silica only if the ampoule contains traces of oxides. For experiments in oxygen- and water-free atmospheres the hydroxide layers adsorbed at the surface must be removed by etching with HF—HNO$_3$ mixtures, immediately followed by short rinsing with metal-free distilled water and subsequent heating at 1000°C for two hours in vacuum. For use at temperatures below 750°C silica glass may be replaced by Vycor† or Suppremax†, below 500°C by Pyrex† or Duran† glass.

Oxide ceramics: Of the pure oxide ceramics, alumina finds the widest application in crystal growth from high-temperature solutions, but the more expensive zircon $ZrSiO_4$, zirconia ZrO_2 and thoria ThO_2 are generally more restant against corrosion. For example, thoria is not measurably attacked by bismuth oxide at 1100° and thus can be used for growth of $Bi_4Ti_3O_{12}$ and similar compounds (Tiche and Spear, 1972). For many applications (melts of metals, sulphides and halides except fluorides) the relatively inexpensive porcelain (mullite, sillimanite), or ceramics based on magnesium silicates, pyrophyllite and Lavite† which can be shaped before the first firing are adequate. For example, glazed porcelain crucibles have been repeatedly used for potassium pyrosulphate melts between 700 and 900°C in order to clean platinum crucibles. According to Jaeger and Krasemann (1952), who investigated the stability of alumina against various agents, borax melts polish the surface of alumina ceramics. However, silicates are not useful in reducing atmospheres at high temperatures. Most ceramic crucibles are sensitive to acids and are either dissolved (BeO, MgO, CaO), or show cracking when reheated after cleaning with acid. In many aspects new types of ceramics which have nearly 100% density and are translucent or even transparent (Al_2O_3, Yttralox† = 90% Y_2O_3, 10% ThO_2) have many advantages (corrosion resistance, vacuum-tight, high maximum working temperature) but are disadvantageous with respect to thermal shock resistance. For special applications single crystalline alumina crucibles and tubes produced by flame fusion or edge-defined film-fed growth might be useful.

Ceramic containers and tubes may be sealed if required either by a high-temperature cement or by welding with a flame or plasma. In principle cheap ceramic containers covered with a thin platinum layer might be used as a replacement for the very expensive platinum crucibles, or corrosion-resistant layers (e.g. of Al_2O_3 or ZrO_2) may be evaporated or plasma-sprayed on to porous and thus shock-resistant crucibles, but such combinations have not yet found wide applications.

Base metals and non-oxide ceramics: Of the base metals nickel, stainless steel, molybdenum, tungsten and tantalum are most frequently used as crucibles, but at temperatures above about 500° oxygen- and water-free atmospheres are required. At such low oxygen pressures as 10^{-3} atm and at high temperatures up to 10 atom % oxygen is dissolved in niobium and tantalum whereas in molybdenum and tungsten the oxygen solubility is smaller by orders of magnitude. At higher oxygen concentrations the metals are completely oxidized. The oxidation behaviour and other properties of

† Trademarks.

refractory metals have been described by Jaffee in High Temperature Technology (1960), and the oxidation of Nb and Ta is discussed by Fromm and Jehn (1972). Molybdenum and tungsten crucibles are suitable for containing melts of pure molten alumina, but are attacked by ruby melts.

In crystal growth from high-temperature solutions Mo, W and Ta crucibles are probably most widely used for growth of chalcogenides of the lower valency states from the corresponding metal melts (e.g. EuS, EuSe, EuTe from Eu) as well as for growth of other refractory compounds (WC, LaB_4) from metallic solutions.

Sawada et al. (1951) and Kawabe and Sawada (1957) used nickel crucibles for the growth of $BaTiO_3$ from Na_2CO_3 solution in a reducing atmosphere and found no attack of the crucible. In air or even in nitrogen, nickel is attacked by carbonate melts. Hauptman et al. (1973) used iron crucibles for growth of Fe_2TiO_4 from $BaO—B_2O_3$ solutions in a nitrogen atmosphere and Garrard et al. (1974), Ni and Mo for growth of fluorides.

Of the non-metal non-oxide refractories graphite is probably the oldest and most popular, and the development of pyrolytic graphite and of glassy carbon has extended the usefulness of carbon crucibles. A characteristic of graphite is that its strength increases with temperature and so is higher than many other refractories (BeO, MgO, ZrO_2, W) above 2000°C. Because of its corrosion resistance, its excellent thermal shock resistance and its relatively low price, graphite is the first choice as crucible material for metals (except those which form carbides), fluorides, and sulphides. Its disadvantages are its reducing action and that it has to be protected from oxidation above about 500°C. The III–V analogue of carbon, boron nitride, shows a slightly better oxidation resistance and a very high electrical resistivity of about $10^7 \, \Omega \, cm$ at 1000°C. BN is resistant to molten nonferrous metals and many salts, and its oxidation in air starts only above about 900°C.

Another III–V analogue is the nitride AlN, and AlN ceramics containing small amounts of silicon carbide decompose above 2200°C and are very resistant towards molten metals. They are not wetted by molten steel and do not react with cryolite melts at 1040°C. In air AlN is oxidized above about 700°C.

Silicon carbide is the most important of the refractory carbides. It has a hardness 9 on the Mohs scale, a high thermal conductivity and low thermal expansion coefficient and is thus very resistant towards thermal shock. Another advantage of SiC is its oxidation resistance up to about 1550° (with the exception of the range below 750°C where "silicon carbide pest" occurs and where the protective SiO_2 layer cannot form so that SiC is oxidized).

Little attention has been given to refractory borides, silicides and sulphides and their application to flux growth although they are expected to exhibit a good corrosion resistance against metals and fused salts, and a reasonable stability because of their high melting points, generally between 2000 and 3000°C. For example, zirconium boride ZrB_2 has a melting point of 3040°C, is oxidation-resistant up to 1000° and is not wetted by many molten non-ferrous metals such as Zn, Al, Pb, Cu, Sn and brass.

B. *Noble metal crucibles.* In crystal growth of oxide compounds and of some fluorides from high-temperature solutions the most common crucible materials are noble metals, and from this group of metals platinum has found by far the widest application. Since platinum has a relatively high price, the initial cost as well as refabrication costs form a significant (if not the largest) portion of the costs of flux growth of oxides. The prices of platinum (and of iridium) fluctuate enormously, as may be seen from Fig. 7.35. In this connection it should be noted that several companies and syndicates ask prices which are 50% and more higher than the free market (U.S.) price. The free market price tends to be a sensitive indicator of demand and supply trends. The bulk of platinum is costed at the fixed producer price, and both prices vary widely. A price of 120 dollars per

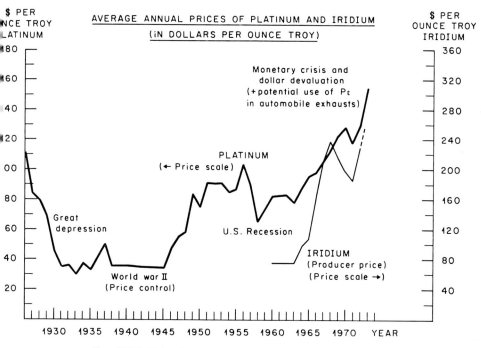

FIG. 7.35. Price fluctuations of platinum and iridium.

ounce platinum may be used as a guideline for the free market price, and 50 dollars per ounce for the producer price. In 1971/1972 there was a shortfall in the free market price of about 40 dollars per ounce until the U.S. government decided not to postpone the Clean Air Act of 1970. The steep increase in the platinum price caused by the potential demand for platinum in automobile exhausts was accelerated by the monetary crisis of 1972/1973. The platinum price may drop significantly if the crisis is settled, if an alternative to platinum in exhausts is adopted or if Russia offers abnormally large quantities of platinum to the Western market as happened at the end of the fifties.

Platinum crucibles for flux growth may be obtained from various sources, either in standard shapes or to special order. Frequently it may be faster and more convenient to order round platinum sheets and have them shaped by a local noble metal spinning shop. A variety of crucible forms used in several laboratories are shown in Fig. 7.36; (a) shows a standard crucible with lid which is commercially available in several sizes; (b) a modification of the crucible rim and lid of Kvapil (1966) and John and Kvapil (1968), which allows sealing by a liquid; (c) a special crucible form used by Scheel (1972) and Scheel and Schulz-DuBois (1971) which is mechanically stable and to which a flat round lid can be welded. After the experiment, crucible and lid are separated by cutting the welded rim, and can be re-used about five times; (d) shows a modification of the lid with a hole of 0.4 mm diameter (Elwell and Morris, unpublished) which is loosely covered by a platinum sheet and which can be sealed after heating up and expansion of the enclosed air by pressing on this sheet from above; (e) the crucible arrangement of Tolksdorf (1968) made from a large diameter platinum tube and with a lid sealed to it having a platinum capillary to release pressure; and (f) shows a diagram of the large crucibles used at Bell Laboratories and at Airtron for pilot plant flux growth. Slightly conical crucibles are advantageous for removing solidified flux and for reshaping which should be done on plastic formers with a plastic or leather hammer. Preferred wall thicknesses are from 0.5 to 1 mm for a crucible volume of 30 to 1000 cm^3, 0.3 mm being the practical minimum. A greater wall thickness increases not only the lifetime of the crucible, but also allows the destruction by beating with a slightly convex hammer of grains formed by recrystallization.

The crucible size is discussed here because it has a dramatic and often restricting effect when expensive noble metals have to be used as crucible materials. According to Block (1930), Goethe had already recognized that increasing the crucible size tends to increase the crystal size, and during the past century crucible sizes up to 50 l were used as discussed in Chapter 2. In recent years platinum crucibles of 4 l and even 8 l were used by various

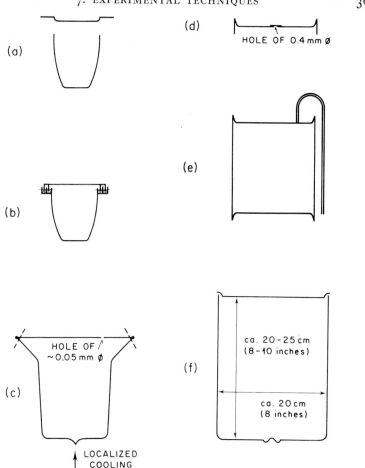

FIG. 7.36. Examples of crucible shapes used in flux growth (not to scale).

companies for the growth of garnets and ruby. However, large crucibles are not always advantageous for a number of reasons.

In Section 6.6.1 the experimental requirements for stable growth are discussed, and for large crucibles extremely low cooling rates for the first weeks or months are required if one or a very few inclusion-free crystals are to be grown. If multinucleation is tolerated or if seed crystals of sufficient area are available "normal" cooling rates (between $0.1° \, h^{-1}$ and $1°C \, h^{-1}$) and duration of experiments can be applied. The effect of the crucible size is demonstrated in Fig. 6.16, which also shows the importance of high solution flow rates at the crystal surfaces in order to achieve stable growth. The larger the crucible the more stirring is necessary in order to prevent convection pockets and thus uncontrolled nucleation, as discussed in more detail in Section 7.2.7. As a general rule stirring is worthwhile for crucibles larger than 100 cm³ and is necessary for crucibles larger than about 500 cm³. Another requirement with large crucibles is that temperature control and programming have to be very precise, and it is estimated that temperature oscillations larger than 0.1°C with periods larger than about one second might have harmful effects on the quality of the grown crystals.

Another obvious disadvantage of large crucibles is their price and the requirements on the size of the furnace and controller, and the total cost of systems based on crucibles more than 1 litre in size is often prohibitive.

An advantage of large crucibles is the low ratio of wetted crucible surface to the volume of solution which determines crucible corrosion and incorporation of crucible material into the growing crystals. The incorporation of, for example, platinum can be significant depending on experimental parameters as was demonstrated by Bradt and Ansell (1967). These authors found 0.1 wt% Pt in $BaTiO_3$ crystals grown from KF when a soaking time of four hours was applied, whereas without soaking only 0.065 wt% Pt was found.

Another advantage is obviously the quantity of crystals which can be grown simultaneously, and large crucibles are required when large-scale production of crystals is applied. Grodkiewicz et al. (1967) obtained a yield of 2.8 kg YIG from 8 l crucibles although about 2/3 of the volume of the YIG crystals contained inclusions. This fraction of included crystals might be reduced if stirring, seed crystals and precise temperature control are applied, as discussed above.

On comparing the advantages and disadvantages of large crucibles the conclusion is reached that an optimum crucible size exists as a compromise. As a rule, crucible sizes of 25–100 cm³ are suggested for crystals of a few mm size, but for crystals larger than 1 cm³ the crucible size should increase by about 100 cm³ per cm³ crystal.

Occasionally an influence of crucible size on the crystallizing phase has been reported and explained by Flicstein and Schieber (1973). If a composition lies near the boundary of stability fields of different phases, then the difference in the evaporation rate in crucibles of different sizes might shift the PbO—PbF_2 ratio (in this case) and thus lead to different phases.

In order to produce the smallest possible surface and thus to minimize corrosion and nucleation sites the crucibles should be mechanically and chemically polished. The latter is performed for example with a potassium pyrosulphate melt at 700 to 800°C (however, Rh-containing crucibles are attacked by pyrosulphate) for one to four hours, and any reshaping and hammering should be followed by chemical polishing.

A short preheating of Pt crucibles to about 1200°C before use prevents excessive welding of new platinum pieces when in contact. Special care in the use and handling of platinum crucibles is necessary in order to prevent crucible failures during the growth experiments (and too frequent reshaping). Platinum crucibles should only be used for ionic melts (fluorides, oxides, silicates) but never for elements and for compounds with strong covalent or metallic bonding. Thus metals, phosphorus, arsenic, antimony, bismuth, silicon, boron and sulphur should never be heated in Pt crucibles. Especially carbon, carbides or organic compounds should not contact hot Pt containers and reducing atmospheres should be prevented under all circumstances. For example, traces of organic compounds or of hydrogen may reduce the heavy-metal compounds like Bi_2O_3, PbO and PbF_2 and may lead to destruction of the crucible. In order to maintain oxidizing conditions an oxidizing agent like PbO_2 (Garton and Wanklyn, 1967) is frequently added, or a buffer like V_2O_5 which is more easily reduced to V_2O_3 than PbO to Pb. Also a reducing atmosphere can be prevented by slowly blowing air into the furnace chamber. As supports stable oxide ceramics should be used since silica and silicates are easily reduced above 1000° to 1200°C, and the resulting SiO (or Si formed by disproportionation of SiO) forms with Pt low-melting eutectics. Thus, only alumina or other stable oxides should be used in conjunction with platinum.

Cleaning of platinum crucibles to remove adhering flux or oxides is done by careful tapping with a plastic hammer, and/or by dissolution with the appropriate hot acid or alkali solution, or by dissolution with a flux. Nitric acid or hydrochloric acid may be used separately, but a mixture of these two acids (aqua regia) will form free chlorine and dissolve platinum (iridium-rich platinum alloys are rather resistant to aqua regia). Adhering oxide crystals, if not easily removed mechanically may be dissolved in melts of potassium pyrosulphate at 700–800°C (SO_3 vapour is formed), in Na_2CO_3—K_2CO_3 melts around 900°, in borax at about 1200°, or in KF or PbF_2 melts at 1200°C. Silicates are easily removed with concentrated hydrofluoric acid. After leaching these melts and after any hammering or reshaping the platinum crucibles should be treated with hot nitric acid for one hour, and after rinsing with sufficient distilled water the crucible should be dried by heat (not with a towel) and kept dust-free until use.

Generally pure platinum is used for growth of oxide compounds from their solutions in molten salts. Alloys of platinum frequently have advantages or are even required for special applications. Thus for growth of $SrTiO_3$ and $BaTiO_3$ from melts containing excess TiO_2, Pt–Ir or Pt–Rh alloys are used because of their higher melting points, and for still higher temperatures iridium might be used. According to Nielsen (1974) iridium crucibles might be used for PbO-free PbF_2 solutions in inert or slightly reducing atmosphere since Ir is not attacked by Pb up to 1600°C.

The disadvantages of Rh and Ir (and Pt–Ir alloys) is that they are more easily corroded than pure platinum by the normal Pb- or Bi- containing solvents and at low temperatures. For example Cloete *et al.* (1969) found in $MgAl_2O_4$ crystals no platinum, but traces of iridium which was probably present as an impurity in the platinum crucible used, and rhodium is frequently found in crystals grown from Pt crucibles. Small rhodium addition (0.5–1%) to the platinum increases its mechanical strength and decreases the tendency to recrystallization. Pt–Au and Pt–Au–Rh alloys are very little wetted by borate and silicate melts, but on the other hand Au and Rh are easily extracted from platinum containers. Iridium evaporation (as IrO_2) might be minimized by rhodium plating.

In summary, for all applications special crucible materials, shapes and sizes may be found which, with the appropriate experimental conditions, might contribute significantly to the size and quality of the flux-grown crystals. Aspects of handling noble metal crucibles are given in booklets and catalogues of the producer companies and in the reviews and papers of Brice and Whiffin (1973), Cockayne (1968), Darling *et al.* (1970), Robertson (1969), Robertson *et al.* (1974a) and Van Uitert (1970). The development of cheaper crucibles would probably extend the application of the flux method.

7.2.4. Separation of crystals from solution

In general crystallization from high-temperature solution is terminated with the solution at a temperature above the eutectic point, and the crystals produced may be separated from the excess solution either at this "final" temperature or following relatively rapid cooling to room temperature. The former alternative is generally preferable when this is convenient, as in top-seeded growth where the crystal may normally be raised out of the solution by a motorized drive mechanism.

When the crystals are grown by spontaneous nucleation, cooling of the crucibles to room temperature prior to removal of the crystals is fairly widely practised because of its relative simplicity. The excess solution is then normally dissolved in some aqueous reagent, a process sometimes referred to as "leaching". The leaching process relies on the differential solubility in this reagent of the crystal and the high-temperature solvent and it is clearly important for easy removal of the crystals that a suitable reagent should be available. Aqueous solvents for the more popular high-temperature solvents are listed in Table 3.13.

The great disadvantage of removing the excess solution at or around room temperature is that the rate of dissolution may be extremely low. A month or even longer may be required to remove, for example, bismuth borates from a crucible of only 100 ml capacity and means of avoiding such delays are clearly desirable. Warming the container used for leaching will promote dissolution and the resulting saving in time can be dramatic especially if leaching is performed at a temperature close to the boiling point of the aqueous reagent. As an example, Elwell and Morris (unpublished) have found that excess $PbO/PbF_2/B_2O_3$ may be dissolved within 1–2 days when the crucible is maintained in fairly dilute nitric acid close to its boiling point. Boiling chips must be added to the acid and the temperature of the acid bath maintained constant by a hot plate or oilbath. The most rapid leaching rates require that the acid be replenished at frequent intervals which depend on the volume of the container used.

Cooling of the crucible and crystals to room temperature suffers from the additional disadvantage that strain may be introduced in the crystals because of the contraction of the solution as it solidifies. This is especially the case when relatively soft crystals like sulphides are grown from molten salts. However, even if crystals are removed while the solution is molten, thermal strains will be present and the crystals should be annealed if freedom from strain is important.

Removal of the excess solution above some critical temperature may be essential if the system is such that re-dissolution of the crystals would occur on further cooling. Even in systems where no such dissolution would occur, the saving in time which is achieved with hot pouring has led to its increasing use. The simplest means of pouring off the excess solution is to remove the crucible from the furnace with tongs and to decant the liquid after removing the crucible lid. This method is possible only when the lid fits loosely onto the crucible, since it is desirable that the crucible should be returned to the furnace as rapidly as possible in order to minimize thermal shock to the crystals; especially when crystals larger than about 1 cm³ are grown cracking may occur on any sudden temperature changes.

An alternative procedure has been suggested by Grodkiewicz et al. (1967) who puncture the base of the crucible from below with a steel spike, without removing the crystal from the furnace. The drainage hole may be re-welded so that the crucible can be used several times before refabrication is necessary.

The seeding technique of Bennett (1968) and Tolksdorf (1968), described in Section 7.1.1, provides a particularly convenient means of removing the crystal from the residual solution by inversion of the crucible. Crucible inversion is applicable to sealed crucibles and may be used for unseeded growth if some mechanism is provided within the furnace. This mechanism may be extremely simple, for example, a wire attached to the base of the crucible which is pulled vertically upwards, provided that the crucible is

N

mounted on a suitable pivot (J. M. Robertson, private communication). A device which may be used for simultaneous hot pouring by inversion of a number of crucibles is shown in Fig. 7.37(a) and (b). The crucibles are mounted in a cemented firebrick block which has the primary advantage of providing a high thermal capacity during growth. At the termination of growth, a clamp mounted on a relatively massive trolley is brought into contact with the block, which may then be removed from the furnace, inverted and returned to the furnace very rapidly with little thermal shock to the crucibles. The crucible lids are crimped tightly to minimize evaporation during growth, but provide imperfect seals so that the excess solution drains away into the surrounding powder and firebrick, which is replaced after every experiment.

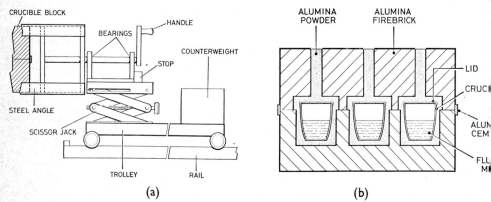

(a)   (b)

FIG. 7.37. Device for simultaneous hot pouring of several crucibles. (a) Mechanical setup. (b) Ceramic block with crucibles (Smith and Wanklyn, 1974).

Completely sealed crucibles permit a greater degree of reproducibility in the crystal growth conditions but inversion and subsequent solidification of the residual solution presents problems in the removal of the lid and of access to the crystals. Scheel (unpublished), using the sealed crucible design shown in Fig. 7.36(c), has found it more convenient to pour off the solution from the crucible by puncturing holes in the lid with a small steel spike after rapid removal of the crucibles from the furnace. The crucible is quickly replaced into a furnace which is then slowly cooled to room temperature. The small thermal shock introduced by this procedure is more than offset by the ease of access to the crystals after cooling to room temperature.

An isothermal technique to remove excess solution has been used by Kawabe and Sawada (1957) who, after growth of $BaTiO_3$ from Na_2CO_3 solution at 900°C, sucked up the still liquid flux through a nickel pipe by a rotary pump, and then cooled the furnace with the crucible to room temperature. In a similar manner for sealed crucibles, pressure from outside could be used to pump the excess solution out of the crucible.

7.2.5. Chemicals

The importance of attention to chemicals cannot be overemphasized and many problems of lack of reproducibility in crystallization from high temperature solution may be related to differences in the starting materials. Most experimenters prefer to buy pure chemicals from specialist suppliers rather than undertake their own purifications but it is well known that different batches of chemicals may vary in their impurity content and in the assay of the substance in question.

In general the chemicals which are supplied can be classified into four broad categories—laboratory or reagent grade with a purity normally quoted in the region of 99%, "analytic grade"—about 99.9%, "spectroscopic grade"—up to 99.999%, and ultra-high purity grades of certain chemicals, supposedly with even higher purities, may be available according to the nature of the material required. The increasing availability of high purity chemicals has been of great value in crystal growth and has been a contributory factor in the preparation of relatively perfect crystals. The experimenter should, however, be aware of the dangers of unquestioning reliance on the manufacturer's analysis. Even where the material is accompanied by the typical result of spectrographic analysis quoting impurity levels in the ppm range, it should be realized that the manufacturer does not guarantee the assay of the material. So, for example, the material may contain volatile impurities such as organic matter, carbon dioxide or water vapour. Moreover, the stoichiometry may differ very considerably from the nominal compostition.

As examples of the types of problem encountered by the authors, a sample of high purity Al_2O_3 from a well-known manufacturer with a stated molecular weight of 101.9 was found by X-ray diffraction to consist mainly of hydrates. Similarly a batch of nickel oxide "NiO" from a different supplier contained about 30% by weight of metallic nickel.

Departures from ideal compositions can increase with storage, particularly once a container has been opened, and particular care is necessary with hygroscopic materials such as B_2O_3, La_2O_3, many halides and even the γ-modification of alumina, especially in powder form. Materials in powder form are particularly liable to attack from the atmosphere on account of their large specific surface area. In several cases it might be necessary to handle such hygroscopic substances or other reactive materials (e.g. metals) in dry-boxes or at least under a dry stream of nitrogen. It is desirable to record in a laboratory notebook the batch number of any new delivery of chemicals, and the dates on which it was received and opened.

Checks of new batches of chemicals by X-ray powder diffraction, and possibly also by chemical analysis, are desirable, and the use of a thermobalance is valuable for detection of departures from stoichiometry of oxides or of the presence of free or bound water. Chemicals should also be checked for the presence of foreign bodies such as hair or other fibres, or pieces of furnace constructional material which may have entered the sample during manufacture. It is obvious that all containers and tools which come in contact with the chemicals or the growth solution must be absolutely dry and clean. The presence of any contaminant will, of course, lead to errors in weighing out the constituents of a growth experiment, and to impurity contamination which in most cases has a bad and sometimes a castastrophic effect as for instance traces of silicon in flux growth of BeO (Austerman, 1965). The positive influence of some impurities on growth has been discussed in Chapter 3, and the effect of impurities on the habit in Chapter 5.

For specialized preparations or when the highest purity is required the availability of facilities for purification is essential. The techniques of synthesis of high-purity substances and of chemical purification have been reviewed by Kröger (1964), by Wilke (1973) and by Jonassen and Weissberger (1963–1968). The synthesis of inorganic compounds in general is the topic of books of Angelici (1969), Brauer (1960/1962), Hecht (1951), Jolly (1970) and Lux (1970) and of the series Inorganic Syntheses (Vols **1**, 1939–**13**, 1972). Purification can be done by *sublimation* or *distillation*, by *heating*, possibly in a vacuum, to evaporate volatile impurities and burn off organic contaminants, and by *recrystallization*. For metals, electrolysis may also be used to separate the metal from its salts and from other metal impurities. These methods all rely on the preferential distribution of impurities between different phases, so that the principles follow the discussion of distribution coefficients outlined in Section 7.1.4. The most important technique for the preparation of high purity metarials for crystal growth is *zone refining*, which employs repeated crystallization from the melt with the impurities concentrated into a thin molten zone which traverses the material. Zone refining or zone melting is the subject of books by Pfann (1966) and Schildknecht (1966), both of whom quote examples of the application of this technique to oxides, salts and metals in addition to the semiconductors and organic materials for which the technique was developed.

Additional purification methods may be used to purify certain materials, in particular by the addition of reagents which react with impurities to form volatile species. As an example, Wanklyn (1969) removed traces of oxides and hydroxide ions from fluorides by heating with ammonium bifluoride $NH_4 . HF_2$ to form volatile ammonium compounds, as by the reaction

$$CoO + NH_4 . HF_2 \rightarrow CoF_2 + NH_4OH\uparrow.$$

The purification of salts is considered by Corbett and Duke (1963) who consider particularly problems of the removal of hydrolysis products. The desirability of purification prior to fusion is mentioned to facilitate the removal of impurities which become more strongly bound in the liquid phase.

Several articles on the purification of semiconductors and of the materials used as dopants are contained in the book of Brooks and Kennedy (1962). The scope of this collection includes the purification of chromium and rare earth metals in addition to semiconducting elements and compounds.

In view of the extensive literature on semiconductor materials, we consider in more detail here some examples of preparation and purification studies of other materials of particular interest.

Boric oxide. The presence of water in B_2O_3 is a well-known problem both in the use of borate fluxes and in the application of this material as an encapsulant in Czochralski growth (Mullin *et al.*, 1965). Chang and Wilcox (1971) found that reagent grade boric oxide contains about 3 wt% of water. Most of this water is removed on heating in a vacuum oven at 260–270°C, but the small traces which remain can be harmful to crystal growth. The removal of moisture from molten B_2O_3 is slow because of its high viscosity but Chang and Wilcox found that drying is accelerated if dry nitrogen is bubbled through the melt. On cooling to room temperature the moisture absorbed by vitreous B_2O_3 was found to be restricted to the surface and could be substantially removed by a vacuum without further heating.

Lead oxide and fluoride. Lead oxide and fluoride are probably the most widely used fluxes and their purification is therefore of particular interest. Schieber (1967) reported that the highest purity PbO could be prepared by precipitating lead carbonate from a solution of lead nitrate by adding ammonium carbonate:

$$PbNO_3 + (NH_4)_2CO_3 \rightarrow PbCO_3\downarrow + 2NH_4NO_3.$$

Lead oxide is produced from the carbonate by heating to 400°C in a clean atmosphere. The preparation of pure lead fluoride is also considered by the same author who proposes that the purest material may be prepared by precipitation on mixing lead acetate and ammonium fluoride solutions:

$$(CH_3COO)_2Pb + 2NH_4F \rightarrow PbF_2\downarrow + 2CH_3COONH_4.$$

Kwestroo and Huizing (1965) also used lead acetate in solution in order to prepare high purity lead oxide by precipitation with ammonia. But it should be said here that PbO and PbF_2 of high purity are now commercially available so that in general purification of these important solvents is not necessary. If for special applications ultrapure PbO or PbF_2 should be required it is suggested to obtain them by zone melting in pure platinum boats or tubes.

A few examples of synthesis and purification of specific solvents and chemicals used in crystal growth from HTS have been reported: $BaCl_2$, $SrCl_2$ (Fong and Yocom, 1964); LiF (Eckstein et al., 1960; Kiyama and Minomura, 1953; Weaver et al., 1963; Thoma et al., 1967); NaCl (Raksani and Voszka, 1969; Lebl and Trnka, 1965); KCl (Capelletti et al., 1968, 1969; Kanzaki and Kido, 1960/1962; Lebl and Trnka, 1965; Butler et al., 1966); Na_2WO_4 (Cockayne and Gates, 1967); V_2O_5 (Jankelevic et al., 1967).

Rare-earth metals present particular problems because of their extremely high reactivity. High purity oxides are commercially available and can be used as the starting material for the elements and hence for other compounds. Alternatively the relatively impure metals available from commercial suppliers may be used as starting material, as in the method of Busch et al. (1971). These authors noted by mass spectrometry that metals claimed to be 99.9% pure actually contain a few wt% of non-metallic impurities, mostly dissolved gases. In their method the dissolved gases are partly desorbed and partly precipitated as non-volatile components by heating the metal to its melting point, followed by distillation. The distillation is carried out in a molybdenum vessel in a sealed UHV system at 3×10^{-7} Torr. Apart from H,C,O, Ca and Ba, impurity levels (including Mo) were less than 1 ppm. Also Habermann et al. (1965) purified the rare-earth metals, namely by reduction of the fluorides followed by distillation. In general, contamination from the crucible is expected to be high and the "cold crucible" of Hukin (1971) was developed to provide a clean container for rare-earth metals. Alternative methods for purification of rare-earth metals are described in the article by Love and Kleber in the collection of Brooks and Kennedy (1962).

The importance of using pure chemicals so that contamination of the crystals is minimized is shown up in a striking fashion be several physical measurements. Sproull (1962) has shown that the thermal conductivity at low temperatures is particularly sensitive to impurities and differences of two orders of magnitude are observed when pure sodium chloride is compared with commercial grade crystals.

Suppliers of chemicals of high purity or of a special pure grade for crystal growth are: Associated Lead Manufacturers Export Company Ltd., London (PbO, PbF_2); The British Drug Houses Ltd., Laboratory Chemicals Division, Poole, England ("Optran"); Eagle-Picher Company, Chemical Division, Cincinnati, Ohio, USA; Koch-Light Laboratories

Ltd., Colnbrook, Bucks, England; Materials Research Corporation, Orangeburg, N.Y.; E. Merck AG, Darmstadt, W-Germany ("Suprapur", "Optipur"); New Metals and Chemicals Ltd., Poole, England; Johnson Matthey Chemicals Ltd., London E.C.1. ("Specpure", "Puratronic", "Spectroflux"); Research Chemical Corporation, Phoenix, Arizona; Molybdenum Corporation of America, 6 Corporate Park Drive, White Plains, N.Y.; MCP Electronics Ltd., Alperton, Wembley, Middlesex, England; Schweizerische Aluminium AG, Neuhausen, Switzerland (Gallium and its compounds); Levy West Laboratories Ltd., Harlow, Essex, England (niobates, tantalates).

7.2.6. Atmosphere control

The importance of atmosphere control has been discussed at several points within this book, particularly in Section 7.1.5 in connection with the preparation of phases having ions in unusual valence states. In general the required phase must be grown in an atmosphere in which it is stable, although in practice some modification of the equilibrium condition may be effected according to the solvent in which the crystal grows. The practical problems of atmosphere control may be considered as threefold: purification of the gases if impurities are likely to enter the crystal or to affect growth; pressure control of a single pure gas phase, and partial pressure control of an active component in a mixture of gases. The vital questions of leak prevention and detection are beyond the scope of this book.

Commercial gases generally are impure, typical impurity concentrations in high-purity oxygen being 0.1% water, and in argon being 20–100 ppm oxygen. The removal of harmful impurities from the atmosphere is of particular importance in high-temperature solution growth in view of the long times required for an experiment and of the crucible lifetime, for example. A variety of methods is available for gas purification, based both on physical and chemical properties. The most commonly used technique is by chemical reaction by flowing the gas over or through a reagent which will react with the impurity but not with the principal constituent. As an example, the concentration of oxygen in nitrogen or argon may be considerably reduced by flow over copper turnings at about 600°C. Other examples of interest to crystal growers are the catalytic removal of oxygen from hydrogen or hydrogen-containing mixtures by using Pt or Pd contacts with removal of the resulting water by freezing or drying agents, and the removal of O_2, N_2, CO and CO_2 from He or Ar by passing the gas over titanium powder at 850°C. Very low oxygen concentrations in inert gases were obtained by Steinmetz et al. (1964) by flowing the gas through aluminium amalgam. The oxygen reacts very readily with the aluminium to form Al_2O_3, which floats on the amalgam, and oxygen partial pressures of 10^{-27} atm may be achieved.

A common practice is the drying of gases, and the efficiency of drying agents increases from $CaCl_2$, $CaSO_4$, H_2SO_4, KOH, $Mg(ClO_4)_2$, BaO to P_2O_5, and especially P_2O_5 distributed in silicagel is a practical drying agent. On the other hand controlled humidities can be produced by bubbling the respective gas through large volumes of specified aqueous solutions as listed in Lange's Handbook of Chemistry (1967).

The principal alternative to chemical methods utilizes differences in the boiling or freezing points of gases, and, for example, water vapour and carbon dioxide may be effectively removed from oxygen at not too high pressures by flow through a reservoir containing liquid air or a similar refrigerant. Hydrogen, which is commonly used as atmosphere in crystal growth from metallic solutions, may be purified by diffusion through solid palladium since its diffusion coefficient is much higher than that of the heavier gases. Methods of purification of gases have been reviewed by Müller and Gnauck (1965) and by Lux (1970).

The total pressure of the gas may be regulated by a device which is based on the operation of a relay by a pressure sensor for the required range. As an example, a device capable of regulating pressures up to 1 atmosphere was described by Oxley and Stockton (1966). Generally pressure regulators for operation both above and below atmospheric pressure are obtained from commercial sources. The use of high vacua is generally to be avoided in HTS growth in view of the volatility of most solvents, and a static or flowing inert atmosphere is normally preferred for syntheses at low oxygen pressures.

Controlled oxygen partial pressures in the range below about 10^{-1} atm may be most readily achieved using gas mixtures such as H_2/CO_2, CO/CO_2 or H_2/H_2O. Data for the variation of oxygen partial pressure with relative concentration and temperature may be obtained from Fig. 3.4. Similarly sulphur partial pressures for mixtures of H_2S with H_2 are available from Fig. 3.5. A fairly detailed account of the importance of partial pressure in material syntheses is available in the book by Reed (1972). Alternative systems for the provision of controlled atmospheres are discussed by Kröger (1964), who treats in great detail crystal imperfections and their relation to the atmosphere.

An unusual example of the use of special atmospheres in crystal growth from high-temperature solution is the preparation of MgO and $LaAlO_3$ crystals doped with ^{17}O by Garton et al. (1972). The high cost of this isotope requires experiments on a small scale and crucibles of 10 ml capacity were used. The isotope-enriched oxygen gas was introduced into an evacuated, sealed platinum tube at room temperature with the tube contained inside a brass cylinder, the pressure within which was adjusted so as to avoid pressure differences which might fracture or distort the platinum tube. The concentration of ^{17}O in the crystals grown was about 10%, a similar value to that in the source powder.

7.2.7. Stirring techniques

Stirring was introduced into the field of crystallization by Wulff (1884) and has become routine in crystal growth from aqueous solutions, both for growth of single crystals and for mass crystallization. Stirring is also applied in other crystal growth techniques (e.g. Czochralski growth, zone melting) whereas in crystal growth from high-temperature solutions it has not yet found widespread application mainly because of experimental problems.

A. Effects of stirring. The main effects of stirring can be divided into two categories, namely the effects due to homogenization of the solution and those due to a high solution flow rate at the growing crystal faces.

A solution homogeneous with respect to solute and impurity concentration and to temperature may result from various stirring techniques. For example, if a stirrer (with seed crystal) is inserted in the solution and rotated in a stationary container, mixing and homogenization occurs between the regions of solution moving with the stirrer and the regions adhering to the container walls. This homogenized solution contains no thermal or solutal convection cells which frequently lead to uncontrolled nucleation. Thus the whole solution is influenced by the presence of any crystal, the supersaturation is not likely to exceed the critical value for nucleation so that spontaneous nucleation can be avoided. Another effect of homgenization is that the growing crystal always sees a homogeneous diffusion field, and thus inhomogeneities in crystals such as striations are prevented or at least reduced to a minimum level. Forced convection, that is a high solution flow rate along crucible walls and crystals decreases the Ostwald-Miers range of metastability, and spontaneous nucleation might occur, especially in highly concentrated and viscous solutions.† Therefore, stirring should be not too vigorous and should be adjusted according to the concentration and to the viscosity in order to avoid spontaneous nucleation. The narrowing of the metastability region leads to a requirement of precise temperature control, which in flux growth we estimate should be to better than $\pm 0.1°C$.

† As an example, in the growth of iodic acid from very concentrated and viscous aqueous solutions, stirring is disadvantageous according to S. Haussühl (private communication). However, Daval (1974) grows large HIO_3 crystals of high quality from stirred solution.

N 2

Forced convection also leads to a decrease in supersaturation, the latter being influenced by the habit of the crystals present, by nucleation at the surface and by the degree of stirring. In general, this decrease in super-saturation corresponds approximately to the narrowing of the metastability range, so that no enhanced spontaneous nucleation is observed, at least at modest stirring rates.

A high solution flow rate along the growing crystal faces has several advantages which have been demonstrated by Scheel and Elwell (1972, 1973) and mentioned in Section 6.6.2. Stirring decreases the diffusion boundary layer and reduces the supersaturation inhomogeneity across the crystal thus allowing a faster maximum stable growth rate than that without stirring. Because of the faster stable growth rates made possible by stirring, the duration of experiments may be decreased (see Fig. 6.16). Stirring is necessary if crystals larger than about 1 cm³ are to be grown economically. The importance of stirring is underlined by the X-ray topograph of Vergnoux *et al.* (1971) which shows growth instability at the event of interruption of stirring, see Fig. 7.38.

The decrease in the boundary layer thickness has an effect on the growth mechanism and on the surface structure of the crystals grown. Surfaces of

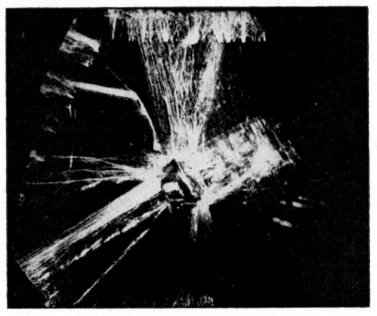

FIG. 7.38. X-ray topograph of a strontium formate crystal showing development of dislocations, probably at inclusions, at the event of interruption of stirring (Vergnoux *et al.*, 1971).

crystals grown from stirred solutions under not too high supersaturation are very flat compared to crystals grown from unstirred solutions. This led Scheel and Schulz-DuBois (1972) to propose stirring in liquid phase epitaxy, for instance by rotation of the substrate(s) in the solution, and Blank and Nielsen (1972) as well as Giess *et al.* (1972) and Ghez and Giess (1973) confirmed by experiment the advantage of stirring. Stirring in LPE has become increasingly widely applied (Vilms and Garrett, 1972; Scheel, 1973).

Through the decreased boundary layer stirring influences the incorporation of impurities or solvent ions as discussed in Section 7.1.3. Accordingly the stirring rate should not be varied too much during an experiment if inhomogeneous incorporation of impurities is to be avoided.

B. Natural convection. In unstirred solutions bulk material transport by natural convection occurs in most growth experiments due to temperature differences (thermal convection) and due to density differences arising from variations in solute concentration (solutal convection). Since natural convection has been discussed in some more detail in Section 6.6.2, here only a few general remarks will be made.

The average convection flow rate for typical high-temperature solutions lies between about 0.01 and 0.1 cm s^{-1} (see Fig. 6.16) and thus is about two orders of magnitude smaller than typical flow rates in stirred aqueous solutions. Problems arise when the convection streams are irregular or when they form convection cells or pockets because in these cases spontaneous nucleation might occur, competing with growth of the crystals already present. Time-dependent convection also leads to striations or even unstable growth. Consequently a steady convection flow should be produced, for instance by heating one crucible wall and cooling the other, or in general by achieving temperature profiles and crucible shapes which favour a steady convection flow. As an alternative, convection may be minimized as in the convection-free cell of Tiller shown in Fig. 6.19, but experimental difficulties may arise when this cell is applied in crystal growth from high-temperature solutions. It is obvious from the discussion of stirring in Section *A*, that if possible smooth and continuous stirring should be applied and that the hydrodynamics in crystal growth from solutions should be dominated by adjustable forced convection.

C. Seed crystals on stirrers. Seed crystals on stirrers are used in crystal growth from aqueous solutions both in small-scale and in large-scale production of monocrystals, thus the advantages of homogeneous solutions and of high solution flow rates can be utilized. However, experimental parameters such as the rotation rate, the position of the seed between the rotation axis and the container walls, the orientation and the fixing of the seed crystals must all be carefully adjusted. Since a small quantity of the solution rotates with the seed crystals and since a sufficient solution flow rate is needed along all faces in order to prevent inclusion veils the rotation should be reversed at least once a minute (Holden, 1949).

On the other hand, Sip and Vanicek (1962) pointed out that abrupt termination and reversal of rotation affect the crystal very unfavourably, and sudden acceleration of the solution might even cause spontaneous nucleation. Therefore Sip and Vanicek proposed that changes in the rotation rate of the seed holder should occur smoothly and that any reversals should be preceded by a pause. As an alternative to attaching the seed crystal to a rotating (or reciprocating) holder it could be fixed to the container, and the solution moved by an independent stirrer.

Seed crystals immersed in solution are not often used in crystal growth from high-temperature solutions. Laudise et al. (1962) applied seeds in an arrangement as shown in Fig. 7.3 in order to grow yttrium iron garnet crystals from $BaO.0.61B_2O_3$ solutions. A crucible of 7.6 cm diameter was used, and the seed holder was rotated at 200 r.p.m. with the direction of rotation reversed every 30 seconds. The dissolution of the feed material was the rate-determining step, as was discussed in detail by Laudise (1963), at least for rotation rates faster than about 50 r.p.m. One would expect that for abrupt changes in the rotation rates as above, uncontrolled nucleation would occur, and this in fact was observed. It seems that stirring is also applied in the commercial synthesis of emeralds but details of the process cannot be found in literature nor obtained from the emerald producers. There is no doubt that in any large-scale commercial crystal synthesis stirring will be applied, and the experience obtained in the growth of crystals from aqueous solutions may be applied in crystal growth from high-temperature solutions. The use of seed crystals on stirrers is generally only proposed when a number of large crystals of the same family have to be grown, in view of the effort required to determine the optimum conditions.

D. *Top-seeded solution growth* (TSSG). This topic has already been discussed with respect to nucleation control in Section 7.1.1, therefore the discussion of TSSG here will be restricted to its hydrodynamic aspects. The principle of TSSG is demonstrated in Fig. 7.39(a) (the whole apparatus is shown in Fig. 7.4). A seed crystal is cooled by an air stream flowing through the seed holder. The solid-liquid interface is below the liquid level so that this method differs from the Czochralski technique where in general the interface is above the melt level. Since in TSSG the crystal dips into the solution, the temperature gradient at the interface is

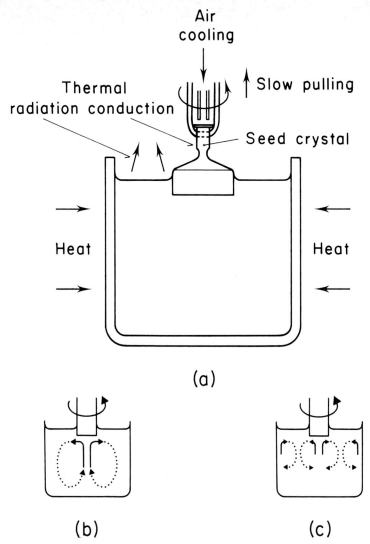

FIG. 7.39. Heat flow (a) and hydrodynamics (b) and (c) of top-seeded solution growth.

less steep than in Czochralski growth, and this encourages the facetting which is normally observed on TSSG-grown crystals.

Due to the similarity of the Czochralski and TSSG techniques the hydrodynamics and the stirring effects are expected to be similar. The effects of seed rotation on Czochralski-grown silicon crystals have been studied by Goss and Adlington (1959), and Robertson (1966) has shown that a large crystal diameter (more than 10 mm) and high crystal rotation rates of more than about 50 r.p.m. are required to produce the upward flow (towards the crystal) as indicated in the examples of Fig. 7.39(b), (c). Cochran (1934) has analysed the flow to a rotating disc (of infinite diameter), and Burton, Prim and Slichter (1953) based their derivation of the boundary layer thickness on Cochran's analysis. An advantage of an axial rotating seed is that the solute flow is, to a first approximation, independent of the crystal radius. Greenspan (1968) extended Ekman's study of the "hydrodynamics" of hurricanes to any fluid medium and demonstrated the existence of the so-called Ekman layer flow. This is a radial flow within a narrow horizontal layer which occurs when there is a difference in rotation rates between the fluid medium and a solid boundary that is approximately perpendicular to the rotation axis (Schulz-DuBois, 1972); see also the discussion in Section 7.2.7E. Carruthers and Nassau (1968) have identified a number of flow patterns as shown in Fig. 7.40 by simulation experiments in which the relative rotation rates of crystals and crucible were varied over a wide range.

In addition to the large variety of flow patterns indicated above thermal convection may lead to temperature oscillations in the melt with the result that it is extremely difficult to grow striation-free crystals by the Czochralski and TSSG techniques. In discussing turbulent-free convection in Czochralski crystal growth Wilcox and Fullmer (1965) agree with Malkus (1954) that fluid baffling is relatively ineffective in reducing turbulent free convection, and propose shielding or an afterheater for this purpose. In contrast Whiffin and Brice (1971) propose, in addition to afterheaters, the use of baffles in order to damp thermal oscillations, as in the convection-free cell of Tiller (1968).

It seems reasonable that a stationary baffle 5 to 15 mm below the rotating crystal will minimize thermal convection not only because of the smaller effective volume of the melt and the reduced (vertical) Rayleigh number (Whiffin and Brice, 1971) but also because of its stirring effect. Also Cockayne et al. (1969) proposed shallow melt depths to reduce temperature fluctuations in the melt and pointed out the influence of convection in the gas above the melt.

Although many of the principles discussed above for Czochralski growth will be valid for TSSG, deviations are expected because of the difference of the height of the interface, the different temperature profile and the facetting of many crystals grown by TSSG.

E. *Accelerated crucible rotation technique* (ACRT). One of the problems of crystal growth from high-temperature solutions using volatile solvents

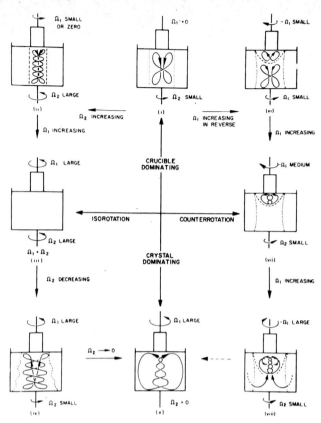

FIG. 7.40. Variation of flow patterns with the relative directions and magnitudes of crystal and crucible rotation (Carruthers and Nassau, 1968).

like PbO and PbF_2 is that the crucible should be sealed in order to prevent solvent evaporation. The high temperatures and the restriction on crucible materials prevent the application of conventional stirring techniques. According to J. W. Nielsen (1974) alternate rotation and counterrotation of crucibles with high-temperature solutions were used by Remeika and Van Uitert at Bell Laboratories and by Lepore at Airtron for initial stirring to promote dissolution of the solute; continuation of this agitation during growth did not allow control of nucleation so that this procedure was abondoned, as by Nelson and Remeika (1964) who used rotation reversal only in the initial stage in order to enhance dissolution of the components but not during crystal growth. However, Scheel demonstrated that a beneficial stirring effect in flux growth could be achieved by periodic acceleration and deceleration of the crucible rotation and by alternating the sense of rotation. This accelerated crucible rotation technique (ACRT) permitted firstly, in a sealed crucible, the restriction of nucleation by localized cooling to one to three crystals and secondly the growth of large inclusion-free crystals (Scheel and Schulz-DuBois, 1971; Scheel, 1972).

There are two typical flow mechanisms which occur during the acceleration and deceleration of container, namely the spiral shearing distortion and the Ekman-layer flow, although on strong deceleration a transient Couette flow might occur.

The *spiral shearing distortion* for an infinite tube (thus neglecting the effect of the crucible base) has been analysed by Schulz-DuBois (1972). This type of flow is conveniently considered by reference to a uniformly rotating cylinder containing two immiscible liquids shown as black and white in the cross-section of Fig. 7.41(a). If the rotation of the tube is

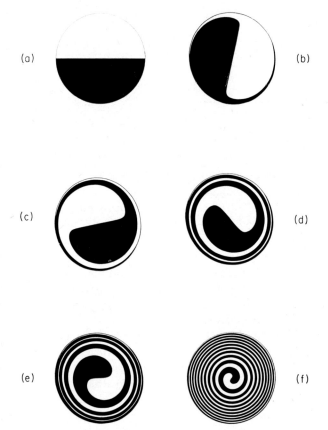

Fig. 7.41. Successive spiral shearing distortion of two liquids in a rotating tube when the rotation is suddenly stopped (after Schulz-DuBois, 1972).

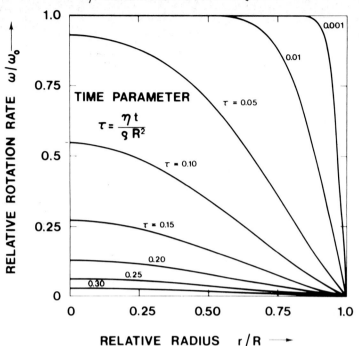

FIG. 7.42. Decrease of relative rotation rate versus relative radius in spiral shearing distortion (Schulz-DuBois, 1972).

suddenly stopped, the rotation rate of the liquid near the wall decreases rapidly due to friction, while the central region of the liquid decelerates much more slowly because of its inertia. Thus shearing occurs between the various ring-like liquid regions, and a spiral develops through successive changes in rotation as demonstrated in Figs 7.41(a)–(f). The decrease of the relative rotation rate ω/ω_0 versus the relative radius r/R is shown in Fig. 7.42, ω_0 being the initial rotation rate. After a time $t=0.1(E\omega_0)^{-1}=0.1\rho R^2/\eta$, the rotation rate at the centre is reduced to about $0.5\omega_0$. For crystal growth by ACRT, especially in tall narrow containers where spiral flow may be the dominant one, the acceleration and deceleration periods should be of this order since for greater periods the changes in the velocity field become increasingly slower. In the above discussion E is the Ekman number $=\eta/\rho\omega_0 R^2$ with η the viscosity and ρ the density of the solution. For a typical melt with a kinematic viscosity $\eta/\rho=10^{-2}$ cm² s⁻¹ and a crucible of diameter $2R=10$ cm the time t has a value of 250 s. However, shorter time intervals are required for the periods of continuous acceleration and deceleration to achieve an optimum and yet smooth stirring effect and for increasing the Ekman flow.

If the liquid is allowed to come to rest after the sudden halt in the rotation, it will have rotated relative to the container by the shear angle

$$\phi(r) = \rho\omega_0(R^2 - r^2)/8\eta = (1 - r^2/R^2)/8E \qquad (7.9)$$

with r being the distance from the axis. From this equation the number N of spiral arms is derived as

$$N = \omega_0\, \rho R^2/16\pi\eta = (16\pi E)^{-1} \qquad (7.10)$$

and the width Δr of the spiral arms is

$$\Delta r = 4\pi\eta/\rho\omega_0\, r = 4\pi ER^2/r. \qquad (7.11)$$

In order to achieve efficient mixing, Δr should be between 0.1 and 1 mm, then homogenization by thermal conduction and by diffusion occurs in less than 10^{-1} seconds. The spiral shearing mechanism as well as the Ekman-layer flow can be clearly recognized in simulation experiments described by Scheel (1972).

A flow mechanism which is of comparable importance to the spiral shearing is the so-called *Ekman-layer flow*. According to Greenspan (1968) Ekman described the rapid flow of air in a relatively thin layer near the surface of the earth in rotating wind systems such as hurricanes. The rapid suction of air occurs in this "Ekman-layer" because there the pressure difference between the high outside the hurricane and the low in the centre is not balanced by centrifugal forces as demonstrated in Figs 7.43(a), (b). A similar Ekman-layer flow occurs in a crucible when the rotation is decelerated. The opposite flow occurs as the crucible is accelerated, due to a thin layer at the crucible bottom of which the liquid is accelerated first and thrown outwards due to centrifugal forces. The Ekman-layer flow for acceleration and deceleration is shown in Figs 7.43(c) and (d) and may be demonstrated with a cup of tea containing tea leaves (Scheel, 1972). The tea leaves at the bottom flow radially outwards, when the tea is stirred, and back towards the centre when the stirring is interrupted.

According to Greenspan (1968) and to Hide and Titman (1967) the thickness d of the Ekman layer may be approximated by

$$d = RE^{1/2} = (\eta/\rho\omega_0)^{1/2}, \qquad (7.12)$$

which is about 0.5 mm for the numerical example given above. The maximum radial velocity in the Ekman layer is

$$v_{max} = \omega_0\, R \qquad (7.13)$$

and easily exceeds 10 cm s^{-1} for the example. Ekman components of the flow velocities in the bulk of the liquid are considerably smaller, about

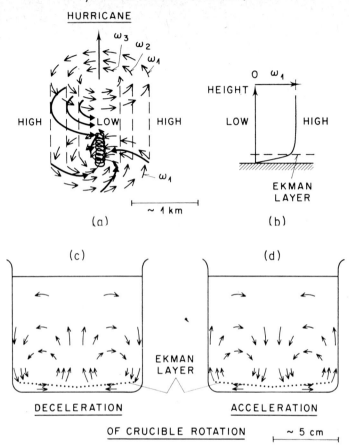

Fig. 7.43. Ekman-layer flow in a hurricane (a), (b) and in a crucible of which rotation is decelerated and accelerated (c), (d).

0.2 cm s^{-1}. Ekman layer flow in a container occurs only during acceleration and deceleration and ceases as soon as uniform rotation is achieved, approximately after the time

$$t = E^{-1/2}\, \omega_0^{-1} = R(\rho/\eta\omega_0)^{1/2}, \qquad (7.14)$$

which for above example corresponds to about 20 s (Schulz-DuBois, 1972). If the whole liquid is accelerated from zero to ω_0 then it has passed once entirely through the Ekman-layer where it experiences acceleration, and thus the Ekman-layer flow is a valuable supplement to the circumferential mixing effect of the spiral shearing.

The high Ekman flow rate given by Eqn 7.13 occurs not only at the crucible base but also on crystals growing there and on any crystal (or baffle) face which is approximately perpendicular to the rotation axis. Thus it also occurs in the top-seeded solution growth (and Czochralski) technique below the rotating crystal if it has a sufficient diameter and rotation rate as discussed in Section 7.2.7D.

Another mechanism of flow near the walls of a rotating cylinder of large diameter is possible when its rotation is decelerated rapidly. This modified, transient type of Couette flow (Chandrasekhar, 1961) occurs in convection cells resulting from interior regions of fluid with higher rotation rates which are driven towards the walls by centrifugal forces during the transient of strong deceleration. It is unlikely that the typical conditions used in ACRT are sufficient to establish the transient Couette flow in the region near the container walls.

The stirring action of ACRT can be adjusted by the rates of acceleration and deceleration, and by the differences of the rotation rates (Scheel, 1972). Typical examples of ACRT cycles are shown in Fig. 7.44(a). The reversal of the sense of rotation as in A is advantageous for homogenizing the interior parts of the liquid which have not come to rest and which otherwise would rotate continuously and thus be affected relatively little by the ACRT mechanism. In order to achieve a smooth change when altering from acceleration to deceleration or the sense of rotation, short periods of constant (or zero) rotation rate might be used as indicated in example C. A slow ACRT stirring action is suggested in the case of relatively concentrated and viscous solutions in order to prevent spontaneous nucleation, but in very viscous systems ACRT would not be of advantage.

In Fig. 7.44(b)–(d) examples of experimental arrangements for flux growth by ACRT are shown, namely for nucleation control by localized cooling (b), for the application of a seed crystal in the slow-cooling technique (c) and for the gradient-transport technique with a seed (d). Similarly, Tolksdorf and Welz (1972) used ACRT in combination with the Bennett–Tolksdorf nucleation control (Section 7.1.1). The crucible is positioned on a rotable pedestal which is connected to a motor by a belt. The rotation cycle of the motor according to Fig. 7.44(a) is obtained, for example, by electromechanical means (motor-driven cam on a motor-speed controlling potentiometer) or by commercially available electronic units with ramp generators.†

ACRT was first applied to the growth of gadolinium aluminate, and the first experiment resulted in a $GdAlO_3$ crystal of 210 g weight shown in Fig. 7.45(a), (b). Only this one crystal nucleated in a 500 cm^3 crucible and represented 67% of the starting material. In successive experiments more large $GdAlO_3$ crystals and large inclusion-free crystals of magnetic garnets have been grown from relatively small crucibles (Scheel and

† Eurotherm Produkte AG, Glattbrugg, Switzerland.

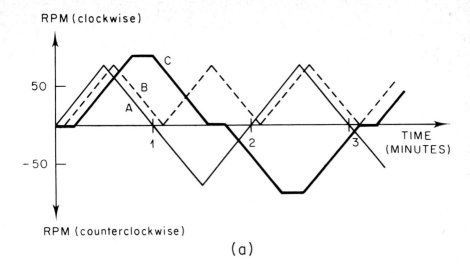

(a)

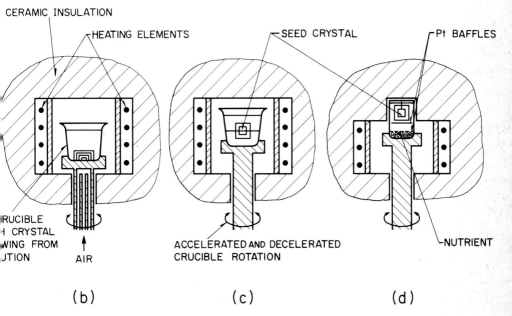

(b)   (c)   (d)

FIG. 7.44. (a) Examples of cycles of accelerated and decelerated crucible rotation. Experimental setup for ACRT with localized cooling (b), with a seed crystal (c), with seed crystal and gradient transport (d) (Scheel, 1972).

(a)

(b)

Fig. 7.45. Example of a large crystal grown by ACRT. (a) Gadolinium aluminate of 210 g weight grown in a 500 cm³ crucible. (b) Same crystal with light reflected inside the crystal at a natural face (Scheel, 1972).

Schulz-DuBois, 1971; Scheel 1972), and experiments in other laboratories have confirmed the value of ACRT-stirring to achieve nucleation control and inclusion-free crystals (Tolksdorf and Welz, 1972; Mateika, 1972). Recently nucleation control and growth of large crystals with the help of ACRT have been achieved by Puttbach at Airtron (Nielsen, 1974) and for growth of 80 g YVO$_4$ crystals at Allied Chemical Corporation (Vichr, 1973).

It is not easy to predict the effect of ACRT when crystals of needle or platy shapes are to be grown but preliminary experiments of Aidelberg *et al.* (1974) indicate a beneficial effect of ACRT in the growth of magneto-plumbite plates.

Striations might occur in systems with large changes in the boundary layer thickness, and obviously changes in the boundary layer thickness do occur in ACRT. But the popular periodic rotation reversal of the seed crystal holder in crystal growth from aqueous and high-temperature solutions also leads to changes in the boundary layer thickness, and to the authors' knowledge no striations due to these hydrodynamic changes have been proven. Only Damen and Robertson (1972) have claimed to have observed striations due to ACRT. However, the periodicity of striations expected from the experimental conditions used is an order of magnitude less than the resolution of X-ray topography. Further studies are required to clarify this question.

Experiences in various laboratories show that ACRT is a powerful stirring technique applicable to laboratory and pilot plant crystal production, and it is obvious that ACRT could be useful in other crystal growth methods, such as the hydrothermal, chemical vapour transport and Bridgman-Stockbarger techniques where stirring is to be achieved in closed containers as well as in Czochralski growth (Scheel, 1972).

An alternative stirring technique has been proposed by Gunn (1971) and consists of circulating the solution around the inside of a container by moving the centre of the container in a horizontal circular path, without rotating the container at all. This technique is frequently used in rinsing out a beaker with water and is applied by chemists during titration as an alternative to magnetic stirring. The circulatory action may be adjusted by varying the angular speed of the centre of the crucible and of the crucible itself; crucible rotation in the same direction as the circular motion of the crucible centre decreases the fluid circulation while counterrotation would enhance it. In contrast to ACRT this action produces a constant solution flow (if it is not interrupted), but it is technically more complicated to apply in crystal growth from HTS.

F. Additional stirring techniques. A variety of stirring mechanisms have been conceived but have not been widely applied in crystal growth from

high-temperature solutions. Kirgintsev and Avvakumov (1965) investigated
the stirring efficiency of the various stirring techniques shown in Fig.
7.46 by measuring the distribution of calcium nitrate impurity in solidi-
fying potassium nitrate. Rotation of a stirrer at high speed and rotation of a
partially filled horizontal tube have been found most effective. Bubbling
gas through the liquid is less effective and vibration of a stirrer at 50 Hz
showed no stirring effect at all. It is interesting to note that Wulff used a
seed crystal and a rotating horizontal cylinder in 1901. Magnetic stirring
is limited to temperatures of about 500°C and is thus not of interest for

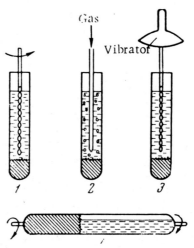

FIG. 7.46. Various stirring techniques studied by Kirgintsev and Avvakumov
(1965).

HTS growth. The same is true for R.F. stirring at about 10 kHz which is
not easily applied to high-temperature solutions in platinum crucibles, and
if R.F. is applied for heating also, then problems arise in obtaining the
required precision of temperature regulation.

7.2.8. High pressure technology

The growth of crystals at pressures in the kilobar region requires a
technology completely different from that for synthesis under modest gas
pressures as described in Section 7.1.5. High pressures are generated by
the action of a hydraulic ram operating on a piston, or by a combination of
several such rams or pistons, and it is now possible to maintain pressures of
200 kbar at temperatures of 2000°C or above. The techniques used in high
pressure technology have been reviewed by Bridgman (1952), Bundy
(1962), Munro (1963) and Rooymans (1972), the latter with emphasis on

chemical syntheses. A bibliography of high pressure techniques is given in the book of Paul and Warschauer (1963).

The simplest arrangement for crystallization under high pressure is that where the sample is located between a piston and a fixed closure within a cylindrical jacket. The maximum pressure that can be attained in this arrangement depends on the bursting pressure of the containing cylinder, which will be around 20 kbar for steel and 60 kbar for tungsten carbide. For higher pressures it is necessary to provide support for the cylinder or to pre-stress the inner wall so that this inner surface is in tension only at pressures in the working range. Composite cylinders are frequently used, and an outer jacket may be shrunk onto the outer surface in order to provide the compression.

In the opposed anvil arrangement, the containing cylinder is eliminated and the specimen is held in a small central region between two anvils which are tapered away from this region at about 10°. This arrangement distributes the stress immediately below the anvil surface over a much larger quantity of material, a device which Bridgman termed the "principle of massive support". The main disadvantage of this arrangement for crystal growth is that the specimen volume is necessarily very small, typically 0.2 mm in thickness and 5 mm in diameter.

Another limitation associated with the thin sample is that thermal conduction to the anvils is relatively high and the associated loss of strength reduces the maximum pressure available at high specimen temperatures. A layer of insulation may be inserted into a recess in the anvil and the pressure limitation is then set by the closeness of approach of the opposing anvils, which is primarily governed by the behaviour of the ring of pyrophyllite which behaves as a fluid under pressure and so forms a gasket enclosing the specimen. A longer compression stroke, and hence a larger specimen volume, is possible when curved anvils are used and optimization of the anvil profiles at the General Electric laboratories led to the "belt" design which is now widely used, for example, in crystallization of diamond and of which the high-pressure region is shown in Fig. 7.47.

This arrangement doubles the axial length of the specimen chamber but the "belt" which governs the movement of the gasket material as the pressure is applied must withstand the bursting pressure of the sample and therefore requires a strong binding ring. The sample is contained in a cylinder with a seal at each end, and the gaskets are normally of pyrophyllite with metal laminations and extrude as the pressure is applied.

Relatively large specimen volumes are also possible with multiple-piston arrangements of which the most popular is the tetrahedral-anvil apparatus (Hall, 1958). Four anvils, each having a triangular end surface, are located at the ends of four rams and enclose a tetrahedron of pyrophyllite

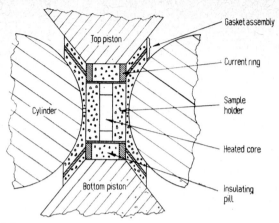

FIG. 7.47. High-pressure region in the belt apparatus used for diamond synthesis (Wentorf, 1966).

some 10% larger in edge dimensions than the anvil faces. The pyrophyllite acts as a pressure-transmitting medium and is bored out to contain the sample and heating element. Normally pressure is applied only to one ram (with the other three mounted in a support ring) which must move in such a way that non-uniform stresses are avoided. Apparatus with six rams acting on a pyrophyllite cube has also been used.

Arrangments for the production of high temperatures combined with high pressures have been reviewed by Strong (1962). External heating of the whole apparatus is possible at pressures below 10 kbar and temperatures less than 600°C. For higher temperatures, the decrease with temperature in the strength of the chamber materials requires that the specimen region only be heated while the anvils and surrounding chamber are kept cool. The main problem is then that of insulation of the specimen together with the prevention of reaction between the pressure transmitting medium and the specimen container, thermocouple and heater coil. Heating is normally effected by passing a current through the specimen or, more usually, through a tube or coil surrounding it. The low resistance of the heating element requires high currents, typically 500–1500 A at 1–3 V. The volume and friction changes caused by increase in temperature tend to reduce the pressure, and compensation for such reductions is difficult because of the absence of a suitable sensing element which may be located close to the specimen.

The measurement of high pressures presents quite severe experimental problems especially at higher values where free-piston and strain gauges are inappropriate. The change in resistivity of manganin wire has been extensively used for pressure measurements but these are inaccurate when pyrophyllite or a similar material is used to transmit pressure, since these media are not perfectly hydrostatic. Calibration of the apparatus using tabulated transformations, such as that of Bi III-V at 77 kbar, is then necessary. The measurement of high pressure has been discussed, for example, by Munro (1963).

Temperature measurement at high pressure is complicated by the effect of pressure on the e.m.f. of the thermocouple. Corrections for Pt/Pt–10% Rh, Pt/Pt–13% Rh, chromel-alumel and iron-constantan thermocouples have been tabulated by Hanneman and Strong (1965). The correction for a Pt/Pt–10% Rh couple at 1300°C and 50 kbar exceeds 50°C. Independent measurements, with some deviations, have been reported by Peters and Ryan (1966) and sources of error in thermocouple measurements are discussed by Hanneman and Strong (1966). In general the measured e.m.f. will vary according to the pressure variation along the thermocouple wire, so that the apparent temperature will change with the design of the cell unless careful precautions are taken.

One problem in crystal growth at high pressures still to be solved in order to obtain large crystals of high quality is stirring. According to the discussion in Chapter 6 and Section 7.2.7 on the effect of stirring on the maximum stable growth rate the duration of experiments (for diamonds of 1 carat, about a week) could be shortened or the crystal size increased if forced convection could be applied. Stirring is particularly required in high pressure systems where natural transport phenomena are retarded by the densification.

7.2.9. Typical procedures for growth of oxides and chalcogenides

In this section detailed instructions are given for the preparation of some representative materials from high-temperature solutions. These "recipes" are intended for beginners in the field as illustrations of the practical procedures which are used. The complexity and the estimated price for apparatus and chemicals increase from the first to the last example; for instance the cost of apparatus and crucibles is about 200 dollars for the first experiment and up to 100 times more for the last experiment while the price of chemicals will vary from 5 dollars to several hundred dollars, depending upon the purity used.

A. *Sodium chromium sulphide*, $NaCrS_2$. 4.5 g coarse chromium powder is placed on the bottom of an unglazed ceramic crucible of about 6 cm diameter, and covered with 10 g sulphur and 60 g $Na_2S \times 9H_2O$. The crucible is closed by a 2 mm thick disc of ceramic wool, which just fits into the crucible, and covered with a ceramic lid. It is placed in the simple

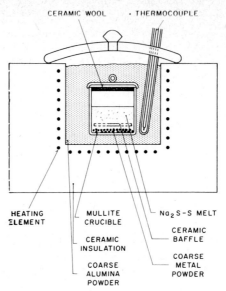

FIG. 7.48. Arrangement of crucible in muffle furnace for crystallization of metal sulphides from alkali polysulphide melts (Scheel, 1974).

muffle furnace shown in Fig. 7.48. The space below, on the sides and above the crucible is filled with coarse alumina powder in order to reduce the oxidation of melt by air. During heating to $1000 \pm 30°C$ (in about three hours) excess sulphur evaporates and removes residual oxygen and water by reaction to SO_2 and H_2S, respectively, and this requires that the furnace is placed in a hood with good ventilation. The temperature is set by a variable transformer, and is reduced by about 25°C per hour by means of a motor-driven variac. At 300°C the power is shut off, and after cooling to room temperature the sodium polysulphide solvent is dissolved in water leaving hexagonal plates of $NaCrS_2$ up to 1 mm thickness and 2 cm diameter. The aqueous solution of polysulphide is poisonous and should be neutralized by an aqueous $ZnCl_2$ solution or should be reacted with $KMnO_4$ solution.

This simple sodium polysulphide process can be used for several other binary and more complex metal sulphides (Scheel, 1974), such as $NaInS_2$, $KCrS_2$, $KFeS_2$, CdS, ZnS, PbS, FeS_2, CoS_2, NiS_2, MoS_2, Cu_3VS_4, CuS, α-MnS, HgS etc., with modified temperature programmes. In order to control nucleation and to achieve stable growth it is necessary to prevent evaporation of sulphur and to apply slower cooling rates. In the simple arrangement shown in Fig. 7.48 a ceramic baffle has been found to reduce the rate of reaction and transport and so to yield larger crystals.

B. *Barium titanate*, $BaTiO_3$. 37 g $BaTiO_3$ and 90 g anhydrous KF are weighed into a 100 cm³ platinum crucible which is covered by a platinum lid and placed into a horizontally loaded resistance furnace at 1100°C. This temperature is held for 8 hours after which the temperature is lowered by 15° h⁻¹ to 900°C. The crucible is then removed from the furnace and the liquid portion poured off (in a well ventilated room). The crucible is then returned rapidly to the furnace and cooled to room temperature; the crystals are freed from the remaining solidified KF by immersing the crucible in hot distilled water. Typical "butterfly" twins, as described in Section 5.5.2, of up to 1 cm size are obtained (Remeika, 1954; Remeika *et al.*, 1966). Undissolved $BaTiO_3$ is required for genesis of the butterfly twins, and the cooling rate has to be adjusted to prevent excessive spontaneous nucleation and to preserve the platy habit. The poisonous flouride vapours require the use of a ventilated hood during growth. A modification of this Remeika process to grow $BaTiO_3$ was published by Bradt and Ansell (1967).

C. *Spinel*, $MgAl_2O_4$. 80.6 g MgO, 204 g Al_2O_3, 1500 g PbF_2 and 10 g B_2O_3 are premelted in a 500 cm³ platinum crucible at 900°C, then a further 600 g PbF_2 is added. The crucible is sealed by a platinum lid which has at its centre a hole of 6 to 9 mm diameter which determines the evaporation rate of the solvent. The crucible is then placed on a zirconia ceramic plate into a silicon carbide muffle furnace which is heated to 1200°C ± <0.5° in 8 hours, and kept at that temperature for two weeks which is sufficient time to evaporate most of the solvent. After cooling to room temperature the crystals are removed mechanically and residual flux is removed by hot dilute nitric acid. Typically octahedra of 1 to 2 cm size may be obtained (Wood and White, 1968). It should be noted that the evaporated PbF_2 is poisonous and heavily corrodes furnace ceramics and heating elements.

D. *Gadolinium aluminate*, $GdAlO_3$. 264 g Gd_2O_3, 120 g Al_2O_3, 840 g PbO, 440 g PbF_2, 48 g B_2O_3 and 12 g PbO_2 are premelted in a new 500 cm³ platinum crucible (previously cleaned and chemically polished with a potassium pyrosulphate melt) of a shape shown in Fig. 7.36(c), then 400 g PbF_2 is added. The platinum lid with a hole of about 0.1 mm is sealed to the crucible by argon-arc welding. The crucible is placed into a Superkanthal muffle furnace and is centred on a ceramic holder which is mounted on a rotatable ceramic tube as shown in Fig. 7.44(b). The temperature of the furnace room is brought to 1290°C and is maintained for 15 hours during which time the accelerated crucible rotation technique (ACRT)† is applied using the cycle of Fig. 7.44(a): C with a maximum

† Furnace and ACRT mechanism are available from Käsermann & Sperisen, Biel/Switzerland and from Eurotherm Schweiz, Glattbrugg ZH, Switzerland.

rotation rate (in both directions) of 90 r.p.m. Localized cooling at the centre of the crucible bottom is now provided by a constant air stream, and the furnace temperature is reduced by 1° per hour to 1200°C, with a maximum ACRT rotation rate of 70 r.p.m. The temperature is raised to 1270°C and held at this temperature overnight, then the temperature is lowered by 0.3°C per hour to 1150°C and then by 0.5°C per hour to 900°C. During the whole cooling time the ACRT action is applied with a maximum rotation rate of 70 r.p.m. and a period of about two minutes. At 900°C the ACRT mechanism is stopped, the crucible is taken out of the furnace, and the excess solution is poured off after two holes have been punched into the lid. Then the hot crucible is quickly placed into a kanthal furnace at 850°C in order to prevent cracking of the (generally) one to three large crystals formed (see Fig. 7.45a, b) and this furnace is then slowly cooled to room temperature. After cutting the crucible rim the residual flux is dissolved in hot dilute nitric acid, and the crystals are mechanically moved from the crucible.

High precision and stability are required of the temperature controller and programmer and the application of the PtRh6/PtRh30 thermopile of Scheel and West (1973) helps to achieve large crystals. Also the air flow for localized cooling has to be well regulated, otherwise nucleation control and continuous stable growth (except for a short dendritic growth period at the beginning) may not be obtained (Scheel and Schulz-DuBois, 1971; Scheel, 1972; Scheel and Elwell, 1972, 1973).

E. *Yttrium iron garnet*, $Y_3Fe_5O_{12}$(YIG). A composition in mole % of 36.3PbO, 27.0PbF_2, 5.4B_2O_3, 20.78Fe_2O_3 and 10.42Y_2O_3 and of approximately 1 kg weight is premelted in a crucible shown in Fig. 7.36(e). A lid containing a narrow S-shaped platinum tube for pressure release is welded to the crucible which must be sealed by welding. A seed crystal is fixed to the centre of the lid. The crucible is mounted into a ceramic arrangement which can be inverted as described by Bennett (1968) and by Tolksdorf (1968) and which additionally can be rotated around a vertical axis in order to achieve stirring by ACRT after the horizontal axial holders have been withdrawn, as discussed by Tolksdorf and Welz (1972). The initial arrangement is shown in Fig. 7.9(a). After heating overnight at a temperature of about 1250°C the crucible base is locally cooled by an air flow of about 160 1 per hour and the temperature is reduced to 1180°C. Then the temperature is slowly reduced by 1°C per hour to 1100°C at which temperature the solution is saturated and spontaneous nucleation occurs (Fig. 7.9b). The crucible is smoothly turned into the position of Fig. 7.9(c) so that the seed crystal is immersed in the saturated solution. The temperature is now reduced at the same rate to 950°C with the cooling air flow maintained on the region now in contact with the seed crystal. The solution is separated from the grown crystals (typically 3 cm in size) by inverting the crucible again into the position of Fig. 7.9(d), and the furnace is cooled to room temperature at 50°C per hour.

During the cooling from 1100° to 950°C, stirring is achieved by the accelerated crucible rotation technique described in detail in the previous example. In addition to the production of YIG crystals this process might be useful when crystals of the same material have to be grown repeatedly to a size larger than 1 cm³ in pilot plant production. The advantages are the Bennett–Tolksdorf nucleation control, the ACRT stirring of Scheel (1972) and the possibility of using the high temperature solution several times, thus saving chemicals.

F. Nickel ferrite, $NiFe_2O_4$. Well-mixed powders of 179.4 g $BaCO_3$ and 43.2 g dry B_2O_3 (to give $BaO \times 0.62B_2O_3$ solvent) are placed in a weighed platinum crucible of 7 cm diameter and 7 cm height which is fitted with a lid and slowly heated to 1000–1100°. An automatic temperature program with a heating and cooling cycle of 24 hours is useful. Several fusions may be required before enough solvent has been added to produce a melt depth of about 3 cm.

After cooling to room temperature the crucible is weighed in order to obtain the mass of borate solvent. NiO and Fe_2O_3 are added in 1:1 molar proportions such that these oxides form 30% of the total weight of solution. This concentration provides several grams of $NiFe_2O_4$ to act as nutrient. The crucible without a lid is placed into the crystal growth furnace in a region such that the base is hotter than the melt surface. A furnace having a horizontal division and with the two sections heated independently is preferable.

The seed crystal of [111] orientation is tied firmly onto a platinum seed holder which can be cooled, rotated and withdrawn at adjustable rates (see for example the apparatus of Belruss *et al.* (1971), but also a commercial Czochralski puller with a modified seed holder would suffice).

The crucible is heated to 1320°C for a day with the seed just above the melt, then cooled to about 1250°C with the temperature difference ΔT across the solution adjusted to 10°C; direct control of this difference by opposed thermocouples is preferable (one advantage of gradient transport is that the exact growth temperature need not be accurately set and the liquidus temperature not accurately known).

The seed crystal is inserted just below the surface of the melt and rotated at 60 r.p.m. Cooling of the seed is desirable by water or air through the seed holder, typically by 5–10 l/min of air. The crystal is observed intermittently, preferably in reflected light. If it grows outward relatively quickly it is raised at about 2 mm/day. The growth speed is regulated by increasing or decreasing ΔT. After a few days the crystal is slowly withdrawn from the solution and cooled over several hours in the furnace or by gradual removal to a cooler region (Smith and Elwell, 1968; see also Linares, 1964; Kestigian, 1967; Whiffin, 1973).

The above six recipes have been tested on several occasions, and the first two procedures are suggested for experimental courses in crystal growth. A collection of crystal growth procedures checked in independent laboratories, similar to "Organic Syntheses" and "Inorganic Syntheses", would be very helpful for many crystal growers and especially for occasional crystal growers such as physicists who require crystals of one or a few materials for research purposes. Such a compilation of "recipes" would prevent much duplication of work in the field of crystal growth.

7.3. Special Techniques, Specific Problems

7.3.1. Crystal growth from HTS at medium and high pressure

The study of high pressure phases and phase transformations has grown rapidly with the development of high pressure apparatus, particularly due to the pioneering work of Bridgman. However, the main effort has been devoted to transformations in the solid state, although an appreciable amount of research has also been carried out on hydrothermal synthesis of minerals at medium pressures and temperatures (500–3000 atm, 200–800°C). The latter work has resulted in the large scale synthesis of quartz and recently of ziron but, as was mentioned in Chapter 1, crystal growth by the hydrothermal method is beyond the scope of this book.

Relatively little crystal growth at medium or high pressure has been effected in non-aqueous solvents, with the notable exception of diamond, which is now produced very widely. The reasons for the neglect of HTS growth under pressure are mainly that:

(a) The experimental effort required is extremely great, and increases very rapidly with the pressure p, and correspondingly with temperature T and solution volume V.
(b) The difficulty of finding solvents in which the solute has an appreciable solubility, but for which corrosion-resistant containers exist, increases with p and T.
(c) The difficulty in controlling the growth parameters to the degree necessary for the synthesis of good quality crystals increases with p and T. At the very high pressures used for diamond synthesis, stirring is still impossible.

In this Section the growth of diamond crystals will be briefly discussed and a few examples will then be given of other crystals which have been grown from HTS at high pressure.

A. Crystallization of diamond. Several tons of diamond are now produced per year for a variety of applications, especially for use as an abrasive. The bulk of the diamond is in the size range from 1 to $100 \mu m$ and the price of synthetic diamond grit is comparable with that of the natural mineral.

Although the synthesis of diamond has attracted the attention of scientists since the last century, a reproducible synthesis was reported only in 1955. The use of metallic solutions at 50–60 kbar and 1400–1600°C is now practised in several centres and has remained the only commercial process, in spite of studies of alternative techniques such as growth from the pure melt (Bundy, 1963), at about 4000 K and 140 kbar, and by metastable epitaxial deposition from the vapour phase at atmospheric pressure (Angus *et al.*, 1968).

The apparatus used for the solution growth experiments of diamond is of the "belt" type (Hall, 1960), the high pressure region of which is shown in Fig. 7.47. Potential solvents have been discussed by Wakatsuki (1966) and especially by Wentorf (1966). Non-metallic solvents such as Cu_2O, CuCl, AgCl, ZnS, CdO, FeS and silicate melts containing OH^- ions were found to produce only graphite, whereas several transition metals such as Fe, Ni, Co, Cr or Mn or alloys of refractory transition metals with Cu produce diamond under the same conditions of temperature and pressure. This observation was attributed by Wentorf to the low solubility ($<1\%$) of carbon in the former group compared with that in the metallic solutions. In addition the nature of the dissolved carbon could play a vital role since Wentorf (1966) showed that carbon has a positive charge in metal solutions, a negative charge in CaC_2 or Li_2C_2 and is neutral in the compound solvents.

The phase diagrams Fe–C, Ni–C and Fe–Ni–C at 57 kbar have been determined by Strong and Hanneman (1967) and Strong and Chrenko (1971) and are shown in Fig. 7.49(a)–(c). In the Fe–C system, crystallization of diamond is restricted to the region between the diamond-graphite equilibrium at 1830 K and the melting point of Fe_3C at 1688 K. The Ni–C eutectic lies at a lower carbon concentration and at a higher temperature than in the Fe–C system. The ternary Fe–Ni–C system has several advantages over the binary systems since the diamond-solution liquidus extends over a wider range of temperature as shown in the 1400°C section. A similar advantage has been reported for the Fe–Al–C system (Strong and Chrenko, 1971). It is interesting that solvents which form compounds with

o

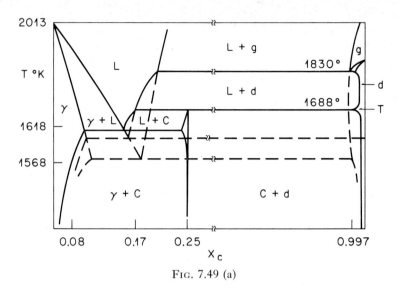

FIG. 7.49 (a)

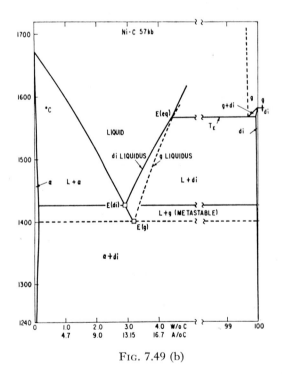

FIG. 7.49 (b)

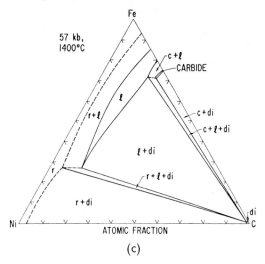

(c)

FIG. 7.49. Phase diagrams at 57 kbar applied to diamond synthesis. (a) Iron-carbon with X_c the fraction of carbon. (b) Nickel-carbon and (c) iron–nickel-carbon (Strong and Hanneman, 1967; Strong and Chrenko 1971).

the solute at lower temperatures (e.g. Fe_3C, Ni_3C) are the most suitable for diamond growth, which supports the principles discussed in Chapter 3.

Use of the phase diagrams has facilitated the growth of gem-quality diamonds up to 5 mm in diameter (0.2 g) by Wentorf (1971). A gradient transport technique was used, as shown in Fig. 7.50, at a constant pressure in the range 55–60 kbar with diamond nutrient at a temperature of about 1450°C and diamond seed crystals at about 1420°C. The use of seeds is essential since kinetic factors otherwise favour graphite crystallization, and

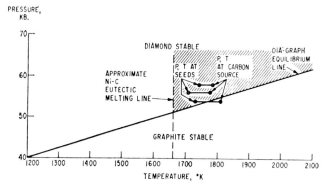

FIG. 7.50. Portion of the pressure-temperature phase diagram of carbon (Wentorf, 1971).

it was found advantageous for the growth of large crystals to locate crystals at the lower end of the pressure cell. Diamond crystallites formed by spontaneous nucleation then tend to rise towards the hotter region and are re-dissolved. With a temperature gradient of about 100°C cm^{-1} the flow of solute to the seed crystals was estimated to be of the order of 10^{-4} g cm^{-2} s^{-1}. If this flow is too rapid unstable growth will occur and graphite may nucleate and grow as large flakes. The maximum stable growth rate was found to be

$$v = \frac{1}{2Yb},$$

where Y is the average crystal diameter in mm and b has a value of 2.5 hr mm^{-2}. Thus the time required to grow a 5 mm diameter crystal is in the region of a week, which is similar to that required for a typical experiment at normal pressure (see Section 6.4).

The growth mechanism is reported to be by spreading of layers nucleated at corners and edges of the crystal (Wentorf, 1971) or at growth spirals of which a typical one published by Strong and Hanneman (1967) is shown in Fig. 7.51. Further observations on surface features and morphology have been reported by Bovenkerk (1961). The habit was observed to depend on the concentration of nitrogen in the solvent; melts containing nitrogen favour the development of {100} and {111} with minor {110} and {113}, while in N-free solutions the {110} and {113} faces, and sometimes {117}, frequently have slower growth rates than the cube and octahedron faces. The dependence of the habit of diamond crystals on the growth temperature and pressure has been discussed by Litvin et al. (1968) and by Bezrukov et al. (1969). Naturally facetted synthetic diamonds of gem quality and weighing about 0.2 g (1 carat) are shown in Fig. 7.52. The nitrogen impurity reduces the overall growth rate by a factor of about 1/3.

The nitrogen content tends to produce a yellow colouration and a resistivity greater than $10^{12}\Omega$ cm, while boron doping leads to a blue colour and a low resistivity of $10-10^6\Omega$ cm. It was recently established (Collins, 1972) that boron rather than aluminium is responsible for the blue colour of natural as well as synthetic diamonds (Chrenko, 1971). Most natural and synthetic diamonds show facet-related incorporation of impurities as discussed in the review of diamond crystal growth by Strong and Wentorf (1972). Although the results of diamond crystal growth are spectacular, the cost of the larger crystals is much higher than that of natural gems. The genesis of the very large naturally occurring diamonds of high quality, such as the Cullinan with a weight of 3616 carats, is still unexplained.

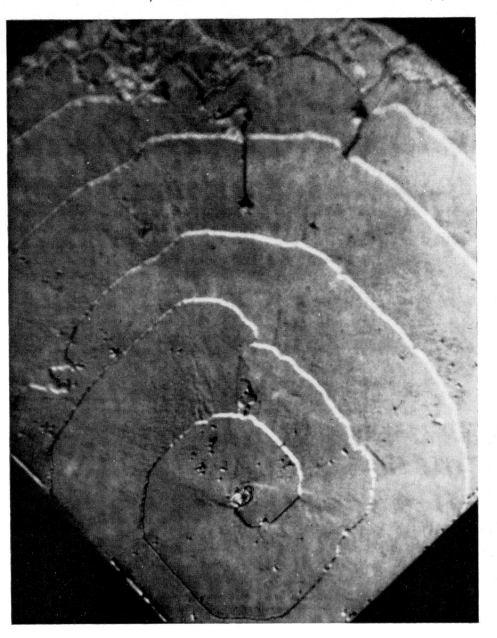

FIG. 7.51. Growth spiral on a (100) face of diamond grown in nickel (Strong and Hanneman, 1967).

FIG. 7.52. Synthetic diamond crystals of gem quality with natural {100} and {111} faces (Courtesy R. H. Wentorf, General Electric Comp.).

B. Other crystals grown at medium and high pressures. The preparation of high pressure phases is of interest to chemists and mineralogists and also offers the materials scientist the opportunity to obtain particularly novel compounds or phases. In addition to the hardest materials diamond and borazon (cubic boron nitride), interesting superconducting compounds have been synthesized as an example of the variety of compounds and modifications the properties of which are related to their highly condensed structure. The crystal chemical aspects of high pressure phases have been reviewed by Neuhaus (1964), Kleber and Wilke (1969), Klement and Jayaraman (1967), Goodenough *et al.* (1972), Rooymans (1972) and by Joubert and Chenavas (1975).

In several high pressure transformations it has been shown that the addition of small amounts of "catalysts", "mineralizers" or "fluxes" has the effect of lowering the transition pressure and temperature or of enhancing the rate of reaction. The transformation of graphite to diamond discussed above will occur only in the presence of the metal solvent and, as a further example, the transformation of quartz into coesite, which occurs without "catalysts" at about 90 kbar and 2000°C, will proceed at 20–35 kbar and at 500 to 750°C with the addition of H_2O, H_3BO_3 or $(NH_4)_2HPO_4$. In reality the additions act as high temperature solvents and thus increase the mobility of the crystal constituents.

The growth of crystals of the high pressure phases is highly desirable for structural studies and for determining the physical properties, but it has been achieved only in relatively few cases. Examples are given below.

Cubic boron nitride ("Borazon", BN): Cubic BN is the III-V analogue of diamond and seems to have a slightly higher hardness than diamond, and also has the advantage of a higher resistance to oxidation. The first successful crystallization of cubic BN was reported by Wentorf (1957, 1961, 1962). He used Li, other alkali and alkaline earth metals and nitrides, Sb, Sn and Pb as solvents at pressures of 45–64 kbar and temperatures of 1500–1900°C and postulated the formation of a $Li_3N \times 3BN$—complex acting as solvent Depending on the growth conditions and on the impurity content small white, yellow or black crystals were prepared. DeVries and Fleischer (1972) determined phase equilibria pertinent to the growth of cubic BN, and Bezrukov *et al.* (1968) and Matecha and Kvapil (1970) studied the perfection and the growth phenomena of high pressure-grown diamond and borazon crystals.

Boron phosphide BP: Boron phosphide decomposes under atmospheric pressure at 1200°C, which is well below its melting point. It has been crystallized by slow cooling from solution in nickel or iron (Stone, 1959) and from solution in black phosphorus at temperatures of 1200–1300°C and a pressure of 15 kbar by Niemyski *et al.* (1967). The growth process appeared to be by isothermal transport of BP in solution with boron located at the base of the graphite container acting as nutrient. Baranov *et al.* (1967) used Cu_3P as solvent and grew crystals up to $4 \times 3 \times 2$ mm in size over a period of one month.

Cobalt diphosphide CoP_2 has been crystallized from germanium at temperatures between 800° and 1200°C at 65 kbar by Donohue (1972). The same author crystallized $MnCoP_4$ from excess phosphorus at 1200°C and 65 kbar.

Crystallization of compounds of high oxidation states at high oxygen pressures: The synthesis of garnets and ferrites free from divalent iron at a few atmospheres oxygen pressure has been mentioned in Section 7.1.5. Although considerable research has been done on "dry" and on hydrothermal systems (see for instance Roy, 1965) high pressure crystal growth to obtain crystals of high valency states does not yet seem to have been given much attention, with the exception of CrO_2.

Ferromagnetic CrO_2 is at atmospheric pressure only stable up to 300°C, but at pressures of 40 kbar the stability field extends to about 1400°C according to DeVries (1967) and Fukunaga and Saito (1968). DeVries (1966) grew CrO_2 layers from CrO_3 by liquid phase epitaxy at high pressure, and Chamberland (1967) grew crystals of CrO_2 up to 0.4 mm size from various fluxes at 60–65 kbar and 500–1000°C, the best flux being Na_2CrO_4.

Crystallization of compounds with high coordination numbers at high pressures: Quite a number of compounds are likely to form denser modifications at high pressures, and the high pressure synthesis of several new compounds has been reported. But here again the limited volume available in high pressure apparatus and the experimental difficulties have prevented a wide activity in this most interesting field. Examples of successful crystal growth experiments are given in the following.

Lithium metaborate $LiBO_2$ has at normal *p–T* conditions a monoclinic chain structure. Marezio and Remeika (1965) reported a tetragonal high pressure phase (γ-$LiBO_2$) with a zinc blende-like structure and were able to grow crystals of this compound from LiCl flux at 15 kbar and 950°C (1966). The solution was held for one hour's growth time in a small tantalum container in a furnace assembly described by Kennedy *et al.* (1962). The solvent was dissolved in methanol, and small colourless crystals were obtained. With a LiCl-rich solution another phase was synthesized which was face-centred cubic with a lattice constant of 12.13 Å and which could be a lithium chloroborate of boracite type although no chemical analysis was made.

In order to study α-Ga_2O_3 as a possible laser host for chromium analogous to the isostructural corundum, crystals of this high pressure modification of Ga_2O_3 have been grown by Remeika and Marezio (1966) although Foster and Stumf (1951) succeeded in the preparation of metastable α-Ga_2O_3 at normal pressure. In a similar experimental arrangement as mentioned for the $LiBO_2$ synthesis Remeika and Marezio used NaOH as solvent, and 44 kbar and 1000°C were applied during one hour. After extraction of the small colourless crystals two kinds of growth habit resembling the hexagonal plates and the rhombohedra of α-Al_2O_3 were observed. Chromium doping resulted in small green rhombohedra of which the difference in colour compared to red ruby was considered to be caused by a difference in crystal field splitting due to the significant size difference between the Al^{3+} and Ga^{3+} ions.

During studies on the effect of high pressures on rare-earth garnets it was found that some of them decomposed to the rare-earth aluminates of perovskite type and the sesquioxides of corundum type (Marezio et al. 1966a). Using this reaction in the presence of a NaOH flux at 70 kbar and 1000°C $YGaO_3$ and $YbGaO_3$ could be crystallized in one hour as small transparent crystallites (Marezio et al. 1966b). Recently Dernier and Maines (1971) reported on the high pressure synthesis and crystal data of the rare earth orthoaluminates.

Further examples of crystal growth at high temperatures and high pressures have been listed by Rooymans (1972), Goodenough et al. (1972) and by Joubert and Chenavas (1974).

7.3.2. Undesirable crystal growth from high temperature solutions

In a number of processes involving high temperature solutions there is undesired crystallization. In metallurgy and in the ceramic and cement industries the development of large crystals or the crystallization of undesired phases might degrade the mechanical and thermal properties of the product, and devitrification in the glass industry makes the products worthless, unless devitrification is achieved intentionally as in the fabrication of Pyroceram†-type ceramics.

A common problem in nuclear-reactor technology occurs in the cooling circuits (when low-melting metals and alloys such as sodium or sodium-potassium are used as coolants) as well as in the recently developed liquid uranium-bismuth alloy fuelled homogeneous reactor. A solid-liquid interaction (corrosion) occurs in the hot region in the reactor, and crystallization of the dissolved container material or of its compounds with the liquid metal occurs in the cooler region (Weeks and Gurinsky, 1958).

Another example of unwanted crystallization is the crystal growth of whiskers by the VLS mechanism which occurs occasionally in electric installations when the appropriate conditions are accidentally fulfilled.

† Trademark of Corning Glass Works, Corning, N.Y.

However, it is not intended to give a broad survey of the topic of this section. The above examples merely illustrate where unwanted crystal growth processes might occur and how important it might be in certain cases to prevent or to control them. In the following discussion a few examples of undesired effects observed in the practice of flux growth are briefly mentioned.

Unwanted crystal growth obviously occurs when the experiment results either in very small crystal or in crystals of an undesired phase. The former problem has been discussed in Sections 3.4.3, 3.6.2., 7.1.1 and

FIG. 7.53. Accidentally formed crystals of lead feldspar (Scheel, 1971).

7.2.7, the latter is a matter of the stability fields as discussed in Section 3.8.5. Another example of unexpected flux growth occurred during the preparation of $ZnAl_2O_4$ crystals from PbO—PbF_2—B_2O_3 solvent: the platinum crucibles covered with a lid (but not sealed) were placed into mullite-type ceramic crucibles with ceramic lids. After the slow-cooling experiment and removal from the furnace the underside of the ceramic lid showed elongated crystals which were identified as solid solutions of the feldspars $KAlSi_3O_8$ and $PbAl_2Si_2O_8$, the first synthesized Pb-containing feldspar crystals (Scheel, 1971), see Fig. 7.53. Wanklyn and Hauptman (1974) reported that silicate crystals grew in solutions for example of Er_2O_3/MoO_3 in PbO/PbF_2 when growth occurred in a silli-manite muffle furnace.

Crystals of up to 5 mm of platinum and of several platinum-lead compounds have been observed in flux growth experiments with corrosive lead- and fluorine-containing melts, especially when the conditions were slightly reducing. In general these crystals are not wanted, especially when they intergrow with the required crystals. Therefore a slightly oxidizing atmosphere or a redox buffer in the melt, such as V_2O_5, are used to minimize flux growth of Pt-containing crystals.

Most experimentalists will have observed examples of unwanted crystal growth from high-temperature solutions. Identification of the crystalline phase, for instance by X-ray powder diffraction, is worthwhile, as in some cases interesting or novel materials may have crystallized. On the other hand an investigation of the crystallized phases may indicate the parameters to be modified in order to prevent the unwanted crystallization.

7.3.3. Unexplored techniques and areas

Nearly all areas of crystal growth from high-temperature solutions are not well understood, thus the whole field might still be said to be "unexplored". However, in the following discussion only examples of those (experimental) topics will be briefly discussed which are or might become of general interest.

A. New materials. A number of materials with interesting properties for research and with potential applications exist which have not yet been grown reproducibly as crystals of sufficient size and quality by crystal growth from melt or from vapour. Several of these could probably be grown from high-temperature solutions, for example BeO†, $BiFeO_3$, SbSI, SiC and TaC. An interesting example of the success of a large effort is the growth of colourless diamond crystals of gem quality and up to 1 carat size as discussed in Section 7.3.1.

Novel materials with interesting properties may still await discovery, because of all possible compounds A_xB_y of elements A and B only approximately 50% are known, and of the compounds $A_xB_yC_z$ the estimate of Lydtin (1972) is that 1% are known, and of quaternary compounds ABCD a small fraction of a percent is known. Thus many compounds are still to be discovered, for instance by investigation of the corresponding phase diagrams or by isomorphous replacement in known structure types. In addition solid solutions might be prepared to optimize the required properties. A class of compounds which has not yet found wide interest is that containing elements in unusual valency states, partially because of experimental difficulties of preparation. Also extreme conditions (pressure, temperature) could produce interesting new materials.

† The reproducible growth of large BeO crystals has recently been reported by S. B. Austerman (private communication).

B. New solvent systems. As in the case of new materials, there are likely to be novel solvent systems for old and new materials which have advantages over the solvents now in use. The approach is to achieve optimum characteristics with respect to solubility and corrosivity, viscosity, metastable region, stable growth, and purity of the crystals grown from them. In developing new solvents the aspects of compound and complex formation as discussed in Chapter 3 might be useful. With increasing knowledge of solutions and of thermodynamic properties of the components the choice of solvents might be made on theoretical grounds.

It is suggested to use the "hydroflux" method for growth of various oxide compounds, hydroflux being a combination of the hydrothermal and the flux techniques. In natural crystallization water is known to be an excellent modifier of solution properties and accordingly one would expect its beneficial influence on certain solvent systems at the medium pressures required. An approach in this direction is the use of highly concentrated salt solutions, of about 50 mole %, in hydrothermal crystal growth.

C. High temperatures, high pressures. Extreme conditions such as very high temperatures and high pressures are difficult to achieve, especially in the volumes required to grow sufficiently large crystals from high-temperature solutions. With the exception of diamond and boron nitride, not much work has been done either at temperatures above 1500°C or at pressures exceeding a few atmospheres. Examples where high temperatures have been required to grow extremely refractory crystals are the growth of SiC, WC and the europium chalcogenides. Recently crystals of LaB_4, which is incongruently melting at around 1800°C, have been prepared by Deacon and Hiscocks (1971) by slow cooling from 1700°C and are reproduced in Fig. 7.54. Although the experimental difficulties increase enormously with temperature and especially with pressure, significant new results are to be expected from studies in this area.

D. Electrolysis; an electric field as a driving force for supersaturation. Crystal growth by electrolysis in molten salts seems a promising area (see Section 7.1.2), especially for the growth of refractory compounds like borides and pnictides and of compounds with defined (lower) valency states. However, the literature indicates that many compounds have been prepared by high-temperature electrolysis, partially on a production scale, but relatively little systematic work has been done on the application of electrolysis to the growth of large high-quality crystals.

Another application of electric fields across ionic solutions might be considered, namely as an easily controlled driving force for supersaturation, even in the absence of electrolytic reduction or oxidation. The growth rate may be controlled by varying the electric current through the

Fig. 7.54. Lanthanum tetraboride crystal of 5.5 mm length grown from lanthanum solution by cooling from 1700° to 1200° (Deacon and Hiscocks, 1971).

solution in an isothermal melt in a uniform temperature zone as was demonstrated by Elwell (unpublished) for the growth of Fe_2O_3 from BaO/B_2O_3. It is possible that an electrical driving force may lead to enhanced stability of growth compared with a thermal driving force. An additional use of an electrical driving force is in the provision of time markers for growth rate determinations, as in the experiments of Lichtensteiger et al. (1971).

E. Growth of special crystal shapes. This topic has been extensively discussed in Chapter 5. Special crystal habits are of advantage for certain applications and physical measurements. For example the $BaTiO_3$ butterfly twins consisting of two thin single-crystalline platelets, as grown by Remeika (1954), were important for a variety of measurements and contributed significantly to the development of ferroelectrics. In the case of $BaTiO_3$ the twin-plane reentrant edge (TPRE) growth on undissolved crystallites was responsible for the formation of platelets of barium titanate which otherwise grows as cubes or similar compact shapes. Other examples of TPRE growth of platelets are alumina Al_2O_3 (Wallace and White, 1967) and beryllia BeO (Austerman, 1964). It is probable that under appropriate conditions many more compounds could be grown as plates by a twin mechanism, and a number of examples of semiconductors grown by TPRE were given by Faust and John (1964).

Crystalline fibres or whiskers of various materials (SiC, Al_2O_3) are applied as components of high-strength materials. Until now most whiskers have been produced from the gas phase via a liquid droplet by the VLS mechanism but they could also be crystallized from bulk high-temperature solutions. For example, crystalline fibres of TiO_2, ZrO_2 and $ZrSiO_4$ have been prepared from borate melts (Russell et al., 1962), and Morgan and Scheffler (1965) showed how these fibres could be separated from the solution.

On the other hand there are a variety of crystals which grow in inconvenient shapes like needles or plates and which are required as equidimensional bulk crystals, such as TiO_2, YVO_4, BeO, SbSI and SiC. There is as yet no simple general solution to this problem. The approach until now has been to modify the solvent, growth temperature and supersaturation, but it might be preferable to control the flow pattern with respect to crystal orientation in order to influence the crystal habit.

F. Liquid phase epitaxy. This topic is discussed in Chapter 8, and here it will be briefly mentioned that LPE is a still unexplored field. One of the reasons is that since the first report on LPE by Nelson (1963) this technique has been little studied as such since most LPE work has been concentrated on fabrication of specific devices. A comparison of LPE and chemical vapour epitaxy has recently been published by Minden (1973), and the first proposal for LPE as a commercial process has been made by Bergh et al. (1973). However, it is a long way from growth on a single substrate to the batch process proposed by Scheel and Schulz-DuBois (see Section 8.4.2) and to a continuous flow process as discussed in Section 8.6.

G. Growth of large crystals. The parameters which are important for growth of large crystals have been discussed in this chapter, and crystals of several cm in diameter and weighing several hundred grams have been obtained of various compounds. However, only two materials are grown commercially as large crystals from HTS, namely magnetic garnets and emerald, the former as inclusion-free crystals 2–5 cm in size or with inclusion-free regions of 1–2 cm thickness. It is expected that large crystals could be grown if required by application of the technology which corresponds to commercial crystal growth from aqueous solutions. Seed crystals could be fixed on rotated seed holders or into rotated solutions, and the precise temperature control and programming as discussed in Chapter 6 and in Section 7.2.2 are required. For volatile solutions accelerated crucible rotation is the alternative stirring technique. It was shown throughout the book that it is now possible to grow large crystals of high purity and perfection, although the effort to achieve this goal varies for different materials.

H. *Mass crystallization*. The large-scale production of small crystals may be divided into two fields. In the first a large excess of solute, of material to be crystallized, is present with only a small amount of solvent whereas in the second field the mass of small crystals is precipitated from previously homogeneous solutions.

An example of the first field is the industrial crystallization of diamond to be used as abrasive: only small amounts of metals or alloys (Fe, Ni, Cr) as solvents are present, and accordingly the term "catalyst" has been frequently used for the metal solvent. In the high-pressure cell, under the temperature and pressure conditions required for diamond synthesis (see Sections 7.2.8 and 7.3.1), graphite or a carbon-containing compound recrystallizes through the thin solvent layer to form diamond.

A similar mechanism, crystal formation or recrystallization through a thin solution layer, occurs in many processes where materials are prepared at high temperatures. Examples are the fabrication of porcelain and cement or the recrystallization of cadmium sulphide phosphors at high temperatures by the addition of a small amount of a salt, usually sodium chloride. Generally the term "mineralizer" is used for the solvent which might remain as a component of the product as in the case of porcelain and cement or which might be subsequently dissolved, as is the NaCl after recrystallization of cadmium sulphide.

Solid state reactions used for synthesis of materials may not always be reactions in the solid state (by diffusion): the formation and crystallization of the new compounds frequently occurs in a liquid layer. This liquid might be a eutectic of the components and impurities which accelerates the rate of reaction significantly. On the other hand, small amounts of a suitable solvent might be deliberately added to the components of the required compound, as in the preparation of CdTe, GaP and other chalcogenides and pnictides at relatively low temperatures by the addition of the corresponding metal iodide (Kwestroo *et al.*, 1969; Kwestroo, 1972). Other examples are the preparation of metallic silicides and germanides by an amalgam method (Mayer *et al.*, 1967), of rare-earth nitrides by the amalgam method by Busch *et al.* (1970), of oxide compounds by the addition of salts (Wilke *et al.*, 1965; Wickham, 1970; Petzold *et al.*, 1971), of several sulphides by the addition of alkali polysulphides (Scheel, 1974), and of carbides by the additon of a metal bath (often 70 wt% Fe, 30 wt% Ni, called "menstruum") as described by Windisch and Nowotny (1972).

Other examples of industrial applications of mass crystallization from high-temperature solutions are the fabrication of Al_2O_3 and SiC whiskers by the VLS mechanism, and the electrolytic preparation of refractory materials (metals, borides, carbides, etc.) from molten salt solutions. In these cases the precipitation of the crystals or layers occurs from homogeneous solutions, the electric field being the driving force for supersaturation. Examples of mass crystallization from high-temperature solutions with solvent evaporation or with slow cooling as driving forces for supersaturation are not known to the authors although there are methods for the preparation of interesting materials which seem easier and better than conventional technology. As an example, tons of CdS phosphor and photoconductor are fabricated by a complicated many-step process (Weisbeck, 1964) and could be crystallized from alkali polysulphide melts (Scheel, 1974) with metallic cadmium or a Cd compound as starting material.

7.4. Summary

In this section no attempt is made to summarize the whole chapter but attention is drawn to a few major points.

The experimental techniques which appear to offer most promise for the growth of large, relatively perfect crystals are slow cooling in sealed crucibles stirred by the accelerated crucible rotation technique, top-seeded solution growth, and the travelling solvent zone method. The two latter methods are restricted to solvents of low volatility and this restricts their applicability. For growth in sealed crucibles, solvents in the system $PbF_2/PbO/B_2O_3$ have been particularly successful for refractory oxides which still form the most important class of materials crystallized from high-temperature solution.

Improvements in crystal size and quality can often be obtained by the use of seeds, and the introduction of the Bennett–Tolksdorf seeding technique is a valuable innovation. The use of temperature cycling and localized cooling to restrict nucleation are particularly worthwhile since the introduction of stirring by the accelerated crucible rotation technique.

Quantitative models are now available for growth by most of the techniques mentioned and the main criteria for the design of apparatus have been established. The necessity to maintain the growth rate below the maximum stable value has been stressed. Temperature stabilization is of particular importance in any experimental method.

Although this chapter has been concerned with the growth of bulk crystals, one prediction which can be made with confidence is that the emphasis will move in the direction of crystal growth of thin layers by liquid phase epitaxy.

References

Abernethy, L. L., Ramsey, T. H. and Ross, J. W. (1961) *J. Appl. Phys.* **32**, 376 S.

Addamiano, A. (1971) *J. Crystal Growth* **11**, 351.

Ahmad, I. and Capsimalis, G. P. (1967) In "Crystal Growth" (H. S. Peiser, ed.) Pergamon, Oxford.

Aidelberg, J., Flicstein, J. and Schieber, M. (1974) *J. Crystal Growth* **21**, 195.

Andrieux, J. L. (1929) Research on the Electrolysis of Metal Oxides dissolved in Anhydrous Boric Oxide and Borate Fluxes, Thesis, Paris.

Andrieux, J. L. and Marion, S. (1953) *Compt. Rend.* **236**, 805.

Andrieux, J. L. and Weiss, G. (1948) *Bull. Soc. Chim.* 596.

Angelici, R. J. (1969) "Synthesis and Technique in Inorganic Chemistry" Saunders, Philadelphia.

Angus, J. C., Will, H. A. and Stanko, W. S. (1968) *J. Appl. Phys.* **39**, 2915.

Arend, H. (1960) *Czech. J. Phys.* B **10**, 971.

Armstrong, W. M., Chaklader, A. C. D. and DeCleene, M. L. A. (1962) *J. Am. Ceram. Soc.* **45**, 407.

Aucoin, T. R., Savage, R. O. and Tauber, A. (1966) *J. Appl. Phys.* **37**, 2908.

Austerman, S. B. (1964) *J. Nucl. Mat.* **14**, 225.

Austerman, S. B. (1965) *J. Am. Ceram. Soc.* **48**, 648.

Austerman, S. B., Myron, S. B. and Wagner, J. W. (1967) *Carbon* **5**, 549.

Baker, H. D., Ryder, E. A. and Baker, N. H. (1953, 1961) "Temperature Measurement in Engineering" Vol. I (1953), II (1961) Wiley, New York.

Ballman, A. A. and Laudise, R. A. (1963) In "The Art and Science of Growing Crystals" (J. J. Gilman, ed.) p. 231. Wiley, New York.

Baranov, B. V., Prochukhan, V. D. and Goryanova, N. A. (1967) *Inorg. Materials USSR* **3**, 1477.

Barns, R. L. and Ellis, W. C. (1965) *J. Appl. Phys.* **36**, 2296.

Bartholomew, R. F. and White, W. B. (1970) *J. Crystal Growth* **6**, 249.

Bartlett, R. W., McCamont, J. W. and Gage, P. R. (1965) *J. Am. Ceram. Soc.* **48**, 551.

Belin, C., Brissot, J. J. and Jesse, R. E. (1972) *J. Crystal Growth* **13/14**, 597.

Bell, R. O., Hemmat, N. and Wald, F. (1970) *Phys. Stat. Sol.* (a) **1**, 375.

Belruss, V., Kalnajs, J., Linz, A. and Folweiler, R. C. (1971) *Mat. Res. Bull.* **6**, 899.

Bennett, G. A. (1968) *J. Crystal Growth* **3/4**, 458.

Bergh, A. A., Saul, R. H. and Paola, C. R. (1973) *J. Electrochem. Soc.* **120**, 1558.

Berkes, J. S., White, W. B. and Roy, R. (1965) *J. Appl. Phys.* **36**, 3276.

Berman, I. and Comer, J. J. (1969) *Mat. Res. Bull.* **4**, 107.

Berman, I. and Ryan, C. E. (1971) *J. Crystal Growth* **9**, 314.

Berry, R. J. (1966) *Metrologia* **2**, No. 2, 80.

Best, R. (1967) *Der Elektroniker* **4**, 203.

Bethe, K. and Welz, F. (1971) *Mat Res. Bull.* **6**, 209.

Bezrukov, G. N., Butuzov, V. P. and Korolev, D. F. (1969) *Growth of Crystals* **7**, 91.

Bezrukov, G. N., Butuzov, V. P., Nikitina, T. P., Fel'dchuk, L. I., Filonenko, N. E. and Khatelishvili, G. V. (1968) *Dokl. Akad. Nauk SSSR* **179**, 1326.

Blank, S. L. and Nielsen, J. W. (1972) *J. Crystal Growth* **17**, 302.

Blank, S. L., Hewitt, B. S., Shick, L. K. and Nielsen, J. W. (1973) Proc. AIP Conf. Mag. and Mag. Mat. No. 10, p. 256.

Block, B. (1930) In "Ullmann's Enzyklopädie der Technischen Chemie", p. 814. 2nd Edition, Berlin-Wien.

Bonner, W. A. and Van Uitert, L. G. (1967) *Mat Res. Bull.* **2**, 131.

Bonner, W. A., Dearborn, E. F. and Van Uitert, L. G. (1965) *Am. Ceram. Soc. Bull.* **44**, 9.

Bonner, W. A., Dearborn, E. F. and Van Uitert, L. G. (1967) In "Crystal Growth" (H. S. Peiser, ed.) 437. Pergamon, Oxford.
Bootsma, G. A. and Gassen, H. J. (1971) *J. Crystal Growth* 10, 223.
Bovee, B. A. (1953) *Am. Ceram. Soc. Bull.* 32, 26.
Bovenkerk, H. P. (1961) *Am Min.* 46, 952.
Bradt, R. C. and Ansell, G. S. (1967) *Mat. Res. Bull.* 2, 585.
Brauer, G. (1960/1962) "Handbuch der präparativen anorganischen Chemie" Enke, Stuttgart.
Brice, J. C. (1965) "The Growth of Crystals from the Melt" North-Holland, Amsterdam.
Brice, J. C. (1973) "The Growth of Crystals from Liquids" North-Holland, Amsterdam.
Brice, J. C. and Whiffin, P. A. C. (1973) *Platinum Metals Rev.* 17, 46.
Bridgman, P. W. (1952) "The Physics of High Pressure" Bell, London.
Brissot, J. J. and Belin, C. (1971) *J. Crystal Growth* 8, 213.
Brixner, L. H. (1959) *J. Amer. Chem. Soc.* 81, 3841.
Brixner, L. H. and Babcock, K. (1968) *Mat. Res. Bull.* 3, 817.
Brixner, L. H. and Bouchard, R. J. (1970) *Mat. Res. Bull.* 5, 61.
Brixner, L. H. and Weiher, J. F. (1970) *J. Solid State Chem.* 2, 55.
Broder, J. D. and Wolff, G. A. (1963) *J. Electrochem. Soc.* 110, 1150.
Brooks, M. S. and Kennedy, J. K. (1962) (editors) "Ultrapurification of Semi-conductor Materials" Macmillan, New York.
Brunst, W. (1957) "Die induktive Wärmebehandlung" Springer, Berlin.
Buckley, H. E. (1951) "Crystal Growth" Chapman and Hall, London.
Bundy, F. P. (1962) In "Modern Very High Pressure Techniques" (R. H. Wentorf, ed.) p. 1. Butterworths, London.
Bundy, F. P. (1963) *J. Chem. Phys.* 38, 631.
Burton, J. A., Prim, R. C. and Slichter, W. P. (1953) *J. Chem. Phys.* 21, 1987.
Busch, G., Kaldis, E., Schaufelberger-Teker, E. and Wachter, P. (1970) Rare Earth Conference Paris May 1969, Coll. Intern. C.N.R.S. I, 359.
Busch, G., Kaldis, E., Muheim, J. and Bischof, R. (1971) *J. Less-Common Metals* 24, 453.
Butler, C. T., Russell, J. R., Quincy jr., R. B. and La Valle, D. E. (1966) Report ORNL-3906.
Caldwell, F. R. (1962) "Thermocouple Materials" NBS Monograph 40, US Government Printing Office, Washington, D.C.
Campbell, J. E. (1959) (editor) "High Temperature Technology" Wiley, New York.
Capelletti, R., Fano, V. and Scalvini, M. (1968) *Ricerca Sci.* 38, 668, 886.
Capelletti, R., Fano, V. and Scalvini, M. (1969) *J. Crystal Growth* 5, 73.
Carruthers, J. R. (1967) *J. Electrochem. Soc.* 114, 959.
Carruthers, J. R. and Nassau, K. (1968) *J. Appl. Phys.* 39, 5205.
Chamberland, B. L. (1967) *Mat. Res. Bull.* 2, 827.
Champion, J. A., Keene, B. J. and Allen, S. (1973) *J. Mat. Sci.* 8, 423.
Chandrasekhar, S. (1961) "Hydrodynamic and Hydromagnetic Stability" Clarendon Press, Oxford.
Chang, C. E. and Wilcox, W. R. (1971) *Mat. Res. Bull.* 6, 1297.
Chase, A. B. and Osmer, J. A. (1967) *J. Am. Ceram. Soc.* 50, 325.
Chicotka, R. J. (1971) *IBM Techn. Discl. Bull.* 13, 3037.
Chrenko, R. M. (1971) *Nature, Phys. Sci.* 229, 165.
Cloete, F. L. D., Ortega, R. F. and White, E. A. D. (1969) *J. Mat. Sci.* 4, 21.
Cobb, C. M. and Wallis, E. B. (1967) Report AD 655 388.

Cochran, W. G. (1934) *Proc. Cambridge Phil. Soc.* **30**, 365.

Cockayne, B. (1968) *Platinum Metals Rev.* **12**, 16.

Cockayne, B. and Gates, M. P. (1967) *J. Mat. Sci.* **2**, 118.

Cockayne, B. Chesswas, M., Plant, J. G. and Vere, A. W. (1969) *J. Mat. Sci.* **4**, 565.

Collins, A. T. (1972) Diamond Res., Suppl. Ind. Diamond Rev. (cited by Strong and Wentorf 1972).

Corbett, J. D. and Duke, F. R. (1963) In "Techniques in Inorganic Chemistry" (H. B. Jonassen and A. Weissberger, eds.) p. 103. Vol. I, Interscience, New York.

Cotton, F. A. (1939–1972) (editor) "Inorganic Syntheses" McGraw-Hill, New York.

Crookes, W. (1912) *Proc. Roy. Soc.* (London) A **86**, 461.

Crystal Growth (1949) *Disc. Faraday Soc.* No. 5.

Cuomo, J. J. and Gambino, R. J. (1970) US Patent 3.498.894. (March 3).

Damen, J. P. M. and Robertson, J. M. (1972) *J. Crystal Growth* **16**, 50.

Daneman, H. L. and Mergner, G. C. (1967) Instrumentation Technology, May and June 1967, Reprint A 1.2102 RP of Leeds and Northrup, Philadelphia.

Darling, A. S., Selman, G. L. and Rushforth, R. (1970) *Platinum Metals Rev.* **14**, 54, 95, 124.

Daval, J. (1974) C.N.R.S. Grenoble, private communication.

Dawson, R. D., Elwell, D. and Brice, J. C. (1974) *J. Crystal Growth* **23**, 65.

Deacon, J. A. and Hiscocks, S. E. R. (1971) *J. Mat. Sci.* **6**, 309.

Derjaguin, B. V., Fedoseev, D. V., Lukyanovich, V. M., Spitzin, B. V., Ryabov, V. A. and Lavrentyev, A. V. (1968) *J. Crystal Growth* **3/4**, 380.

Dernier, P. D. and Maines, R. G. (1971) *Mat Res. Bull.* **6**, 433.

DeVries, R. C. (1966) *Mat. Res. Bull.* **1**, 83.

DeVries, R. C. (1967) *Mat. Res. Bull.* **2**, 999.

DeVries, R. C. and Fleischer, J. F. (1972) *J. Crystal Growth* **13/14**, 88.

DiBenedetto, B. and Cronan, C. J. (1968) *J. Am. Ceram. Soc.* **51**, 364.

Dike, P. H. (1958) "Thermoelectric Thermometry" 3rd Edition, Leeds and Northrup, Philadelphia.

Dike, P. H., Gray, W. T. and Schroyer, F. K. (1966) Optical Pyrometry, Leeds and Northrup Techn. Publ. A. 1.4000.

Donohue, P. C. (1972) *Mat Res. Bull.* **7**, 943.

Dugger, C. O. (1966) *J. Electrochem. Soc.* **113**, 306.

Dugger, C. O. (1967) In "Crystal Growth" (H. S. Peiser, ed.) p. 493. Pergamon, Oxford.

Duncan, C. S., Hopkins, R. H. and Mazelsky, R. (1971) *J. Crystal Growth* **11**, 50.

Eckstein, J., Holas, M., Jindra, J., Uchytilova, A. and Wachtl, Z. (1960) *Czech. J. Physics* B **10**, 247.

Egli, P. H. and Johnson, L. R. (1963) In "The Art and Science of Growing Crystals" (J. J. Gilman, ed.) p. 194. Wiley, New York.

Ellis, W. C., Frosch, C. J. and Zetterstrom, R. B. (1968) *J. Crystal Growth* **2**, 61.

Elmer, T. H. (1953) *Am. Ceram. Soc. Bull.* **32**, 23.

Epstein, D. J. (1970) Report AD 715'312.

Espig, H. (1960) *Chem. Tech.* **12**, 6, 327.

Euler, J. and Ludwig, R. (1960) "Arbeitsmethoden der optischen Pyrometrie" Braun, Karlsruhe.

Fajans, K. (1921) *Naturwiss.* **9**, 729.

Fajans, K. and Karagunis, G. (1930) *Z. Angew. Chem.* **43**, 1046.

Faust jr., J. W. and John, H. F. (1964) *J. Phys. Chem. Solids* **25**, 1407.

Ferretti, A., Kunnmann, W. and Wold, A. (1962) *J. Appl. Phys.* **34**, 388.
Filby, J. D. and Nielsen, S. (1966) *Microelectronics and Reliability* **5**, 11.
Finch, C. B. and Clark, G. W. (1965) *J. Appl. Phys.* **36**, 2143.
Finch, D. I. (1966) General Principles of Thermoelectric Thermometry, Techn. Publ. D 1.1000, Leeds and Northrup, Philadelphia.
Fitzer, E. (1956) Plansee Proceed., 2nd Seminary, Reutte, Tirol, p. 56.
Flicstein, J. and Schieber, M. (1973) *J. Crystal Growth* **18**, 265.
Flörke, O. W. (1955) *Ber. Dt. Keram. Ges.* **32**, 369.
Foguel, M. and Grajower, R. (1971) *J. Crystal Growth* **11**, 280.
Fong, F. K. and Yocom, P. N. (1964) *J. Chem. Phys.* **41**, 1383.
Foster, L. M. and Stumpf, H. C. (1951) *J. Amer. Chem. Soc.* **73**, 1590.
Fromm, E. and Jehn, H. (1972) *Metallurg. Transact.* **3**, 1685.
Frosch, C. J. (1967) In "Crystal Growth" (H. S. Peiser, ed.) 305. Pergamon, Oxford.
Fukunaga, O. and Saito, S. (1968) *J. Am. Ceram. Soc.* **51**, 362.
Garrard, B. J., Wanklyn, B. M. and Smith, S. H. (1974) *J. Crystal Growth* **22**, 169.
Gärtner, K. J., Rittinghaus, K. F., Seeger, A. and Uelhoff, W. (1972) *J. Crystal Growth* **13/14**, 619.
Garton, G. and Wanklyn, B. M. (1967) *J. Crystal Growth* **1**, 164.
Garton, G., Hann, B. F., Wanklyn, B. M. and Smith, S. H. (1972) *J. Crystal Growth* **12**, 66.
Gasson, D. B. (1965) *J. Scient. Instr.* **42**, 114.
Geary, D. A. and Hough, J. M. (1968) *J. Crystal Growth* **2**, 113.
Ghez, R. and Giess, E. A. (1973) *Mat. Res. Bull.* **8**, 31.
Giess, E. A. (1966) *J. Am. Ceram. Soc.* **49**, 104.
Giess, E. A., Kuptsis, J. D. and White, E. A. D. (1972) *J. Crystal Growth* **16**, 36.
Gilman, J. J. (1963) "The Art and Science of Growing Crystals" Wiley, New York.
Givargizov, E. I. and Sheftal, N. N. (1971) *J. Crystal Growth* **9**, 326.
Goodenough, J. B., Kafalas, J. A. and Longo, J. M. (1972) In "Preparative Methods in Solid State Chemistry" (P. Hagenmuller, ed.) p. 1. Academic Press, New York.
Goodrum, J. W. (1970) *J. Crystal Growth* **7**, 254.
Goodwin, A. R., Gordon, J. and Dobson, C. D., (1968) *J. Phys.* D **1**, 115.
Goss, A. J. and Adlington, R. E. (1959) *Marconi Rev.* **22**, 18.
Gray, W. T. and Finch, D. I. (1971) *Physics Today* **24**, 9, 32.
Greenspan, H. P. (1968) "The Theory of Rotating Fluids" Cambridge Univ. Press.
Griffiths, L. B. and Mlavsky, A. I. (1964) *J. Electrochem. Soc.* **111**, 305.
Grodkiewicz, W. H. and Nitti, D. J. (1966) *J. Am. Ceram. Soc.* **49**, 576.
Grodkiewicz, W. H., Dearborn, E. F. and Van Uitert, L. G. (1967) In "Crystal Growth" (H. S. Peiser, ed.) 441. Pergamon, Oxford.
Guldberg, C. and Waage, P. (1867) "Etudes sur les affinités chimiques" Christiana.
Gunn, J. B. (1971) private communication, see also *IBM Techn. Discl. Bull.* **15**, 1050 (1972).
Habermann, C. E., Daane, A. H. and Palmer, P. E. (1965) *Trans. AIME* **233**, 1038.
Hall, H. T. (1958) *Rev. Sci. Instrum.* **29**, 267.
Hall, H. T. (1960) *Rev. Sci. Instrum.* **31**, 125.
Hall, R. N. (1952) *Phys. Rev.* **88**, 135.
Hall, R. N. (1953) *J. Phys. Chem.* **57**, 836.
Hamaker, R. W. and White, W. B. (1968) *J. Appl. Phys.* **39**, 1758.
Hamaker, R. W. and White, W. B. (1969) *J. Electrochem. Soc.* **116**, 478.
Hanneman, R. E. and Strong, H. M. (1965) *J. Appl. Phys.* **36**, 523.

Hanneman, R. E. and Strong, H. M. (1966) *J. Appl. Phys.* **37,** 612.
Harbaugh, W. E. and Eastman, G. Y. (1970) *Heating, Piping and Air Cond.* **42,** 10, 92.
Harrison, T. R. (1960) "Radiation Pyrometry and its Underlying Principles of Radiant Heat Transfer" Wiley, New York.
Hart, P. B. (1965) Brit. Pat. Applicat. 21979/62 (quoted by White 1965).
Hauptman, Z., Wanklyn, B. M. and Smith, S. H. (1973) *J. Mat. Sci.* **8,** 1695.
Haussühl, S. (1964) *Neues Jahrb. Min. Abh.* **101,** 343.
Hayes, A. and Chipman, J. (1939) *Trans. AIME* **135,** 85.
Hecht, H. (1951) "Präparative anorganische Chemie" Springer, Berlin.
Hein, C. C. (1956) US Patent 2'747'196.
Helferich, R. L. and Zanis, C. A. (1973) Report AD 754'092.
Hemmat, N., Lampoŕt, C. B., Menna, A. A. and Wolff, G. A. (1970) Solution Growth of Electronic Compounds and Their Solid Solutions. Preprint volume, Mat. Engng. and Sciences Div., Biennial Conf., Febr. 15–18, 1970, Atlanta, Amer. Inst. of Chem. Engs., 112–121 (see also AD 722,785).
Hide, R. and Titman, C. W. (1967) *J. Fluid Mech.* **29,** 39.
High Temperature Technology (1960) Proc. Int. Symp. 1959, McGraw-Hill, New York.
Hintzmann, W. and Müller-Vogt, G. (1969) *J. Crystal Growth* **5,** 274.
Hippel, A. von *et al.* (1963) *MIT Techn. Report* **178,** 45.
Hoffmann, W. and Scheel, H. J. (1969) *Z. Krist.* **129,** 396.
Holden, A. N. (1949) *Disc. Faraday Soc.* **5,** 312.
Holden, A. and Singer, P. (1960) "Crystals and Crystal Growing" Heinemann, London.
Holonyak, N., Wolfe, C. M. and Moore J. S. (1965) *Appl. Phys. Lett.* **6,** 64.
Hukin, D. (1971) Paper at ICCG 3, Marseille.
Hurle, D. T. J. (1966) *Phil. Mag.* **13,** 305.
Hurle, D. T. J. (1967) In "Crystal Growth" (H. S. Peiser, ed.) p. 659. Pergamon, Oxford.
Hurle, D. T. J., Mullin, J. B. and Pike, E. R. (1964) *Phil. Mag.* **9,** 423.
Hurle, D. T. J., Mullin, J. B. and Pike, E. R. (1967) *J. Mat. Sci.* **2,** 46.
Hurle, D. T. J., Jakeman, E. and Pike, E. R. (1968) *J. Crystal Growth* **3/4,** 633.
Hurst, J. J. and Linz, A. (1971) *Mat. Res. Bull.* **6,** 163.
IPTS-68 (1969) Internat. Practical Temperature Scale of 1968, in *Metrologia* **5,** No. 2, 35.
Jaeger, G. and Krasemann, R. (1952) *Werkstoffe u. Korrosion* **3,** 401.
Jankelevic, G., Vinarov, I. V., Seka, I. A. and Frajman, I. B. (1967) *Ukrain. Chim. Z.* **33,** 1040 (russ.).
Janssen, G.A.M., Robertson, J.M. and Verheijke, M.L. (1973) *Mat. Res. Bull.* **8,** 59.
Janz, G. J. (1967) "Molten Salts Handbook", p. 388. Academic Press, New York–London.
John, V. and Kvapil, J. (1968) *Kristall u. Technik* **3,** 59.
Jolly, W. (1970) "The Synthesis and Characterization of Inorganic Compounds" Prentice-Hall, Englewood Cliffs, N.J.
Jonassen, H. B. and Weissberger, A. (1963–1968) (editors) "Technique of Inorganic Chemistry" Vols. I–VII, Interscience, New York.
Joubert, J. C. and Chenavas, J. (1975) to be published.
Kamenetskaya, D. S. (1967) *Sov. Phys.-Cryst.* **12,** 1; (1968) In "Growth and Perfection of Metallic Crystals" (D. E. Ovsienko, ed.) p. 245. Consultants Bureaux, New York.

Kaneko, K., Ayabe, M., Dosen, M., Morizane, K., Usui, S. and Watanabe, N. (1973) *Proc. IEEE* **61**, 884; (1974), *J. Electrochem. Soc.* 121, 556.

Kanzaki, H. and Kido, K. (1960) *J. Phys. Soc. Japan* **15**, 529; (1962), *J. Appl. Phys.* **33**, 482.

Kawabe, K. and Sawada, S. (1957) *J. Phys. Soc. Japan* **12**, 218.

Keezer, R. C. and Wood, C. (1966) *Appl. Phys. Lett.* **8**, 139.

Kelman, L. R., Wilkinson, W. D. and Yaggee, F. L. (1950) U.S. A.E.C. Report ANL-4417.

Kennedy, G. C., Jayaraman, A. and Newton, R.C. (1962) *Phys. Rev.* **126**, 1363.

Kennedy, T. N., Hakim, R. and Mackenzie, J. D. (1967) *Mat. Res. Bull.* **2**, 193.

Kestigian, M. (1967) *J. Am Ceram. Soc.* **50**, 165.

Khamskii, E. V. (1969) "Crystallization from Solutions" Consultants Bureau, New York.

Kim, K. M., Witt, A. F. and Gatos, H. C. (1972) *J. Electrochem. Soc.* **119**, 1218.

Kingery, W. D. (1959) "Property Measurement at High Temperatures" Wiley, New York.

Kirgintsev, A. N. and Avvakumov, E. G. (1965) *Sov. Phys.-Cryst.* **10**, 375.

Kiyama, R. and Minomura, S. (1953) *Rev. Phys. Chem. Japan* **23**, 10.

Kjems, J. K., Shirane, G., Müller, K. A. and Scheel, H. J. (1973) *Phys. Rev.* B **8**, 1119.

Kleber, W. and Wilke, K. T. (1969) *Kristall u. Technik* **4**, 165.

Kleinknecht, H. P. (1966) *J. Appl. Phys.* **37**, 2116.

Klement, W. and Jayaraman, A. (1967) *Progr. Solid State Chem.* **3**, 289.

Knippenberg, W. F. (1963) *Philips Res. Repts.* **18**, 161.

Knippenberg, W. F. and Verspui, G. (1966) *Philips Res. Repts.* **21**, 113.

Kostkowski, H. J. (1960) In "High Temperature Technology" Intern. Symp. 1959, p. 33. McGraw-Hill, New York.

Kostkowski, H. J. and Lee, R. D. (1962) "Theory and Methods of Optical Pyrometry" NBS Monograph 41, US Government Printing Office, Washington, D.C.

Kröger, F. A. (1964) "The Chemistry of Imperfect Crystals" North-Holland, Amsterdam.

Kunnmann, W. (1971) In "Preparation and Properties of Solid State Materials" (R. A. Lefever, ed.) 1. Dekker, New York.

Kunnmann, W. and Ferretti, A. (1964) *Rev. Sci. Instrum.* **35**, 465.

Kunnmann, W., Ferretti, A. and Wold, A. (1963) *J. Appl. Phys.* **34**, 1264.

Kvapil, J. (1966) *Kristall u. Technik* **1**, 97.

Kwestroo, W. (1972) In "Preparative Methods in Solid State Chemistry" (P. Hagenmuller, ed.) p. 563. Academic Press, New York.

Kwestroo, W. and Huizing, A. (1965) *J. inorg. nucl. Chem.* **27**, 1951.

Kwestroo, W., Huizing, A. and deJonge, J. (1969) *Mat. Res. Bull.* **4**, 817.

Lange, N. O. (1967) "Handbook of Chemistry" 10th Edition, p. 1432. McGraw-Hill, New York.

Laudise, R. A. (1963) In "The Art and Science of Growing Crystals" (J. J. Gilman, ed.) Wiley, New York.

Laudise, R. A., Linares, R. C. and Dearborn, E. F. (1962) *J. Appl. Phys.* **33**, 1362.

Lebl, M. and Trnka, J. (1965) *Z. Physik* **186**, 128.

Levin, E. M., Rynders, G. F. and Dzimian, R. J. (1952) Growing Beryllium Oxide Crystals, NBS Report No. 1916.

Lichtensteiger, M., Witt, A. F. and Gatos, H. C. (1971) *J. Electrochem. Soc.* **118**, 1013.

Lierde, W. van and Bressers, J. (1966) *J. Appl. Phys.* **37**, 444.
Linares, R. C. (1960) *J. Phys. Chem.* **64**, 941.
Linares, R. C. (1962) *J. Am. Ceram. Soc.* **45**, 307.
Linares, R. C. (1964) *J. Appl. Phys.* **35**, 433.
Linares, R. C. (1965) *J. Am. Ceram. Soc.* **48**, 68.
Linares, R. C. (1967) *Am. Min.* **52**, 1554.
Linares, R. C. (1968) *Trans. Met. Soc. AIME* **242**, 441.
Linz, A., Belruss, V. and Naiman, C. S. (1965) *Electrochem. Soc.* Spring Meeting, San Francisco, Extended Abstracts **2**, 87.
Litvin, Y. A., Lisoyvan, V. I., Sobolev, Y. V. and Butuzov, V. P. (1968) *Dokl. Akad. Nauk SSSR* **183**, 144.
Lord, G. W. and Moss, R. H. (1970) *J. Phys. E: Scientif. Instr.* **3**, 177.
Lux, H. (1970) "Anorganisch-chemische Experimentierkunst" 3rd Edition. Barth, Leipzig.
Luzhnaya, N. P. (1968) *J. Crystal Growth* **3/4**, 97.
Lydtin, H. (1972) private communication.
Lyons, V. J. (1965) US Patent 3.198.606 (Aug. 3, 1965).
Makram, H. (1966) *Czech. J. Physics* B **17**, 387.
Makram, H. and Krishnan, R. (1969) In "Crystal Growth" (H. S. Peiser, ed.) p. 467. Pergamon, Oxford.
Makram, H., Touron, L. and Loriers, J. (1968) *J. Crystal Growth* **3/4**, 452.
Malkus, M. V. R. (1954) *Proc. Roy. Soc. (London)* A **225**, 185.
Manzel, M. (1967) *Kristall u. Technik* **2**, 61.
Marezio, M. and Remeika, J. P. (1965) *J. Phys. Chem. Solids* **26**, 2083.
Marezio, M. and Remeika, J. P. (1966) *J. Chem Phys.* **44**, 3348.
Marezio, M., Remeika, J. P. and Jayaraman, A. (1966a) *J. Chem. Phys.* **45**, 1821.
Marezio, M., Remeika, J. P. and Dernier, P. D. (1966b) *Mat. Res. Bull.* **1**, 247.
Margrave, J. L. (1965) (editor) "Refractory Materials, a Series of Monographs" Academic Press, New York–London.
Mason, D. R. and Cook, J. S. (1961) *J. Appl. Phys.* **32**, 475.
Matecha, J. and Kvapil, J. (1970) *J. Crystal Growth* **6**, 199.
Mateika, D. (1971) *J. Crystal Growth* **9**, 249.
Mateika, D. (1972) *J. Crystal Growth* **13/14**, 698.
May, E. (1950) "Industrial High-Frequency Electric Power" Wiley, New York.
Mayer, I., Shidlovsky, I. and Yanir, E. (1967) *J. Less-Common Metals* **12**, 46.
McLaren, E. H. (1957) *Can. J. Phys.* **35**, 78.
McWhan, D. B. and Remeika, J. P. (1970) *Phys. Rev.* B **2**, 3734.
Meyer, A. (1970) *J. Less-Common Metals* **20**, 353.
Miller, C. E. (1958) *J. Appl. Phys.* **29**, 233.
Millett, E. J., Brice, J. C., Whiffin, P. A. C. and Whipps, P. W. (1967) In "Crystal Growth" (H. S. Peiser, ed.) p. 673. Pergamon, Oxford.
Minden, H. T. (1973) *Solid State Technol.* (January, 1973).
Mlavsky, A. I. and Weinstein, M. (1963) *J. Appl. Phys.* **34**, 2885.
Monchamp, R. R., Belt, R. and Nielsen, J. W. (1967) In "Crystal Growth" (H. S. Peiser, ed.) p. 463. Pergamon, Oxford.
Morgan, W. L. and Scheffler, L. F. (1965) US Patent 3.224.843.
Morozewicz, J. (1899) *Tschermaks Min. Mitt.* **18**, 1, 105, 225.
Morris, D. F. C. (1968) *Structure and Bonding* **4**, 63; **6**, 157.
Müller, G. and Gnauck, G. (1965) "Reinste Gase" VEB Deutscher Verlag der Wissenschaften, Berlin.
Mullin, J. W. (1972) "Crystallization" 2nd Edition. Butterworths, London.

Mullin, J. B., Straughan, B. W. and Brickell, W. S. (1965) *J. Phys. Chem. Solids* **26**, 782.

Munro, D. C. (1963) In "High Pressure Physics and Chemistry" (R. S. Bradley, ed.) Vol. 1, p. 11. Academic Press, London.

Nelson, D. F. and Remeika, J. P. (1964) *J. Appl. Phys.* **35**, 522.

Nelson, H. (1963) *RCA Review* **24**, 603.

Neuhaus, A. (1956) *Chem. Ing. Techn.* **155**, 350.

Neuhaus, A. (1964) *Chimia* **18**, 93.

Newkirk, H. W. and Smith, D. K. (1965) *Am. Min.* **50**, 22, 44.

Nielsen, J. W. (1964) *Electronics* (Nov. 30, 1964).

Nielsen, J. W. (1974) private communication.

Nielsen, J. W. and Dearborn, E. F. (1960) *J. Phys. Chem.* **64**, 1762.

Nielsen, J. W., Lepore, D. A. and Leo, D. C. (1967) In "Crystal Growth" (H. S. Peiser, ed.) p. 457. Pergamon, Oxford.

Niemyski, T., Mierzejewska-Appenheimer, S. and Majewski, J. (1967) In "Crystal Growth" (H. S. Peiser, ed.) p. 585. Pergamon, Oxford.

O'Kane, D. F., Kwap, T. W., Gulitz, L. and Bednowitz, A. L. (1972) *J. Crystal Growth* **13/14**, 624.

Oppelt, W. (1964) "Kleines Handbuch technischer Regelvorgänge" 4th Edition. Verlag Chemie, Weinheim.

Otto, C. A. (1958) "Electric Furnaces" London.

Oxley, C. E. and Stockton, J. (1966) *J. Scient. Instr.* **43**, 767.

Parker, H. S. and Brower, W. S. (1967) In "Crystal Growth" (H. S. Peiser, ed.) p. 489. Pergamon, Oxford.

Patel, A. R. and Bhat, H. L. (1971) *J. Crystal Growth* **11**, 166.

Paul, W. and Warschauer, D. M. (1963) "Solids under Pressure" McGraw-Hill, New York.

Perloff, D. S. and Wold, A. (1967) In "Crystal Growth" (H. S. Peiser, ed.) p. 361. Pergamon, Oxford.

Perry, F. W. (1967) In "Crystal Growth" (H. S. Peiser, ed.) p. 483. Pergamon, Oxford.

Peters, E. T. and Ryan, J. J. (1966) *J. Appl. Phys.* **37**, 933.

Petzold, D. R., Schultze, D. and Wilke, K. T. (1971) *Z. Anorg. Allg. Chem.* **386**, 288.

Pfann, W. G. (1955) *Trans. AIME* **263**, 961.

Pfann, W. G. (1958) In "Liquid Metals and Solidification" (B. Chalmers, ed.) *Amer. Soc. f. Metals*, Cleveland, Ohio.

Pfann, W. G. (1966) "Zone Melting" 2nd Edition. Wiley, New York.

Pfann, W. G., Benson, K. E. and Wernick, J. H. (1957) *J. Electron.* **2**, 597.

Plaskett, T. S. (1969) *J. Electrochem. Soc.* **116**, 1722.

Plaskett, T. S., Blum, S. E. and Foster, L. M. (1967) *J. Electrochem. Soc.* **114**, 1304.

Pohl, D. and Scheel, H. J. (1975) to be published.

Poiblaud, G. and Jacob, G. (1973) *Mat. Res. Bull.* **8**, 845.

Pointon, A. J. and Robertson, J. M. (1967) *J. Mat. Sci.* **2**, 293.

Poland, D. E., Green, J. W. and Margrave, J. L. (1961) Corrected Optical Pyrometer Readings, NBS Monograph 30, US Government Printing Office, Washington, D.C.

Rabenau, A. (1963/1964) *Philips' Techn. Rundschau* **25**, 365.

Raksani, K. and Voszka, R. (1969) *Kristall u. Technik* **4**, 227.

Rea, J. R. and Kostiner, E. (1971) *J. Crystal Growth* **11**, 110.

Reed, T. B., Fahey, R. E. and Honig, J. M. (1967) *Mat. Res. Bull.* **2**, 561.

Reed, T. B. (1972) "Free Energy of Binary Compounds" MIT Press.
Reed, T. B. (1973) Solid State Research Report, Lincoln Lab., No. 1, 25.
Remeika, J. P. (1954) *J. Amer. Chem. Soc.* **76**, 940.
Remeika, J. P. (1958) U.S. Pat. **2**, 852, 400.
Remeika, J. P. (1963) U.S. Patent 3'079'240 (Febr. 26, 1963).
Remeika, J. P. (1970) Paper presented at Int. Conf. on Ferrites, Kyoto.
Remeika, J. P. and Marezio, M. (1966) *Appl. Phys. Lett.* **8**, 87.
Remeika, J. P. and Spencer, E. G. (1964) *J. Appl. Phys.* **35**, 2803.
Remeika, J. P., Dodd, D. M. and DiDomenico, M. (1966) *J. Appl. Phys.* **37**, 5004.
Reynolds, G. F. and Guggenheim, H. J. (1961) *J. Phys. Chem.* **65**, 1655.
Robertson, D. S. (1966) *Brit. J. Appl. Phys.* **17**, 1047.
Robertson, G. D. and Scholl, R. F. (1971) *Rev. Sci. Instrum.* **42**, 882.
Robertson, J. M. (1969) *Engelhard Industries Techn. Bull.* **10**, 77.
Robertson, J. M. and Neate, B. W. (1972) *J. Crystal Growth* **13/14**, 576.
Robertson, J. M. (1974) private communication.
Robertson, J. M., Smith, S. H. and Elwell, D. (1969) *J. Crystal Growth* **5**, 189.
Robertson, J. M., Kamminga, W. and Rooymans, C. J. M. (1975) to be published.
Robertson, J. M., Wittekoek, S., Popma, T. J. A. and Bongers, P. F. (1973) *Appl. Phys.* **2**, 219.
Robertson, J. M., Damen, J. P. M., Jonker, H. D., Van Hout, M. J. G., Kamminga, W. and Voermans, A. B. (1974a) *Platin. Metals Rev.* **18**, 15.
Robertson, J. M., Van Hout, M. J. G., Verplanke, J. C. and Brice, J. C. (1974b) *Mat. Res. Bull.* **9**, 555.
Rodot, H., Hruby, A. and Schneider, M. (1968) *J. Crystal Growth* **3/4**, 305.
Roeser, W. F. and Lonberger, S. T. (1958) Methods of Testing Thermocouples and Thermocouple Materials, NBS Circular 590, US Government Printing Office, Washington, D.C.
Rogers, D. B., Ferretti, A. and Kunnmann, W. (1966) *J. Phys. Chem. Solids* **27**, 1445.
Rooymans, C. J. M. (1972) In "Preparative Methods in Solid State Chemistry" (P. Hagenmuller, ed.) p. 72. Academic Press, New York.
Roy, R. (1965) *Bull. Soc. Chim. Franc.* 1065.
Roy, R. (1966) *Mat. Res. Bull.* **1**, 299.
Roy, R. and White, W. B. (1968) *J. Crystal Growth* **3/4**, 33.
Rubin, L. G. Temperature-Concepts, Scales and Measurement Techniques, Report A 1.0001 RP, Leeds and Northrup, Philadelphia.
Russell, R. G., Morgan, W. L. and Scheffler, L. F. (1962), US Patent 3.065.091.
Sawada, S., Nomura, S. and Fujii, S. (1951) *Rept. Inst. Sci. Techn., Univ. Tokyo* **5**, 7.
Schäfer, H. (1964) "Chemical Transport Reactions" Academic Press, New York.
Scheel, H. J. (1968) *J. Crystal Growth* **2**, 411.
Scheel, H. J. (1971) *Z. Krist.* **133**, 264.
Scheel, H. J. (1972) *J. Crystal Growth* **13/14**, 560.
Scheel, H. J. (1973) Swiss Patent 541.353 (Sept. 15, 1973), US Patent pending.
Scheel, H. J. (1974) *J. Crystal Growth* **24/25**, 669.
Scheel, H. J. and Schulz-DuBois, E. O. (1971) *J. Crystal Growth* **8**, 304.
Scheel, H. J. and Elwell, D. (1972) *J. Crystal Growth* **12**, 153.
Scheel, H. J. and Schulz-DuBois, E. O. (1971) *IBM Techn. Discl. Bull.* **14**, 2850.
Scheel, H. J. and Elwell, D. (1973) *J. Electrochem. Soc.* **120**, 818.
Scheel, H. J. and West, C. H. (1973) *J. Phys. E: Scientif. Instr.* **6**, 1178.
Schieber, M. (1967) *Kristall u. Technik* **2**, 55.

Schildknecht, H. (1966) "Zone melting" Academic Press, New York.

Schönherr, E. (1971) *J. Crystal Growth* **9**, 346.

Schulz-DuBois, E. O. (1972) *J. Crystal Growth* **12**, 81.

Seidenstricker, R. G. (1966) *J. Electrochem. Soc.* **113**, 152.

Senhouse, L. S., Paolis, M. V. de, Loomis, T. C. (1966) *Appl. Phys. Lett.* **8**, 173.

Shafer, M. W., Torrance, J. B. and Penney, T. (1972) *J. Phys. Chem. Sol.* **33**, 2251.

Shenker, H., Lauritzen jr., J. I., Corruccini, R. J. and Lonberger, S. T. (1955) NBS Circular No. 561, US Government Printing Office, Washington, D.C.

Shick, L. K. and Von Neida, A. R. (1971) US Patent 3'627'498.

Sickafus, E. N. and Barker, D. B. (1967) *J. Crystal Growth* **1**, 93.

Simpson, P. G. (1960) "Induction Heating Coil and System Design" McGraw-Hill, New York.

Sip, V. and Vanicek, V. (1962) *Growth of Crystals* **3**, 191.

Slichter, W. P. and Burton, J. A. (1958) In "Transistor Technology" (H. E. Bridgers, J. H. Scaff and J. N. Shive, eds.) p. 107. Van Nostrand, Princeton.

Smakula, A. (1962) "Einkristalle" Springer, Berlin.

Smith, C. L. and Murrill, P. W. (1966) *ISA Journal* **13**, 5, 50.

Smith, S. H. and Elwell, D. (1968) *J. Crystal Growth* **3/4**, 471.

Smith, S. H. and Wanklyn, B. (1974) *J. Crystal Growth* **21**, 23.

Sproull, R. L. (1962) *Scient. Amer.* (December 1962).

Stambaugh, E. P., Miller, J. E. and Hines, R. C. (1961) "Metallurgy of Elemental and Compound Semiconductors" p. 317. Interscience, New York.

Stansel, N. R. (1949) "Induction Heating" McGraw-Hill, New York.

Steininger, J. and England, R. E. (1968) *Trans. AIME* **242**, 436.

Steininger, J. and Reed, T. B. (1972) *J. Crystal Growth* **13/14**, 106.

Steininger, J. and Strauss, A. J. (1972) *J. Crystal Growth* **13/14**, 657.

Stimson, H. F. (1961) Internat. Practical Temp. Scale of 1948—Text Revision of 1960, NBS Monograph 37, US Government Printing Office, Washington, D.C.

Stone, B. D. (1959) US Patent 3.009.780 (June 20, 1959).

Stringfellow, G. B. and Greene, P. E. (1971) *J. Electrochem. Soc.* **118**, 805.

Strong, H. M. (1962) In "Modern Very High Pressure Techniques" (R. H. Wentorf, ed.) p. 93. Butterworths, London.

Strong, H. M. and Hanneman, R. E. (1967) *J. Chem. Phys.* **46**, 3668.

Strong, H. M. and Chrenko, R. M. (1971) *J. Phys. Chem.* **75**, 1838.

Strong, H. M. and Wentorf, jr., H. R. (1972) *Naturwiss.* **59**, 1.

Tarjan, I. and Matrai, M. (1972) "Laboratory Manual on Crystal Growth" Akademiai Kiado, Budapest.

"Temperature, its Measurement and Control in Science and Industry" Vol. 1 (1941), Vol. 2 (1955), Vol. 3, Parts I and II (1962) Reinhold, New York.

Thoma, R. E., Goss, R. G. and Weaver, C. F. (1967) In "Crystal Growth" (H. S. Peiser, ed.) p. 147. Pergamon Press, Oxford.

Thurmond, C. D. and Struthers, J. D. (1957) *J. Phys. Chem. Solids* **57**, 831.

Tiche, D. and Spear, K. E. (1972) Report AD 736.639.

Tiller, W. A. (1963) *J. Appl. Phys.* **34**, 2757.

Tiller, W. A. (1968) *J. Crystal Growth* **2**, 69.

Timofeeva, V. A. and Zalesskii, A. V. (1959) *Growth of Crystals* **2**, 69.

Timofeeva, V. A. and Kvapil, J. (1966) *Sov. Phys. Cryst.* **11**, 263.

Timofeeva, V. A., Lukyanova, N. I., Guseva, I. N. and Lider, V. V. (1969) *Sov. Phys.-Cryst.* **13**, 747.

Tolksdorf, W. (1968) *J. Crystal Growth* **3/4**, 463.

Tolksdorf, W. and Welz, F. (1972) *J. Crystal Growth* **13/14**, 566.
Tolksdorf, W. (1973) *J. Crystal Growth* **18**, 57.
Tolksdorf, W. (1974a) private communication.
Tolksdorf, W. (1974b) *Acta Electron.* **17**, 57.
Tolksdorf, W. and Welz, F. (1973) *J. Crystal Growth* **20**, 47.
Trainor, A. and Bartlett, B. E. (1961) *Solid State Electron.* **2**, 106.
Triebwasser, S. (1959) *Phys. Rev.* **114**, 63.
Trumbore, F. A., Porbansky, E. M. and Tartaglia, A. A. (1959)) *J. Phys. Chem. Solids* **11**, 239.
Tsushima, K. (1966) *J. Appl. Phys.* **37**, 443.
Tyco Laboratories Inc., Waltham, Mass. (1971) Report AD 722.785.
Van Hook, A. (1961) "Crystallization, Theory and Practice" Reinhold, New York.
Van Uitert, L. G. (1970) *Platinum Metals Rev.* **14**, 118.
Van Uitert, L. G., Bonner, W. A., Grodkiewicz, W. H., Pietroski, L. and Zydzik, G. J. (1970) *Mat. Res. Bull.* **5**, 825.
Vergnoux, A. M., Riera, M., Ribet, J. L. and Ribet, M. (1971) *J. Crystal Growth* **10**, 202.
Vichr, M. (1973) private communication.
Vidal, C. R. and Haller, F. B. (1971) *Rev. Sci. Instrum.* **42**, 1779.
Vilms, J. and Garrett, J. P. (1972) *Sol. State. Electr.* **15**, 443.
Vines, R. F. (1941) "The Platinum Metals and their Alloys" Internat. Nickel Comp., New York.
Von Neida, A. R. and Shick, L. K. (1969) *J. Appl. Phys.* **40**, 1013.
Von Philipsborn, H. (1967) *J. Appl. Phys.* **38**, 955; (1967) *Helv. Phys. Acta* **40**, 810.
Von Philipsborn, H. (1969) *J. Crystal Growth* **5**, 135.
Von Philipsborn, H. (1971) *J. Crystal Growth* **9**, 296.
Wagner, C. (1950) *J. Phys. and Coll. Chem.* **53**, 1030.
Wagner, C. (1954) *Trans. AIME* **200**, 154.
Wagner, R. S. and Ellis, W. C. (1964) *Appl. Phys. Lett.* **4**, 89.
Wagner, R. S. and Ellis, W. C. (1965) *Trans. AIME* **233**, 1053.
Wagner, R. S. (1967) In "Crystal Growth" (H. S. Peiser, ed.) p. 347. Pergamon, Oxford.
Wagner, R. S. (1968) *J. Crystal Growth* **3/4**, 159.
Wakatsuki, M. (1966) *Japan. J. Appl. Phys.* **5**, 337.
Wald, F., Bell, R. O. and Menna, A. A. (1971) Interim Techn. Rept., Contact No. AT(30-1)-4202, Tyco Labs., Waltham, Mass.
Wallace, C. A. and White, E. A. D. (1967) In "Crystal Growth" (H. S. Peiser, ed.) p. 431. Pergamon, Oxford.
Wanklyn, B. M. (1969) *J. Crystal Growth* **5**, 279.
Wanklyn, B. M. (1972) *J. Mat. Sci.* **7**, 813.
Wanklyn, B. M. and Hauptman, Z. (1974) *J. Mat. Sci.* **9**, 1078.
Weaver, C. F., Ross, R. G. and Thoma, R. E. (1963) *J. Appl. Phys.* **34**, 1827.
Webster, F. W. and White, E. A. D. (1969) *J. Crystal Growth* **5**, 167.
Weeks, J. R. and Gurinsky, D. H. (1958) In "Liquid Metals and Solidification" p. 106. Amer. Soc. for Metals, Cleveland/Ohio.
Weinstein, M. and Mlavsky, A. I. (1964) *J. Appl. Phys.* **35**, 1892.
Weisbeck, R. (1964) *Chem. Ing. Techn.* **36**, 442.
Wemple, S. H. (1964) *MIT Techn. Rept.* **425**, (see also the Thesis, MIT, 1963).
Wentorf jr., R. H. (1957) *J. Chem. Phys.* **26**, 956.
Wentorf jr., R. H. (1961) *J. Chem. Phys.* **34**, 809 (see also US Patent 3'192'015).
Wentorf jr., R. H. (1962) *J. Chem. Phys.* **36**, 1990.

Wentorf jr., R H. (1966) *Ber. Bunsenges. Phys. Chem.* **70,** 975.
Wentorf, jr., R. H. (1971) *J. Chem. Phys.* **75,** 1833.
Wernick, J. H. (1956). *J. Chem. Phys.* **25,** 47.
Wernick, J. H. (1957) *J. Metals* **9,** 1169.
Whiffin, P. A. C. and Brice, J. C. (1971) *J. Crystal Growth* **10,** 91.
Whiffin, P. A. C. (1973) Mullard Research Labs. Report No. 2848.
Whipps, P. W. (1972) *J. Crystal Growth* **12,** 120.
White, E. A. D. (1965) *Brit. J. Appl. Phys.* **16,** 1415.
White, E. A. D. and Brightwell, J. W. (1965) *Chem. and Ind.* 1662.
Wickham, D. G. (1970) Proceed. Intern. Conf. Ferrites (Kyoto), 105.
Widmer, R. (1971) *J. Crystal Growth* **8,** 216.
Wilcox, W. R. (1968) *Ind. Eng. Chem.* **60,** 13.
Wilcox, W. R. (1970) *J. Crystal Growth* **7,** 203.
Wilcox, W. R. and Fullmer, L. D. (1965) *J. Appl. Phys.* **36,** 2201.
Wilcox, W. R. and Fullmer, L. D. (1966) *J. Am. Ceram. Soc.* **49,** 415.
Wilke, K. T. (1956) *Fortschritte d. Mineral.* **34,** 85.
Wilke, K. T. (1963) "Methoden der Kristallzüchtung", Verlag H. Deutsch, Frankfurt and VEB Deutscher Verlag der Wissenschaften, Berlin.
Wilke, K. T. (1973) "Kristallzüchtung" VEB Deutscher Verlag der Wissenschaften, Berlin.
Wilke, K. T., Töpfer, K. and Schultze, D. (1965) *Z. Phys. Chem.* **230,** 112.
Windisch, S. and Nowotny, H. (1972) In "Preparative Methods in Solid State Chemistry" (P. Hagenmuller, ed.) p. 533. Academic Press, New York.
Wirkus, C. D. and Wilder, D. R. (1966) *J. Am. Ceram. Soc.* **49,** 173.
Witt, A. F. and Gatos, H. C. (1968) *J. Electrochem. Soc.* **115,** 70.
Witt, A. F. and Gatos, H. C. (1969) *J. Electrochem. Soc.* **116,** 511.
Wöhler, F. (1823) *Ann. Chim. Phys.* **29,** 43.
Wold, A., Kunnmann, W., Arnott, R. J. and Ferretti, A. (1964) *Inorg. Chem.* **3,** 545.
Wolfe, R., Sturge, M. D., Merritt, F. R. and Van Uitert, L. G. (1971) *Phys. Rev. Lett.* **26,** 1570.
Wolff, G. A. (1965) In "Adsorption et Croissance Cristalline" p. 722. Colloq. Intern. CNRS, No. 152, Paris.
Wolff, G. A. and Labelle, H. E. (1965) *J. Am. Ceram. Soc.* **48,** 441.
Wolff, G. A. and Mlavsky, A. I. (1965) In "Adsorption et Croissance Cristalline" p. 711. Colloq. Intern. CNRS, No. 152, Paris.
Wolff, G. A. and Mlavsky, A. I. (1974) In "Crystal Growth, Theory and Techniques" Vol. 1 (C. H. L. Goodman, ed.) 193. Plenum Press, London–New York.
Wolff, G. A., Herbert, R. A., Broder, J. D. (1958) Proc. Intern. Coll. on Semiconductors and Phosphors.
Wolff, G. A., LaBelle, H. E. and Das, B. N. (1968) *Trans. AIME* **242,** 436.
Wolff, G. A., Das, B. N., Lamport, C. B., Mlavsky, A. I. and Trickett, E. A. (1969) *Mat Res. Bull.* **4,** 567.
Wood, C. and Van Pelt, B. (1972) *Rev. Sci. Instrum.* **43,** 1374.
Wood, J. D. C. and White, E. A. D. (1968) *J. Crystal Growth* **3/4,** 480.
Woodall, J. M. (1971) *J. Electrochem. Soc.* **118,** 150.
Wulff, L. (1884) German Patent DRP 33.190.
Ziegler, J. G. and Nichols, N. B. (1942) *Trans. ASME* **64,** 759.

8. Liquid Phase Epitaxy

8.1. Introduction to Liquid Phase Epitaxy (LPE)

The development of electronic technology has been increasingly in the direction of devices utilizing a thin layer of material. Such planar devices offer advantages not only of ease of access for the input and retrieval of information in some form but also in terms of the material requirements. Many materials of interest are difficult to prepare as bulk crystals without serious inhomogeneities, strain, inclusions or other defects. The preparation of thin crystalline films therefore makes possible a range of devices which would not be possible (or economically feasible) if only bulk crystals were considered. Epitaxial chemical vapour deposition has already reached the stage where computer-controlled plants are used for commercial device production, and the realization of a similar stage in liquid phase epitaxy (LPE) appears to be close at hand.

In this chapter the methods and mechanisms of LPE will be reviewed, with particular reference to the two types of device which appear to offer the greatest current promise for commercial production. Although several devices have been proposed and investigated, the great majority of the effort to date has been concentrated on two types of material—the III–V semiconductors for applications as light-emitting diodes, including lasers, and magnetically anisotropic garnet materials which exhibit cylindrical "bubble" domain structures.

LPE normally involves the deposition of epitaxial films, typically several μm in thickness, on carefully prepared substrates of similar structure. The substrate material should approximately match the material to be deposited both in lattice parameter and in linear thermal expansion coefficient, otherwise strains are introduced into the film and dislocations or cracks occur if the mismatch is excessive. A distinction is made between *homoepitaxy*, in which case the film is of the same composition as the substrate except possibly for the nature or concentration of some dopant (as in *p–n* junction fabrication) and *heteroepitaxy* where the film and substrate differ markedly in composition.

The widespread use of LPE, especially for devices based on relatively thick layers ($>1 \mu m$), depends upon its convenience and on the high quality of the deposited film compared with that which can be obtained from the melt or from the vapour phase. An important advantage of LPE films over those grown from the melt is in the low concentration of point defects, which may be particularly troublesome in semiconductor devices which require a low residual impurity concentration. In addition, as shown by Vilms and Garrett (1972) for the growth of GaAs, certain impurities in the source material are removed from the gallium solution by evaporation into the stream of hydrogen gas flowing over the solution. It is possible that similar purification processes are operative in other solutions in addition to the normal impurity rejection by the growing crystal.

The LPE process involves the introduction of a carefully prepared substrate into a supersaturated solution, or into a nearly saturated solution in which supersaturation is created after a short period during which etching may occur. After growth of the layer has occurred, the substrate is removed and it is necessary to ensure that the surface is free from any drops of excess solution.

There are essentially two ways of promoting growth on the substrate crystal: either the substrate may be immersed in the solution, or the solution may be transported into the region of the crucible in which the substrate is located. There are, however, many variations on these basic processes and we shall examine the experimental methods in some detail. A comparison will be given in this chapter between the various techniques used by different workers and an attempt will be made to establish the important factors which determine the quality of the films, and to suggest the direction of future trends in this rapidly changing field.

A complementary approach has been adopted in a review of liquid phase epitaxy by Dawson (1972), who considers the application to specific III–V semiconductors and solid solutions. A recent review of LPE of magnetic bubble materials has been given by Ghez and Giess (1974).

8.2. Light-Emitting Diodes

The methods used for film growth by LPE have been evolved in many cases to meet a particular device requirement and it is desirable to consider the principal applications of LPE films prior to a detailed discussion of the experimental techniques. Of the major applications of devices which utilize LPE films, that with largest potential market is probably for semi-conducting diodes which emit visible light. These light-emitting diodes (LED's) have the advantages of cheapness, low power consumption, reliability and ease of fabrication into complex arrays, and are particularly attractive for numerical display and similar applications. Consideration has been given to their use for TV displays and for general illumination although these are some way from realization. Reviews of the preparation and properties of LED's have been given by Casey and Trumbore (1970), Thomas (1971), and, along with a comparison of chemical vapour de-position and LPE, by Minden (1973).

The excitation of electromagnetic radiation relies on the injection under forward bias of minority carriers, predominantly of electrons into the p-type region. The fraction of minority carriers that recombine with emission of radiation to the total that recombine is termed the *internal quantum efficiency* of the diode. A more meaningful quantity, however, is the *external quantum efficiency* which is lower because of the loss of photons by absorption within the material.

The wavelength λ of the emitted radiation depends upon the energy band gap W_G of the semiconductor according to the relation $\lambda \simeq hc/W_G$, where h is Planck's constant and c the velocity of light. For the emission of visible light, the band gap must exceed 1.8 eV, which corresponds to a wavelength of about 7000 Å. This requirement excludes germanium and silicon and, although the II–VI semiconductors have band gaps in the required region, they will also be excluded from this chapter since attempts to make a stable p–n junction have so far been unsuccessful. The III–V compounds which remain may be divided into two categories according to their band structure—in *direct gap* materials such as GaAs a photon is absorbed by the crystal with the creation of an electron and a hole, while in *indirect gap* materials like GaP the excitation of carriers involves a lattice phonon for momentum conservation.

Emission of radiation from direct gap materials is a relatively simple process in which the injected electrons recombine directly with holes, with the emission of a photon. The energy of the photon will be the same as the band gap except for a decrease with increase in the doping level due to the formation of a "tail" on the valence band by the merging of impurity levels with the band edge. The major problem with direct gap semi-conductors is self-absorption by the material of the radiation emitted, since the characteristic wavelength for absorption is the same as that for emission. As an illustration, Archer and Kerps (1967) reported an internal quantum efficiency for GaAs of nearly 50% but the external quantum efficiency was only 1%. Rupprecht *et al.* (1966) produced GaAs diodes of external efficiency 4% with silicon as the dominant impurity, and attributed this high value to the close compensation of the dopant. This value may be considerably improved by the use of a domed structure, and Ashley and Strack (1969) achieved values as high as 20% for a domed GaAs diode. Since the band gap of GaAs is only 1.4 eV, the radiation emitted is in the infra-red.

Because of the high efficiency of GaAs devices, LED structures have been made in which a diode is covered by a layer of phosphor which transforms some of the radiation into the visible region. An alternative approach is to alloy the GaAs with an element which results in an increase in the bandgap. Table 8.1 lists some of the materials which are currently

TABLE 8.1. Materials for Light-Emitting Diodes (after Thomas 1971)

Direct gap		Indirect gap		GaAs + phosphor	
Alloy	W_{max}(eV)	Compound	W_G(eV)	Phosphor	W_{ph}(eV)
$GaAs_{1-x}P_x$	2.0	GaP	2.3	YOCl(Yb, Er)	1.9
$Ga_{1-x}Al_xAs$	1.9	SiC	3.0	YF_3(Yb, Er)	2.3
$Ga_{1-x}In_xP$	2.2			YF_3(Yb, Tm)	2.6

of interest for LED's. $Ga_{1-x}Al_xAs$ layers are readily grown by LPE on GaAs substrates since the lattice parameters of GaAs and AlAs differ by only 0.009 Å. One problem which is encountered in the production of such layers is the large distribution coefficient of Al which normally leads to a decreasing aluminium concentration during growth.

The maximum direct gap of 2.2 eV in the $Ga_{1-x}In_xP$ system makes possible the emission of a range of wavelengths from red to green. The red-emitting $GaAs_{1-x}P_x$ alloys are widely used commercially but the diodes are normally prepared from the vapour phase.

In GaP and other indirect gap materials the emission of radiation is a much more complex process which involves an interaction with *exciton* centres. The emission of red light depends upon the presence on adjacent lattice sites of zinc (or Cd) and oxygen atoms which form a neutral Zn—O complex. An electron trapped by the complex can decay by pair emission with a hole bound to the Zn acceptor or by radiative (exciton) decay with the bound hole. The great advantage of indirect gap materials is that absorption at the gap energy is very small so that self-absorption by the material is low. The external quantum efficiency is not so high as in Si-doped GaAs but the value of 7% reported by Saul et al. (1969) still appears to be the highest for any electroluminescent device emitting in the visible region.

Nitrogen in GaP also acts as a shallow electron trap due to the difference in atomic number between nitrogen and phosphorus, although both are trivalent. The trapping energy in this case is only about 0.01 eV. When electrons are injected into a nitrogen-doped p-type region, they are trapped by the nitrogen and can attract holes to form a bound exciton, which decays with the emission of green light. The efficiency of nitrogen-doped diodes is relatively low, about 0.6%, but the diodes appear roughly as bright as the red-emitting diodes because of the greater sensitivity of the eye to green light. Lorim et al. (1973) proposed that higher efficiencies could be achieved by overcompensation as a means of producing the junction; this has the advantage that the wafers need not be removed from the solution during growth.

Silicon carbide is included in Table 8.1 since it could be used for the emission of a wide range of wave lengths, including the ultraviolet region. The bandgap varies according to the polytype and yellow emission has been reported for the hexagonal 6H form. However the difficulty of fabrication of epilayers and of substrate crystals has prevented the widespread development of practical devices.

In Fig. 8.1 a comparison is given of the light output from various LED structures, according to the review of Thomas (1971). The red GaP diodes are seen to require the lowest power for displays and are therefore preferred at present but other materials are capable of development by a greater factor, and rapid changes are likely in the immediate future. InP LED'S have been studied recently by Williams et al. (1973), who report a quantum efficiency of 1.5% at 300°K.

Lasers. The emission of high intensity electromagnetic radiation from junction diodes led to early attempts to fabricate laser diodes, and Nelson (1963) reported the first semiconductor lasers to be made by LPE. The diodes were made by cleaving the wafers to provide plane-parallel ends perpendicular to a flat p–n junction obtained by deposition of n-GaAs on a p-GaAs substrate. The threshold current for these devices was lower by a factor 4 than for diffused diodes of the same geometry.

Subsequent attempts to reduce the threshold current density $\mathcal{J}_{th}$ in order to achieve continuous laser action at room temperature were reported by Dousmanis et al. (1964) and by Rupprecht (1967). The threshold of the GaAs devices made by the former group was about 4×10^4 A cm^{-2} at 300 K, while Rupprecht achieved a value of 2.6×10^4 A cm^{-2} at the same temperature, using a modified laser structure. Silicon, which exhibits a temperature-dependent amphoteric doping behaviour, was the principal dopant for both n- and p-type material.

Still lower threshold currents may be achieved if heterojunctions are used so that there is a sharp discontinuity in band structure and refractive

P

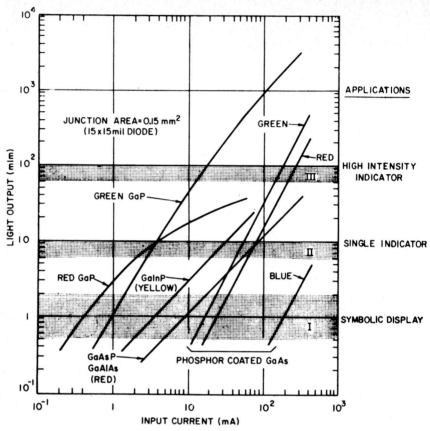

FIG. 8.1. Light output versus current for various types of LED. The values quoted are the highest reported (Thomas, 1971).

index at the p–n junction. The use of heterojunction diodes was suggested by Krömer (1963), and Rupprecht *et al.* (1967) reported the fabrication of LED'S using $Ga_{1-x}Al_xAs$ deposited on GaAs. Single heterojunction (SH) lasers with thresholds as low as 1.0×10^4 A cm^{-2} were made by Hayashi *et al.* (1969).

A comparison of the homo- and single hetero-junction laser diode structures is given in Fig. 8.2. Electrons injected into the p-GaAs layer of Fig. 8.2(b) are repelled by the step in the conduction band at the boundary with the $Ga_{1-x}Al_xAs$ layer, while hole injection is prevented by the difference in the effective bandgap of highly doped n-GaAs and the compensated p-GaAs. Recombination therefore occurs between the p–n junction and the hetero-layer boundary, and the threshold current depends

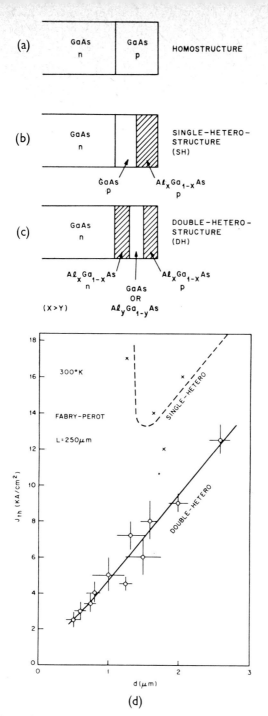

FIG. 8.2. Schematic cross-section of (a) homostructure, (b) single heterostructure and (c) double heterostructure lasers, (d) threshold current density *versus* film thickness *d* for SH and DH diodes for a Fabry–Perot device of length 250 μm (Hayashi *et al.*, 1971).

very strongly on the width of the p-GaAs layer. A value of 2 μm was found to be optimum, with an increase leading to a lower light intensity for a given current and a decrease causing a higher $\mathcal{J}_{th}$ because of hole injection.

The threshold current density may be still further reduced by the use of double heterostructure (DH) diodes as shown in Fig. 8.2(c). Alferov *et al.* (1969b, c) reported a room temperature threshold current of 4.3×10^3 A cm^{-2} and Casey *et al.* (1974) subsequently a chieved values around 650 A cm^{-2} for a square totally internally reflecting diode. The discontinuity in the bandgap at a p–p junction in the $Ga_{1-x}Al_xAs$ system is entirely in the conduction band (Alferov *et al.*, 1969a) so that electrons are blocked by this potential barrier. Hole injection is prevented by the residual potential barrier in the valence band and low values of $\mathcal{J}_{th}$ are observed, particularly if the thickness of the GaAs layer is below 1 μm. A comparison between $\mathcal{J}_{th}$ values for SH and DH structures is shown in Fig. 8.2(d). Fabry–Perot lasers on diamond heat sinks have been operated continuously at room temperature, giving about 20 mW maximum power at a current of 1 A, corresponding to a current density of about 3000 A cm^{-2}. A review of the development of DH lasers has been given by Hayashi *et al.* (1971). One of the problems in the application of DH lasers for high density data transmission via glass fibres is the lifetime of the devices, which at present does not exceed 100 hours.

It is expected that even lower thresholds will be possible with a one-dimensional superlattice structure in which the composition varies periodically with a spacing of 100–200 Å between layers (Esaki and Tsu, 1970). Such a structure has been achieved by vapour growth and techniques for its realization by LPE will be discussed in Section 8.4. This structure may also lead to novel high frequency devices.

8.3. Bubble-Domain Devices

It was established by Bobeck and coworkers (Bobeck, 1967; Bobeck *et al.*, 1969) that magnetically anisotropic materials of low magnetization will support highly mobile cylindrical or "bubble" domains which are of great interest for memory applications. Garnet materials appear to be most suitable for this application since both the bubble diameter and the mobility have acceptable values (Bobeck *et al.*, 1970; van Uitert *et al.*, 1970). The growth of garnet films and the factors affecting the choice of composition have been reviewed by Robertson (1973) and by Stein (1974).

The bubble diameter b is related to the saturation magnetization M_s, magnetic exchange energy W_{ex} and uniaxial magnetocrystalline anisotropy K_u by an expression due to Thiele (1969)

$$b \simeq W_{ex}^{1/2} K_u^{1/2} M_s^{-2}.$$

$4\pi M_s$ is typically about 150 gauss at room temperature, the reduction in magnetization compared with the undoped iron garnets normally being achieved by replacing some tetrahedral Fe by Ga or Al. The magnetization should be substantially independent of temperature and this requirement is often fulfilled by selecting a composition in which room temperature lies midway between the Curie temperature and the compensation temperature.

The uniaxial magnetic anisotropy arises either from stress-induced anisotropy due to mismatch between the lattice parameters of the film and the substrate or from growth-induced anisotropy. According to Rosencwaig and Tabor (1971) and Callen (1971), the latter effect is due to ordering of the ions occupying the dodecahedral sites in the garnet structure and their models explain most of the observed data at least qualitatively. However, Isherwood (1968) demonstrated that even undoped yttrium iron garnet exhibits a departure from cubic symmetry and Stacy and Tolksdorf (1972) reported a noncubic magnetic anisotropy in YIG. The latter observation was explained by Stacy and Rooymans (1971) in terms of ordering of defects such as impurity ions or vacancies and it is possible that other non-annealable mechanisms may contribute to the anisotropy.

The stress σ in an epitaxial film is given (Besser et al., 1971, 1972; Carruthers, 1972) by

$$\sigma = \frac{E}{(1-\mu)}(1-\eta)\left(\frac{a_s - a_f}{a_f}\right) + \eta(\alpha_s - \alpha_f)\,\Delta T$$

where E is Young's modulus and μ Poisson's ratio of the film, a_s and a_f the room temperature lattice parameter of substrate and film respectively, α_s and α_f their expansion coefficients, ΔT the temperature difference between the growth temperature and ambient, and η the fractional stress relief. The lattice parameter mismatch between the film and the substrate may be made negligibly small but stress arises from the difference in expansion coefficient between the gallium garnet substrate ($\sim 9.2 \times 10^{-6}\,°C^{-1}$) and the iron garnets ($\sim 10.4 \times 10^{-6}\,°C^{-1}$). In the absence of stress relief the resulting tension would be in the region of 3×10^8 Nm^{-2} or 45,000 p.s.i. (Blank and Nielsen, 1972). This value is close to the fracture limit of many oxide materials and cracking may be expected if the mismatch in a or α has a larger value. For garnet films grown on {111} substrates, the anisotropy resulting from the stress is given by

$$K_u = -\tfrac{3}{2}\,\sigma\lambda_{111}$$

so that the anisotropy may be varied by selecting rare earth ions having the appropriate magnetostriction coefficient λ_{111}. The values of this parameter together with other useful data for the choice of a garnet composition are given in the paper of van Uitert et al. (1970). Current practice

(Roberson, 1973) is to minimize stress-induced anisotropy and to choose a composition in which a rare earth ion of large ionic radius such as Y, Gd or La is combined with one of small ionic radius such as Tm, Yb or Lu. Ions in a spectroscopic S-state are preferred since the bubble mobilities are found to be higher than in materials in which there is an orbital contribution to the magnetic moment. Bubble mobilities of the order of 10^3 cm s^{-1} Oe^{-1} are required for devices.

Another factor which might influence the choice of film composition is the requirement of a good mismatch of lattice parameter and expansion coefficient with the substrate. However a variety of non-magnetic garnets are available as high quality crystals for use as substrates and it is often more convenient to change the substrate rather than the film composition to achieve a suitable match.

The variation in stress-induced anisotropy for films grown on substrates of different lattice parameter has been studied by Giess and Cronemeyer (1973), and the change in anisotropy due to the dependence on temperature of the misfit was investigated by Stacy et al. (1973).

8.4. Experimental Techniques

8.4.1. Tipping

The development of liquid phase epitaxy, first reported by Nelson in 1961, was stimulated by the paper of Nelson (1963) which describes the epitaxial growth of germanium and gallium arsenide. The apparatus used is shown diagrammatically in Fig. 8.3(a). The substrate wafer is held tightly against the base of a graphite boat which is located in the centre of a uniform temperature zone of a furnace, inside a quartz glass tube through which hydrogen gas is flowed.

Initially the substrate wafer is well clear of the solution and the furnace is heated to form a solution of, for example, GaAs in Sn. The power is then switched off and the furnace tipped so that the wafer is covered by the solution. Initially the solution is not quite saturated and some dissolution of the substrate material occurs. On continued cooling, the solution becomes saturated and eventually supersaturated so that epitaxial growth occurs on the GaAs substrate. After deposition of a suitable layer, typically 20 μm in thickness, the furnace is tilted back to its original position and the graphite boat immediately removed from the furnace and any remaining solution wiped off the epitaxial layer. The type of temperature programme used by Nelson is shown in Fig. 8.3(b). The cooling rate will depend on the volume of solution used, the linear growth rate being in the region of 200 Å s^{-1} as for bulk crystals. A period of only 20 minutes is required for the growth of a layer of sufficient thickness for device fabrication compared

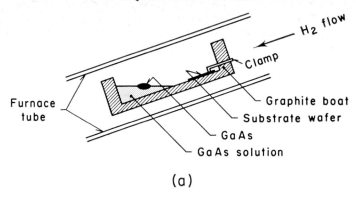

(a)

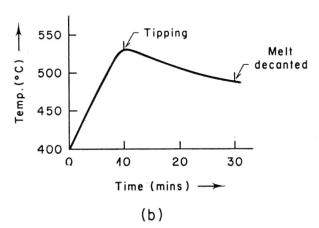

(b)

FIG. 8.3. (a) Apparatus for epilayer growth by tipping, (b) temperature programme (Nelson, 1963).

with many days for the growth of a bulk crystal. Nelson prepared GaAs
laser diodes from Sn or Ga solution and Ge tunnel diodes using indium as
solvent.

Burns (1968) used a similar arrangement to prepare n-GaSb from Ga
solution but with a different type of temperature programme. In this case
the temperature changes were carried out rapidly in steps of 2–7°C, with
long holding periods after each change. The advantage of this procedure
would appear to be that etch-back of the substrate, equilibration and
growth occur under substantially isothermal conditions but a comparison
of the resulting films with those grown by the type of programme used by
Nelson was not given.

Trumbore *et al.* (1967) drew attention to the importance of the method of doping for the properties of a *p–n* junction, considering in particular the preparation of (Zn + O)-doped GaP films. They used an open-tube system with a flow of hydrogen and nitrogen, with a separate Zn source to provide Zn vapour to compensate for evaporation losses from the solution. A two-zone furnace was used in order to regulate the partial pressure of zinc in the region containing the solution. Direct flow of gas over the solution was avoided in order to reduce the loss of Zn and Ga_2O by evaporation. The principal disadvantage of this system is the poor control of the oxygen content of the solution.

Tipping in a closed tube system was reported by Shih *et al.* (1968) and by Allen and Henderson (1968). Ladany (1969) used a gas flow but with a covered boat in order to reduce dopant evaporation. Saul *et al.* (1969) achieved the very high value of 7.2% for the efficiency of a GaP light emitting diode by performing a 2-step anneal in the closed tube tipping furnace. The importance of doping in relation to the performance of light-emitting diodes will be discussed further in Section 8.6.

The first successful preparation of magnetic garnet films of high quality was described by Linares (1968) using a method very similar to that of Nelson (1963) but with a platinum boat. Gadolinium gallium garnet was found to be the best substrate material, mainly because of the lattice parameter match to the magnetic garnets at the growth temperature. Irregular surfaces developed on (111) substrates if the growth rate exceeded 100 Å s^{-1}.

Orthoferrite films were grown by Shick and Nielsen (1971), who used a platinum screen to prevent undissolved particles of orthoferrite from drifting into the growth region. The substrate, fastened to a platinum sheet, was mounted in a vertical position at the end of the boat since this position was found to give a more uniform temperature distribution across the wafer. Best results were obtained with the substrate held some 5–10°C below the temperature of the solution prior to tipping, in order to prevent partial dissolution of the substrate. The solvent was $50PbO:1B_2O_3$ and the linear growth rate only 10 Å s^{-1}. Growth was not possible on any non-magnetic substrate due to the relatively rapid dissolution of these materials. In all hetero-epitaxial deposition it is important to realize that the solution is not saturated with respect to the substrate material, and kinetic factors determine whether epitaxial layer growth will occur in preference to substrate dissolution.

Shick *et al.* (1971) described the growth of magnetic garnet films on gadolinium gallium garnet using the same technique and discussed the characteristics of the resulting materials in relation to their application in bubble-domain devices.

Recently it appears that tipping has become less popular for the growth of garnet films than the dipping techniques which will be described in the next section. A notable exception is the work of Plaskett (1972) who reported the highly reproducible growth of films of good quality. An important problem in the achievement of uniform, homogeneous films is the avoidance of temperature gradients across the substrate, and Plaskett has mounted the substrates in such a way that the surface is horizontal during growth, rather than inclined to the surface of the liquid at the tipping angle.

In the growth of semiconductors by tipping, Donahue and Minden (1970) introduced rotary crucibles in which the tilting is replaced by a rotation of 180° or 360°. The principle of this technique is analogous to that used for nucleation control in flux growth of bulk crystals by Bennett and Tolksdorf which is described in Chapter 7. The great advantage of this

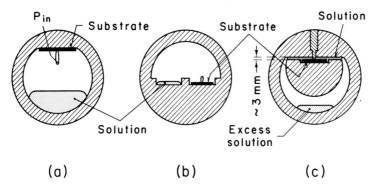

(a) (b) (c)

FIG. 8.4. Apparatus for epilayer growth using small volume of solution (Donohue and Minden, 1970).

technique is that only the crucible is moved, rather than the whole furnace. The various boats used by Donahue and Minden are shown in cross-section in Fig. 8.4. In the simplest version, depicted in Fig. 8.4(a), the substrate wafer is held against a flat portion of the inner wall of the cylindrical graphite crucible by an L-shaped pin which is anchored in the base. The solution, which has a relatively small volume, is made to flow over the substrate by inverting the crucible.

The crucible shown in Fig. 8.4(b) is an alternative version in which the substrate is held in one of two wells, the second of which contains the solution. A 360° rotation is necessary in this case to bring the solution into contact with the substrate. Rotation in the opposite direction is used to decant the melt after growth of the film by slow cooling.

In the crucibles described so far, the wafer is normally located further from the axis of the furnace than the solution, so that the temperature gradient is in the wrong direction for stability according to the supersaturation gradient criterion. This is remedied in the version shown in Fig. 8.4(c) in which the substrate wafer is held in a recess so that it is closer to the furnace axis than the solution. This design has the additional advantage that only a very thin layer of solution is in contact with the wafer. The solution is held in place by surface tension in a 3 mm gap between the substrate and the crucible wall. In all the experiments of Donahue and Minden, 5% of Zn was used in place of the equivalent amount of gallium; this was found to enhance wetting of the substrate without appreciable substitution in the lattice. The design shown in Fig. 8.4(c) was found to give the best quality films, but all the films showed an irregular surface structure.

Rotatable crucibles were also used by Stringfellow and Greene (1971), who required to grow films under isothermal conditions. They encountered particular difficulties in attempting to grow $InAs_{1-x}Sb_x$ solid solutions by polythermal methods, because of the large distribution coefficient of As and the relatively strong dependence of the lattice parameter on the As concentration. As with the growth of bulk crystals of solid solutions, a steady state method is required in which a temperature gradient is maintained across the solution with nutrient material at the hotter and the substrate wafer at the cooler end. Because of the large distribution coefficient of the arsenic, InAs was used as substrate material since the rate of depletion of Sb from the melt was slight.

The apparatus is shown in Fig. 8.5. The constituents of the solution

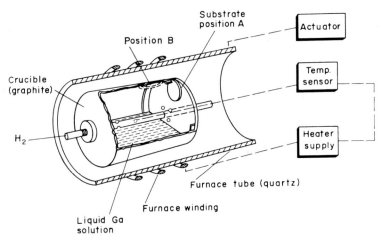

FIG. 8.5. Rotary crucible (Vilms and Garrett, 1972).

are placed in the crucible and held at the growth temperature for several hours in order to achieve equilibration. The substrate and source are then simultaneously introduced into the solution by rotating the crucible through 180° about the axis of the furnace. Rotation to the original position is used to terminate growth after a suitable period. A temperature difference between source and substrate of 7–15°C was found to be optimum, with higher values resulting in the incorporation of solvent inclusions. No evidence was found of substrate dissolution prior to growth. With an InAs substrate, the composition of the epitaxial layer varied from that of the substrate to that of the solid solution within the first 4 μm of growth, and remained constant over the 80 μm of subsequent growth. The greatest Sb concentration in the epitaxial layer was $x = 0.25$, beyond which the lattice parameter mismatch with InAs is excessive.

A steady state method of epitaxial growth had been used previously by Panish et al. (1966), who utilized a radial temperature gradient to transport GaAs from nutrient material at the perimeter of the solution to a central seed mounted onto a silica heat sink.

Vilms and Garrett (1972) have given a detailed account of the growth of GaAs using the type of rotary crucible shown in Fig. 8.5, describing in particular the impurity concentration of doped and undoped layers. It was found that small air leaks in the gas supply are a major cause of impurities in the epilayers. In a leak-tight system the concentration of $1-2 \times 10^{14}$ cm^{-2} was achieved by growth at temperatures below 700°C. The rate of growth, about 4×10^{-7} cm s^{-1}, was reproducible to within 12%, and the variation in thickness across a film was in the region of $\pm 20\%$. The average doping density was reproducible to within $\pm 6\%$. A comparison was not given between wafers mounted horizontally and those mounted vertically.

One disadvantage of the methods described so far is that the solution is stationary during growth so that stirring occurs only by thermal and solutal convection. Controlled stirring could be achieved in principle in the crucible of Fig. 8.5 by alternate rotation in opposite directions about its axis by, say, 10–15°. As discussed extensively in Chapters 6 and 7, growth in an unstirred solution may result in more rapid growth at the edges of the substrate and the maximum stable growth rate will be much less than that in a stirred solution. An apparatus for the simultaneous growth of epilayers on several substrates under stirring has been proposed by Scheel and Schulz-DuBois (1972) and is shown in Fig. 8.6. The wafers are located in recesses in a shallow crucible which can be tilted by about 15°. The solution is allowed to homogenize in an outer reservoir A and the crucible is then tilted so that the solution runs into the central chamber where it is retained by a dam which separates this chamber from the reservoir. Growth occurs by programmed cooling with the crucible in its

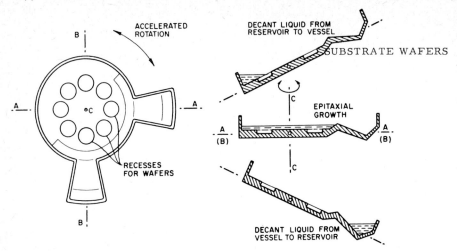

FIG. 8.6. Crucible for epilayer growth of several wafers in stirred solution (Scheel and Schulz-DuBois, 1972).

horizontal position, with stirring by accelerated rotation (see Chapter 7). When the epilayer has been grown, the solution is decanted from the central compartment by tilting the crucible about either the A- or B-axis. The use of more than one reservoir permits multiple layers to be deposited. A simpler means of stirring, by rotation of a horizontal substrate, was also mentioned by Scheel and Schulz-DuBois (1972), and suitable substrate holders are shown in the next section. In the apparatus of Lien and Bestel (1973) the solution is forced from an inner chamber to an outer one containing eight substrates by centrifugal force as the crucible is rotated at 500–1000 rpm.

8.4.2. Dipping

The dipping method, in which the substrate is immersed into a supersaturated solution, appears to have been first reported by Linares et al. (1965), who grew garnet films by gradient transport onto YAG and GdGaG substrates. A similar technique was used by Rupprecht (1967) who required to deposit n-GaAs on a p-type substrate. The advantage of dipping is that the time of growth is short and well-defined, and no movement of the crucible is necessary.

Deitch (1970) advocated a "quick dip" method for the growth of GaAs from Ga solution, with the substrate temperature some $50°$ lower than that of the solution. The initial growth is then very rapid, but the large temperature gradient at the substrate-solution interface has a stabilizing effect. In subsequent studies of the same system by Crossley and Small (1972), smoother layers and better reproducibility were obtained when the substrate was undercooled by only 5–10°C, and most workers have preferred to avoid a large temperature difference between substrate and solution.

Levinstein *et al.* (1971) pointed out that the particular advantage of the dipping technique for LPE growth of magnetic garnet films is that the solution in a $50PbO : 1B_2O_3$ flux may be supercooled by up to 120°C and that the supercooled melt can be stable with respect to garnet nucleation. Growth may occur under isothermal conditions and several films can be grown by dipping of successive substrates with the solution remaining at constant temperature. In these experiments the substrate was held in a vertical position and maintained just above the melt for 5 min. before dipping in order to attain a temperature close to that of the solution. The growth rate for a supercooling of 30°C was 70 Å s^{-1}. The wafer was slowly withdrawn and the solution found not to wet the film during withdrawal. The density of etchable defects was less than 10 cm^{-2} and a value of zero found in some cases. The uniformity of the film was ± 2 to 5% over an area of 0.6 cm^2.

More detailed investigations of a similar system were carried out by Blank and Nielsen (1972) and Blank *et al.* (1972) and a cross-section of their apparatus is shown in Fig. 8.7. The crucible has a volume of about 80 ml and is located in a constant temperature region about 15 cm in length. The constituents of the solution are initially melted in a larger crucible and the solution poured into the crucible to ensure uniformity and to produce only small garnet crystallites, so that redissolution is rapid on heating to the growth temperature.

In order to deposit garnet films it is necessary to select a region of the phase diagram such that garnet is the only stable phase. The general features of the pseudoternary phase diagram flux —R_2O_3—Fe_2O_3 are shown in Fig. 8.8, and the composition range found by Blank and Nielsen (1972) to be optimum for the growth of their garnet films is summarized in Table 8.2. The distribution coefficient for gallium in yttrium iron garnet grown from an unstirred solution was found to vary from 2.26 at 1077°C to 2.09 at 879°C and was substantially independent of the garnet composition. These values are slightly higher than those of Giess *et al.* (1972) or Janssen *et al.* (1973) but there is general agreement on the tendency of the distribution coefficient to decrease with decreasing temperature. The actual values observed will depend on the growth rate.

The investigations of Blank and Nielsen revealed some very unusual features of the behaviour of the supersaturated solution. In the region where the orthoferrite is the stable phase, the *maximum* supercooling is only about 5°C and the addition of only about 0.2% by weight of the total melt of Fe_2O_3 will increase this value to 30–60°C if it displaces the com-

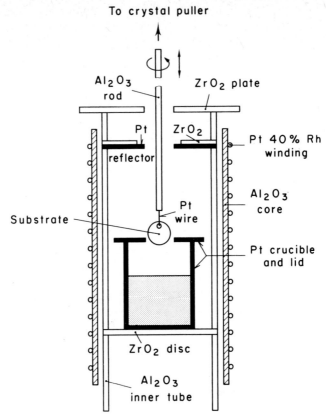

FIG. 8.7. Furnace for garnet film growth on vertical rotating substrate (Blank and Nielsen, 1972).

position into the garnet stable region. By selecting a composition close to the phase boundary, it is possible to produce a solution saturated with respect to orthoferrite but supersaturated with respect to garnet, so that a garnet film may be grown epitaxially in the presence of orthoferrite crystallites, but without the spontaneous nucleation of any garnet. The time over which a supercooled melt is stable increases with the distance from the orthoferrite boundary.

As stressed by Nielsen (1972), high growth temperatures must be avoided in order to prevent appreciable dissolution of the substrate. A lower growth temperature will also reduce the rate of solvent evaporation and hence the attack on the substrate prior to immersion.

Blank and Nielsen also found that there exists a critical mismatch between the lattice parameters of a garnet film and the non-magnetic

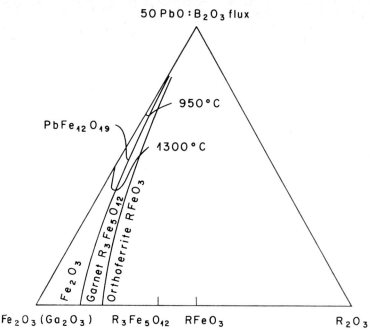

FIG. 8.8. Pseudo-ternary phase diagram showing stability fields of garnet and orthoferrite phase (Blank and Nielsen, 1972).

garnet substrate. Nucleation did not occur even in a strongly super-saturated solution if the mismatch between the substrate and the stable solute phase exceeded 0.19 Å. This value is of course much larger than that for cracking of the films, which Tolksdorf *et al.* (1972) reported to be 0.015 Å.

If devices based on LPE films are to be commercially viable, holders suitable for the simultaneous growth of several films under stirring are necessary, and designs discussed by Scheel and Schulz-DuBois (1972) are shown diagrammatically in Fig. 8.9. In the version shown in Fig. 8.9(a),

TABLE 8.2. Solution Composition for LPE Growth of Magnetic Garnets

Ratio	Recommended values (Blank *et al.*, 1972)
$Fe_2O_3 : R_2O_3$	14–25
$Fe_2O_3 : Ga_2O_3$ (or Al_2O_3)	6–16
$PbO : B_2O_3$	15–6
Garnet : Total solution	~0.08
Er : Eu in $Er_{3-x}Eu_xFe_{4.3}Ga_{0.7}O_{12}$	1.7–3

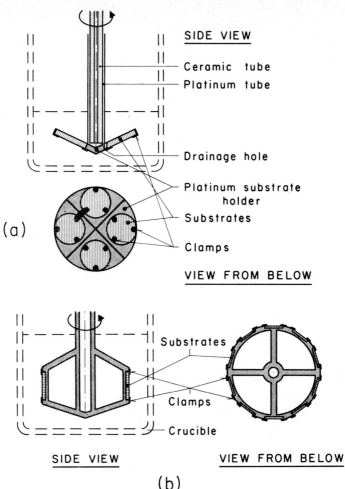

Fig. 8.9. Rotating substrate holder (a) with slightly inclined substrates, (b) with vertical substrates (after Scheel and Schulz-DuBois, 1972).

the wafers are held at a small angle to the horizontal in order to facilitate wetting by preventing air bubble formation during dipping and to make possible complete drainage of the solution. If the clamps are flat, the substrates may be withdrawn from the solution without adhering droplets. The clamps should be of minimum thickness in order to avoid irregular flow patterns. A drainage hole is provided to prevent trapping of the solution in the hollow region above the substrates. The rotable substrate holder should be made of a platinum alloy containing not more than 5% of rhodium (for increased strength) in the case of oxide layer growth. For semiconductor growth, the substrate holder would be made from graphite. Stirring has been used in the apparatus of Lorim *et al.* (1973) who used two vertically mounted GaP wafers on a holder moved over 30° in an oscillatory motion.

Figure 8.9(b) shows an alternative version in which the substrates are mounted in a vertical position. The turbulence introduced on rotation will be minimized if several wafers are supported around the periphery of the holder so that its circumference is almost circular. The advantages of the simultaneous preparation of a batch of devices instead of single ones are obvious for applications in research and industry.†

Giess *et al.* (1972) used a single horizontal rotating substrate dipped into a supercooled melt in order to grow bubble garnet films. The wafer is held against a flat platinum base by a number of platinum tabs bent over the edges of the wafer.

Prior to immersion in the solution, the wafer is heated to a temperature close to that of the liquid. During this preheating it is necessary to protect the substrate from attack by the flux vapour. Rotation is maintained during growth but is stopped as the wafer is removed; when the wafer has cleared the melt surface, rapid rotation at about 1000 r.p.m. is used in order to remove any droplets of solution which may be adhering to the surface. The holder may be withdrawn from the furnace extremely rapidly and cooled to room temperature in about 3 min, apparently without cracking or similar damage to the film. The concentration of lead from the flux was in the range 0.2–3.0 wt%, increasing with the growth rate. Film thickness uniformity was within 2×10^{-5} cm, or 2% for a 10 μm film, and the composition was found to be uniform across the film.

Several films can be produced from the same melt, the depletion of solute due to film growth being offset to some extent by solvent evaporation. The process is now controlled by an automated system which permits accurate control of the time during which growth occurs. The operation of apparatus over extended periods and the resulting reproducibility of film parameters have been discussed by Stein (1974).

Hiskes and Burmeister (1972) proposed the use of a $BaO/BaF_2/B_2O_3$ flux, which has a much lower volatility than the PbO/B_2O_3 flux. An additional advantage of the barium flux is that the Ba concentration in the garnet films was some two orders of magnitude lower than the Pb concentration found when the lead flux was employed. The magnetic properties and defect densities were comparable to those of films grown from PbO/B_2O_3, but the lattice parameter and optical absorption exhibited marked differences (Hiskes and Burmeister, 1972). The high surface tension of these fluxes is

† The simultaneous growth of films on several horizontal substrates mounted in a vertical stack has been achieved in various laboratories.

an important disadvantage (Elwell *et al.*, 1974) since the solution is difficult to remove from the film after it has been raised out of the liquid.

8.4.3. Sliders

Sliders are boats, normally of graphite, in which a small volume of solution is transported by the lateral movement of a graphite plate. The initial development of sliders resulted from the requirement to remove all traces of excess solution from a semiconductor film following growth.

The earliest form of slider appears to be that of Rosztocy (1968), mentioned in the review of Casey and Trumbore (1970), the operation of which is illustrated in Fig. 8.10. The initial homogenization of the solution and the epilayer growth is by a conventional tipping procedure, as shown in Fig. 8.10(a) and (b), except that the substrate is held in a recess with its surface level with the base of the boat. When the boat is tilted to terminate growth, as in (c), the movement of the sliding section wipes the solution off the surface of the film and confines the residual solution in a region well removed from the wafer. Panish *et al.* (1969) described a slider used simply to remove any oxide layer from the surface of the solution prior to the introduction of the seed.

Sliders also have the advantage that they are particularly suitable for the deposition of multiple films as required for single or double heterostructure lasers. In the version described by Panish *et al.* (1971), tipping is used to

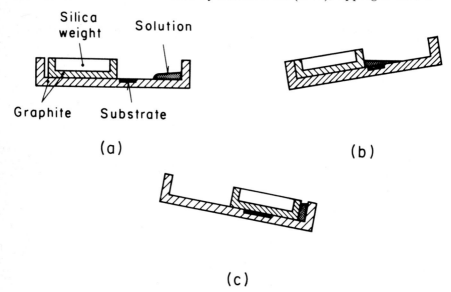

FIG. 8.10. Slider used to move solution in tipping process (Rosztocy, 1968).

move successively three solutions containing $Ga_{1-x}Al_xAs$ with different dopants over a single substrate so that p-type layers of different composition are deposited on an n-type substrate. The time of contact can be extremely short so that very thin layers may be deposited. In the alternative design of Hayashi *et al.* (1970) a push rod is used so that the boat remains in a horizontal position during growth. Continuous cooling at $1-3°C/min$ was used for the deposition of multiple $Ga_{1-x}Al_xAs$ layers and films of thickness $1-3$ μm (apart from the initial layer) were deposited in a time of only 15 seconds.

A sliding shutter was employed by André *et al.* (1972) in the LPE growth of $Ga_{1-x}Al_xAs$. The shutter covered the wafer prior to growth of the epilayer and was used only as a means of initiating growth.

Blum and Shih (1972) described an improved form of slider which is illustrated in Fig. 8.11. The solution is contained in a cavity in plate 2, which is held in position by the weight of plate 1. After equilibration with the source material, the solution is moved to a position over the substrate by displacement of plate 2, and an epitaxial layer is grown by cooling the furnace. When a layer of suitable thickness has been grown the solution is returned to its original position and any solution on the surface of the layer is wiped off during the displacement. Suitable dopants for subsequent layers may be introduced into the solution by displacement of plate 1.

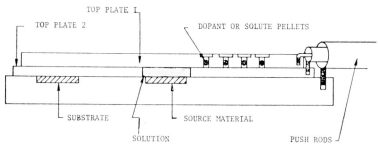

FIG. 8.11. Apparatus with two sliding plates (Blum and Shih, 1972).

The slider is located in a vertical temperature gradient with the substrate cooler than the solution, but the temperature is constant in the horizontal direction. In the growth of $Ga_{1-x}Al_xAs$ films on a GaAs substrate the variation in thickness across the film was $\pm 6\%$, and the reproducibility between different runs was within $\pm 8\%$. A modified version of this apparatus was used to prepare four wafers each 2 cm² in area on one wafer. High-quality light-emitting diodes were made by depositing a p-type $Ga_{1-x}Al_xAs$ layer on a p-type GaAs substrate prior to deposition of the n-type layer to form the diode.

A further development was described by Lockwood and Ettenberg (1972) in which a very small volume of solution containing the appropriate dopant is used for each layer. The GaAs source material floats in each compartment on a layer of solution about 0.2–1 g in weight and 1 mm in thickness. A quartz glass rod and graphite spacer are used to cover the solution, and so this design should have the advantage of low volatilization as well as the stability expected from the use of a thin layer of solution.

Bergh *et al.* (1973) have proposed design features of a commercial system for LPE growth using slider arrangements and thin solvent "aliquots". The quality of the substrates and the amount of dopant rather than the degree of constitutional supercooling were found to affect the usefulness of the layer.

8.4.4. Other methods

Woodall (1971, 1972) has developed apparatus for the growth of very thin layers of $Ga_{1-x}Al_xAs$ for application in heterojunction lasers and which avoid the use of mechanical wiping on termination of growth. The wiping may result in damage to the epilayer if too high a pressure is applied, or to imperfect wiping of residual solution if the pressure is insufficient.

In the earlier version (Woodall, 1971) growth is terminated by rapid isothermal transfer of the substrate from the solution, or into a solution of different composition, by rotation of a central cylindrical substrate holder. In the outer region of the graphite crucible are located two annular regions containing respectively the solution used for deposition of the epilayer and a second solution which will normally contain a low concentration of solute if a single film is to be grown. Alternatively the second solution may be a saturated solution of different composition, and continued slow cooling with periodic transfer then used to prepare multiple heterojunction films. One disadvantage of this apparatus is that a small amount of solution is transferred with the wafer as it is rotated from one solution to the other.

In the later version (Woodall, 1972), which is shown in Fig. 8.12, the substrate is supported horizontally in a recess in part b, which is situated immediately below part c which contains a cavity of cross section slightly larger than that of the substrate. Part a is an outer crucible which contains b, c and also part d in which are located separate cavities containing the solutions. The location of the GaAs source material is shown at e, while f is a quartz glass tube used for the addition of a dopant during the growth cycle.

In operation, b and c may be rotated from one melt to the other. If the temperature is lowered continuously with the wafer in contact with one or other of the melts, growth occurs as in the conventional tipping process. Alternatively the substrate may be rotated to a position between the

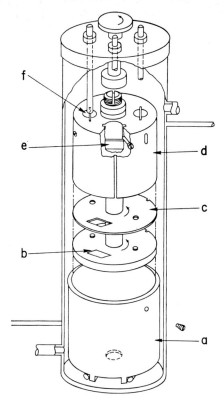

Fig. 8.12. Apparatus for substrate transfer between two solutions by rotating of central holder (Woodall, 1972).

solution chambers and cooling promoted so that deposition occurs from a very small volume of solution. A third possibility is that the substrate is rotated alternately between two melts of different composition. Growth occurs as a small volume of Ga/Al/As solution, which is transferred with the wafer, is mixed with a second solution in equilibrium to produce a new supersaturated solution. This technique is termed *isothermal solution mixing* by Woodall and its principle may be understood with reference to Fig. 8.13. This diagram shows the portion of the Ga/Al/As phase diagram in the Ga-rich corner. Points A and B represent two compositions on the same isotherm at which solid and liquid phases are in equilibrium. If liquids of these compositions are mixed, the composition will lie along the straight line joining A and B and the resulting composition will be in the supersaturated region, below the equilibrium liquidus curve.

   In the preparation of very thin layers, the second method appears to offer the greatest advantages and Woodall was able to prepare linear

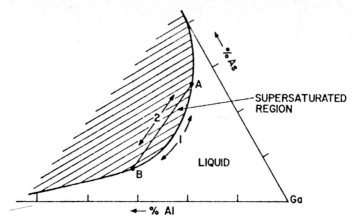

FIG. 8.13. Phase diagram of Ga/Al/As at Ga-rich corner showing creation of supersaturation by isothermal solution mixing.

superlattice structures of periods as low as 0.1 μm by successive deposition from two solutions of different composition. An example of such a super-lattice structure is given in Fig. 8.14, which shows a scanning electron micrograph of a section through a $Ga_{1-x}Al_xAs$ layer. The volume of melt in contact with the layer may be conveniently varied by changing the thickness of the spacer C, which may be used as a means of varying the film thickness.

The deposition of thin layers using a small volume of solution was also

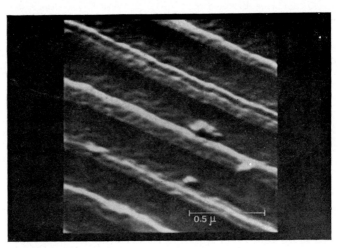

FIG. 8.14. Scanning electron micrograph of superlattice structure of $Ga_{1-x}Al_xAs$ films (Woodall, 1972).

investigated by Panish and Sumski (1971) using the simple method shown diagrammatically in Fig. 8.15. The pair of wafers are separated by etching one wafer to produce small feet or shims of the desired thickness. The wafers are heated by a carbon strip heater having water-cooled electrodes and any oxide on the outer surface is removed by a flow of hydrogen when the temperature is above 800°C. The solution is introduced by capillary flow from a small drop on the end of a micromanipulator. A solution is formed by contact between the drop and the GaAs rod but wetting is

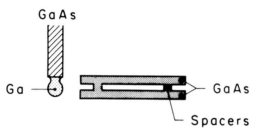

FIG. 8.15. Apparatus for epilayer growth by temperature gradient transport in a thin layer of solution (Panish and Sumski, 1971).

found to occur only if the solution is unsaturated. Growth occurs by transport from the hotter to the cooler wafer in the temperature gradient and abrupt junctions may be obtained.

Popov and Pamplin (1972) described a method for the deposition of thin layers using a VLS method with a thin zone of solvent which is made to move across the substrate by the travelling solvent method.

8.5. Growth-Rate Controlling Mechanisms in LPE

In addition to their practical significance, LPE experiments have also provided some of the most interesting measurements of growth rates under carefully controlled conditions. LPE growth of semiconductors to date appears to have been carried out mainly in unstirred crucibles and at rather high supersaturation, so that it is reasonable to assume that the growth rate will be controlled by solute diffusion. Mitsuhata (1970) gave expressions for the growth and etch-back in terms of the solubility data and temperature programme assuming a homogeneous solution and his results suggest that volume diffusion is rate-determining. Diffusive transport has been assumed to be the rate-controlling factor by Small and Barnes (1969), by Minden (1970) and by Ghez (1973). Crossley and Small (1971) applied numerical methods to calculate the rate of etch-back and growth and interpreted data for both tipping and dipping experiments from a number of laboratories. Boundary layer diffusion was found to give the best fit to the experimental data for the vertical dipping geometry, with growth rates of 50–300 Å s^{-1} and boundary layer thicknesses in the region of 0.5 mm. With the horizontal tipping geometry bulk diffusion gave a better fit, and in both geometries allowance was necessary for nucleation within the solution, particularly at the higher cooling rates. The authors concluded that solutal convection had a major effect on the dipping systems. With the relatively short growth periods, limited melt volumes and rapid cooling rates used in LPE, steady state conditions will not be realized in the majority of experiments. The instantaneous growth rate will therefore depend upon the previous temperature programme, and average growth rates estimated from the total layer thickness and the duration of cooling may differ appreciably from the maximum value. Examples of the expected variation with time of the deposition rate are quoted in the above references and by Tiller and Kang (1968). Moon and Kinoshita (1974) find that the film thickness varies as A dT/dt $t^{3/2} + B$, where A and B are constants; the term B depends on the initial transient. The diffusion coefficient for GaAs in Ga was found to be about 3×10^{-5} cm^2 s^{-1} at 700–900°C.

Since it is difficult to obtain information on the interface kinetic mechanism by growth rate measurements in LPE growth of semiconductors, the main evidence for the kinetic process is provided by observations of the surface features. Most of the epilayer growth described in the literature is on substrates of high dislocation density and the resulting films frequently exhibit irregularities such as corrugations and cusps, although very flat areas may be present on the same films. It appears likely that the substrate should be considered as a rough surface with nucleation occurring randomly rather than at screw dislocations. Evidence for this conclusion is provided, for example, by Vilms and Garrett (1972), who found that cusps on the surface did not correspond to dislocations in the substrate. The surface features observed under normal conditions of film deposition will, however, depend on the substrate perfection and flat films are to be expected on films free from etchable defects.

Of vital importance to the performance of semiconductor junction devices is the profile of the dopant and problems might be expected from compound formation or complexing between the dopant and the solvent. However, simple theoretical models appear to be appropriate for estimates of the dopant distribution. As an example, André et al. (1972) were able to account for the aluminium distribution in Ga$_{1-x}$Al$_x$As layers by the assumption of a regular solution in the gallium solvent and an ideal solid solution between GaAs and AlAs. Complications will arise if a dopant undergoes appreciable diffusion in the solid at the growth temperature, as in the case of Zn in GaP (Peaker et al., 1972). It should also be noted that Jordan et al. (1973) reported that Te doping in GaP was found to be lower than expected from the equilibrium value, and an interface limitation was postulated to explain results for oxygen doping.

Growth-rate measurements on magnetic garnet films by LPE have been reported by Giess *et al.* (1972), Blank and Nielsen (1972), Blank *et al.* (1973) and Knight *et al.* (1974). In each case growth occurred by dipping into supercooled melts and the growth rate was determined as a function of the rotation rate, or of time in unstirred melts. The measured growth rates depend strongly on the substrate rotation rate ω, and the data of Giess *et al.* (1972), Blank *et al.* (1973) and Knight *et al.* (1974) indicate a linear dependence of the growth rate v on $\omega^{1/2}$, provided that a correction is made for the growth prior to the establishment of steady state conditions. The influence of the interface kinetic mechanism is not appreciable except at rotation rates above 160 r.p.m. where the rate of solute flow by diffusion becomes relatively rapid. Using the Burton, Prim and Slichter (1953) approximation for the boundary layer thickness, the data of Blank *et al.* give a solute diffusion coefficient between 9×10^{-7} and 6×10^{-6} cm² s⁻¹ for different temperatures and supercooling. A plot of $v\omega^{-1/2}$ *versus* $v^{1/m}$ (Eqn 4.21) gave a more nearly linear variation for $m=2$ than for $m=1$ but the range of values of v and ω is insufficient for this result to be interpreted with confidence, and additional uncertainty is introduced by the effect of the initial transient. The data for zero rotation rate (Fig. 2 of Blank *et al.*) indicate a value of $D = 7 \times 10^{-6}$ to 3×10^{-5} cm² s⁻¹ if the Carlson (1959) expression is used for the boundary layer thickness, with the solution flow velocity estimated at 0.1–1 cm s⁻¹.

Ghez and Giess (1973) extended the data of Giess *et al.* (1972) and performed a detailed analysis of results in the transient regime and in the steady state. Their data also indicate that the growth rate is primarily limited by solute diffusion although evidence is presented for a kinetic limitation of the growth rate. A correction to the observed data is required because of the effect of a non-uniform lead impurity on the refractive index, so that a film of thickness 0.62 μm (as determined by SEM) had an apparent thickness of 0.74 μm by optical measurement. The authors subtracted a value of 0.12 μm from all measured thicknesses, but this procedure is not valid for very thin films, and a more accurate interpretation is likely if the data points are multiplied by the ratio 0.62/0.74. With this amendment a plot of $v\omega^{-1/2}$ *versus* $v^{1/m}$ from Fig. 1 of Ghez and Giess (1973) gives a linear variation for $m=1$, but also gives an approximately linear relation for $m=2$. The analysis of Ghez and Giess assumes a first order reaction, but more data are required before the order of the kinetic process can be determined with confidence. Additional evidence for a partial kinetic limitation of the growth rate is obtained from a plot for growth at zero substrate rotation of the thickness h *versus* $t^{1/2}$, which has a small negative intercept on the $t=0$ axis. With the correction mentioned above, the intercept has a value of about 0.06 μm. This value is probably significant in view of the care taken in these experiments to achieve a reproducible process with accurate determination of the time for which the substrate is immersed.

With the modified correction for the refractive index, and assuming a first order kinetic process with $v=C(n_i-n_e)$, Ghez and Giess' data for rotating substrate gives $D=6\times10^{-6}$ cm s^{-1} and $C=3\times10^{-2}$ cm^4 g^{-1} s^{-1}. (The supersaturation is taken to be 0.032 gcm^{-3}, in agreement with the phase diagram of Blank and Nielsen, 1972.) The unstirred melt data give $D=3\times10^{-6}$ cm^2 s^{-1} and $K=1\times10^{-3}$ cm s^{-1}. The mean value of C corresponds to a growth rate of 20,000 Å s^{-1} for a 1% supersaturation at the interface, which is very rapid compared with growth on a habit face.

Knight *et al.* (1974) introduce a term linear in time to the expression for the film thickness to correct for growth due to convection in the unstirred case and find a value of $(5\pm1)\times10^{-7}$ cm^2 s^{-1} for the diffusion coefficient at about 850°C. This value for D is in good agreement with that obtained for growth on a rotating substrate, where the correction due to growth prior to establishment of the boundary layer is relatively minor.

The data quoted are in substantial agreement in that they demonstrate the dominant effect of solute diffusion. The discrepancies in the calculated values of the diffusion coefficient probably reflect the inevitable problem of correcting for the initial transient, in addition to differences in interpretation. The values found for the diffusion coefficient are rather low in relation to values for alkali halides but this is not unexpected for the diffusion of complex ions. Knight *et al.* (1973) estimate the radius of the diffusing species to be about 2.8 Å.

It is clear from the above data that the interface kinetic process is extremely rapid and nucleation may be expected to occur uniformly over the whole area of substrate. LPE deposition of garnet films differs from the growth of bulk garnets in that the growth normally occurs in the fast $\langle111\rangle$ directions rather than on the $\{110\}$ or $\{211\}$ habit faces. As in LPE growth of semiconductors, it appears that screw dislocations are not necessary for growth on $\{111\}$ substrates. The energy required for the generation of dislocations is, in fact, very high in the garnets and substrates are often dislocation-free (Brandle *et al.*, 1972). Defects in the substrates which give rise to etch pits are normally inclusions, for example of iridium from the crucibles used for Czochralski growth, but Plaskett (1972) claims that such defective regions do not give rise to preferential local growth.

The use of time markers in growth rate studies has been applied to LPE growth by Kumagawa *et al.* (1973) and this technique is likely to lead to improved understanding of the growth process.

The incorporation of impurities such as Pb and the composition of solid solution garnet films are governed by the same principles which are described for the growth of bulk crystals in Section 7.1.3. An additional factor which may affect the composition as well as the growth rate is the lattice parameter mismatch between film and substrate and different considerations will apply in many cases according to whether the film is in tension or in compression. (See, for example, Plaskett et al., 1972.) Giess et al. (1972) noted a tendency for the effective distribution coefficient of gallium to change with time at constant growth rate and constant temperature, and this may be due to the transition from hetero-epitaxial growth in the initial stage to homoepitaxial growth as the thickness of the deposited film increases. A rapid variation in lead concentration across the films has been confirmed by Robertson et al. (1974). A gradation in lattice parameter or other properties across the film-substrate boundary is to be expected on some scale, even if this is difficult to detect, for all epilayers and this variation must be considered in interpreting results on epitaxial film growth.

8.6. Important Factors and Future Trends

In this section an attempt is made to specify those aspects which are of greatest importance in the preparation of high quality, uniform, homogeneous epilayers with a high degree of reproducibility. As with all statements on crystal growth processes, generalizations are dangerous. Also there may be overriding economic factors which influence a decision in favour of some particular technique or procedure.

A factor which is stressed throughout the literature of LPE is the importance of cleanliness in the preparation of the substrate and holder. As an example of the necessary precautions, Giess et al. (1972) clean the substrate and holder, after mounting, in Alconox† alkaline solution for 15 min. at 80°C. The holder is then rinsed in deionized water, boiled in deionized water for 15 min. and finally rinsed with a jet of ethanol. According to Blank and Nielsen (1972), defects arising through handling of the substrate are amongst the most troublesome in garnet film preparation, and transfer of the wafer from the polishing stage to the furnace should occur under clean conditions which may be obtained by the use of laminar air flow cabinets.

Contamination from the crucible must be avoided in semiconductor preparation where low impurity concentrations are vital for device performance. This problem has been considered particularly by Wolfe and Stillman (1971), who refer to studies using both graphite and silica crucibles for the growth of GaAs. Silicon contamination in the latter case was studied in detail by Hicks and Greene (1971).

† Proprietary name, Alconox Inc., New York.

Similarly, cleanliness in the growth atmosphere is essential. The observation by Vilms and Garrett (1972) of contamination of semi-conductor films resulting from air leaks has already been mentioned in Section 8.4.1. Dean *et al.* (1968) found that silicon contamination of GaP from Ga solution could be reduced by adding water to the hydrogen gas stream.

Substrate preparation is also of fundamental importance. In the case of semiconductors, many surface treatments have been used but a standard procedure has been adopted by most workers which comprises lapping, mechanical polishing and chemical polishing with a solution of 1% bromine in methanol. However, Vilms and Garrett followed this procedure by successively immersing the substrate in concentrated HCl, rinsing in water, drying in nitrogen, immersing in concentrated H_2SO_4 followed by H_2SO_4 with H_2O_2, and finally rinsing in hot water and drying.

GaAs and GaP substrates differ from the garnet substrates used for bubble films in that they often contain a high density of dislocations, introduced either during growth or during the subsequent cutting and polishing. The surface damage may be removed by the chemical polish but the residual dislocations will affect the layer growth and probably the device performance. Ladany *et al.* (1969) noted a reduction in the disloca-tion density from 5×10^5 cm^{-2} in the substrate to 5×10^4 cm^{-2} in the epi-layer, and found a dependence of the efficiency of light emission on the dislocation density. Saul (1971) demonstrated a reduction in the dislocation density by etching back the wafer during growth (in addition to an initial etch-back of the substrate) since the crystal-solution interface provides a source of point defects which are required for dislocation mobility. It is not clear, however, how far this reduction in dislocation density may be taken by repeated growth and dissolution cycling. Queisser (1972) has proposed that it is the movement of layers along the surface in the earliest stages of growth which leads to the annihilation or termination of dis-locations.

Generally etch-back of the substrate is to be avoided in heteroepitaxial growth since it will lead to a graded film composition. It would appear very desirable in LPE growth of the III–V semiconductors to improve the growth and handling of substrate crystals so that dislocation-free wafers could be produced for routine device fabrication, and the etch-back stage eliminated. According to Rozgonyi and Iizuka (1973), it is decorated dislocations which are important in affecting luminescent properties and these occur only when another type of defect (called a saucer or S-pit) is present.

In the growth of magnetic garnet films it is clearly established that substrate dissolution is undesirable, which is one reason for the choice of supercooled melts with growth at relatively high supersaturation. Substrate preparation normally consists of mechanical polishing using diamond grit, followed by Syton† polishing. Robertson *et al.* (1973a) have suggested an alternative procedure which involves the deposition on the sawn slice of a homoepitaxial layer following an initial etch-back of the slice in the high temperature solution. A saw blade of good quality with internal cutting edge is required for slicing the wafer but this procedure is otherwise very rapid and convenient compared with lapping and polishing since the substrate can be easily transferred between adjacent furnaces for deposition of the magnetic garnet layer. These advantages may, however, be offset by a reduction in film uniformity (Nielsen and Blank, private communication). Glass (1972) has shown that LPE films greatly enhance surface scratches in the substrate.

Attack of the substrate by the flux vapour is possible if it is held for several minutes above the solution, particularly if the solvent has a relatively high volatility, and this etching results in a deterioration in film uniformity. Giess *et al.* (1972) found it necessary to pre-heat the substrate in a shielded enclosure prior to immersion in the solution, rather than in the furnace immediately above the solution, otherwise some etching by the vapour was observed.

One question which is not normally discussed in the literature is the accuracy of substrate orientation. The sensitivity of the growth rate and surface topography to orientation has not been established and slight misorientations in cutting and polishing may account for some of the scatter in experiments aimed at assessment of the reproducibility of a technique.

Reproducibility, film uniformity and homogeneity are extremely important criteria, particularly as regards device production, and these parameters will have a very strong influence on the technique which is preferred for large-scale production. Of the results reported in Section 8.4, dipping has given good results for garnet films, particularly if rotation is used, and the sliders have given good results in terms of reproducibility and thickness uniformity for semiconductor epilayers. An important reservation which must be expressed about slider techniques is that homogeneity is unlikely to be achieved across any film grown by slow cooling. If homogeneity is of importance to the device performance a steady-state method is desirable, and growth on a rotating substrate by temperature gradient transport (see Section 7.1.2) is recommended.

Since nucleation occurs uniformly across the surface of a film, it is particularly important to ensure homogeneity both of temperature and of solute distribution over the surface, otherwise more rapid growth is likely

† Proprietary name, Monsanto Inc.

to occur at the film edges (as discussed in Chapter 6). Stirring the solution will lead to more uniform layers, so long as turbulence with non-uniform flow patterns is avoided. If a single substrate is used, a horizontal position is more favourable but has the disadvantage that solution droplets tend to adhere to the surface after removal from the liquid. These may be removed by rapid rotation of the wafer following withdrawal, but draining of excess solution will occur more readily in the substrate holders shown in Fig. 8.9.

Assuming that a uniform layer of solution is achieved during growth, the major problem remaining is that of avoiding a supersaturation gradient. The possible techniques are to use a very thin layer of solvent or to apply a stabilizing temperature gradient. Longo *et al.* (1972) have reported major improvements in film quality on mounting the substrate in contact with a heat sink through which cooling is achieved by a flow of air.

Future trends. The requirement of reproducibility and close control over the properties of the epilayers will require increasingly precise control over substrate preparation, growth temperature and duration and indeed of all parameters which affect the film growth. The very large potential market for such devices as low current LED display arrays will ensure continuing interest and development in this field. Additional semiconductor devices such as solar cells (Hovel and Woodall, 1973) will be produced by LPE.

Practical production of any device will require techniques for the preparation of multiple films and for the deposition on large area substrates. Experimental techniques must be concentrated on the development of batch processes for the simultaneous production of many films, once conditions for growth of a single film have been established. Large-scale production will require the development of apparatus for automatic production of devices, preferably by a continuous flow process, and design proposals for such a system have been described by Bergh *et al.* (1973).

Further development is necessary of techniques for the reproducible preparation of very thin layers, of thickness in the region of 100 Å, for devices such as heterojunction lasers. In addition to the devices mentioned above, the growth of high quality films by LPE is certain to lead to the development of completely new devices, and investigations are taking place at present of optical devices and of novel magnetic applications of epitaxial films. For example, thermomagnetic recording has been discussed by Krumme *et al.* (1972), and a review of Bi-substituted garnet films for modulation and storage of information has been given by Robertson *et al.* (1973b). Integrated thin film optical systems have been discussed by Tien *et al.* (1972), and laser action in $Y_3Al_5O_{12}$ films by van der Ziel *et al.* (1973).

References

Alferov, Z. I., Andreev, V. M., Korolkov, V. I., Portnoi, E. L. and Tretyakov, D. N. (1969a) *Sov. Phys.-Semicond.* **2**, 843.

Alferov, Z. I., Andreev, V. M., Korolkov, V. I., Portnoi, E. L. and Tretyakov, D. N. (1969b) *Sov. Phys.-Semicond.* **2**, 1289.

Alferov, Z. I., Andreev, V. A., Portnoi, E. L. and Trukan, M. K. (1969c) *Fiz. Tekh. Poluprov* **3**, 1328.

Allen, H. A. and Henderson, G. A. (1968) *J. Appl. Phys.* **39**, 2977.

André, E., LeDuc, J. M. and Mahien, M. (1972) *J. Crystal Growth* **13, 14**, 663.

Archer, R. J. and Kerps, D. (1967) Proc. 1966 Int. Symp. GaAs (Inst. of Phys., London), 103.

Ashley, K. L. and Strack, H. A. (1969) Proc. 1968 Int. Symp. GaAs (Inst. of Phys., London), 123.

Bergh, A. A., Saul, R. H. and Paola, C. R. (1973) *J. Electrochem. Soc.* **120**, 1558.

Besser, P. J., Mee, J. E., Elkins, P. E. and Heinz, D. M. (1971) *Mat. Res. Bull.* **6**, 1111.

Besser, P. J., Mee, J. E., Glass, H. L., Heinz, D. M., Austerman, S. B., Elkins, P.E., Hamilton, T. N. and Whitecomb, E. C. (1972) Proc. 17th Conf. Magn. and Magn. Mat. Chicago (AIP Conf. Proc. No. 5), p. 125.

Blank, S. L. and Nielsen, J. W. (1972) *J. Crystal Growth* **17**, 302.

Blank, S. L., Hewitt, B. S., Shick, L. K. and Nielsen, J. W. (1972) A.J.P. Conf. Proc. No. 10, 256.

Blum, J. M. and Shih, K. K. (1972) *J. Appl. Phys.* **43**, 1394.

Bobeck, A. H. (1967) *Bell Syst. Techn. J.* **46**, 1901.

Bobeck, A. H., Fischer, R. F., Perneski, A. J., Remeika, J. P. and Van Uitert, L. G. (1969) *IEEE Trans. Magn.* **5**, 544.

Bobeck, A. H., Spencer, E. G., Van Uitert, L. G., Abrahams, S. C., Barns, R. L., Grodkiewicz, W. H., Sherwood, R. C., Schmidt, P. H., Smith, D. H. and Walters, E. M. (1970) *Appl. Phys. Lett.* **17**, 131.

Brandle, C. D., Miller, D. C. and Nielsen, J. W. (1972) *J. Crystal Growth* **12**, 195.

Burns, J. W. (1968) *Trans. Met. Soc. AIME* **242**, 432.

Burton, J. A., Prim, R. C. and Slichter, W. P. (1953) *J. Chem. Phys.* **21**, 1987.

Callen, H. (1971) *Appl. Phys. Lett.* **18**, 311.

Carruthers, J. R. (1972) *J. Crystal Growth* **16**, 45.

Casey, H. C. and Trumbore, F. A. (1970) *Mat. Sci. and Engng.* **6**, 69.

Casey, H. C., Panish, M. B., Shlosser, W. O. and Paoli, T. L. (1974) *J. Appl. Phys.* **45**, 322.

Crossley, I. and Small, M. B. (1971) *J. Crystal Growth* **11**, 157.

Crossley, I. and Small, M. B. (1972) *J. Crystal Growth* **15**, 175.

Dawson, L. R. (1972) Prog. Solid State Chem. (H. Reiss and J. O. McCaldin, eds.) **7**, 117.

Dean, P. J., Frosch, C. J. and Henry, C. H. (1968) *J. Appl. Phys.* **39**, 5631.

Deitch, R. H. (1970) *J. Crystal Growth* **7**, 69.

Donahue, J. A. and Minden, H. T. (1970) *J. Crystal Growth* **7**, 221.

Dousmanis, G. C., Nelson, H. and Staebler, D. L. (1964) *Appl. Phys. Lett.* **5**, 174.

Elwell, D., Capper, P. and Lawrence, C. M. (1974) *J. Crystal Growth* **24/25**, 651.

Esaki, L. and Tsu, R. (1970) *IBM J. Res. Develop.* **14**, 61.

Ghez, R. (1973) *J. Crystal Growth* **19**, 153.

Ghez, R. and Giess, E. A. (1973) *Mat. Res. Bull.* **8**, 31.

Ghez, R. and Giess, E. A. (1974) In "Epitaxial Growth" (J. W. Matthews, ed.) Academic Press, New York.

Giess, E. A. and Cronemeyer, D. C. (1973) *Appl. Phys. Lett.* **22**, 601.

Giess, E. A., Kuptsis, J. D. and White, E. A. D. (1972) *J. Crystal Growth* **16**, 36.

Glass, H. L. (1972) *Mat. Res. Bull.* **7**, 385.

Hayashi, I., Panish, M. B. and Foy, P. W. (1969) *IEEE J. Quant. Electron.* **5**, 211.

Hayashi, I., Panish, M. B., Foy, P. W. and Sumski, S. (1970) *Appl. Phys. Lett.* **17**, 109.

Hayashi, I., Panish, M. B. and Reinhart, F. K. (1971) *J. Appl. Phys.* **42**, 1929.

Hicks, H. G. B. and Greene, P. D. (1971) Proc. 1970 Conf. GaAs, Aachen (Institute of Physics, London) p. 92.

Hiskes, R. and Burmeister, R. A. (1972) Proc. AIP Conf. Mag. and Mag. Mat., p. 304.

Hovel, H. J. and Woodall, J. M. (1973) *J. Electrochem, Soc.* **120**, 1246.

Isherwood, B. J. (1968) *J. Appl. Cryst.* **1**, 299.

Janssen, G. A. M., Robertson, J. M. and Verheijke, M. L. (1973) *Mat. Res. Bull.* **8**, 59.

Jordan, A. S., Trumbore, F. A., Wolfstirn, K. B., Kowalchick, M. and Roccasecca, D. D. (1973) *J. Electrochem. Soc.* **120**, 791.

Knight, S., Hewitt, B. S., Rode, D. L. and Blank, S. L. (1974) *Mat. Res. Bull.* **9**, 895.

Krömer, J. (1963) *Proc. IEEE* **51**, 1782.

Krumme, J. P., Verweel, J., Haberkamp, J., Tolksdorf, W., Bartels, G. and Espinosa, G. P. (1972) *Appl. Phys. Lett.* **20**, 451.

Kumagawa, M., Witt, A. F., Lichtensteiger, M. and Gatos, H. C. (1973) *J. Electrochem. Soc.* **120**, 583.

Ladany, I. (1969) *J. Electrochem. Soc.* **116**, 993.

Ladany, I., McFarlane, S. H. and Bass, S. J. (1969) *J. Appl. Phys.* **40**, 4984.

Levinstein, H. J., Licht, S., Landorf, R. W. and Blank, S. L. (1971) *Appl. Phys. Lett.* **19**, 486.

Lien, S. Y. and Bestel, J. L. (1973) *J. Electrochem. Soc.* **120**, 1571.

Linares, R. C. (1968) *J. Crystal Growth* **3/4**, 443.

Linares, R. C., McGraw, R. B. and Schroeder, J. B. (1965) *J. Appl. Phys.* **36**, 2884.

Lockwood, J. F. and Ettenberg, M. (1972) *J. Crystal Growth* **15**, 81.

Longo, J. T., Harris, J. S., Gertner, E. R. and Chu, J. C. (1972) *J. Crystal Growth* **15**, 107.

Lorim, O. G., Hackett, W. H. and Bachrach, R. Z. (1973) *J. Electrochem. Soc.* **120**, 1424.

Minden, H. (1970) *J. Crystal Growth* **6**, 228.

Minden, H. T. (1973) *Solid State Technol.* (January) 31.

Mitsuhata, T. (1970) *Jap. J. Appl. Phys.* **9**, 90.

Moon, R. L. and Kinoshita, J. (1974) *J. Crystal Growth* **21**, 149.

Nelson, H. (1961) Epitaxial Crystal Growth from the Liquid Phase, Solid State Device Conference, Stanford Univ., June 26, 1961.

Nelson, H. (1963), *RCA Review* **24**, 603.

Nielsen, J. W. (1972) Invited paper at the IVth All Union Conference on Crystal Growth, Tashkhadzor USSR (to be published in *Growth of Crystals*).

Panish, M. B. and Sumski, S. (1971) *J. Crystal Growth* **11**, 101.

Panish, M. B. and Ilegems, S. M. (1972) Prog. Sol. State Chem. (H. Reiss and J. O. McCaldin, eds.) **7**, 39.

Panish, M. B., Queisser, H. J., Derick, L. and Sumski, S. (1966) *Solid State Electron.* **9**, 311.

Panish, M. B., Hayashi, I. and Sumski, S. (1969) *IEEE J. Quant. Electron.* **5**, 210.

Panish, M. B., Sumski, S. and Hayashi, I. (1971) *Met. Trans.* **2**, 795.

Peaker, A. R., Sudlow, P. D. and Mottram, A. (1972) *J. Crystal Growth* **13/14**, 651.

Plaskett, T. S. (1972) Paper at AACG Conference on Crystal Growth, Princeton.

Plaskett, T. S., Klokholm, E., Hu, H. L. and O'Kane, D. F. (1972) AIP Conf. Proc. 10, p. 319.

Popov, V. P. and Pamplin, B. R. (1972) *J. Crystal Growth* **15**, 129.

Queisser, H. J. (1972) *J. Crystal Growth* **17**, 169.

Robertson, J. M. (1973) Philips Research Laboratories Report M.S. 8324.

Robertson, J. M., van Hout, M. J. G., Janssen M. M. and Stacy W. T. (1973a) *J. Crystal Growth* **18**, 294.

Robertson, J. M. Wittekoek, S., Popma, T. J. A. and Bongers, P. F. (1973b) *Appl. Phys.* **2**, 219.

Robertson, J. M., van Hout, M. J. G. and Verplanke, J. C. (1974) *Mat. Res. Bull.* **9**, 555.

Rosencwaig, A. and Tabor, W. J. (1971) *J. Appl. Phys.* **42**, 1643.

Rosztocy, F. E. (1968) *Electrochem. Soc. Electron. Div. Abstracts* **17**, 516.

Rozgonyi, G. A. and Iizuka, T. (1973) *J. Electrochem. Soc.* **120**, 673.

Rupprecht, H. (1967) Proc. 1966 Int. Symp. GaAs (Institute of Physics, London) p. 57.

Rupprecht, H., Woodall, J. M., Konnerth, K. and Pettit, D. G. (1966). *Appl. Phys. Lett.* **9**, 22.

Rupprecht, H., Woodall, J. M. and Pettit, G. D. (1967) *Appl. Phys. Lett.* **11**, 81.

Saul, R. H. (1971) *J. Electrochem. Soc.* **118**, 793.

Saul, R. H., Armstrong, J. and Hackett, Jr., W. H. (1969) *Appl. Phys. Lett.* **15**, 229.

Scheel, H. J. and Schulz-DuBois, E. O. (1972) *IBM Techn. Discl. Bull.* **14**, 2850.

Shick, L. K. and Nielsen, J. W. (1971) *J. Appl. Phys.* **42**, 1554.

Shick, L. K., Nielsen, J. W., Bobeck, A. H., Kurtzig, A. J., Michaelis, P. C. and Reekstin, J. P. (1971) *Appl. Phys. Lett.* **18**, 89.

Shih, K. K., Lorenz, M. R. and Foster, L. M. (1968) *J. Appl. Phys.* **39**, 2747.

Small, M. B. and Barnes, J. F. (1969) *J. Crystal Growth* **5**, 9.

Stacy, W. T. and Rooymans, C. J. M. (1971) *Solid State Commun.* **9**, 2005.

Stacy, W. T. and Tolksdorf, W. (1972) Proc. 17th Conf. Magnet. and Magnet. Mat., Chicago, Nov. 1971 (Amer. Inst. Phys. Conf. Proc. No. 5), p. 185.

Stacy, W. T., Janssen, M. M., Robertson, J. M. and van Hout, M. J. G. (1973) Proc. 18th Conf. Mag. and Mag. Mat., Denver (AIP Conf. Proc. No. 10), p. 314.

Stein, B. F. (1974) *Proc. 19th AI C Conf. Mag. and Mag. Met.*

Stringfellow, G. B. and Greene, P. E. (1971) *J. Electrochem. Soc.* **118**, 805.

Thiele, A. (1969) *Bell. Syst. Techn. J.* **48**, 3287.

Thomas, D. G. (1971) *IEEE Trans.* **Ed-18**, 621.

Tien, P. K., Martin, R. J., Blank, S. L., Wemple, S. H. and Varnerin, L. J. (1972) *Appl. Phys. Lett.* **21**, 207.

Tiller, W. A. and Kang, C. (1968) *J. Crystal Growth* **2**, 345.

Tolksdorf, W., Bartels, G., Espinosa, G. P., Holst, P., Mateika, D. and Welz, F. (1972) *J. Crystal Growth* **17**, 322.

Trumbore, F. A., Kowalchik, M. and White, H. G. (1967) *J. Appl. Phys.* **38**, 1987.

Van Uitert, L. G., Bonner, W. A., Grodkiewicz, W. H., Pitroski, L. and Zydzik, G. J. (1970) *Mat. Res. Bull.* **5**, 825.

Van der Ziel, J. P., Bonner, W. A., Kopf, L., Singh, S. and van Uitert, L. G. (1973) *Appl. Phys. Lett.* **22**, 656.

Vilms, J. and Garrett, J. P. (1972) *Solid State Electr.* **15**, 443.

Williams, E. W., Porteous, P., Astles, M. G. and Dean, P. J. (1973) *J. Electrochem. Soc.* **120**, 1757.

Wolfe, C. M. and Stillman, G. E. (1971) Proc. 1970 Conf. GaAs, Aachen (Institute of Physics, London) p. 3.

Woodall, J. M. (1971) *J. Electrochem. Soc.* **118**, 150.

Woodall, J. M. (1972) *J. Crystal Growth* **13/14**, 32.

9. Characterization

9.1. Necessity of Crystal Characterization

It is obvious that there are no ideal crystals in reality and all crystals grown by any technique contain some defects, impurities and inhomogeneities. Most of the physical properties are sensitive to the deviation from ideality, therefore generally the characterization of the grown crystals is a necessity. The results of a careful characterization should be published along with any crystal growth results and with any physical measurements on the crystals. However, in the words of Holtzberg (1970) "There are too many examples of very sophisticated measurements on poorly characterized materials, leading to complex and incorrect theoretical analyses". Schieber recently mentioned in a review talk on Trends in Materials Research that it will be up to our children and grandchildren to repeat a number of physical measurements we are doing now, but on highly characterized crystals. Several years ago Roy (1965) tried to draw attention to the characterization problem, but still nowadays papers on measurements of physical properties of solid-state materials are published which are of little value because of the partial or complete absence of characterization.

Frequently in physical publications the only "characterization" given of the crystal used is its source (name of company or of the crystal grower), but this is insufficient except when reference can be given to a detailed published characterization of that crystal or of the corresponding batch of crystals.

A full characterization of a material or a crystal is very time-consuming and needs a large instrumental effort, which is why characterization by both crystal growers and crystal consumers is often left to the other party. Ideally characterization should be performed by a specialist group, but such groups are available only in a very few universities and in a few large government and industrial research laboratories, since the instrumentation for the modern physical characterization techniques is very expensive.

The reliability, precision and speed of chemical, structural and defect characterization by physical techniques can be enhanced significantly by computer-based laboratory automation. For instance at IBM research laboratories at Yorktown Heights (New York) and San Jose (California) a medium-size computer controls on-line a number of characterization experiments (simultaneously with other physical measurements and with crystal-growth experiments), processes all data and presents the results as tables, plots or as graphs on display units.

Who is responsible for characterization? Reporting on a workshop discussion on magnetic materials, Wolf (1970) proposed that 75% of the responsibility for crystal characterization should be that of the "crystal chemist" and 25% that of the "consumer". This division of responsibility is perhaps too closely defined and the ratio should rather depend on whether it is the crystal grower or the consumer who most benefits from the crystal synthesis and who publishes results of either crystal growth or physical measurements.

Not only the physicist, but also the crystal grower should be most interested in characterization of the crystals grown. It is important to correlate the quality of the crystal with the growth technique and the growth parameters. The demand for crystals of the highest quality is increasing, and only systematic characterization enables the crystal grower to optimize the growth parameters in order to obtain better crystals.

A great stimulus to careful crystal growth and characterization work would be if the scientists responsible for these activities were to participate in some of the "physical" publications, since frequently the effort required to grow and characterize crystals is much larger than that for the physical measurements. In several research establishments joint publications of physicists, crystal growers and analysts have become routine, and this is to the benefit of all participants and of the results.

There are several definitions of "characterization",† but in the following discussion the definition is meant to include the full chemical and physical description of a material which is obtained from a whole spectrum of techniques. A single crystal may be characterized by a description of its chemical composition, of its structure, of its defects, and of the spatial distribution of these three features. A full characterization should also include determination of electronic and excited states of the chemical constituents of the material. In addition, the results of many physical measurements can assist in a full description of a specific crystal. However,

† Frequently the term "appraisal" or "evaluation" is used with a similar meaning.

physical properties alone of a given crystal cannot sufficiently characterize a crystal. Even a set of various physical properties will not allow the unequivocal identification of a material except for elements and some simple compounds. Chemical and crystallographic data are obviously preferable for the identification of a material.

An excellent review on characterization and on the potential, the sensitivity and the accuracy of the various characterization techniques has been compiled by a Committee on Characterization of Materials, Materials Advisory Board, National Academy of Sciences, Washington D.C., 1967 (Report AD 649 941). It is hoped that the recommendations in this report will be increasingly observed. On the other hand, a revised edition taking into account the more recent developments in characterization techniques would be beneficial if widely distributed.

A book on characterization of the semiconductor materials Ge, Si and III–V compounds has been written by Kane and Larrabee (1970), and many aspects of characterization are reviewed in Vol. 1 of *Treatise on Solid State Chemistry* edited by Hannay (1974).

In the following sections the factors necessary for a description of the chemical composition, structure and defects of crystals will be reviewed and the corresponding characterization techniques described. Further sections deal with the growth history as it can be deduced from characterization techniques, and with the determination and removal of inclusions. The final section is devoted to a proposal of a standard characterization procedure which, possibly in a modified form, could be taken as a requirement for experimental physics and crystal-growth papers to be accepted for publication in appropriate journals of high standard.

In this chapter we mention well-known techniques only briefly since they are described in many textbooks, and concentrate on new techniques and on recent developments in classical methods as well as on a few potentially important characterization techniques. This chapter is relatively long but we believe a detailed treatment is justified by the great importance of the (so often neglected) characterization.

9.2. Chemical Composition and Homogeneity

The determination of the chemical composition of a (single-phase) crystal may be separated into three problems, which may require different techniques. The first is a measurement of the concentration of the major constituents, with an accuracy which is sufficiently high to permit any significant departure from stoichiometry to be detected. Secondly, any minor impurities which are present in "trace" quantities must be detected and their concentration measured. The third problem is the determination of any inhomogeneities or striations of the major constituents or of the traces. Frequently the determination of the valence state of the ions in polar compounds might be of interest and represents a further problem particularly when elements with multiple valence states, such as many transition and rare earth elements, are involved.

Another approach to the classification of characterization techniques is by the sample preparation necessary. Table 9.1 lists the various techniques,

TABLE 9.1. Characterization of Chemical Composition

Sample preparation	Analytical technique	
MATERIAL (CRYSTAL)→	*Nondestructive methods*	
↓		
Sampling, grinding, mixing, surface cleaning	electron probe microanalysis X-ray fluorescence analysis neutron activation analysis optical emission spectroscopy with laser excitation (nearly nondestructive)	
↓		
SAMPLE POWDER→	*Direct methods (powders)* optical arc and spark emission spectroscopy spark source mass spectrometry (for traces only)	
↓		
Dissolving in acids or fused salts, preparation of defined solution		
↓		
SAMPLE SOLUTION→	*Direct methods (solutions)* flame emission spectroscopy atomic absorption colorimetry	
↓	titrimetry ⎱ if no interference gravimetry ⎰ by other constituents coulometry	
Separation of constituents	polarography	
↓		
SOLUTIONS OF MAJOR→ CONSTITUENTS	*Determination of major constituents* colorimetry titrimetry gravimetry coulometry polarography	
↓		
Separation (and enrichment, pre-concentrating) of minor constituents or traces		
↓		
PRECIPITATES OR SOLUTIONS→ OF TRACES	*Trace determination* colorimetry titrimetry polarography coulometry	

INCREASING COST OF APPARATUS (general trend)

separated into classes of non-destructive, direct and non-direct methods. Non-destructive and direct methods generally require less time per analysis than the non-direct methods, whereas the cost of the instrumentation for nondestructive and direct methods is generally very high.

The various chemical and physical techniques for the determination of composition and of inhomogeneities are reviewed in the Committee on Characterization Report (1967). Further general information on this subject may be obtained from the books of Kolthoff and Elving (1959–1972), Wilson and Wilson (1959–1971), Berl (1960), Wainerdi and Uken (1971) and Maxwell (1968).

The most important techniques which may be applied in chemical analysis are listed in Table 9.2, together with the normal limits of sensitivity and average precision. The lower limit of quantitative analysis by several techniques is higher by one or two orders of magnitude than the sensitivity, because a barely detectable signal cannot be measured with the necessary precision. The values quoted in Table 9.2 are approximate and can vary with the element to be detected, and they often depend strongly on the other major elements present in the crystal. The sensitivity and accuracy of most analytical techniques are dependent on the type of instrument and on the analyst, as was found by test analyses of standard samples sent for analysis to various laboratories. A reliable analytical service is obviously a necessity when the characterization is to be correlated with the physical properties of a given crystal or material.

9.2.1. Analysis for major constituents

The nature and the concentration of the major constituents have to be determined when new crystalline phases are synthesized or when solid solutions are grown. In most cases, however, the major constituents and their ratio are fixed by the structure type of the material and by the requirement of electrical neutrality in the crystal. Also the stoichiometry of many crystals grown from high temperature solutions is very good since the growth temperature is relatively low and does not vary much (relative to the melting point). Many crystals of the spinel, garnet, perovskite, etc., type are often stoichiometric or show very little deviation from stoichiometry. In these cases, when no novel phases are to be expected, identification by X-ray techniques and often by the habit is sufficient, unless solid solutions are grown and unless the distribution of the cations on the various sites and the distribution of their valency states (e.g. in spinels and in garnets) is not clear. From the many analytical techniques only those which have a precision of better than about 1% are of interest for analysis of the major constituents. Nevertheless less accurate techniques are frequently used.

TABLE 9.2. Comparison of Chemical Analytical Methods with respect to Sensitivity and Precision

Method	Detection limit (p.p.m.)	Precision %	Elements detected	Destructive or Non-destructive	Standards†	Min. sample weight mg.	Determination of sample weight necessary
Colorimetry	10^{-2}–10^2	0.1–10	most	D	a	10(0.1)	yes
Titrimetry	$\sim$10–10^3	0.01–1.0	many	D	c	10(0.1)	yes
Gravimetry	10^3–10^5	0.01–1.0	many	D	c	100(10)	yes
Coulometry	major phase	0.001–0.005	most	D	c	10	yes
Polarography	10^{-2}–10^2	0.1–10	most	D	a	10	yes
Opt. emission spectroscopy	10^{-2}–10^2	1–10	most	D	b	10	yes
Atomic absorption	10^{-2}–10^2	1–10	most	D	a	10	yes
X-ray fluorescence	10–2.10^2	0.1–2	Atomic No. >5 – 11	N	b	100(10)	no
Electron microprobe	10^3–10^4	0.5–5	>5 – 11	N	b	10^{-3}	no
Mass spectrometry	10^{-2}–1	>5	all	D	a, b	10(0.1)	yes
Neutron activation	10^{-3}–10^{-1}	2–10	many	N	b (c)	10(0.1)	yes
Radioactive tracer	10^{-3}–10^{-1}	$\sim$1	many	N	b (c)	10(0.1)	no

† a. standard solutions necessary, b. internal standards or standard samples necessary, c. no standards necessary (but of advantage for checking).

Of the methods used for the determination of the major constituents, the classical method of *wet chemistry* is still the most widely used, although its importance has declined rapidly as new physical techniques have been developed. Wet analysis involves the dissolution of the crystal (or its powder) in acids, fused salts or in conventional fused salt mixtures†, separation when necessary of the various chemical constituents, and the quantitative determination of the elements by gravimetric, titrimetric or colorimetric methods (see for instance Kolthoff and Elving, 1961, 1959–1972; Ringbom, 1963; Gordon *et al.*, 1959; Vogel, 1966; Seel, 1970; Sandell, 1959; Boltz, 1958; Schwarzenbach, 1957 and Kodama, 1963). The results of wet chemical analysis are sometimes unreliable, when interfering elements are present. Some elements, such as boron and oxygen, are difficult or impossible to determine by this method. Apart from problems of interference by other elements, the main disadvantage of wet chemical analysis is the amount of time required for a quantitative determination. Except for simple routine analyses, the complexity of wet analysis is such that considerable experience is required before reliable results are obtained. Perhaps for this reason, relatively few examples of complete chemical analyses of HTS-grown crystals have been reported, although there are many examples of results of trace analysis or analysis of a single component.

The value of careful chemical analysis in HTS growth has been demonstrated by investigations such as that of Arend and Novak (1966) and of Arend *et al.* (1969) on barium titanate crystals of the butterfly twin morphology. In one example, the composition of a crystal of nominal formula $BaTiO_3$ was found to be

$$Ba_{0.99}K_{0.01}Ti_{0.99}Fe_{0.004}Pt_{0.006}O_{2.988}F_{0.01}\square_{0.002},$$

where $\square$ denotes an anion vacancy. Of particular significance are the rather high concentration of platinum from the crucible material and the replacement of an appreciable amount of O^{2-} by F^- from the KF flux. That the fluorine ions are present substitutionally rather than in inclusions is indicated by the changes in properties of the crystals with time at ambient temperature, due to hydrolysis. The observation of F^- substituted for O^{-2} is of considerable importance since fluoride solvents are often used for crystal growth of oxides. The probability of fluorine substitution will

† Rapid and complete decomposition of a variety of oxide components is achieved according to Biskupski (1965) by fusing the powdered sample at 850°C in a platinum crucible with 2 g B_2O_3 and 3 g LiF, and by dissolving the product in H_2SO_4. According to Thilo *et al.* (1955) alumina and ruby powder dissolve rapidly in a fused mixture of Na_2CO_3, $Na_2B_4O_7$ and $KClO_3$ in the weight ratio 5 : 4 : 1. Fusion techniques in chemical analysis have been reviewed by Bock (1972) and by Dolezal *et al.* (1968), and pre-molten fluxes of high purity are commercially available.

depend critically on the presence of charge compensating ions which can enter the lattice simultaneously.

Coulometry (Lingane, 1958; Taylor *et al.*, 1965; Stock, 1965, is a very precise analytical technique for major and minor constituents. It is essentially titration with electrons and therefore replaces the classical titrimetric methods. In this quantitative electrolytic technique a constant electrolysis current is applied, which is very convenient, or the electrolysis potential is controlled very precisely, the latter technique being more selective. The sensitivity of coulometric techniques approaches the p.p.m. range, and the precision is of the order of 0.001 to 0.005%. Its disadvantage is that solutions have to be prepared as in the wet chemical techniques, and the necessary steps are time-consuming and might reduce the accuracy if impurities are introduced.

Polarography (Kolthoff and Lingane, 1952; Breyer and Bauer, 1963; Neeb, 1969) is used for the determination of most elements in solution when an electrolytic oxidation or reduction process can be used. Several elements in minor or major concentration can be measured simultaneously, and sensitivities are often below the p.p.m. range. The precision of this technique for traces is of the order of 20%, whereas for favourable concentration ranges it can be better than 1%. A disadvantage of polarography is that the sample has to be dissolved prior to analysis, and the same is true for optical (flame) emission and atomic absorption spectrometry discussed in the following section.

Optical emission spectrometry is used mainly for trace analysis but can be used for major constituents. Excitation of the spectrum is achieved by various techniques according to the nature of the specimen. The flame (Mavrodineanu and Boiteux, 1965) and d.c. arc were used in most early experiments and the book of Ahrens and Taylor (1961) is based exclusively on the d.c. arc. More recently the a.c. arc, a.c. spark (with a higher voltage) and combinations of arc and spark have been used as sources, and lasers have been used to initiate combustion of insulating samples. The important criterion for accurate quantitative analysis is that combustion of the specimen must be complete and spark sources are normally considered more satisfactory, especially for metallic samples (Harrison *et al.*, 1963; Harvey, 1964; Nachtrieb, 1950). The main advantage of lasers over conventional methods of excitation is that a small volume of a crystal can be vaporized selectively and the spectrum of the resulting vapour excited by means of an auxiliary spark source. The region examined may be smaller than 50μ in diameter and 25μ deep. By this means the distribution of major constituents and of trace elements can be determined without slicing up the crystals (Moenke and Moenke, 1966; Snetsinger and Keil, 1967; Blackburn *et al.*, 1968).

Q2

Dispersion of the emitted radiation is achieved by prisms or gratings. The requirements of sensitivity and resolution are diametrically opposed and Cooper *et al.* (1969) describe the simultaneous use of three types of spectrograph, each receiving radiation from an arc. Medium- and large-prism spectrographs were used for elements having lines well resolved from those of other elements present, while a grating spectrograph was used for transition elements and the actinides because of their complex spectra.

The spectra are recorded either photographically or with photoelectric counting systems. Photographic emulsions require careful calibration and their relative inconvenience has led to gradual replacement by direct-recording techniques. For the highest accuracy it is necessary to compare the spectra for the sample under test with that of a standard material.

The limit of detection of optical emission spectroscopy varies between 0.3% for potassium and 0.4 p.p.m. for magnesium, but for many elements lies between 1 and 100 p.p.m. Optical emission spectroscopy is the subject of books by Herzberg (1944), Ahrens and Taylor (1961), Clark (1960), Harrison *et al.* (1963) and Slavin (1971). The emission spectra can be evaluated with the help of computers (Thompson *et al.*, 1969; Helz *et al.*, 1969).

Atomic absorption spectrometry is a related but newer method which permits the concentration of a large number of elements to be determined using relatively simple and inexpensive apparatus. The precision depends upon the element considered but can be as high as $\pm 0.5\%$. The principle relies upon the absorption of light at a characteristic wavelength by atoms in a vapour produced from the sample. The sample is normally dissolved and the resulting solution vaporized by spraying through an aerosol into a flame or heated graphite tube. A spectrometer selects the wavelength corresponding to the strongest spectral line of the element and the decrease in intensity on passing through the vapour is measured.

The absorption is related to the concentration n of the element in the vapour by the usual law

$$\ln \frac{I}{I_o} = \mu n l,$$

where I_o and I are the intensity of the radiation before and after passing through the flame respectively, l the path length and μ the absorption coefficient. The latter parameter is normally found by calibration using solutions of known concentration.

A number of light sources are required to cover a wide range of elements. As with emission spectroscopy, the accuracy is expected to be higher for elements with relatively simple spectra. The accuracy may be affected by

interference from other elements present, and possible causes of interference must be considered in the interpretation of analytical data. A
detailed account of the advantages and limitations of atomic absorption
spectrometry has been given by Slavin (1968), and other books on the
topic have been written by Ramirez-Munoz (1968) and Angino and
Billings (1967).

Perhaps the best example of the use of atomic absorption spectrometry
in the appraisal of HTS-grown crystals is the study of the lead concentration in various rare-earth orthoferrites by Remeika and Kometani (1968).
Their analyses for Pb were considered to be accurate to within 2% of the
amount present, which varied from 0.08% to 13.4% by weight of the
material. The high concentration of lead in the heavier rare-earth orthoferrites would have a major effect on the properties of these materials if it
were incorporated in the lattice (see Chapter 3).

X-ray fluorescence analysis has been used increasingly as an alternative
analytical method and has advantages of directness and convenience over
wet chemistry. As may be seen from Table 9.2, it is one of relatively few
analytical methods which may be non-destructive, so that analysis can be
performed on the actual crystal or crystalline layer used for a subsequent
experiment. The crystal is placed in a strong beam of "white" X-rays
produced by a heavy-metal anode, and each excitable element in the crystal
will emit X-rays of its characteristic wavelengths. This fluorescent radiation
is analysed by a counter spectrometer employing a crystal such as sodium
chloride, and the intensity of radiation at a characteristic wavelength is
proportional to the concentration of the particular element. The precision
which can be obtained in a single analysis is of the order of 5 to 10%, but
by routine procedures it may be reduced to about 0.5%. Such precision
can be achieved by forming a composite sample containing some standard
material in addition to that analysed so that errors due to a difference in
density between the sample and the calibration standard can be eliminated.
X-ray fluorescence is often carried out with non-dispersive systems,
usually attached to a scanning electron microscope. Reviews of X-ray
fluorescence spectroscopy have been given by Birks (1969), Adler (1966),
Jenkins and De Vries (1967), and Müller (1972).

Electron probe microanalysis (EPMA) has been increasingly used, particularly to study variations in the concentration of an element in the region
near the surface of a crystal. It is probably the most powerful method for
the investigation of compositional gradients and striations, although
optical methods may be preferable if the crystal contains only atoms of
low atomic number. As in X-ray fluorescence, the analysis depends upon
the emission by an element of its characteristic X-ray spectrum, but
the excitation in EPMA is produced by a beam of electrons. The great

advantage of electron excitation is that the beam can be focused, and a spatial resolution of less than 1 μm is possible with EPMA compared with >50 μm for X-ray fluorescence.

A schematic diagram of an electron probe microanalyser is shown in Fig. 9.1. The idea was patented by Hillier (1947) but the first instrument was constructed by Castaing (1951). The specimen is mounted on a stage which can be displaced or rotated, and is observed through a microscope so that it can be continuously viewed in order to select a particular region for analysis. An area of the specimen is usually scanned and an image is simultaneously observed at a number of wavelengths corresponding to the constituent elements. The instrument is particularly powerful if combined with a scanning electron microscope and commercial instruments with both facilities are available.

As with X-ray fluorescence analysis, the range of elements which can be detected depends upon the X-ray spectrometer and detector, which are difficult to build for very long wavelengths. The range of elements which can be detected is slowly increasing and instruments could be built to detect all the elements except possibly helium and hydrogen, but in current commercial instruments the limit is at atomic number 5 (boron).

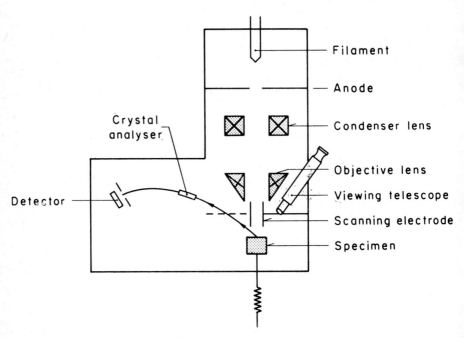

FIG. 9.1. Electron microprobe analyzer (diagrammatic).

The beam causes local heating of the sample which must therefore be stable in a vacuum at temperatures up to a few hundred degrees. This problem is encountered only with insulating samples and can be overcome by coating the surface with a thin layer of carbon or aluminium. This layer also serves to prevent charging of the sample and so stabilizes the electron beam.

X-ray intensities can be determined to better than 1% but corrections are necessary in order to achieve comparable accuracy in the concentration values. The principal corrections are for absorption of the X-rays and for excitation of fluorescent radiation from other elements in the specimen. In addition, the stopping power of electrons and the probability of back-scatter vary with the atomic number of the element and with the energy of the incident electrons. The theory and application of these corrections have been discussed in reviews by Castaing (1960) and Heinrich (1966, 1967). Generally the precision is not better than 5%.

Frazer et al. (1971) have proposed a procedure for quantitative analysis by microprobe in which two or more beam energies are employed and the intensities extrapolated to the values at the critical excitation potential. At this potential the absorption, fluorescence and back-scatter corrections become negligible and the resulting values require correction only for the electron retardation.

The use of EPMA in investigating the homogeneity of dopants has been reported for instance by Timofeeva et al. (1969a) (Cr and Nd in Al_2O_3). This method was also used by Sobon et al. (1967) to determine the Pb concentration in garnets. Monographs on electron probe microanalysis include those of Birks (1963), McKinley (1966), Reuter (1971), Heinrich (1968), Russ (1970) and Andersen (1973).

Electron spectroscopy has been developed to a valuable analytical technique by Siegbahn and coworkers (1967). Several terms are found in the literature, for example ESCA (electron spectroscopy for chemical analysis), IEE (induced electron emission), UPS (UV-photoemission spectroscopy), XPS (X-ray photoelectron spectroscopy), SXS (soft X-ray emission spectroscopy) and INS (ion neutralization spectroscopy).

Preisinger et al. (1971) proposed the classification of electron spectroscopical techniques according to the stimulating process:

Ion electron spectroscopy	IES
Electron electron spectroscopy	EES
X-ray electron spectroscopy	XES
UV electron spectroscopy	UVES

UV irradiation (3–50 eV) results in ionizing interactions with the bonding electrons whereas X-rays interact with the inner electron shells. The latter process is therefore useful for chemical analysis whereas the former process gives information on the electronic structure of semiconductors, metals and recently also of insulators. Photons of uniform energy produce photo-electrons and Auger electrons simultaneously. Electron spectroscopy allows the determination of oxidation states of elements and also the analysis of elements in the surface layers (Chang, 1971; McDonald, 1970; Benning-hoven, 1973) especially of the light elements (second period from Li to F).

Mass spectrometry is, with some exceptions such as the O_2 determination mentioned below, only valuable for analysis of minor elements (below 1%) because of its inaccuracy. Mass spectrometry relies on the variation with mass of the path of an ion in a magnetic field and/or an electric field. The major limitation is that it can be applied only to gaseous ions and so the material must be vaporized by a suitable ion source. At present the type of instrument most suitable for studies of solids is the spark-source mass spectrometer. Very small quantities of materials are required, and a large number of elements may be determined simultaneously. Monographs on mass spectrometry include those of Ahearn (1966), McDowell (1963), Kienitz (1968) and Roboz (1968).

The spark is produced by a potential of 20–100 kV at about 1 MHz and ionizes all elements with approximately the same sensitivity. The majority of ions are singly ionized, but doubly and triply ionized atoms are also produced. The major disadvantage is, however, that the ions have a wide range of energy and so the analysing system must separate and focus ions of different mass independent of their initial energy and direction. A suitable spectrometer is the double-focusing system of Mattauch and Herzog (1934), in which electric and magnetic fields of critical values are applied in series to the beam of ions. The spectra are recorded on photographic plates.

Oxygen (in oxygen-containing compounds) is one of the most difficult elements to determine but can be analyzed by mass spectrometry if it is extracted from the material by (a) reaction with agents such as mercuric cyanide or silver cyanide to yield CO_2 (Anbar and Guttmann, 1959; Shakhashiri and Gordon, 1966), (b) reduction with carbon at high tem-peratures to yield CO (Schwander, 1953; Dontsova, 1959) or (c) oxidation by fluorine or halogen fluorides to yield O_2. The latter oxygen extraction technique seems to be suitable for many oxides and oxide compounds and has been developed by Clayton and Mayeda (1963). 5 to 30 mg of the material is reacted for 12 hours at 450–700°C with bromine pentafluoride BrF_5, a colourless liquid at room temperature. As an example the reaction between orthoclase and BrF_5 may be written as

$$5KAlSi_3O_8 + 16BrF_5 \rightarrow 5KF + 5AlF_3 + 15SiF_4 + 8Br_2 + 20O_2.$$

The extracted oxygen is separated by a liquid nitrogen trap and is determined as such or after its transformation to CO_2. The quantitative extraction of oxygen from a variety of materials and even from refractory oxides has been achieved at high temperatures with carbon as reducing agent by Kraus (1972) and by Paesold et al. (1967), and the latter were able to determine oxygen in a variety of oxides with a standard deviation of 1 to 1.5%. However, this precision and a high sensitivity were only possible by IR absorption analysis of the CO (formed during the extraction) with specialized apparatus.

The main problem in quantitative mass spectrometric analysis is to establish the correspondence between the elemental composition of the ion beam and that of the sample. Measurements may be made over different recording periods, the general principle being to compare the measured intensity with that of an added standard in the sample. With photographic recording techniques, the precision is usually worse than 10%. The main cause of error is variability of the spark source, and recent developments have been concentrated on improving the means of ion production. The best accuracy claimed is still in the region of 5%, and it is hoped that higher precision will be achieved in the near future.

The analysis of thin layers and of solids as a function of depth can be achieved by sputtering and the determination of the positive (or negative) ions in the very sensitive *ion microprobe* (Socha, 1971), by secondary ion mass spectrometry (SIM) described by Herzog and Viehböck (1949), Honig (1958), Benninghoven (1969), Benninghoven and Storp (1971), and Benninghoven (1973), or of the sputtered (by rf glow discharge) neutral particles by mass spectrometry (Coburn and Kay 1971). Lasers might be useful ionizing sources in the future (Fenner and Daly, 1968). Also the energy distribution of the back-scattered ions allows qualitative and quantitative analysis of the uppermost surface layer, and there is no interference from the continuum of ion energies associated with scattering from atoms beneath the surface. Thus the differentiation between the S- and the Cd-face of the polar CdS is possible (Smith, 1971).

9.2.2. Deviation from stoichiometry

Often crystals are nonstoichiometric and contain vacancies of either anions or cations, and the equilibrium number of such defects increases with temperature (see Section 9.5). The concentration of vacancies may also depend on the atmosphere employed during growth and on the nature of the solvent. Vacancies are also influenced by the kinetics of the growth process, and their concentration thus depends on the growth rate and on other growth parameters. Vacancies are often constituents of larger defects and their distribution can be random or they can be ordered, thus forming crystallographic superstructures (see Rabenau, 1970).

A crystal showing a deviation from stoichiometry can be regarded as a solid solution of the stoichiometric crystal with a compound of the next higher or lower valence state of the metal ion, or with a constituent element or with a constituent component without change of valence state. So CdS with sulphur vacancies can be regarded as a solid solution of CdS with Cd, and FeO with oxygen excess can be described as a solid solution of FeO with Fe_2O_3. Many crystalline phases show a large range of composition at high temperatures (near the melting point), whereas at low temperatures the composition range of that phase can be small, even immeasurably small. For example, the width of the Ni_3S_2 phase is close to zero below 550°C and extends from $Ni_{2.56}S_2$ to $Ni_{3.68}S_2$ at 640°C (Huber and Liné, 1963).

Nonstoichiometry is a field of increasing importance and is treated in the books of Rabenau (1970), Wadsley (1964), Mandelkorn (1964), Brebrick (1969), Kröger (1964), Eyring and O'Keeffe (1970), and in several volumes of "Progress in Solid State Chemistry" (H. Reiss, editor).

Large deviations from stoichiometry (larger than 0.1–1%) are indicated in carefully determined phase diagrams and are determined by several of the more precise analytical methods described in the previous section. In the following discussion, methods will be described for the determination of small deviations from stoichiometry. This is possible by special chemical techniques, by thermogravimetry, by exact determination of density, lattice constants and other physical properties, or by determination of the abnormal valence state by physical techniques such as optical absorption, electron spin resonance, Mössbauer spectroscopy, X-ray spectroscopy and electron emission spectroscopy.

The chemical determination of the concentration of ions of abnormal valance state can be achieved by an appropriate oxidation or reduction reaction, by titrimetry, or by colorimetry. As examples, excess metal in ZnO and BaO is analysed by dissolving the materials in acids and by measuring the hydrogen formed (Berdennikova, 1932; Libowitz, 1953).

Metal excess can also be analyzed by oxidizing the metal and by titrating back the excess of the oxidizing agent (Alsopp and Roberts, 1957) and by similar methods (Novak and Arend, 1964; Kleinert and Funke, 1960), or by coulometric titration (Engell, 1956). Hildisch (1968) developed a technique which allows determination of 1 p.p.m. excess metal in a sample of 10–20 mg CdS by evolution of hydrogen and its determination by gas chromatography. Gruehn (1966) has developed a micro-technique for the quantitative determination of low oxidation states of metals. The oxides (or other compounds) are dissolved in molten KOH and thereby oxidized to the normal valence state and, according to the concentration of metal of low valence state, hydrogen is produced according to

$$(\mathrm{Nb}_m^{4+} \mathrm{Nb}_{1-m}^{5+})\mathrm{O}_x + m\mathrm{KOH} \rightarrow m\mathrm{KNb}^{5+}\mathrm{O}_3 + \frac{m}{2}\mathrm{H}_2.$$

This reaction is quantitative, and the hydrogen can be determined. The precision of the measured x values for the case of $\mathrm{Nb}_2\mathrm{O}_{5-x}$ was ± 0.002.

Thermogravimetry (Duval, 1963; Garn, 1965) is a useful method for analyzing nonstoichiometry. The change of weight of a heated sample during oxidation or reduction reactions to a definite state (stoichiometric oxide of normal valency, for example) is a measure of the deviation from stoichiometry. Another approach by TGA (thermogravimetric analysis) is to produce controlled (by weight) nonstoichiometric compounds under given temperature and atmospheric conditions, check the chemical composition by chemical analysis (or TGA) and identify these compounds by X-ray data. Bartholomew and White (1970) crystallized several "non-stoichiometric" titanium oxides from solution and most probably identified them by X-ray data (comparison with earlier studies). Thermogravimetry has become quite popular since high precision instrumentation has become available and it has, for example, been of value in proving the existence of a variety of nonstoichiometric oxides of the rare earths (Wiedemann, 1964) and in analysis of the oxidation states and defect structure of ferrites (Reijnen, 1970).

Several properties are sensitive to deviations from stoichiometry, and especially the density often varies considerably with the composition. Nonstoichiometry can often be studied by precision lattice parameter measurements as was shown with gallium arsenide by Willoughby *et al.* (1971). The correlation of physical properties such as resistivity, thermoelectric power or Hall effect with the deviation from stoichiometry is often possible, and these properties may be extremely sensitive, but the effect of other defects, of trace impurities and compensation must not be overlooked in these cases.

Nonstoichiometry in lithium niobate LiNbO_3 could be correlated with the $^{93}\mathrm{Nb}$ NMR linewidth and to the Curie temperature by Carruthers *et al.* (1971), whereas Scott and Burns (1972) used measured Raman linewidths and frequencies to determine nonstoichiometry in LiNbO_3 and LiTaO_3. These techniques allowed determination of the subsolidus of the corresponding phase diagrams. Various models have been proposed in order to understand these deviations from stoichiometry (Nassau and Lines, 1970).

Specific physical techniques can be used in several cases to determine abnormal valance states. Many oxides (e.g. TiO_2) and oxide compounds (titanates, niobates, tantalates, molybdates, tungstates) which show little absorption in the visible are deeply coloured (blue, brown, black) in oxygen-deficient states and so can be examined by *absorption spectroscopy*.

Electron paramagnetic resonance (EPR, or ESR for the equivalent term electron spin resonance) can give details of the presence of certain transition-metal ions (with unfilled shells) and of the local structure but generally the paramagnetic ions have to be in very dilute concentrations in diamagnetic host crystals for ESR absorption to be detectable (Ayscough, 1967). As an example, Ti^{3+} has been studied in TiO_2 by Chester (1961). A useful technique for the determination of unusual valence states is *Mössbauer spectroscopy* (see Greenwood in Rabenau, 1970). For example, Mullen (1963) detected Fe^+ and Fe^{2+} in NaCl, Gallagher *et al.* (1964) studied the oxidation of $SrFeO_{2.5}$ to the perovskite $SrFeO_3$ containing Fe^{4+}, and Hannaford *et al.* (1965) investigated Sn^{2+} defects in neutron-radiation damaged $Mg_2Sn^{4+}O_4$ by chemical isomer shift. The application of the Mössbauer technique is somewhat restricted since, from the 70 isotopes of 40 elements which fulfill the conditions for recoilless emission of γ-rays, only [57]Fe and [119]Sn have been widely studied. A few publications mention [121]Sb, [125]Te, [151]Eu, [166]Er, [170]Yb and [197]Au in the study of nonstoichiometric phases.

The use of electron spectroscopy for the determination of the valence state of elements is mentioned in Section 9.2.1.

9.2.3. Analysis for minor constituents and of traces

For trace analysis the sensitivity of the method is generally more important than the accuracy. Because of their high sensitivity most of the techniques mentioned in the Sections 9.2.1 and 9.2.2 can also be applied to the analysis of minor constituents (compare Table 9.2). Minor constituents can be dopants added deliberately or may originate from impure starting materials, from the solvent or from the crucible, and impurities can enter the growth system and the crystal via the furnace atmosphere. Frequently a wide range of impurities is found in the grown crystals, often in an unexpectedly high concentration. Therefore careful selection of chemicals as starting materials (with accompanied batch analysis) and careful experiments in sealed systems are necessary if crystals of high purity are to be grown. Impure solvents are particularly troublesome because of their relatively high concentration in a typical solution growth experiment. Fortunately the growing crystal rejects many impurities according to their low effective distribution coefficients, especially when the crystal can grow slowly enough as is discussed in connection with the Burton-Prim-Slichter equation (Section 9.2.4). Trace analysis has been treated, for example, in the books of Meinke and Scribner (1967), Alimarin and Petrikova (1962), Sandell (1959), Tölg (1970) and Koch and Koch-Dedic (1964). Qualitative determination of trace elements is the subject of the books of Feigl (1970) and Vogel (1966).

The most important techniques for trace analysis (determination of impurity concentrations of less than 100 p.p.m.) are *optical emission spectroscopy*, *atomic absorption*, *mass spectrometry*, *neutron activation analysis*, and the relatively simple *titrimetry* and *colorimetry*.

Optical emission spectrography is very convenient since it is particularly rapid, requires only a small sample, and can be used for most elements. Errors may arise through interference, particularly for elements having very complex spectra. The limit of sensitivity for arc and spark emission spectrography for the chemical elements has been tabulated by Addink (1957) and by Morrison and Skogerboe (1965).

Neutron activation analysis (Guinn and Lukers, 1965; De Soete *et al.*, 1971; Lyon, 1964; Kruger, 1971; De Voe, 1969; Siegbahn, 1965; Taylor, 1964; Bowen and Gibbons, 1963) is based on the principle that, if a material is bombarded with neutrons, the induced radioactivity will depend in nature, energy and intensity on the composition of the material. The effect of the bombardment on a stable isotope will normally be to produce a radio-isotope, which will decay with the emission of its characteristic radiation. As the bombardment is continued, a steady condition will eventually be reached where the rate of decay of the radioactive species is equal to the rate of production from the stable isotope. The decay rate is then proportional to the number of nuclei of that isotope. If shorter irradiation times are used, the activity can be calculated in terms of the steady value and the half life of the radio-isotope.

"Thermal" neutrons are normally used as the bombarding particles since most elements have a high capture cross-section for neutrons at relatively low energies, and high neutron fluxes are readily available from nuclear reactors. Alternatively high-energy neutron sources may be used, in instruments which are smaller and easier to erect in an analytical laboratory. These sources normally contain a target of a cooled metal tritide (^{3}H) which is bombarded by deuterons (^{2}H) accelerated to an energy of 100–200 keV. The reaction which generates neutrons is

$$^3_1H + {}^2_1H \rightarrow {}^4_2He + n.$$

Analysis is normally performed by measuring the γ-ray emission by a scintillation counter and pulse height analyser, with data evaluation by computer. The sensitivity is about 10^{-8} g or 0.01 p.p.m. for a 1 g sample. The precision of the determination is typically 1–3% if suitable care is taken. The majority of elements have more than one stable isotope and so a correction is required for the fraction in the element of the isotope used for the analysis. Cloete *et al.* (1969) performed neutron activation analysis of $MgAl_2O_4$ crystals grown from PbF_2 solution at 1200–1250°C in 100 cm³ platinum crucibles and found 13 p.p.m. iridium (present in platinum in small amounts), but no platinum. A remarkable result was that the authors found by absorption spectroscopy less than 10 p.p.m. lead in the spinel crystals.

Even higher sensitivity than with neutron activation is sometimes possible with *tracer methods* in which a known concentration of a radio-isotope is added to the solution prior to crystallization and the activity of the crystal measured after growth. This determination is based on the assumption that the fraction of the radio-isotope in the crystal is the same as that in the melt. Although the apparatus required for such a determination is relatively simple, the use of radioactive tracer methods in high temperature solution growth has been reported in the literature only recently when their application to a variety of problems was described by Janssen *et al.* (1973).

Comparable sensitivity is sometimes possible with *mass spectrometry*, the limit being around 10^{-3} ppm, with 10^{-1}–10^{-3} ppm typical for most elements in any matrix. Wolfe *et al.* (1970) used mass spectrometry to measure impurity concentrations at the level of 10^{13} cm^{-3} in gallium arsenide. The range of sensitivity between different elements with mass spectrometry is less than with most methods, being normally within a factor 3. The other chief advantage is that different areas of the sample may be studied using selective volatilization but the spectrometer is expensive and the accuracy relatively low. In addition, both mass spectrometry and neutron activation analysis are time consuming compared with optical emission spectrometry. Trace analysis much below the p.p.m. level is achieved by correlation of impurity concentration with sensitive physical properties, for example with the resistivity of metals and semiconductors. Luminescent properties are also extremely sensitive to the smallest impurity levels (see for example Meinke and Scribner, 1967). Care is necessary in these cases since they are often not specific or are easily interfered with by other impurities. Other physical techniques based on photon spectroscopy (such as *nuclear magnetic resonance, electron spin resonance,* and *Mössbauer spectrometry*) or on electron spectroscopical methods (see Section 9.2.2) are useful in certain cases of composition or trace determination, but their main potential lies in the study of structure and of defects, and they will be discussed further in connection with these latter aspects.

Linares *et al.* (1965) have shown that a variety of rare-earth ions could be detected in concentrations as low as 0.02 to 1 p.p.m. by *optical fluorescence* which was excited by X-rays (probably not directly but by photo-electrons which then preferentially transfer their energy to impurity atoms by inelastic scattering).

Impurities such as oxygen, nitrogen and hydrogen in the p.p.m. range in metals and semiconductors have an appreciable effect on the material properties. The detection of such volatile traces is possible with modern apparatus by vacuum fusion as developed by Thompson *et al.* (1937), with determination of the gases by infrared absorption (CO) and other physical properties (Kraus, 1972; Paesold *et al.*, 1967) or by mass spectrometry.

Frequently preconcentration techniques such as zone melting, chromatography, electrophoresis, extraction, and precipitation are used to increase the sensitivity of the determination, and also coprecipitation and mixing techniques can be helpful in this respect.

Other methods for trace analysis are based on *chromatography, ion exchange chromatography* and *electrophoresis*. In these techniques a mixture of ions or molecules is separated during its travel with a solvent phase (a) along a second (immiscible) liquid layer absorbed on thin films or columns of cellulose, silica gel, alumina, organic polymers etc. (= chromatography), or (b) along ion-exchange resins which are reversibly replacing an equivalent amount of other ions, or (c) in an electric field (= electrophoresis). The determination is performed directly on the dried and fixed systems by specific colouring agents (paper chromatography, thin-film chromatography, or electrophoresis), or the separated chemical constituents are extracted and determined by other techniques. Advantages of these techniques are the possibility of separation of a variety of ions and the high sensitivity which can be lower than 1 ppm, also the apparatus is simple. Disadvantages are the inaccuracy (5–20%), the necessity of preparing solutions, and the high level of experience necessary. Monographs on the above methods have been written by Lederer and Lederer (1960) and by Lederer (1971).

9.2.4. Inhomogeneities and their detection

Inhomogeneities of the crystal composition, either of the major or minor constituents, occur on all scales. Inclusions are discussed in Section 9.4.6, inhomogeneities on an atomic scale in connection with defects in Section 9.4.1, and structural inhomogeneities in Section 9.3. Methods of producing homogeneously doped crystals and of homogeneous solid solutions have been described in Section 7.1.3.

Origin of inhomogeneities. According to the Burton-Prim-Slichter equation (1953) the effective distribution coefficient is

$$k_{\text{eff}} = \frac{k_o}{k_o + (1 - k_o) \exp - (v\delta/D)},$$

where k_o is the equilibrium distribution coefficient, v the growth rate, δ the boundary layer thickness and D the diffusion coefficient. Thus inhomogeneities are produced when any of the parameters k_o, v, δ or D

change during the course of a crystallization experiment. k_o is a function of the temperature according to the ideal solution concept (Thurmond, 1959). Also temperature-dependent equilibria in the solution and the enrichment (or depletion) of impurities in the solution during growth will affect k_o. The growth rate v is strongly dependent on the supersaturation and hence on temperature changes. In the many cases where the rate of crystallization depends on adsorption on specific faces or on the crystallographic direction, an orientation-dependent impurity incorporation is to be expected, and an example of such zonal growth is shown in Fig. 9.2. The change with temperature in the diffusion coefficient D is expected to be relatively minor, whereas the boundary layer thickness δ is more likely to cause inhomogeneities since it is sensitive to any changes in the hydrodynamics in the solution. However the most important causes of inhomogeneities in flux grown crystals are temperature variations and oscillations, the latter leading to banded growth (*striations*), and extended variations of composition are due to changes in k_o.

It is possible that the majority of striations are due to temperature

Fig. 9.2. Orientation-dependent impurity incorporation in zircon (Scheel, unpublished).

fluctuations which arise in the solution because of unstable convection. Temperature oscillations due to convective motion in fluids are well known but their importance in crystal growth was not appreciated until comparatively recently. Wilcox and Fulmer (1965) demonstrated that fluctuations in the dopant concentration in calcium fluoride crystals grown from the melt could be correlated with temperature oscillations within the melt, and many examples of the relation between such oscillations and striations in melt-grown crystals have now been reported (Hurle, 1966, 1967; Cockayne and Gates, 1967; Witt and Gatos, 1968). Striations due to variations in the Ta/Nb ratio in crystals of $KTa_{1-x}Nb_xO_3$ grown by top-seeded solution growth were attributed by Whiffin (1973) to temperature oscillations which were measured as $\pm 1°C$ in amplitude.

The incidence of temperature oscillations is discussed in Section 6.6.2 where reference is made to "overstability" which is a probable cause of such oscillations in high temperature solutions. Figure 9.3 shows temperature oscillations at different depths in a solution of nickel ferrite in

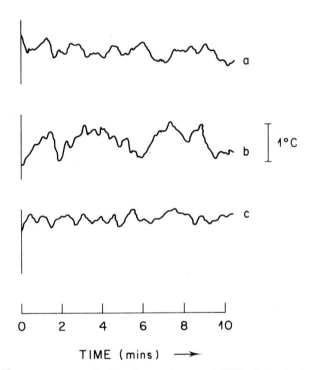

FIG. 9.3. Temperature oscillations in a solution of $NiFe_2O_4$ in $BaO \cdot 0.62B_2O_3$ (Smith and Elwell, 1968). (a) Thermocouple junction just below melt surface. (b) Junction 4 mm below surface. (c) as (b), but crucible rotated at 10 r.p.m.

$BaO \cdot 0.62B_2O_3$ at 1200°C. The melt depth in this case was only 3 cm but oscillations of amplitude 0.5°C were measured at a point 4 mm below the surface in a temperature gradient of about 10°C/cm. The amplitude could be reduced somewhat by rotating the crucible, as shown in Fig. 9.3(c).

Interesting observations in Czochralski growth of tellurium doped InSb were made by Kim, Witt and Gatos (1972). They found by "time markers" that only under thermal (convective) stability were the macroscopic and microscopic growth rates identical. Thermal stability was achieved by a stabilizing temperature gradient and by making the depth of liquid small. Otherwise oscillatory thermal instability or turbulent convection occurred leading to back melting and to microscopic growth rates more than 20 times greater than the average macroscopic growth rate. The striations produced by the temperature and the corresponding growth rate fluctuations could be made visible by selective etching and optical techniques. Time markers would be useful also in similar experiments on the origin of inhomogeneities in flux-grown crystals, and in effect such experiments are in progress according to Damen and Robertson (1972) and White (1971). The use of time markers in vapour growth was reported by Kroko (1966).

The incidence of oscillatory behaviour in high-temperature solutions can be reduced by the use of shallow melts or small temperature gradients as mentioned above. Since the depth of solution appears in Eqns (9.2) and (9.3) (which are valid for the critical range only, see Section 6.6.2) as l^3, its effect is particularly marked. Brice *et al.* (1971) have shown that the incidence of oscillations may be reduced by the use of baffles which lower the effective depth of the melt and it is possible that similar arrangements will be necessary in HTS-growth for the growth of crystals which are particularly prone to striations. The tendency to form striations may also be reduced by using a growth rate which is well below the maximum stable value.

Other sources of growth-rate oscillations are also possible. If solute transport occurs only by diffusion, then a solution of the diffusion equations for $n(x, t)$ can give oscillatory behaviour in terms of the distance x from the interface but not in terms of the time. An oscillatory interface kinetic process has not been proposed although some periodically varying behaviour might be envisaged in the bunching of layers which leads to the formation of macrosteps. The possibility of striations due to an oscillating growth rate in an unstable supersaturation gradient was proposed by Landau (1958) and Petrov (1956).

Self-excited oscillatory motion *along* the interface near the breakdown of stability was predicted by Mullins and Sekerka (1964), and this theory was extended to growth from stirred melts by Delves (1968, 1971). A detailed discussion of these problems is given in Chapter 6.

Wilcox and Chase (1967) considered the effect of strain produced by incorporated impurities on the distribution coefficient. The authors developed an expression for the effective distribution coefficients for the case where no plastic deformation occurs. Although no experimental proof has yet been given, it seems plausible that a certain fraction of inhomogeneous impurity incorporation can be attributed to this effect.

Techniques for determination of inhomogeneities. Inhomogeneities in a bulk crystal cannot be measured quantitatively at present, but qualitative observations can frequently be made. Crystals often show effects due to variation in the refractive index, light absorption, etc., and crystals may show strain due to different lattice parameters and differing thermal expansion coefficients of the inhomogeneous regions. Since interferometry is extremely sensitive to fluctuations in optical density it is frequently used to prove qualitatively the high quality of crystals. Qualitative indication of inhomogeneous impurity incorporation is readily obtained when radioactive impurities are used (Landau, 1958) and occasionally selective etching can be used to display striations (Witt, 1967). Frequently inhomogeneities are the cause of line-broadening of X-ray diffraction patterns (see Section 9.4.2).

Quantitative determinations of inhomogeneities are performed by various techniques which allow local analysis of the surfaces of bulk crystals, of crystal sections, or of layers grown by liquid phase epitaxy. Examples of physical parameters (Curie temperature, magnetization, conductivity, optical absorption, etc.) which have been used to characterize inhomogeneities in flux-grown crystals are given later in this section.

Techniques for the analysis of inhomogeneities in crystals are based on narrow beams of electrons, photons or particles which strike the surface and excite electrons, X-rays or photons of characteristic energies or which cause local evaporation of the material. The possible techniques are summarized in Table 9.3. Many of the methods listed are applicable only to special problems, while several techniques are in the early stages of development.

Electron probe microanalysis is the most powerful and most widely used technique for the analysis of inhomogeneities and is discussed in Section 9.2.1. Here it should be added that the region which is excited and which emits X-rays has a larger diameter than the electron beam and a certain depth since the locally produced X-rays of short wavelengths excite other elements (secondary emission). Future developments in electron probe microanalysis are proceeding in the direction of narrowing the electron beam and of building sensitive X-ray detectors which allow a fast quantitative analysis of the whole X-ray spectrum.

TABLE 9.3. Techniques for Analysis of Local Inhomogeneities

Exciting beam	Emitted radiation or particles	Measuring technique	Local resolution	Sensitivity	Precision	References
X-rays (focussed or collimated)	X-ray diffraction	identification, cell parameters (see Sec. 9.4.2)	~50 μm†	~5%	—	See Section 9.2.4.
	X-rays (emitted)	X-ray fluorescence analysis	~100 μm†	~0.1%	~0.1–1% ~2%	See Section 9.2.1. Linares et al., 1965
	photons (luminescence)	luminescence spectroscopy	~100 μm†	~0.1–1 p.p.m.†	~10%	Chang, 1971;
	electrons	electron spectroscopy	~100 μm†			Preisinger et al., 1971
UV	photons (luminescence)	luminescence spectroscopy	~10 μm‡		~10%	Chang, 1971; Preisinger et al., 1971;
	electrons	electron spectroscopy (ESCA)	~10 μm‡			Siegbahn et al., 1967
Laser	material vapour	optical emission spectroscopy	~50 μm‡	~1–10³ p.p.m.	~1–10%	Snetsinger and Keil, 1967; Blackburn et al., 1968
		atomic absorption (?)	~50 μm‡	<1 p.p.m.	>5%	Fenner and Daly, 1968
		mass spectrometry	~50 μm‡			
Electrons	X-rays	electron probe microanalysis	~1 μm	~0.1–1%	0.5–5%	See Section 9.2.1
	photons (luminescence)	electron spectroscopy (Auger electr. spectr.)	~1 μm	0.1–1%		Chang, 1971
	electrons		~1 μm			
	electrons (diffraction)	identification, cell parameters	<0.1 μm	—	—	See Section 9.4.3
			<0.1 μm		~5%	See Section 9.4.3
Neutrons, Protons and Alpha particles	neutrons and other backscattered particles	energy distribution	200Å in depth for B in silicon, <1 mm	3 p.p.m.		Ziegler et al., 1972; Ziegler and Baglin, 1971; Mitchell et al., 1971; Davies et al., 1967
Ions (sputtering)	electrons	electron spectroscopy	50Å in depth			Coburn and Kay, 1972; Gupta and Tsui, 1970
	positive ions	ion microprobe			>5%	Socha, 1971
	material particles	mass spectrometry				Coburn and Kay, 1971
	reflected ions	ion scattering spectroscopy				Smith, 1971

† Expansion to <5 μm should be feasible

The analysis of impurities as a function of depth has been reviewed by Coburn and Kay (1972) who distinguish three categories of techniques: (a) non-destructive, (b) microsectioning with observation of the remaining material, and (c) microsectioning with observation of the material removed. For (b) and (c) chemical etching or preferably etching by sputtering under clean conditions has been used. The measuring techniques are included in Table 9.3.

Inhomogeneities observed in flux-grown crystals. It seems to be widely accepted that striations in crystals grown from high-temperature solutions arise from temperature fluctuations which are caused by insufficient temperature control or by convective instability. Striations have been observed in numerous cases, therefore in the following only a few examples will be given.

Striations corresponding to variations in the chromium concentration were found in the ruby crystals grown by White and Brightwell (1965). These were believed to arise because of preferred adsorption on different faces, giving rise to a series of sharp boundaries in some crystals cut parallel to the *c*-axis.

Chase (1968) found a variety of striations in indium oxide crystals. The bands in the innermost zone were very clear and decreased in width with distance from the growth centre. In the intermediate zone very fine striations were observed with wider bands, about 1 mm across, superimposed. The fine striations were absent in the outermost zone but less well-defined variations in colour were observed. The various types of striation could be readily correlated with the mode of growth. The most marked striations, in the inner zone, were produced at the time of most rapid growth following nucleation. In the intermediate zone, growth occurred by nucleation of layers at corners and edges of the crystals and the fine bands are presumably related to the period between the passage of successive layers. The bands of relatively minor intensity were associated with a more stable mode of growth at screw dislocations giving rise to growth hillocks. The striations could all be correlated with fluctuations in the furnace temperature resulting from an imperfect controller.

The very regular, fine bands of striations in a dysprosium orthoferrite crystal as shown in Fig. 9.4 have been attributed by Wanklyn (1975) to an on–off temperature controller. Banding may be a periodic array of inclusions caused by periodic unstable growth as demonstrated in Fig. 9.5. A careful distinction should be made if possible between striations (banding due to varying composition or impurity incorporation) and banding of inclusions. Both phenomena may be of similar origin, namely the fluctuations in growth rate caused by changes in the temperature or hydro-

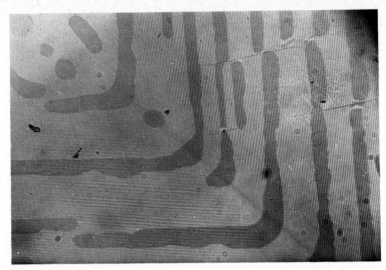

FIG. 9.4. Striations (the fine lines) in a $DyFeO_3$ crystal caused by an on–off temperature controller (courtesy B. M. Wanklyn).

FIG. 9.5. Bands of inclusions in $GdAlO_3$ parallel to as-grown faces (Scheel and Elwell, 1972).

dynamics. In several cases "striations" have been reported which could have been periodic layers of tiny inclusions due to unstable growth as discussed above. This effect is to be expected when the growth rate is near the maximum stable value. Then growth rate variations due to temperature oscillations will cause alternatively stable and unstable growth, or the

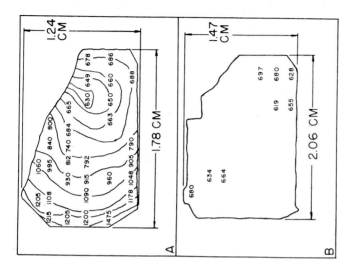

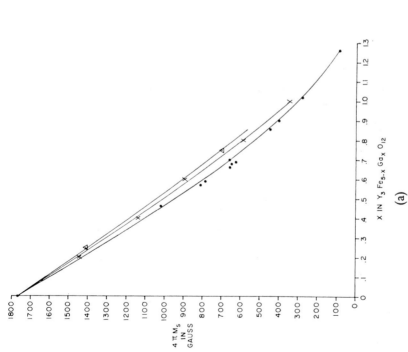

FIG. 9.6. (a) $4\pi M_s$ *versus* composition in $Y_3Fe_{5-x}Ga_xO_{12}$. (b) $4\pi M_s$ variations in crystal slabs of $Y_3Fe_{5-x}Ga_xO_{12}$ solid solutions (Nielsen *et al.* 1967).

growth rate variations may be self-exciting as described by Landau (1958).

Lefever *et al.* (1961) found a banded appearance of finely divided inclusions of 100 μm diameter in yttrium iron garnet. These bands appeared to be associated with the presence of divalent iron in the crystals since their incidence could be reduced by growing the crystals in an oxygen atmosphere or by reducing the silicon concentration in the melt. Silicon enters the garnet lattice readily as Si^{4+} and so produces Fe^{2+} ions by charge compensation.

Giess *et al.* (1970) observed striations with a periodicity of about 2 to 60 μm in rare-earth orthoferrite crystals but their origin has not been found. The lead concentration was found to vary from 1.10% in the darker bands to 0.72% in the lighter regions. These bands must therefore be formed during growth. It appears likely that the lead would be present in finely divided inclusions but the authors were unable to distinguish whether or not the impurity had entered the crystal lattice.

Few examples have been published where physical measurements were made to demonstrate the presence of inhomogeneities in crystals. In the interesting paper of Nielsen *et al.* (1967) local saturation magnetization ($4\pi M_s$) measurements on flux-grown solid solutions of $Y_3Fe_{5-x}Ga_xO_{12}$

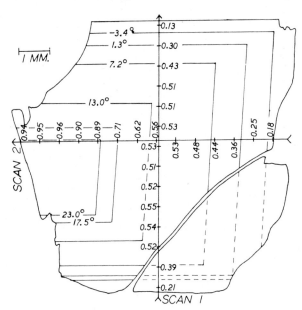

FIG. 9.7. Outline of the boundary of the surface of Rh-doped Fe_2O_3 crystal. The uncorrected Rh_2O_3 dopings are shown on the scan paths. The straight lines indicate the antiferromagnetic–weakly ferromagnetic boundaries for several temperatures (Morrish and Eaton, 1971).

could be correlated with compositional inhomogeneity using the data of Fig. 9.6(a). The authors were able to correlate the $4\pi M_s$ variation and therefore the compositional inhomogeneity with the weight loss of the solution during the experiment: a 20% weight loss (mainly by PbF_2 evaporation) resulted in a $4\pi M_s$ variation of 845 gauss along the crystal section, whereas in another experiment a 4.2% weight loss resulted in a $4\pi M_s$ variation of only 260 gauss. Typical examples of measured saturation magnetization values for various crystal sections are shown in Fig. 9.6(b), where A is an example of a crystal from a run with high weight loss and B is typical of a crystal grown in a tightly closed crucible.

The magnetic (Morin) transition temperature T_M of pure haematite at about 263°K is very sensitive to doping; Al^{3+}, Ga^{3+}, Ti^{4+} and Sn^{4+} all lower T_M whereas Rh doping raises T_M. Morrish and Eaton (1971) studied rhodium-doped haematite crystals and found a correlation between the doping level and the magnetic transition temperatures. Their results as measured by electron microprobe are shown along scan 1 and scan 2 in Fig. 9.7 for a doping level of about 0.5 mole%Rh. The isotherms of the magnetic transition were parallel to the growth front, from left to right. In the first stage much rhodium was incorporated and T_M is about 20°C, whereas the outer portion of the crystal contains little Rh and therefore shows T_M values below 0°C.

9.3. Structural Aspects and their Determination

Most physical phenomena are sensitive to the structural aspects of crystals, to the three-dimensional periodic arrangement of atoms, ions, and molecules and to the array of bonds between these species. However, only those techniques which use diffraction of a radiation with a wavelength similar to the spacings within the crystals can give a picture of the structure as will be discussed in connection with the diffraction of X-rays, electrons, or neutrons. Most physical phenomena only indicate the anisotropy (or isotropy of cubic crystals) and the symmetry of the crystals. From optical phenomena and optical measurements symmetry, orientation of the crystal, and occasionally other structural details can be very conveniently deduced as discussed below. The characterization of thin films by optical and other methods has been described by Heavens (1965).

9.3.1. Optical techniques

Microscopy. Routine examination of crystals is most frequently performed by optical microscopy, particularly if the crystals are transparent. It is considered necessary in most crystal growth laboratories to have permanent access to a polarizing microscope, preferably with a universal stage and a camera attachment.

Microscopic techniques have been reviewed by a number of authors, for example Chamot and Mason (1958), Burri (1959), Winterbottom and McLean (1960) and Schaeffer (1966). The principal applications of microscopy in the study of crystals grown from high temperature solution may be classified as follows:

(a) Identification and examination of optical properties,
(b) studies of surface features,
(c) investigations of defects.

The use of optical methods in crystallography has been described by Winchell (1937), Bloss (1961), Wood (1963), Gay (1967), Rath (1969) and Hartshorne and Stuart (1969). A complete optical examination of a transparent crystal would include the following stages:

(i) Examination in unpolarized light and observations of colour, habit, edge angles, cleavage, refractive index and optical dispersion (variation of refractive index with the wavelength of light). Precautions necessary for precision measurements of refractive index are discussed by Hafner and Rood (1967), and Lawless and DeVries (1964) described the refinement of refractive index measurements of thin samples.

(ii) Use of crossed polarizers to distinguish between isotropic and anisotropic crystals and to look for twinning and strain. Measurement of the extinction angle and determination of birefringence (difference in refractive index for ordinary and extraordinary ray).

(iii) Determination of the principal refractive indices of anisotropic crystals and a study of pleochroism (variation of absorption of light with direction of vibration of polarized light).

(iv) Observation in convergent polarizing light between crossed polars, using the microscope as a *conoscope*. The advantage of convergent rather than parallel light is that the resulting image or *interference figure* depends upon the optical character in many crystallographic directions rather than in a single direction. The converging beam is produced by the condenser lens and a real image is formed by the objective. Each point in the image corresponds to a definite direction of light through the crystal and the image may be examined through the analyzer directly or by use of an auxiliary lens, the Bertrand lens, to bring the image into the focal plane of the eyepiece. The image may be used to classify anisotropic crystals as uniaxial or biaxial and to determine the positive or negative character and the angle between the two optic axes of biaxial crystals. For various optical measurements and for determining the orientation of twins, of lamellae due to unmixing, and of inclusions, the universal stage of Fedoroff and the spindle stage are useful or necessary (Reinhart, 1931; Fairbairn and Podolsky, 1951; Wilcox, 1959; Emmons, 1943).

Suitable modifications to the optical microscope permit several additional investigations. *Ultramicroscopy* is a technique in which a narrow beam of light, usually from a laser, is incident on the crystal in a direction at right angles to the axis of the microscope so that light scattered from inclusions and other defects is observed. This technique was used by Newkirk and Smith (1967) to study inclusions in BeO, and particles less than 1000 Å in diameter were detected.

Phase contrast microscopy (Zernicke, 1938; Sunagawa, 1967) and *interference microscopy* (Tolansky, 1943, 1970; Nomarski and Weill, 1954) are modifications of optical microscopy in which the image is extremely sensitive to the surface topography, and these techniques are particularly useful for the study of surface features such as growth spirals and layers. A height resolution of about 4 Å may be achieved under favourable conditions. The great advantage of Tolansky's multiple beam interference method is that the interference fringes are narrowed and sharpened relative to those in the classical Newton's rings experiment and so the height resolution is perhaps 100 times that possible in the latter experiment.

Infrared microscopy (Sherman and Black, 1970; Sunshine and Goldsmith, 1972) is useful for materials which absorb in the visible but are partially transparent in the infrared, such as many chalcogenides and ferrites. Valuable information on the magnetic domain structure (which is sensitive to defects) of the magnetic bubble domain materials and devices may be obtained using polarized infrared light, and defects in several semiconductors may be studied easily, especially in thin films.

Infrared absorption analysis may sometimes be usefully applied to the identification of inorganic materials or impurities (Lawson, 1961; Moenke, 1962; Nakanoto, 1963; Kendall, 1966; Harrick, 1967).

Crystal handling. In the case of materials of low transparency, interior features of the crystal may be examined if it is cut into thin sections. Thin crystal slices with as little damage as possible are also a necessity for X-ray topography as will be discussed in Section 9.4.3. The method used for sectioning will depend on the material and will normally involve *cleaving*, *sawing, chemical or electro-chemical machining* or *spark-cutting*. The latter can be used only for electrically conducting materials and it produces surface damage as does mechanical sawing. The most suitable saw blades are normally thin metallic discs impregnated with diamond grit which are rotated at high speeds. The main disadvantage of such saws is that the amount of damage will increase with ageing of the blade as the number of irregularities increases. In general the damage is less for an annular blade impregnated along its inner edge and supported along its perimeter than for a blade supported at its centre and with a peripheral cutting edge,

R

and even less damage is reported for the use of a diamond-impregnated wire saw. Successful cleaving depends upon the existence of a cleavage plane within the crystal, and relatively few flux-grown crystals of practical interest will cleave easily. Chemical machining, as with an acid saw, produces the least damage but the cutting rate is often prohibitively slow.

It is normally assumed that cutting will result in some damage to the crystal and that the damaged layer must be removed if it is required to examine the "as grown" crystal characteristics. The slice is first *mechanically polished* on some abrasive paper to give a flat surface. It is sometimes desirable to mount the crystal in a cylinder of transparent plastic to facilitate handling and to prevent rounding of the edges during polishing. Progressively finer grades of abrasive paper are used to give a good polish and the final stage is normally performed with fine powder such as alumina, magnesia, chromic oxide or jeweller's rouge. The powders are dispersed on napless cloth of cotton or silk, with velvet of fairly deep pile used for the final polish. Alternatively the whole polishing process may be effected using diamond pastes, containing diamond grit of various diameters down to about 0.1 μm. A jig for holding specimens for automatic polishing using diamond paste and soft metal lapping plates has been described by Bennett and Wilson (1966). The crystal in this case is held in position by wax and polishing is effected by rotating the lapping plate at 10–60 r.p.m. while the jig traverses this plate on a reciprocating arm.

A process which is currently very popular for a final polish involves the use of "Syton".† This is a silica abrasive in an alkaline colloidal suspension, which gives a good surface finish on, for example, refractory oxides.

Metal slices may be thinned by electrolysis, and this *electrolytic polishing* has the advantage that there is no mechanical disturbance. The principal disadvantage is that it frequently does not result in a smooth surface, since selective attack tends to occur in the region of any tiny cracks, pores or inclusions.

An alternative technique is to combine mechanical polishing with chemical attack, for example by impregnating a polishing cloth with a chemical reagent.

Since any mechanical polishing will normally result in some surface damage and the generation of dislocations, the final stage in the preparation of thin samples is normally *chemical polishing* in which a uniform layer of crystal is removed by the action of a solvent. Fused salts have been used quite commonly for this purpose but suffer from the disadvantage that the specimen cannot be observed during the polishing process. Shick (1971) has proposed the use of phosphoric acid at 380–425°C as a chemical polish for orthoferrites. The polishing rate in his experiments was about

† Trade mark, Monsanto Company.

10 μm/min compared with 3–6 μm/hr for Syton, and an improved surface finish was obtained. Shick found rotation at about 40 r.p.m. to give a surface of optimum smoothness and describes a jig for this purpose. In his experiments layers of orthoferrite initially 0.1 mm in thickness were reduced to half this value and the resulting improvement in the sharpness of the domain pattern is evidence for the complete removal of the strained outer layers. Phosphoric acid was also used by Basterfield (1969) as a chemical polish for yttrium iron garnet, while Reisman *et al.* (1971) preferred mixtures of phosphoric acid and sulphuric acid for spinel and sapphire.

Certain solvents have the property of causing preferential attack at strained regions rather than removing a layer uniformly. The preferential attack results since atoms in these regions are less tightly bound because of the strain energy. Solvents which possess this property are unsuitable for use in chemical polishing but may be used as etchants for measurement of the concentration of those defects which give rise to the strain. The most likely cause of *etch-pits* are dislocations and counting of etch-pits is the most convenient method of studying the concentration and distribution of dislocations in crystals. Etch-pit counts may, however, underestimate the dislocation density if selective etching of edge or screw dislocations occurs. Reviews of etching for the study of dislocations have been given by Regel *et al.* (1960) and Amelinckx (1964); some typical etchants are also listed by Laudise (1970). Examples of dislocation densities determined by etching are listed in Table 9.4.

Examples of optical appraisal studies. A large number of appraisal studies of crystals grown from high-temperature solution have appeared in literature, and we shall consider only a few typical examples.

Lefever *et al.* (1961) examined polished sections of yttrium iron garnet crystals by transmitted light and were able to observe the dendritic core, layers of inclusions and striations. Janowski *et al.* (1965) studied the types of inclusion in ruby crystals by transmission microscopy and attempted to relate their distribution to that of dislocations revealed by etching the same crystals. They were thus able to show that dislocations were sometimes produced by solvent inclusions. In addition, twin planes were observed and were found to be associated with groups of etch pits only infrequently, the more common result of etching being to produce a shallow groove along the twin boundary.

Nelson and Remeika (1964) used a Twyman-Green interferometer to investigate the optical homogeneity and departure from face flatness of ruby crystals used for laser generation. The variation in optical path across the crystal was about $\lambda/10$ for red light. A conoscope was used to look for strain but this was barely detectable.

TABLE 9.4. Dislocation Density of HTS-Grown Crystals

Material	Solvent/flux	Count	Face	Method	Reference
Al_2O_3	Various	0 (over 1–2 cm^2)	—	etch	Linares, 1965b
Al_2O_3	PbO/PbF_2	10^2	{001}	etch	Stephens and Alford, 1964
TiO_2	Alkali borates	4×10^4	—	etch?	Berkes et al., 1965
$BaTiO_3$	KF?	10^2–10^4	{001}	etch	Waku, 1962
$Y_3Al_5O_{12}$; $Y_3Ga_5O_{12}$	—	10^2–10^4	{110} {211}	X-ray topography/etch	Belt, 1969
ThO_2	$Li_2W_2O_7$	$10^{\pm 2}$	{111}	X-ray rocking curve	Finch and Clark, 1965
Al_2O_3	PbF_2	10^2–10^4	{001}	etch	White and Brightwell, 1965
$Y_3Al_5O_{12}$	$PbF_2/PbO/B_2O_3$	6×10^3	{110}	etch	Timofeeva et al., 1969a
Al_2O_3	Bi_2O_3/PbF_2; Bi_2O_3/BiF_3	1–10	{001}	etch	Janowski et al., 1965
Al_2O_3	PbO/B_2O_3	1×10^4	—	etch	Sahagian and Schieber, 1969
Al_2O_3	PbF_2	10^2–10^4	—	etch	Champion and Clemence, 1967
$MgAl_2O_4$	PbF_2	50–200	{111}	X-ray topography	Wang and McFarlane, 1968
$MgAl_2O_4$	PbF_2	0–200	{111}	X-ray topography	Wang and Zanzucchi, 1971
Graphite	Fe, Ni	10^7 / 10^{3-5}	{001} other planes	etch } etch	Austerman et al., 1967
$BaTiO_3$	TiO_2	10	{100}	etch	Belruss et al., 1971

Detailed optical tests on rubies grown by various methods were performed by Bradford *et al.* (1964). A conoscopic investigation was used to look for variations in the optic axis, based on the disturbance of the normal "isogyre" figure which occurs when the crystal is not perfectly uniaxial. (An isogyre is characterized by dark arms extending in the directions of polarization of the polarizers, with a set of dark circles filling the otherwise bright areas.) On the flux-grown ruby the disturbance of the figure was found to be very slight, and was attributed to strain near the edges of the crystal.

The tests also included *shadowgraphs* produced simply by passing light from a zirconium arc through a filter and the crystal onto a distant film. The resulting photographs show up defects in the crystal as variations in the intensity of the image. The shadowgraphs included some taken with crossed polarizers, which may be expected to reveal contrast due to photo-elastic effects and possibly also due to variations in the optic axis. *Small angle scattering* of a laser beam was also used as an indication of general optical quality; the degree of scattering depends upon the number and nature of scattering centres in the crystal. The quality of the flux grown crystals was generally good, but all crystals suffered from variations in the distribution of the chromium dopant.

Newkirk and Smith (1965) and Austerman *et al.* (1965) examined the surface features and inclusions in crystals of BeO. Oblique illumination was used to reveal the central projection associated with the inversion twin (see Section 5.5.3) and the height of this projection was measured by an interference method.

A fairly detailed microscopic study of defects in yttrium aluminium garnet was reported by Timofeeva *et al.* (1969). Observations were made by transmitted, reflected and by scattered light, and polarized light was used to show the stress distribution. Refractive index variations were also monitored using a Michelson interferometer. These observations permitted a distinction to be made between inclusions arising during growth and those which occurred during the subsequent cooling, since the latter were not accompanied by high stress. Many inclusions were in the form of small particles of size less than 1 μm, which were best revealed by ultramicroscopy. Crystalline inclusions of Al_2O_3, Nd_2O_3 and Y_2O_3 were also observed and found to give rise to large stress fields.

Photoelastic stress patterns associated with dislocations were observed by Bond and Andrus (1956) and in yttrium gallium garnet by Prescott and Basterfield (1967). The dislocations in this material were found to be introduced and propagated during the growth process. An example of the use of this technique is given in Fig. 9.8, which shows the pattern in the region of the growth centre on a (110) face of $Y_3Ga_5O_{12}$. The difference in

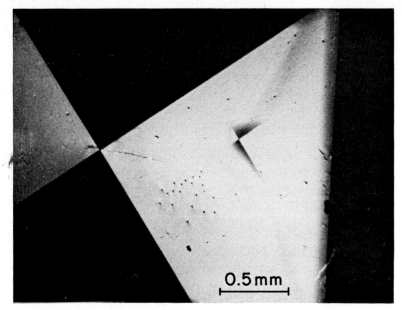

FIG. 9.8. Stress birefringent pattern on (110) growth face of $Y_3Ga_5O_{12}$ (Prescott and Basterfield, 1969).

strain direction in adjacent quadrants is clearly indicated, as is the presence of a secondary growth centre in the right-hand quadrant.

Spectroscopy. The increasing use of crystals in lasers has encouraged the use of *optical spectroscopy* as a means of characterization. A good example is the study of optical emission spectra of Cr^{3+} in ruby by Linares (1965b). The room temperature fluorescence linewidth was found to be the same in crystals grown by the Verneuil flame fusion method as in crystals grown by various fluxes. At 77 K, however, rubies grown from fluxes free from fluorine exhibited a linewidth of less than 0.5 cm^{-1}, comparable with the best flame fusion ruby, while crystals grown from PbF_2 or BaF_2/AlF_3 fluxes exhibited linewidths greater than 1.5 cm^{-1}. The emission studies also revealed fluorescence due to contamination from the crucible material, whether platinum, iridium or rhodium was used. The same author (Linares, 1967) also reported spectroscopic studies of CeO_2 and ThO_2.

Wang and Zanzucchi (1971) made emission, absorption, reflection and transmission measurements on $MgAl_2O_4$ crystals grown from a PbF_2 flux and also on crystals grown by the Czochralski and flame fusion methods. All crystals were found to be transparent between 0.3 μm and 7.2 μm, but the flux-grown crystals exhibited a much sharper absorption edge below

0.28 μm which was attributed to Fe impurity. Reflection measurements were made in the range 10–40 μm in which bands occur due to lattice vibrations associated with the octahedral and tetrahedral sites in the spinel lattice. The reflectivity of the flux-grown crystals was rather lower than that of the Czochralski or flame fusion crystals, again probably because of chemical impurities or inclusions. The emission linewidth of all the synthetic crystals was much broader than that of natural $MgAl_2O_4$, an observation which rules out the possible application of the synthetic spinel as a laser host material.

Infrared absorption spectroscopy was used by Wickersheim *et al.* (1960) to determine the concentration of silicon in yttrium iron garnets.

9.3.2. X-ray methods

X-ray diffraction is by far the most popular method for the identification of substances and for the investigation of crystal structure and degree of crystal perfection. Many diffraction techniques have been developed, for a wide range of problems. A simple approach to X-ray diffraction is provided by the Bragg (1913) condition

$$n\lambda = 2d_{hkl} \sin \theta_{hkl},$$

where n is the order of the diffracted beam, θ_{hkl} the angle between the incident X-ray beam and the atomic planes hkl which give rise to the diffraction peak considered, d_{hkl} the spacing between such planes and λ the X-ray wavelength. Alternatively it may be convenient to consider the reciprocal lattice P_{hkl} which may be envisaged by rewriting the Bragg condition in the form (for $n = 1$)

$$\sin \theta_{hkl} = \frac{1/d_{hkl}}{2/\lambda} = \frac{P_{hkl}}{2/\lambda}.$$

Diffraction occurs when a reciprocal lattice point passes through the surface of the sphere of reflection from the hkl planes. The relation between the various diffraction techniques and the reciprocal lattice concept have been treated by Azaroff (1968). Azaroff also gives a clear overview of the application of X-ray diffraction techniques to crystal structure analysis, which is discussed in detail by Buerger (1960, 1962), Lipson and Cochran (1966), Guinier (1962), Barrett and Massalski (1966), and by Stout and Jensen (1968).

Identification of inorganic substances by X-ray diffraction patterns. The observation of a characteristic diffraction pattern provides a convenient means of identification of a crystal, to demonstrate the incidence of twinning or polycrystallinity, or to determine imperfections such as stacking fault densities (Warren, 1969). Single crystals are normally necessary for the identification of new compounds since indexing of powder patterns is difficult unless the symmetry is high. If only small single crystals of less than 1 mm are available (minimum size 0.05 mm, optimum size 0.1–0.2 mm) information on the crystal symmetry (point group and space group) and on the unit-cell geometry may be obtained from X-ray diffraction data with the single-crystal diffractometer (Arndt and Willis, 1966), the Buerger precession camera (Buerger, 1964), or the Weissenberg camera. Of the two latter, the precession camera is preferable since it gives photographs of the undistorted reciprocal lattice of the crystal, although the Weissenberg camera presents the complete (but distorted) reciprocal lattice. With this information and some knowledge of the elements present in the compound, of the density etc. it should be possible to identify any known compound using "Crystal Data" of Donnay et al. (1963, 1972) or "Crystal Structures" of Wyckoff (1963–1968).

One of the easiest and yet most reliable techniques to identify crystalline phases is by X-ray diffraction of powders. Generally about 10 mg powder of the unknown material is available so that the more convenient powder diffraction patterns obtained, for example by the Guinier-DeWolff camera, can be used for identification. The d-values of the lines can be directly measured by means of a ruler, or the distances (angles) of the diffraction lines from the primary X-ray beam are measured and converted into the d-values. With these d-values of the 3 to 8 strongest lines a reliable identification is normally possible by use of the Powder Diffraction File (1973), especially with some additional chemical or physical knowledge of the material. In not too complex cases even the constituents of mixtures can be identified and their approximate concentrations estimated.

Since the identification of crystals of known compounds from their powder diagrams is generally much easier than from single crystal data, Gandolfi (1967) has developed a camera which allows a powder pattern to be obtained from a small single crystal. This technique is very useful when crystals cannot be powdered due to their value or rarity.

Precision measurement of lattice parameters. The exact lattice constants may be used as a measure of the composition of the crystal, since changes in lattice dimensions are caused by substitution of ions which differ in size from those of the host crystal. For accurate determination of lattice constants, powder samples (optimum crystallite size of the order of 5 μm) are normally preferred. X-ray powder techniques have been reviewed by Klug and Alexander (1954), by Peiser et al. (1955) and by Azaroff and Buerger (1958), and the precise determination of lattice constants by Bond's method and the refinement of the lattice constants by a least squares computer programme has been described by Barns (1967) and by Segmüller (1970).

For the Debye-Scherrer method a powder camera of large diameter, 14.5 cm or more, is necessary to give good resolution. In order to achieve the highest accuracy, corrections are necessary for the film position and shrinkage, eccentricity of the specimen, absorption by the specimen and divergence of the X-ray beam. Temperature stabilization is also necessary for very high accuracy. The best accuracy which can be achieved is normally not better than a few parts in 10^5, so that the uncertainty in the determination is in the fourth decimal place if sufficient care is taken.

Counter diffractometers have considerable advantages over film cameras and can give somewhat higher precision. The counter used to detect the X-rays is normally located at a greater distance from the sample and improved resolution is possible by the use of *Soller slits* which limit the horizontal and vertical spread of the beam. In addition, use is made of the tendency of a divergent beam of X-rays to come to a "focus" after diffraction by the specimen in the normal *Bragg-Brentano* arrangement. If the counting arrangement is automated, the profile of a diffraction peak can be plotted very accurately. In this way Baker *et al.* (1966) were able to measure *changes* in lattice parameter of 1 part in 10^7, using on-line computer control.

The geometrical arrangement of Seeman (1919) and Bohlin (1920) in which the diffracted X-rays converge to a point can also be used to give improved resolution in film cameras. Particularly high resolution can be obtained if this *parafocussing* geometry is used together with a curved-crystal monochromator as in the *Guinier* (1937) camera. Such cameras, for instance the Guinier-De Wolff and the Jagodzinski camera, have an accuracy approaching 1 part in 10^5 and, under ideal conditions, even one order of magnitude better, especially when good reference samples with well-defined lattice constants (Si, NaCl, Al, etc.) are used. Another advantage of the two focussing camera types mentioned is that three or four samples can be analyzed simultaneously.

As an example of the changes in lattice parameter which may be expected, Wang and McFarlane (1968) found that the lattice parameter of 7 batches of $MgAl_2O_4$ crystals varied from 8.0797 to 8.0848 Å, with impurities in the range 10–100 p.p.m. The crystal of highest purity had a lattice parameter of 8.0830 Å.

It is clearly preferable if measurements are performed on single crystals since local variations in lattice parameter can be studied and the possibility is avoided of changes brought about by strain in the process of powdering. Isherwood (1968) has applied the *double diffraction* technique (Isherwood and Wallace, 1966) to the precise determination of the lattice parameter

R 2

of yttrium iron garnet. The precision of any one measurement was
± 0.0004 Å, and the lattice parameter of crystals of various impurity levels
and grown from a $PbO/PbF_2/B_2O_3$ flux was found to vary between
12.3752 and 12.3800 Å. This variation appeared to result from a change in
the lead content of the crystals rather than the level of rare-earth impurities. An interesting result of this investigation was the observation
of a departure from cubic symmetry caused by a deformation in the growth
direction, for both $\langle 110 \rangle$ and $\langle 211 \rangle$. A reversion to cubic symmetry
occurred on light abrasion of the crystal surfaces. This anisotropy is of
interest in the application of garnets in bubble domain devices, and is
discussed in Chapter 8.

Willoughby *et al.* (1971) correlated non-stoichiometry of gallium
arsenide with precisely measured lattice parameters.

Orientation of crystals. In order to orient crystals for cutting or for physical
measurements, optical techniques are useful when the crystal is transparent
(Bunn, 1961; Wood, 1963). Greater convenience and precision are obtained
when the crystal is oriented by X-ray diffraction. Although other single-
crystal diffraction techniques are frequently used for this purpose the Laue
back-reflection geometry is the most convenient and popular technique
for crystal orientation (Barrett and Massalski, 1966; Wood, 1963). From
Laue photographs, the symmetry may be easily obtained as well, whereas
indexing of the Laue spots in order to obtain some structural information
has become unnecessary in view of the development of the elegant single
crystal techniques and especially of the powerful computer-automated
single crystal diffractometers (Ahmed, 1969).

9.3.3. Electron and neutron diffraction

X-rays interact with the electron shell whereas an electron beam interacts
more strongly (by a factor of about 10^4) with the nucleus of the atom so
that the absorption by the sample is much stronger and only very small
crystals ($<1\ \mu m$) or very thin layers (<1000 Å) can be examined by transmitted electron diffraction. Strongly absorbing samples can only be
examined by electron back reflection. Generally electron diffraction is
performed in electron microscopes with the necessary accessories. Normally
the lattice constants may be determined to no better than 1%. Electron
diffraction is especially powerful for the detection of impurities and of
unmixing, with a sensitivity of about 0.1% depending on the distribution.
Also for the identification of the constituents of complex mixtures and for
indexing X-ray powder diagrams electron diffraction might be of help
since it produces an undistorted pattern of the reciprocal lattice which in
X-ray diffraction is only achieved by, for example, the precession method

for the study of single crystals. Another application of electron diffraction is in the observation of magnetic and ferroelectric domain boundaries, although the application of electron diffraction to structure determination is limited. The wavelength of the electron beam can be adjusted by the applied voltage, and in the case of long wavelength (of the order of 1 Å) electrons one speaks of Low Energy Electron Diffraction (LEED) which is used for surface studies (see Section 9.4). Electron diffraction is the topic of the reviews of Pinsker (1953) and of Vainshtain (1960, 1964), and is treated also in the books on electron microscopy.

In *neutron diffraction* (Bacon, 1962) the neutrons interact with the nucleus and no Coulomb interactions occur. Because of the complexity of the instrumentation and of the limited applicability of neutron diffraction it is only used for structure determination when other techniques such as X-ray and electron diffraction are not applicable. For example, mono-chromatized thermal (slow) neutrons are used for the determination of hydrogen positions in H-containing compounds, and for the cation distribution of elements with similar atomic weights but with different scattering amplitudes for neutrons (e.g. Fe, Co).

Since neutrons possess magnetic dipole moments they interact with electronic magnetic moments or electronic magnetic fields. Thus the magnetic scattering amplitude depends on the electronic structure of the atoms, and the magnetic structures of many ferromagnetic, ferrimagnetic and anti-ferromagnetic materials have been analyzed (Bertaut, 1963; De Gennes, 1963; Forsyth, 1970).

9.3.4. Various physical techniques

In addition to the classical methods for the study of crystal structures or structural aspects, several physical techniques have been used with increased emphasis in order to obtain "indirectly" information on the structure and on the relative positions of atoms in crystals. For example, the point group can be determined by optical dielectric constants, acoustic measurements, magnetostriction and electrostriction. The local symmetry around specific atoms or ions and the local point group can be derived by electron spin resonance (ESR, EPR), nuclear magnetic resonance (NMR), optical absorption spectroscopy and by Mössbauer spectroscopy, whereas magnetic symmetry properties of crystals are studied by neutron diffrac-tion, magnetic measurements (of susceptibility, saturation magnetization, magnetic anisotropy and magnetostriction), NMR, Mössbauer spectro-scopy, and the Kerr and Faraday effects. By several of the above-mentioned techniques it is possible to analyze the cation distribution in, for instance, spinels or garnets and so to study order–disorder phenomena.

In several cases critical temperatures and pressures at which crystallographic, magnetic or electronic *phase transitions* occur can help in the characterization. Such critical data are obtained with the polarizing microscope, by X-ray, neutron or electron diffraction, by calorimetry (differential thermal analysis DTA or measurement of specific heat), by measurement of the magnetic or dielectric susceptibility, by determination of the resistivity or the thermoelectric power, of dilation and elastic constants, or by EPR, NMR and Mössbauer spectroscopy under the appropriate conditions. Only the most important or popular phenomena which are sensitive to phase transitions have been mentioned and many other physical properties and phenomena are sensitive to phase changes. Several of the physical techniques are discussed briefly in connection with defects (Section 9.4.5).

9.3.5. Determination of material constants

A complete description of a material contains its physical properties, thermodynamic data, and chemical properties. These properties are related to the structure, but most of them are too unspecific or cannot be measured with the necessary accuracy in order to obtain information on the structure. The exceptions have been mentioned in previous sections. Many properties are, of course, very sensitive to specific deviations from the ideal structure or composition. For example, by precise determination of the density a deviation from stoichiometry or the presence of vacancies can be detected, and mechanical properties such as hardness, ductility, tensile strength or elasticity are sensitive to dislocations and other imperfections. The melting point, vapour pressure and other thermodynamic data are useful and often necessary if an optimum choice of a crystal growth technique and crystal growth parameters has to be made. Also chemical corrosion behaviour, the thermal expansion coefficient and many other properties not mentioned above might be of value to potential crystal users and to other crystal growers and should be published when they are measured. Most of such measurements, however, are only meaningful when the material is sufficiently characterized by its chemical composition, by its structure and by inhomogeneities and defects.

9.4. Defects in Crystals and their Determination

9.4.1. The nature of defects in crystals

In Table 9.5 are listed the more important defects which may be found in crystals grown from high-temperature solution (and in crystals generally). The list is not exhaustive and the classification not perfectly rigorous, but no system of classification is ideal. The defects could alternatively be grouped into chemical defects, such as foreign atoms and nonstoichiometry,

TABLE 9.5. Defects in Crystals

Point Defects
    Schottky defect (vacancy)
    Frenkel defect (interstitial and compensating vacancy)
    Interstitial atom
    Foreign (substitutional) atom
    Colour centre
Line Defects
    Dislocations
Planar Defects
    Low angle boundaries
    Stacking faults
    Twin boundaries
Bulk Defects
    Mosaic structure
    Compositional inhomogeneity, striations
    Inclusions (solid, liquid or gas)

---

and physical defects such as dislocations and twin boundaries. Defects in general have been treated by Van Bueren (1960) and Bollmann (1970).

Point defects. Vacancies are unoccupied sites in the crystal lattice. The majority of vacancies in alkali halides are *Schottky defects* which are in thermodynamic equilibrium with the lattice and arise because of thermal fluctuations. The concentration of Schottky defects increases exponentially with temperature according to $n_v/N \approx \exp - (\Delta G_v/kT)$ with n_v/N the ratio of vacancies to atoms and with ΔG_v the heat of formation of a vacancy. The vacancy concentration becomes particularly significant as the melting point is approached. Cation vacancies are compensated by anion vacancies in charge-compensated systems. Another type of vacancy is compensated by an interstitial atom in the neighbourhood of the vacancy, and this is the *Frenkel defect*. The number of Frenkel defects is given by

$$n = (NN')^{1/2} \cdot \exp\left(-\Delta G_F/2kT\right)$$

with N' the number of possible interstitial sites and ΔG_F the heat of formation of a Frenkel defect. *Nonstoichiometry* can be produced by a high concentration of vacancies. The crystal then contains an excess of cations or anions rather than exhibiting a simple cation–anion ratio. Compounds of the transition metals are particularly prone to departures from stoichiometry according to their stability range and, for example, the Fe : O ratio in "FeO" can show large departures from unity depending on the partial pressure of oxygen in the growth atmosphere and to the growth temperature.

Frenkel defects are formed during growth or during heat treatment by the migration of atoms from their normal lattice sites to interstitial positions. The concentration of such defects, as of Schottky defects, is clearly lower in crystals grown from HTS than in crystals grown from the melt.

Foreign atoms will be present in any crystal to a greater or lesser extent according to the effective distribution coefficients, to the purity of the chemicals used and to the solvent from which the crystal was grown. The impurity atoms will usually be present substitutionally in a site normally occupied by the atoms in the host lattice, but they may be in interstitial sites. The dependence of the impurity concentration on the choice of solvent has been discussed in Chapter 3.

Colour centres are light-absorbing point defects (F-, V-centres etc.), which may be present in (nonstoichiometric) compounds. For example, if sodium chloride contains an excess of sodium compared with the chlorine, the excess sodium will be present as an atom rather than as a Na^+ ion. The valence electron of the sodium atom, located in a Cl^- vacancy, will have a characteristic absorption in the visible range and so will produce coloration of the crystal. Colour centres may also be introduced by the presence of certain impurities, or by bombardment of the crystal by ionizing radiations. Complex centres may be formed by groups of simple colour centres or by the trapping of electrons by pairs of ions. An *exciton* is a neutral, mobile excited state of a crystal, usually a bound electron-hole pair. The term is sometimes used more generally for any local excited state.

Monographs on point defects have been written, for example by Van Bueren (1960), Kröger (1964), Schulmann and Compton (1962), Markham (1966), Fowler (1968) and by Crawford and Slifkin (1972).

Dislocations. Strain in the crystal will cause displacements of atoms from their equilibrium positions. The magnitude of such displacements will depend on the severity of the strain, which may result in dislocations and other defects. A large number of books and reviews on the theory, the effects and the detection of dislocations have been published. As examples the books of Van Bueren (1960), Cottrell (1953), Friedel (1964), Hirth and Lothe (1968), Nabarro (1967), Read (1953), Rosenfield *et al.* (1968) and Simmons *et al.* (1970) (see also Discussions Faraday Society (1964) on the theory or general aspects of dislocations) are cited, whereas monographs on dislocation detection techniques will be mentioned under electron microscopy, etching, X-ray topography etc.

Dislocations occur at regions in the crystal where adjacent planes of atoms fail to meet perfectly. They may be classified as *edge* or *screw* dislocations according to whether the lattice displacement is perpendicular or parallel to the slip direction. The importance of screw dislocations for the nucleation of layers on the surface of the crystal is discussed in Chapter 4.

Measurement of the dislocation density (the number of dislocations per unit area of crystal) gives a valuable indication of the perfection of the crystal. In Table 9.5 are listed a number of determinations of the dislocation density in HTS-grown crystals. The normal value is between 10 and 10^4 cm^{-2} which is lower than the average for melt-grown crystals. Examples are quoted where crystals have been grown dislocation-free or containing large regions without measurable dislocations. Evidence from a large number of investigators indicates that dislocations are normally generated either on nucleation or seeding, or by solvent inclusions. Figure 9.9 shows etch pits which correspond to dislocations generated by a solvent inclusion in a tantalum carbide crystal (Rowcliffe and Warren, 1970). The relation between inclusions and dislocations will be discussed in more detail in section 9.4.6. In addition, Wagner (1967) has noted that silicon whiskers produced by the VLS method are dislocation-free. The dislocation density of epitaxial films of III–V semiconducting materials depends primarily on that of the substrate crystal but a reduction by typically a factor 3–10 is normally observed in the regrown layer.

Decorated dislocations occur when the dislocations act as nucleation centres for the precipitation of some solid phase. The decoration may

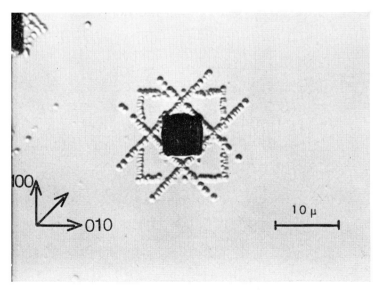

FIG. 9.9. Etch pits around a solvent inclusion in TaC (Rowcliffe and Warren, 1970).

occur during growth of the crystal or may be deliberately produced subsequently as a means of revealing the presence and distribution of the dislocations.

Misorientation. Strain may also result in the separation of the crystal into many crystallites each of which is misoriented with respect to neighbouring regions by a small angle, typically a few minutes of arc. The crystal is then said to exhibit a *mosaic structure* with the relatively perfect mosaic blocks, about 500–5000 units cells in diameter, separated by the small misorientation boundaries. If the crystallites and misorientations are larger, the terminology is changed and the crystallites are referred to as *grains*. The *low-angle grain-boundaries* between such grains may be formed from a number of dislocations, as illustrated in Fig. 9.10. The dislocations arise since some planes of atoms must terminate at the boundary and an edge dislocation will run through the crystal normal to such terminations, as shown in the diagram.

The rather simple situation shown in Fig. 9.10 rarely occurs in practice

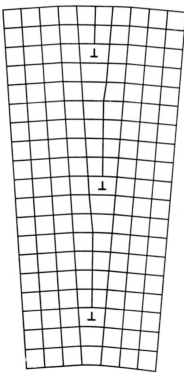

FIG. 9.10. Simple tilt boundary with edge dislocations denoted by the symbol ⊥.

since the misorientation between neighbouring grains may have a component normal to the plane of the diagram. The misorientation may be resolved into a *tilt* component in the plane of the grain surface (as in Fig. 9.10) and a *twist* component normal to this surface. Twist boundaries will generate screw dislocations, whereas tilt boundaries generate only edge dislocations. A detailed discussion of grain boundaries has been given by Read (1953), Nabarro (1969) and by Chaudhari and Matthews (1972). Table 9.6 shows data by a number of investigators on the misorientation

TABLE 9.6. Lattice Misorientation in HTS-Grown Crystals

Material	Solvent	Mis-orientation	Method	Reference
Al_2O_3	PbO/B_2O_3; PbO/PbF_2	22″	X-ray double crystal	Berkes *et al.*, 1965
ThO_2	$Li_2W_2O_7$	3′	X-ray linewidth	Finch and Clark, 1965
Al_2O_3	PbF_2	<2′	X-ray topography	White and Brightwell, 1965
Al_2O_3	PbF_2	20″	X-ray topography	Champion, 1969
$KNbO_3$	K_2CO_3	10′	Neutron diffraction	Hurst and Linz, 1971
$LaAlO_3$	$PbO/PbF_2/V_2O_5$	6″	Neutron diffraction	Kjems *et al.*, 1973

measured in some HTS grown crystals. The best value which has been realized in practice is seen to be about 10 seconds of arc, although an improvement by perhaps a factor 3 might be expected for crystals grown under very stable conditions.

Twinning. Twin boundaries occur between the components of a composite crystal when these components are related in a simple crystallographic manner. Of particular interest to the crystal grower are *growth twins* which are formed during growth rather than during subsequent deformation of the crystal. Examples of the influence of twinning on the growth mechanism have been discussed in Chapter 5 and particularly in Section 4.14.

Stacking faults. Stacking faults are irregularities in the sequence of atomic layers which cause departures from the arrangement in a perfect lattice. For example, if in a perfect lattice the repeat units are stacked in a regular sequence $\nabla\Delta\nabla\Delta\nabla\Delta$ ... and so on, the insertion of an extra plane would result in a sequence such as $\nabla\Delta\Delta\nabla\Delta\nabla\Delta$.... The relative ease of formation of such faults depends strongly on the type of structure. Stacking faults may be determined by X-ray diffraction (see Section 9.3.2) and other techniques, and have been reported in a number of HTS-grown crystals, for example by Belt (1967), Cook and Nye (1967) and Wang and McFarlane (1968).

Inclusions are regions within the crystal occupied by some phase other than that of the host crystal. There are a great variety of types of inclusion which may form either during growth or during subsequent cooling to room temperature. Inclusions which are formed on cooling due to a decreasing solubility in the crystal of some solid are referred to as *precipitates*. *Foreign particles* of various kinds may be trapped by the growing crystal if they are present in the solution. Inclusions of platinum metal are quite common in crystals grown from fluxed melts, particularly if the flux has a fairly high solubility for platinum. *Gas* and *liquid* inclusions are comparatively rare in solution-grown crystals but are observed frequently in naturally occurring minerals.

The most serious defects in HTS-grown crystals are often *solvent inclusions*, which tend to form whenever growth becomes unstable (see Chapter 6). Solvent inclusions may be classified as follows (Wilcox, 1968; Elwell, 1975):

(a) *Filled-in dendrites.* The initial growth following spontaneous nucleation is frequently "dendritic" with arms extending along the fast growth directions and a very high concentration of inclusions may be trapped as facets form and join up the ends of the dendrite arms. Such inclusions have been very widely observed, for example by Doughty and White (1960) and Giess (1962) (see Fig. 4.28a).

(b) *Veils.* These are thin sheets of small inclusions which may have a variety of origins. The mechanism of formation of *starvation* veils was described by Carlson (1959). These are due to a low supersaturation at the centre of a crystal face such that layers spread from corners and edges of the face and overgrow the central region, trapping inclusions there. Veils may also arise due to cracks which develop during growth and which enclose a film of solvent as the crack heals by subsequent growth. The film then tends to break up into a number of tiny droplets. The formation of such veils has been filmed by Powers (1970) during observations of sucrose growth from aqueous solution.

(c) *Ghosts.* These are oriented veils, parallel to a natural face of the crystal. They form at various stages during growth and so trace out the development of the crystal, hence the name. Ghosts, or less regular veils, may form when growth occurs erratically in a periodic sequence of relatively rapid and unstable growth followed by an interval of stable growth at lower supersaturation.

(d) *Clouds.* Inclusions are sometimes observed not in bands but in localized "clouds" or aggregates of very fine particles, of say, less than 1 μm diameter. Such clouds have been observed, for example, by Timofeeva *et al.* (1969b).

(e) *Fjords* are narrow channels of inclusions which are trapped between fine projections, a good example being shown in Fig. 6.13. The projections of the crystal may be in a direction perpendicular to the growth direction, if growth is erratic as described in the previous section, or normal to the crystal surface. The tendency of an interface to develop an instability which exhibits a regular periodicity was described in Section 6.3.

The nature and concentration of solvent inclusions depend on the conditions of growth, and these inclusions can be eliminated completely if growth occurs under stable conditions. These conditions have been discussed extensively in Chapters 6 and 7.

9.4.2. Optical methods

Optical techniques as described in Section 9.3.1 may be applied to the detection of many of the defects mentioned above. The handling of crystals for optical studies has been described in Section 9.3.1 so that in the following only a brief summary of the potential of optical methods for the detection of defects will be given.

Optical absorption spectroscopy allows the detection of a variety of point defects. Optically active vibrational modes near atomic point defects show IR isotope effects, whereas optically inactive vibrational modes show the Raman effect for the case where a sufficient concentration of point defects (10^{19} cm^{-3}) is present. Luminescence (fluorescence, phosphorescence, thermoluminescence) and photoconduction are relatively sensitive ($\sim 10^{10}$ cm^{-3}) to specific point defects. Optical microscopy is useful for the detection of decorated dislocations or of etch pits and allows in favourable cases the detection of 1 line defect per square micron (spatial resolution 1 μm). Twinning and low-angle grain-boundaries may be seen in the polarization microscope or by optical reflection on the as-grown or on the etched faces. Optical techniques for the detection of inclusions are presented in Section 9.4.6.

9.4.3. X-ray topography

A simple means of assessment of the perfection of a crystal is provided by the size of the spots in a normal X-ray single crystal diffraction pattern or *Laue photograph*. If mosaic structure causes a spread in the lattice parameter over a range of values Δd, the spots will have an angular width given by differentiation of the Bragg condition as

$$\Delta\theta = -\tan\theta \, \frac{\Delta d}{d}.$$

The width $\Delta\theta$ is greatest for high values of $\tan\theta$, and so the degree of imperfection is best determined from back reflection photographs. The directional variation of the imperfection may be assessed by rotating the crystal, so that back reflection photographs are obtained for a number of orientations. An extensive discussion of the use of conventional X-ray crystallography for appraisal studies has been given by Guinier (1962).

Laue photographs may also be used to demonstrate the existence of twinning and the presence of precipitates or inclusions, but only when these are present in concentrations of about 2–5%. The sensitivity to defects is, however, inferior to that obtained by X-ray diffraction topography which has been particularly developed for crystal quality appraisal studies.

X-ray topography is the examination by X-ray diffraction of the surface or bulk of a crystal, either of a particular region or of the whole crystal. The term embraces a variety of techniques which are distinguished from studies using, for example, the size of the Laue spots in that the region of the crystal which produces each part of the image can be identified. An X-ray topograph is thus a representation of the crystal which is sensitive to imperfections of the crystal lattice.

The techniques of X-ray topography have been reviewed by Barrett (1967), Bonse, Hart and Newkirk (1966) and Lang (1970) and the arrangements of source, crystal and film which have been most widely used are shown in Fig. 9.11.

The simplest arrangement is probably that due to Schulz (1954), shown in Fig. 9.11(a). A beam of white X-rays diverging from a point source is diffracted by a crystal inclined at about 25° to the beam. The image is a series of Laue spots and is about the same size and shape as the crystal, and faults in the crystal cause gaps or overlap regions in the image. White radiation is used so that there is no major variation in intensity due to the different angle of incidence of the beam on various regions of the crystal. The exposure time with this technique can be very short, of the order of minutes, but the sensitivity and resolution are poor. A similar technique using transmitted radiation was proposed by Guinier and Tennevin (1949).

Figure 9.11(b) shows the arrangement for the Berg-Barrett technique (Barrett, 1945) which uses monochromatic radiation from a collimated line source. The photographic plate in this case is mounted very close to the crystal, in order to minimize doubling of the image due to K_{α_1} and K_{α_2} radiation and to obtain maximum resolution. Typically, the source to specimen distance is 30–50 cm and the specimen to film distance is about 1 mm. Newkirk (1959) was able to achieve a resolution of about 1 μm and so resolve single dislocations. The area of specimen irradiated is about 1 mm² and the size of the image about 5–10 mm², depending on the angle of incidence of the beam. Relatively soft radiation may be used and this

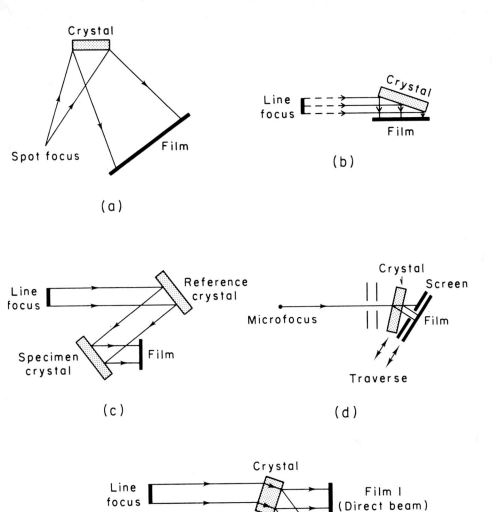

FIG. 9.11. X-ray topographic techniques: (a) Schulz (1954). (b) Berg-Barrett (Barrett, 1945). (c) Double-crystal spectrometer. (d) Lang (1957). (e) Transmission (Borrmann *et al.*, 1958).

has the additional advantage that the photographic emulsion may be thin, so that it is not essential for the beam to strike the film normally. Other topographic methods require nuclear emulsions, and dense emulsions may be necessary to achieve convenient exposure times.

The Berg-Barrett technique is the most widely used reflection method and is particularly suitable for the rapid examination of crystals in order to distinguish good quality crystals from those containing a high concentration of defects. A similar method may be used for transmitted diffracted radiation (Barth and Hosemann, 1958).

The double-crystal spectrometer (Bond and Andrus, 1952; Bonse and Kappler, 1958) shown in Fig. 9.11(c) uses Bragg reflection at a relatively perfect crystal C_1 and then from the specimen C_2 before the radiation reaches the film F. Similar materials are used for C_1 and C_2 so that the result is substantially independent of the spread $\Delta\lambda$ of the radiation used. The intensity pattern of F is very sensitive to local misorientations in the crystal C_2, and changes in interplanar spacing of about 1 part in 10^8 can be detected.

The most popular method of X-ray topography is probably that due to Lang (1957), shown in Fig. 9.11(d). A narrow beam of characteristic radiation is diffracted by the crystal and a screen S is set so that only the diffracted beam reaches the photographic plate F. The orientation of the crystal slice is such that the diffracting planes are approximately normal to the faces of the slice. A topograph of the whole slice may be obtained by traversing the sample and film simultaneously across the stationary slit.

The maximum resolution of the Lang technique approaches the theoretical limit, typically a few μm, which is related to the effect on the diffracted image of the region around the highly strained "core" of the dislocation. The area of which X-ray topography will detect the dislocation depends on the width of the rocking curve for the perfect crystal and so on the wavelength and order of the reflection (Authier, 1967, 1970). The resolution is clearly much less than that possible with transmission electron microscopy but this disadvantage is more than compensated by the greater penetrating power of X-rays and the relative absence of damage produced by the beam itself. The attainment of optimum resolution requires a highly collimated beam, produced either by a microfocus generator or by use of a long source-to-crystal distance. The beam must be sufficiently narrow that it acts as its own collimator, with only the image due to say K_{α_1} radiation being recorded on the plate.

A phenomenon which has also been used in transmission topography is that of anomalous transmission (Borrmann et al., 1958). In a perfect crystal the incident and diffracted waves form a standing wave pattern fitting onto the atomic planes. The energy transport of the wave field depends on the position of the antinodes of this pattern and anomalously low absorption will occur if the antinodes lie between the planes. A perfect crystal will therefore transmit X-rays at a thickness which would absorb nearly all the energy in the absence of the interference between the incident and diffracted beams. Any imperfection in the lattice causes a reduction in the intensity of both the diffracted beam and the beam which is transmitted in the direction of incidence, and both these transmitted beams may be used to obtain topographs, see Fig. 9.11(e).

Contrast in the image obtained in X-ray topography is caused by a local departure from the Bragg condition and by its gradient (Authier, 1966). Even when the distortions are not so great, "dynamical contrast" due to bending of the diffracted beam and to a shift of energy between incident and diffracted beams may be observed in reflection and transmission methods, respectively. The contrast in dynamical images will differ from that in the direct images and the theory of image contrast in X-ray topographs is rather complex. A review of this topic has been given by Authier (1970).

Table 9.7, which is taken from the review of Bonse et al. (1966), gives a comparison between the various techniques of X-ray topography. The table summarizes the practical advantages and disadvantages of the techniques described above and includes information on the types of defect for which each is most suitable.

Many examples of X-ray topographic studies of HTS-grown crystals have appeared in the literature, and these have often provided data which have been invaluable for an understanding of the growth mechanism.

Austerman et al. (1965) obtained both Berg-Barrett and Lang topographs of BeO crystals grown from a lithium molybdate flux. These topographs showed the boundary of the twin core and confirmed that the twin boundary intercepts the growth face at the base of a conical feature, so forming a re-entrant angle. Parallel bands were observed crossing the crystal at rather irregular intervals along the core axis. Radial bands were also noted on the oxygen-rich side of a transverse section, but not on the Be-rich side, suggesting that these are caused by a stress field associated with the twin core. The other notable feature was a screw dislocation forming a helix of pitch 200–500 Å along the axis of the crystal.

Wallace and White (1967) studied the growth of Al_2O_3 crystals from PbF_2 flux by Lang topography, and clearly demonstrated the existence of twinning, the importance of which to the habit of alumina crystals was discussed in Chapter 5. Striations arising from imperfect temperature control were also observed. Champion (1969) used the Berg-Barrett technique to compare the perfection of Al_2O_3 crystals with those grown by the Verneuil method, and established that the misorientation in the flux-grown crystals was less than 20 seconds of arc.

TABLE 9.7. Comparison of X-ray Topographic Techniques

Technique	Schulz; Guinier and Tennevin	Berg-Barrett	Double crystal	Wide beam transmission (Barth and Hosemann)		Scanning transmission (Lang)	
apparatus	simple	simple	complicated	$\mu_0 t > 10$ simple	$\mu_0 t < 10$† simple	$\mu_0 t \sim 3$ complicated	$\mu_0 t < 1$† complicated
exposure time	10–25 hr	~1 hr	~1 hr	~10 hr	~1 hr	10–30 hr	2–10 hr
defect for which technique is most suitable	grain misorientation, subgrains	subgrains, dislocations	subgrains, dislocations, stacking faults	dislocations	subgrains, dislocations, stacking faults	dislocations	subgrains, dislocations, stacking faults
best resolution	50 μm	1 μm	1 μm	1 μm	1 μm	1 μm	1 μm
sensitivity to deformations	low	low	high	high	low	high	low
sensitive to sense of deformations	tilts: yes inhomogeneous deform: no	subgrains: yes dislocations: no	yes	yes	no	yes	no
thickness t of specimen contributing to topograph	$S \leqslant 5$ μm $G \& T$ 50–100 μm	⩽ 5 μm	⩽ 5 μm (ref.) ⩽ 300 μm (transm.)	1–5 mm	0–2 mm	0.1→5 mm	0–2 mm
dislocation image width	—	1–5 μm	up to 150 μm	⩾ 50 μm	~5 μm	up to 150 μm	1→10 μm
upper limit of dislocation density	—	5×10^6	10^5	5×10^3	5×10^6	5×10^3	5×10^6

† μ_0 is the absorption coefficient of the material.

Belt (1967) examined lithium ferrite crystals grown from a PbO/B_2O_3 flux by means of Lang topography. Highly perfect regions were found, together with defective areas mainly associated with flux inclusions on the {111} habit planes. Additional contrast observed in the crystals was attributed to strain or impurities in {110} planes and to stacking faults on {111}. The same author (Belt, 1969) made a careful study of defects in $Y_3Fe_5O_{12}$ and $Y_3Ga_5O_{12}$ crystals by etching and topography and showed that etch pits are not necessarily associated with dislocations. The crystals exhibited microscopic parallel bands, located principally on the fast growing {100} and {111} planes. Defects clearly identifiable as dislocations were also present and possible Burgers vectors determined. The origin of the bands was difficult to ascertain, but impurities and strain arising from thermal fluctuations were thought to be responsible for these defects.

Basterfield and Prescott (1967) used Lang topography to investigate the domain structure in $Tb_3Fe_5O_{12}$, but did not comment in any detail on the perfection of the crystals studied. Stacy and Tolksdorf (1972) compared X-ray topographs of magnetic garnet crystals with the domain structure detected optically by means of Faraday rotation. The domain pattern was found to be correlated with strain which was responsible for a non-cubic magnetic anisotropy.

$MgAl_2O_4$ was studied by Wang and McFarlane (1968) using the Lang method. The topographs showed dislocations, in low concentration, and striations leading to strain along the $\langle 111 \rangle$ directions, normal to the growth facets. Precipitates surrounded by an isotropic strain field were also observed, together with stacking faults.

The Borrmann technique of anomalous transmission topography has been comparatively little used in appraisal studies but a good example of its use was described by Wolff and Das (1966). They employed this technique to confirm the improvement in perfection of the GaSb layer grown by the travelling solvent method compared with that of the seed crystal. The regrown material had a dislocation density of 3×10^3 cm^{-2} and the topographs showed that the elimination of 70% or more of the dislocations present in the seed was due to the formation of closed loops or half loops, or a transformation to propagation in a direction perpendicular to the growth direction. This elimination occurred in the first 100 μm of regrown material. Examples of X-ray topographs of crystals grown from solution are given in Section 9.5.

9.4.4. Electron microscopy, LEED

Continuous advances in *transmission electron microscopy* have led to its increasing use, particularly for studies requiring high magnification.

Lattice planes with spacings of about 3 Å have been resolved, and techniques for the resolution of heavy atoms in a rigid matrix are being developed. The main limitation of electron microscopy arises from sample preparation. Because of the low penetrating power of electrons, samples for 100 kV electron microscopy have to be thinner than 500 to 5000 Å depending on the material. In the recently developed high-voltage electron microscopes with accelerating voltages up to several MV, thicker samples up to a few microns can be used but resolution decreases with sample thickness. However, high-voltage electron microscopy is useful for the study of precipitates, although its application in the study of mixed phases is limited by the difficulty of preparation of thin samples.

Sample preparation techniques are described in Section 9.3.1, and special thinning techniques (chemically by a jet of etching solution or physically by ion bombardment at low angles) have been developed. Literature on electron microscopy is extensive and includes the books of Heidenreich (1964), Thomas (1962), Amelinckx (1964), Nicolson et al. (1965), Hawkes (1972), Thomas et al. (1972) and the review of Phillips and Lifshin (1971). Murr (1970) has described the applications of electron optics in materials science. In the following, examples of applications of electron microscopy in the characterization of HTS-grown crystals are given.

Lefever et al. (1961) used transmission electron microscopy to investigate the nature of striations in yttrium iron garnet. They were able to demonstrate that the bands contained a relatively opaque material in the form of fine particles, about 0.1 μm in diameter. Electron micrographs of *small particles* of $BaTiO_3$ were used by Nielsen et al. (1962) in an investigation of the nucleation on such particles of the "butterfly twin" crystals.

An alternative to the use of thinned samples is to make a thin-film replica of the crystal surface either in the as-grown or etched condition. Replication imposes a limit on the resolution of features and introduces uncertainties in interpretation. Resolution with the etch pit method for the determination of dislocation densities is, however, much higher than is possible with the optical microscope (Boswell, 1957). Replication was used by Lefever et al. (1961) in their study of banding in YIG and they were able to use the resulting high magnification to show the form of the precipitate in the impurity bands. Newkirk et al. (1967) studied surface features at twin boundaries of BeO crystals by a replica technique.

Cook and Nye (1967) examined replicas of the surface features on crystals of hexagonal ferrites as well as of cleaved and etched surfaces. They were able to observe stacking faults and also the nature of the inclusions. A systematic study of the stacking faults and the layered structure of hexagonal ferrites by combined X-ray diffraction and electron microscopy has been reported by Kohn *et al.* (1967), and Savage and Tauber (1967) describe the growth of complex hexaferrites and X-ray identification of the layer types.

The electron microscope with appropriate accessories may also be used in the investigation of crystal perfection by electron diffraction, particularly in the identification of dislocations and stacking faults. The diffraction pattern for a crystal at known orientation is compared either with a computed micrograph (Humble, 1970) or an optical transform (Taylor and Lipson, 1964). Differences between ideal and observed patterns may be analyzed to identify the types of defect.

Scanning electron microscope. The scanning electron microscope (SEM) differs strongly in concept from the transmission electron microscope, and has a resolution intermediate between that of the latter and that of an optical microscope, namely of about 100 Å at present. In the normal mode of operation, electrons accelerated by a potential of 5–50 kV are directed onto the sample, and scanning coils cause the beam to move across the specimen surface in a square raster. The secondary electrons which are emitted from the specimen strike a collector electrode and the resulting current is amplified and used to modulate the brightness of a corresponding spot which is displayed on a cathode-ray tube. The time associated with emission and collection of the secondary electrons is negligible compared with the time of the scan, and so the number of secondary electrons collected from any point on the specimen is determined only by the "brightness" of that point. Image contrast can, in fact, arise from a number of factors, particularly surface topography, atomic number, electrical conductivity, specimen orientation and electric or magnetic fields. The scanning electron microscope has no imaging lens and magnification depends only on the ratio of the sizes of the raster on the cathode-ray tube and that on the specimen. The minimum spot size, and hence the maximum magnification, is determined by aberration in the electron lenses. Practical limitations in resolution are set by the number of lines in the raster, the time of recording and noise in the instrument, and magnifications greater than 20,000 times are rarely used.

The great advantage of the scanning electron microscope is its very large depth of focus, which permits convenient observation of relatively large features on a crystal surface. (The instrument is therefore complementary to the interference microscope which has high sensitivity to surface topography.) The other advantage is that the SEM can be used in a large number of modes, and alternative images obtained, for example from reflected electrons or from monitoring the current through the specimen. Reviews of the principles, modes of operation and applications of SEM have been given by Kammlott (1971), Booker (1970), Reimer and Pfeffer-korn (1973), Thornton (1968) and by Oatley *et al.* (1965), while Minkoff (1967) reviewed the applications of SEM in materials science.

Since the image formed using secondary electrons is obtained from the first 10–100 atomic layers, the SEM is used mainly in the study of surface features rather than of defects inside the crystal (see, for example, Elwell and Neate, 1971).

Accessories to electron microscopes allow analysis of the characteristic X-rays produced by the electron beam so that electron micrographs and elemental analysis by *electron probe microanalysis* can be made from the same spot on the sample. Such a combination of techniques is very valuable since under high magnifications the localization of special features on the sample is extremely difficult.

Another way of obtaining information on composition and the type of bonding is to analyze the energy distribution of secondary electrons using special accessories to the electron microscopes, as described in Section 9.2.2.

Photo-emission electron microscopy (PhEEM). As described in Section 9.2.2, the energy distribution of the emitted electrons excited by photons can be used for chemical analysis. By a geometrical arrangement of electron lenses corresponding to that of an electron microscope the emitted electrons give an enlarged picture of the sample surface. The principle is shown in Fig. 9.12. Contrast in the image is given by the specific electron emission of various (chemical) materials, but the electron emission is also dependent on the crystallographic orientation and on defects. Topographic contrast locally distorts the electrical field, and therefore carefully polished samples should be used. In order to prevent contamination, surface cleaning, for example by a short ion bombardment, a high vacuum, and heating the sample to temperatures over 250°C, are necessary or advantageous. The photo-emission electron microscope has only been developed recently (Wegmann, 1969, 1970), but its successful application to various characterization problems in conducting samples and in insulators has been demonstrated (Weber, 1972; Wegmann, 1972).

A composition of PhEEM with SEM has been given by Bode *et al.* (1971), and its use for quantitative measurements is discussed by Wegmann and Dannöhl (1971). Lateral resolution in PhEEM is of the order of 150 Å, and the electrons originate from a maximum depth of about 100 Å. Examples of photo-electron emission micrographs are shown in Fig. 9.13. In (a) a corner is shown of a polished natural crystal of SnO_2, and the intensity changes are due to a variation in the Fe content. The contrast in (b) which shows the surface of a graphite crystal, is mainly topographic

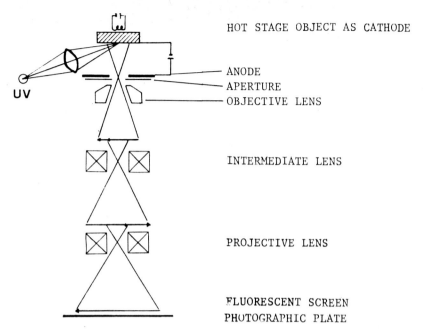

HOT STAGE OBJECT AS CATHODE

ANODE
APERTURE
OBJECTIVE LENS

UV

INTERMEDIATE LENS

PROJECTIVE LENS

FLUORESCENT SCREEN
PHOTOGRAPHIC PLATE

FIG. 9.12. Photo-emission electron microscope (diagrammatic).

and this photograph demonstrates the potential of PhEEM for studies of surface features.

Field-emission microscopy and *field-ion-emission microscopy* developed by Müller allow resolution of single heavy atoms and are used for the study of defects but their application is limited to refractory metals and alloys. The latter technique has been reviewed by Bowkett and Smith (1970) and by Müller and Tien Tzou Tsong (1969).

Low-energy electron diffraction (LEED). LEED provides structural information on the first two or three atomic levels. This implies that very special preparation techniques and ultra-high vacuum have to be used. Frequently the interpretation of LEED patterns is difficult because of surface damage introduced during sample handling or by electron impact. So LEED and the recently developed reflection-mode high-energy electron diffraction (RHEED) are mainly used for the study of adsorption layers, surface reactions and crystallite reorientation in surface layers (Somorjai, 1969; Marcus, 1969; Jona, 1970; Estrup and McRae, 1971). Very little application of electron-diffraction techniques for characterization of flux-grown crystals and layers has been reported.

(a)

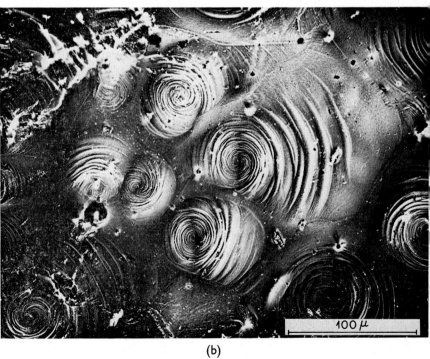

(b)

FIG. 9.13. Photo-electron emission micrographs. (a) Cassiterite SnO_2, showing variation in Fe content. (b) Graphite, showing growth hillocks (Wegmann, 1972).

9.4.5. Miscellaneous physical techniques

The assessment of crystal perfection is often made by a wide variety of physical measurements other than the optical and more general techniques described above. These measurements are often specific for the type of material considered and the information obtained is supplementary to that obtained by such methods as X-ray topography, and should not be considered as a substitute for a general study of crystal perfection. Table 9.8 lists the property measurement techniques as classified by Laudise *et al.* (1971).

However, one powerful technique for the detection of impurities (or ions possessing an unfilled electronic shell) and for the detection of colour centres should be particularly mentioned here, namely, *electron paramagnetic resonance* EPR (or ESR for electron spin resonance). Since ions of more than half of the elements of the periodic table have incomplete shells (e.g., the transition metal ions) EPR absorption can be used to identify a number of these ions when they are diluted in a diamagnetic

TABLE 9.8. Property Measurement Techniques

1. Electric field
   Resistivity
   Dielectric constant

2. Magnetic field
   Magnetic susceptibility
   Coercive force
   Uniaxial anisotropy

3. Gravitational field
   Density

4. Optical field
   Refractive index
   Birefringence
   Spectroscopic absorption (bandgap)

5. Thermal field
   Thermal conductivity
   Heat capacity and heat of transition
   Thermal expansion

6. Stress field
   Elastic moduli
   Mode of permanent deformation

7. Multiple fields
   Coefficients
      Thermoelectric, photoelectric, photoelastic, elasto-optic, piezoelectric, pyro-electric, electro-optic, acousto-optic, magneto-optic, lifetime of carriers, mobility of carriers, residual resistance ratio, electroluminescence, photoluminescence
   Transitions
      Superconducting temperature, Curie temperature, phase-matching temperature, spin reorientation temperature
   Resonance
      Nuclear magnetic, paramagnetic, Mössbauer, cyclotron, nuclear double resonance

crystal. The g-factor (magnitude of splitting of the groundstate level by an external magnetic field), the ligand field splitting, the nuclear hyperfine splitting and the spin-lattice relaxation time are characteristic of a paramagnetic ion and its local symmetry. Due to the high sensitivity of EPR the transition metal ions can be detected in the concentration range of p.p.m. to a few percent. Limitations of the application of EPR are due to: (1) a high symmetry which is useful for interpretation of the spectra, although recently EPR spectra of paramagnetic ions in monoclinic and triclinic crystals have been interpreted; (2) strain in the crystals; (3) the presence of other paramagnetic ions in high concentrations.

EPR has been reviewed by Ayscough (1967), Abragam and Bleaney (1970), Low (1960, 1963) and Pake (1962). It is surprising that, although EPR has been widely used to investigate structural aspects around the incorporated ions, phase transitions (Müller, 1971) and qualitatively to detect the presence of certain ions, EPR has not been widely used for quantitative trace determination despite the fact that Burns proposed this in 1964.

The detection of Pb^{3+} in ThO_2 grown from PbF_2—B_2O_3 flux by Scheel (unpublished) is of interest: Röhrig and Schneider (1969) found in crystals containing 1% lead a large hyperfine interaction for the isotope ^{207}Pb at 77 K, and Pb^{3+}—F^- centres would account for the weak satellite lines near the Pb^{3+} lines.

Wertheim et al. (1971) have reviewed the determination of the electronic structure of point defects by spin resonance and by Mössbauer spectroscopy.

Ultrasonic attenuation and dispersion might become interesting as a nondestructive characterization technique since ultrasonic waves are not bound to optical transmission, electrical conductivity and so forth, so that all kinds of crystals and crystalline layers can be tested. Another advantage is the high sensitivity to defects of phonon scattering. The main difficulties are, experimentally, to obtain reliable and reproducible coupling to the transducer and, theoretically, to correlate the phonon spectra with the types of defect. Ultrasonic techniques for characterization are reviewed in the books of Herzfeld and Litovitz (1959), Filipczynski et al. (1966), and Sharpe (1970).

The importance of ferrimagnetic oxides has been mentioned at several points in the text and a valuable measurement for such materials is the *saturation magnetization*. This property will be affected by solvent inclusions, which contribute to the mass but not to the magnetization, and by impurities, the effect of which will depend on the Bohr magneton number of the ion and on its location in the crystal lattice. Nielsen et al. (1967) took samples from various regions of Ga-doped yttrium iron garnet

crystals and plotted their results in the form of contours of equal magnetization on a representation of the crystal surfaces (see Fig. 9.6b). The
resulting plot illustrates dramatically the variation in gallium concentration
across the crystal face.

The local concentration of a non-magnetic impurity such as gallium
will also affect the *Curie temperature* and the difference in Curie temperature between different parts of a crystal will be low if the crystals are
relatively homogeneous. Linares (1965a) plotted histograms showing the
spread in Curie temperature among doped garnet crystals in batches
grown respectively by slow cooling and thermal gradient transport (see
Fig. 9.14). The spread was found to be much less in the crystals grown by
gradient transport, as may be expected since the growth is isothermal;
the relatively large variation in crystals grown by slow cooling arises from
the high effective distribution coefficient of gallium.

A parameter which is of considerable importance in studies of magnetic
oxides of iron is the concentration of Fe^{2+} or Fe^{4+} ions. This concentration

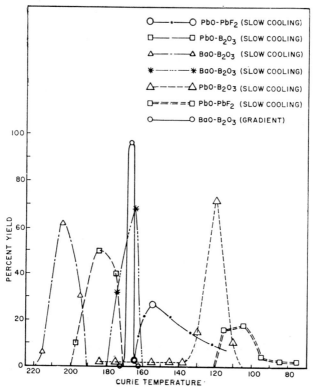

FIG. 9.14. Yield of yttrium iron garnet vs. Curie temperature (Linares, 1965a).

is often small compared with that of Fe^{3+} and is difficult to determine to high precision. The problem of determining the Fe^{2+} concentration in yttrium iron garnet was discussed by Robertson and Elwell (1969) who considered several measurements in addition to wet chemical analysis: electrical conductivity, thermoelectric power, infra-red absorption, saturation magnetization, ferrimagnetic resonance linewidth and magneto-crystalline anisotropy. The electrical conductivity is very sensitive to the Fe^{2+} concentration but also varies rapidly with temperature and with the concentration of impurities. Thermoelectric power is less sensitive to impurities and appears to be one of the most reliable methods for concentrations greater than about 2%, but calibration is required using samples of known Fe^{2+} concentration. Infra-red absorption curves have been plotted by Wood and Remeika (1966) for both Si- and Ca-doped yttrium iron garnet containing Fe^{2+} and Fe^{4+} ions, respectively. The absorption shows a sharp edge for wavelengths just above 1 μm and the absorption coefficient at a fixed wavelength will give the Fe^{2+} or Fe^{4+} concentration; a wavelength of 1.2 μm was used by Nassau (1968), who formulated a model of Fe^{2+}—Fe^{4+} equilibrium. Ferrimagnetic resonance linewidth is frequently used as a measure of the quality of magnetic oxide crystals, since the linewidth increases with the Fe^{2+} concentration and with the concentration of inclusions. It is not normally possible to separate these two main contributions to the linewidth, but a more reliable estimate of the Fe^{2+} concentration can be obtained from the increase in linewidth at low temperatures. For a quantitative measurement the anisotropy of ferrimagnetic resonance or a torque magnetometer may be used to determine the temperature dependence of the magnetocrystalline anisotropy constant K_1. The Fe^{2+} concentration is determined from the contribution to K_1 due to the ferrous ions, and is accurate only if the energy level splitting of the lowest doublet of the Fe^{2+} ions is known.

Measurements of ferrimagnetic resonance linewidth and anisotropy have also been used extensively in other investigations, notably by Makram and co-workers (Makram and Krishnan, 1967; Makram, 1968; Makram et al., 1968).

The properties of HTS-grown dielectric materials have also been investigated intensively as a means of crystal characterization. As an example, Giess et al. (1969) measured the spread in the ferroelectric Curie temperature of different barium sodium niobate crystals as a means of assessing homogeneity.

Reference has been made in Chapter 1 to examples of measurements which demonstrate the higher quality of semiconductor layers deposited by liquid phase epitaxy compared with that of the substrate crystals. In the original paper of Nelson (1963), diodes formed by LPE were reported to have improved performance as tunnel diodes over those produced from the melt, and diodes used as lasers had lower thresholds. Shih and Blum (1971) reported a high efficiency of light emission from $Ga_{1-x}Al_xAs$ diodes for display devices.

9.4.6. Determination and removal of inclusions

The origin of inclusion formation and the theoretical and experimental conditions to prevent unstable growth are discussed in Chapter 6, and Figs 6.6, 6.12 and 6.13 show typical examples of inclusions found in flux-grown crystals. The theory of stability of plane interfaces predicts a periodicity of inclusions of the order of 10–30 μm, and inclusions of that periodicity have been found. Depending on the growth system and the experimental parameters inclusions of submicroscopic size as well as bulk inclusions are formed.

Macroscopic inclusions are clearly detectable by the naked eye. In the following, a few techniques will be briefly mentioned which are used or which could be used for the detection and analysis of microscopic and submicroscopic inclusions. *Optical microscopy* is of course the simplest way of detecting inclusions of a size larger than 1 μm in transparent crystals, either in the as-grown crystals or in thin sections. Special microscopic techniques for examining inclusions in natural gemstones for identification purposes have been described by Gübelin (1953). Newkirk and Smith (1967) used *ultramicroscopy* where light scattered from a light source (here a helium-neon gas laser) perpendicular or at least diagonal to the microscope axis is examined and photographed. Thus inclusions of a size less than 1000 Å can be detected, but it is difficult to determine the exact size distribution and the morphology of the inclusions by examination of the scattered light. *Light scattering* is reviewed in the books of Van de Halst (1957) and of Stacey (1956). Orientation-dependent light scattering was observed from a variety of rotating cylinders of synthetic crystals by Guseva *et al.* (1971).

Several materials (e.g. ferrites, semiconductors) opaque to visible light have transmission windows in the infrared so that they can be examined by *infrared microscopy* (Sherman and Black, 1970; Sunshine and Goldsmith, 1972). Ferroelectric and magnetic domains are sensitive to inclusions, probably influenced by the strain produced by the inclusions. This strain is partially released by the formation of dislocations, so that *X-ray topographs* may indicate the presence of inclusions. *Ultrasonic dispersion*, as mentioned in Section 9.4.5, is extremely sensitive to tiny inclusions.

For the study of submicroscopic inclusions *transmission electron microscopy*, *replica electron microscopy* and *scanning electron microscopy* are valuable techniques when the problem of preparation of appropriate samples is solved.

Some of the techniques suitable for the analysis of chemical inhomogeneities are also valuable for the detection and analysis of inclusions (e.g. *electron microprobe analysis*), see Section 9.2.4. A semiquantitative determination of the inclusion content of a crystal is possible by grinding the crystal, extracting the inclusions and determining the total content of matter included in the extract.

X-ray radiography has been used by Doughty and White (1960) to examine bulk crystals for major defects and inclusions. While this technique is very insensitive, it will reveal major flaws in opaque crystals, such as the dentritic core frequently present in crystals grown by spontaneous nucleation. Radiography is relatively simple and may be used to examine rapidly a batch of crystals in order to eliminate those containing gross defects prior to subsequent cutting and polishing of sections for the application intended.

It was pointed out by Chase and Wilcox (1966) that the migration of a solvent zone in a temperature gradient may be used to remove solvent inclusions from HTS-grown crystals. The main objective of inclusion removal is to permit more accurate determinations of such properties as density, optical absorption coefficient or the concentration of substitutional impurities. However this technique may also be used as a simple means of determining the inclusion concentration (Elwell *et al.*, 1972). The crystal is heated in a strong gradient at a temperature well above the melting point of the solvent for a time sufficiently long to allow all the solvent to migrate to the hotter surface. After cooling the crystal to room temperature, the high temperature solvent may be removed by dissolution in nitric acid or some alternative reagent, and the resulting change in weight of the crystal will be a measure of the initial concentration of solvent inclusions.

Wilcox (1968) has given a review of the topic of *inclusion removal* from crystals grown from both high temperature and aqueous solutions. The rate of migration of inclusions is given approximately by

$$v_R = \frac{D(\mathrm{d}T/\mathrm{d}z)}{m(1 - w_o)} \frac{\rho_{sn}}{\rho_c},$$

where $\mathrm{d}T/\mathrm{d}z$ is the temperature gradient, D the solute diffusion coefficient, m the slope of the liquidus, w_o the solute concentration (expressed as a weight fraction) and ρ_c and ρ_{sn} the density of crystal and solution, respectively. Some examples of removal rate using this equation are quoted in Table 9.9 together with experimental data where these are available.

Agreement between theory and experiment is as good as can be expected since the theory leading to the expression for v_R is highly simplified. In

TABLE 9.9. Rate of Movement v_R of Solvent Inclusions in Crystals

Crystal	Solvent	Temperature °C	dT/dz °C/cm	w_0 wt fraction	m °C/wt fraction	D cm² s⁻¹	ρ g cm⁻³	ρ_{sn} g cm⁻³	v_R mm hr⁻¹ Theory	v_R mm hr⁻¹ Expt.	Reference
Al₂O₃	PbF₂	1200	100	0.13	6700	5×10^{-5}†	4.0	7.6	0.06	—	Wilcox (1968)
BaTiO₃	KF	1200	100	0.42	670	5×10^{-5}†	5.0	3.5	0.3	—	
CaCO₃	Li₂CO₃	690	400	0.48	1500	1×10^{-5}†	2.71	2.2†	0.015	0.2	Bélin et al. (1972)
GaAs	Ga	900	750	0.21	1300	1×10^{-4}	4.22	5.0†	3.1	0.46	Mlavsky and Weinstein (1963)
Ge	Al	667	31	0.74	1100	1.7×10^{-4}	3.5	3.0†	0.6	0.76–1.8	Wernick (1957)
InAs	In	800	500	0.39	530	2×10^{-4}	5.67	6.0†	1.2	0.7	Kleinknecht (1966)
NiFe₂O₄	BaO/B₂O₃	1200	20	0.10	1000	2×10^{-5}	5.37	4.55	0.014	0.05	Elwell et al. (1972)
SiC	Si	2000	100	0.012	1.4×10^4	2×10^{-4}	2.96	2.15	0.03	—	Wilcox (1968)
ThO₂	PbF/Bi₂O₃	1200	400	0.006	6000	5×10^{-5}	10.0	8.2	0.12	0.1	Wilcox (1968); Chase and Wilcox (1966)

† Denotes an estimated quantity.

practice the travel rate may depend on the size of inclusions and on the interface kinetics (Tiller, 1963). The data reported in the table on the semiconductors GaAs, Ge, InAs and SiC were taken on crystal-growth experiments by the travelling solvent method and were used to obtain values of the diffusion coefficient by an equation related to that given above for v_R.

The rather large discrepancy between experimental and theoretical values of v_R for GaAs and InAs indicates a difference of interpretation between Wilcox (1968) and the original authors. It may be that the values either of the temperature gradient or of the diffusion coefficient have been overestimated in these examples.

Disagreement between theory and practice in the experiments of Bélin et al. (1972) is to be expected since the crystal was rotated so that growth was probably not limited by diffusion in that case. For a more detailed discussion of inclusion removal, reference may be made to Wilcox (1968).

9.5. The Growth History of a Crystal as deduced from Characterization

Reference has been made at various parts in the text, especially in Chapter 4, to the mode of growth of crystals following nucleation or seeding. It is instructive here to consider the information to be obtained from characterization, particularly from X-ray topography, on the growth history and on the generation and propagation of dislocations and other defects. Detailed studies have in general been made of crystals grown from aqueous solutions but the evidence available suggests that similar results will be obtained on crystals grown from high-temperature solution.

Emara et al. (1969) investigated the distribution of dislocations in potash alum grown by spontaneous nucleation. The majority of dislocations are generated in the early stages of growth and radiate as bundles or sometimes as single dislocations, in a direction roughly perpendicular to the crystal faces (see Fig. 6.8). Additional dislocations are generated at growth accidents occurring at a later stage, usually at an inclusion because of imperfect lattice closure. The number of dislocations remains approximately constant as growth proceeds, so that the dislocation density decreases.

Veils of solution were observed in these crystals, and those in the faster-growing sectors gave rise to a high degree of strain. "Clouds" of inclusions in the form of small bubbles were also observed. Some bubbles act as dislocation sources but in other instances the dislocations terminate on bubbles so that the lattice heals when closing around a bubble of solvent. It was frequently observed that the small crystal faces have a relatively high density of dislocations, indicating that a high density of dislocation leads to more rapid growth.

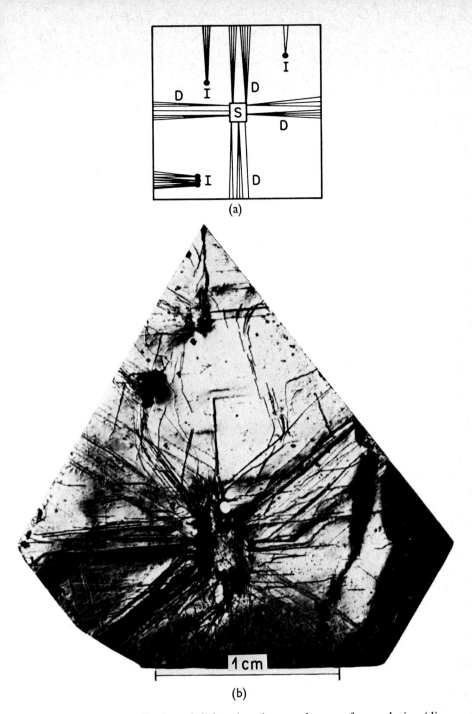

FIG. 9.15. (a) Distribution of dislocations in crystal grown from solution (diagrammatic, after Authier, 1972). S = seed, D = dislocation bundles, I = inclusions. (b) X-ray topograph (Authier, 1972) of TGS, showing dislocations generated at the seed and at inclusions.

Generally similar observations were reported by Ikeno *et al.* (1968) who investigated crystals of Mn-doped NaCl. They also noted additional dislocations which appeared to be generated at the position where the crystal was in contact with the beaker. These authors stress the observation that the crystals appeared to grow in six parts, each specified by one of the {110} growth surfaces and separated by a diagonal surface passing through the nucleus.

In a review of the applications of X-ray topography in crystal-growth studies, Authier (1972) considers seeded growth from aqueous solution. The majority of the dislocations originate at the seed and are propagated normal to the crystal faces, so that distribution of the dislocated regions is very similar to that in unseeded crystals. Figure 9.15(a) shows schematically the location of dislocations in a crystal of cubic habit. An actual topograph made by Mrs. A. Izrael is shown in Fig. 9.15(b) which shows a crystal of triglycine sulphate. In this case the dislocations propagate in six main bundles and large areas of crystal are substantially dislocation-free. Two regions can be seen where large solvent inclusions have resulted in the generation of further bands of dislocations. The mechanism of generation of dislocations at inclusions has been discussed by Matthews (1972).

Klapper (1971) made a topographic study of benzyl crystals grown from solution in xylol, and also found dislocation bundles radiating from the nucleus or from inclusions. He demonstrated that the direction of dislocations corresponds to a minimum of potential energy per unit length. Thus dislocations deviating 10° from the growth direction were found in benzyl, and Fig. 9.16(a) shows a topograph of a crystal of lithium formate hydrate of which the explanation is given in Fig. 9.16(b). The preferred orientation of dislocations can be calculated from the elastic constants of the material according to Klapper (1972, 1973).

An interesting series of topographs was obtained by Vergnoux *et al.* (1971) on strontium formate and sulphur crystals. Their observations confirm what appears to be a general tendency in solution growth for the dislocations to propagate in bundles (see Fig. 6.8). Some of the dislocations were found to originate in the seed rather than in the region of initial growth on the seed. As indicated by other observations of surface features, a tendency was noted for the number of dislocations to decrease during growth. During slow growth of sulphur crystals, the dislocation density on the (100) face was observed to fall from 200 cm^{-2} after 8 days to 40 after 32 days and to 1 after 58 days. However, any disturbance in the growth process is unfavourable and results in a production of new dislocations. As an example, an interruption in stirring led to a remarkable increase in dislocation density (see Fig. 7.38).

LPE studies have shown that the number of dislocations in the crystal

(a)

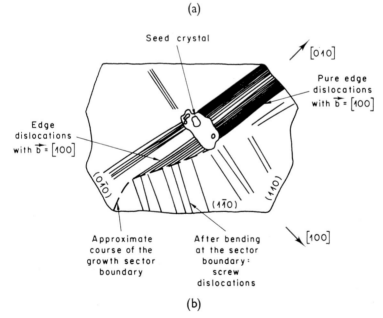

Seed crystal

[010]

Pure edge
dislocations
with $\vec{b} = [100]$

Edge
dislocations
with $\vec{b} = [100]$

(010)

(1$\bar{1}$0)

(110)

[100]

Approximate
course of the
growth sector
boundary

After bending
at the sector
boundary:
screw
dislocations

(b)

FIG. 9.16. (a) X-ray topograph of lithium formate hydrate crystal. (b) Interpretation of Fig. 9.16(a). After the formation of screw dislocations at the growth sector
boundary, enhanced growth of the (1$\bar{1}$0) face is observed (Klapper, 1973).

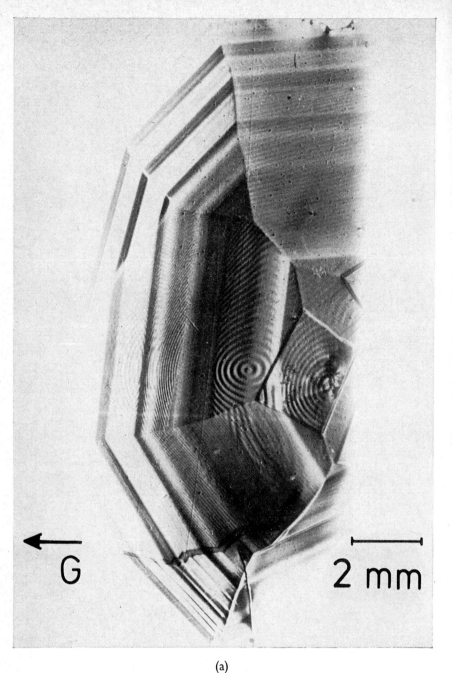

(a)

FIG. 9.17. X-ray topographs of (a) (111) slice of $Y_3Fe_{4.45}Ga_{0.55}O_{12}$, grown by Pavlik, showing "growth ripples"; (b) enlargement of dominant feature shown in (a); (c) $Y_3Fe_4Ga_1O_{12}$, grown by W. Tolksdorf, showing striations and inclusions generated by dislocations (Courtesy Dr. W. T. Stacy, Phillips Research Laboratories).

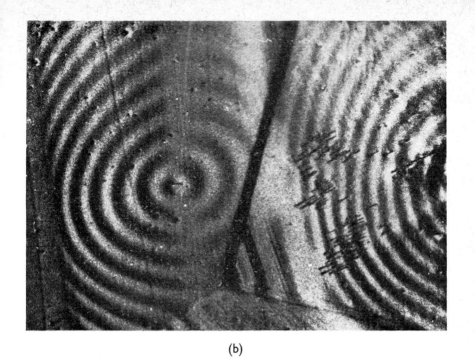

(b)

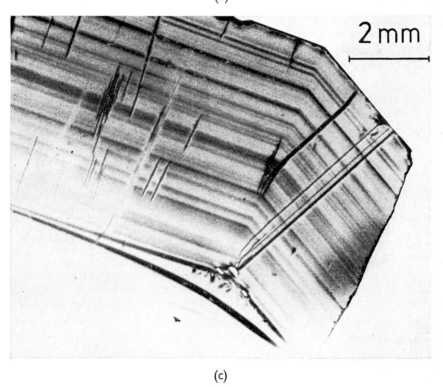

(c)

should ideally be less than that in the seed, and so more perfect crystals than those described above could be grown by careful seeding. A change in growth direction through 90° may also be effective in reducing the number of dislocations produced by an imperfect seed, as proposed by Schroeder and Linares (1966).

Relatively few topographic investigations of flux-grown crystals have been reported in the literature, but some particularly interesting topographs by W. T. Stacy are shown in Fig. 9.17. The "growth ripple" of Fig. 9.17(a) is formed by a dislocation surrounded by concentric regions of alternating strain. The dislocation, seen more clearly in Fig. 9.17(b), runs perpendicular to these bands and is associated with a growth hillock on the (111) face. Fig. 9.17(c) shows typical striations and the generation of dislocations at inclusions incorporated during a period of rapid growth following partial dissolution.

9.6. Proposed Standard for Routine Characterization of Crystals

As outlined at the beginning of this chapter there is no doubt that much material published in the literature is of dubious validity because insufficient effort has been made to characterize the materials, both in terms of their composition and structural perfection. As far as we are aware, only one proposal has been made, by Roy (1965), to specify the minimum data desirable for the characterization of a crystal on which results are presented for publication. An analogous set of recommendations for the acceptance of new minerals (including synthetic crystals) has been proposed by Donnay and Fleischer (1970).

In Table 9.10 we have outlined a proposal for a set of data which should be presented when measurements on crystals are reported or in publications on the synthesis of a novel material in single-crystal form. The experimental techniques which may be used to obtain these data are also indicated. In addition, in papers on physical measurements on crystals either details of the growth technique or a reference to the corresponding crystal-growth publication containing the details should be given. It is proposed to present the characterization data and the crystal-growth data in the form of one or two tables in order to keep publications short. A few details on any heat treatment following growth and on sample preparation (surface damage!) should be given.

Data of flux-grown crystals approximating those of Table 9.10 are quite rare in literature. Among the best characterized crystals are ThO_2 (Finch and Clark, 1965) and BeO (Newkirk and Smith, 1965, 1967; Newkirk et al., 1967; Austerman et al., 1964). Compilation of such characterization data, when done with the necessary care, is of course very time-consuming and needs a large effort on apparatus. It would be very valuable if the national

TABLE 9.10. Data Required for Crystal Characterization

Parameter	Experimental methods
Nominal chemical formula	
Chemical analysis for major constituents, deviation from stoichiometry	Chemical and physical techniques for chemical analysis
Trace analysis	Emission spectroscopy, activation analysis, colorimetry, etc.
Homogeneity of major and minor constituents	Electron microprobe analysis; local variation of T_c etc.
Colour	In several cases the optical absorption spectrum might be of interest
Crystal habit, cleavage, twinning, refractive index, optical activity	Optical goniometry, microscopy
Size and weight of single crystals, size of inclusion-free regions, density	Three-dimensional data on size
X-ray identification	X-ray focussing powder cameras, X-ray diffractometer, Gandolfi camera for small crystals
For new crystals: space group (+ structure determination)	Single crystal X-ray diffraction techniques, automatic single crystal diffractometer
Precision determination of lattice constants	Focussing X-ray powder techniques, automatic single crystal diffractometer
Low-angle boundaries, degree of misorientation	Laue X-ray patterns, X-ray topography
Quantity and distribution of inclusions, cracks, twinning, strain	Microscopy, ultramicroscopy, X-ray topography, etc.
Dislocation density, distribution of dislocations	Etching, X-ray topography
For metals, alloys, semiconductors: Resistivity	4-probe
For magnetic materials: saturation magnetization	Magnetic balance, vibrating sample magnetometer

research institutions were to support or to maintain specialized laboratories for characterization. Ideally the characterization specialist should participate in the research programme along with physics and crystal-growth specialists, with the characterization service provided free of charge to collaborating establishments.

A greater emphasis on characterization would probably reduce the rate of publication but would be highly beneficial in the long run for all interested parties.

References

Abragam, A. and Bleaney, B. (1970) "Electron Paramagnetic Resonance of Transition Ions" Clarendon Press, Oxford.

Addink, N. W. H. (1957) *Spectrochim. Acta* **11**, 168.

Adler, I. (1966) "X-Ray Emission Spectrography in Geology" Elsevier, Amsterdam.

Ahearn, A. J. (1966) (editor) "Mass Spectroscopy of Solids" Elsevier, Amsterdam.

Ahmed, F. R. (1969) (editor) "Crystallographic Computing" Munksgaard, Copenhagen.

Ahrens, L. H. and Taylor, S. R. (1961) "Spectrochemical Analysis" 2nd edition. Addison-Wesley, Reading, Mass.

Alimarin, I. P. and Petrikova, M. N. (1962) "Anorganische Ultramikroanalyse" Deutscher Verlag der Wissenschaften, Berlin.

Alsopp, H. J. and Roberts, J. P. (1957) *Nature* **180**, 603; *Analyst* **82**, 474.

Amelinckx, S. (1964) "The Direct Observation of Dislocations." Supplement of Solid State Physics (F. Seitz and D. Turnbull, eds.).

Anbar, M. and Guttman, S. (1959) *Internat. J. Appl. Radiation Isotopes* **5**, 233.

Andersen, C. A. (1973) (editor) "Microprobe Analysis" Wiley-Interscience, New York.

Anderson, E. E., Cunningham, J. R. Jr., McDuffle, G. E. Jr. and Ayers, W. E. (1959) Polycrystalline Ferrimagnetic Garnets, Navord Report No. 6741, U.S. Naval Ordnance Laboratory, White Oak, Maryland.

Angino, E. E. and Billings, G. K. (1967) "Atomic Absorption Spectrometry in Geology" Elsevier, Amsterdam.

Arend, H. and Novak, J. (1966) *Kristall u. Technik* **1**, 93.

Arend, H., Novak, J. and Coufova, P. (1969) *Growth of Crystals* **7**, 244.

Arndt, U. W. and Willis, B. T. M. (1966) "Single Crystal Diffractometry" Cambridge University Press, New York.

Austerman, S. B., Newkirk, J. B. and Smith, D. K. (1965) NRL Rept. 6080.

Austerman, S. B., Myron, S. M. and Wagner, J. W. (1967) *Carbon* **5**, 549.

Authier, A. (1966) *J. de Physique* **27**, 57.

Authier, A. (1967) *Adv. in X-ray Analysis* **10**, 9.

Authier, A. (1970) In "Modern Diffraction and Imaging Techniques in Material Science" (S. Amelinckx, R. Gevers, G. Remaut and I. Van Landuyt, eds.) p. 481. North-Holland, Amsterdam.

Authier, A. (1972) *J. Crystal Growth* **13/14**, 34.

Ayscough, P. B. (1967) "Electron Spin Resonance in Chemistry" Methuen, London.

Azaroff, L. V. (1968) "Elements of X-ray Crystallography". McGraw-Hill, New York.

Azaroff, L. V. and Buerger, M. J. (1958) "The Powder Method in X-ray Crystallography" McGraw-Hill, New York.

Bacon, G. E. (1962) "Neutron Diffraction" Oxford University Press.

Baker, T. W., George, J. D., Bellamy, R. A. and Causer, R. (1966) *Nature* **210**, 720.

Barns, R. L. (1967) *Mat. Res. Bull.* **2**, 273.

Barrett, C. S. (1945) *Trans. AIME* **161**, 15.

Barrett, C. S. (1967) "Experimental Methods of Materials Research" (H. Herman, ed.) Interscience, New York.

Barrett, C. S. and Massalski, T. B. (1966) "Structure of Metals" McGraw-Hill, New York.

Barth, H. and Hosemann, R. (1958) *Z. Naturf.* **13a,** 792.

Bartholomew, R. F. and White, W. B. (1970) *J. Crystal Growth* **6,** 249.

Basterfield, J. (1969) *J. Phys.* D **2,** 1159.

Basterfield, J. and Prescott, M. J. (1967) *J. Appl. Phys.* **38,** 3190.

Bélin, C., Brissot, J. J. and Jesse, R. E. (1972) *J. Crystal Growth* **13/14,** 597.

Belruss, V., Kalnajs, J., Linz, A. and Folweiler, R. C. (1971) *Mat. Res. Bull.* **6,** 899.

Belt, R. F. (1967) *Mat. Res. Bull.* **2,** 919.

Belt, R. F. (1969) *J. Appl. Phys.* **40,** 1644.

Bennett, G. A. and Wilson, R. B. (1966) *J. Sci. Instrum.* **43,** 669.

Benninghoven, A. (1969) *Z. Phys.* **230,** 403.

Benninghoven, A. (1973) *Appl. Phys.* **1,** 3.

Benninghoven, A. and Storp, S. (1971) *Z. Angew. Chem.* **31,** 31.

Berdennikova, T. P. (1932) *Phys. J. Soviet Union* **2,** 77 (ref. from Kröger, 1964).

Berkes, J. S., White, W. B. and Roy, R. (1965) *J. Appl. Phys.* **36,** 3276.

Berl, W. G. (1960) "Physical Methods in Chemical Analysis" Academic Press, New York–London.

Bertaut, E. F. (1963) "Magnetism" Vol. 3 (G. T. Rado and H. Suhl, eds.) Academic Press, New York–London.

Birks, L. S. (1963) "Electron Probe Microanalysis" Wiley-Interscience, New York.

Birks, L. S. (1969) "X-ray Spectrochemical Analysis" 2nd edition. Interscience, New York.

Blackburn, W. H., Pelletier, Y. J. A. and Dennen, W. H. (1968) *Appl. Spectroscopy,* April, 278.

Bloss, F. D. (1961) "An Introduction to the Methods of Optical Crystallography" Holt, Rinehart and Winston, New York.

Bock, R. (1972) "Aufschlussmethoden der anorgan. u. organ. Chemie" Verlag Chemie, Weinheim.

Bode, M., Pfefferkorn, G., Schur, K. and Wegmann, L. (1971) *J. Microscopy* **95,** 323.

Bohlin, H. (1920) *Ann. Physik* **61,** 421.

Bollmann, W. (1970) "Crystal Defects and Crystalline Interfaces" Springer, Berlin.

Boltz, D. F. (1958) (editor) "Colorimetric Determination of Nonmetals" Wiley-Interscience, New York.

Bond, W. L. and Andrus, J. (1952) *Am. Min.* **37,** 622.

Bond, W. L. and Andrus, J. (1956) *Phys. Rev.* **101,** 1211.

Bonse, U. and Kappler, E. (1958) *Z. Naturf.* **13a,** 348.

Bonse, U. K., Hart, M. and Newkirk, J. B. (1966) In "Encyclopaedic Dictionary of Physics" Supp. 1, p. 391. Pergamon, Oxford.

Booker, G. R. (1970) In "Modern Diffraction and Imaging Techniques in Materials Science" (S. Amelinckx, R. Gevers, G. Remaut and J. van Landuyt, eds.) p. 553. North Holland, Amsterdam.

Borrmann, G., Hartwig, W. and Irmler, H. (1958) *Z. Naturf.* **13a,** 423.

Boswell, F. W. C. (1957) *Metal Prog.* **72,** 92.

Bowen, H. J. and Gibbons, D. (1963) "Radioactivation Analysis" Clarendon Press, Oxford.

Bowkett, K. M. and Smith, D. A. (1970) "Field-Ion Microscopy" North-Holland, Amsterdam.

Bradford, J. N., Ekardt, R. C. and Tucker, J. W. (1964) NRL Rept. 6080.

Bragg, W. L. (1913) *Proc. Cambridge Phil. Soc.* **17**, 43.

Brebrick, R. F. (1969) International Conference on Non-Stoichiometry, Tempe, U.S.A.

Breyer, B. and Bauer, H. H. (1963) "Alternating Current Polarography and Tensammetry" Wiley-Interscience, New York.

Brice, J. C., Hill, O. F., Whiffin, P. A. C. and Wilkinson, J. A. (1971) *J. Crystal Growth* **10**, 133.

Buerger, M. J., (1960) "Crystal Structure Analysis" Wiley, New York.

Buerger, M. J. (1962) "X-Ray Crystallography" Wiley, New York.

Buerger, M. J. (1964) "The Precession Method" Wiley, New York.

Bunn, C. W. (1961) "Chemical Crystallography". Oxford Univ. Press

Burns, G. (1964) *Phys. Rev.* **135**, A 479.

Burri, C. (1959) "Das Polarisationsmikroskop" Birkhäuser, Bale.

Burton, J. A., Prim, R. C. and Slichter, W. P. (1953) *J. Chem. Phys.* **21**, 1987.

Carlson, A. E. (1959) In "Growth and Perfection of Crystals" (R. H. Doremus, B. W. Roberts and D. Turnbull, eds.) p. 421. Wiley, New York.

Carruthers, J. R., Peterson, G. E. and Grasso, M. (1971) *J. Appl. Phys.* **42**, 1846.

Castaing, R. (1951) Ph.D. Thesis, University of Paris.

Castaing, R. (1960) "Advances in Electronics and Electron Physics" Vol. 8, 317 (L. Marton, ed.) Academic Press, New York–London.

Chamot, E. M. and Mason, C. W. (1958) "Handbook of Chemical Microscopy" Vol. 1, 3rd edition. Wiley, New York.

Champion, J. A. (1969) *Trans. Brit. Ceram. Soc.* **68**, 91, 1969.

Champion, J. A. and Clemence, M. Y. (1967) *J. Mat. Sci.* **2**, 153.

Chang, C. C. (1971) *Surface Sci.* **25**, 53.

Chase, A. B. (1968) *J. Am. Ceram. Soc.* **51**, 507.

Chase, A. B. and Wilcox, W. R. (1966) *J. Electrochem. Soc.* **113**, 299.

Chaudhari, P. and Matthews, J. W. (1972) (editors) "Grain Boundaries and Inter-faces" North-Holland, Amsterdam (*Surface Sci.* **31**, 1–622).

Chester, P. F. (1961) *J. Appl. Phys.* **32**, 866.

Clark, G. L. (1960) (editor) "The Encyclopedia of Spectroscopy" Reinhold, New York.

Clayton, R. N. and Mayeda, T. K. (1963) *Geochim. Cosmochim. Acta* **27**, 43.

Cloete, F. L. D., Ortega, R. F. and White, E. A. D. (1969) *J. Mat. Sci.* **4**, 11.

Coburn, J. W. and Kay, E. (1971) *Appl. Phys. Letters* **19**, 350.

Coburn, J. W. and Kay, E. (1972) to be published (IBM Research Report RJ951).

Cockayne, B. and Gates, M. P. (1967) *J. Mat. Sci.* **2**, 118.

Committee on Characterization of Materials (1967). Report AD 649, 941.

Cook, C. F. and Nye, W. F. (1967) *Mat. Res. Bull.* **2**, 1.

Cooper, B. S., Gerrard, P. W., Sear, L. G. and Wiggins, G. M. (1969). Paper presented at ACCG Conference on Crystal Growth.

Cottrell, A. H. (1953) "Dislocations and Plastic Flow in Crystals" Clarendon Press, Oxford.

Crawford, J. H. and Slifkin, L. M. (1972) (editors) "Point Defects in Solids" (Vol. 1: General and Ionic Crystals, Vol. 2: Defects in Semiconductors, Vol. 3: Defects in Metals) Plenum Press, New York.

Damen, J. P. M. and Robertson, J. M. (1972) *J. Crystal Growth* **16**, 50.

Davies, J. A., Denhartog, J., Erikson, L. and Mayer, J. W. (1967) *Can. J. Phys.* **45**, 4053.

De Gennes, P. G. (1963) In "Magnetism" Vol. 3 (G. T. Rado and H. Suhl, eds.) Academic Press, New York–London.

Delves, R. T. (1968) *J. Crystal Growth* **3/4**, 562.

Delves, R. T. (1971) *J. Crystal Growth*, **8**, 13.

De Soete, D., Gijbels, R. and Hoste, J. (1971) "Neutron Activation Analysis" Wiley-Interscience, New York.

DeVoe, J. R. (1969) (editor) "Modern Trends in Activation Analysis". National Bureau of Standards, Washington (Special publication 312, 2 volumes).

Discussions Faraday Society (1964) "Dislocations in Solids" (No. 38).

Dolezal, J., Povondra, P. and Sulzek, Z. (1968) "Decomposition Techniques in Inorganic Analysis" Iliffe Books, London.

Donnay, G. and Fleischer, M. (1970) *Am. Min.* **55**, 1017.

Donnay, J. D. H. *et al.* (1963) "Crystal Data, Determinative Tables", second edition, Amer. Crystallographic Assoc., ACA Monograph 5.

Donnay, J. D. H. and Ondik, H. M. (1972) "Crystal Data, Determinative Tables", third edition (to be published by Joint Committee on Powder Diffraction Standards, Swarthmore, Pennsylv.).

Dontsova, E. I. (1959) *Geokhimiya* (Translation), 824.

Doughty, D. C. and White, E. A. D. (1960) *Acta Cryst.* **13**, 761.

Duval, C. (1963) "Inorganic Thermogravimetric Analysis" 2nd edition. Elsevier, Amsterdam.

Elwell, D. (1975) In "Crystal Growth" (B. R. Pamplin, ed.) Pergamon, Oxford.

Elwell, D. and Neate, B. W. (1971) *J. Mat. Sci.* **6**, 1499.

Elwell, D., Morris, A. W. and Neate, B. W. (1972) *J. Crystal Growth* **16**, 67.

Emara, S. H., Lawn, B. R. and Lang, A. R. (1969) *Phil. Mag.* **19**, 7.

Emmons, R. C. (1943) "The Universal Stage", Geol. Soc. Amer. Mem. **8.**

Engell, H. J. (1956) *Z. Electrochem.* **60**, 905.

Estrup, P. J. and McRae, E. G. (1971) *Surface Sci.* **25**, 1.

Eyring, L. and O'Keeffe, M. (1970) (editors) "The Chemistry of Extended Defects in Non-metallic Solids" North-Holland, Amsterdam.

Fairbairn, H. W. and Podolsky, T. (1951) *Am. Min.* **36**, 823.

Feigl, F. (1970) "Spot Tests in Inorganic Analysis" 5th edition. Elsevier, Amsterdam.

Fenner, N. C. and Daly, N. R. (1968) *J. Mat. Sci.* **3**, 259.

Finch, C. B. and Clark, G. W. (1965) *J. Appl. Phys.* **36**, 2143.

Filipczynski, L., Pawlowski, Z. and Wehr, J. (1966) "Ultrasonic Methods of Testing Materials" Butterworths, London.

Forsyth, J. B. (1970) In "Thermal Neutron Scattering" (B. T. M. Willis, ed.) Oxford University Press.

Fowler, W. B. (1968) (editor) "Physics of Color Centers" Academic Press, New York–London.

Frazer, J. Z., Fujita, H. and Fitzgerald, R. W. (1971) *Mat. Res. Bull.* **6**, 711.

Friedel, J. (1964) "Dislocations" Pergamon, Oxford.

Gallagher, P. K., McChesney, J. B. and Buchanan, D. N. E. (1964) *J. Chem. Phys.* **41**, 2429.

Gandolfi, G. (1967) *Miner. Petr. Acta* **13**, 67.

Garn, P. D. (1965) "Thermoanalytical Methods of Investigation" Academic Press, New York–London.

Gay, P. (1967) "An Introduction to Crystal Optics" Longmans, London.

Giess, E. A. (1962) *J. Am. Ceram. Soc.* **45**, 12.

Giess, E. A., Scott, B. A., O'Kane, D. F., Olson, B. L., Burns, G. and Smith, A. W. (1969) *Mat. Res. Bull.* **4,** 741.

Giess, E. A., Cronemeyer, D. C., Rosier, L. L. and Kuptsis, J. D. (1970) *Mat. Res. Bull.* **5,** 495.

Gordon, L., Salutsky, M. L. and Willard, H. H. (1959) "Precipitation from Homogeneous Solution" Wiley, New York.

Gruehn, R. (1966) *Zeitschr. Analyt. Chem.* **221,** 146.

Gübelin, E. J. (1953) "Inclusions as a Means of Gemstone Identification" Gemmolog. Soc. of Amer., Los Angeles.

Guinier, P. (1937) *Compt. Rend.* **204,** 1115.

Guinier, A. (1962) "X-Ray Diffraction" Freeman, New York.

Guinier, A. and Tennevin, J. (1949) *Acta Cryst.* **2,** 133.

Guinn, V. P. and Lukers, H. R. Jr. (1965) In "Trace Analysis" (G. H. Morrison, ed.) p. 325. Interscience, New York.

Gupta, D. and Tsui, R. T. C. (1970) *Appl. Phys. Lett.* **17,** 294.

Guseva, I. N., Bagdasarov, Kh. S. and Rakhshtadt, U. A. (1971) Paper presented at the Intern. Conf. Crystal Growth, Marseille.

Hafner, H. C. and Rood, J. L. (1967) *Mat. Res. Bull.* **2,** 303.

Hannaford, P., Howard, C. J. and Wignall, J. W. G. (1965) *Phys. Letters* **19,** 257.

Hannay, N. B. (1974) (editor) "Treatise on Solid State Chemistry" Vol. 1. The Chemical Structure of Solids. Plenum Press, New York.

Harrison, G. R., Lord, R. C. and Loofbourow, J. R. (1963) "Practical Spectroscopy" Prentice-Hall, Englewood Cliffs, N.J.

Harrick, N. J. (1967) "Internal Reflection Spectroscopy" Interscience, New York.

Hartshorne, N. H. and Stuart, A. (1969) "Practical Optical Crystallography" 2nd edition. Arnold, London.

Harvey, C. E. (1964) "Semiquantitative Spectrochemistry" Applied Research Laboratories, Glendale, California.

Hawkes, P. W. (1972) "Electron Optics and Electron Microscopy" Taylor and Francis, London.

Heavens, O. S. (1965) "Optical Properties of Thin Films" Dover, New York.

Heidenreich, R. D. (1964) "Fundamentals of Transmission Electron Microscopy" Interscience, New York.

Heinrich, K. F. J. (1966) "The Electron Microprobe" Wiley, New York.

Heinrich, K. F. J. (1967) In "Experimental Methods of Materials Research" (H. Herman, ed.) p. 145. Wiley, New York.

Heinrich, K. F. J. (1968) "Quantitative Electron Probe Microanalysis" (National Bureau of Standards Spec. Publ. No. 298).

Helz, A. W., Walthall, F. G. and Berman, S. (1969) *Appl. Spectroscopy* **23,** 508.

Herzberg, G. (1944) "Atomic Spectra and Atomic Structure" Dover, New York.

Herzfeld, K. F. and Litovitz, T. A. (1959) "Absorption and Dispersion of Ultrasonic Waves" Academic Press, New York.

Herzog, R. F. K. and Viehböck, R. P. (1949) *Phys. Rev.* **76,** 855.

Hildisch, L. (1968) *J. Crystal Growth* 3/4, 131.

Hillier, J. (1947) U.S. Pat. 2,418,029.

Hirth, J. P. and Lothe, J. (1968) "Theory of Dislocations" McGraw-Hill, New York.

Holtzberg, F. (1970) *J. Appl. Phys.* **41,** 1283.

Honig, R. E. (1958) *J. Appl. Phys.* **29,** 549.

Huber, M. and Liné, G. (1963), *C.R. Acad. Sci.* **256**, 3118.

Humble, P. (1970) In "Modern Diffraction and Imaging Techniques in Material Science" (S. Amelinckx, R. Gevers, G. Remaut and J. van Landuyt, eds.) p. 99. North-Holland, Amsterdam.

Hurle, D. T. J., (1966) *Phil. Mag.* **13**, 305.

Hurle, D. T. J. (1967) In "Crystal Growth" (H. S. Peiser, ed.) Pergamon Press, Oxford.

Hurst, J. J. and Linz, A. (1971) *Mat Res. Bull.* **6**, 163.

Ikeno, S., Maruyana, H. and Kato, N. (1968) *J. Crystal Growth*, **3/4**, 683.

Isherwood, B. J. (1968) *J. Appl. Cryst.* **1**, 299.

Isherwood, B. J. and Wallace, C. A. (1966) *Nature* **212**, 173.

Janowski, K. R., Chase, A. B. and Stofel, E. J. (1965) *Trans. AIME* **233**, 2087.

Janssen, G. A. M., Robertson, J. M. and Verheijke, M. L. (1973) *Mat. Res. Bull.* **8**, 59.

Jenkins, R. and DeVries, J. L. (1967) "Practical X-Ray Spectrometry" Springer, New York.

Jona, F. (1970) *IBM J. Res. Develop.* **14**, 444.

Kammlott, G. W. (1971) *Surface Sci.* **25**, 120.

Kane, P. F. and Larrabee, G. B. (1970) "Characterization of Semiconductor Materials" McGraw-Hill, New York.

Kendall, D. N. (1966) (editor) "Applied Infrared Spectrometry" Reinhold, New York.

Kienitz, H. (1968) (editor) "Massenspektrometrie" Verlag Chemie, Weinheim, Bergstr.

Kim, K. M., Witt, A. F. and Gatos, H. C. (1972) *J. Electrochem. Soc.* **119**, 1218.

Kjems, J. K., Shirane, G., Müller, K. A. and Scheel, H. J. (1973) *Phys. Rev.* **B 8**, 1119.

Klapper, H. (1971) *J. Crystal Growth* **10**, 13.

Klapper, H. (1972) *Phys. Stat. Sol.* **14**, 99, 443.

Klapper, H. (1973) *Z. Naturforschg.* **28a**, 614.

Kleinert, P. and Funke, A. (1960) *Naturwiss.* **47**, 106.

Kleinknecht, H. P. (1966) *J. Appl. Phys.* **37**, 2116.

Klug, H. P. and Alexander, L. E. (1954) "X-Ray Procedures for Polycrystalline and Amorphous Materials" Wiley, New York (2nd edition 1974).

Koch, O. G. and Koch-Dedic, G. A. (1964) "Handbuch der Spurenanalyse" Springer, Berlin.

Kodama, K. (1963) "Methods of Quantitative Inorganic Analysis: An Encyclopedia of Gravimetric, Titrimetric and Colorimetric Methods" Wiley-Interscience, New York.

Kohn, J. A., Eckart, D. W. and Cook, C. F. Jr. (1967) *Mat. Res. Bull.* **2**, 55.

Kolthoff, I. and Lingane, J. J. (1952) "Polarography" Interscience, New York.

Kolthoff, I. M. and Elving, P. I. (1959–1972) "Treatise on Analytical Chemistry" (Series) Interscience, New York.

Kolthoff, I. M. and Elving, P. J. (1961) "Treatise on Analytical Chemistry" Part I, Theory and Practice, Vol. 1. Wiley, New York.

Kraus, T. (1972) *Neue Hütte* **17**, 369.

Kröger, F. A. (1964) "The Chemistry of Imperfect Crystals" North-Holland, Amsterdam.

Kroko, L. J. (1966) *J. Electrochem. Soc.* **113**, 801.

Kruger, P. (1971) "Principles of Activation Analysis" Wiley-Interscience, New York.

Landau, A. I. (1958) *Phys. Met. Metallogr.* **6,** 132.

Lang, A. R. (1957) *Acta Met.* **5,** 358.

Lang, A. R. (1970) In "Modern Diffraction and Imaging Techniques in Materials Science" (S. Amelinckx, R. Gevers, G. Remaut and I. Van Landuyt, eds.) p. 407. North-Holland, Amsterdam.

Laudise, R. A. (1970) "The Growth of Single Crystals" Prentice-Hall, Englewood Cliffs, New York.

Laudise, R. A., Carruthers, J. R. and Jackson, K. A. (1971) *Ann. Rev. Mat. Sci.* **1,** 253.

Lawless, W. N. and DeVries, R. C. (1964) *J. Opt. Soc. Amer.* **54,** 1225.

Lawson, K. E. (1961), "Infrared Absorption of Inorganic Substances" Reinhold, New York.

Lederer, M. (1971) (editor) "First CAMAG Intern. Symp. on Advanced Thin-Layer Chromatography" (*J. Chromatogr.* **63,** No. 1) Elsevier, Amsterdam.

Lederer, E. and Lederer, M. (1960) "Chromatography" 2nd edition. Elsevier, Amsterdam.

Lefever, R. A., Chase, A. B. and Torpy, J. W. (1961) *J. Am. Ceram. Soc.* **44,** 141.

Libowitz, G. G. (1953) *J. Amer. Chem. Soc.* **75,** 1501.

Linares, R. C. (1965a) *J. Amer. Chem. Soc.* **48,** 68.

Linares, R. C. (1965b) *J. Phys. Chem. Solids* **26,** 1817.

Linares, R. C. (1967) *J. Phys. Chem. Solids* **28,** 1285.

Linares, R. C., Schroeder, J. B. and Hurlbut, L. A. (1965) *Spectrochim. Acta* **21,** 1915.

Lingane, J. J. (1958) "Electroanalytical Chemistry" 2nd edition. Interscience, New York.

Lipson, H. and Cochran, W. (1966) "The Determination of Crystal Structures" 2nd edition. Bell and Sons, London.

Low, W. (1960) "Paramagnetic Resonance in Solids" Academic Press, New York–London.

Low, W. (1963) (editor) "Paramagnetic Resonance" Academic Press, New York.

Lyon, W. S. (1964) "Guide to Activation Analysis" Van Nostrand, New York.

Makram, H. (1968) *J. Crystal Growth* **34,** 447.

Makram, H. and Krishnan, R. (1967) *J. Phys. Chem. Solids Supp.* **1,** 467.

Makram, H., Touron, L. and Loriers, J. (1968) *J. Crystal Growth* **3/4,** 452.

Mandelkorn, L. (1964) "Non-stoichiometric Compounds" Academic Press, New York–London.

Marcus, R. B. (1969) In "Physical Measurement and Analysis of Thin Films" p. 105. Plenum Press, New York.

Markham, J. J. (1966) "F centers in Alkali Halides" Academic Press, New York–London.

Mattauch, J. and Herzog, R. (1934) *Z. Physik* **89,** 786.

Matthews, J. W. (1972) *Phys. Stat. Sol.* **15,** 607.

Mavrodineanu, R. and Boiteux, H. (1965) "Flame Spectroscopy" Wiley, New York.

Maxwell, J. A. (1968) "Rock and Mineral Analysis" Interscience, New York.

McDonald, N. C. (1970) *Appl. Phys. Letters* **16,** 76.

McDowell, C. A. (1963) (editor) "Mass Spectrometry" McGraw-Hill, New York.

McKinley, J. D. (1966) (editor) "The Electron Microprobe" Wiley-Interscience, New York.

Meinke, W. M. and Scribner, B. F. (1967) (editors) "Trace Characterization, Chemical and Physical" National Bureau of Standards Monograph 100, Washington D.C.

Minkoff, I. (1967) *J. Mat. Sci.* **2**, 388.

Mitchell, I. V., Kamoshida, M. and Mayer, J. W. (1971) *J. Appl. Phys.* **42**, 4378.

Mlavsky, A. E. and Weinstein, M. (1963) *J. Appl. Phys.* **34**, 2885.

Moenke, H. (1962) "Mineralspektren" Akademie Verlag, Berlin.

Moenke, H. and Moenke-Blankenburg, L. (1966) "Einführung in die Laser-Mikro-Emissionsspektralanalyse" Akad. Verlagsgesellsch. Geest and Portig, Leipzig.

Morrish, A. H. and Eaton, J. A. (1971) *J. Appl. Phys.* **42**, 1495.

Morrison, G. H. and Skogerboe, R. K. (1965) In "Trace Analysis" (G. H. Morrison, ed.) p. 1. Interscience, New York.

Müller, E. W., and Tien Tzon Tsong (1969) "Field-Ion Microscopy" Elsevier, New York.

Müller, K. A. (1971) In "Structural Phase Transitions and Soft Modes" (E. J. Samuelsen, E. Andersen and J. Feder, eds.) Universetetsforlaget, Oslo.

Müller, R. O. (1972) "Spectrochemical Analysis by X-Ray Fluorescence" Plenum Press, New York.

Mullen, J. G. (1963) *Phys. Rev.* **131**, 1415.

Mullins, W. W. and Sekerka, R. F. (1964) *J. Appl. Phys.* **35**, 444.

Murr, L. E. (1970) "Electron Optical Applications in Materials Science" McGraw-Hill, New York.

Nabarro, F. R. N. (1967) "Theory of Crystal Dislocations" Oxford University Press.

Nachtrieb, N. H. (1950) "Principles and Practice of Spectrochemical Analysis" McGraw-Hill, New York.

Nakanoto, K. (1963) "Infrared Spectra of Inorganic and Coordination Compounds" Wiley, New York.

Nassau, K. (1968) *J. Crystal Growth* **2**, 215.

Nassau, K. and Lines, M. E. (1970) *J. Appl. Phys.* **41**, 533.

Neeb, R. (1969) "Inverse Polarographie und Voltammetrie, Neuere Verfahren zur Spurenanalyse" Verlag Chemie, Weinheim/Bergstr.

Nelson, D. F. and Remeika, J. P. (1964) *J. Appl. Phys.* **35**, 522.

Nelson, H. (1963) *RCA Review* **24**, 603.

Newkirk, J. B. (1959) *Trans AIME* **215**, 483.

Newkirk, H. W. and Smith, D. K. (1965) *Am. Min.* **50**, 44.

Newkirk, H. W. and Smith, D. K. (1967) *J. Appl. Phys.* **38**, 2705.

Newkirk, H. W., Smith, D. K., Meieran, E. S. and Mattern, D. A. (1967) *J. Mat. Sci.* **2**, 194.

Nicolson, R. B., Kirsch, P. B., Howie, A., Pashley, D. W. and Whelan, H. J. (1965) "Electron Microscopy of Thin Crystals" Butterworth, London.

Nielsen, J. W., Linares, R. C. and Koonce, S. E. (1962) *J. Am. Ceram. Soc.* **45**, 12.

Nielsen, J. W., Lepore, D. A. and Leo, D. C. (1967) In "Crystal Growth" (H. S. Peiser, ed.) p. 457. Pergamon, Oxford.

Nomarski, G. and Weill, A. R. (1954) *Bull. Soc. franç. Minér. Crist.* **77**, 840.

Novak, J. and Arend, H. (1964) *J. Am. Ceram. Soc.* **47**, 530.

Oatley, C. W., Nixon, W. C. and Pease, R. F. W. (1965) In "Advances in Electronics and Electron Physics" Vol. 21, p. 181. Academic Press, New York-London.

Paesold, G., Müller, K. and Kieffer, R. (1967) *Z. Analyt. Chem.* **232,** 31.

Pake, G. E. (1962) "Paramagnetic Resonance" Benjamin, New York.

Peiser, H. S., Rooksby, H. P. and Wilson, A. J. C. (1955) "X-Ray Diffraction by Polycrystalline Materials" Chapman and Hall, London.

Petrov, D. A. (1956) *Zh. fiz. Khim.* **30,** 50.

Phillips, V. A. and Lifshin, E. (1971) In "Annual Review of Materials Science" (R. A. Huggins, ed.) Annual Reviews, Palo Alto, California.

Pinsker, Z. G. (1953) "Electron Diffraction" Butterworths, London.

Powder Diffraction File (1973), published regularly as cards, books, microfiche, magnetic tapes etc. by the Joint Committee on Powder Diffraction Standards, Swarthmore, Pennsylvania.

Powers, H. E. C. (1970) *Sugar Tech. Rev.* **1,** 86.

Preisinger, A., Viehböck, F. P. and Weissmann, W. (1971) In "Ergebnisse der Hochvakuumtechnik und der Physik dünner Schichten" (M. Auwärter, ed.) Vol. II. Wiss. Verlagsgesellschaft, Stuttgart.

Prescott, M. J. and Basterfield, J. (1967) *J. Mat. Sci.* **2,** 583.

Rabenau, A. (1970) (editor) "Problems of Nonstoichiometry" North-Holland, Amsterdam.

Ramirez-Munoz, J. (1968) "Atomic Absorption Spectroscopy and Analysis by Atomic Absorption Flame Photometry" Elsevier, Amsterdam.

Rath, R. (1969) "Theoretische Grundlagen der allgemeinen Kristalldiagnose im durchfallenden Licht" Springer, Berlin.

Read, W. T. (1953) "Dislocations in Crystals" McGraw-Hill, New York.

Regel, V. R., Urusorskaya, A. A. and Kolomiichuk, V. N. (1960) *Sov. Phys. Cryst.* **4,** 895.

Reijnen, P. J. L. (1970) *Philips Tech. Rev.* **31,** 24.

Reimer, L. and Pfefferkorn, G. (1973) "Raster-Elektronenmikroskopie" Springer, Berlin.

Reinhart, M. (1931) "Universaldrehtischmethoden" Wepf, Basel.

Reisman, A., Berkenblit, M., Cuomo, J. and Chan, S. A. (1971) *J. Electrochem. Soc.* **118,** 1653.

Remeika, J. P. and Kometani, T. Y. (1968) *Mat. Res. Bull.* **3,** 895.

Reuter, W. (1971) *Surface Science* **25,** 80.

Ringbom, A. (1963) "Complexation in Analytical Chemistry" Interscience, New York.

Robertson, J. M. and Elwell, D. (1969) *J. Phys. C.* **2,** 14.

Roboz, J. (1968) "Introduction to Mass Spectrometry" Wiley-Interscience, New York.

Röhrig, R. and Schneider, J. (1969) *Phys. Letters* **30A,** 371.

Rosenfield, A. R., Hahn, G. T., Bement, A. L. and Jaffee, R. I. (1968) "Dislocation Dynamics" McGraw-Hill, New York.

Rowcliffe, D. J. and Warren, W. J. (1970) *J. Mat. Sci.* **5,** 345.

Roy, R. (1965) *Physics Today* **18,** 71.

Russ, J. C. (1970) "Energy Dispersion X-Ray Analysis", Amer. Soc. for Testing and Materials, Philadelphia.

Sahagian, C. S. and Schieber, M. S. (1969) *Growth of Crystals* **7,** 183.

Sandell, E. B. (1959) "Colorimetric Determination of Traces of Metals" 3rd edition, Interscience, New York.

Savage, R. O. and Tauber, A. (1967) *Mat. Res. Bull.* **2,** 469.

Schaeffer, H. F. (1966) "Microscopy for Chemists" Dover, New York.

Scheel, H. J. and Elwell, D. (1972) *J. Crystal Growth* **12**, 153.

Schroeder, J. B. and Linares, R. C. (1966) *Prog. Ceram. Sci.* **4**, 195.

Schulmann, J. H. and Compton, W. D. (1962) "Colour Centers in Solids" Pergamon Press, Oxford.

Schulz, L. G. (1954) *J. Metals* **200**, 1082.

Schwander, H. (1953) *Geochim. Cosmochim. Acta* **4**, 261.

Schwarzenbach, G. (1957) "Die Komplexometrische Titration" Ferdinand Enke, Stuttgart.

Scott, B. A. and Burns, G. (1972) *J. Amer. Ceram. Soc.* **55**, 225.

Seel, F. (1970) "Grundlagen der analytischen Chemie" 5th edition. Verlag Chemie, Weinheim/Bergstr.

Seeman, H. (1919) *Ann. Physik* **59**, 455.

Segmüller, A. (1970) *Advances in X-Ray Analysis* **13**, 455.

Shakhashiri, B. Z. and Gordon, G. (1966) *Talanta* **13**, 142.

Sharpe, R. S. (1970) "Research Techniques in Nondestructive Testing" Academic Press, London–New York.

Sherman, B. and Black, J. F. (1970) *Appl. Optics* **9**, 802.

Shick, L. K. (1971) *J. Electrochem. Soc.* **118**, 179.

Shih, K. K. and Blum, J. M. (1971) *J. Electrochem. Soc.* **118**, 1631.

Siegbahn, K. (1965) "Alpha-, Beta-, and Gamma-Ray Spectroscopy" North-Holland, Amsterdam.

Siegbahn, K. *et al.* (1967) "ESCA, Atomic, Molecular and Solid State Structure studied by Means of Electron Spectroscopy". *Nova Acta Regiae Societatis Scientarum Upsaliensis* Ser. IV, **20**, 1–282 (Uppsala, Almquist and Wiksells Boktryckeri AB).

Simmons, J. A., DeWit, R. and Bullough, R. (1970) "Fundamental Aspects of Dislocation Theory", National Bureau of Standards, Washington D.C. (Spec. Publ. No. 317).

Slavin, M. (1971) "Emission Spectrochemical Analysis" Wiley-Interscience, New York.

Slavin, W. (1968) "Atomic Absorption Spectroscopy" Interscience, New York.

Smith, D. P. (1971) *Surface Science* **25**, 171.

Smith, S. H. and Elwell, D. (1968) *J. Crystal Growth* **3/4**, 471.

Snetsinger, K. G. and Keil, K. (1967) *Am. Min.* **52**, 1842.

Sobon, L. E., Wickersheim, K. A., Robinson, J. C. and Mitchell, M. J. (1967) *J. Appl. Phys.* **38**, 1021.

Socha, A. J. (1971) *Surface Science* **25**, 147.

Somorjai, G. A. (1969) (editor) "The Structure and Chemistry of Solid Surfaces" Wiley, New York.

Stacy, K. A. (1956) "Light Scattering in Physical Chemistry" Academic Press, New York–London.

Stacy, W. T. and Tolksdorf, W. (1972) AIF Conf. Proceed. No. 5, *Magn. and Magnet. Mat.* (1972), 185.

Stephens, D. L. and Alford, W. J. (1964) *J. Am. Ceram. Soc.* **47**, 81.

Stock, J. T. (1965) "Amperometric Titrations" Wiley-Interscience, New York.

Stout, G. H. and Jensen, L. H. (1968) "X-ray Structure Determination" Macmillan, New York.

Sunagawa, I. (1967) *J. Crystal Growth* **1**, 102.

Sunshine, R. A. and Goldsmith, N. (1972) *RCA Review* **33**, 383.

Taylor, C. A. and Lipson, H. (1964) "Optical Transforms" Bell, London.

Taylor, D. (1964) "Neutron Irradiation and Activation Analysis" Van Nostrand, Princeton N.J.

Taylor, J. K., Maienthal, E. J. and Marinenko, G., (1965) "Electrochemical Methods in Trace Analysis—Physical Methods" (G. H. Morrison, ed.) Interscience, New York.

Thilo, E., Jander, J. and Seemann, H. (1955) *Z. anorg. allg. Chemie* **279**, 16.

Thomas, G. (1962) "Transmission Electron Microscopy of Metals" Wiley, New York.

Thomas, G. *et al.* (1972) "Electron Microscopy and Structure of Materials" Univ. of Calif. Press.

Thompson, G., Paine, K. and Manheim, F. (1969) *Appl. Spectroscopy* **23**, 264.

Thompson, J. G., Vacher, H. C. and Bright, H. A. (1937) *J. Research NBS* **18**, 259.

Thornton, P. R. (1968) "Scanning Electron Microscopy" Chapman and Hall, London.

Thurmond, C. D. (1959) In "Semiconductors" (N. B. Hannay, ed.) p. 145. Reinhold, New York.

Tiller, W. A. (1963) *J. Appl. Phys.* **34**, 2757.

Timofeeva, V. A., Lukyanova, N. I., Guseva, I. N. and Lider, V. V. (1969a) *Soviet Phys. Crystallog.* **13**, 747.

Timofeeva, V. A., Guseva, I. N. and Melankholin, N. M. (1969b) *Growth of Crystals* **7**, 247.

Tölg, G. (1970) "Ultramicro Elemental Analysis" Wiley, New York.

Tolansky, S. (1943) *Nature* **152**, 722.

Tolansky, S. (1970) "Multiple-Beam Interference Microscopy of Metals" Academic Press, London–New York.

Vainshtain, B. K. (1960) Application of Electron Diffraction to the Study of the Chemical Bond in Crystals, *Quart. Rev.* **14**, 105.

Vainshtain, B. K. (1964) "Structure Analysis by Electron Diffraction". Pergamon, Oxford.

Van Bueren, H. G. (1960) "Imperfections in Crystals" North-Holland, Amsterdam.

Van de Halst, H. C. (1957) "Light Scattering by Small Particles" Wiley, New York.

Vergnoux, A. M., Riera, M., Ribet, J. L. and Ribet, M. (1971) *J. Crystal Growth* **10**, 202.

Vogel, A. I. (1966) "A Textbook of Quantitative Inorganic Analysis" 3rd edition. Longmans, London.

Wadsley, A. D. (1964) "Non-Stoichiometric Compounds" Academic Press, New York–London.

Wagner, J. S. (1967) *J. Appl. Phys.* **38**, 1554.

Wainerdi, R. E. and Uken, E. A. (1971) "Modern Methods of Geochemical Analysis". Plenum, New York.

Waku, S. (1962) *J. Phys. Soc. Japan* **17**, 1068.

Wallace, C. A. and White, E. A. D. (1967) *J. Phys. Chem. Sol. Supp.* **1**, 431.

Wang, C. C. and McFarlane, S. H. (1968) *J. Crystal Growth* **3/4**, 485.

Wang, C. C. and Zanzucchi, P. J. (1971) *J. Electrochem. Soc.* **118**, 586.

Wanklyn, B. M. (1975) In "Crystal Growth" (B. R. Pamplin, ed.) Pergamon, Oxford.

Warren, B. E. (1969) "X-ray Diffraction" Addison-Wesley, Reading, Mass.

Weber, L. (1972) *Schweizer. Min. Petrogr. Mitt.* **52**, 349.

Wegmann, L. (1969) "Photoemissions—Elektronenmikroskopie" in "Handbuch der zerstörungsfreien Materialprüfung" (E. A. W. Müller, ed.) Oldenbourg, München.

Wegmann, L. (1970) *Mikroskopie* **26**, 99.

Wegmann, L. (1972) *J. Microscopy* **96**, 1.

Wegmann, L. and Dannöhl, H. D. (1971) *Beitr. el. mikrosk. Direktabb. Oberfl.* **4/2**, 221.

Wernick, J. H. (1957) *Trans AIME* **209**, 1169.

Wertheim, G. K., Hausmann, A. and Sander, W. (1971) "The Electronic Structure of Point Defects" North-Holland, Amsterdam.

Whiffin, P. A. C. (1973) Mullard Research Report No. 2848.

White, E. A. D. (1971) private communication.

White, E. A. D. and Brightwell, J. W. (1965) *Chem. and Ind.* 1662.

Wickersheim, K. A., Lefever, R. A. and Hanking, B. M. (1960) *J. Chem. Phys.* **32**, 271.

Wiedemann, H. G. (1964) *Chem.-Ing.-Techn.* **36**, 1105.

Wilcox, R. E. (1959) *Am. Min.* **44**, 1272.

Wilcox, W. D. and Fulmer, L. D. (1965) *J. Appl. Phys.* **36**, 2201.

Wilcox, W. R. (1968) *Ind. Eng. Chem.* **60**, 13.

Wilcox, W. R. and Chase, A. B. (1967) In "Crystal Growth" (H. S. Peiser, ed.) p. 617. Pergamon, Oxford.

Willoughby, A. F. W., Driscoll, C. M. H. and Bellamy, B. A. (1971) *J. Mat. Sci.* **6**, 1389.

Wilson, C. L. and Wilson, D. W. (1959–1971) "Comprehensive Analytical Chemistry" (Series) Elsevier, Amsterdam–New York.

Winchell, A. N. (1937) "Elements of Optical Mineralogy" Wiley, New York.

Winterbottom, A. B. and McLean, D. (1960) In "The Physical Examination of Metals" (B. Chalmers and A. G. Quarrell, eds.) 2nd edition, p. 1. Arnold, London.

Witt, A. F. (1967) *J. Electrochem. Soc.* **114**, 298.

Witt, A. F. and Gatos, H. C. (1968) *J. Electrochem. Soc.* **115**, 70.

Wolf, W. P. (1970) *J. Appl. Phys.* **41**, 1281.

Wolfe, C. M., Stillman, G. E. and Owens, E. B. (1970) *J. Electrochem. Soc.* **117**, 129.

Wolff, G. A. and Das, B. N. (1966) *J. Electrochem. Soc.* **113**, 299.

Wood, D. L. and Remeika, J. P. (1966) *J. Appl. Phys.* **37**, 1232.

Wood, W. A. (1963) "Crystal Orientation Manual" Columbia Univ. Press.

Wyckoff, R. W. G. (1963–1968) "Crystal Structures" 2nd edition, Vols. 1–4. Interscience, New York.

Zernicke, F. (1938) *Physica* **5**, 785.

Ziegler, J. F. and Baglin, J. E. E. (1971) *J. Appl. Phys.* **42**, 2031.

Ziegler, J. F., Cole, G. W. and Baglin, J. E. E. (1972) *J. Appl. Phys.* **43**, 3809.

10. Crystals Grown from High-Temperature Solutions

In this chapter a list is given of the crystals which have been prepared from high-temperature solutions during the 20th century. This tabulation is intended to provide the crystal grower with an extensive list of references to previous work, which should minimize duplication and permit a choice to be made of a suitable solvent in those cases where successful growth experiments have been reported. The table will also indicate to "crystal consumers" a possible source of crystals.

Previous compilations of crystal syntheses have been made by Laurent (1969), by Wanklyn (1974) and by Wilke (1973), and this chapter includes these earlier tabulations except for references which could not be traced by the authors.

In the table the materials crystallized are listed in the order of the element which is conventionally used as the first symbol in the formula for the compound. For example, zinc ferrite is normally written as $ZnFe_2O_4$ rather than Fe_2ZnO_4 and we have adhered to this normal practice. Solid solutions are listed under the first element alphabetically except where the concentration of this element is very low. Where elements are added in very low concentrations, these are denoted in brackets after the host material, so that $Al_2O_3(Cr)$ denotes chromium-doped alumina crystals; some materials are listed alphabetically by the first authors' names. The second column lists the solvent used in each case and in general no attempt is made to specify an exact composition in those cases where binary or ternary solvent systems have been used. The table is not intended to provide sufficient information for a crystal growth experiment to be attempted but is rather a guide to the literature and reference should be made to the original papers for details of the compositions employed.

In those cases where an extensive investigation was made of several solvent systems for growth of a given solute, the optimum system has been selected according to the original author's findings. As an additional restriction on the length of the table, an exhaustive list was not given of the many experiments in which, for example, GaAs or GaP films were grown under rather similar conditions from Ga solution for device applications or detailed measurements of parameters. Chapter 8 contains many references to this topic and to experiments on the growth of magnetic garnet films, in addition to the references to previous reviews.

Column 3 of the table lists the technique used in the experiment, and the abbreviations used are explained at the head of the table. In column 4 is listed the temperature or temperature range employed, while column 5 gives the value of the parameter which controls the growth rate. In a slow cooling experiment, the parameter listed is the cooling rate in degree/hour, while for an evaporation experiment it is the duration in hours of the experiment which is given since values of the solvent evaporation rate are not normally quoted. The temperature difference across the melt is quoted in the case of gradient transport experiments, and the current is quoted for electrolytic growth.

The size in mm of the largest crystals grown is listed in column 6, or alternatively the largest dimension where only one is given. In some examples of growth under steady state conditions, the growth rate is quoted in mm/day. Some workers prefer to state the crystal weight in g rather than the linear dimensions, and this value has been included in such cases. The reference to the source publication is given in the final column.

Ommisions from the Table and additions of recent data are tabulated in the Appendix. Although an attempt has been made to make the Table and the Appendix comprehensive, some omissions will have occurred. The authors would be grateful for notification of such omissions, or of improvements realized in more recent work, for inclusion in future editions.

Abbreviations used:

ACRT	Accelerated crucible rotation technique		THM	Travelling heater method	
CR	Chemical reaction		TR	(Gradient) transport	
EL	Electrolysis		TSM	Travelling solvent method	
EV	Evaporation		TSSG	Top-seeded solution growth	
HPS	High-pressure solution growth		VLS	Vapour-liquid-solid	
LPE	Liquid phase epitaxy		VLRS	Vapour-liquid-reaction-solid	
SC	Slow cooling		X	Seed crystal	

Crystal	Solvent	Technique	Temp. range	Cooling rate duration	Crystal size (mm)	Reference
$AgNiF_3$	AgF	SC	700–435	0.8	3	Gluck et al. (1974)
Al	$KNO_3/NaNO_3$	—	600°	—	$150 \times 4 \times 0.5$	Makarevich (1965)
AlAs	Al	SC	950–	1.3	small	Faust and John (1964)
$Al_xGa_{1-x}As$	Ga	LPE/SC	800–775	120	10 µm	Sugiyama and Kawakimi (1971)
AlB_2	Al	CR	975	348 h	$14 \times 10 \times 0.01$	Sirtl and Woerner (1972)
Al_3Er	Al	SC	900–640	4; 85	~ 2	Meyer (1970)
AlF_3	$PbCl_2/PbF_2$	SC	930–650	2	13	Wanklyn (1969)
AlN	Ca_3N_2	SC	1610–1520	3	1×0.3	Dugger (1974)
$Al_2O_3(Cr)$	PbF_2	SC/EV	1400–1000		15	Adams et al. (1966)
Al_2O_3	Na_3AlF_6	SC	1040–960	1.5	$12 \times 12 \times 5$	Arlet et al. (1967)
Al_2O_3	PbO/B_2O_3, PbF_2/Bi_2O_3	SC	$\sim$ 1200–800	1.5	$15 \times 10 \times 0.2$	Barks and Roy (1967)
Al_2O_3	PbF_2	SC	1320–950	1–2	$20 \times 20 \times 3$	Bibr et al. (1966)
Al_2O_3	PbF_2	SC	1300–1100	1–5	30	Butcher and White (1965)
Al_2O_3	$Bi_2O_3/PbF_2/La_2O_3$	SC	1250–1000	2–8	~ 10	Chase (1966)
Al_2O_3	$PbF_2/B_2O_3/La_2O_3$	SC	1250–1000	2–8		Chase (1966)
Al_2O_3	PbF_2/Bi_2O_3	SC	1250–1000	2–10	$6 \times 6 \times 4$	Chase and Osmer (1967)
Al_2O_3	Bi_2O_3/V_2O_5	SC	1300–900	5	$5 \times 5 \times 5$	Garton et al. (1972b)
Al_2O_3	PbF_2	SC	1250–?	1–2	3 g	Giess (1964)
Al_2O_3	PbF_2/B_2O_3	EV	—			Grodkiewicz and Nitti (1966)
$Al_2O_3(Ga)$	PbF_2	SC	1200–900	3	10	Hart and White, see White (1966)
Al_2O_3	Fluorides	SC	1200–700		$6 \times 6 \times 1$	Izvekov et al. (1968)
Al_2O_3	Na_2O/TiO_2	CR/EV	1550	4 h	$\sim 4 \times 4 \times 1$	Jones (1971)
Al_2O_3	$Pb_5O_4F_6$	SC	1260–1000	3	$10 \times 10 \times 5$	Linares (1962c)
Al_2O_3	PbO/B_2O_3	SC	1250–950	0.25–5	$\Big\}$ 1.3/day	Linares (1965)
Al_2O_3	PbO/PbF_2	SC	1250–950	0.25–5		

Al₂O₃(Cr)	PbO/B₂O₃	SC	1300–600	4	~7×7×4	Nelson and Remeika (1964)
Al₂O₃	PbO/B₂O₃	SC	1300–915	2	30 (plate)	Remeika (1963)
Al₂O₃	PbF₂	SC/EV	1250–1300	2–3 mm/h	5×5×5	Timofeeva (1968)
Al₂O₃(Fe, Cr)	PbF₂	EV	1100–1300	20 h	4×5×0.2	Tsushima (1966)
Al₂O₃	Na₂O/WO₃	TR(X)	~1200	$\Delta T=50°$	6–8	Voronkova et al. (1968a)
Al₂O₃	SrO/WO₃	SC	1500–1100	16	1.5–2	Voronkova et al. (1968a)
Al₂O₃	PbF₂	SC	1300–1100	1–5	30	Wallace and White (1967)
Al₂O₃	PbF₂	—	—		30 (plate)	White (1961)
Al₂O₃(Cr)	PbF₂	SC	1400–900	1	30 (plate)	White and Brightwell (1965)
Al₂O₃(Cr)	PbF₂	TR/SC	~1100	0.5	40×40×12	White and Brightwell (1965)
AlP	Al	SC	1350		3×3×0.5	Sonomura and Miyaauchi (1969)
AlSb	Al	SC	960–	0.7	small	Faust and John (1964)
AlSb	Al	SC			small	Wolff et al. (1954)
Al₂(WO₄)₃	Na₂O/WO₃	SC	1150–700	5–20	1–2	Voronkova et al. (1968b)
As	Tl	THM	457	0.5 mm/day	20×9	Tester et al. (1971)
B	Pt	SC	1200–830	—	—	Horn (1959)
BN	Li₃N	HPS/SC	1750–1610	55 kb	small	DeVries and Fleischer (1972)
BN	Mg	HPS	1500–1800	66–75 kb	~0.5	Matecha and Kvapil (1970)
BN	Mg₃N₂, Ca₃N₂	HPS/SC	1750–1300	55 kb	small	Wentorf (1960)
BN	Li₃N	HPS/SC	1750–1600	49 kb	2	Wentorf (1965)
B₆P	Ni	VLRS/TSSG	1300	1200 psi	30	Burmeister and Green (1967)
BP	BPO₄	HPS	850	1 month	small	Ananthanarayanan et al. (1973)
BP	Cu₃P	TR	1100–450	$\Delta T=5$	4×3×2	Baranov et al. (1967)
BP	Ni/P	TR	1200	~15 kb	4	Chu et al. (1973)
BP	Ni	HPS	~1250–400		5×2×2	Iwami et al. (1971)
BP	P	HPS	1000–1500		small	Niemyski et al. (1967)
BP	FePB₂	SC	—			Rundquist (1962)
BaAl₁₂O₁₉(Pb)		SC	—		3×1×1	Stone and Hill (1959)
BaAl₁₂O₁₉(Fe)	Pb₂OF₂	SC	1200–1000	3	3×3×3	Linares (1962c)
BaB₂O₄	PbF₂	EV	1260	100 h	2×2×0.1	Tsushima (1966)
BaCl₂·BaF₂	BaCl₂	CR	1000, 1200	2 h	small	Brixner and Babcock (1968)
Ba₂Co_xZn₂₋_xFe₁₂O₂₂	NaCl	SC	950	50	3×2×0.1	Patel and Singh (1969)
BaCrO₄	NaFeO₂	SC	1350–900	2–5	small	Agapora and Perekalina (1968)
BaCrO₄	BaCl₂	CR	1000, 1200	2 h	3	Brixner and Babcock (1968)
BaCrO₄	LiCl	SC	800–	0.7		Packter and Roy (1971)

Crystal	Solvent	Technique	Temp. range	Cooling rate duration	Crystal size (mm)	Reference
BaF_2	LiF	TSSG/SC	880–860	—	~1 cm	Neuhaus et al. (1967)
$BaFe_{12}O_{19}$	$BaO/PbO/B_2O_3$	SC/ACRT	1200–1300	1.6–10	15 × 2.5	Aidelberg et al. (1974)
$BaFe_{12}O_{19}$	$Fe_2O_3 + BaF_2$	CR	1250	50	1	Brixner (1959)
$BaFe_{12}O_{19}$	Na_2CO_3	SC	1200–900	4.5	5.6 g	Gambino and Leonhard (1961)
$BaFe_{12}O_{19}$	Na_2CO_3	SC	1350–900	20	20 mg	Mones and Barks (1958)
$BaFe_6O_{10}$	$BaCl_2$	SC	1250–		0.015	Hamilton (1963)
$BaFe_{12-x}Sc_xO_{19}$	$NaFeO_2$	SC	1250–940	1–3	—	Perekalina and Cheparin (1968)
$BaFe_{12-2x}Ir_xZn_xO_{19}$	Bi_2O_3	SC	1300–1000	1.5	6 × 6 × 1	Tauber et al. (1963)
$BaFe_{12-2x}Zn_xTi_xO_{19}$	$BaO/Na_2O/B_2O_3$	SC(X)	—			Dixon et al. (1971)
$BaFeO_{3-x}$	$Ba(OH)_2$	HPS	600	1 kb		Takeda et al. (1974)
$BaGe_4O_9$	GeO_2	TSSG/SC	1369	0.5	20.1	Belruss et al. (1971)
$Ba_2MgGe_2O_7$	BaO/GeO_2	TSSG/SC	1345–	1	22 g	Belruss et al. (1971)
$Ba_3MgTa_2O_9$	BaF_2/MgF_2	CR/SC	1315–970	3.5	72 g	Weaver and Li (1969)
$Ba_5(MnO_4)_3Cl$	$BaCl_2$	CR	1000, 1200	2 h	1–4	Brixner and Babcock (1968)
$BaMoO_4$	Li_2MoO_4	TSSG/TR	1040–1140	0.5 mm/h	small	Chen (1973)
$BaMoO_4$	$LiCl, BaCl_2$	SC	1000–	0.7	60 × 15	Packter and Roy (1971)
$BaNb_2O_6$	BaB_2O_4	SC	1450–1350	2	4	Galasso et al. (1968)
$BaNb_2O_6, Ba_5Nb_4O_{15}$	$BaCl_2$	CR	1000, 1200	2 h	small	Brixner and Babcock (1968)
BaO	Ba	EV	900	60–80 h	small	Libowitz (1953)
BaO	$Ba(OH)$	CR/EV	510	30 days	2	Lynch and Lander (1959)
$Ba_5(PO_4)_3Cl$	$BaCl_2$	CR	1000, 1200	2 h	small	Brixner and Babcock (1968)
$BaPbO_3$	$BaCl_2$	CR	1000–	0.7	small	Brixner and Babcock (1968)
$BaSO_4$	$LiCl, BaCl_2$	SC	900–700	10	1.4	Packter and Roy (1971)
$BaSO_4$	$NaCl$	SC/EV	1000–700	65	6 × 3	Patel and Bhat (1971)
$BaSO_4$	$NaCl$	SC	1000–700	5	~1	Patel and Koshy (1968)
$BaSO_4$	$LiCl, NaCl$	SC	1000–800	5	~8	Wilke (1962)
$BaSb_2O_6$	$BaCl_2$	CR	1000, 1200	2 h	small	Brixner and Babcock (1968)
$BaSi_2O_5$	$BaCl_2$	CR	1000, 1200	2 h	small	Brixner and Babcock (1968)
$BaSnB_2O_6$	BaB_2O_4	SC	1300–	5–50	small	Schultze et al. (1971)
$BaTa_2O_6$	BaB_2O_4	SC	1500–1100	1–2	few mm	Layden (1967)
$Ba_5Ta_4O_{15}$	PbO				0.1	Shannon and Katz (1970)
Ba_2TiO_4	$BaCl_2/BaCO_3$	SC	1340–	fast	—	Bland (1961)
$BaTi_3O_7$	$BaCl_2$	CR	1000, 1200	2 h	small	Brixner and Babcock (1968)
$BaTiO_3$	$BaCl_2$	EV	1200–1300		4 × 4 × 0.5	Arend (1960)

BaTiO$_3$	TiO$_2$	…C	…	…	10-2	…eev et al. (19…)
BaTiO$_3$	Na$_2$CO$_3$/K$_2$CO$_3$	TR/CR	800-900	3 days	—	Belyaev (1962)
BaTiO$_3$	BaCl$_2$	SC	1200-750	—	—	Belyaev et al. (1951)
BaTiO$_3$	BaCl$_2$	SC/EV	—	—	8	Benes et al. (1955)
(Ba, Sr)TiO$_3$	TiO$_2$	TSSG/SC	1500-1340	0.5	2	Bethe and Welz (1971)
BaTiO$_3$	BaCl$_2$	SC	—	—	—	Blattner et al. (1947)
Ba$_4$Ti$_{12}$PtO$_{10}$	BaCl$_2$	SC	1220-	14 days	~3 × 3 × 3	Blattner et al. (1949)
BaTi$_{0.75}$Pt$_{0.25}$O$_3$	Na$_2$CO$_3$/K$_2$CO$_3$	SC	1110-20	5 days	4 × 2	Blattner et al. (1949)
Ba$_{1-x}$Pb$_x$TiO$_3$	KF	SC	940-	14 days	3 × 3	Blattner et al. (1949)
BaTi$_{1-x}$Sn$_x$O$_3$	KF	SC	1200-400	—	5	Bogdanov et al. (1960)
BaTiO$_3$	KF	SC	1200-400	—	—	Bogdanov et al. (1960)
(Ba, Pb)TiO$_3$	KF	SC	1120-975	12	twins	Bradt and Ansell (1967)
BaTiO$_3$(Ni)	KF	SC	1100-1000-8000	20-50	5 × 0.5	Cherepanov (1962)
BaTiO$_3$	KF	SC	1090-	20	0.03-0.2	Coufova and Novak (1969)
BaTiO$_3$	KF	EV	1000	—	film	DeVries (1962)
BaTiO$_3$	KF	SC	950-	10	5 × 5 × 5	Eustache (1957)
BaTiO$_3$	BaCl$_2$	SC	1300-1030	10-15	12 × 0.1	Feltz and Langbein (1971)
BaTiO$_3$	KF	SC	1150-850	—	~6 × 6 × 0.5	Gavrilyachenko et al. (1968)
BaTiO$_3$	BaCl$_2$	EV	1400-1480	—	—	Gliki et al. (1956)
BaTiO$_3$	BaB$_2$O$_4$	SC	1150-950	3	2	Goto and Cross (1969)
Ba$_{1-x}$Sr$_x$TiO$_3$	Na$_2$CO$_3$	SC	1200-900	20	10 × 10 × 0.3	Kawabe and Sawada (1957)
BaSn$_x$Ti$_{1-x}$O$_3$	KF	EV/SC	~1100-850	—	1.5	Khodakov et al. (1956, 1958)
BaTiO$_3$(Ni)	KF	SC	—	11-15	12 (plate)	Khodakov and Sholokhovich (1960)
BaTiO$_3$	NaF	SC	1160-→	30	5 × 5 × 0.1	Kudzin (1962)
BaTiO$_3$	BaCl$_2$	SC	—	—	—	Kudzin et al. (1964)
BaTiO$_3$	KF	SC	1200-800	25	—	Matthias (1948)
(Ba, Pb)TiO$_3$	BaB$_2$O$_4$/PbB$_2$O$_4$	TSSG/SC	1125-1000	0.2	—	Nielsen et al. (1962)
(Ba, Pb)TiO$_3$	KF	SC	1200-950	13	~15 × 8	Perry (1967)
BaTiO$_3$	KF	SC	1050-900	—	twins	Perry et al. (1967)
BaTiO$_3$(Fe)	KF	SC	1200-900	—	—	Remeika (1953)
BaTiO$_3$(Fe)	KF	SC	1200-850	20	—	Remeika (1954)
BaTiO$_3$(Bi)	KF	SC	1190-855	15	34 × 24 × 0.4	Remeika (1958a)
BaTiO$_3$	KF	SC	1100-900	60	15 × 15 × 0.2	Remeika et al. (1966)
BaTiO$_3$	KF	SC	1200-830	30-50	—	Sasaki (1965)
BaTiO$_3$	KF+BaO	SC	1200	50	6	Sasaki (1965)
BaTiO$_3$	KF	SC	1150-900	5	—	Sasaki (1968)
BaTiO$_3$	KF	SC	1180-850	—	—	Sasaki and Kurokawa (1965)
Ba(Ti, Sn)O$_3$	KF	SC	1200-900	11	11 × 0.4	Sholokhovich and Khodakov (1962)

Crystal	Solvent	Technique	Temp. range	Cooling rate duration	Crystal size (mm)	Reference
$Ba_{1-x}Pb_xTiO_3$	KF	SC	1000→	—	—	Sholokhovich et al. (1956)
$BaTiO_3$	KF	SC	1200-800	14-18	16 × 3	Sholokhovich et al. (1968)
$BaTiO_3$	$BaCl_2$	SC	1400	20, 4	2	Timofeeva (1959)
BaB_2O_6	BaB_2O_4	SC	1040-	5-50	small	Schultze et al. (1971)
$BaTiB_2O_6$	$BaCl_2$	CR	1000, 1200	2 h	small	Brixner and Babcock (1968)
$Ba_5(VO_4)_3Cl$	$BaCl_2$	CR	1000, 1200	2 h	small	Brixner and Babcock (1968)
$BaWO_4$	LiCl	SC	700-800	—	small	Packter and Roy (1973)
$BaWO_4$	$LiCl, BaCl_2$	SC	1000-	0.7	4	Packter and Roy (1971)
$BaWO_4$	NaCl	SC	1000-750	2	3	Patel and Arora (1973)
$BaWO_4$	$Na_2W_2O_7$	SC	1100-1250→700	2.5	10 × 5 × 5	Van Uitert and Soden (1960)
$Ba_w(Zn, Mn, Ni)_xFe_yO_z$	$BaB_2O_4/NaFeO_2$	SC	1375-950	0.5-4	3 × 0.5	Savage and Tauber (1967)
$BaY_2Mg_2Ge_3O_{12}$	BaO/GeO_2	TSSG/SC	1570-	1	48 g	Belruss et al. (1971)
$Ba_2Zn_2Al_xFe_{12-x}O_{22}$	BaO/B_2O_3	SC	1350-800	2-3	5 × 3 × 1	Agapora et al. (1969)
$Ba_2Zn_2Fe_{12}O_{22}$	$Ba_2B_2O_5$	TSSG/SC	1200	0.1-0.2	10 g	Aucoin et al. (1966)
$Ba_2Zn_2Fe_{12}O_{22}$	BaO/B_2O_3	SC/TR	1250→	0.25	15	Gendelev and Zvereva (1971)
$Ba_4Zn_2Fe_{36}O_{60}$	$NaFeO_2$	SC	~1300→	~4	—	Kerecman et al. (1968)
$Ba_4Zn_2Fe_{36}O_{60}$	BaO/B_2O_3	SC	1300-1000	0.5	—	Kerecman et al. (1969)
$Ba_2Zn_2Fe_{12}O_{22}$	$NaFeO_2$	SC	1250(-1375)→	0.75-4	12 × 12 × 6	Savage and Tauber (1964)
$Ba_2Zn_2Fe_{12}O_{22}(Mn)$	$Ba_2B_2O_5$	SC	1250-1050	1	6 × 6 × 3	Savage et al. (1965)
$Ba_2Zn_2Fe_{12}O_{22}$	$Ba_2B_2O_5$	SC/TSSG	1300-900	0.5-5	—	Shinoyama and Suemune (1970)
$Ba_4Zn_2Fe_{12}O_{22}$	$Na_4Fe_2O_4$	SC	1250-1050	0.75-4	12 × 12 × 3	Tauber et al. (1963)
$Ba_2Zn_2Fe_{12}O_{22}$	BaO/B_2O_3	SC	1300-1000	0.5-2	6 × 6 × 3	Tauber et al. (1964)
$BaZrB_2O_6$	BaB_2O_4	SC	1300-	5-50	small	Schultze et al. (1971)
$BeAl_2O_4$	$PbF_2/PbO/SiO_2/B_2O_3$	SC	1300-700	0.5	44	Bonner and Van Uitert (1968)
$BeAl_2O_4$	PbO	SC	1375-	12.5	2 × 2 × 2	Farrell and Fang (1964)
$Be_3Al_2Si_6O_{18}$	Li_2O/MoO_3	TR(X)	1100-800	2-5	8	Ashida (1968)
$Be_3Al_2Si_6O_{18}$	V_2O_5	SC	600-1200	—	0.7/day	Ballman et al. (1966)
$Be_3Al_2Si_6O_{18}$	Li_2O/MoO_3	TR(X)	975-790	4	1	Lefever et al. (1962)
$Be_3Al_2Si_6O_{18}$	V_2O_5	TR(X)	1050	$\Delta T = 50$	~10 × 6 × 6	Linares et al. (1962)
$Be_3Al_2Si_6O_{18}$	PbO/V_2O_5	SC	1250-900	1	5 mm³	Linares (1967b)
$Be_3Al_2Si_6O_{18}$	PbO/V_2O_5	TR	1000	$\Delta T = 50$	25 × 25 × 10	Linares (1967b)
BeO	Li_2O/MoO_3	SC	1400-	1.5-25	5	Austerman (1963)
BeO	Li_2O/MoO_3	TR	~1150	$\Delta T = 30$	2	Austerman (1964)
BeO	KOH	—	400-500	—	0.3 × 0.06	Levin et al. (1957)

BeO	Li_2O/MoO_3	EV	1165	5 days	1–5	Osmer and Chase (1972)
Be_2SiO_4	$Na_2Mo_3O_{10}$	TR	900–1000	ΔT=40–80	2 mm/day	Ballman and Laudise (1965)
Be_2SiO_4	Li_2O/MoO_3	EV	1165	5 days	1–5	Osmer and Chase (1972)
$Bi_{3-x}Ca_{2x}Fe_{5-x}V_xO_{12}$	PbO/Bi_2O_3	SC	1200–1000	3–10	5	Espinosa and Geller (1964)
$Bi_{3-x}Ca_{2x}Fe_{5-x}V_xO_{12}$	PbO/Bi_2O_3	SC	1200–1050	2	12	Hodges et al. (1967)
$Bi_{3-x}Ca_{2x}Fe_{5-x}V_xO_{12}$	PbO/B_2O_3	SC	1200–1050	0.5–1	8	Krishman (1969)
$Bi_{3-x}Ca_{2x}Fe_{5-x}V_xO_{12}$	PbO/Bi_2O_3	SC	1200–1000	1	35 × 30, 60 g	Suzuki et al. (1971)
$BiFeO_3$	Bi_2O_3	SC +EV	860(820)–760	1.5	1–2	Teague et al. (1970)
$Bi_2Fe_4O_9$	Bi_2O_3	EV(?)	850	—	0.1–0.3	Koizumi et al. (1964)
$Bi_2Mo_2O_9$	MoO_3	TSSG/TR	670	0.5–0.7 mm/h	30 × 10	Chen (1973)
$Bi_2Mo_3O_{12}$	MoO_3	TSSG/TR	~650	0.25–0.5 mm/h	50 × 20	Chen (1973)
$Bi_{0.5}Na_{0.5}TiO_3$	NaF	SC	1300–850	10–15	3 × 3 × 2	Homma and Wada (1971)
$Bi_2Sn_2O_7$	PbF_2	SC	1280–900	6	2 × 2 × 2	Wanklyn (1972)
$Bi_2Sn_2O_7$	KF	SC	1100–850	6	1 × 1 × 1	Wanklyn (1972)
$Bi_4Ti_3O_{12}$	Bi_2O_3	SC	1100–940	10	sheets 100 μm	Cummins and Cross (1968)
$Bi_4Ti_3O_{12}$	Bi_2O_3	SC	—	—	small	Van Landuyt et al. (1969)
$Bi_4Ti_3O_{12}$	Bi_2O_3	SC	1200–	—	sheets	Van Uitert and Egerton (1961)
C(diamond)	—	—	—	—	small	Bezrukov et al. (1966, 1969)
C(diamond)	Fe, Ni	—	—	—	1	Bundy et al. (1955)
C(diamond)	—	—	—	—		Litvin and Butuzov (1968)
C(diamond)	—	—	—	—		Lundblad (1964)
C(diamond)	Ni	HPS	1500–1800	60–75 kb	0.5	Matecha and Kvapil (1970)
C(diamond)	Fe, Ni	TR	1550–1400	57 kb	5 × 4 × 4	Strong and Chrenko (1971)
C(diamond)	Ni	TR(X)	1450	54 kb	0.5	Strong and Hanneman (1967)
C(diamond)	Fe, Ni	TR/SC	1500–1400	55–60 kb	~6	Strong and Wentorf (1972)
C(graphite)	Fe, Ni	SC	1500–	1–2.5	3	Austerman et al. (1967)
C(graphite)	Fe, Ni	TR	1200–1500	—		Austerman et al. (1967)
C(graphite)	Fe, etc.		review			Austerman (1968)
C(graphite)	Fe, Ni	SC	1500–700	1–2.5	~2	Austerman et al. (1968)
C(graphite)	Fe, Ni	TR	1200–1400	ΔT=25–50	30 × 0.5 × 0.5	Austerman et al. (1968)
C(graphite)	Si/Al/Na	SC	—	—	small	Minkoff (1968)
C(graphite)	Fe	SC	1900–1500	5	30 × 0.06	Sumiyoshi et al. (1968)
$Ca_3Al_{10}O_{18}$	$CaCl_2$	CR	1000, 1200	2 h	small	Brixner and Babcock (1968)
$CaAl_{12}O_{19}(Pb)$	Pb_2OF_2	SC	1260–1000	3	3 × 3 × 3	Linares (1962c)
CaB_6	B_2O_3	VLS	1810	—	0.1	Rea and Kostiner (1971)

Crystal	Solvent	Technique	Temp. range	Cooling rate duration	Crystal size (mm)	Reference
CaB_2O_4	LiCl	SC/HPS	1000–	20 kb, 30	small	Marezio et al. (1969)
$CaCO_3$	Li_2CO_3	TSM/TR	~750	5 mm/day	30 × 10	Belin et al. (1972)
$CaCO_3$	Li_2CO_3	TSM/TR	700–800	5 mm/day	30 × 10	Brissot and Belin (1971)
$CaCO_3$	Li_2CO_3	SC	800–650	0.5–1	10 × 10 × 1	Nester and Schroeder (1967)
$CaCrO_4$	$CaCl_2$	CR	1000, 1200	2 h	small	Brixner and Babcock (1968)
$CaCrO_4$	LiCl	SC	800–	0.7	0.6	Packter and Roy (1971)
$Ca_3Cr_2(SiO_4)_3$	$Na_2O/K_2O/B_2O_3$	EV	1000	72 h	0.2	Lowell et al. (1971)
CaF_2	NaCl/NaF	SC	890–760	15	~1	Leckebusch and Recker (1972)
$CaFe_2O_4$	$CaCl_2$	CR	1000, 1200	2 h	small	Brixner and Babcock (1968)
$Ca_2Fe_2O_5$	PbF_2/PbO	EV	1245	11 days	4 × 3 (plate)	Wanklyn (1972)
$CaGe_2O_5$	GeO_2	TSSG/SC	1208–	0.5	25 g	Belruss et al. (1971)
$CaMgSi_2O_6$	PbO	SC	1350–900	2	2	Grodkiewicz and Van Uitert (1963)
$CaMn_2O_4$	$CaCl_2$	CR	1000, 1200	2 h	5	Brixner and Babcock (1968)
$CaMo_{1-x}W_xO_4$	NaCl	SC	1200–500	15–18	1–4	Kukui et al. (1967)
$CaMoO_4$	LiCl, $CaCl_2$	SC	1000–	0.7	2	Packter and Roy (1971)
$CaMoO_4$	Li_2SO_4	TSM	~835–960	0.2 mm/day	2	Parker and Brower (1967)
$Ca_2NaMg_2V_3O_{12}$	V_2O_5; NaV_3O_8	SC	1220–850	2	2–10	Havlicek et al. (1971)
$CaNb_2O_6$	$CaF_2/MgF_2/Nb_2O_5$	CR/SC	1315–915	3–5	5 × 5 × 3	Weaver and Li (1969)
$Ca_2Nb_2O_7$	$CaCl_2$	CR	1000, 1200	2 h	3	Brixner and Babcock (1968)
CaO	NaCl	CR	950	1–3 h	~1	Sakamoto and Setoguchi (1964)
Ca_2PO_4Cl	$CaCl_2$	CR	~1000	—	small	Brixner and Babcock (1968)
$Ca_5(PO_4)_3Cl$	$CaCl_2$	SC	1280–1060	2–4	3–4	Prener (1967)
$Ca_5(PO_4)_3F$	CaF_2	SC	1375–1220	2–4	~6	Prener (1967)
$CaRuO_3$	$CaCl_2$	SC	1260–800	2	2 × 1	Bouchard and Gillson (1972)
$CaSO_4$	LiCl, $CaCl_2$	SC	1000–	0.7	1	Packter and Roy (1971)
$CaSO_4$	Na_2SO_4	SC	1000–800	5	11 × 7 × 2	Wilke (1962, 1968)
β-$CaSiO_3$	NaCl	SC	1200–800	5	6 × 0.03 × 0.03	Setoguchi and Sakamoto (1967)
Ca_2SiO_4	$CaCl_2$	CR	1000, 1200	2 h	small	Brixner and Babcock (1968)
$CaSnO_3$	$CaCl_2$	EV	800–900	—	—	Smith and Welch (1960)
$CaSnB_2O_6$	CaB_2O_4	SC	1350–	5–50	small	Schultze et al. (1971)
$CaTiO_3$	$BaCl_2/CaCl_2$	SC	1150–	—	1	Kay and Bailey (1957)
$CaWO_4$	LiCl	SC	~700	—	1	Anikin (1958)
$CaWO_4$	Fluorides	SC	1200–700	—	4 × 3 × 0.5	Ivrekov et al. (1968)

Ca₆WO₉	Na₂WO₄/K₂Cr₂O₇	SC	1250–	9	5 × 1 × 1	Nassau and Mills (1962)
Ca₂WO₅ · CaCl₂	LiCl/CaCl₂	SC	~1000–500	—	20 × 5	Barta et al. (1971)
CaY₂Mg₂Ge₃O₁₂	CaO/GeO₂/B₂O₃	TSSG/SC	1491–	0.4	115 g	Belruss et al. (1971)
CaZrB₂O₆	CaB₂O₄	SC	1150–	5–50	small	Schultze et al. (1971)
CdCr₂S₄	CrCl₃	VLRS	900–850	—	3–6	Von Neida and Shick (1969)
CdCr₂Se₄	CdCl₂	SC	—	2–5	0.8	Berger and Pinch (1967)
CdCr₂Se₄	CrCl₃(+Pt)	VLRS	—	—	2	Eastman and Shafer (1967)
CdCr₂Se₄	CdCl₂	SC	—	2–5	1	Harbeke and Pinch (1966)
CdCr₂Se₄	PbCl₂/CdCl₂	SC	—	slow	3	Kuse (1970)
CdCr₂Se₄	CdCl₂	SC	900–500	2.4	3	Larson and Sleight (1968)
CdCr₂S₄₋ₓSeₓ		—	—	—	1	Pickardt et al. (1970)
CdCr₂Se₄	CrCl₂/CrSe	VLS	850–1200	7	4	Shick and Von Neida (1971)
CdCr₂Se₄	CrCl₃(+Pt)	VLRS	875–825	1–4	1–4	Von Neida and Shick (1969)
CdCr₂Se₄	CrCl₃(+Pt)	VLRS	700–	2–3	2–3	Von Philipsborn (1967, 1969)
CdFe₂O₄	PbO	SC	1300–	~5	<10	Remeika (1958b)
CdGeP₂	Sn	SC	–750	1–8 h	—	Spring-Thorpe and Pamplin (1968)
CdIn₂Te₄	In₂Te₃	TR	800–	40	—	Mason and Cook (1961)
Cd₂Nb₂O₇	NaF	SC	1050–700	2	~2	Jona et al. (1955b)
Cd₂Nb₂O₇	CdF₂	SC	1225–700	—	10 × 10 × 1	Kestigian (1963)
Cd₂PbO₄	PbO	HPS/SC	750,400	5–100	50 μm	Keester and White (1970)
CdS	CdCl₂	SC	1000–	8–9	1	Bibr et al. (1966)
CdS	CdCl₂	SC	900–	38	5 × 0.2	Bidnaya et al. (1962)
CdS	Ga, In, Sn, Tl	SC(X)	1180–	83	0.7 × 0.01 × 0.01	Harsy (1967)
CdS	CdCl₂/NaCl	SC	650–400	2.5 mm/day	4 × 4 × 4	Haworth and Lake (1965)
CdS	Cd	SC	~940	—	small	Hemmat and Weinstein (1967)
CdS	CdCl₂	SC	1200–700	—	~4	Izvekov et al. (1968)
CdS	Bi, Cd, Sn	CR/SC/EV	—	~25	15 × 1 × 1	Rubenstein (1968)
CdS	Na₂Sₓ	CR/SC/EV	700–300	~25	4 × 4 × 0.1	Scheel (1972b, 1974)
CdS	Na₂Sₓ	SC	760–300	5	4 × 1.5	Scheel (1972b, 1974)
CdS	CdCl₂	SC/TR	800–600	—	~4	Sysoev et al. (1966)
CdSe	Bi, Cd, Sn	TSM	—	—	20	Rubenstein (1968)
CdSe	Se	SC	1050–1100	20	—	Steininger (1968)
CdSiAs₂	Sn	SC	–750	—	23 × 3.5	Spring-Thorpe and Pamplin (1968)
CdSiP₂	Bi, Sn, Sb	SC	—	—	~10	Borshchevskii et al. (1967)
CdSiP₂	Sn	SC	1200–	—	30 (needles)	Buehler and Wernick (1971)
CdSiP₂	Sn	SC	1200–600	4–16	10	Mughal et al. (1969)
CdSiP₂	Sn	SC	–750	—		Spring-Thorpe and Pamplin (1968)

Crystal	Solvent	Technique	Temp. range	Cooling rate duration	Crystal size (mm)	Reference
$CdSnB_2O_6$	CdB_2O_4	SC	1200–	5–50	small	Schultze et al. (1971)
$CdSnP_2$	Sn	SC	1200–	2–3	~10	Buehler and Wernick (1971)
$CdSnP_2$	Sn	SC	730–	2–3	$8 \times 2 \times 0.2$	Buehler et al. (1971)
$CdSnP_2$	Cd	SC	—	—	small	Loshakova et al. (1966)
$CdSnP_2$	Sn	SC	–750	—	—	Spring-Thorpe and Pamplin (1968)
CdTe	CdJ_2	CR	600	6 h	small	Kwestroo et al. (1969)
CdTe	Cd	Moving Boat	1025–970	~1 h	10	Lorenz (1962)
CdTe	Bi, Cd, Sn	SC/TR			~4	Rubenstein (1968)
(Cd, Hg)Te	Te	THM	~750	0.2–0.3 mm/h	~10	Ueda et al. (1972)
$CdTiO_3$	$NaCl/NaBO_2/Na_2CO_3$	SC	900–	fast	0.1	Kay and Miles (1957)
$CdWO_4$	LiCl	SC	~700	—	1	Anikin (1958)
$CdWO_4$	Na_2SO_4, Na_2O/WO_3	SC	1000–	—	small	Schultze et al. (1967)
$CdWO_4$	$Na_2W_2O_7$	SC	1100–1250–700	2.5	$10 \times 5 \times 5$	Van Uitert and Soden (1960)
$CdZrB_2O_6$	CdB_2O_4	SC	1040–	5–50	small	Schultze et al. (1971)
Ce-compounds see also R-						
$CeAlO_3$	KF	SC	1300–840	4–5	~0.05	Zonn (1965)
CeF_3	SrF_2, BaF_2 etc.	SC	~1000–900	3		Myakishev and Klokman (1966)
CeO_2	PbF_2	SC	1280–850	3	2.5	Baker et al. (1969)
CeO_2	$Li_2W_2O_7$	TR	1070–1100	$\Delta T 25$–30	2	Finch and Clark (1966)
CeO_2	PbF_2/B_2O_3	EV			1–2	Grodkiewicz and Nitti (1966)
CeO_2	$Na_2B_4O_7$	EV				Harari et al. (1967)
CeO_2	PbF_2/B_2O_3	SC	1325–	1	3	Linares (1967c)
CeO_2	Li_2O/MoO_3	TR(X)	970	$\Delta T 100$	$10 \times 4 \times 4$	Linares (1967c)
CeO_2	PbF_2	EV	1200	17 days	$8 \times 4 \times 4$	Wanklyn (1969)
CeO_2	PbF_2/MgF_2	SC/EV	1200–800	1.2–3	3–5	Wanklyn (1969)
CeO_{2-x}	NaCl	CR	1000	50–80	small	Wilke (1964)
CeO_2	PbO/PbF_2	SC	1300–	3–4	$5 \times 4 \times 3$	Zonn and Joffe (1969)
$CoCr_2O_4$	Na_2O/WO_3	SC	1450–1000	3	—	Kunnmann et al. (1965)
$CcCr_2S_4$	$CrCl_3(+Pt)$	VLRS	906–856	—	4	Shick and Von Neida (1969)
CoF_2	KCl	SC	1050–400	4	$20 \times 2 \times 2$	Garrard et al. (1974)
CoF_2	$PbCl_2$	SC	930–650	3	~5	Wanklyn (1969)
$CoFe_2O_4$	PbO—B_2O_3	SC	1320–950	1–2	~8	Bibr et al. (1966)
$CoFe_2O_4$	$NaFeO_2$	SC	1590–1350	6–10	$25 \times 8 \times 8$	Ferretti et al. (1963)

$CoFe_2O_4$	$Na_2B_4O_7$	TR(X)	1200–1300	—	7	Timofeeva and Zalesskii (1959)
$Co_{2.2}Mn_{0.8}O_4$	$NaKSO_4$	CR	~1000	20 h	small	Petzold et al. (1971)
$Co_{1.9}Mn_{1.1}O_4$	$NaKSO_4$	CR	~1000	20 h	small	Petzold et al. (1971)
$CoMn_2O_4$	PbF_2	EV	1150	70 h	$2 \times 2 \times 2$	Tsushima (1966)
$CoNb_2O_6$	Na_2O/WO_3	SC	1260–700	3	$10 \times 1.5 \times 1.5$	Wanklyn (1972)
CoO	$NaCl$	CR	various	various	small	Packter (1968)
CoO	Na_2SO_4, KF	SC/CR	1000	50–80	small	Wilke (1964)
CoS_2	Na_2S_x	SC	700–300	25	2	Scheel (1974)
CoS_2	$PbCl_2$	SC	750–400	3.5	small	Wilke et al. (1967)
$CoSnB_2O_6$	CoB_2O_4	SC	1180–	5–50	small	Schultze et al. (1971)
$Co_{1+x}V_{2-x}O_4$	Na_2O/WO_3	EL	~900	40 mA/cm²	10	Rogers et al. (1966)
$CoWO_4$	Na_2SO_4, Na_2WO_4	SC/CR	1000–	1.5	small	Schultze et al. (1967)
$CoWO_4$	Na_2O/WO_3	SC	1250–600	—	$4 \times 4 \times 4$	Wanklyn (1972)
$CrBO_3$	B_2O_3	HPS	900, 1 h	60–65 kb	small	Chamberland (1967)
CrF_3	$PbCl_2/PbF_2$	SC	930–650	3	2	Wanklyn (1969)
Cr_2O_3	Li_2O/B_2O_3	SC	1300–1050	5–25	~$10 \times 8 \times 0.2$	Barks and Roy (1967)
CrO_2	PbO/PbF_2, Na_2CrO_4, K_2CrO_4, $K_2Cr_2O_7$	HPS	900, 1 h	60–65 kb	small	Chamberland (1967)
CrO_2	CrO_3	HPS/LPE	~425	1.3	2 μm layer	DeVries (1966)
Cr_2O_3	Bi_2O_3/V_2O_5	SC	1345–750	—	5	Garton et al. (1972b)
Cr_2O_3	KOH	HPS	900	36 kb	small	Porta et al. (1972)
Cr_2O_3	$K_2B_4O_7/PbO$	EV	1220	3 weeks	$8 \times 8 \times 0.3$	Scheel (unpublished)
Cr_2O_3	Na_2S_x	SC/CR	1150–400	5	$6 \times 6 \times 0.2$	Scheel (1974)
Cr_2O_3	PbF_2/Bi_2O_3	EV	1300	150 h	$5 \times 5 \times 1$	Tsushima (1967)
Cr_2O_3	$NaCl$	CR/SC	1000	50–80	small	Wilke (1964)
$CsCoCrF_6$	$CsCl/CoCl_2$	—	850–950	3.5	3	Nouet et al. (1971)
$CsKCrF_6$	PbF_2	SC	1055–500	—	2×1	Garton and Wanklyn (1967a)
$CsZnCrF_6$	$CsCl/ZnCl_2$	—	800–	—	3	Nouet et al. (1971)
Cu	Tl	SC	1000–	0.2–10	small	Liaw and Faust (1971, 1972)
$CuCl$	KCl	THM	360	2	35×8	Kvapil and Perner (1971)
$CuCl$	$SrCl_2$, $PbCl_2$	TR	345–390	$\Delta T5$–12	$25 \times 15 \times 7$	Parker and Pinnell (1970)
$CuCl$	KCl, $BaCl_2/SrCl_2$	THM	~330	0.2 mm/h	40×7	Perner (1969)
$CuCl$	KCl	TSSG	~400	0.1–0.3 mm/h	$50 \times 20 \times 20$	Wilcox and Corley (1967)
$CuCl$	KCl	SC	400–	2	$10 \times 10 \times 5$	Wilcox and Corley (1967)
$Cu_2Cr_2O_4$, $CuCr_2O_4$	Cu_2O	HPS	1000, 2 h	60–65 kb	small	Chamberland (1967)
$CuCr_2Se_{4-x}Br_x$	CuBr	SC	—	—	10	Sleight and Jarret (1969)

Crystal	Solvent	Technique	Temp. range	Cooling rate duration	Crystal size (mm)	Reference
$CuFe_2O_4$	PbO/PbF_2	SC	950–500	11–13	2	Miyada et al. (1964)
$CuFe_2O_4$	PbO/B_2O_3	SC	980–500	5–10	3	Petrakovskii et al. (1969)
$CuGa_{1-x}In_xS_2$	In	SC	1150–300	4	$5 \times 5 \times 0.1$	Yamamoto and Miyauchi (1972)
$CuMoO_4$	MoO_3	SC	790–	2–10 mm/h	6	Nassau and Abrahams (1968)
CuO	KF	CR/SC	~1000	50–80	small	Wilke (1964)
CuP_2	Sn	SC	—	20	$1\ cm^2$	Goryunova et al. (1968a)
CuS	Na_2S_x	SC/EV	~600–300	25	$2 \times 2 \times 0.1$	Scheel (1974)
$Cu_5V_2O_{10}$	KVO_3	SC	1000–500	10	—	Shannon and Calvo (1973)
Cu_3VS_4	Na_2S_x	SC/EV	~650–300	~25	1	Scheel (1974)
Dy-compounds see also R-						
$DyAlO_3$	PbO	EV	1260°	10 days	~2	Garton and Wanklyn (1967b)
$DyPO_4$	$Pb_2P_2O_7$	SC	1330–930	0.6–0.9	$15 \times 2 \times 2$	Smith and Wanklyn (1974)
$DyVO_4$	$Pb_2V_2O_7/NaVO_3/Na_2B_4O_7$	SC	1300–1040	2	$10 \times 5 \times 0.5$	Garton and Wanklyn (1969)
$DyVO_4$	PbO/V_2O_5	SC	1330–950	1	$20 \times 1.5 \times 1.5$	Garton et al. (1972b)
$DyVO_4$	$Pb_2V_2O_7$	SC	1330–930	0.6–0.9	$15 \times 2 \times 2$	Smith and Wanklyn (1974)
Er-compounds see also R-						
$Er_3Al_5O_{12}(Tm)$	$PbO/PbF_2/B_2O_3$	SC	1300–950	1	117 g	Van Uitert et al. (1965)
ErB_2	Er	CR	1750	20 min.	$1 \times 1 \times 0.01$	Castellano (1972)
$ErMnO_3$	$PbF_2/PbO/B_2O_3$	SC	1280–900	1	$20\ mm^2$	Wanklyn (1972)
ErOF	$PbF_2/PbO/B_2O_3$	SC	1255–940	1	$20 \times 5 \times 1$	Wanklyn (1972)
Er_2SiO_5	$Li_2Mo_2O_7$	—	1100	—	—	Harris and Finch (1965)
Eu-compounds see also R-						
EuB_6	$B_2O_3(?)$	VLS	~1800	—	small	Rea and Kostiner (1971)
$EuFeO_3$	PbO	SC	1250–1000	6	$5 \times 5 \times 1$	Drofenik et al. (1973)
$Eu_3Fe_5O_{12}$	PbO/B_2O_3	SC	1280–900	—	60 mg	Schieber (1967)
$Eu_3Ga_5O_{12}$	PbF_2/PbO	SC	1240–900	—	1.8 g	Schieber (1964)
$Eu_3Ga_5O_{12}$	PbF_2/PbO	SC	1250–	—	—	Schieber and Holmes (1965)
$Eu_3Ga_5O_{12}$	PbO/B_2O_3; PbO/PbF_2	SC	1280–900	—	0.4 g	Schieber (1967)
EuO	Eu $(+Ta)$	SC	2000–1550	25	~10	Guerci and Shafer (1966)
EuO	Eu	SC	2000–1400	7–100	~10	Shafer et al. (1972)
EuO	Eu	SC	2200–	5–50	60 g	Reed and Fahey (1971)

FeBO$_3$	Bi$_2$O$_3$/MoO$_3$, PbMoO$_4$	SC	1000–750	1–2	5	Chadwick (1972)
FeBO$_3$	B$_2$O$_3$	CR	670–860	—	small	Joubert et al. (1968)
FeBO$_3$	PbO/PbF$_2$/B$_2$O$_3$	SC	1100–700	9	5 × 5 × 2	Lecraw et al. (1969)
FeBO$_3$	Bi$_2$O$_3$/B$_2$O$_3$	SC	850–600	0.5	8	Makram et al. (1972)
(Fe$_{0.9}$Ga$_{0.1}$)BO$_3$	Bi$_2$O$_3$/B$_2$O$_3$	SC	1145–835	1.1	—	Bernal et al. (1963)
FeCr$_2$S$_4$	CrCl$_3$(+Pt)	VLRS	906–865	—	4	Shick and Von Neida (1969)
FeF$_2$	KCl	SC/EV	980–500	5	10 × 1 × 1	Garrard et al. (1974)
FeNbO$_4$	Na$_2$O/WO$_3$	SC	1250–600	1.5	3 × 2 × 2	Wanklyn (1972)
Fe$_2$O$_3$	Li$_2$O/B$_2$O$_3$	SC	1200–1050	2	1.5	Anderson and Schieber (1963)
Fe$_2$O$_3$	Na$_2$B$_4$O$_7$, K$_2$B$_4$O$_7$	SC	1200–880	5–80	10, 300 mg	Barks and Roy (1967)
Fe$_2$O$_3$	Na$_2$B$_4$O$_7$	SC	1100–700	45	6 × 6	Barks et al. (1964)
Fe$_2$O$_3$	Na$_2$B$_4$O$_7$	SC	1260–1040	0.5–15	6 × 6 × 0.4	Chase and Morse (1973)
Fe$_2$O$_3$	Bi$_2$O$_3$	SC	1320–900	0.5–2	20	Curry et al. (1965)
Fe$_2$O$_3$	Bi$_2$O$_3$/B$_2$O$_3$	SC	1200–500	5	—	Flanders and Remeika (1965)
Fe$_2$O$_3$	PbO/B$_2$O$_3$	SC	1300–	10	—	Flanders and Remeika (1965)
Fe$_2$O$_3$	Bi$_2$O$_3$/V$_2$O$_5$	SC	1300–900	5	7	Garton et al. (1972b)
Fe$_2$O$_3$	PbO/V$_2$O$_5$	SC	1345–750	1.3	30	Garton et al. (1972b)
Fe$_2$O$_3$	Na$_2$B$_4$O$_7$, Bi$_2$O$_3$/B$_2$O$_3$	SC	1250–800	1	7 × 3, 3 g	Nielsen (1969)
Fe$_2$O$_3$	Bi$_2$O$_3$/V$_2$O$_5$/Li$_2$B$_4$O$_7$	SC	1220–1000	2	6 × 5 × 2	Scheel (unpublished)
Fe$_2$O$_3$	Li$_2$O/B$_2$O$_3$	SC	1150–800	—	5 mg	Schieber (1964)
Fe$_2$O$_3$	PbO/Bi$_2$O$_3$/B$_2$O$_3$	SC	1300–800	2–3	15 × 15 × 3	Schieber (1966)
Fe$_2$O$_3$	Na$_2$B$_4$O$_7$/CuO	SC	1250–800	1.5	80 mg	Tasaki and Iida (1963)
Fe$_2$O$_3$	Na$_2$B$_4$O$_7$	SC	1300–	10	3 × 3 × 0.5	Vichr (1966)
Fe$_2$O$_3$	Pb$_2$V$_2$O$_7$	SC	—	—	25	Wanklyn (1970)
Fe$_2$O$_3$	Pb$_2$P$_2$O$_7$/MgO	CR/SC	1310–900	4.3	5	Wickham (1962)
Fe$_2$O$_3$	Na$_2$SO$_4$, NaCl, LiCl, KF	EV?	~1000	50–80	~1	Wilke (1964)
Fe$_2$O$_3$	Na$_2$SO$_4$/NaCl	SC/HPS	1200	—	4	Wilke (1968)
Fe$_3$O$_4$	Na$_2$B$_4$O$_7$	SC	1300–1000	2	2	Vichr and Makram (1969)
FeS	Fe	SC	~1150–	15–20	20 × 10	Takahashi et al. (1973)
FeS$_2$	Na$_2$S$_x$	SC/EV	500–300	25	3 × 3 × 2	Scheel (1974)
FeS$_2$	FeCl$_3$, FeBr$_3$	CR/TR	750	—	4	Wilke (1968)
FeS$_2$	PbCl$_2$	SC	750–400	3.5	~3	Wilke et al. (1967)
Fe$_2$TiO$_4$	BaO/B$_2$O$_3$	SC	1200–900	3	4	Hauptman et al. (1973)
Fe$_2$TiO$_5$	PbO/V$_2$O$_5$	SC	1330–950	1	25 × 8	Garton et al. (1972b)
Fe$_2$TiO$_5$	Pb$_2$V$_2$O$_7$	SC	—	—	20 × 3 × 1	Wanklyn (1970)

Crystal	Solvent	Technique	Temp. range	Cooling rate duration	Crystal size (mm)	Reference
$FeVO_4$	V_2O_5	SC	1130–700	10	$4 \times 1.5 \times 1.5$	Levinson and Wanklyn (1971)
$FeWO_4$	$Na_2W_2O_7$	SC	1000–	—	small	Schultze et al. (1967)
$Ga_xIn_{1-x}As$	Ga/In	LPE/SC	$\gtrsim 750$	—	—	Antypas (1970)
$GaAs_{1-x}P_x$	Ga	SC	1100–600	3–10	~1	Badzian et al. (1969)
$GaAs$	Ga, Au, Pd, Pt	VLS(X)	980–830	4 h	whisker	Barns and Ellis (1965)
$GaAs_{1-x}P_x$	Ga	TSSG	1215	1 mm/h	36×21	Cerrina et al. (1973)
$GaAs$	Ga	SC	1050–	1.3	small	Faust and John (1964)
$GaAs$	Ga	TR	500–900	$\Delta T 25$–200	—	Lyons (1965)
$GaAs$	Ga	LPE/SC	800–	45–300	—	Miki and Otsubo (1971)
$GaAs$	Ga	TSM/TR(X)	850	40–80°/mm	—	Mlavsky and Weinstein (1963)
$GaAs(Si, Ge)$	Ga	LPE/SC	800–400	1800–2400	15–30 μm	Moriizumi and Takahashi (1969)
$GaAs$	Ga	VLS	550–680	—	~40 μm	Nickl and Just (1971)
$GaAs_xP_{1-x}$	Ga	LPE/SC	~1100–	120–300	—	Panish (1969)
$GaAs$	Ga	TR(X)	800	4–5 days	~20 × 2	Panish et al. (1966)
$GaAs$	Ga	TR	900–1000	—	—	Queisser and Panish (1967)
$GaAs$	Ga	LPE/SC	900–875	100	~15 μm layer	Rosztoczy et al. (1970)
$GaAs$	Ga	LPE/SC	~840	~35	1 mm	Rupprecht (1966)
$GaAs_xP_{1-x}$	Ga	LPE/SC	955–860	24	2 × 0.02	Shih (1970)
$Ga_xAl_{1-x}As$	Ga	LPE/SC	955–860	24	60–80 μm	Shih and Blum (1971)
$GaAs(Si)$	Ga	LPE	700	1	—	Spitzer and Panish (1969)
$Ga_xIn_{1-x}As$	In/Ga	LPE/SC	700–400	60 or 900	—	Takahashi et al. (1971)
$GaAs$	Ga	SC	~910–		small	Wolff et al. (1954)
$Ga_xAl_{1-x}As$	Ga	LPE/SC	1300–900	30	~200 μm	Woodall et al. (1969)
$GaFeO_3$	PbO/V_2O_5	SC	1000–600	5	10 × 3 × 3	Garton et al. (1972b)
$GaFeO_3$	Bi_2O_3/BiF_3	SC	1125–500	—	12 × 6 × 6	Linares (1962c)
$Ga_{2-x}Fe_xO_3$	Bi_2O_3/B_2O_3	SC	1300–800	4–7	10	Remeika (1960)
$Ga_{2-x}Fe_xO_3$	$PbO/Bi_2O_3/B_2O_3$	SC	1300–800	2–3	—	Schieber (1966)
$Ga_{2-x}Fe_xO_3$	Bi_2O_3/B_2O_3	SC	1250–1000	2–3	—	Schieber (1966)
$\beta\text{-}Ga_2O_3$	PbF_2/Bi_2O_3	SC	1300–900	4	10 × 5 × 3	Chase and Osmer (1967)
$\beta\text{-}Ga_2O_3$	Bi_2O_3/V_2O_5	SC	1200–900	5	6 × 2 × 1	Garton et al. (1972b)
$\beta\text{-}Ga_2O_3$	PbF_2	SC	1225–872	3	10	Hart and White, see White (1966)
$\beta\text{-}Ga_2O_3$	PbF_2	SC		4	16, 0.2 g	Katz and Roy (1966)

T 2

β-Ga₂O₃	MoO₃	SC	1050-70	—	50 mg	Schieber (...)
β-Ga₂O₃	PbO/PbF₂	SC	1390-900	—	0.1 g	Schieber (1964)
β-Ga₂O₃	PbF₂, PbO/PbF₂	EV	1200-	1	3-4	Timofeeva (1968)
β-Ga₂O₃	Na₂O/WO₃	SC	1350-900	10	2-4	Voronkova et al. (1968a)
GaP	Ga	TR	1140-820	—	8 × 5 × 2	Besselere and LeDuc (1968)
GaₓIn₁₋ₓP	In/Ga	LPE	900-1168	$\Delta T 5$-20	—	Blom and Plaskett (1971)
GaP	Ga	TSM	1100-100	<6 day	25 × 9 × 9	Broder and Wolff (1963)
GaP	NaPO₃, NaF, Ga₂O₃	EL	800	50 mA/cm²	~10 μm/h	Cuomo and Gambino (1968)
GaP	Ga	VLS	800-950	20 mm/h	20 × 2	Ellis et al. (1968)
GaP	Ga	SC	1050-	0.7	small	Faust and John (1964)
GaP	Sn	SC	1050-	0.5	small	Faust and John (1964)
GaₓIn₁₋ₓP	In/Ga	LPE/SC	875-800	120-180		Hakki (1971)
GaP	Cu, Zn	VLS(X)	950	22 days	whiskers	Holonyak et al. (1965)
GaP	GaJ₃	CR	250-500	12 h	small	Kwestroo et al. (1969)
Ga₁₋ₓInₓP	In, Sn	TR	1000	ΔT~10-20	5	Macksey et al. (1972)
GaₓAl₁₋ₓP	Ga	SC	1150-900	10	"few"	Merz and Lynch (1967)
GaₓAl₁₋ₓP	Ga	TR	1080	50-70 h	—	Merz and Lynch (1967)
GaₓIn₁₋ₓP	Ga/In	SC	1150-	100	—	Okuno et al. (1971)
GaP	Ga	VLRS	1200-1000			Plaskett (1969)
GaP	Ga	TSM	1160	4/day	34 × 14 × 14	Plaskett et al. (1967)
GaP	Ga	VLRS	1000	1 mm/day	~20 × 20	Poiblaud and Jacob (1973)
GaP	Ga	TR	~1050-800	10 μm/h	1 cm³	Rodot et al. (1968)
GaP(Zn, O)	Ga	LPE/SC	1060-600	30-1080	—	Saul and Hackett (1970)
GaP	Ga	SC	1140-800	570	—	Shiraki (1969)
GaP(N, Zr)	Ga	SC	1200-800	40	~10	Thomas and Lynch (1967)
GaP	Ga	SC	1275/1000-750	3-5	5 × 5 × 0.5	Thomas et al. (1964)
GaP(Te)	Ga	SC	1200-	5-10	1 cm³	Toyama et al. (1969)
GaP	Sn	SC	1085-784			Trumbore et al. (1964)
GaP	Bi	SC	1200-400		5 × 5 × 0.2	Ugai and Gukov (1965)
GaP	Ga	TSSG	1430-1200	7-13 mm/h	25 g	Von Neida et al. (1972)
GaP	Ga	TSM	850	ΔT~200	10 × 10 × 2	Weinstein and Mlavsky (1964)
GaP	Ga	TR(X)	850	0.5/h	10 × 10 × 2	Weinstein and Mlavsky (1964)
GaP	Ga	SC	—		—	Wolff et al. (1954)
GaS	Ga	SC	~1100-1200	~20	5 × 5 × 0.2	Harsy (1968)
GaₓIn₁₋ₓSb	In or Pb	TSM	420	—	9 thick	Hamaker and White (1969)
GaSb	Ga, Pb, Sb, Sn	SC	650-	0.7	small	Faust and John (1964)
GaₓIn₁₋ₓSb	Ga, In, Sb	LPE	~400	10	200 μm	Plaskett and Woods (1971)

Crystal	Solvent	Technique	Temp. range	Cooling rate duration	Crystal size (mm)	Reference
GaSb	Sb	TR	~650	1.6, 7.2	—	Reid et al. (1966)
GaSb	Ga	SC	—	—	—	Wolff et al. (1954)
Ga₂Se₃	Ga	SC	~1200–	~20	~5 × 5 × 0.2	Harsy (1968)
Gd-compounds see also R-						
GdAlO₃	Bi₂O₃/B₂O₃	EV	1200	17 days	4 mm³	Airtron
GdAlO₃(Nd)	Bi₂O₃/PbF₂/B₂O₃	SC/ACRT	1300–950	0.3	20 × 10 × 10	Courtens and Scheel (1974)
GdAlO₃	PbO/PbF₂	EV	1200	17 days	4 × 4 × 3	Garton and Wanklyn (1967b)
GdAlO₃	Bi₂O₃/B₂O₃	SC	1340–840	0.7–1.7	~5 × 4 × 4	Garton and Wanklyn (1967b)
GdAlO₃	PbO/PbF₂	SC	1260–1000	3	6 × 6 × 3	Linares (1962c)
GdAlO₃	Bi₂O₃/B₂O₃	SC	1340–840	1.7	—	Remeika (1956)
GdAlO₃	PbO/PbF₂/B₂O₃	SC/ACRT	1300–900	0.3–0.6	35 × 30 × 25 (210 g)	Scheel and Schulz-DuBois (1971); Scheel (1972a)
Gd₃Al₅O₁₂	PbO/PbF₂	SC	1245–	0.6	7.3 g	Manabe and Egashira (1971)
Gd₃Al₅O₁₂	PbO/PbF₂/B₂O₃	SC	1300–950	0.3	30 × 20 × 10	Scheel (unpublished)
GdCo₅	Al, Ga	SC	1000–		10	Gambino and Ruf (1971)
Gd₃Fe₅O₁₂	PbO/B₂O₃	SC	~1300–	5	—	Giess (1962a)
Gd₃Fe₅O₁₂	Bi₂O₃/B₂O₃	SC	1300–925	2	10	Remeika (1963)
Gd₃Fe₅O₁₂	Fe₂O₃	SC	1520–1440	—	~1	Smith et al. (1959)
GdGaO₃	see RGaO₃					
Cd₃Ga₅O₁₂	PbO/B₂O₃	SC	1300–1000	2	7.5	Remeika (1963)
Gd₃Ga₅O₁₂	PbF₂/PbO	SC	1240–900	—	1.2 g	Schieber (1964)
Gd₂GeMoO₈	PbO/MoO₃	SC	1290–900	2–3	2	Swithenby et al. (1974)
Gd₂GeWO₈	PbO/WO₃	SC	1280–900	4	2	Swithenby et al. (1974)
GdMnO₃	PbO/PbF₂	SC	1260–950	2	2 × 2 × 1	Wanklyn (1972)
GdMnO₃	PbF₂	SC	1290–1180	1	3 × 3 × 2	Wanklyn (1972)
Gd₂(MoO₄)₃	MoO₃	TSSG/SC	1184–	0.1	12 g	Belruss et al. (1971)
GdPO₄	Pb₂P₂O₇	SC	1370–950	1	7 × 1.5 × 0.6	Wanklyn (1972)
Gd₂Si₂O₇	Bi₂O₃	—	—	—	2 × 1 × 0.5	Bondar et al. (1971)
GdVO₄	see also RVO₄					
GdVO₄	Pb₂V₂O₇/NaVO₃/Na₂B₄O₇	SC	1300–1000	2	6 × 1.9 × 1.6	Garton and Wanklyn (1969)
GdVO₄	Bi₂O₃/V₂O₅	SC	1320–950	1	7.5 × 2 × 2	Garton et al. (1972b)
GdVO₄	Pb₂V₂O₇	SC	1150–950	2 (Oscill.)	7 × 3 × 0.5	Hintzmann and Müller-Vogt (1969)
GdVO₄	Pb₂V₂O₇	SC	1250–950	0.3	7 × 3 × 2	Scheel (unpublished)
Ge	several metals	SC				

Compound	Flux/solvent	Method	Temp.		Size	Reference
Ge	Bi, Bi/Sn	LPE	900–700	20–40	—	Kijima et al. (1971)
Ge	In, In/Ga	TR(X)	~600	$\Delta T \sim 50$	3 μm/min.	Laugier et al. (1967)
Ge	Sn	TR(X)	~650	$\Delta T \sim 50$	0.2 layer	Laugier et al. (1967)
Ge(As)	Pb/Sn	TSSG/EV	—	18–19 h	34 g	Trumbore and Porbansky (1960)
GeO$_2$	As	SC	1100–950	≲1	$3 \times 2 \times 2$	Finch and Clark (1968)
GeO$_2$	Li$_2$O/MoO$_3$ or Li$_2$O/WO$_3$	TSSG/SC	1050–975	2	10×3	Goodrum (1970)
GeO$_2$	Li$_2$O, Na$_2$O	SC	1100–700	3	1–2	Hart and White, see White (1966)
GeO$_2$	Li$_2$O/MoO$_3$	TSSG	~1025	0.04	$5 \times 1 \times 1$	Swets (1971)
GeO$_2$	Na$_2$O	TSSG	1050–950	0.5–1	$\sim 10 \times 3 \times 2$	Goodrum (1972)
HfB$_2$	Li$_2$O/WO$_3$	SC	1700–400	5	~3	Nakano et al. (1971)
HfO$_2$	Fe	SC	1250–1000	4	$3 \times 2 \times 1$	Chase and Osmer (1966b)
HfO$_2$	PbF$_2$	EV	—	—	—	Grodkiewicz and Nitti (1966)
HfO$_2$	PbF$_2$/B$_2$O$_3$	EV	—	—	—	Grodkiewicz and Nitti (1966)
HfO$_2$	BiF$_3$	EV	1250–690	2–4	$5 \times 5 \times 1.5$	Harari et al. (1967)
HfO$_2$	Na$_2$B$_4$O$_7$	SC	1300–950	3	6×0.2	Zonn and Joffe (1969)
HgS	Bi$_2$O$_3$/PbF$_2$	SC	630–450	10	5	Cruceanu and Nistor (1969)
HgS	Hg	SC	350–230	1.5–6.5	1–2	Garner and White (1970)
Hg(Se, Te)	Na$_2$S$_4$	SC	650–	6	~6	Cruceanu and Nistor (1969)
HgSe	Hg	SC	650–	~10	1–2	Nistor and Cruceanu (1969)
(Hg, Zn)Te	Hg	SC	650–	6	small	Cruceanu and Nistor (1969)
HgTe	Hg	SC	650–	0.5	~3	Dziuba (1971)
HgTe	Hg, Te	SC	—	~6	~10	Nistor and Cruceanu (1969)
(Hg, Cd)Te	Hg	THM	~750	0.2–0.3 mm/h		Ueda et al. (1972)
Ho-compounds see also R-	Te					
HoFeO$_3$	PbO/PbF$_2$/B$_2$O$_3$	SC	1275–900	4	$40 \times 15 \times 3$	Feigelson (1971)
HoFeO$_3$	PbO/B$_2$O$_3$	SC	1300–850	2	$15 \times 5 \times 5$	Remeika (1963)
HoFeO$_3$(Cr)	PbO/PbF$_2$/B$_2$O$_3$	SC	1300–900	1	0.5 cm^3	Van Uitert et al. (1970)
HoPO$_4$	Pb$_2$P$_2$O$_7$	SC	1370–950	1	$20 \times 2 \times 2$	Wanklyn (1972)
InAs	In	SC	800–	0.7	small	Faust and John (1964)
InAs	In/Cd	SC	—	10–100	$7 \times 5 \times 1$	Koppel (1966)
InAs$_x$Sb$_{1-x}$	In	LPE	720–520	ΔT1.5–14	0.17–15 μm/min.	Stringfellow and Greene (1971)
InAs	In	LPE	480–300	1200	20 μm	Tamura et al. (1971)
InAs	In	SC	—	—	—	Wolff et al. (1954)
In$_2$O$_3$	PbO/B$_2$O$_3$	SC	1250–900	2.3–10	—	Chase (1968)

Crystal	Solvent	Technique	Temp. range	Cooling rate duration	Crystal size (mm)	Reference
In_2O_3	PbO/B_2O_3	SC	1250–1000	4–6	—	Chase and Wilcox (1967)
$In_2O_3(Mg)$	PbO/B_2O_3	SC	1250–900	2.3–4	$2 \times 2 \times 1$	Chase and Tippins (1967)
In_2O_3	PbO/B_2O_3	SC	1250–900	2.3–4	$5 \times 5 \times 3$	Chase and Tippins (1967)
In_2O_3	PbO/B_2O_3	—	—	—	$2 \times 2 \times 0.5$	DeWit (1972)
In_2O_3	PbF_2	SC	1200–900	3	7	Hart and White, see White (1966)
In_2O_3	PbO/B_2O_3	SC	1200–500	10	$10 \times 10 \times 1$	Remeika and Spencer (1964)
InP	In	LPE/SC	~600–	60	~20 μm layer	Astles et al. (1973)
InP	In	SC	970–	0.7	small	Faust and John (1964)
InP	In	SC	—	10	~5	Röder et al. (1970)
In'	In	SC	—	—	—	Wolff et al. (1954)
InS	In	SC	~1200–	~20	—	Harsy (1968)
In_2S_3	As_2S_3	SC	850–300	2	$5 \times 5 \times 0.2$	Diehl and Nitsche (1973)
InSb	In, In + Zn	SC	480	0.7	$20 \times 20 \times 1$	Faust and John (1964)
InSb	Hg	SC	325	0.7	small	Faust and John (1964)
InSb	In	SC	—	—	small	Wolff et al. (1954)
K_3AlF_6	KCl	SC	1055–500	3.5	$2 \times 2 \times 1$	Garton and Wanklyn (1967a)
$K(Na)Ba_2Nb_5O_{15}$	K_2CO_3/Na_2CO_3	SC	1500–1300	2	~3	Giess et al. (1969)
KCl	AgCl	TSM	400–500	820 Å/sec	—	Rashkovich (1969)
$KCoF_3$	KCl	SC	1050–400	4	$10 \times 10 \times 10$	Garrard et al. (1974)
$KCoF_3$	KCl	SC	1100–650	3.5	small	Wanklyn (1969)
K_3CrF_6	PbF_2	SC	1055–500	3.5	~2	Garton and Wanklyn (1967a)
K_3CrF_6	KCl	SC	1140–880	0.7	~2	Garton and Wanklyn (1967a)
$KCrS_2$	K_2S_x	SC/EV	~1000–300	~25	$2 \times 2 \times 0.1$	Scheel (1974)
KEr_3F_{10}	MnF_2	TSSG/SC		0.5–2	10 g	Gabbe and Pierce (1974)
$KFeF_3$	KCl	SC/EV	850–510	6	6	Garrard et al. (1974)
KFe_5O_8	KF	CR/SC	1000–746	10	$2 \times 1 \times 0.05$	Roth and Romanczuk (1969)
$KFe_{12}O_{19}$	KF	SC	1200–900	3	5	Hart and White, see White (1966)
$KFeS_2$	Na_2S_x/K_2S_x	SC/EV	~600–300	25	$10 \times 0.2 \times 0.2$	Scheel (1974)
$K_2Ga_2O_{10}$	KF	SC	1200–900	3	1	Hart and White, see White (1966)
$K_2Ge_4O_9$	KF/GeO_2	CR/SC	1100–800	3–5	—	Weaver and Li (1969)
$K_3Li_2Ta_xNb_{5-x}O_{15}$	K_2CO_3/Li_2CO_3	TSSG/SC	1250–1000	1	$12 \times 6 \times 5$	Fukuda et al. (1970)
$KMgF_4$	KF	TSSG/SC	845–798	0.8 mm/h	8	Neuhaus et al. (1967)
$KMg_2LiSi_4O_{12}F_2$	PbO/B_2O_3	SC	1050–800			

Compound	Flux	Method	Temp (°C)	Rate	Size	Reference
K_2NaAlF_6	KCl/NaCl	SC	905–700	3.5	~2	Garton and Wanklyn (1967a)
K_2NaCrF_6	PbF_2	SC	975–600	3.5	~2	Garton and Wanklyn (1967a)
K_2NaCrF_6	KCl/NaCl	SC	975–600	3.5	1.5×1.5×1	Garton and Wanklyn (1967a)
K_2NaGaF_6	PbF_2	SC	1060–550	3.5	17 g	Garton and Wanklyn (1967a)
$(K, Na)TaO_3$	K_2O	TSSG/SC	~1345–	0.5	30×30×15	Belruss et al. (1971)
$KNbO_3$	K_2CO_3	SC	~1045–	0.5–1	2 cm³	Fukuda and Uematsu (1972)
$KNbO_3$	K_2CO_3	TSSG/SC	1085–	0.5	10	Hurst and Linz (1971)
$KNbO_3$	KF, KCl	SC	—	3	12–15 g	Matthias and Remeika (1951)
$KNbO_3$	K_2CO_3	SC(X)	1070–1046	1–3 mm/h	17×15×15	Miller (1958)
$KNbO_3$	K_2CO_3/Li_2CO_3	SC(X)	1050–	5	10×10×0.25	Timofeeva and Bychkov (1966)
$K_4Nb_6O_{17}$	KCl	SC	1250–	2	8	Hirano and Fukuda (1968)
$KNiF_3$		SC	1100–700	5	5	Wanklyn (1969)
$K_3Pb_3Nb_5ZnO_{18}$	PbO	TSSG/SC	1200–800	0.5	20 g	Kojima and Nomura (1973)
$KTa_{0.35}Nb_{0.65}O_3$	K_2O	—	~1265–	—	~7	Belruss et al. (1971)
$K(Ta, Nb)O_3$	K_2O	TSSG/SC	1260	—	25, 1300 g	Salnikov and Shelarkin (1967)
$K(Ta, Nb)O_3$	K_2O/K_2CO_3	TSSG/TR	~1400–1000	—	40×40×30	Bonner et al. (1967)
$K(Ta, Nb)O_3$	K_2O/K_2CO_3	TSSG/TR	—	—	12×8×5	Gentile and Andres (1967)
$KTaO_3$	K_2O	TSSG/SC	~1200	0.5	33 g	Whipps (1972)
$KTaO_3$	K_2CO_3/Li_2CO_3	SC	1346–	2	280 g	Belruss et al. (1971)
$KTaO_3$	K_2CO_3	SC	1400–1000	10	3×2×1	Bonner et al. (1967)
$K_2Ta_4O_{11}$	MoO_3	EV	1250–600; 950–1400	—	—	Demurov and Venevtsev (1971); Ovechkin et al. (1968)
$K_2Ti_6O_{13}$	KCl/KF	SC	1000–850	6–18	1.5×1 μm fibers	Roth et al. (1973)
KV_2F_6	KCl	SC	980–500	5	6	Garrard et al. (1974)
La-compounds see also R-						
$LaAlO_3(^{17}O)$	PbO/PbF_2	EV	1280	6 days	3	Garton et al. (1972a)
$LaAlO_3$	$PbO/PbF_2/B_2O_3/V_2O_5$	SC	1300–980	0.5	11×9×6	Kjems et al. (1973)
$LaAlO_3(Mn)$	$Pb_7O_2F_{10}$	EV	1200	100 h	3×3×0.4	Tsushima (1966)
$LaAlO_3$	PbO/PbF_2	SC	1200–800	1.5–2	10	Zonn and Joffe (1968)
LaB_4	La	VLS	1700–1250	16	5–8	Deacon and Hiscocks (1971)
LaB_6	B_2O_3(?)		~1800		0.05	Rea and Kostiner (1971)
$LaCoO_3$	PbO/PbF_2	EV	1250	6 days	2×1×1	Wanklyn (1972)
$LaCrO_3$	PbF_2	EV	1380	50 h	3×3×3	Tsushima (1967)
$LaFeO_3$	PbO/B_2O_3	SC	1300–850	2	5×1×1	Remeika (1963)
$LaNbO_3$	PbF_2	SC	1200–900	3	5	Hart and White, see White (1966)
$La(Y)Nb(Ta)O_4$	KF	SC/EV	1250–850	4.2	—	Wolten and Chase (1967)

Crystal	Solvent	Technique	Temp. range	Cooling rate duration	Crystal size (mm)	Reference
La_2O_3	NaCl	CR/SC	~1000	50–80	small	Wilke (1964)
LaOCl	$LaCl_3$	CR	1050	2 h	3×3 plates	Wanklyn (1972)
LaOCl	$MgSO_4 + LaCl_3$	CR	1050	2 h	3×3 plates	Wanklyn (1972)
$La_{1-x}Pb_xMnO_3$	PbO/PbF_2	—	—	—	—	Morrish (1970)
$La_{0.7}Pb_{0.3}MnO_3$	PbF_2	—	—	—	~3	Nielsen and Dearborn (1960)
LaS_{2-x}	Na_2S_x	—	—	~25	~1	Scheel (1974)
La_2SiO_5	KF	SC/EV	1300–850	1–2.5	$4 \times 3 \times 2$	Bondar et al. (1968)
$La_2Ti_2O_7$	$PbF_2/PbO/B_2O_3$	SC	1230	3 days	4×5 plate	Wanklyn (1972)
$LiAlO_2$	PbO/B_2O_3	EV	1300–840	2.5–10	6	Schwarzer and Neels (1971)
$LiAlO_2$	PbO/B_2O_3	SC/EV	1140–	2.5	~6	Schwarzer and Neels (1971)
$LiAl_5O_8$	PbF_2/PbO	SC/EV(X)	1300–800	7–10	2–3	Anselm et al. (1966)
γ-$LiBO_2$	LiCl	SC	950–	44 kb	small	Marezio and Remeika (1966)
$LiBaF_3$	LiF	TSSG/SC	821–770	35 h	$30 \times 30 \times 30$	Neuhaus et al. (1967)
LiCuGe	LiCl/LiF	—	—	—	—	Oleksiv and Kripyakevich (1967)
Li_2CuGe	LiCl/LiF	—	—	—	—	Oleksiv and Kripyakevich (1967)
$LiFe_5O_8$	Li_2O/B_2O_3	SC	1100–750	2	1.5	Anderson and Schieber (1963)
$LiFeO_2$	Li_2O/B_2O_3	SC	1090–800	2	1.5	Anderson and Schieber (1963)
$LiFe_5O_8$	$Li_2O/B_2O_3/PbO$	SC	1000–600	4	$4 \times 4 \times 4$	Arai and Tsuya (1972)
$LiFe_5O_8$	PbO/PbF_2	SC	1300–	1		Folen (1962)
$LiFe_5O_8$	Li_2CO_3	LPE/EV	1000–1400	5–20 h	$10 \times 10 \times 0.05$	Gambino (1967)
$LiFe_5O_8$	PbO/B_2O_3	SC	1060–600	0.8	$10 \times 10 \times 10$	Pointon and Robertson (1967)
$LiFe_{5-x}Ga_xO_8$	PbO/B_2O_3	SC	1050–300	4	10	Petrakovskii et al. (1969)
$LiFe_5O_8$	PbO/B_2O_3	SC	800–300	2		Remeika and Comstock (1964)
$LiFeO_2$	Li_2O/B_2O_3	SC	1050–700	—	2 mg	Schieber (1964)
$LiFe_5O_8$	$LiF/Li_2O/B_2O_3$	SC	1050–700	—	5 mg	Schieber (1964)
$LiFe_5O_8$(Zn, Co)	PbO/B_2O_3	SC	1150–400	6–2	12	Seleznev et al. (1971)
$LiFe_5O_8$	PbO/B_2O_3	SC	1060–	—	20 g	Spencer et al. (1968)
$LiFe_5O_8$	Li_2SO_4/Na_2SO_4	CR	800	1 h	small	Wickham (1962)
$LiFeO_2$	$CaSO_4$	CR	—	—	small	Wickham (1962)
$LiGaO_2$	PbO/B_2O_3	SC	1300–500	5		Remeika and Ballman (1964)
$LiGaO_2$	PbO/B_2O_3	SC/EV	1300–840	2.5–10	12	Schwarzer and Neels (1971)
$LiGaO_2$	PbO/B_2O_3	SC(X)	1140–	2.5	1.5 g	Schwarzer and Neels (1971)
(Li, Ni)O	Na_2SO_4	—	—	—	small	Wilke et al. (1965)

Compound	Flux	Method	Temp.	Time/rate	Size	Reference
Li_xWO_3	Li_2WO_3					
$LiRMo_2O_8$	$Li_2Mo_2O_7$	LPE	1050	—	3 μm	Clark et al. (1972)
Lu-compounds see also R-						
$LuCrO_3$	$PbF_2/PbO/B_2O_3$	EV	1260	9 days	3 × 2 × 2	Wanklyn (1972)
$MgAl_2O_4$	BaF_2/MgF_2	CR/SC	1650–1510	5	4 × 3	Dugger (1966)
$MgAl_2O_4$	BaF_2/MgF_2	CR/SC	1640–1500	7	12 × 8	Dugger (1967)
$MgAl_2O_4$	Pb_2OF_2	SC	1260–1000	3	3 × 3 × 3	Linares (1962c)
$MgAl_2O_4$	PbF_2	EV	1200–1175	5 days	10 × 10 × 10	Robertson and Taylor (1968)
$MgAl_2O_4$	$PbO/PbF_2/B_2O_3$	SC	1300–900	2–7	3	Tabata and Okuda (1971)
$MgAl_2O_4$	PbF_2/B_2O_3	EV	1250	500 h	25	Wang and Zanzucchi (1971)
$MgAl_2O_4$	PbF_2/B_2O_3	EV	1250		25	Wang and McFarlane (1968)
$MgAl_2O_4$	PbF_2/B_2O_3	EV	1250–1220	8 days	25	Wood and White (1968)
$MgAl_2O_4$	PbO/B_2O_3	SC	1300–850	2	2.5	Remeika (1963)
$MgAlFeO_4$		TR	600–800		2 × 0.3 × 0.2	Scheel (1968)
$MgAl_2Si_3O_{10}$	Li_2O/WO_3	SC	1120–600	4.5–45	2	Kleber et al. (1967)
$Mg_3B_2O_6$	$Na_2O/B_2O_3, PbO/B_2O_3$	SC	1350–900	2	small	Grodkiewicz and Van Uitert (1963)
$MgCa_3Si_2O_8$	PbO	SC	920–		0.5	Nakatani et al. (1970)
$MgCu_2$ etc.	$MgCl_2$	SC	1300–	2	—	Kunnmann et al. (1965)
$MgFe_2O_4$	Na_2O/WO_3	SC	1300–850	2	5	Remeika (1958b, 1963)
$MgFe_2O_4$	PbO/B_2O_3	SC	1310–900	5	2–4	Rozhdestvenskaya et al. (1962)
$MgFe_2O_4$	PbO	SC	1300–900	6	2–4	Rozhdestvenskaya et al. (1962)
$MgFe_2O_4$	PbO/B_2O_3	SC	1100		—	Viting (1965)
$MgFe_2O_4$	PbF_2	EV	1310–900	4.3	3	Wickham (1962)
$MgFe_2O_4$	$Pb_2P_2O_7$	SC	1250–850	0.5	1–2 g	Giess (1962b)
$MgGa_2O_4$	Pb_2OF_2	SC	–750		—	Spring-Thorpe and Pamplin (1968)
$MgGeP_2$	Sn	SC		—		Von Dreele et al. (1970)
$Mg_{14}Ge_5O_{24}$	PbO	SC	1300–900	4	1	Kostiner and Bless (1971)
$Mg_{28}Ge_{7.5}O_{38}F_{10}$	PbF_2	SC	1250–	4	small	Bless et al. (1971)
$MgMoO_4$	LiCl	SC	800–	0.7	1	Packter and Roy (1971)
$MgO(^{17}O)$	PbF_2	EV	1280	5 days	5 × 3 × 2	Garton et al. (1972a)
MgO	KOH	SC	600–200	5–40	1	Kleber et al. (1967)
MgO	PbF_2	SC			~3	Nielsen and Dearborn (1958, 1960)
MgO	$Na_2O/WO_3/P_2O_5$	TR	1300–1200	6 days	5 × 2 × 2	Vora and Zupp (1970)
MgO	PbF_2	EV	1230	0.3 mm/day	1 cm³	Webster and White (1969)
MgP_4	Bi/Pb/Sn	SC	1100–450	1	20 × 3 × 0.1	Gibinski et al. (1974)
Mg_2SiO_4	PbO	SC	1350–900	2	small	Grodkiewicz and Van Uitert (1963)

Crystal	Solvent	Technique	Temp. range	Cooling rate duration	Crystal size (mm)	Reference
Mg_2SiO_4	Li_2O/MoO_3	SC	1500–1100	5	10	Vutien et al. (1972)
$MgSnB_2O_6$	MgB_2O_4	SC	1240–	5–50	small	Schultze et al. (1971)
$MgWO_4$	LiCl	SC	700–800	—	small	Packter and Roy (1971, 1973)
$MgWO_4$	Na_2SO_4, Na_2O/WO_3	SC	1000–	—	small	Schultze et al. (1967)
$MgWO_4$	$Na_2W_2O_7$	SC	1100–1250–700	2.5	10 × 5 × 5	Van Uitert and Soden (1960)
$MnAl_6$	Al	CR	658–710	—	—	Barber (1964)
MnBi	Bi	TR	500–300	—	—	Ellis et al. (1957)
$MnCr_2O_4$	Bi_2O_3	SC	1300–800	1.5		Mikami (1971)
$Mn_xCr_{3-x}O_4$	PbO/PbF_2	SC	1220–850	1	3	Nevriva (1971)
$MnCr_2O_4$	Na_2O/WO_3	SC	1250–	5	4	Kunnmann et al. (1965)
$MnCr_2O_4$	BiF_3	SC	1100–	5	5	Tsushima et al. (1968)
$MnFe_2O_4$	$PbO—B_2O_3$	SC	1320–950	1–2	8	Bibr et al. (1966)
$Mn_{2-x}Fe_xO_3$	Na_2O/WO_3	EL	725	90 mA	2–3	Barks and Kostiner (1966)
$Mn_xFe_{3-x}O_4$	Bi_2O_3/V_2O_5	SC	1150–950	0.5	10, 0.5 g	Nevriva and Holba (1972)
$MnFe_2O_4$	$Na_2B_4O_7$	TR(X)	1200–1300	—	6–7	Timofeeva and Zalesskii (1959)
$MnGa_6$	Ga				1	Girgis and Schulz (1971)
Mn_5Ge_3	Hg	CR/SC	700–800	5 h, ~200	small	Mayer et al. (1967)
$MnNb_2O_6$	Na_2O/WO_3	SC	1260–700	3	4 × 1 × 1	Wanklyn (1972)
Mn_2O_3(Al, Fe)	$PbF_2/PbCl_2$	SC/TR	800–850	—	11 × 9	Espinosa (1971)
MnO	PbF_2	SC	1050–950	10–25	3	Linares (1962c)
Mn_3O_4	$Na_2B_4O_7$	SC	1050–900	1	14 × 1 × 1	Nielsen (1969)
Mn_2O_3	PbF_2	SC/CR		—	~3	Nielsen and Dearborn (1960)
Mn_3O_4	$PbF_2/PbO/B_2O_3$	CR/SC	1280–950	1.2	7 × 5 × 5	Wanklyn (1972)
Mn_2O_3	LiCl, NaCl, Na_2SO_4	CR/SC/EV	~1000	50–80	small	Wilke (1964)
α-MnS	Na_2S_x	SC	~700–300	~25	<1	Scheel (1974)
MnS_2	$PbCl_2$	CR/SC	750–400	3.5	small	Wilke (1967)
Mn_5Si_3	Hg	SC	550	20 h, ~200	small	Mayer et al. (1967)
$MnSnB_2O_6$	MnB_2O_4	SC	1110–	5–50	50 × 15 × 15	Schultze et al. (1971)
MnTe	Te	ACRT/SC	1018–920	0.4	3 × 3 × 3	Mateika (1972)
$MnWO_4$	Na_2O/WO_3	SC/CR	1250–600	1.5	small	Wanklyn (1972)
$MnWO_4$	Na_2SO_4, $Na_2W_2O_7$	SC	1000–	—	7 × 3 × 2	Schultze et al. (1967)
MoO_2	Na_2O/MoO_3	EL	675	0.75 V, 20 mA	—	Perloff and Wold (1967)
MoO_2	Na_3MoO_4	EL	675	—		Wold (1964)
MoO_3	Na_3AlF_6	THM	700–645	0.5/min.		Vidorovich and Marychev (1964)

Compound	Solvent	Method	Temperature		Size	Reference
NaBa₂Nb₅O₁₅	Na₂CO₃	SC	1000–300	~25	20 × 20 × 0.2	Scheel (1974)
NaCrS₂	Na₂Sₓ	CR/SC/EV	1350–	20	10 × 1 × 1	Reid et al. (1968)
NaFeTiO₄	Fe₂O₃	SC	1050–700	—	1.2 g	Schieber (1964)
Na₀.₅Gd₀.₅MoO₄	Na₂O/MoO₃	SC	1350–800	10	—	Chukhlantsev and Poleskev (1965)
Na₂HfSiO₅	Na₂SiO₃	SC	~750–300	~25	15 × 15 × 0.2	Scheel (1974)
NaInS₂	Na₂Sₓ	CR/SC/EV	1350–1040	0.5	2 × 2 × 3	Summergrad and Banks (1957)
Na₀.₅La₀.₅Fe₁₂O₁₉	Na₂CO₃	SC	550	40	—	Wold et al. (1964)
NaMo₆O₁₇	Na₂O/MoO₃	EL	1300–200	—	1	Cross and Nicholson (1955)
NaNbO₃	NaF/NaHCO₃	SC	1300–	—	—	Gliki and Timofeeva (1960)
NaNbO₃	NaF	SC	—	—	10 × 10 × 3	Matthias and Remeika (1951)
NaNbO₃	NaF	SC	—	—	~2 mm	Miller et al. (1962)
NaNbO₃(V)	Na₂O/V₂O₅	SC	1300–	5–30	6 × 6 × 3	Wanklyn (1972)
NaNbO₃	Na₂O/WO₃	SC	1260–950	2	10 × 1 × 1	Reid et al. (1968)
NaScTiO₄	Na₂O/TiO₂	SC	1150–850	~200	1	Kay and Miles (1957)
NaTaO₃	NaBO₂/Na₂CO₃	SC	1250–	—	—	Roth et al. (1973)
Na₂Ta₄O₁₁	MoO₃	EV	950–1400	~1000°C	—	Fredlein and Damjanovic (1972)
NaₓWO₃	Na₂O/WO₃	EL	750–803	15–50 mA	3 × 0.1 × 0.1	Scott et al. (1970)
NaₓWO₃	Na₂CO₃	EL	800	10–75 mA	—	Shanks et al. (1962)
NaₓWO₃	Na₂O/WO₃	EL	750–900	1–1.5 V, 30–50 mA	30 × 30 × 30	Shanks (1972)
NaₓWO₃	Na₂O/WO₃	EL	750–900	0.5–3 V, 10–20 mA	~10	Weller et al. (1970)
NaₓWO₃	Na₂O/WO₃	EL	750–900	—	10 × 10 × 10	Weller and Grandits (1972)
NaₓWO₃	Na₂O/WO₃	TSSG/EL	700–900	—	<4	Wilke (1968)
NaₓWO₃	Na₂O/WO₃	EL	720–800	10	0.1–8	Chukhlantsev and Poleskev (1965)
Na₂ZrSiO₅	Na₂SiO₃	SC	700	—	—	
Nb₃Sn	Sn	SC	1350–800	—	—	Hanak and Johnson (1969)
Nd-compounds see also R-						
NdAlO₃	PbO/PbF₂	SC	1300–800	4.5	~5	Zonn and Joffe (1968)
NdAlO₃	PbO/PbF₂	SC	1280–910	0.7	~3	Garton and Wanklyn (1967b)
Nd₃Fe₅₋ₓScₓO₁₂	3PbO/4PbF₂	SC	1260–950	5	6 × 6	Loriers et al. (1971)
Nd₃Ga₅O₁₂	PbF₂/PbO	SC	1240–900	—	0.3 g	Schieber (1964)
NdNa₅W₄O₁₆	Na₂WO₄	SC	1100–700	5	0.1	Hong and Dwight (1974)
NdNbO₄	PbF₂	SC	1200–800	5	5	Lou and White (1967)
Ni	NiCl₂	CR	1000	5–60 min.	1–10 μm	Lefever (1961)
Ni	NiCl₂ + KCl + Al	CR	800	—	small	Lefever (1962)
NiAl₃	Al	CR	730	—	—	Barber (1964)
NiF₂	KCl, PbCl₂/PbF₂	SC	930–650	3	4 × 1.5 × 1.5	Wanklyn (1969)
NiFe₂O₄	PbO—B₂O₃	SC	1320–950	1–2	~8	Bibr et al. (1966)
NiFe₂O₄	Na₂Mo₂O₇	SC	1200–700	3	3	Damay and Heindl (1965)

Crystal	Solvent	Technique	Temp. range	Cooling rate duration	Crystal size (mm)	Reference
$NiFe_2O_4$	$BaO/Bi_2O_3/B_2O_3$	SC	1320–1150	1	5	Elwell et al. (1972)
$NiFe_2O_4$	$Na_2B_2O_7$	SC	1330–1250	2	2	Galt et al. (1950)
$Ni(Zn)Fe_2O_4$	$NaCO_3$	LPE/EV	1000–1400	5–20 h	$10 \times 10 \times 0.05$	Gambino (1967)
$NiFe_2O_4$	Bi_2O_3/V_2O_5	SC	1350–950	1	3	Garton et al. (1972b)
$NiFe_2O_4$	PbO/PbF_2	—	—	—	6	Gendelev et al. (1964)
$NiFe_2O_4$	$Na_2Fe_2O_4$	TR	1350	14 days	~10	Kunmann et al. (1963)
$NiFe_2O_4$	PbF_2/PbO	SC	1220–900	3–4	10	Makram and Krishnan (1962)
$NiFe_2O_4$	PbO/PbF_2	SC	1220–900	3	10	Makram and Krishnan (1963)
$Ni(Zn, Co)Fe_2O_4$	PbO/B_2O_3	SC	1280–	4	4	Manzel et al. (1963)
$Ni(Zn)Fe_2O_4$	PbO/PbF_2	SC	1220–1150	2	4	Manzel (1967)
$NiFe_2O_4$	$NaBO_2/NaF$	SC	1100–	0.3	15	Neuhaus and Liebertz (1962)
$NiFe_2O_4$	PbO	SC	1330–850	10	5–7	Nowicki (1962)
$NiFe_2O_4$	BaO/B_2O_3	SC	1300–1187	0.5	3	Quon et al. (1970)
$NiFe_2O_4$	PbO	SC	1300–900	10	—	Remeika (1954, 1958b)
$NiFe_2O_4$	PbO/B_2O_3	SC	1300–850	2	5	Remeika (1963)
$NiFe_2O_4$	$BaO/Bi_2O_3/B_2O_3$	SC	1350–900	0.8	5	Robertson et al. (1969)
$NiFe_2O_4$	PbO	SC	1280–900	5.4	2–3	Sekizawa and Sekizawa (1962)
$NiFe_2O_4$	BaO/B_2O_3	TR/TSSG	1230		$10 \times 8 \times 8$	Smith and Elwell (1968)
$Ni_{1+x}Ga_xFe_{2-2x}O_4$	PbO/B_2O_3	SC	1250–900	1	—	Krishnan (1968b)
$NiMn_2O_4$	Bi_2O_3/B_2O_3	SC	1280–860	1	6	Makram (1967)
$NiNb_2O_6$	Na_2O/WO_3	SC	1260–700	3	$2.5 \times 0.7 \times 0.7$	Wanklyn (1972)
NiO	$NaCl, PbF_2$	SC			small	Packter (1968)
NiO	PbF_2	SC	1200–900	3	0.5	Hart and White, see White (1966)
NiO	PbF_2	SC	1290–800	2	6	Hill and Wanklyn (1968)
NiO	KF, LiCl, Na_2SO_4	CR/SC	~1000	50–80	small	Wilke (1964)
$NiO(Li)$	Na_2SO_4	—	—	—	small	Wilke et al. (1965)
NiS_2	Na_2S_x	CR/SC/EV	800–300	~25	$1 \times 1 \times 1$	Scheel (1974)
NiS_2	$PbCl_2$	SC	750–400	3.5	~2	Wilke et al. (1967)
Ni_2Si	Hg	CR/SC	500	10 h, ~200	small	Mayer et al. (1967)
$NiTiO_3$	PbO/V_2O_5	SC	1320–950	1	$3 \times 3 \times 3$	Garton et al. (1972b)
$NiTiO_3$	$Pb_2V_2O_7$	SC	—	—	10	Wanklyn (1970)
$NiWO_4$	$Na_2WO_4 + NiCl_2 + NaCl$	SC/CR	900–	—	—	Keeling (1957)
$NiWO_4$	$Na_2SO_4, Na_2O/WO_3$	SC	1000–	—	small	Schultze et al. (1967)
$NiWO_4$	Na_2O/WO_3	SC/CR	1250–600	1.5	$6 \times 6 \times 6$	Wanklyn (1972)
NbO_2	PbF_2/B_2O_3	TR/SC	1250–1250			

NpO_2	PbF_2/B_2O_3	EV	1250–1350	4–10 days	3	[…] and Clark (1970)
NpO_2	Li_2O/MoO_3	TR	1315–1310	—	2	Finch and Clark (1970)
NpO_2	Li_2O/MoO_3	EV	1250–1350	4–10 days	3	Finch and Clark (1970)
$PbAl_{12}O_{19}$	$PbO/PbF_2/La_2O_3$	SC	1250–1000	4	2	Chase and Wolten (1965)
$PbAl_{12}O_{19}$	$PbO/PbF_2/B_2O_3$	SC		1	1	Comer et al. (1967)
$PbAl_{12}O_{19}$	Pb_2OF_2	SC	1260–1000	3	3×3×3	Linares (1962c)
$PbAl_2Si_2O_8$	PbO/PbF_2	SC/CR	1250–900	0.5–1	15×2×0.1	Scheel (1971)
$Pb_3CdNb_2O_9$	PbO	SC	1200–800	5–10	10	Ichinose et al. (1971)
Pb_2CoWO_6	PbO/WO_3	SC	1200–800	5		Bokov et al. (1965)
$Pb_3M(Nb,Ta)_2O_9$ with $M=Co, Mg, Ni, Zn$	PbO	SC	1300–800	30–100	1–2	Bokov and Mylnikova (1961a)
$PbCrO_3$	PbO	HPS	900	60–65 kb	small	Chamberland (1967)
$PbCrO_3$	PbO	HPS	1150	65 kb	0.25	DeVries and Roth (1968)
$PbFe_{12}O_{22}$	PbO/PbF_2	SC	1350–	1–3	20	Komatsu and Sunagawa (1964)
$PbFe_{12}O_{22}$	—	SC	1390–900		0.8 g	Schieber (1964)
$Pb(Fe,Al)_{12}O_{19}$	$PbO/PbF_2/B_2O_3$	SC	1300–950	0.5–5	25×20	Van Uitert et al. (1970b)
Pb_2FeTaO_6	PbO	SC	1200–900	5–10	1×1×1	Nomura et al. (1968)
$PbGa_{12}O_{19}$	$PbF_2/Bi_2O_3/La_2O_3$	SC	1250–1000	4	2	Chase and Wolten (1965)
$Pb_3MgNb_2O_9$	PbO	SC	1200–	20–40		Bokov and Mylnikova (1961b)
$Pb_3MgNb_2O_9$	$PbO/MgO/B_2O_3$	TSSG/SC	1150	0.5	10×10	Bonner and Van Uitert (1967)
$Pb_3MgNb_2O_9$	PbO	SC	1200–800	10–60	0.5	Mylnikova and Bokov (1962)
$Pb(NO_3)_2$	$KNO_3/NaNO_3$ or $KNO_3/LiNO_3$	EV	140–250	2 days	5–10	Egghart (1967)
$Pb_3NiNb_2O_9$	PbO	SC	1200–	20–40		Bokov and Mylnikova (1961b)
$Pb_3NiNb_2O_9$	PbO	SC	1200–800	10–60	0.5	Mylnikova and Bokov (1962)
PbS	Na_2S_x	CR/SC/EV	650–300	25	1×1×1	Scheel (1974)
$PbSO_4$	Na_2SO_4	SC/EV	900–700	10	1	Patel and Bhat (1971)
$Pb_xSr_{1-x}TiO_3$	$PbO/SrO/B_2O_3$	THM/TR	~1100	0.1–10/day	20×10×10	Dibenedetto and Cronan (1968)
$PbTa_2O_6$	$Pb_2V_2O_7$	TSSG/SC	1186	1.5, 4.0	5×2×2	Bruton and White (1973)
$PbTa_2O_6$	Bi_2O_3/B_2O_3	TSSG/SC	1054	1.5, 4.0	4	Bruton and White (1973)
$Pb_3Ta_4O_{13}$	PbF_2/Ta_2O_5	CR/SC	1100–800	3–5		Weaver and Li (1969)
$PbTe$	Pb	LPE/SC		12–60		Wagner and Thompson (1970)
$PbTiO_3$	PbO/V_2O_5	SC	~1200–800	—	1	Belyaev and Nesterova (1952)
$Pb(Ti,Zr)O_3$	KF/PbF_2; PbO/PbF_2	SC/EV	1200–800	20		Fushimi and Ikeda (1964)
$Pb(Ti,Zr)O_3$	KF/PbF_2	SC	1150–950	7	0.1×0.08×0.03	Fushimi and Ikeda (1965)
$Pb(Ti,Zr)O_3$	PbF_2/KF	SC	1000–700	40	~0.5	Ikeda and Fushimi (1962)
$PbTiO_3$	KF	SC	920–600	60	9×7×0.05	Kobayashi (1958)

Crystal	Solvent	Technique	Temp. range	Cooling rate duration	Crystal size (mm)	Reference
$PbTiO_3$	$PbCl_2$	SC	~900–500	5–7	1	Nomura and Sawada (1952)
$PbTiO_3$	PbO	SC	1100–700	5	$5 \times 3 \times 1$	Remeika and Glass (1970)
$PbTiO_3$	PbO	SC	1300–1040	50	$3 \times 1.5 \times 0.2$	Rogers (1952)
$PbTiO_3$	PbO/B_2O_3	SC	~1000–850	—	2–3	Sholokhovich (1958)
$PbTiO_3$	Na_2SiO_3	SC	~1000–850	—	—	Sholokhovich and Berkova (1955)
$PbTiO_3$	PbO/B_2O_3	SC	960–600	5–20	4×0.1	Sholokhovich and Khodakov (1962)
$Pb(Ti, Zr)O_3$	various	SC	~1250–550	5–20	—	Tsuzuki et al. (1968)
$Pb(Ti, Zr)O_3$	$KF/PbF_2/Pb_3(PO_4)_2$	SC	1140–700	3–40	—	Tsuzuki et al. (1969)
$Pb(Ti, Zr)O_3$	$KF/PbF_2/Pb_3(PO_4)_2$	SC	1115–800	5	1–2.5	Tsuzuki et al. (1973)
$PbTiP_2O_8$	$Pb_2P_2O_7$	SC/CR	1280–910	2	$4 \times 1.5 \times 0.5$	Wanklyn (1972)
$PbWO_4$	Na_2WO_4	SC	1000–	—	small	Schultze et al. (1967)
$Pb_3ZnNb_2O_9$	$PbO/Na_2B_4O_7$	—	—	—	$0.8 \times 0.8 \times 0.5$	Berezhnoi et al. (1968)
$Pb_3ZnNb_2O_9$	PbO/B_2O_3	SC?	—	—	$2 \times 0.5 \times 0.1$	Krainik et al. (1971)
$Pb_3ZnNb_2O_9$	PbO	SC	1200–	5–10	8	Yokomizo et al. (1970)
$PbZrO_3$	PbF_2	SC	1250–	50	0.3	Jona et al. (1955a)
$PbZrO_3$	$PbF_2/PbO/B_2O_3$	SC	1280–800	2–3	1–3	Scott and Burns (1972)
Pr-compounds see also R-						
$PrAlO_3$	PbO/PbF_2	SC	1285–960	3.5	2	Garton and Wanklyn (1967b)
$PrAlO_3$	PbO/PbF_2	SC	1300–800	4.5	~5	Zonn and Joffe (1968)
$PrAlO_3$	PbF_2	EV	850	—	—	Vutien and Anthony, after Laurent (1969)
$Pr_3(Fe, Sc)_5O_{12}$	$3PbO/4PbF_2$	SC	1260–950	5	1–2	Loriers et al. (1971)
PrO_{2-x}	PbO	SC	1260–1000	3	1–2	Feigelson (1971)
Pt_2Si	Hg	CR/SC	450–550	~200	small	Mayer et al. (1967)
PuO_2	$Li_2Mo_2O_7$	TR	1300–1270	1 mm/weak	$3 \times 3 \times 2$	Finch and Clark (1972)
R-compounds see also individual rare earths and yttrium!						
$RAlO_3(R = Dy—Lu)$	$NaOH$	HPS	1200	32.5 kb, 4 h	small	Dernier and Maines (1971)
$RAlO_3$	Pb_2OF_2	SC	1260–1000	3	$6 \times 6 \times 3$	Linares (1962c)
$RAlO_3$	PbO	SC	1300–850	30	~1	Remeika (1956)
$RAlO_3$	$PbO/PbF_2/B_2O_3$	SC	1290–800	2	3–6	Wanklyn (1969)
$RAlO_3$	Bi_2O_3/B_2O_3	SC	1315–815	2	3–6	Wanklyn (1969)
$R_3Al_5O_{12}$	PbO/PbF_2	SC	1250–1000	4.2	2–10	Chase and Lefever (1960)
$R_3Al_5O_{12}(Cr)$	$PbO/PbF_2/B_2O_3$	SC	1300–950	0.25–2	—	Kestigian and Holloway (1967)
$R_3Al_5O_{12}$	$PbF_2, PbF_4/PbO$	EV	1350			

Compound	Flux	Method	Temperature	Time	Size	Reference
$RAl_3(BO_3)_4$		SC				
$RAsO_4$	$Pb_2As_2O_7$	SC/EV	1300–925	1–4	$7 \times 2 \times 0.2$	Feigelson (1967)
$RAsO_4$	$Pb_2As_2O_7$	SC	1250–	2	$6.5 \times 1 \times 1$	Hintzmann and Müller-Vogt (1969)
RB_4	Al	SC	1550–1000	10 min.	1	Fisk et al. (1972)
RB_4	R	SC	1700–1000	10 min.	small	Fisk et al. (1972)
RBO_3	PbO, $PbO/PbF_2/MoO_3$	SC	1280–800	1–6	~ 8	Wanklyn (1973b)
$RCoO_3$	PbO	SC	1300–850	30	~ 1	Remeika (1956)
$RCrO_3$	PbO, Bi_2O_3	SC	1300–850	30	~ 1	Remeika (1956)
$RCrO_3$	PbF_2/B_2O_3	EV	1240	10 days	~ 5	Wanklyn (1969)
$RCr_3(BO_4)_3$	K_2SO_4/MoO_3	SC	1150–900	2	10	Ballman (1962)
RF_3	BeF_3	SC	1100–		small	Van Uitert et al. (1969)
$RFeO_3$	$PbO/PbF_2/B_2O_3$	SC	1300–950	0.5–1	2–18	Gendelev and Titova (1971)
$RFeO_3$	$PbO/PbF_2/B_2O_3$	SC	1300–850	2–4	10	Giess et al. (1970)
$RFeO_3$	PbO	SC	1300–850	30	~ 1	Remeika (1956)
$ReFeO_3$	PbO/B_2O_3	SC	1300–950	2	$10 \times 5 \times 2$	Remeika and Kometani (1968)
$RFeO_3(Bi)$	$PbO/Bi_2O_3/B_2O_3$	SC	1300–950	1–4	~ 10	Remeika et al. (1969)
$RFeO_3$	PbO/B_2O_3	LPE/SC	1150–1100	2–20	$4\ \mu m/h$	Shick and Nielsen (1971)
$R_3Fe_5O_{12}$	$PbO/PbF_2/B_2O_3$	SC	1290–850	2	5	Wanklyn (1969)
$R_3Fe_5O_{12}$	PbO; Bi_2O_3	SC	1350–930	1	6 g	Nielsen and Dearborn (1958), Nielsen (1958)
$R_3Fe_5O_{12}$	$PbO/PbF_2/B_2O_3$	SC	1300–950	0.5–0.8	50	Van Uitert et al. (1970c)
$RGaO_3$	PbO	SC	1300–850	30	~ 1	Remeika (1956)
$RGaO_3$	NaOH	HPS	1000	72 kb, 200	small	Marezio et al. (1966)
$RGaO_3$	NaOH	HPS	1000	72 kb	small	Marezio et al. (1968a)
$R_3Ga_5O_{12}$	PbO/B_2O_3	SC	1300–	1–5	small	Marezio et al. (1968a)
$R_3Ga_5O_{12}$	PbO/PbF_2	SC	1250–1000	4.2	2–10	Chase and Lefever (1960)
$R_3Ga_5O_{12}$	PbO/PbF_2	SC	1250	100 h	$8 \times 8 \times 8$	Tsushima (1967)
RGe_2	Hg	CR/SC	450–550	~ 200	small	Mayer et al. (1967)
R_2GeO_5	PbO/GeO_2	SC	1270–800	5	8	Wanklyn (1973a)
$R_2Ge_2O_7$	PbF_2	EV	1270	1 week	8	Wanklyn (1973a)
R_2GeMoO_8	PbO/MoO_3	SC	1290–900	2–3	2	Swithenby et al. (1974)
R_2GeWO_8	PbO/WO_3	SC	1280–900	4	2	Swithenby et al. (1974)
$RLiF_4$	LiF	TSSG/SC	800–900–	0.5–2	$20 \times 20 \times 100$	Gabbe and Harmer (1968)
$RMnO_3$	Bi_2O_3	SC	1400–1000	—	1	Bertaut et al. (1963)
$RMnO_3$	Bi_2O_3	EV	1450	—	1	Bertaut et al. (1963)
$RMnO_3$	$PbF_2/PbO/B_2O_3$	SC	1280–900	2	Plates $\sim 20\ mm^2$	Wanklyn (1972)
$RMnO_3$	Bi_2O_3	SC	1200–	—	0.5	Yakel et al. (1963)
RMn_2O_5	$PbO/PbF_2/B_2O_3$	SC	1280–900	2	—	Schieber et al. (1972a)

Crystal	Solvent	Technique	Temp. range	Cooling rate duration	Crystal size (mm)	Reference
RMn_2O_5	$PbF_2/PbO/B_2O_3$	SC	1280–900	1.2	7 × 5 × 5	Wanklyn (1972)
RMn_x	Hg	CR	700–1100	—	small	Kirchmayr (1965)
RN	Hg	CR/SC	~320	—	small	Busch et al. (1970)
RN	Hg	CR/EL	1000	—	small	Magyar (1968)
$R_{0.5}Na_{0.5}MoO_4$	Na_2O/MoO_3	SC	1250–850	1.1–3.2	1.2 g	Schieber and Holmes (1964)
$RNbO_4$	Bi_2O_3/B_2O_3	SC	1340–1050	3	5	Garton and Wanklyn (1968)
R_3NbO_7	PbO/PbF_2	SC	1285–950	3	1	Garton and Wanklyn (1968)
ROF	$PbO/PbF_2/Al_2O_3$	SC	1280–845	3		Garton and Wanklyn (1968)
$R_{2/3}Pb_{1/3}MnO_3$	Pb_2OF_2	SC	1150–		1–5	Janes and Bodnar (1971)
RPO_4	$Pb_2P_2O_7$	SC	1300–975	4	25 × 4 × 0.5	Feigelson (1964)
RPO_4	$Pb_2P_2O_7$	SC	1150–950	2	15 × 1.5 × 1	Hintzmann and Müller-Vogt (1969)
RPO_4	$Pb_2P_2O_7$	SC	1330–930	0.6–0.9	15 × 2 × 2	Smith and Wanklyn (1974)
RPO_4(Tb—Er)	$Pb_2P_2O_7$	SC	1320–950	1	20 × 2 × 2	Wanklyn (1972)
RPO_4(La—Gd)	$Pb_2P_2O_7$	SC	1330–1050	1	3 × 0.5 × 0.2	Wanklyn (1972)
$RScO_3$	PbO	SC	1300–850	30	~1	Remeika (1956)
RSi_2	Hg	CR/SC	450–550	~200	small	Mayer et al. (1967)
$R_2Ti_2O_7$	PbF_2	EV	1235	1 week	3	Garton and Wanklyn (1968)
R_2TiO_5	PbO	SC	1280–940	3	3	Garton and Wanklyn (1968)
RVO_4	V_2O_5	TR	1000		10	Brixner and Abramson (1965)
RVO_4	$Pb_2V_2O_7$	SC	1300–950	2–4	6 × 0.5 × 0.1	Feigelson (1968)
RVO_4	$Pb_2V_2O_7$	SC	1250–	2	7 × 3 × 0.5	Hintzmann and Müller-Vogt (1969)
RVO_4	$Pb_2V_2O_7$	SC	1330–930	0.6–0.9	15 × 2 × 2	Smith and Wanklyn (1974)
Rb_2CoCl_4	$CoCl_2$	SC	538–		~0.5 cm³	Makovsky et al. (1974)
Rb_3CoCl_5	$CoCl_2$	SC	506–		~0.5 cm³	Makovsky et al. (1974)
$RbCoCrF_6$	$RbCl/CoCl_2$	—	950–		8	Nouet et al. (1971)
$RbCoVF_6$	$RbCl/CoCl_2$	—	950–		4	Nouet et al. (1971)
Rb_2KGaF_6	RbCl/KCl	SC	1060–550	3.5	~2	Garton and Wanklyn (1967a)
Rb_2KGaF_6	PbF_2	SC	1060–550	3.5	~2	Garton and Wanklyn (1967a)
$Rb_3Mn_2Cl_4$	RbCl	SC	475–		25 × 14	Makovsky et al. (1971)
Rb_2MnCl_4	RbCl	SC	462–		25 × 14	Makovsky et al. (1971)
Rb_2MnF_4	RbF	TSSG/SC	600–700–	5–10	25–30 g	Gabbe and Linz (1974)
Rb_2MnF_4	$RbF + YF_3$	TSSG/SC	600–700–	5–10	25–30 g	Gabbe and Linz (1974)
$RbMnCrF_6$	$RbCl/MnCl_2$	—	900–		2	Nouet et al. (1971)
$RbNiF_3$	RbCl	EV/CR	800–900	—	15 × 5 × 5	Syrnikov (1971)

Compound	Additive/Solvent	Method	Temp.	Rate/Time	Size	Reference
RbₓWO₃			555			
RbZnCrF₆		—	650–	—	1	Nouet et al. (1971)
SbSI	SbI₃	SC	450–	12	10 × 5 × 0.5	Mori and Tamura (1964)
SbSI	SbI₃	SC	475–350	0.4–0.75	30 × 6 × 6	Nassau et al. (1970)
ScN	Hg	CR/SC	~320	—	small	Busch et al. (1970)
Sc₂O₃	PbO/PbF₂	SC	1260–1000	3	2	Feigelson (1971)
Sc₂TiO₅	Na₂WO₄	SC	1460–1000	2	—	Ito (1971)
Se	Tl₂Se₃	VLS	185	25 μm/h	0.5 × 0.03	Keezer and Wood (1966)
Si	several metals	SC	800–1000–	0.15, 0.7, 1.5	small	Faust et al. (1968)
Si	Al, Ag, Au, Zn	SC	980–1050–	0.7	small	Faust and John (1964)
Si	Ga, Sn	SC	500	0.15, 1.5	small	Faust and John (1964)
Si	Au	VLS(X)	950–1150	60 h	15 μm layer	Filby and Nielsen (1966)
Si	Au, Ni	VLS(X)			Whisker	James and Lewis (1965)
Si	Ga, In	SC	—		small	Keck (1953)
Si	Au	VLS	~1050		15 × 0.3 × 0.3	Wagner and Doherty (1966)
Si	Au	VLS(X)	950		0.2 Whisker	Wagner and Ellis (1964)
Si	Au etc.	VLS	~750–950	1 μm/min.	0.2 Whisker	Wagner and Ellis (1965)
SiC	Si	TR	~1600		15 × 0.1 × 0.1	Bartlett et al. (1967)
SiC	Si	VLS	1400–1450		5 × 1 × 1	Berman and Ryan (1971)
SiC	Si	TR/SC	1450–2600		1	Beckmann (1963)
SiC	Si	SC	1800–1400		0.2	Bibr et al. (1966)
SiC	Cr	THM	1800	~6 μm/min.	small	Gillessen and Von Münch (1973)
SiC	Si	TSM	~1750		10 × 3 × 3	Griffiths and Mlavsky (1964)
SiC	Si/Al	TSSG	~1665		4	Halden (1960)
SiC	Cr	LPE	1600		0.1	Hall (1958)
SiC	Cr	TR	1750–1800		12	Kalnin and Tairov (1966)
SiC	Cr	TSM/LPE	1750		210 μm/h	Knippenberg and Verspui (1966)
SiC	Co/Si	TSM	~1650		small	Kumagawa et al. (1970)
SiC	Si	TSSG	~1600–2000	8–24 h	10 × 2 × 0.2	Marshall (1969)
SiC	Si	SC	2000	50°/cm	1 g	Marshall (1969)
SiC	Cr	TR	1500		2 × 0.5 × 0.5	Nelson et al. (1966)
SiC		TR	—		small	Shaskov and Shushlebina (1964)
SiC	Cr	SC	1900–1640	26	few mm	Wolff et al. (1969)
SiC	Cr	TSM	1650–1900	3 mm/day	5 × 5 × 2	Wolff et al. (1969)
SiC	Cr/Si, Cr/Ta	TSM	1400–1700	~0.1/h	4 × 0.3	Wright (1965)
Si₃N₄	Si	TR	1650	6–12 h		Inomata and Yamane (1974)
SiO₂	Li₂O/MoO₃	EV	1165	5 days	0.5	Osmer and Chase (1972)

Crystal	Solvent	Technique	Temp. range	Cooling rate duration	Crystal size (mm)	Reference
SiO_2(Coesite)	$Na_2SiO_3/(NH_4)_2HPO_4$	HPS	750	35 kb, 15 h	small	Robertson et al. (1957)
SiP_2	Sn/Mg	SC	–750	—	$5 \times 0.5 \times 0.1$	Spring-Thorpe and Pamplin (1968)
SiP_2	Mg/Sn	SC	1150–400	10	$20 \times 2 \times 2$	Spring-Thorpe (1969)
Sm-compounds see also R-						
$SmAlO_3$	PbO/PbF_2	SC	1290–960	2.3	2	Garton and Wanklyn (1967b)
$SmAlO_3$	PbO/PbF_2	SC	1200–800	1.5	~10	Zonn and Joffe (1968)
SmB_6	Al	SC	1500–1000		3	Sturgeon et al. (1974)
$Sm_{1-x}Dy_xFeO_3$	$PbO/PbF_2/B_2O_3$	SC	1300–900	0.5		Pierce et al. (1969)
$SmFeO_3$	PbF_2/PbO	SC	1280–900		0.1 g	Schieber (1964)
$(Sm, Tb)FeO_3$	$PbO/PbF_2/B_2O_3$	SC	1300–900	0.5	15	Van Uitert et al. (1970a)
$Sm_3Fe_5O_{12}$	PbF_2/PbO	SC	1240–900		0.25 g	Schieber (1964)
$Sm_3Ga_5O_{12}$	PbF_2/PbO	SC	1240–900		85 mg	Schieber (1964)
$Sm_3Ga_5O_{12}$	PbF_2/PbO	SC	1250–		—	Schieber and Holmes (1965)
$(Sm, Tb)FeO_3$	$PbO/PbF_2/B_2O_3$	SC/EV	1300–970	0.5	—	Quon et al. (1971)
SnO_2	PbF_2	SC	1200–900	3	1	Hart and White, see White (1966)
SnO_2	Cu_2O	EV?	1250	1 week	$10 \times 1 \times 1$	Kunkle and Kohnke (1965)
$SnZnAs_2$	Sn	SC	800–	1.3	small	Faust and John (1964)
$SrAl_{12}O_{19}$(Pb)	Pb_2OF_2	SC	1260–1000	3	$3 \times 3 \times 3$	Linares (1962c)
$SrCrO_4$	LiCl	SC	800–	0.7	1	Packter and Roy (1971)
$SrEu_2Fe_2O_7$	PbO	SC	1250–1000	6	$1 \times 1 \times 0.4$	Drofenik et al. (1973)
Sr_2EuFeO_5	SrO(?)	SC	1560–1540		0.4×0.3	Drofenik et al. (1974)
$SrFe_{12}O_{19}$	$Fe_2O_3 + SrCl_2$	CR	1250	50	1	Brixner (1959), Brixner and Weiher (1970)
$SrFe_{6-x}Al_xO_{19}$	SrO/B_2O_3	SC	1350–1000	6–35	$10 \times 10 \times 0.5$	Goto and Takahashi (1973)
$SrFe_{12}O_{19}$	Na_2O	SC				Shick and Buessen (1969), see Gambino and Leonhard (1961)
$Sr_2KTa_5O_{15}$	$K_3AlF_6/SrCl_2$	SC	1350–900	5	$8 \times 4 \times 4$	Sugai and Wada (1971)
$Sr_2MgGe_2O_7$	GeO_2	TSSG/SC	1468–	0.5	12 g	Belruss et al. (1971)
$SrMoO_4$	$LiCl, SrCl_2$	SC	800–	0.7	4	Packter and Roy (1971)
$SrNb_2O_7$	$SrCl_2$	CR	1000, 1200	2 h	small	Brixner and Weiher (1970)
$Sr_5(PO_4)_3Cl$	$SrCl_2$	CR	1000, 1200	2 h	10	Brixner and Weiher (1970)
$SrRuO_3$	$SrCl_2$	SC	1260–800	2	$2 \times 2 \times 1$	Bouchard and Gillson (1972)
$SrSO_4$	$LiCl, SrCl_2$	SC	1000–	0.7	1	Packter and Roy (1971)
$SrSO_4$	NaCl	SC	)50–25	~38	6×2	Patel and Bhat (1971)
S-SO	M-Cl					

Compound	Flux	Method	Temperature	Time	Size	Reference
SrSO₄	LiCl, NaCl, Li₂SO₄	SC				
SrSnO₃	SrCl₂	CR	1000, 1200	2 h	small	Brixner and Weiher (1970)
SrSnO₃	SrCl₂	EV	800–900	—	—	Smith and Welch (1960)
SrSnB₂O₆	SrB₂O₄	SC	1340–	5–50	small	Schultze et al. (1971)
Sr₂Ta₂O₇	SrCl₂	CR	1000, 1200	2 h	small	Brixner and Weiher (1970)
SrTiO₃	TiO₂	TSSG/SC	1530–	0.5	7.5 g	Belruss et al. (1971)
SrTiO₃	SrCl₂	CR	1000, 1200	2 h	small	Brixner and Weiher (1970)
SrTiO₃	Fluorides	SC	1200–700	—	15 × 0.2 × 0.2	Izvekov et al. (1968)
SrTiO₃	SiO₂	SC	1410–900	6	2.5	Robbins (1968)
SrTiO₃	KF/LiF	SC	1200–770	2.7	7 × 7 × 7	Sugai et al. (1968)
SrTiO₃	KF/K₂MoO₄	SC	1200–770	2.7	4 × 4 × 4	Sugai et al. (1968)
Sr₂VO₄Br	SrBr₂	CR	800, 1000	2 h	1	Brixner and Bouchard (1970)
Sr₃VO₄Cl	SrCl₂	CR	1000, 1200	2 h	1	Brixner and Bouchard (1970)
SrWO₄	SrCl₂	CR	1000, 1200	2 h	small	Brixner and Weiher (1970)
SrWO₄	LiCl	SC	700–800	—	small	Packter and Roy (1971, 1973)
SrWO₄	Na₂W₂O₇	SC	1000–	—	small	Schultze et al. (1967)
SrWO₄	Na₂W₂O₇	SC	1100–1250–700	2.5	10 × 5 × 5	Van Uitert and Soden (1960)
SrZrO₃	KF	SC	1250–	7	—	Nemeth and Tinklepaugh (1966)
SrZrB₂O₆	SrB₂O₄	SC	1300–	5–50	small	Schultze et al. (1971)
TaC	Al, Fe	SC	1500–2100–	—	—	Robbins (1959)
TaC	Fe	SC	2200–1450	10	2	Rowcliffe and Warren (1970)
Tb-compounds see also R-						
TbAlO₃	PbO/PbF₂	EV	1207	21 days	2	Garton and Wanklyn (1967b)
TbFeO₃(Cr)	PbF₂/B₂O₃	EV	1300	5 days	5	Van Uitert et al. (1970)
TbNbO₄	Bi₂O₃/V₂O₅	SC	1330–900	1	5	Garton et al. (1972c)
ThO₂₋ₓ	PbO	SC	1260–1000	3	1–2	Feigelson (1971)
ThO₂	PbF₂/Bi₂O₃	SC	1260–950	4	3.5–6	Chase and Osmer (1964)
ThO₂	PbF₂/Bi₂O₃	SC	1250–1000	2–4	6 × 4 × 3	Chase and Osmer (1967)
ThO₂	Li₂W₂O₇/B₂O₃	TR/EV(X)	~1250	ΔT=50–75	10 × 3	Finch and Clark (1965)
ThO₂	PbO/V₂O₅	SC	1330–1000	1	2.5 × 2.5	Garton et al. (1972b)
ThO₂	PbF₂/B₂O₃	EV				Grodkiewicz and Nitti (1966)
ThO₂	Na₂B₄O₇	EV			1–2	Harari et al. (1967)
ThO₂	Li₂O/MoO₃	SC	1100–700	3	2	Hart and White, see White (1966)
ThO₂	NaF/PbF₂/B₂O₃	SC	1300–1000		11 × 3 × 3	Linares (1967c)
ThO₂	NaF/B₂O₃	TR(X)	920	ΔT=75	0.2/day	Linares (1967c)
ThSiO₄	Li₂O/MoO₃	SC	1225–900	2–4	2–7	Chase and Osmer (1966a)
ThSiO₄	Li₂W₂O₇, Li₂Mo₂O₇, Na₂W₂O₇	TR	1200	10–20°/cm	7 × 5 × 2	Finch et al. (1964)

Crystal	Solvent	Technique	Temp. range	Cooling rate duration	Crystal size (mm)	Reference
$ThTi_2O_6$	$Na_2B_4O_7$	EV	—	—	5–8	Harari et al. (1967)
TiB_2	Fe	SC	—	—	—	Bernard (1963)
TiB_2	Fe	SC	1700–1400	5	~3	Nakano et al. (1971)
TiB_2	Al	SC	1450–25	—	~2	Higashi and Atoda (1970)
TiO_2	$Na_2B_4O_7/LiF$	SC	1200–800	0.5–1	~3	Anikin et al. (1965)
TiO_2	Na_2O/B_2O_3	EV	1200	—	$15 \times 0.5 \times 0.5$	Berkes et al. (1964)
TiO_2	$Na_2B_6O_{10}$	SC/EV	1300–1000	22	$30 \times 1.5 \times 0.5$	Berkes et al. (1965)
TiO_2	PbO/V_2O_5	SC	1330–1000	1	$20 \times 1.5 \times 1.5$	Garton et al. (1972b)
TiO_2	PbF_2/B_2O_3	EV	—	—	—	Grodkiewicz and Nitti (1966)
TiO_2	$Na_2B_4O_7$	EV	1200	—	5–8	Harari et al. (1967)
TiO_2	Li_2O/MoO_3, Na_2O/MoO_3, K_2O/MoO_3	SC	1100–700	3	$10 \times 3 \times 3$	Hart and White, see White (1966)
TiO_2	Fluorides	SC	1200–700	—	$10 \times 1 \times 1$	Izvekov et al. (1968)
TiO_2	NaCl					Packter (1968)
TiO_2	Li_2O/WO_3	SC	1250–800	5	$20 \times 1.5 \times 1.5$	Sugai et al. (1967)
TiO_2	Na_2O/B_2O_3	SC	1370–980	1	100	Russell et al. (1962)
TiO_2	Na_2SO_4	CR/SC	~1000	50–80	small	Wilke (1964)
Ti_3O_5	$Na_2B_4O_7/B_2O_3$	CR/EV	1360–1250		$5 \times 0.5 \times 0.2$	Bartholomew and White (1970), Roy and White (1972)
Ti_4O_7	$Na_2B_4O_7/B_2O_3$	CR/EV	1200–1150		$5 \times 0.5 \times 0.2$	Bartholomew and White (1970), Roy and White (1972)
Ti_5O_9, Ti_6O_{11}, Ti_8O_{15}, $Ti_{10}O_{19}$	$Na_2B_4O_7/B_2O_3$	CR/EV	1350–1150		small	Bartholomew and White (1970), Roy and White (1972)
Tm-compounds see also R-						
TmB_2	Tm	CR	1750	20 min.	$1 \times 1 \times 0.01$	Castellano (1972)
$TmFeO_3$	Bi_2O_3	SC	1300–950	0.5–1	3×12	Gendelev and Titova (1971)
$Tm_3Ga_5O_{12}$	PbO/B_2O_3, PbO/PbF_2	SC	1280–900	—	0.1 g	Schieber (1967)
UO_2	$Na_2B_4O_7(+H_2)$	EV?	1100	8 h	0.02	Bard (1957)
VGa_5	Ga	SC	900–	~20	0.05	Girgis et al. (1966)
VO_2	V_2O_5	CR/EV	1215	30–50 h	~17	Aramaki and Roy (1968)
VO_2	V_2O_5	SC/CR	1230–	—	small	Kimizuka et al. (1970)
VO_2	V_2O_5	CR/EV	900	2 days	1	Kitahiro and Watanabe (1967)

Compound	Flux	Method	Temperature	Time/rate	Size	Reference
VO₂	V₂O₅	CR/EV	1000	7 days	—	Sasaki and Watanabe (1966)
VO₂	V₂O₅	CR/SC	1000–750	4	10 × 2 × 2	Sobon and Greene (1966)
V₂O₃(Cr)	KF/V₂O₅	CR/SC	930–940	90 h, 40	6.5 × 2	Foguel and Grajower (1971)
V₂O₃	KF/V₂O₅	CR/SC	950	72 h	5 × 1	McWhan and Remeika (1970)
V₂O₃	V₂O₅	CR				Takei and Koide (1966)
WC	Co	TSSG/SC	~1700–	0.15	10	Gerk and Gilman (1968)
WC	Co	SC	1600–1500	60–180	5	Takahashi and Freise (1965)
WO₃	Na₃AlF₆	THM	870–780	0.5/min.	—	Vigdorovich and Marychev (1964)

Y-compounds see also R-

Compound	Flux	Method	Temperature	Time/rate	Size	Reference
YAlO₃	PbO/PbF₂	EV	1260	10 days	2	Garton and Wanklyn (1967b)
Y₃Al₅O₁₂(Fe)	PbO/PbF₂	SC	1200–950	5	20	Gendelev and Titova (1968)
Y₃Al₅O₁₂(Nd)	PbF₂/PbO/B₂O₃	SC	1300–950	0.4–1	15	Ashida (1968)
Y₃Al₅O₁₂	BaO × 0.6B₂O₃	SC	1125–835	—	1 × 1 × 1	Bakradze and Kusnetsova (1969)
Y₃Al₅O₁₂	PbO/B₂O₃	TSSG/SC	1463–	0.5	8 g	Carlo (1973)
Y₃Al₅O₁₂	PbO/PbF₂	SC	1250–1000	4	10	Chase and Osmer (1969)
Y₃Al₅O₁₂(Si)	PbO/PbF₂	SC	1250–1000	4	~10	Chase and Osmer (1972)
Y₃Al₅O₁₂(Nd)	PbO/PbF₂/B₂O₃	SC	1300–950	0.5	150 g	Grodkiewicz et al. (1967)
Y₃Al₅O₁₂(Cr)	PbO/PbF₂/B₂O₃	SC	1300–950	0.25–2		Kestigian and Holloway (1967)
Y₃Al₅O₁₂	PbF₂/PbO	SC	1150–800	4.3	1.5 g	Lefever et al. (1961)
Y₃Al₅O₁₂	PbO/PbF₂	SC/EV	1150–	4.3	—	Lefever and Chase (1962)
Y₃Al₅O₁₂	PbO/B₂O₃	SC	1250–950	1	10 × 7 × 7	Linares (1962a)
Y₃Al₅O₁₂	Pb₂OF₂	SC	1260–1000	1	2.5 g	Linares (1962a)
Y₃Al₅₋ₓGaₓO₁₂	PbO/B₂O₃	SC	1300–	3–5	large	Marezio et al. (1968b)
Y₃Al₅O₁₂(RE)	PbF₂/B₂O₃	SC	1300–850	>1	7 × 6 × 6	Monchamp et al. (1967)
Y₃Al₅O₁₂	PbO/PbF₂/B₂O₃	SC	1290		25	Timofeeva and Kvapil (1966)
Y₃Al₅O₁₂(Nd + Cr)	PbO/PbF₂/B₂O₃	SC	1300–1000	1	60 g	Timofeeva (1967, 1968)
Y₃Al₅O₁₂(Nd)	PbO/PbF₂/B₂O₃	SC	1300–950		small	Van Uitert et al. (1965)
YAu	Hg	CR	700–1100		0.4	Kirchmayr (1965)
(Y, Bi)₃(Ca, Fe, Si)₅O₁₂	PbO/Bi₂O₃/MoO₃/B₂O₃	SC	1320–	2.3		Schieber et al. (1968)
YCo	Hg	CR	700–1100	—	small	Kirchmayr (1965)
YCrO₃	PbF₂/B₂O₃	EV	—	—	—	Grodkiewicz and Nitti (1966)
YCrO₃	PbF₂	EV	1350	50 h	8 × 5 × 5	Tsushima (1967)
YCu	Hg	CR	700–1100	—	small	Kirchmayr (1965)
Y₃₋ₓErₓFe₅O₁₂	PbO/B₂O₃	SC	1300–1000	2	2.5	Remeika (1963)
Y₁₋ₓEuₓVO₄	NaVO₃	SC	1200–	—	—	Van Uitert et al. (1962)

Crystal	Solvent	Technique	Temp. range	Cooling rate duration	Crystal size (mm)	Reference
YFe	Hg	CR	700–1100	—	small	Kirchmayr (1965)
$YFeO_3$	$BaO/BaF_2/B_2O_3$	SC	1175–1050	1	$4 \times 3 \times 3$	Hiskes et al. (1972)
$YFeO_3$	$PbO/PbF_2/B_2O_3$	SC	1300–950	0.5–1.7	—	Quon et al. (1971)
$YFeO_3$	PbO/B_2O_3	SC	1300–850	2	$15 \times 5 \times 5$	Remeika (1963)
$Y_3Fe_5O_{12}$	PbF_2/PbO	SC	1250–	2	9.2 g	Barry and Roberts (1961)
$Y_3Fe_5O_{12}$	$PbO/PbF_2/B_2O_3$	SC/TSSG	1200–1000	0.5	25; 57 g	Bennett (1968)
$Y_3Fe_5O_{12}$	PbO/B_2O_3	SC	1320–950	1–2	10	Bibr et al. (1966)
$Y_3Fe_5O_{12}(Si)$	PbO/PbF_2	SC	1260–1010	0.65	—	Broese et al. (1967)
$Y_3Fe_5O_{12}$	$BaO/B_2O_3/BaF_2$	TSSG/SC	1194–	0.3	17.5	Carlo (1973)
$Y_3Fe_5O_{12}$	$PbF_2/PbO/B_2O_3$	SC	1390–	—	—	Chabria (1967)
$Y_3Fe_5O_{12}(Si)$	PbO/PbF_2	SC	1250–1000	4	~10	Chase and Osmer (1972)
$Y_3Fe_5O_{12}$	$PbO/PbF_2/B_2O_3$	SC	1300–900	0.5	250 g	Grodkiewicz et al. (1967)
$(Y, Ca)_3(Fe, Ir, Zn)_5O_{12}$	$PbO/PbF_2/B_2O_3$	SC, ACRT	1120–1000	0.5	8	Hansen et al. (1973)
$Y_3Fe_5O_{12}$	PbO/PbF_2	SC	1250–980	0.5	$15 \times 15 \times 10$	John and Kvapil (1968)
$Y_3Fe_5O_{12}$	PbO/B_2O_3	SC	1275–1000	1	—	Jonker (1974)
$Y_3Fe_5O_{12}$	BaO/B_2O_3	TR(X), TSSG	1080	$\Delta T = 25$	$44 \times 10 \times 10$	Kestigian (1967)
$Y_3Fe_5O_{12}$	PbO/PbF_2	SC	1350–	1–3	—	Komatsu and Sunagawa (1964)
$Y_3Fe_{5-2x}Ni_xGe_xO_{12}$	$PbO/PbF_2/B_2O_3$	SC	1250–900	1.5–2	—	Krishnan (1966)
$Y_3Fe_5O_{12}(Cu)$	$PbO/PbF_2/B_2O_3$	SC	1220–950	1–2	—	Krishnan (1968a)
$Y_3Fe_{5-x}Cr_xO_{12}$	$PbO/PbF_2/B_2O_3$	SC	1330/1240–1000	0.5–1	6–18	Krishnan (1972)
$Y_3Fe_5O_{12}$	Na_2O/WO_3	SC	1300–	2	—	Kunnmann et al. (1965)
$Y_3Fe_5O_{12}$	PbO/PbF_2	SC	1250–1000	~1	$3 \times 3 \times 3$	Kvapil (1966)
$Y_3Fe_5O_{12}$	BaO/B_2O_3	TR(X)	1170	$\Delta T = 40$	—	Laudise et al. (1962)
$Y_3Fe_5O_{12}$	BaO/B_2O_3	SC	1180–1000	1	0.5 g	Linares (1962b)
$Y_3Fe_5O_{12}$	BaO/B_2O_3	TR(X)	1100	$\Delta T = 50$, 1.3/day	~10	Linares (1964)
$Y_3Fe_5O_{12}$	BaO/B_2O_3	LPE	1050	4–20 sec.	~400 Å	Linares et al. (1965)
$Y_3Fe_5O_{12}(Al, Ga)$	PbO/PbF_2	SC	1250–1000	1	—	Linares (1965a)
$Y_3Fe_5O_{12}(Al, Ga)$	PbO/B_2O_3	SC	1250–1000	1	—	Linares (1965a)
$Y_3Fe_5O_{12}(Al, Ga)$	BaO/B_2O_3	SC	1250–1000	1	—	Linares (1965a)
$Y_3Fe_5O_{12}(Al, Ga)$	BaO/B_2O_3	TR	1100		—	Linares (1965a)
$Y_3Fe_5O_{12}$	PbO/B_2O_3	LPE/SC	1100–1000	25–100 Å/sec.	0.5–100 μm	Linares (1968b)
$Y_3Fe_5O_{12}$	PbO/PbF_2	SC	1320–950	—	—	Makram (1968)
$Y_3Fe_5O_{12}$	PbF_2/PbO	SC	1250–	3–4	10–15	Makram and Krishnan (1964, 1967)
$Y_3Fe_5O_{12}(Ga)$	$PbO/PbF_2/B_2O_3$	HPS/SC	1270–950	2.1		

(continued)	PbO, B₂O₃				8	
$Y_3Fe_5O_{12}$	PbO	SC	1350–780	1–5	—	Nielsen (1960b)
$Y_3Fe_5O_{12}$	$4PbF_2/3PbO$	SC	1260–950	0.5–5	19 g	Nielsen (1960a)
$Y_3Fe_5O_{12}$	$3PbF_2/2PbO$	SC	1260–1040	0.5	300 g	Nielsen (1964)
$Y_3Fe_{5-x}Ga_xO_{12}$	PbO/PbF_2	SC	1270–1050	~1	~20	Nielsen et al. (1967)
$Y_3Fe_{5-2x}Co_xGe_xO_{12}$	PbO	SC	1280–900	1.8	—	Okada et al. (1963)
$Y_3Fe_5O_{12}$	PbO	SC	1350–1030	2.4	—	Porter et al. (1958)
$Y_3Fe_5O_{12}$	PbO/B_2O_3	SC	1300–1000	2	5	Remeika (1963)
$Y_3Fe_5O_{12}$	$PbO/PbF_2/B_2O_3$	SC/HPS	1330–1060	0.75	12	Robertson and Neate (1972)
$Y_3Fe_{5-x}In_xO_{12}$	PbO/PbF_2	SC	—	—	—	Rubinstein et al. (1965)
$Y_3Fe_5O_{12}(Ca, Si, Bi)$	PbF_2/PbO	SC	1280–900	—	1.8 g	Schieber (1964)
$Y_3Fe_5O_{12}$	$PbO/B_2O_3/MoO_3/Bi_2O_3$	SC	—	1.8	—	Schieber et al. (1972)
$Y_3Fe_5O_{12}$	PbO/PbF_2	SC	1270–950	4	10–14 g	Suzuishi and Ito (1967)
$Y_3Fe_5O_{12}$	PbO/B_2O_3	SC	1350–900	0.5	12	Titova (1959, 1962)
$Y_3Fe_5O_{12}$	PbO/PbF_2	SC	1180–950	—	30, 49 g	Tolksdorf (1968)
$Y_3Fe_5O_{12}$	PbO/PbF_2	ACRT	—	1	60 × 60, 250 g	Tolksdorf (1974b)
$Y_3Fe_5O_{12}$	BaO/B_2O_3	TSSG	1000	$\Delta T = 10°$	15 × 15 × 6	Tolksdorf (1974a)
$Y_3Fe_5O_{12}$	BaO/B_2O_3	TSM	—	50 μm/h	3	Tolksdorf (1974a)
$Y_3Fe_{5-x}Ga_xO_{12}$	$PbO/PbF_2/B_2O_3$	ACRT/SC	1100–1010	1	30 × 30	Tolksdorf and Welz (1972)
$Y_3Fe_5O_{12}$	PbO/PbF_2	TR(X)/ACRT	1070	1	~22 × 20, 14 g	Tolksdorf and Welz (1973)
$Y_3Ga_5O_{12}$	$PbO/YF_3/B_2O_3$	SC/CR	1300–900	4	1	Watanabe et al. (1970)
$Y_3Ga_5O_{12}$	PbO/PbF_2	SC	1250–1000	4	7 × 7 × 5	Chase and Osmer (1967)
$Y_3Ga_5O_{12}$	PbO/PbF_2	SC	1250–1000	0.5–5	10	Chase and Osmer (1969, 1972)
$Y_3Ga_{4.95}Fe_{0.05}O_{12}$	Pb_2OF_2	SC	1260–950	2	—	Nielsen (1960a)
$Y_3Ga_5O_{12}(Nd)$	PbO/B_2O_3	SC	1300–1000	0.9, 1.6	2.5	Remeika (1963)
$Y_3Ga_5O_{12}$	PbO/PbF_2	SC	1250–900	3	0.5–2 g	Suzuishi et al. (1968)
$YMnO_3$	Bi_2O_3	SC	1400–	—	Platelets 50 μm	Bokov et al. (1964)
$YMnO_3$	Bi_2O_3	SC	1200–	3	—	Kohn and Tasaki (1965)
YN	Hg	CR/SC	320	—	small	Busch et al. (1970)
YNi	Hg	CR	700–1100	—	small	Kirchmayr (1965)
YOF	PbF_2	SC	1340–900	3	3 × 2	Garton and Wanklyn (1968)
Y_2O_3	PbF_2	SC	1200–900	3	5	Hart and White, see White (1966)
$Y_2SiO_5, Y_4Si_3O_{12}$	$Li_2Mo_2O_7$	—	1100	—	—	Harris and Finch (1965)
Y_2SiO_5	KF	SC	1300–850	1–2.5	4 × 3 × 2	Bondar et al. (1968)
$Y_2Ti_2O_7$	$TiO_2/BaO/B_2O_3$	TSSG/SC	1330–	0.5	14 g	Belruss et al. (1971)
YVO_4	V_2O_5	SC	1200–900	3	2	Hart and White, see White (1966)
YVO_4	V_2O_5	TSSG/SC	1050–900	50 μm/h	15 × 6	Loriers and Vichr (1972)
YVO_4	$NaVO_3/Na_2B_4O_7$	TR(X)	—	4 days	15 × 3 × 3	Phillips and Pressley (1967)

Crystal	Solvent	Technique	Temp. range	Cooling rate duration	Crystal size (mm)	Reference
YWO_4Cl	YCl_3	SC	900–700	3	0.4	Yocom and Smith (1973)
Yb-compounds see also R-						
YbB_6	$B_2O_3(?)$	VLS	—	—	0.6	Rea and Kostiner (1971)
$YbCrO_3$	$PbF_2/B_2O_3/PbO_2$	EV	1260	9 days	3 × 3 × 2	Wanklyn (1972)
$YbFeO_3$	$PbF_2/PbO/B_2O_3$	SC	1300–1100	3	12 × 9	Damen and Robertson (1972)
$YbFeO_3$	$PbO/PbF_2/B_2O_3$	SC	1250–900	4	30 × 15 × 3	Feigelson (1971)
$YbFeO_3$	$PbO/PbF_2/B_2O_3$	SC/EV	1300–950	0.5	—	Quon et al. (1971)
$YbFeO_3$	PbO/B_2O_3	SC	1300–1000	2	2.5	Remeika (1963)
$Yb_3Fe_5O_{12}$	PbO/PbF_2	SC	1280–900		0.4 g	Schieber (1967)
$Yb_3Fe_2Ga_3O_{12}$	$PbO/PbF_2/B_2O_3$	SC	1180–	2.3	3	Schieber et al. (1968)
$Yb_3(Fe, Ga)_5O_{12}$	$PbO/MoO_3/B_2O_3/Bi_2O_3$	SC				Schieber et al. (1972b)
$Yb_3Fe_5O_{12}(Ca, Si)$	PbO/B_2O_3	SC	1280–900		60 mg	Schieber (1967)
$Yb_3Ga_5O_{12}$	$YbCl_3$	SC	1050–	8	1 × 1 × 0.1	Brandt and Diehl (1974)
$YbOCl$	KF	SC	1300–850	1–2.5	4 × 3 × 2	Bondar et al. (1968)
$Yb_2Si_2O_7$	$YbCl_3$	SC	1050–	8	1.5 × 1.5 × 0.5	Diehl and Brandt (1974)
$Yb_3Si_2O_8Cl$						
$ZnAl_2O_4$	$PbF_2/PbO/SiO_2/B_2O_3$	SC	1350–700	0.5–2	12	Bonner and Van Uitert (1968)
$Zn(Al, Ga)_2O_4$	$PbF_2/Bi_2O_3/B_2O_3$	SC	1250–1000	4	7 × 7 × 5	Chase and Osmer (1967)
$ZnAl_2O_4$	PbF_2	SC	1250→	1–2	10	Giess (1964)
$ZnAl_2O_4$	Pb_2OF_2	SC	1260–1000	3	3 × 3 × 3	Linares (1962c)
$ZnAl_2O_4(Mn)$	PbF_2	EV	1200	24 h	1 × 1 × 1	Tsushima (1966)
$Zn_2Ba_2Fe_{12}O_{22}$	$Ba_2B_2O_5$	TSSG/TR	1000	$\Delta T = 10°$	~20 × 16 × 4	Tolksdorf (1973)
$ZnCr_2O_4$	Na_2O/WO_3	SC	1250–	5	3	Kunnmann et al. (1965)
$ZnCr_2Se_4$	$CrCl_3$	VLRS	950–900		8	Miyatani et al. (1968)
$ZnCr_2Se_4$	$CrCl_3$	VLRS	950–900	—	10	Von Neida and Shick (1969)
$ZnFe_2O_4$	$BaO/Bi_2O_3/B_2O_3$	SC	1320–1150	1	<10	Elwell et al. (1972)
$ZnFe_2O_4$	PbO	SC	1300–	~5		Remeika (1958b)
$ZnGeP_2$	Sn	SC	1100–600	4–16	small	Mughal et al. (1969)
$ZnGeP_2$	Bi	SC			6 × 1 × 0.5	Samogyi and Bertoti (1972)
$ZnGeP_2$	Sn	SC	~750	—		Spring-Thorpe and Pamplin (1968)
$ZnMn_2O_4$, $Zn_{1.25}Mn_{1.75}O_4$	$NaKSO_4$	CR	~1000	20 h	small	Petzold et al. (1971)
$ZnMn_2O_4$	PbF_2	EV	1100	120 h	2 × 2 × 2	Tsushima (1967)
ZnO	PbF_2	SC	1250–900	4	12 × 10 × 7	Chase and Osmer (1967)
ZnO	PbF_4	TR	900	3.5/day		

ZnO	PbF₂	SC	1150–800	1–10	30 × 30	Nielsen and Dearborn (1960)
ZnO	PbF₂	SC	1160–	—	10 × 10 × 1	Timofeeva (1968)
ZnO	V₂O₅, P₂O₅	SC	1330–980	1	20 × 0.3	Wanklyn (1970)
ZnRh₂O₄	PbO	SC	1330–880	3.5	1.5 × 1.5 × 1.5	Arlett (1968)
ZnS	Ga	SC	—	—		Bertoti et al. (1965)
ZnS	Ga, In, Sn, Tl	SC	1180	38	~5 × 0.2	Harsy (1967, 1968)
ZnS	ZnF₂	SC	1000–800	1	12 × 6 × 0.9	Linares (1962d)
ZnS	PbCl₂	SC/TR	800–500	1–10	5 × 5 × 1.2	Linares (1968a)
ZnS	NaCl	EV/SC	1070–1040	10–20	10 × 0.2 × 0.2	Mita (1961)
ZnS	KCl, etc.	SC	1200–900	2–10	10 × 1 × 1	Mita (1962)
ZnS	KCl, KJ/ZnCl₂	TR	845	30 days	10	Parker and Pinnell (1968)
ZnS	Bi, Sn	SC	—	—	~4	Rubenstein (1968)
ZnS	Na₂Sₓ	CR/SC/EV	600–300	~25	1 × 1 × 1	Scheel (1974)
ZnSb	Sb	SC	546–505	1	—	Eisner et al. (1961)
ZnSb	Sb	TR	560–575	0.3 μm/sec.		Eisner et al. (1961)
Zn₇Sb₂O₁₂	PbF₂	SC	1150–800	1–5	6	Linares and Mills (1962)
ZnSe	Zn	SC	950–	1.3	small	Faust and John (1964)
ZnSe	Ga, In	SC	~1200–	~20	~1	Harsy (1968)
ZnSe	Bi, Sn	SC	—	—	~4	Rubenstein (1968)
ZnSe	Ga, In	SC	—	4	3.5	Wagner and Lorenz (1966)
ZnSeₓTe₁₋ₓ	Te	TSM	500	1–5 days	35 × 20 × 10	Steininger and England (1968)
ZnSiAs₂	Sn	SC	1000–	5	1	Gentile and Stafsudd (1974)
ZnSiAs₂	Sn	SC	–750	—		Spring-Thorpe and Pamplin (1968)
ZnSiP₂	Bi, Sb, Sn	SC		20	25 × 3.5	Borshchevskii et al. (1967)
ZnSiP₂	Sn	SC	1300–		~2	Buehler and Wernick (1971)
ZnSiP₂	Sn	SC	1100–600	4–16	25 × 5 × 2	Mughal et al. (1969)
ZnSiP₂	Sn	SC	1120–680	7.5	17 × 2.5 × 0.3	Mughal and Ray (1973)
ZnSnAs₂	Sn	SC	–750	—	10	Spring-Thorpe and Pamplin (1968)
ZnSnAs₂	Sn	SC	800–	1.3	small	Faust and John (1964)
ZnSnP₂	Sn	SC	1000–	5	1	Gentile and Stafsudd (1974)
ZnSnP₂	Sn	SC	870–	—	3 × 1.5 × 0.5	Loshakova et al. (1966)
ZnSnP₂	Sn	SC	910–350	60	small	Mughal et al. (1969)
ZnSnP₂	Sn	SC	900–600	10	4 × 4 × 0.3	Rubenstein and Ure (1968)
ZnSnSb₂	Sn	SC	–750	—	4 × 4 × 0.5	Spring-Thorpe and Pamplin (1968)
ZnTe	Bi	LPE/SC	~350–250	3–100	30 μm	Goryunova et al. (1968b)
ZnTe	Zn	SC	800–600	30–150	—	Fujita et al. (1971)
ZnTe	Zn	SC	1200–500	12–60	4 × 4 × 0.2	Fuke et al. (1971)

Crystal	Solvent	Technique	Temp. range	Cooling rate duration	Crystal size (mm)	Reference
ZnTe	Te	TR	900	10°/cm	$3 \times 3 \times 3$	Fuke et al. (1971)
ZnTe	Ga	SC	$\sim$1200–	$\sim$20	$\sim$1	Harsy (1968)
ZnTe	Bi, Sn, Zn	SC	–	–	$\sim$4	Rubenstein (1968)
ZnTe	In	LPE/SC	500–480	120	10 μm	Tamura et al. (1971)
ZnTe	Ga, In	SC	–	4	3.5	Wagner and Lorenz (1966)
ZnV_2O_6	KCl	HPS	1030	45 kb, 1 h	0.1	Gondrand et al. (1974)
$ZnWO_4$	$Na_2W_2O_7$	SC	1000–	–	small	Schultze et al. (1967)
$ZnWO_4$	$Na_2W_2O_7$	SC	1100–1250–700	2.5	$10 \times 5 \times 5$	Van Uitert and Soden (1960)
ZrB_2	Fe	SC	1700–1400	5	$\sim$3	Nakano et al. (1971)
ZrO_2	PbF_2	EV/SC	–	–	–	Anthony and Vutien (1965)
ZrO_2	PbF_2	SC	1040–800	3	$8 \times 5 \times 1$	Chase and Osmer (1966b)
ZrO_2	PbF_2	SC	1040–800	4	$4 \times 4 \times 1$	Chase and Osmer (1967)
ZrO_2	$Na_2B_4O_7$/KF	SC	–	–	3	Fujiki and Ono (1972)
ZrO_2	$Na_2B_4O_7$	EV	1120	2–4	5–8	Harari et al. (1967)
ZrO_2	$Na_2B_4O_7$	SC	1250–	5	1	Kleber et al. (1966)
ZrO_2	Na_2O/B_2O_3	SC	1450–930	3	25	Russel et al. (1962)
ZrO_2	Na_2SO_4	CR/SC	$\sim$1000	50–80	small	Wilke (1964)
$ZrSiO_4$	$Li_2Mo_3O_{10}$, $Na_2Mo_3O_{10}$	SC	1400–900	2	2 mm/day	Ballman and Laudise (1965)
$ZrSiO_4$	$Na_2Mo_3O_{10}$	TR	900–1000	$\Delta T = 40$–80		Ballman and Laudise (1965)
$ZrSiO_4$	Li_2O/MoO_3	SC	1250–1000	2	2–7	Chase and Osmer (1966a)
$ZrSiO_4$	Li_2O/WO_3	SC	1374–969	2.3	$3 \times 3 \times 3$	Dharmarajam et al. (1972)
$ZrSiO_4$	Li_2O/V_2O_5	SC	1366–907	1.3	4	Dharmarajam et al. (1972)

References

Adams, I., Nielsen, J. W. and Story, M. S. (1966) *J. Appl. Phys.* **37**, 832.

Agapora, N. N. and Perekalina, T. M. (1968) *Sov. Phys.–Solid State* **9**, 1825.

Agapora, N. N., Sizov, V. A. and Yamzin, I. I. (1969) *Sov. Phys.–Solid State* **10**, 2258.

Aidelberg, J., Flicstein, J. and Schieber, M. (1974) *J. Crystal Growth* **21**, 195.

Airtron, US Office Naval Research, Report NONR-4616(00), ARPA 306-62.

Ananthanarayanan, K. P., Mohanty, C. and Gielisse, P. J. (1973) *J. Crystal Growth* **20**, 63.

Anderson, J. C. and Schieber, M. (1963) *J. Phys. Chem.* **67**, 1838.

Anikin, I. N. (1958) *Growth of Crystals* **1**, 259.

Anikin, I. N., Naumova, I. I. and Rumgantseva, G. V. (1965) *Sov. Phys.–Cryst.* **10**, 172.

Anselm, L. N., Bir, G. L., Mylnikova, I. E. and Petrov, M. P. (1966) *Sov. Phys.–Solid State* **8**, 812.

Anthony, A. M. and Vutien, L. (1965) *Compt. Rend.* **260**, 1383.

Antypas, G. S. (1970) *J. Electrochem. Soc.* **117**, 1393.

Arai, K. I. and Tsuya, N. (1972) *J. Phys. Soc. Japan* **33**, 1581.

Aramaki, S. and Roy, R. (1968) *J. Mat Sci.* **3**, 643.

Arend, H. (1960) *Silikaty* **4**, 29; *Czech. J. Physics* B **10**, 971.

Arend, H. and Novak, J. (1966) *Kristall u. Technik* **1**, 93.

Arend, H., Novak, J. and Coufova, P. (1969) *Growth of Crystals* **7**, 244.

Arlett, R. H. (1968) *J. Am. Ceram. Soc.* **51**, 292.

Arlett, R. H., Robbins, M. and Harkart, P. G. (1967) *J. Am. Ceram. Soc.* **50**, 58.

Ashida, S. (1968) *Kobutsugaku Zasshi* **8**, 407.

Astles, M. G., Smith, F. G. H. and Williams, E. W. (1973) *J. Electrochem. Soc.* **120**, 1750.

Aucoin, T. R., Savage, R. O. and Tauber, A. (1966) *J. Appl. Phys.* **37**, 2908.

Austerman, S. B. (1963) *J. Am. Ceram. Soc.* **46**, 6.

Austerman, S. B. (1964) *J. Nucl. Mat.* **14**, 225.

Austerman, S. B. (1968) *Chem and Phys. of Carbon* **4**, 137.

Austerman, S. B. and Gehman, W. G. (1966) *J. Mat Sci.* **1**, 249.

Austerman, S. B. and Smith, D. K. (1967) *J. Appl. Phys.* **38**, 2705.

Austerman, S. B., Berlincourt, D. A. and Krueger, H. H. A. (1963) *J. Appl. Phys.* **34**, 339.

Austerman, S. B., Myron, S. M. and Wagner, J. W. (1967) *Carbon* **5**, 549.

Austerman, S. B., Wagner, J. W. and Myron, S. M. (1968) US-AEC Report NAA-SR-12486.

Badzian, A. R., Wisniewska, K., Widaj, B., Krukowska-Fulde, B. and Niemyski, T. (1969) *J. Crystal Growth* **5**, 222.

Baker, J. M., Copland, G. M. and Wanklyn, B. M. (1969) *J. Phys. C* **2**, 862.

Bakradze, R. V. and Kusnetsova, G. P. (1969) *Inorg. Mater. (USSR)* **5**, 1257.

Ballman, A. H. (1962) *Am. Min.* **47**, 1380.

Ballman, A. and Laudise, R. A. (1965) *J. Am. Ceram. Soc.* **48**, 130.

Ballman, A. A., Linares, R. C. and Van Uitert, L. G. (1966) US Patent 3.234.135.

Baranov, B. V., Prochukhan, V. D. and Goryunova, N. A. (1967) *Inorg. Mater. (USSR)* **3**, 1477.

Barber, D. J. (1964) *J. Appl. Phys.* **35**, 398.

Bard, R. J. (1957) Los Alamos Report LA-2076.

Barks, E. and Kostiner, E. (1966) *J. Appl. Phys.* **37**, 1423.

u

Barks, R. E. and Roy, D. M. (1967) In "Crystal Growth" (H. S. Peiser, ed.) p. 497. Pergamon, Oxford.

Barks, R. E., Roy, D. M. and White, W. B. (1964) *Bull. Am. Ceram. Soc.* **43**, 255.

Barns, R. L. and Ellis, W. C. (1965) *J. Appl. Phys.* **36**, 2296.

Barry, D. and Roberts, R. W. (1961) *J. Appl. Phys.* **32**, 1405.

Barta, C., Schultze, D., Wilke, K.-Th. and Zemlicka, J. (1971) *Z. Anorg. Allg. Chem.* **380**, 41.

Bartholomew, R. F. and White, W. B. (1970) *J. Crystal Growth* **6**, 249.

Bartlett, R. W., Nelson, W. E. and Halden, F. A. (1967) *J. Electrochem. Soc.* **114**, 1149.

Beckmann, G. E. J. (1963) *J. Electrochem. Soc.* **110**, 84.

Belin, C., Brissot, J. J. and Jesse, R. E. (1972) *J. Crystal Growth* **13/14**, 597.

Belruss, V., Kalnajs, J., Linz, A. and Folweiler, R. C. (1971) *Mat. Res. Bull.* **6**, 899, and private communication (July 1974).

Belt, R. F. (1964) *J. Appl. Phys.* **35**, 3063.

Belt, R. F. (1967) *J. Am. Ceram. Soc.* **50**, 588.

Belt, R. F. (1969) *J. Appl. Phys.* **40**, 1664.

Belyaev, I. N. (1962) *Growth of Crystals* **3**.

Belyaev, I. N. and Nesterova, A. K. (1952) *J. Gen. Chem. USSR* **22**, 469.

Belyaev, I. N., Novosiltsev, N. V., Khodakov, A. L. and Fesenko, E. G. (1951) *Dokl. Akad. Sci. USSR* **78**, 875.

Benes, J., Bednarova, V. and Safrata, S. (1955) *Czech. J. Physics* **5**, 564.

Bennett, G. A. (1968) *J. Crystal Growth* **3/4**, 458.

Berezhnoi, A. A., Bukhman, V. N., Kudinova, L. T. and Mylnikova, I. E. (1968) *Sov. Phys.–Solid State* **10**, 192.

Berger, S. B. and Pinch, H. L. (1967) *J. Appl. Phys.* **38**, 949.

Berkes, J. S. and White, W. B. (1965) *J. Appl. Phys.* **36**, 3276.

Berkes, J. S., White, W. B. and Roy, R. (1964) *Bull. Am. Ceram. Soc.* **43**, 255.

Berkes, J. S., White, W. B. and Roy, R. (1965) *J. Appl. Phys.* **36**, 3276.

Berman, I. and Ryan, C. E. (1971) *J. Crystal Growth* **9**, 314.

Bernal, I., Struck, C. W. and White, J. G. (1963) *Acta Cryst.* **16**, 849.

Bernard, A. (1963) US Patent 3.096.149.

Bertaut, E. F., Forrat, F. and Fang, P. (1963) *Compt. Rend.* **256**, 1958.

Bertoti, I. (1970) *J. Mat. Sci.* **5**, 1073.

Bertoti, J., Lendvay, E., Fahrkas-Jahnke, M., Harsy, M. and Kovacs, P. (1965) *Phys. stat. sol.* **12**, K1.

Besselere, J. P. and LeDuc, J. M. (1968) *Mat. Res. Bull.* **3**, 797.

Bethe, K. and Welz, F. (1971) *Mat. Res. Bull.* **6**, 209.

Bezrukov, V. A., Bezrukov, G. N., Butuzov, V. P., Varagin, V. S., Vorozejkin, K. F., Kirova, N. F. and Litvin, Yu.A. (1966) *Zap. Vsesojuzn. mineral. Obsc.* **95**, 3.

Bezrukov, G. N., Butuzov, V. P. and Korolev, D. F. (1969) *Growth of Crystals* **7**, 91.

Bibr, B., Kvapil, J. and Smid, J. (1966) *Techn. Digest* **31**, 163.

Bidnaya, D. S., Obukhovskii, Y. A. and Sysoev, L. A. (1962) *J. Inorg. Chem. USSR* **7**, 1391.

Bland, J. A. (1961) *Acta Cryst.* **14**, 875.

Blattner, H., Matthias, B. and Merz, W. (1947) *Helv. Phys. Acta* **20**, 225.

Blattner, H., Känzig, W. and Merz, W. (1949) *Helv. Phys. Acta* **22**, 35.

Bless, P. W., Von Dreele, R. B., Kostiner, E. and Hughes, R. E. (1971) *J. Solid State Chem.* **4**,

Blom, G. and Plaskett, T. S. (1971) *J. Electrochem. Soc.* **118,** 1831.
Bogdanov, S. V., Kovalenko, G. M. and Cherepanov, A. M. (1960) *Bull. Acad. Sci. USSR Phys. Ser.* **24,** 1236.
Bokov V. A. and Mylnikova, I. E. (1961a) *Sov. Phys.–Solid State* **2,** 2428.
Bokov, V. A. and Mylnikova, I. E. (1961b) *Sov. Phys.–Solid State* **3,** 613.
Bokov, V. A., Smolenskii, G. A., Kizhaev, S. A. and Mylnikova, I. E. (1964) *Sov. Phys.–Solid State* **5,** 2646.
Bokov, V. A., Kizhaev, S. A., Mylnikova, I. E. and Tutov, A. G. (1965) *Sov. Phys.–Solid State* **6,** 2419.
Bondar, I. A., Koroleva, L. N. and Toropov, N. A. (1968) *Growth of Crystals* **6A,** 101; see also *Mat. Res. Bull.* **2** (1967) 479.
Bondar, I. A., Udalov, Y. P. and Kalinin, A. I. (1971) *Inorg. Mater. (USSR)* **7,** 1304.
Bonner, W. A. and Van Uitert, L. G. (1967) *Mat. Res. Bull.* **2,** 131.
Bonner, W. A. and Van Uitert, L. G. (1968) US Patent 3.370.963.
Bonner, W. A., Dearborn, E. F. and Van Uitert, L. G. (1967) In "Crystal Growth" (H. S. Peiser, ed.) p. 437. Pergamon, Oxford.
Borshchevskii, A. S., Goryunova, N. A., Kesamanly, F. P. and Nasledov, D. N. (1967) *Phys. stat. sol.* **21,** 9.
Bouchard, R. J. and Gillson, J. L. (1972) *Mat. Res. Bull.* **7,** 873.
Bradt, R. C. and Ansell, G. S. (1967) *Mat. Res. Bull.* **2,** 585.
Brandt, G. and Diehl, R. (1974) *Mat. Res. Bull.* **9,** 411.
Brissot, J. J. and Belin, C. (1971) *J. Crystal Growth* **8,** 213.
Brixner, L. H. (1959) *J. Amer. Chem. Soc.* **81,** 3841.
Brixner, L. H. and Abramson, E. (1965) *J. Electrochem. Soc.* **112,** 70.
Brixner, L. H. and Babcock, K. (1968) *Mat. Res. Bull.* **3,** 817.
Brixner, L. H. and Bouchard, R. J. (1970) *Mat. Res. Bull.* **5,** 61.
Brixner, L. H. and Weiher, J. F. (1970) *J. Solid State Chem.* **2,** 55.
Broder, J. D. and Wolff, G. A. (1963) *J. Electrochem. Soc.* **110,** 1150.
Broese van Groenou, A., Page, J. L. and Pearson, R. F. (1967) *J. Phys. Chem. Solids* **28,** 1017.
Bruton, T. M. and White, E. A. D. (1973) *J. Crystal Growth* **19,** 341.
Buehler, E. and Wernick, J. H. (1971) *J. Crystal Growth* **8,** 324.
Buehler, E., Wernick, J. H. and Shay, J. L. (1971) *Mat. Res. Bull.* **6,** 303.
Bundy, F. P., Hall, H. T., Strong, H. M. and Wentorf, R. H. (1955) *Nature* **176,** 51.
Burmeister, R. A. and Green, P. E. (1967) *Trans AIME* **239,** 408.
Busch, G., Kaldis, E., Schaufelberger-Teker, E. and Wachter, P. (1970) Rare Earth Conference 1969, Coll. Intern. CNRS No. 180, Vol. I, p. 359.
Butcher, M. M. and White, E. A. D. (1965) *J. Am. Ceram. Soc.* **48,** 492.
Carlo, J. T. (1973) Ph.D. Thesis, MIT, Dept. Electr. Engng.
Castellano, R. N. (1972) *Mat. Res. Bull.* **7,** 261.
Cerrina, F., Margadonna, D. and Perfetti, P. (1973) *J. Crystal Growth* **18,** 202.
Chabria, J. R. (1967) *Ceram. Ind.* **89,** 52.
Chadwick, J. (1972) Mullard Research Report 2843.
Chamberland, B. L. (1967) *Mat. Res. Bull.* **2,** 827.
Chase, A. B. (1966) *J. Am. Ceram. Soc.* **49,** 233.
Chase, A. B. (1968) *J. Am. Ceram. Soc.* **51,** 507.
Chase, A. B. and Lefever, R. (1960) *Am. Min.* **45,** 1126.
Chase, A. B. and Morse, F. L. (1973) *J. Crystal Growth* **19,** 18.

Chase, A. B. and Osmer, J. A. (1964) *Am. Min.* **49**, 1469.
Chase, A. B. and Osmer, J. A. (1966a) *J. Electrochem. Soc.* **113**, 198.
Chase, A. B. and Osmer, J. A. (1966b) *Am. Min.* **51**, 1808.
Chase, A. B. and Osmer, J. A. (1967) *J. Am. Ceram. Soc.* **50**, 325.
Chase, A. B. and Osmer, J. A. (1969) *J. Crystal Growth* **5**, 239.
Chase, A. B. and Osmer, J. A. (1972) *J. Crystal Growth* **15**, 249.
Chase, A. B. and Tippins, H. H. (1967) *J. Appl. Phys.* **38**, 2469.
Chase, A. B. and Wilcox, W. R. (1967) *J. Am. Ceram. Soc.* **50**, 332.
Chase, A. B. and Wolten G. M. (1965) *J. Am. Ceram. Soc.* **48**, 276.
Chen, T. (1973) *J. Crystal Growth* **20**, 29.
Cherepanov, A. M. (1962) *Growth of Crystals* **3**, 324.
Chu, T. L., Jackson, J. M. and Smeltzer, R. K. (1973) *J. Electrochem. Soc.* **120**, 802.
Chuklantsev, V. G. and Poleskev, Y. M. (1965) *Izv. Vyss. Ubechn. Zavedenii Khim; Khim. Tech.* **8**, 357.
Clark, G. W., Finch, C. B., Harris, L. A. and Yakel, H. L. (1972) *J. Crystal Growth* **16**, 110.
Comer, J. J., Croft, W. J., Kestigian, M. and Carter, J. R. (1967) *Mat. Res. Bull.* **2**, 293.
Coufova, P. and Novak, J. (1969) *Sov. Phys.-Cryst.* **14**, 61.
Courtens, E. L. and Scheel, H. J. (1974) *IBM Techn. Discl. Bull.* **16**, 2451.
Cross, L. E. and Nicholson, B. J. (1955) *Phil. Mag.* **46**, 453.
Cruceanu, E. and Nistor, N. (1969) *J. Crystal Growth* **5**, 206; *Growth of Crystals* **7**, 230.
Cummins, S. E. and Cross, L. E. (1968) *J. Appl. Phys.* **39**, 2268.
Cuomo, J. J. and Gambino, R. J. (1968) *J. Electrochem. Soc.* **115**, 755.
Curry, N. A., Johnston, G. B., Besser, J. B. and Morish, A. H. (1965) *Phil. Mag.* **12**, 221.
Damay, F. M. G. and Heindl, R. W. (1965) French Patent 1.469.158.
Damen, J. P. M. and Robertson, J. M. (1972) *J. Crystal Growth* **16**, 50.
Deacon, J. A. and Hiscocks, S. E. R. (1971) *J. Mat. Sci.* **6**, 309.
Demurov, D. G. and Venevtsev, Y. N. (1971) *Sov. Phys.–Solid State* **13**, 553.
Dernier, P. D. and Maines, R. G. (1971) *Mat. Res. Bull.* **6**, 433.
DeVries, R. C. (1962) *J. Am. Ceram. Soc.* **45**, 225.
DeVries, R. C. (1966) *Mat. Res. Bull.* **1**, 83.
DeVries, R. C. and Fleischer, J. F. (1972) *J. Crystal Growth* **13/14**, 88.
DeVries, R. C. and Roth, W. L. (1968) *J. Am. Ceram. Soc.* **51**, 72.
DeWit, J. H. (1972) *J. Crystal Growth* **12**, 183.
Dharmarajan, R., Belt, R. F. and Puttbach, R. C. (1972) *J. Crystal Growth* **13/14**, 535.
Dibenedetto, B. and Cronan, C. J. (1968) *J. Am. Ceram. Soc.* **51**, 364.
Diehl, R. and Brandt, G. (1974) *Mat. Res. Bull.* **9**, 421.
Diehl, R. and Nitsche, R. (1973) *J. Crystal Growth* **20**, 38.
Dixon, S., Aucoin, T. R. and Savage, R. O. (1971) *J. Appl. Phys.* **42**, 1732.
Donnally, J. P. and Milnes, A. G. (1966) *J. Electrochem. Soc.* **113**, 297.
Drofenik, M., Kolar, D. and Golic, L. (1973) *J. Crystal Growth* **20**, 75.
Drofenik, M., Golic, L. and Kolar, D. (1974) *J. Crystal Growth* **21**, 170, 305.
Dugger, C. O. (1966) *J. Electrochem. Soc.* **113**, 306.
Dugger, C. O. (1967) In "Crystal Growth" (H. S. Peiser, ed.) p. 493. Pergamon, Oxford.
Dugger, C. O. (1974) *Mat. Res. Bull.* **9**, 331.

Dziuba, E. Z. (1971) *J. Crystal Growth* **8**, 221.
Eastman, D. E. and Shafer, M. W. (1967) *J. Appl. Phys.* **38**, 4761.
Egghart, H. C. (1967) *Inorg. Chem.* **6**, 2121.
Eisner, R. L., Mazelsky, R. and Tiller, W. A. (1961) *J. Appl. Phys.* **32**, 1833.
Ellis, W. C., Williams, H. J. and Sherwood, R. C. (1957) *J. Appl. Phys.* **28**, 1215.
Ellis, W. C., Frosch, C. J. and Zetterstrom, R. B. (1968) *J. Crystal Growth* **2**, 61.
Elwell, D., Morris, A. W. and Neate, B. W. (1972) *J. Crystal Growth* **16**, 67.
Espinosa, G. P. (1971) *J. Crystal Growth* **10**, 323.
Espinosa, G. P. and Geller, S. (1964) *J. Appl. Phys.* **35**, 2551.
Eustache, H. (1957) *Compt. Rend.* **244**, 1029.
Farrell, E. F. and Fang, J. H. (1964) *J. Am. Ceram. Soc.* **47**, 274.
Faust, J. W. and John, H. F. (1964) *J. Phys. Chem. Solids* **25**, 1407.
Faust, J. W., John, H. F. and Pritchard, C. (1968) *J. Crystal Growth* **3/4**, 321.
Feigelson, R. S. (1964) *J. Am. Ceram. Soc.* **47**, 257.
Feigelson, R. S. (1967) *J. Am. Ceram. Soc.* **50**, 433.
Feigelson, R. S. (1968) *J. Am. Ceram. Soc.* **51**, 538.
Feigelson, R. S. (1971) private communication.
Feltz, A. and Langbein, H. (1971) *Kristall u. Technik* **6**, 359.
Ferretti, A., Kunnmann, W. and Wold, A. (1963) *J. Appl. Phys.* **34**, 388.
Fesenko, E. G. and Kolesova, R. V. (1960) *Sov. Phys.-Cryst.* **4**, 54.
Filby, J. D. and Nielsen, S. (1966) *Brit. J. Appl. Phys.* **17**, 81.
Finch, C. B. and Clark, G. W. (1965) *J. Appl. Phys.* **36**, 2143.
Finch, C. B. and Clark, G. W. (1966) *J. Appl. Phys.* **37**, 3910.
Finch, C. B. and Clark, G. W. (1968) *Am. Min.* **53**, 1394.
Finch, C. B. and Clark, G. W. (1970) *J. Crystal Growth* **6**, 245.
Finch, C. B. and Clark, G. W. (1972) *J. Crystal Growth* **12**, 181.
Finch, C. B., Harris, L. A. and Clark, G. W. (1964) *Am. Min.* **49**, 782.
Fisk, Z., Cooper, A. S., Schmidt, P. H. and Castellano, R. N. (1972) *Mat. Res. Bull.* **7**, 285.
Flanders, P. J. and Remeika, J. P. (1965) *Phil. Mag.* **11**, 1271.
Foguel, M. and Grajower, R. (1971) *J. Crystal Growth* **11**, 280.
Folen, V. J. (1962) *J. Appl. Phys.* **33**, 1084.
Fredlein, R. A. and Damjanovic, A. (1972) *Solid State Chem.* **4**, 94.
Fujiki, Y. and Ono, A. (1972) *Yogyo Kyokai Shi.* **80**, 506.
Fujita, S., Itoh, K., Arai, S. and Sakaguchi, T. (1971) *Japan J. Appl. Phys.* **10**, 516.
Fuke, S., Washiyama, M., Otsuka, K. and Aoki, M. (1971) *Japan J. Appl. Phys.* **10**, 687.
Fukuda, T. and Uematsu, Y. (1972) *Japan J. Appl. Phys.* **11**, 163.
Fukuda, T., Hirano, H. and Koide, S. (1970) *J. Crystal Growth* **6**, 293.
Fushimi, S. and Ikeda, T. (1964) *Japan J. Appl. Phys.* **3**, 171.
Fushimi, S. and Ikeda, T. (1965) *J. Phys. Soc. Japan* **20**, 2007.
Gabbe, D. and Harmer, A. L. (1968) *J. Crystal Growth* **3/4**, 544.
Gabbe, D. and Linz, A. (1974) private communication.
Gabbe, D. and Pierce, J. (1974) private communication.
Galasso, F., Layden, G. and Ganung, G. (1968) *Mat. Res. Bull.* **3**, 397.
Galt, J. K., Matthias, B. T. and Remeika, J. P. (1950) *Phys. Rev.* **79**, 391.
Gambino, R. J. (1967) *J. Appl. Phys.* **38**, 1129.
Gambino, R. J. and Leonhard, E. (1961) *J. Am. Ceram. Soc.* **44**, 221.
Gambino, R. J. and Ruf, R. R. (1971) *IBM Techn. Discl. Bull.* **13**, 3082.
Garner, R. W. and White, W. B. (1970) *J. Crystal Growth* **7**, 343.

Garrard, B. J., Wanklyn, B. M. and Smith, S. H. (1974) *J. Crystal Growth* **22**, 169.

Garton, G. and Wanklyn, B. M. (1967a) *J. Crystal Growth* **1**, 49.

Garton, G. and Wanklyn, B. M. (1967b) *J. Crystal Growth* **1**, 164.

Garton, G. and Wanklyn, B. M. (1968) *J. Mat. Sci.* **3**, 395.

Garton, G. and Wanklyn, B. M. (1969) Proc. Rare Earth Conf., Grenoble, p. 343.

Garton, G., Hann, B. F., Wanklyn, B. M. and Smith, S. H. (1972a) *J. Crystal Growth* **12**, 66.

Garton, G., Smith, S. H. and Wanklyn, B. M. (1972b) *J. Crystal Growth* **13/14**, 588.

Gavrilyachenko, V. G., Dudkevich, V. P. and Fesenko, E. G. (1968) *Sov. Phys.-Cryst.* **13**, 277.

Gendelev, S. S., Lopovok, B. L. and Rubinstein, B. E. (1964) *Sov. Phys.–Solid State* **5**, 2223.

Gendelev, S. S. and Titova, A. G. (1968) *Growth of Crystals* **6A**, 90.

Gendelev, S. S. and Titova, A. G. (1971) *Bull. Acad. Sci. USSR Phys. Ser.* **6**, 1131.

Gendelev, S. S. and Zvereva, R. I. (1971) *Bull. Acad. Sci. USSR Phys. Ser.* **6**, 1142.

Gentile, A. L. and Anders, F. H. (1967) *Mat. Res. Bull.* **2**, 853.

Gentile, A. L. and Stafsudd, O. M. (1974) *Mat. Res. Bull.* **9**, 105.

Gerk, A. P. and Gilman, J. J. (1968) *J. Appl. Phys.* **39**, 4497.

Gibinski, T., Cisowska, E., Zdanowicz, W., Henkie, Z. and Wojakowski, A. (1974) *Kristall u. Technik* **9**, 161.

Giess, E. A. (1962a) *J. Am. Ceram. Soc.* **45**, 53.

Giess, E. A. (1962b) *J. Appl. Phys.* **33**, 2143.

Giess, E. A. (1964) *J. Am. Ceram. Soc.* **47**, 388.

Giess, E. A., Scott, B. A., O'Kane, D. F., Olson, B. L., Burns, G. and Smith, A. W. (1969) *Mat. Res. Bull.* **4**, 741.

Giess, E. A., Cronemeyer, D. C., Rosier, L. L. and Kuptsis, J. D. (1970) *Mat. Res. Bull.* **5**, 495.

Gillessen, K. and Von Münch, W. (1973) *J. Crystal Growth* **19**, 263.

Girgis, K., Laves, F. and Reinmann, R. (1966) *Naturwiss.* **53**, 610.

Girgis, K. and Schulz, H. (1971) *Naturwiss.* **58**, 95.

Gliki, N. V. and Timofeeva, V. A. (1960) *Sov. Phys.-Cryst.* **5**, 96.

Gliki, N. V., Plateneva, I. A. and Timofeeva, V. A. (1956) *Sov. Phys.-Cryst.* **1**, 478.

Gluck, R. M., Lee, T. H. and Smith, F. T. J. (1974) *Mat. Res. Bull.* **9**, 305.

Gondrand, M., Collomb, A., Joubert, J. C. and Shannon, R. D. (1974) *J. Solid State Chem.* **11**, 1.

Goodrum, J. W. (1970) *J. Crystal Growth* **7**, 254.

Goodrum, J. W. (1972) *J. Crystal Growth* **13/14**, 604.

Goryunova, N. A., Baranov, B. V., Grigoreva, V. S., Kradinova, L. V., Maksimova, V. A. and Prochukhan, V. D. (1968a) *Izv. Akad. Nauk USSR Inorg. Mat.* **4**, 931.

Goryunova, N. A., Orlov, V. M., Sokolova, V. I., Shpenkov, G. P. and Tsuetkova, E. V. (1968b) *Phys. stat. sol.* **25**, 513.

Goto, Y. and Cross, L. E. (1969) *Yogyo Kyokai Shi.* **77**, 355.

Goto, Y. and Takahashi, K. (1973) *Jap. J. Appl. Phys.* **12**, 948.

Griffiths, L. B. and Mlavsky, A. I. (1964) *J. Electrochem. Soc.* **111**, 805.

Grodkiewicz, W. H. and Van Uitert, L. G. (1963) *J. Am. Ceram. Soc.* **46**, 356.

Grodkiewicz, W. H. and Nitti, D. J. (1966) *J. Am. Ceram. Soc.* **49**, 576.

Grodkiewicz, W. H., Dearborn, E. F. and Van Uitert, L. G. (1967) In "Crystal Growth" (H. S. Peiser, ed.) p. 441. Pergamon, Oxford.

Guerci, C. F. and Shafer, M. W. (1966) *J. Appl. Phys.* **37**, 1406.

Hakki, B. W. (1971) *J. Electrochem. Soc.* **118,** 1469.
Halden, F. A. (1960) In "Silicon Carbide" (J. R. O'Connor and J. Smiltens, eds.) p. 115. Pergamon, London.
Hall, R. N. (1958) *J. Appl. Phys.* **29,** 914.
Hamaker, R. W. and White, W. B. (1969) *J. Electrochem. Soc.* **116,** 478.
Hamilton, P. M. (1963) US Patent 3.115.469.
Hanak, J. J. and Johnson, D. E. (1969) Paper at Amer. Conf. Crystal Growth, NBS, Gaithersburg, Md.
Hansen, P., Schuldt, J. and Tolksdorf, W. (1973) *Phys. Rev.* B **8,** 4274.
Harari, A., Théry, J. and Collongues, R. (1967) *Rev. Int. Hautes Temp. et Refract.* **4,** 207.
Harbeke, G. and Pinch, H. (1966) *Phys. Rev. Lett.* **17,** 1090.
Harris, L. A. and Finch, C. B. (1965) *Am. Min.* **50,** 1493.
Harsy, M. (1967) *Kristall u. Technik* **2,** 3, 447.
Harsy, M. (1968) *Mat. Res. Bull.* **3,** 483.
Hauptman, Z., Wanklyn, B. W. and Smith, S. H. (1973) *J. Mat. Sci.* **8,** 1695.
Havlicek, V., Novak, P. and Vichr, M. (1971) *Phys. stat. sol.* **44,** K21.
Haworth, D. T. and Lake, D. P. (1965) *Chem. Comm.* **21,** 553.
Hemmat, N. and Weinstein, M. (1967) *J. Electrochem. Soc.* **114,** 403.
Higashi, I. and Atoda, T. (1970) *J. Crystal Growth* **7,** 251.
Hill, G. J. and Wanklyn, B. M. (1968) *J. Crystal Growth* **3/4,** 475.
Hintzmann, W. and Müller-Vogt, G. (1969) *J. Crystal Growth* **5,** 274.
Hirano, H. and Fukuda, T. (1968) *Japan J. Appl. Phys.* **7,** 1413.
Hiskes, S. R., Felmlee, T. L. and Burmeister, R. A. (1972) *J. Electron. Mat.* **1,** 458.
Hodges, L. R., Wilson, W. R., Rodrigue, G. P. and Harrison, G. R. (1967) *J. Appl. Phys.* **38,** 1127.
Holonyak, N., Wolfe, C. M. and Moore, J. J. (1965) *Appl. Phys. Lett.* **6,** 64.
Homma, K. and Wada, M. (1971) *Rec. El. & Comm. Eng. Convers. Tohoku Univ.* **40,** 50.
Hong, H. Y. P. and Dwight, K. (1974) *Mat. Res. Bull.* **9,** 775.
Horn, F. H. (1959) *J. Electrochem. Soc.* **106,** 905.
Hurst, J. J. and Linz, A. (1971) *Mat. Res. Bull.* **6,** 163.
Ichinose, N., Takahashi, T. and Yokomizo, Y. (1971) *J. Phys. Soc. Japan* **31,** 1848.
Ikeda, T. and Fushimi, S. (1962) *J. Phys. Soc. Japan* **17,** 1202.
Inomata, Y. and Yamane, T. (1974) *J. Crystal Growth* **21,** 317.
Ito, J. (1971) *Am. Min.* **56,** 1105.
Iwami, M., Fujita, N. and Kawabe, K. (1971) *Japan J. Appl. Phys.* **10,** 1746.
Izvekov, V. N., Sysoev, L. A., Obukhovskii, Ya. A. and Birman, B. I. (1968) *Growth of Crystals* **6A,** 106.
James, D. W. F. and Lewis, C. (1965) *Brit. J. Appl. Phys.* **16,** 1089.
Janes, D. L. and Bodnar, R. E. (1971) *J. Appl. Phys.* **42,** 1500.
John, V. and Kvapil, J. (1968) *Kristall u. Technik* **3,** 59.
Jona, F., Shirane, G. and Pepinsky, R. (1955a) *Phys. Rev.* **97,** 1584.
Jona, F., Shirane, G. and Pepinsky, R. (1955b) *Phys. Rev.* **98,** 903.
Jones, C. M. (1971) *J. Am. Ceram. Soc.* **54,** 347.
Jonker, H. D. (1974) to be published.
Joubert, J. C., Shirk, T., White, W. B. and Roy, R. (1968) *Mat. Res. Bull.* **3,** 671.
Kalnin, A. A. and Tairov, Y. M. (1966) *Izv. Leningrad Elektrotech. Inst.* **61,** 26.
Katz, G. and Roy, R. (1966) *J. Am. Ceram. Soc.* **49,** 168.
Kawabe, K. and Sawada, S. (1957) *J. Phys. Soc. Japan* **12,** 218.

Kay, H. F. and Bailey, P. C. (1957) *Acta Cryst.* **10**, 219.

Kay, H. F. and Miles, J. L. (1957) *Acta Cryst.* **10**, 213.

Keck, P. (1953) *Phys. Rev.* **90**, 369 (abstract only).

Keeling, R. O. (1957) *Acta Cryst.* **10**, 209.

Keester, K. L. and White, W. B. (1970) *J. Am. Ceram. Soc.* **53**, 39.

Keezer, R. C. and Wood, C. (1966) *Appl. Phys. Lett.* **8**, 139.

Kerecman, A. J., Tauber, A., Aucoin, T. R. and Savage, R. O. (1968) *J. Appl. Phys.* **39**, 726.

Kerecman, A. J., Aucoin, T. R. and Dattilo, W. P. (1969) *J. Appl. Phys.* **40**, 1416.

Kestigian, M. (1963) *J. Am. Ceram. Soc.* **46**, 563.

Kestigian, M. (1967) *J. Am. Ceram. Soc.* **50**, 165.

Kestigian, M. and Holloway, W. W. (1967) In "Crystal Growth" (H. S. Peiser, ed.) p. 451. Pergamon, Oxford.

Khodakov, A. L. and Sholokhovich, M. L. (1960) *Bull. Acad. Sci. USSR* **24**, 1240.

Khodakov, A. L., Sholokhovich, M. L., Fesenko, E. G. and Kramanov, O. P. (1956) *Dokl. Akad. Nauk* **108**, 379; (1958) *Growth of Crystals* **1**, 232.

Kijima, K., Miyamoto, N. and Nishizawa, J. I. (1971) *J. Appl. Phys.* **42**, 486.

Kimizuka, N., Saeki, M. and Nakahira, M. (1970) *Mat. Res. Bull.* **5**, 403.

Kirchmayr, H. R. (1965) *Z. Metallkunde* **56**, 767.

Kitahiro, I. and Watanabe, A. (1967) *Japan J. Appl. Phys.* **6**, 1023.

Kjems, J. K., Shirane, G., Müller, K. A. and Scheel, H. J. (1973) *Phys. Rev.* B **8**, 1119.

Kleber, W. and Mlodoch, R. (1966) *Kristall u. Technik* **1**, 249.

Kleber, W., Ickert, L. and Doerschel, J. (1966) *Kristall u. Technik* **1**, 237.

Kleber, W., Fehling, W. and Röhl, M. (1967) *Kristall u. Technik* **2**, 489.

Knippenberg, W. F. and Verspui, G. (1966) *Philips Res. Repts.* **21**, 113.

Kobayashi, J. (1958) *J. Appl. Phys.* **29**, 866.

Kohn, K. and Tasaki, A. (1965) *J. Phys. Soc. Japan* **20**, 1273.

Koizumi, H., Niizeki, N. and Ikeda, T. (1964) *Japan J. Appl. Phys.* **3**, 495.

Kojima, F. and Nomura, S. (1973) *J. Phys. Soc. Japan* **35**, 624.

Komatsu, H. and Sunagawa, I. (1964) *Min. J. Japan* **4**, 203.

Koppel, K. D. (1966) *Inorg. Mater.* (*USSR*) **2**, 1527.

Kostiner, E. and Bless, P. W. (1971) *J. Electrochem. Soc.* **118**, 351.

Krainik, N. N., Gokhberg, L. S. and Mylnikova, I. E. (1971) *Sov. Phys.–Solid State* **12**, 1885.

Kressel, H., Nelson, H., McFarlane, S. H., Abrahams, M. S., LeFur, P. and Buiocchi, C. J. (1969) *J. Appl. Phys.* **40**, 3587.

Krishnan, R. (1966) *J. Am. Ceram. Soc.* **49**, 678.

Krishnan, R. (1968a) *Phys. stat. sol.* **26**, K47.

Krishnan, R. (1968b) *J. Appl. Phys.* **39**, 1340.

Krishnan, R. (1969) *Phys. stat. sol.* **35**, K63.

Krishnan, R. (1972) *J. Crystal Growth* **13/14**, 582.

Kudzin, A. Y. (1962) *Sov. Phys.–Solid State* **4**, 1369.

Kudzin, A. Y., Guskina, L. G. and Petrushkevich, I. S. (1964) *Sov. Phys.–Solid State* **6**, 73.

Kukui, A. L., Parkhomovskii and Budko, V. L. (1967) *Zap. Vses. Min. Obshchest.* **96**, 306.

Kumagawa, M., Ozaki, M. and Yamada, S. (1970) *Japan J. Appl. Phys.* **9**, 1422.

Kunkle, H. F. and Kohnke, E. E. (1965) *J. Appl. Phys.* **36**, 1489.

Kunnmann, W., Wold, A. and Banks, E. (1962) *J. Appl.* **33**, 1364.

Kunnmann, W., Ferretti, A. and Wold, A. (1963) *J. Appl. Phys.* **34,** 1264.
Kunnmann, W., Ferretti, A., Arnott, R. J. and Rogers, D. B. (1965) *J. Phys. Chem. Solids* **26,** 311.
Kuse, D. (1970) *IBM J. Res. Develop.* **14,** 315.
Kvapil, J. (1966) *Kristall u. Technik* **1,** 97.
Kvapil, J. and Perner, B. (1971) *J. Crystal Growth* **8,** 162.
Kwestroo, W., Huizing, A. and DeJonge, J. (1969) *Mat. Res. Bull.* **4,** 817.
Larson, G. H. and Sleight, A. W. (1968) *Phys. Lett.* **28A,** 203.
Laudise, R. A., Linares, R. C. and Dearborn, E. F. (1962) *J. Appl. Phys. Suppl.* **33,** 1362.
Laugier, A., Garrand, M. and Mesnard, G. (1967) *Bull. Soc. Franc. Min. Crist.* **90,** 176.
Laurant, Y. (1969) *Rev. chim. min.* **6,** 1145.
Layden, G. K. (1967) *Mat. Res. Bull.* **2,** 533.
Leckebusch, R. and Recker, K. (1972) *J. Crystal Growth* **13/14,** 276.
Lecraw, R. C., Wolfe, R. and Nielsen, J. W. (1969) *Appl. Phys. Lett.* **14,** 352.
Lefever, R. A. (1961) *J. Electrochem. Soc.* **108,** 107.
Lefever, R. A. (1962) *J. Appl. Phys.* **33,** 3596.
Lefever, R. A. and Chase, A. B. (1962) *J. Am. Ceram. Soc.* **45,** 32.
Lefever, R. A., Torpy, J. W. and Chase, A. B. (1961) *J. Appl. Phys.* **32,** 962.
Lefever, R. A., Chase, A. B. and Sobon, L.E. (1962) *Am. Min.* **47,** 1450.
Lepore, D. A., Belt, R. F. and Nielsen, J. W. (1967) *J. Appl. Phys.* **38,** 1421.
Levin, E. M., Rynders, G. F. and Dzimian, R. J. (1952) NBS Report 1916.
Levinson, L. M. and Wanklyn, B. M. (1971) *J. Solid State Chem.* **3,** 131.
Liaw, H. M. and Faust, J. W. (1971) *J. Crystal Growth* **10,** 302.
Liaw, H. M. and Faust, J. W. (1972) *J. Crystal Growth* **13/14,** 772.
Libowitz, G. G. (1953) *J. Am. Chem. Soc.* **75,** 1601.
Linares, R. C. (1962a) *J. Am. Ceram. Soc.* **45,** 119.
Linares, R. C. (1962b) *J. Am. Ceram. Soc.* **45,** 307.
Linares, R. C. (1962c) *J. Appl. Phys.* **33,** 1747.
Linares, R. C. (1962d) In "Metallurgy of Advanced Electronic Materials" (G. E. Brock, ed.) p. 329. Interscience, New York.
Linares, R. C. (1964) *J. Appl. Phys.* **35,** 433.
Linares, R. C. (1965a) *J. Am. Ceram. Soc.* **48,** 68.
Linares, R. C. (1965b) *J. Phys. Chem. Solids* **26,** 1817.
Linares, R. C. (1967a) *Am. Min.* **52,** 1211.
Linares, R. C. (1967b) *Am. Min.* **52,** 1554.
Linares, R. C. (1967c) *J. Phys. Chem. Solids* **28,** 1285.
Linares, R. C. (1968a) *Trans. Met. Soc. AIME* **242,** 441.
Linares, R. C. (1968b) *J. Crystal Growth* **3/4,** 443.
Linares, R. C. and Mills, A. D. (1962) *Acta Cryst.* **15,** 1048.
Linares, R. C., Ballman, A. A. and Van Uitert, L. G. (1962) *J. Appl. Phys.* **33,** 3309.
Linares, R. C., McGraw, R. B. and Schroeder, J. B. (1965) *J. Appl. Phys.* **36,** 2884.
Litvin, Yu.A. and Butuzov, V. P. (1968) *Dokl. Akad. Nauk SSSR* **181,** 1123
Lorenz, M. R. (1962) *J. Appl. Phys.* **33,** 3304.
Lorenz, M. R., Woods, J. F. and Gambino, R. J. (1967) *J. Phys. Chem. Solids* **28,** 403.
Loriers, J. and Vichr, M. (1972) *J. Crystal Growth* **13/14,** 593.
Loriers, J., Vichr, M. and Makram, H. (1971) *J. Crystal Growth* **8,** 69.
Loshakova, G. V., Plechko, R. L., Vaipolin, A. A., Pawlov, B. V., Valov, Y. A. and Goryunova, N. A. (1966) *Izv. Akad. Nauk SSR Neorgan. Mat.* **2,** 1702.

Lou, F. H. and White, E. A. D. (1967) *J. Mat. Sci.* **2**, 97.

Lowell, J., Naurotsky, A. and Holloway, J. R. (1971) *J. Am. Ceram. Soc.* **54**, 466.

Lundblad, E. (1964) *Wire Products* **39**, 1188.

Lynch, R. T. and Lander, J. J. (1959) *J. Appl. Phys.* **30**, 1614.

Lyons, V. J. (1965) US Patent 3.198.606.

Macksey, H. M., Holonyak, N., Dupuis, R. D. and Campbell, J. C. (1972) *J. Appl. Phys.* **43**, 1334.

Magyar, B. (1968) *Inorg. Chem.* **7**, 1457.

Makarevich, A. I. (1965) *Dokl. Akad. Nauk Belorussk SSSR* **2**, 94.

Makovsky, J., Zodkevitz, A. and Kalman, Z. H. (1971) *J. Crystal Growth* **11**, 99.

Makovsky, J., Horowitz, A. and Gazit, D. (1974) *J. Crystal Growth* **22**, 241.

Makram, H. (1967) *J. Crystal Growth* **1**, 325.

Makram, H. (1968) *J. Crystal Growth* **3/4**, 447.

Makram, H. and Krishnan, R. (1962) *J. Phys. Rad.* **23**, 1000.

Makram, H. and Krishnan, R. (1963) *Compt. Rend.* **257**, 624.

Makram, H. and Krishnan, R. (1964) *J. Phys.* (*France*) **25**, 343.

Makram, H. and Krishnan, R. (1967) In "Crystal Growth" (H. S. Peiser, ed.) p. 467. Pergamon, Oxford.

Makram, H., Touron, L. and Loriers, J. (1968) *J. Crystal Growth* **3/4**, 452.

Makram, H., Touron, L. and Loriers, J. (1972) *J. Crystal Growth* **13/14**, 585.

Manabe, T. and Egashira, K. (1971) *Mat. Res. Bull.* **6**, 1167.

Manzel, M. (1967) *Kristall u. Technik* **2**, 61.

Manzel, M., Voigt, F. and Kleinert, P. (1963) *Phys. stat. sol.* **3**, 1392.

Marezio, M. and Remeika, J. P. (1966) *J. Chem. Phys.* **44**, 3348.

Marezio, M., Remeika, J. P. and Dernier, P. D. (1966) *Mat. Res. Bull.* **1**, 247.

Marezio, M., Remeika, J. P. and Dernier, P. D. (1968a) *Inorg. Chem.* **7**, 1337.

Marezio, M., Remeika, J. P. and Dernier, P. D. (1968b) *Acta Cryst.* **B24**, 1670.

Marezio, M., Remeika, J. P. and Dernier, P. D. (1969) *Acta Cryst.* **B25**, 955.

Marshall, R. C. (1969) *Mat. Res. Bull. Suppl.* **4**, S73.

Mason, D. R. and Cook, J. S. (1961) *J. Appl. Phys.* **32**, 475.

Matecha, J. and Kvapil, J. (1970) *J. Crystal Growth* **6**, 199.

Mateika, D. (1972) *J. Crystal Growth* **13/14**, 698.

Matthias, B. T. (1948) *Phys. Rev.* **73**, 808.

Matthias, B. T. and Remeika, J. P. (1951) *Phys. Rev.* **82**, 727.

Mayer, I., Shidlovsky, I. and Yamir, E. (1967) *J. Less–Common Metals* **12**, 46.

McWhan, D. B. and Remeika, J. P. (1970) *Phys. Rev.* B **2**, 3734.

Merz, J. L. and Lynch, R. T. (1967) *J. Appl. Phys.* **38**, 1988.

Meyer, A. (1970) *J. Less–Common Metals* **20**, 353.

Mikami, I. (1971) private communication.

Miki, H. and Otsubo, M. (1971) *Japan J. Appl. Phys.* **10**, 509.

Miller, C. E. (1958) *J. Appl. Phys.* **29**, 233.

Miller, R. C., Wood, E. A., Remeika, J. P. and Savage, A. (1962) *J. Appl. Phys.* **33**, 1623.

Minkoff, I. (1968) *J. Crystal Growth* **3/4**, 328.

Mita, Y. (1961) *J. Phys. Soc. Japan* **16**, 1484.

Mita, Y. (1962) *J. Phys. Soc. Japan* **17**, 784.

Miyada, T., Matsuo, Y. and Miyahara, S. (1964) *J. Phys. Soc. Japan* **19**, 1747.

Miyatani, K., Wada, Y. and Okamoto, F. (1968) *J. Phys. Soc. Japan* **25**, 369.

Mlavsky, A. I. and Weinstein, M. (1963) *J. Appl. Phys.* **34**, 2885.

Monchamp, R. R., Belt, R. and Nielsen, J. W. (1967) In "Crystal Growth" (H. S. Peiser, ed.) p. 463. Pergamon, Oxford.

Mones, A. H. and Barks, E. (1958) *J. Phys. Chem. Solids* **4**, 217.

Moore, R. E. (1961) US Patent 3.011.868.

Mori, T. and Tamura, H. (1964) *J. Phys. Soc. Japan* **19**, 1247.

Moriizumi, T. and Takahashi, K. (1969) *Japan J. Appl. Phys.* **8**, 348.

Morrish, A. H. (1970) Proc. Int. Conf. Ferrites Kyoto, p. 574.

Mughal, S. A. and Ray, B. (1973) *J. Mat. Sci.* **8**, 1171.

Mughal, S. A., Payne, A. J. and Ray, B. (1969) *J. Mat. Sci.* **4**, 895.

Myakishev, K. G. and Klokman, V. R. (1966) *Soviet Radiochemistry* **8**, 361.

Mylnikova, I. E. and Bokov, V. A. (1962) *Growth of Crystals* **3**, 309.

Nakano, K., Hiyashi, H. and Imura, T. (1971) *Japan J. Appl. Phys.* **10**, 513.

Nakatani, I., Kitano, Y., Korekado, M. and Komura, Y. (1970) *Japan J. Appl. Phys.* **9**, 842.

Nassau, K. and Abrahams, S. C. (1968) *J. Crystal Growth* **2**, 136.

Nassau, K. and Mills, A. D. (1962) *Acta Cryst.* **15**, 808.

Nassau, K., Shiever, J. W. and Kowalchik, M. (1970) *J. Crystal Growth* **7**, 237.

Nelson, D. F. and Remeika, J. P. (1964) *J. Appl. Phys.* **35**, 522.

Nelson, W. E., Halden, F. A. and Rosengreen, A. (1966) *J. Appl. Phys.* **37**, 333.

Nemeth, J. and Tinklepaugh, J. R. (1966) *Bull. Amer. Ceram. Soc.* (abstract only).

Nester, J. F. and Schroeder, J. B. (1967) *Am. Min.* **52**, 276.

Neuhaus, A. and Liebertz, J. (1962) *Chem. Ing. Techn.* **34**, 813.

Neuhans, A., Holz, H. G. and Klein, H. D. (1967) *Z. Phys. Chem. (Neue Folge)* **53**, 163.

Nevriva, M. (1971) *Kristall u. Technik* **6**, 517.

Nevriva, M. and Holba, P. (1972) *Kristall u. Technik* **7**, 1195.

Newkirk, H. W. and Smith, D. K. (1965) *Am. Min.* **50**, 22, 44.

Nickl, J. J. and Just, W. (1971) *J. Crystal Growth* **11**, 11.

Nielsen, J. W. (1958) *J. Appl. Phys.* **29**, 390.

Nielsen, J. W. (1960a) *J. Appl. Phys.* **31**, 51S.

Nielsen, J. W. (1960b) US Patent 2.957.827.

Nielsen, J. W. (1964) *Electronics* (November 30).

Nielsen, J. W. and Dearborn, E. F. (1958) *J. Phys. Chem. Solids* **5**, 202.

Nielsen, J. W. and Dearborn, E. F. (1960) *J. Phys. Chem.* **64**, 1762.

Nielsen, J. W., Linares, R. C. and Koonce, S. E. (1962) *J. Am. Ceram. Soc.* **45**, 12.

Nielsen, J. W., Lepore, D. A. and Leo, D. C. (1967) In "Crystal Growth" (H. S. Peiser, ed.) p. 457. Pergamon, Oxford.

Nielsen, O. V. (1969) *J. Crystal Growth* **5**, 398.

Niemyski, T., Mierzejewska-Appenheimer, S. and Majewski, J. (1967) In "Crystal Growth" (H. S. Peiser, ed.) p. 585. Pergamon, Oxford.

Nistor, N. and Cruceanu, E. (1969) *Kristall u. Technik* **4**, 337.

Nomura, S. and Sawada, S. (1952) *Rept. Inst. Sci. Tech. Univ. Tokyo* **6**, 191.

Nomura, S., Takabayashi, H. and Nakagawa, T. (1968) *Japan. J. Appl. Phys.* **7**, 600.

Nouet, J., Jacobini, C., Ferey, G., Gérard, J. Y. and DePape, R. (1971) *J. Crystal Growth* **8**, 94.

Nowicki, L. (1962) *Acta Phys. Polon.* **22**, 287.

Okada, T., Sakizawa, H. and Iida, S. (1963) *J. Phys. Soc. Japan* **18**, 981.

Okuno, Y., Suto, K. and Nishizawa, J. (1971) *Japan J. Appl. Phys.* **10**, 388.

Oleksiv, G. I. and Kripyakevich, P. I. (1967) *Visn. Lviv. Derz. Univ. Ser. Khim.* **9**, 25.

Osmer, J. A. and Chase, A. B. (1972) *J. Am. Ceram. Soc.* **55**, 292.
Ovechkin, E. K., Kraslinova, N. P., Volova, L. M., Shutova, A. T. and Kovalenko, L. S. (1968) *Bull. Acad. Sci. USSR Inorg. Mat.* **4**, 1141.
Packter, A. (1968) *Kristall u. Technik* **3**, 51.
Packter, A. and Roy, B. N. (1971) *Kristall u. Technik* **6**, 39.
Packter, A. and Roy, B. N. (1973) *J. Crystal Growth* **18**, 86.
Panish, M. B. (1969) *J. Phys. Chem. Solids* **30**, 1083.
Panish, M. B., Queisser, H. J., Derick, L. and Sumski, S. (1966) *Solid State Electronics* **9**, 311.
Parker, H. S. and Brower, W. S. (1967) In "Crystal Growth" (H. S. Peiser, ed.) p. 489. Pergamon, Oxford.
Parker, S. G. and Pinnell, J. E. (1968) *J. Crystal Growth* **3/4**, 490.
Parker, S. G. and Pinnell, J. E. (1970) *J. Electrochem. Soc.* **117**, 107.
Patel, A. R. and Arora, S. K. (1973) *J. Crystal Growth* **18**, 175.
Patel, A. R. and Bhat, H. L. (1971) *J. Crystal Growth* **8**, 153; **11**, 166.
Patel, A. R. and Koshy, J. (1968) *J. Crystal Growth* **2**, 128.
Patel, A. R. and Singh, R. P. (1969) *J. Crystal Growth* **5**, 70.
Perekalina, T. M. and Cheparin, V. P. (1968) *Sov. Phys.–Solid State* **9**, 2524.
Perloff, D. S., Vlasse, M. and Wold, A. (1969) *J. Phys. Chem. Solids* **30**, 1071.
Perloff, D. S. and Wold, A. (1967) *J. Phys. Chem. Solids Suppl.* **1**, 361.
Perner, B. (1969) *J. Crystal Growth* **6**, 86.
Perry, F. W. (1967) In "Crystal Growth" (H. S. Peiser, ed.) p. 483. Pergamon, Oxford.
Perry, F. W., Hutchins, G. A. and Cross, L. E. (1967) *Mat. Res. Bull.* **2**, 409.
Petrakovskii, G. A., Sablina, K. A. and Smokotin, E. M. (1969) *Sov. Phys.–Solid State* **10**, 2005.
Petzold, D. R., Schultze, D. and Wilke, K.-Th. (1971) *Z. Anorg. Allg. Chem.* **386**, 288; see also *Kristall u. Technik* **6**, 53.
Phillips, W. and Pressley, R. J. (1967) *Bull. Am. Ceram. Soc.* **46**, 366.
Pierce, R. D., Wolfe, R. and Van Uitert, L. G. (1969) *J. Appl. Phys.* **40**, 1241.
Pickardt, J., Riedel, E. and Reuter, B. (1970) *Z. Anorg. Allg. Chem.* **373**, 15.
Plaskett, T. S. (1969) *J. Electrochem. Soc.* **116**, 1722.
Plaskett, T. S. and Woods, J. F. (1971) *J. Crystal Growth* **11**, 341.
Plaskett, T. S., Blum, S. E. and Foster, L. M. (1967) *J. Electrochem. Soc.* **114**, 1303.
Poiblaud, G. and Jacob, G. (1973) *Mat. Res. Bull.* **8**, 845.
Pointon, A. J. and Robertson, J. M. (1967) *J. Mat. Sci.* **2**, 293.
Porta, P., Marezio, M., Remeika, J. P. and Dernier, P. D. (1972) *Mat. Res. Bull.* **7**, 157.
Porter, C. S., Spencer, E. G. and Le Craw, R. C. (1958) *J. Appl. Phys.* **29**, 495.
Prener, J. S. (1967) *J. Electrochem. Soc.* **114**, 77.
Queisser, H. J. and Panish, M. B. (1967) *J. Phys. Chem. Solids* **28**, 1177.
Quon, H. H., Eastwood, H. K. and Potvin, A. J. (1970) *J. Can. Ceram. Soc.* **39**, 27.
Quon, H. H., Potvin, A. and Entwistle, S. D. (1971) *Mat. Res. Bull.* **6**, 1175.
Rashkovich, L. N. (1969) *Growth of Crystals* **7**, 280.
Rea, J. R. and Kostiner, E. (1971) *J. Crystal Growth* **11**, 110.
Reed, T. B. and Fahey, R. E. (1971) *J. Crystal Growth* **8**, 337.
Reid, F. J., Baxter, R. D. and Miller, S. E. (1966) *J. Electrochem. Soc.* **113**, 713.
Reid, A. F., Wadsley, A. D. and Sienko, M. J. (1968) *Inorg. Chem.* **7**, 112.
Remeika, J. P. (1954) *J. Amer. Chem. Soc.* **76**, 940.
Remeika, J. P. (1956) *J. Amer. Chem. Soc.* **78**, 4259.

Remeika, J. P. (1958a) US Patent 2.852.400.
Remeika, J. P. (1958b) US Patent 2.848.310.
Remeika, J. P. (1960) *J. Appl. Phys.* **31**, 263 S.
Remeika, J. P. (1963) US Patent 3.075.831, 3.079.240, 3.110.674.
Remeika, J. P. and Ballman, A. A. (1964) *Appl. Phys. Lett.* **5**, 180.
Remeika, J. P. and Comstock, R. L. (1964) *J. Appl. Phys.* **35**, 3320.
Remeika, J. P. and Kometani, T. Y. (1968) *Mat. Res. Bull.* **3**, 895.
Remeika, J. P. and Marezio, M. (1966) *Appl. Phys. Lett.* **8**, 87.
Remeika, J. P. and Glass, A. M. (1970) *Mat. Res. Bull.* **5**, 37.
Remeika, J. P. and Spencer, E. G. (1964) *J. Appl. Phys.* **35**, 2803.
Remeika, J. P., Dodd, D. M. and DiDomenica, M. (1966) *J. Appl. Phys.* **37**, 5004.
Remeika, J. P., Gyorgy, E. M. and Wood, D. L. (1969) *Mat. Res. Bull.* **4**, 51.
Robbins, C. R. (1968) *J. Crystal Growth* **2**, 402.
Robbins, D. A. (1959) Nat. Phys. Lab. Symp. 9, Vol. II, 7 B, p. 2.
Robertson, D. S. and Cockayne, B. (1966) *J. Appl. Phys.* **37**, 927.
Robertson, E. C., Birch, F. and MacDonald, G. J. F. (1957) *Amer. J. Sci.* **255**, 115.
Robertson, J. M. and Taylor, R. G. (1968) *J. Crystal Growth* **2**, 171.
Robertson, J. M., Smith, S. H. and Elwell, D. (1969) *J. Crystal Growth* **5**, 189.
Robertson, J. M. and Neate, B. W. (1972) *J. Crystal Growth* **13/14**, 576.
Röder, O., Heim, U. and Pilkuhn, M. H. (1970) *J. Phys. Chem. Solids* **31**, 2625.
Rodot, H., Hruby, A. and Schneider, M. (1968) *J. Crystal Growth* **3/4**, 305.
Rogers, D. B., Ferretti, A. and Kunnmann, W. (1966) *J. Phys. Chem. Solids* **27**, 1445.
Rogers, H. H. (1952) Techn. Report 56, Lab. Ins. Res. MIT.
Rostoczy, F. E., Ermanis, F., Hayashi, I. and Schwarz, B. (1970) *J. Appl. Phys.* **41**, 264.
Roth, R. S., Parker, H. S., Brewer, W. S. and Waring, J. L. (1973) In "Fast Ion Transport in Solids, Solid State Batteries and Devices" (W. Van Gool, ed.) p. 217. North-Holland, Amsterdam.
Roth, W. L. and Romanczuk, R. J. (1969) *J. Electrochem. Soc.* **116**, 975.
Rowcliffe, D. J. and Warren, W. J. (1970) *J. Mat. Sci.* **5**, 345.
Roy, R. and White, W. B. (1972) *J. Crystal Growth* **13/14**, 78.
Rozhdestvenskaya, M. V., Romanovskaya, O. M. and Yureva, E. K. (1962) *Sov. Phys.–Solid State* **3**, 1698.
Rubenstein, M. (1968) *J. Crystal Growth* **3/4**, 309.
Rubenstein, M. and Ure, R. W. (1968) *J. Phys. Chem. Solids* **29**, 551.
Rubinstein, B. E., Titova, A. G. and Lapovok, B. L. (1965) *Sov. Phys.–Solid State* **6**, 2834.
Rundquist, S. (1962) *Acta Chem. Scand.* **16**, 1.
Rupprecht, H. (1966) Proc. Int. Symp. GaAs, Reading, p. 57.
Russell, R. G., Morgan, W. L. and Scheffler, L. F. (1962) US Patent 3.065.091.
Sakamoto, C. and Setoguchi, M. (1964) *Gypsum and Lime (Japan)* **73**, 234.
Salnikov, V. D. and Shelarkin, V. I. (1967) *Tr. Vses. NAUK Issled. Inst. Khim. Reakt. Osobo Chist. Khim. Veschestv.* **30**, 415.
Samogyi, K. and Bertoti, J. (1972) *Jap. J. Appl. Phys.* **11**, 103.
Sasaki, H. (1965) *J. Phys. Soc. Japan* **20**, 264.
Sasaki, H. (1965, 1967) Brit. Patent 1.095.805.
Sasaki, H. (1968) US Patent 3.409.412.
Sasaki, H. and Kurokawa, E. (1965) *J. Am. Ceram. Soc.* **48**, 171.
Sasaki, H. and Watanabe, A. (1964) *J. Phys. Soc. Japan* **19**, 1748.

Saul, R. H. and Hackett, W. H. (1970) *J. Electrochem. Soc.* **117,** 921.
Savage, R. O. and Tauber, A. (1964) *J. Am. Ceram. Soc.* **47,** 13.
Savage, R. O. and Tauber, A. (1967) *Mat. Res. Bull.* **2,** 469.
Savage, R. O., Dixon, S. and Tauber, A. (1965) *J. Appl. Phys.* **36,** 873.
Scheel, H. J. (1968) *J. Crystal Growth* **2,** 411.
Scheel, H. J. (1971) *Z. Krist.* **133,** 264.
Scheel, H. J. (1972a) *J. Crystal Growth* **13/14,** 560.
Scheel, H. J. (1972b) Swiss Patent 530.939.
Scheel, H. J. (1974) *J. Crystal Growth* **24/25,** 669.
Scheel, H. J. and Schulz-DuBois, E. O. (1971) *J. Crystal Growth* **8,** 304.
Schieber, M. (1964) *J. Am. Ceram. Soc.* **47,** 537.
Schieber, M. (1966) *J. Appl. Phys.* **37,** 4588.
Schieber, M. (1967) *Kristall u. Technik* **2,** 55.
Schieber, M. and Holmes, L. (1964) *J. Appl. Phys.* **35,** 1004.
Schieber, M. and Holmes, L. (1965) *J. Appl. Phys.* **36,** 1159.
Schieber, M., Grill, A. and Shidlovsky, I. (1968) *J. Crystal Growth* **3/4,** 467.
Schieber, M., Grill, A., Nowik, I., Wanklyn, B. M., Sherwood, R. C. and Van Uitert, L. G. (1972a) *J. Appl. Phys.* **43,** 1864.
Schieber, M., Grill, A. and Avigal, Y. (1972b) *J. Crystal Growth* **13/14,** 579.
Schultze, D., Wilke, K.-Th. and Waligora, C. (1967) *Z. Anorg. Allg. Chem.* **352,** 184.
Schultze, D., Wilke, K.-Th. and Waligora, C. (1971) *Z. Anorg. Allg. Chem.* **380,** 37.
Schwarzer, H. and Neels, H. (1971) *Kristall u. Technik* **6,** 639.
Scott, B. A. and Burns, G. (1972) *J. Am. Ceram. Soc.* **55,** 331.
Scott, J. F., Leheny, R. F., Remeika, J. P. and Sweedler, A. R. (1970) *Phys. Rev.* B **2,** 3883.
Sekizawa, H. and Sekizawa, K. (1962) *J. Phys. Soc. Japan* **17,** 359.
Seleznev, V. N., Pukhov, I. K., Dronin, A. I. and Shapolov, V. A. (1971) *Sov. Phys.–Solid State* **12,** 683.
Setoguchi, M. and Sakamoto, C. (1967) *J. Ceram. Assoc. Japan* **75,** 325.
Shafer, M. W., Torrance, J. B. and Penney, T. (1972) *J. Phys. Chem. Solids* **33,** 2251.
Shanks, H. R., Sidles, P. H. and Danielson, G. C. (1962) In "Non-Stoichiometric Compounds" Am. Chem. Soc., p. 237.
Shanks, H. R. (1972) *J. Crystal Growth* **13/14,** 433.
Shannon, J. and Katz, L. (1970) *Acta Cryst.* **B26,** 102.
Shannon, R. D. and Calvo, C. (1973) *Acta Cryst.* **B29,** 1338.
Shaskov, Y. M. and Shushlebina, N. Y. (1964) *Sov. Phys.–Solid State* **6,** 1134.
Sherwood, R. C., Remeika, J. P. and Williams, H. J. (1959) *J. Appl. Phys.* **30,** 217.
Shick, B. T. and Buessen, W. R. (1969) *J. Appl. Phys.* **40,** 1294.
Shick, L. K. and Nielsen, J. W. (1971) *J. Appl. Phys.* **42,** 1554.
Shick, L. K. and Von Neida, A. R. (1969) *J. Crystal Growth* **5,** 313.
Shick, L. K. and Von Neida, A. R. (1971) US Patent 3.627.498.
Shih, K. K. (1970) *J. Electrochem. Soc.* **117,** 387.
Shih, K. K. and Blum, J. M. (1971) *J. Electrochem. Soc.* **118,** 1631.
Shinoyama, S. and Suemune, Y. (1970) Proc. Int. Conf. Ferrites Kyoto, p. 346.
Shiraki, H. (1969) *Japan J. Appl. Phys.* **8,** 279.
Sholokhovich, M. L. (1958) *J. Inorg. Chem.* **3,** 207.
Sholokhovich, M. L. and Berkova, G. V. (1955) *J. Gen. Chem. USSR* **25,** 1201.

Sholokhovich, M. L. and Khodakov, A. L. (1962) *Growth of Crystals* **3**, 328.
Sholokhovich, M. L., Berberova, L. M. and Varicheva, V. I. (1968) *Growth of Crystals* **6A**, 85.
Sholokhovich, M. L., Fesenko, E. G., Kramanov, O. P. and Khodakov, A. L. (1956) *Dokl. Akad. Sci. USSR* **5**, 1025.
Sirtl, E. and Woerner, L. M. (1972) *J. Crystal Growth* **16**, 215.
Sleight, A. W. and Jarret, H. S. (1969) *J. Phys. Chem. Solids* **29**, 868.
Smith, S. H. and Elwell, D. (1968) *J. Crystal Growth* **3/4**, 471.
Smith, S. H. and Wanklyn, B. M. (1974) *J. Crystal Growth* **21**, 23.
Smith, A. J. and Welch, A. J. E. (1960) *Acta Cryst.* **13**, 653.
Smith, W. V., Overmeyer, J. and Calhoun, B. A. (1959) *IBM J. Res. Develop.* **3**, 153.
Sobon, L. E. and Greene, P. E. (1966) *J. Am. Ceram. Soc.* **49**, 106.
Somogyi, K. and Bertoti, I. (1972) *Japan J. Appl. Phys.* **11**, 103.
Sonomura, H. and Miyaauchi, T. (1969) *Japan J. Appl. Phys.* **8**, 1263.
Spencer, E. G., Lepore, D. A. and Nielsen, J. W. (1968) *J. Appl. Phys.* **39**, 732.
Spitzer, W. G. and Panish, M. B. (1969) *J. Appl. Phys.* **40**, 4200.
Spring-Thorpe, A. J. and Pamplin, B. R. (1968) *J. Crystal Growth* **3/4**, 313.
Spring-Thorpe, A. J. (1969) *Mat. Res. Bull.* **4**, 125.
Steininger, J. (1968) *Mat. Res. Bull.* **3**, 595.
Steininger, J. and England, R. E. (1968) *Trans. Met. Soc. AIME* **242**, 444.
Stone, B. and Hill, D. (1959) *Bull. Am. Phys. Soc.* **4**, 408; *Phys. Rev. Lett.* **4**, (1960) 282.
Stringfellow, G. P. and Greene, P. E. (1971) *J. Electrochem. Soc.* **118**, 805.
Strong, H. M. and Chrenko, R. M. (1971) *J. Phys. Chem.* **75**, 1838.
Strong, H. M. and Hanneman, R. E. (1967) *J. Chem. Phys.* **46**, 3668; In "Crystal Growth" (H. S. Peiser, ed.) p. 580. Pergamon, Oxford.
Strong, H. M. and Wentorf, R. H. (1972) *Naturwiss.* **59**, 1.
Sturgeon, G. D., Mercurio, J. P., Etourneau, J. and Hagenmuller, P. (1974) *Mat. Res. Bull.* **9**, 117.
Sugai, T. and Wada, M. (1971) *Japan J. Appl. Phys.* **10**, 955.
Sugai, T., Hasegawa, S. and O'Hara, G. (1967) *Japan J. Appl. Phys.* **6**, 901.
Sugai, T., Hasegawa, S. and O'Hara, G. (1968) *Japan J. Appl. Phys.* **7**, 358.
Sugiyama, K. and Kawakami, T. (1971) *Japan J. Appl. Phys.* **10**, 1007.
Sumiyoshi, Y., Ito, N. and Noda, T. (1968) *J. Crystal Growth* **3/4**, 327.
Summergrad, R. N. and Banks, E. (1957) *J. Phys. Chem. Solids.* **2**, 312.
Suzuishi, M. and Ito, H. (1967) *Japan J. Appl. Phys.* **6**, 653.
Suzuishi, M., Ito, H. and Yamaguchi, T. (1968) *Japan J. Appl. Phys.* **7**, 183.
Suzuki, K., Nunomura, K. and Koyama, N. (1971) Ann. Meetg. Jap. Soc. Electron. Commun., p. 362.
Swets, D. E. (1971) *J. Crystal Growth* **8**, 311.
Swithenby, S. I., Wanklyn, B. M. and Wells, M. R. (1974) *J. Mat. Sci.* **9**, 845.
Syrnikov, P. P. (1971) *Bull. Acad. Sci. USSR Phys. Ser.* **6**, 1154.
Sysoev L. A., Obukhovskii, Ya.A. and Bidnaya, D. S. (1966) *Growth of Crystals* **4**, 130.
Tabata, H. and Okuda, H. (1971) *Yogyo Kyokai Shi.* **79**, 15.
Takahashi, T. and Freise, E. J. (1965) *Phil. Mag.* **12**, 1.
Takahashi, T., Morizumi, T. and Shirose, S. (1971) *J. Electrochem. Soc.* **118**, 1639.
Takahashi, T., Yamada, O. and Ametani, K. (1973) *J. Crystal Growth* **20**, 89.
Takeda, Y., Shimada, M., Kanamaru, F., Koizumi, M. and Yamamoto, N. (1974) *Mat. Res. Bull.* **9**, 537.

Takei, H. and Koide, S. (1966) *J. Phys. Soc. Japan* **21**, 1010.

Tamura, T., Moriizumi, T. and Takahashi, K. (1971) *Japan J. Appl. Phys.* **10**, 813.

Tasaki, A. and Iida, S. (1963) *J. Phys. Soc. Japan* **18**, 1148.

Tauber, A., Kohn, J. A. and Savage, R. O. (1963) *J. Appl. Phys.* **34**, 1265.

Tauber, A., Dixon, S. and Savage, R. O. (1964) *J. Appl. Phys.* **35**, 1008.

Teague, J. R., Gerson, R. and James, W. J. (1970) *Solid State Comm.* **8**, 1073.

Tester, J. W., Reid, R. C. and Wolff, G. A. (1971) *Mat. Res. Bull.* **6**, 1265.

Thomas, D. G. and Lynch, R. T. (1967) *J. Phys. Chem. Solids* **28**, 433.

Thomas, D. G., Gershenzon, M. and Trumbore, F. Z. (1964) *Phys. Rev.* **133**, A 269.

Timofeeva, V. A. (1959) *Growth of Crystals* **2**, 73.

Timofeeva, V. A. (1967) In "Crystal Growth" (H. S. Peiser, ed.) p. 445. Pergamon, Oxford.

Timofeeva, V. A. (1968) *J. Crystal Growth* **3/4**, 496; *Growth of Crystals* **6A**, 79.

Timofeeva, V. A. and Bychkov, V. Z. (1966) *Growth of Crystals* **4**, 114.

Timofeeva, V. A. and Kvapil, I. (1966) *Sov. Phys.-Cryst.* **11**, 263.

Timofeeva, V. A. and Zalesskii, A. V. (1959) *Growth of Crystals* **2**, 69.

Titova, A. G. (1959) *Sov. Phys.–Solid State* **1**, 1871.

Titova, A. G. (1962) *Growth of Crystals* **3**, 306–308.

Tolksdorf, W. (1968) *J. Crystal Growth* **3/4**, 463.

Tolksdorf, W. (1973) *J. Crystal Growth* **18**, 57.

Tolksdorf, W. (1974a) *Acta Electron.* **17**, 57.

Tolksdorf, W. (1974b) private communication.

Tolksdorf, W. and Welz, F. (1972) *J. Crystal Growth* **13/14**, 566.

Tolksdorf, W. and Welz, F. (1973) *J. Crystal Growth* **20**, 47.

Toyama, M., Naito, M. and Kasami, A. (1969) *Japan J. Appl. Phys.* **8**, 358.

Trumbore, F. A. and Porbansky, F. M. (1960) *J. Appl. Phys.* **31**, 2068.

Trumbore, F. A., Kowalchik, M., White, H. G., Logan, R. A. and Luke, C. L. (1964) *J. Electrochem. Soc.* **111**, 748.

Tsushima, K. (1966) *J. Appl. Phys.* **37**, 443.

Tsushima, K. (1967) *Tech. J. Jap. Broadcasting Corp.* **19**, 1.

Tsushima, K., Kino, Y. and Funahashi, S. (1968) *J. Appl. Phys.* **39**, 626.

Tsuzuki, K., Sakata, K., Hagewa, S. and Ohara, G. (1968) *Japan J. Appl. Phys.* **7**, 953.

Tsuzuki, K., Sakata, K. and Wada, M. (1969) *Japan J. Appl. Phys.* **8**, 816.

Tsuzuki, K., Sakata, K., Ohara, G. and Wada, M. (1973) *Japan J. Appl. Phys.* **12**, 1500.

Ueda, R., Ohtsuki, O., Shinohara, K. and Ueda, Y. (1972) *J. Crystal Growth* **13/14**, 668.

Ugai, Ya.A. and Gukov, O.Ya. (1965) *Zh. Neorgan. Khim.* **1**, 857.

Van Landuyt, J., Remaut, G. and Amelinckx, S. (1969) *Mat. Res. Bull.* **4**, 329.

Van Uitert, L. G. and Egerton, L. (1961) *J. Appl. Phys.* **32**, 959.

Van Uitert, L. G. and Soden, R. R. (1960) *J. Appl. Phys.* **31**, 328.

Van Uitert, L. G., Pictroski, L. and Grodkiewicz, W. H. (1969) *Mat. Res. Bull.* **4**, 777.

Van Uitert, L. G., Sherwood, R. C., Bonner, W. A., Grodkiewicz, W. H., Pictroski, L. and Zydzik, G. (1970a) *Mat. Res. Bull.* **5**, 153.

Van Uitert, L. G., Smith, D. H., Bonner, W. A., Grodkiewicz, W. H. and Zydzik, G. J. (1970b) *Mat. Res. Bull.* **5**, 455.

Van Uitert, L. G., Bonner, W. A., Grodkiewicz, W. H., Pictroski, L. and Zydzik, G. J. (1970c) *Mat. Res. Bull.* **5**, 825.

Van Uitert, L. G., Linares, R. C., Soden, R. R. and Ballman, A. A. (1962) *J. Chem. Phys.* **36**, 702.

Van Uitert, L. G., Grodkiewicz, W. H. and Dearborn, E. F. (1965) *J. Am. Ceram. Soc.* **48**, 105.

Vichr, M. (1966) *Kristall u. Technik* **1**, 581.

Vichr, M. and Makram, H. (1969) *J. Crystal Growth* **5**, 77.

Vigdorovich, V. N. and Marychev, V. V. (1964) *Sov. Phys. Doklady* **159**, 157.

Viting, L. M. (1965) *Vestn. Mosk. Univ. Ser. II Khim.* **20**, 54.

Von Dreele, R. B., Bless, P. W., Kostiner, E. and Hughes, R. E. (1970) *J. Solid State Chem.* **2**, 612.

Von Neida, A. R. and Shick, L. K. (1969) *J. Appl. Phys.* **40**, 1013.

Von Neida, A. R., Oster, L. J. and Nielsen, J. W. (1972) *J. Crystal Growth* **13/14**, 647.

Von Philipsborn, H. (1967) *J. Appl. Phys.* **38**, 955; *Helv. Phys. Acta* **40**, 810.

Von Philipsborn, H. (1969) *J. Crystal Growth* **5**, 135.

Vora, H. and Zupp, R. R. (1970) *Mat. Res. Bull.* **5**, 977.

Voronkova, V. I., Yanovskii, V. K. and Koptsik, V. A. (1968a) *Sov. Phys. Doklady* **12**, 994.

Voronkova, V. I., Yanovskii, V. K. and Koptsik, V. A. (1968b) *Moscow Univ. Phys. Astron.* **23**, 109.

Vutien, L. and Anthony, A. M. (unpublished), see Laurent, Y. (1969).

Vutien, L., Grandin De L'Eprevier, A., Gabis, V. and Anthony, A. M. (1972) *J. Crystal Growth* **13/14**, 610.

Wagner, J. W. and Thompson, A. G. (1970) *J. Electrochem. Soc.* **117**, 936.

Wagner, P. and Lorenz, M. R. (1966) *J. Phys. Chem. Solids* **27**, 1749.

Wagner, R. S. and Doherty, C. J. (1966) *J. Electrochem. Soc.* **113**, 1300.

Wagner, R. S. and Ellis, W. C. (1964) *Appl. Phys. Lett.* **4**, 89.

Wagner, R. S. and Ellis, W. C. (1965) *Trans. Met. Soc. AIME* **233**, 1053.

Wallace, C. A. and White, E. A. D. (1967) In "Crystal Growth" (H. S. Peiser, ed.) p. 431. Pergamon, Oxford.

Wang, C. C. and McFarlane, S. H. (1968) *J. Crystal Growth* **3/4**, 485.

Wang, C. C. and Zanzucchi, P. J. (1971) *J. Electrochem. Soc.* **118**, 586.

Wanklyn, B. M. (1969) *J. Crystal Growth* **5**, 219, 279, 323.

Wanklyn, B. M. (1970) *J. Crystal Growth* **7**, 107, 368.

Wanklyn, B. M. (1972) *J. Mat. Sci.* **7**, 813.

Wanklyn, B. M. (1973a) *J. Mat. Sci.* **8**, 649.

Wanklyn, B. M. (1973b) *J. Mat. Sci.* **8**, 1055.

Wanklyn, B. M. (1974) In "Crystal Growth" (B. R. Pamplin, ed.). Pergamon, Oxford.

Watanabe, M., Gando, T. and Hashimoto, T. (1970) Proc. Int. Conf. Ferrites Kyoto, p. 493.

Weaver, E. A. and Li, C. T. (1969) *J. Am. Ceram. Soc.* **52**, 335.

Webster, F. W. and White, E. A. D. (1969) *J. Crystal Growth* **5**, 167.

Weinstein, M. and Mlavsky, A. I. (1964) *J. Appl. Phys.* **35**, 1892.

Weller, P. F. and Grandits, D. M. (1972) *J. Crystal Growth* **12**, 63.

Weller, P. F., Taylor, B. E. and Mohler, R. L. (1970) *Mat. Res. Bull.* **5**, 465.

Wentorf, R. H. (1960) US Patent 2.947.617; *J. Chem. Phys.* **36**, 1990 (1962).

Wentorf, R. H. (1965) US Patent 3.192.015.

Whipps, P. W. (1972) *J. Crystal Growth* **12**, 120.

White, E. A. D. (1961) *Nature* **191**, 901.

White, E. A. D. (1966) *J. Mat. Sci.* **1,** 199.
White, E. A. D. and Brightwell, J. W. (1965) *Chem. and Ind.* 1662.
Wickersheim, K. A., Lefever, R. A. and Hanking, B. M. (1960) *J. Chem. Phys.* **32,** 271.
Wickham, D. G. (1962) *J. Appl. Phys.* **33,** 3597.
Wilcox, W. R. and Corley, R. A. (1967) *Mat. Res. Bull.* **2,** 571.
Wilke, K.-Th. (1962) *Ber. Geol. Ges.* **7,** 500.
Wilke, K.-Th. (1964) *Z. Anorg. Allg. Chem.* **330,** 164.
Wilke, K.-Th. (1968) *Growth of Crystals* **6A,** 71.
Wilke, K.-Th. (1973) "Kristallzüchtung". VEB Dt. Verlag der Wiss., Berlin.
Wilke, K.-Th., Töpfer, K. and Schultze, D. (1965) *Z. Physik. Chem.* **230,** 112.
Wilke, K.-Th., Schultze, D. and Töpfer, K. (1967) *J. Crystal Growth* **1,** 41.
Wold, A., Kunnmann, W., Arnott, R. J. and Ferretti, A. (1964) *Inorg. Chem.* **3,** 545.
Wolff, G. A. and Labelle, H. E. (1965) *J. Am. Ceram. Soc.* **48,** 441.
Wolff, G. A., Keck, P. H. and Broder, J. D. (1954) *Phys. Rev.* **94,** 753 (abstract only).
Wolff, G. A., Das, B. N., Lamport, C. B., Mlavsky, A. I. and Trickett, E. A. (1969) *Mat. Res. Bull. Suppl.* **4,** S67.
Wolten, G. M. and Chase, A. B. (1967) *Am. Min.* **52,** 1536.
Wood, D. L., Kolb, E. D. and Remeika, J. P. (1968) *J. Appl. Phys.* **39,** 1139.
Wood, J. D. and White, E. A. D. (1968) *J. Crystal Growth* **3/4,** 480.
Woodall, J. M., Rupprecht, H. and Reuter, W. (1969) *J. Electrochem. Soc.* **116,** 899.
Wright, M. A. (1965) *J. Electrochem. Soc.* **112,** 1114.
Yakel, H. L., Koehler, W. C., Bertaut, E. F. and Forrat, F. (1963) *Acta Cryst.* **16,** 957.
Yamamoto, N. and Miyauchi, T. (1972) *Japan J. Appl. Phys.* **11,** 1383.
Yocom, P. N. and Smith, R. T. (1973) *Mat. Res. Bull.* **8,** 1293.
Yokomizo, Y., Takahashi, T. and Nomura, S. (1970) *J. Phys. Soc. Japan* **28,** 1279.
Zonn, Z. N. (1965) *Inorg. Mater. (USSR)* **1,** 1034.
Zonn, Z. N. and Joffe, V. A. (1968) *Growth of Crystals* **6A,** 112.
Zonn, Z. N. and Joffe, V. A. (1969) *Growth of Crystals* **8,** 63.

Appendix to Chapter 10

Abbreviations used:

ACRT – Accelerated crucible rotation technique
CR   – Chemical reaction
EL   – Electrolysis
EV   – Evaporation
HPS   – High-pressure solution growth
LPE   – Liquid phase epitaxy
SC   – Slow cooling
THM – Travelling heater method
TR   – (Gradient) transport
TSM   – Travelling solvent method
TSSG – Top-seeded solution growth
VLS   – Vapour-liquid-solid
VLR S – Vapour-liquid-reaction-solid
X   – Seed crystal

Crystal	Solvent	Technique	Temp. range	Cooling rate duration	Crystal size (mm)	Reference
Ag_3AsS_3	silicone oil	SC	250	—	0.01–2	Weil et al. (1954)
Al_2O_3	Na_3AlF_6	TR/X	1120–1000	50–200 h	$20 \times 20 \times 5$	Watanabe and Sumiyoshi (1974)
As_4S_4	silicone oil	SC	250	—	0.01–2	Weil et al. (1954)
Ba_2CoWO_6	$BaCl_2$	SC(?)	—	0.5	small	Voorhoeve et al. (1974)
Ba_2FeNbO_6	$BaCl_2$	SC(?)	—	0.5	small	Voorhoeve et al. (1974)
$BaFe_{12}O_{19}$	NaCl/KCl	CR/SC	1000–1050	30 min.	$<1\,\mu m$	Arendt (1973)
$Ba_xSr_{1-x}Nb_2O_6$	$BaO/SrO/B_2O_3$	SC	1350–700	1	2–6 mm	Whipps (1972)
$BaMoO_4$	LiCl/KCl	SC	—	—	<10	Potkin (1972)
Ba_2NiTeO_6	$BaCl_2$	SC	1150–800	5	small prism	Köhl et al. (1972)
$BaTiO_3(Sr)$	KF	SC	—	—	—	Goulpeau et al. (1973)
$Ba_2TiSi_2O_8$	TiO_2	SC	1425–1000	3	5	Robbins (1970)
$BaWO_4$	LiCl/NaCl	EV	950–670	140 h	$\sim 5 \times 4 \times 1$	Patel and Arora (1974)
$BaWO_4$	LiCl/KCl	SC	—	—	<10	Potkin (1972)
$BaWO_4$	NaCl, KCl	SC	950–650	2–50	~3	Voigt and Neels (1971)
$BeAl_2O_4$	$PbO/PbF_2/B_2O_3$	SC	1270–800	3	$\sim 10 \times 3 \times 1$	Tabata et al. (1974)
$Be_3Al_2Si_6O_{18}(Cr)$	V_2O_5 etc.	TR/X	1050	$\Delta T = 1{-}42$	~5	Ushio and Sumiyoshi (1972)
$Be_3Al_2Si_6O_{18}(Cr)$	V_2O_5	TR/X	1050	$\Delta T \sim 15°$	~10	Ushio and Sumiyoshi (1973a)
$Bi_{40}Fe_2O_{63}$	Bi_2O_3	TSSG/SC	~780	0.5	$15 \times 10 \times 10$	Bruton et al. (1974)
$Bi_{40}Ga_2O_{63}$	Bi_2O_3	TSSG/SC	—	—	$15 \times 15 \times 10$	Bruton et al. (1974)
$Bi_{12}TiO_{20}$	Bi_2O_3	TSSG/SC	~860	—	~100 g	Bruton et al. (1974)
$Bi_4Ti_3O_{12}$	Bi_2O_3	SC	1150–850	0.88	$70 \times 70 \times 1$	Morrison et al. (1970)
$Bi_{12}TiO_{20}$	Bi_2O_3	SC	110–850	0.75	$15 \times 10 \times 4$	Morrison (1971)
Ca_9BO_3Cl	$CaCl_2$	SC	820–	2–5	<20	Majling et al. (1974)
$CaCO_3$	Li_2CO_3	TSSG/HPS	~900°	—	$\sim 10 \times 10$	Balascio and White (1974)
$CaFe_xMn_{1-x}O_3$	$CaCl_2$	SC	1280–900	5–10	$3 \times 3 \times 3$	Banks et al. (1972)
$CaMoO_4$	LiCl/KCl	SC	—	—	<10	Potkin (1972)
$Ca_3Sc_2Ge_3O_{12}$	$Pb_2V_2O_7$	SC	1250–1000	—	—	Havlicek et al. (1974)
$CaWO_4$	LiCl/KCl	SC	—	—	<10	Potkin (1972)
$CaWO_4$	NaCl, KCl	LPE	950–650	2–50	~3	Voigt and Neels (1971)
$CdSnP_2$	Sn	LPE/SC	—	—	—	Bachmann et al. (1974)
$CdSnP_2$	Sn	LPE/SC	510–	10	—	Shay et al. (1974)
$CdTe$	Te, $CdCl_2$	THM	675	7 mm/day	—	Taguchi et al. (1974)

Compound	Flux/solvent	Method	Temperature	ΔT/rate	Size	Reference
$Cs_2Pt_4S_6$	Cs_2CO_3	SC	—	—	small	Rüdorff et al. (1968)
$DyAlO_3$	PbO/PbF_2	SC/EV	1290–	—	1	Flicstein and Schieber (1973)
$Dy_3Al_5O_{12}$	PbO/PbF_2	SC/EV	1290–	—	2	Flicstein and Schieber (1973)
Dy_2SiO_5	PbF_2	VLS/SC	1280–	—	3×3	Wanklyn and Hauptman (1974)
Dy_4SiO_8	PbF_2	VLS/SC	1280–	—	10×1	Wanklyn and Hauptman (1974)
$DyVO_4$	$Pb_2V_2O_7$	TSSG/TR	1200	$\Delta T = 2$–5/cm	$6 \times 3 \times 3$	Smith et al. (1974)
$Er_3Fe_5O_{12}$	PbF_2/BaF_2	SC	1300–900	4–5	—	Perekalina et al. (1972)
Er_2SiO_5	PbF_2/PbO	VLS/SC	1285–	—	—	Wanklyn and Hauptman (1974)
$(Eu, Y)_3Fe_5O_{12}$	PbO/B_2O_3	LPE	738–830	—	0.19–0.58 μm/min.	Plaskett et al. (1973)
EuS	$LiCl/KCl$	SC/CR	500	—	1×0.2	Koutaissoff (1964)
$GaAs$	Ga	LPE/SC	725–	30–120	—	Rosztocy and Kinoshita (1974)
$GaAs_{1-x}P_x$	Ga	VLS/TSM	950–1150	$\Delta T \sim 8$–40/cm	$28 \times 20 \times 20$	Saito and Seki (1974)
$GaAs_{1-x}Sb_x(Al)$	Ga	LPE/SC	760–700	60	10–15 μm	Antypas and Moon (1974)
$Ga_xIn_{1-x}Sb$	In	LPE/SC	410–280	10–200	—	Joullié et al. (1974)
GaP	Ga	VLRS	1330–1024	$\Delta T = 7$–46	0.1–0.7 mm/day	Kaneko et al. (1973)
$(Gd, Y, Yb)_3$ $(Fe, Ga)_5O_{12}$	$BaO/B_2O_3/BaF_2$	LPE	~ 1000	—	0.1–0.5 μm/min.	Hiskes and Burmeister (1973)
$Gd_{0.19}WO_3$	$RCl_3 + WO_3$	EL	1200	<60 mA	10	Collins and Ostertag (1966)
HfB_2	Fe	SC	1720–	5	2×2	Nakano et al. (1974)
$In_{1-x}Ga_xP$	In	LPE/TR	800	$\Delta T \sim 10$	50–75 μm	Macksey et al. (1973)
InP	In	LPE/SC	720–560	6–13	5–120 μm	Hess et al. (1974)
IrS_2	K_2CO_3	SC	—	—	small	Rüdorff et al. (1968)
$K_9Bi_3Zn_2Nb_{10}O_{36}$	Bi_2O_3	SC	1130–800	5–10	<10	Nomura and Kojima (1973)
$KNbO_3$	K_2CO_3	TSSG/SC	—	0.5	$30 \times 30 \times 5$	Fukuda et al. (1974)
$KNb_{1-x}Ta_xO_3$	K_2CO_3	TSSG/TR	1100	$\Delta T \sim 0$–9	14×14	Whiffin and Brice (1974)
$K_2Pd_3PtS_6$	K_2CO_3	SC			small	Rüdorff et al. (1968)
$K_2Pd_3SnS_6$	K_2CO_3	SC			small	Rüdorff et al. (1968)
$K_2Pd_3TiS_6$	K_2CO_3	SC			small	Rüdorff et al. (1968)
$K_2Pt_4S_6$	K_2CO_3	SC			small	Rüdorff et al. (1968)
$K_2Pt_3IrS_6$	K_2CO_3	SC			small	Rüdorff et al. (1968)
$K_2Pt_3SnS_6$	K_2CO_3	SC			small	Rüdorff et al. (1968)

Crystal	Solvent	Technique	Temp. range	Cooling rate duration	Crystal size (mm)	Reference
$K_2Pt_3TiS_6$	K_2CO_3	SC	900–1000	1 h	small	Rüdorff et al. (1968)
$K_{1-x}SbO_{3-x}F_x$	KF	EV			0.5 mm	Brower et al. (1974)
LaB_6	Al	SC	1500–	30	1	Aita et al. (1974)
LaP_5O_{14}	$H_4P_2O_7$	EV	750	—	10 × 3	Miller et al. (1974)
$LiFe_5O_8(Ru)$	PbO/B_2O_3	SC	1050–600	1		Jonker (1974)
Li_2MgSiO_4	Li_2MoO_4	SC	1420–900	3.3–7.6	8 × 8 × 1	Setoguchi and Sakamoto (1974)
Li_2ZnSiO_4	Li_2MoO_4	SC	1420–900	3.3–7.6	2.5 × 2.5	Setoguchi and Sakamoto (1974)
$Mg_{8.5}As_3O_{16}$	PbO/As_2O_5	SC	1225–600	8	small	Bless and Kostiner (1973)
$MgFeBO_4$	$Na_2B_4O_7$	SC	—	—		Mikov et al. (1973)
$Mg_2SiO_4 \cdot MgF_2$	PbF_2	VLS/SC	1280–	—	6	Wanklyn and Hauptman (1974)
MnB_4O_7	Bi_2O_3/B_2O_3	SC	1060–600	1.5	2	Abrahams et al. (1974)
$MnBi$	Bi	TSSG	435°, <355°		~60 × 4 × 4	Chen (1974)
$MnSi$	Cu	SC	1200–500	10	2	Johnson et al. (1973)
$NaNbO_3$	$NaBO_2$	TSSG/TR	1105	$\Delta T \sim 6$–36	160–400 Å s^{-1}	Dawson et al. (1974)
$Nb_2Z_6O_{17}$	BaO/V_2O_5	SC	1200–		small	Galy and Roth (1973)
NdP_5O_{14}	$H_3PO_4, H_4P_2O_7$	EV	550	7 days	3 × 3 × 1	Danielmeyer et al. (1974)
NdP_5O_{14}	$H_4P_2O_7$	EV	750		5 × 1	Miller et al. (1974)
Ni_2FeBO_5	$Na_2B_4O_7$	SC			small	Mikov et al. (1973)
$PbTiO_3$	PbO/B_2O_3	SC	1000–800	20	12 × 12 × 1	Fesenko et al. (1972)
$R_3Al_5O_{12}$	PbF_2/PbO	VLS/SC	1280–1150	1	—	Wanklyn and Garton (1974)
$RCrO_3$	PbF_2/Bi_2O_3	EV	1230	—	—	Subba Rao et al. (1971)
$RFeO_3$	$PbO/PbF_2/B_2O_3$	SC	1280–950	1.25–5		Akaba (1974)
R_2MoO_6	PbF_2/PbO	SC	1270–1160	0.7	5 × 2	Wanklyn (1974)
$Rb_2Pt_4S_6$	Rb_2CO_3	SC			small	Rüdorff et al. (1968)
$Rb_2Pt_3SnS_6$	Rb_2CO_3	SC			small	Rüdorff et al. (1968)
$Sm_{0.09}WO_3$	$SmCl_3 + WO_3$	EL	1200	<60 mA	5	Collins and Ostertag (1966)
$SrBi_2Ta_9O_9(Ba)$	Bi_2O_3	SC	1250–900	2	1 × 1 × 0.05	Newnham et al. (1973)
$SrFe_xMn_{1-x}O_3$	$SrCl_2$	SC	1330–900	5–10	1 × 1 × 1	Banks et al. (1972)
$SrFe_{12}O_{19}$	$NaCl/KCl$	CR/SC	1000–1050	30 min.	<1 μm	...et (1972)

$SrMoO_4$	LiCl/KCl	SC	—	—	<10	...rotkin (1972)
$SrTiO_3$	KF	SC	1000–	3–4	~10	Kojima and Kochi (1968)
$SrWO_4$	LiCl/NaCl	EV	950–730	40–100 h	~4×3×3	Patel and Arora (1974)
$SrWO_4$	LiCl/KCl	SC	—	—	<10	Potkin (1972)
$SrWO_4$	NaCl, KCl	SC	950–650	2–50	~3	Voigt and Neels (1971)
$TbVO_4$	$Pb_2V_2O_7$	TSSG/TR	1200	$\Delta T \sim 2$–5/cm	~3×3×3	Smith et al. (1974)
TiB_2	Fe, Co, Ni	SC	1700–1600	5	5×5	Nakano et al. (1974)
$TmVO_4$	$Pb_2V_2O_7$	TSSG/TR	1200	$\Delta T \sim 2$–5/cm	~6×4×4	Smith et al. (1974)
$Tm_{0.1}WO_3$	$TmCl_3 + WO_3$	EL	1200	<60 mA/cm²	few	Collins and Ostertag (1966)
$Y_3Al_5O_{12}$	$PbO/PbF_2/B_2O_3$	SC	1250–	2	3–10	Komatsu et al. (1974)
$Y_3Al_{5-x}Cr_xO_{12}$	$PbO/PbF_2/B_2O_3$	SC	1350–1050	—	—	Takasu and Shimanuki (1974)
$Y_3Fe_5O_{12}$(Ca, Te)	BaO/B_2O_3	LPE	1058	—	—	Davies and White (1974)
$Y_3Fe_5O_{12}$	$PbO/PbF_2/B_2O_3$	SC	1160–960	0.3–1.4	50 g	Dominé-Bergès et al. (1973)
$Y_3Fe_5O_{12}$	PbO/B_2O_3	TR	965	$\Delta T \sim 5$	—	Görnert and Hergt (1973)
$Y_3Fe_5O_{12}$	$PbO{\cdot}0.2B_2O_3$	SC/ACRT	1250–1200	0.4	10×10×10	Görnert and Hergt (1973)
$Y_3Fe_5O_{12}$(Pb, Si)	PbO/B_2O_3	LPE	960–780	—	~5 μm	Jonker (1975)
$Y_3Fe_5O_{12}$(Sm, Ga)	PbO/B_2O_3	HPS/SC	1300–950	3	—	Larsen and Robertson (1974)
$Y_3Fe_5O_{12}$(Bi)	BaO/B_2O_3	LPE	1020–980	—	—	Mroczkowski (1974)
$Y_3Fe_5O_{12}$(La, Ga)	PbO/B_2O_3	LPE	965	$\Delta T \sim 10$	20×20×5 μm	Robertson et al. (1973)
$(Y, R)_3(Fe, Ga)_5O_{12}$	$BaO/BaF_2/B_2O_3$	LPE	1050	—	—	Robertson et al. (1975)
$Y_3Fe_5O_{12}$	review	—	—	—	—	Suemune and Inone (1974)
						Tolksdorf (1974)
$Y_2Pt_2O_7$	PbO/PbO_2	HPS/SC	1290–950	1.5	6	Ostorero and Makram (1974)
ZnO	KOH	TR	480	—	needles	Kashyap (1973)
ZnO	$Na_2B_4O_7$	TSSG	—	—	2	Nevyantseva et al. (1969)
ZnS	$BaZnS_3$	SC	1300–	—	3	Malur (1966)
$ZnSnSb_2$	Sn	SC	427–360	0.44	—	Scott (1973)
$ZnTe$	Bi, Zn	LPE/SC	1050–	30–300	—	Fujita et al. (1973)
ZrB_2	Fe	SC	1700–	5	2	Nakano et al. (1974)
ZrO_2	PbF_2, V_2O_5	TR/EV	1050, 1010	$\Delta T = 40, 50$	5×5×2	Fujiki and Suzuki (1974)
ZrO_2	PbF_2	TR/EV	970	$\Delta T \sim 50$	5×3×1	Fujiki et al. (1972)
ZrO_2	Na_3AlF_6	SC/EV	1080–	1–10	~1	Ushio and Sumiyoshi (1973b)

References

Abrahams, S. C., Bernstein, J. L., Gibart, P., Robbins, M. and Sherwood, R. C. (1974) *J. Chem. Phys.* **60**, 1899.

Aita, T., Kawabe, U. and Honda, Y. (1974) *Japan J. Appl. Phys.* **13**, 391.

Akaba, R. (1974) *J. Crystal Growth* **24/25**, 537.

Antypas, G. A. and Moon, R. L. (1974) *J. Electrochem. Soc.* **121**, 416.

Arendt, R. H. (1973) *J. Solid State Chem.* **8**, 339.

Bachmann, K. J., Buehler, E., Shay, J. L. and Kammlott, G. W. (1974) *J. Electron. Mat.* **3**, 451.

Balascio, J. F. and White, W. B. (1974) *J. Crystal Growth* **23**, 101.

Banks, E., Berkooz, O. and Nakagawa, T. (1972) *Natl. Bur. Stand. Spec. Publ.* No. **364**, 265.

Bless, P. W. and Kostiner, E. (1973) *J. Solid State Chem.* **6**, 80.

Brower, W. S., Minor, D. B., Parker, H. S., Roth, R. S. and Waring, J. L. (1974) *Mat. Res. Bull.* **9**, 1045.

Bruton, T. M., Brice, J. C., Hill, O. F. and Whiffin, P. A. C. (1974) *J. Crystal Growth* **23**, 21.

Chen, T. (1974) *J. Crystal Growth* **24/25**, 454.

Collins, C. V. and Ostertag, W. (1966) *J. Amer. Chem. Soc.* **88**, 3171.

Danielmeyer, H. G., Jeser, J. P., Schönherr, E. and Stetter, W. (1974) *J. Crystal Growth* **22**, 298.

Davies, J. E. and White, E. A. D., (1974) *J. Mat. Sci.* **9**, 1374.

Dawson, R. D., Elwell, D. and Brice, J. C. (1974) *J. Crystal Growth* **23**, 65.

Dominé-Bergès, M., Loriers, J., Makram, H. and Villers, G. (1973) *Compt. Rend.* **277**, 1113.

Donohue, P. C. (1972) *Mat. Res. Bull.* **7**, 943.

Fesenko, E. G., Gavrilyachenko, V. G., Grigoreva, E. A. and Feronov, A. D. (1972) *Sov. Phys.-Cryst.* **17**, 122.

Flicstein, J. and Schieber, M. (1973) *J. Crystal Growth* **18**, 265.

Fujiki, Y. and Suzuki, Y. (1974) *J. Crystal Growth* **24/25**, 661.

Fujiki, Y., Suzuki, Y. and Ono, A. (1972) *J. Japan. Assoc. Min. Petr. Econ. Geol.* **67**, 20.

Fujita, S., Moriai, F., Arai, S. and Sakaguichi, T. (1973) *Japan J. Appl. Phys.* **12**, 1841.

Fukuda, T., Uematsu, Y. and Ito, T. (1974) *J. Crystal Growth* **24/25**, 450.

Galy, J. and Roth, R. S. (1973) *J. Solid State Chem.* **7**, 277.

Görnert, P. and Hergt, R. (1973) *Phys. stat. sol. (a)* **20**, 577.

Goulpeau, L., Pilet, J. C. and LeMontagner, S. (1973) *Bull soc. sci. Bretagne* **48**, 183.

Havlicek, V., Novak, P. and Mill, B. V. (1974) *Phys. stat. sol. (b)* **64**, K19.

Hess, K., Stath, N. and Benz, K. W. (1974) *J. Electrochem. Soc.* **121**, 1208.

Hiskes, R. and Burmeister, R. A. (1973) *Amer. Inst. Phys. Conf. Proc.* **10**, 304.

Johnson, V., Finley, A. and Wold, A. (1973) *Inorg. Synth.* **14**, 182.

Jonker, H. D. (1974) *J. Solid State Chem.* **10**, 116.

Jonker, H. D. (1975) *J. Crystal Growth*, to be published.

Joullié, A., Aulombard, R. and Bougnot, G. (1974) *J. Crystal Growth* **24/25**, 276.

Kaneko, K., Ayabe, M., Dosen, M., Morizane, K., Usui, S. and Watanabe, N. (1973) *Proc. IEEE* **61**, 884; (see also (1974) *J. Electrochem. Soc.* **121**, 556).

Kashyap, S. C. (1973) *J. Appl. Phys.* **44**, 4381.

Köhl, P., Müller, U. and Reinen, D. (1972) *Z. Anorg. Allg. Chem.* **392**, 124.

Kojima, H. and Kochi, A. (1968) *Yamanashi Daigaku Kogakuby Kenkyu Hokoku* **19**, 123.

Komatsu, H., Homma, S., Kimura, S., Miyazawa, Y, and Shindo, I. (1974) *J. Crystal Growth* **24/25**, 633.

Koutaissoff, A. (1964) Ph.D. Thesis No. 3572, ETH Zurich, p. 15.

Larsen, P. K. and Robertson, J. M. (1974) *J. Appl. Phys.* **45**, 2867.

Macksey, H. M., Lee, M. H., Holonyak, N., Hitchers, W. R., Dupuis, R. D. and Campbell, J. C. (1973) *J. Appl. Phys.* **44**, 5035.

Majling, J., Figusch, V., Corba, J. and Hanic, F. (1974) *J. Appl. Cryst.* **7**, 402.

Malur, J. (1966) *Kristall u. Technik* **1**, 261.

Mikov, V. T., Apostolov, A. V., Taraleshkova, V. S. and Andreevska, V. G. (1973) *Compt. Rend. Acad. Bulg. Sci.* **26**, 859.

Miller, D. C., Shick, L. K. and Brandle, C. D. (1974) *J. Crystal Gowth* **23**, 313.

Morrison, A. D. (1971) *Ferroelectrics* **2**, 59.

Morrison, A. D., Lewis, F. A. and Miller, A. (1970) *Ferroelectrics* **1**, 75.

Mroczkowski, S. (1974) *J. Crystal Growth* **24/25**, 683.

Nakano, K., Hayashi, H. and Imura, T. (1974) *J. Crystal Growth* **24/25**, 679.

Nevyantseva, R. R., Kidyarov, B. I., Stroitelev, S. A. and Nikolaev, I. V. (1969) *Inorg. Mater. (USSR)* **5**, 1806.

Newnham, R. E., Wolfe, R. W., Horsey, R. S. and Diaz-Colon, F. A. (1973) *Mat. Res. Bull.* **8**, 1183.

Nomura, S. and Kojima, F. (1973) *Japan J. Appl. Phys.* **12**, 205.

Ostorero, J. and Makram, H. (1974) *J. Crystal Growth* **24/25**, 677.

Patel, A. R. and Arora, S. K. (1974) *J. Crystal Growth* **23**, 95.

Perekalina, T. M., Fonton, S. S., Magakova, Yu. G. and Voskanyan, R. A. (1972) *Sov. Phys.-Solid State* **13**, 2693.

Plaskett, T. S., Klokholm, E., Hu, H. C. and O'Kane, D. F. (1973) *Amer. Inst. Phys. Conf. Proc.* **10**, 319.

Potkin, L. I. (1972) *Rost Kristallov* **9**, 90 (russ.).

Robbins, C. R. (1970) *J. Res. Natl. Bur. Stand.* **74A**, 229.

Robertson, J. M., Wittekoek, S., Popma, T. J. A. and Bongers, P. F. (1973) *Appl. Phys.* **2**, 219.

Robertson, J. M., Tolksdorf, W. and Jonker, H. D. (1975) *J. Crystal Growth*, to be published.

Rosztocy, F. E. and Kinoshita, J. (1974) *J. Electrochem. Soc.* **121**, 439.

Rüdorff, W., Stössel, A. and Schmidt, V. (1968) *Z. Anorg. Allg. Chem.* **357**, 264.

Saito, T. and Seki, Y. (1974) *J. Crystal Growth* **23**, 217.

Scott, W. (1973) *J. Appl. Phys.* **44**, 5165.

Setoguchi, M. and Sakamoto, C. (1974) *J. Crystal Growth* **24/25**, 674.

Shay, J. L., Bachmann, K. J. and Buehler, E. (1974) *J. Appl. Phys.* **45**, 1302; *J. Crystal Growth* **24/25**, 260.

Smith, S. H., Garton, G. and Tanner, B. K. (1974) *J. Crystal Growth* **23**, 335.

Subba Rao, G. V., Wanklyn, B. M. and Rao, C. N. R. (1971) *J. Phys. Chem. Solids* **32**, 345.

Suemune, Y. and Inone, N. (1974) *Japan J. Appl. Phys.* **13**, 204; *J. Crystal Growth* **24/25**, 646.

Sugaya, T. and Watanabe, O. (1971) *Trans .Jap. Inst. Met.* **12**, 301.

Tabata, H., Ishii, E. and Okuda, H. (1974) *J. Crystal Growth* **24/25**, 656.

Taguchi, T., Shirafuji, J. and Inuishi, Y. (1974) *Japan J. Appl. Phys.* **13**, 1169.

Takasu, S. and Shimanuki, S. (1974) *J. Crystal Growth* **24/25**, 641.

Tolksdorf, W. (1974) *Acta Electron.* **17,** 57.

Ushio, M. and Sumiyoshi, Y. (1972) *J. Chem. Soc. Japan* No. 9, 1648.

Ushio, M. and Sumiyoshi, Y. (1973a) *J. Chem. Soc. Japan* No. 3, 506; No. 5, 941.

Ushio, M. and Sumiyoshi, Y. (1973b) *J. Chem. Soc. Japan* No. 12, 2301.

Voigt, D. O. and Neels, H. (1971)*Kristall u. Technik* **6,** 651.

Voorhoeve, R. J. F., Trimble, L. E. and Khattak, C. P. (1974) *Mat. Res. Bull.* **9,** 655.

Wanklyn, B. M. (1974) *J. Mat. Sci.* **9,** 1279.

Wanklyn, B. M. and Garton, G. (1974) *J. Mat. Sci.* **9,** 1378.

Wanklyn, B. M. and Hauptman, Z. (1974) *J. Mat. Sci.* **9,** 1078.

Watanabe, K. and Sumiyoshi, Y. (1974) *J. Crystal Growth* **24/25,** 666.

Weil, R., Hocart, R. and Monier, J. C. (1954) *Bull. Soc. Franc. Min.* **77,** 1084.

Whiffin, P. A. C. and Brice, J. C. (1974) *J. Crystal Growth* **23,** 25.

Whipps, P. W. (1972) *Solid State Chem.* **4,** 281.

Symbols and Abbreviations

(unless otherwise defined)

A_m	surface area occupied per molecule, cm²
a	cell edge, cm
ACRT	accelerated crucible rotation technique, see Ch. 7, 10
b	ratio of mean displacement on surface to diffusion mean free path $(y_s/\varLambda)$, see Ch. 4
BCF	Burton–Cabrera–Frank theory, see Ch. 4
C	constant in BCF equation
c	subscript : crystal
CR	chemical reaction, see Ch. 7, 10
CVD	chemical vapour deposition
D	diffusion coefficient, cm² s⁻¹
D_s	surface diffusion coefficient, cm² s⁻¹
EL	electrolytic growth, see Ch. 7, 10
EV	evaporation technique, see Ch. 7, 10
F faces	flat faces, see Ch. 5
HPS	high-pressure solutions
HTS	high-temperature solutions
I	nucleation rate, s⁻¹
j_s	surface diffusion (solute) flux, g cm⁻² s⁻¹
j_v	volume diffusion (solute) flux, g cm⁻³ s⁻¹
K	Kelvin
k_{eff}	interface distribution coefficient
k_0	equilibrium distribution coefficient
K faces	kinked faces, Ch. 5
L	latent heat, J mole⁻¹
LPE	liquid phase epitaxy
N_c	number of molecules in critical nucleus
n	solute concentration, g cm⁻³
n_e	equilibrium solute concentration, g cm⁻³
n_{sn}	solute concentration in bulk of solution, g cm⁻³
PBC	periodic bond chain concept of Hartman and Perdok, see Ch. 5
R	universal gas constant, J K⁻¹

R	Rayleigh number
r^*	radius of critical nucleus, cm
S_r	surface roughness
s	subscript : surface
SC	slow cooling technique, see Ch. 7, 10
sn	subscript : solution
st	subscript : step
S faces	stepped faces, Ch. 5
T	temperature, K
T_e	equilibrium temperature, K
THM	travelling heater method, see Ch. 7, 10
TPRE	twin-plane re-entrant edge growth mechnaism, see Ch. 5
TR	(temperature gradient) transport technique, see Ch. 7, 10
TSM	travelling solvent technique, see Ch. 7, 10
TSSG	top-seeded solution growth, see Ch. 7, 10
TSZM	travelling-solvent-zone method, see Ch. 7, 10
u	solution flow velocity, cm s^{-1}
V_M	molar volume, cm^3 mole^{-1}
v	linear growth rate, cm s^{-1}
v_{st}	step velocity, cm s^{-1}
VLS	vapour-liquid-solid growth, see Ch. 7, 10
VLRS	vapour-liquid-reaction-solid growth, see Ch. 7, 10
W_D	activation energy for volume diffusion, J mole^{-1}
W	surface energy per atom, J atom^{-1}
W_B	binding energy per molecule, J molecule^{-1}
X	seed crystal, see Ch. 10
x_0	mean kink separation, cm
y_0	step separation, cm
y_s	mean displacement on surface, cm
Z^*	collision frequency, s^{-1}

coordinate system with directions x, y, z

( )	one particular crystallographic plane
{ }	general type of crystallographic planes
[ ]	crystallographic direction
$\langle\ \rangle$	general type of crystallographic direction
α	critical angle (twin, grains)
β	step boundary condition parameter, see Ch. 4
Γ	Gibbs–Thompson capillarity constant $= V_M \gamma / RT$
γ	interfacial surface energy, J cm^{-2}
γ'	Temkin parameter

γ_e	edge energy of nucleus, J cm^{-1}
δ	boundary layer for solute diffusion, cm
ϵ	spiral interaction parameter
η	viscosity, poise
κ	thermal conductivity, J cm^{-1} s^{-1} K^{-1}
Λ	adsorption mean free path, cm
ν	kinematic viscosity, η/ρ_{sn}
ρ_c	crystal density, g cm^{-3}
ρ_{sn}	solution density, g cm^{-3}
σ	relative supersaturation
σ_1	constant in BCF equation
σ_s	surface supersaturation
τ_{deads}	deadsorption time, s
ϕ	angle (between twins, grains)
ϕ	heat of crystallization, J mole^{-1}
ϕ_m	heat of crystallization per molecule, J molecule^{-1}
ψ	$\sigma - \sigma_s$
Ω	volume of growth unit, cm^3
ω	angular velocity, rad s^{-1}

Index